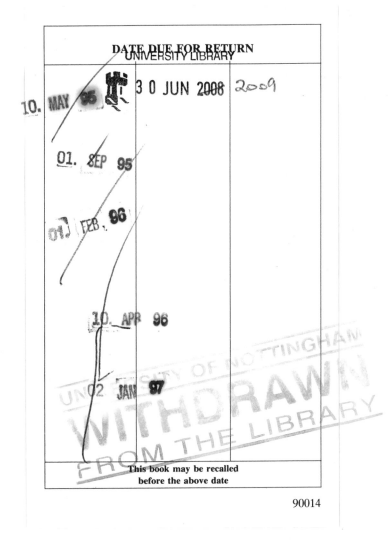

Other Volumes in PROGRESS IN BRAIN RESEARCH

Volume 81: The Central Neural Organization of Cardiovascular Control, by J. Ciriello, M.M. Caverson and C. Polosa (Eds.) — 1989, ISBN 0–444–81224–5.

Volume 82: Neural Transplantation: From Molecular Basis to Clinical Applications by S. Dunnett and S.-J. Richards (Eds.) — 1990, ISBN 0–444–81137–0

Volume 83: Understanding the Brain through the Hippocampus, by J. Storm-Mathison, J. Zimmer and O.P. Ottersen (Eds.) — 1990, ISBN 0–444–81149–4.

Volume 84: Cholinergic Neurotransmission: Functional and Clinical Aspects, by S.-M. Aquilonius and P.-G. Gillberg (Eds.) — 1990, ISBN 0–444–81148–6.

Volume 85: The Prefrontal Cortex: Its Structure, Function and Pathology, by H.B.M. Uylings, C.G. van Eden, J.P.C. de Bruin, M.A. Corner and M.G.P. Feenstra (Eds.) — 1991, ISBN 0–444–81124–9.

Volume 86: Molecular and Cellular Mechanisms of Neuronal Plasticity in Normal Aging and Alzheimer's Disease, by P.D. Coleman, G.A. Higgins and C.H. Phelps (Eds.) — 1990, ISBN 0–444–81121–4.

Volume 87: Role of the Forebrain in Sensation and Behavior, by G. Holstege (Ed.) — 1991, ISBN 0–444–81181–8.

Volume 88: Neurology of the Locus Coeruleus, by C.D. Barnes and O. Pompeiano (Eds.) — 1991, ISBN 0–444–81394–2.

Volume 89: Protein Kinase C and its Substrates: Role in Neuronal Growth and Plasticity, by W.H. Gispen and A. Routtenberg (Eds.) — 1991, ISBN 0–444–81436–1.

Volume 90: GABA in the Retina and Central Visual System, by R.R. Mize, R.E. Marc and A.M. Sillito (Eds.) — 1992, ISBN 0–444–81446–9.

Volume 91: Circumventricular Organs and Brain Fluid Environment, by A. Ermisch, R. Landgraf and H.-J Rühle (Eds.) — 1992, ISBN 0–444–81419–1.

Volume 92: The Peptidergic Neuron, by J. Joosse, R.M. Buijs and F.J.H. Tilders (Eds.) — 1992, ISBN 0–444–81457–4.

Volume 93: The Human Hypothalamus in Health and Disease, by D.F. Swaab, M.A. Hofman, M. Mirmiran, R. David and F.W. van Leeuwen (Eds.) — 1992, ISBN 0–444–89538–8.

Volume 94: Neuronal-Astrocyte Interactions: Implications for Normal and Pathological CNS Function, by A.C.H. Yu, L. Hertz, M.D. Norenberg, E. Syková and S. Waxma (Eds.) — 1992, ISBN 0–444–89537–x.

Volume 95: The Visually Responsive Neuron: From Basic Neurophysiology to Behavior, by T.P. Hicks, S. Molotchnikoff and T. Ono (Eds.) — 1993, ISBN 0–444–89492–6.

Volume 96: Neurobiology of Ischemic Brain Damage, by K. Kogure, K.-A. Hossmann and B.K. Siesjö (Eds.) — 1993, ISBN 0–444–89603–1.

Volume 97: Natural and Artificial Control of Hearing and Balance, by J.H.J. Allum, D.J. Allum-Mecklenburg, F.P. Harris and R. Probst (Eds.) — 1993, ISBN 0–444–81252–0.

Volume 98: Cholinergic Function and Dysfunction, by A.C. Cuello (Ed.) — 1993, ISBN 0–444–89717–8.

Volume 99: Chemical Signalling in the Basal Ganglia, by G.W. Arbuthnott and P.C. Emson (Eds.) — 1993, ISBN 0–444–81562–7.

Volume 100: Neuroscience: From the Molecular to the Cognitive, by F.E. Bloom (Ed.) — 1994. ISBN 0–444–01678–x.

Volume 101: Biological Function of Gangliosides, by L. Svennerholm et al. (Eds.) — 1994, ISBN 0–444–01658–5.

PROGRESS IN BRAIN RESEARCH

VOLUME 102

THE SELF-ORGANIZING BRAIN: FROM GROWTH CONES TO FUNCTIONAL NETWORKS

Proceedings of the 18th International Summer School of Brain Research, held at the University of Amsterdam and the Academic Medical Center (The Netherlands) from 23 to 27 August 1993

EDITED BY

J. VAN PELT, M.A. CORNER, H.B.M. UYLINGS

Netherlands Institute for Brain Research, Amsterdam, The Netherlands

and F. H. LOPES DA SILVA

Department of Experimental Zoology, University of Amsterdam, The Netherlands

ELSEVIER
AMSTERDAM – LAUSANNE – NEW YORK – OXFORD – SHANNON – TOKYO
1994

ISBN 0-444-81819-7 (volume)
ISBN 0-444-80104-9 (series)

Published by:
Elsevier Science BV.
P.O. Box 211
1000 AE Amsterdam
The Netherlands

Printed in The Netherlands on acid-free paper

List of Contributors

M. Abeles, Department of Physiology, School of Medicine, The Hebrew University, P.O. Box 12272, Jerusalem 91120, Israel.

M.F. Bear, Department of Neuroscience, Institute for Brain and Neural Systems, Brown University, Providence, RI 12912, USA.

Y. Ben-Ari, INSERM U-29, 123 Blvd de Port-Royal, Paris 75014, France.

H. Bergman, Department of Physiology, School of Medicine, the Hebrew University, P.O. Box 12272, Jerusalem 91120, Israel.

M.M. Black, Department of Anatomy and Cell Biology, Temple University School of Medicine, 3400 N. Broad Street, Philadelphia, PA 19140, USA.

J.-P. Bourgeois, Molecular Neurobiology Laboratory, Pasteur Institute, Paris, France.

J. Burrill, Department of Biology, University of Michigan, Ann Arbor, MI 48109, USA.

B.W. Connors, Department of Neuroscience, Brown University, Providence, RI 02912, USA.

M. Constantine-Paton, Department of Biology, Yale University, Kline Biology Tower, New Haven, CT 06511, USA.

L.N. Cooper, Department of Physics, Institute for Brain and Neural Systems, Brown University, Providence, RI 02912, USA.

M.A. Corner, Graduate School Neurosciences, and Netherlands Institute for Brain Research, Meibergdreef 33, 1105 AZ Amsterdam, The Netherlands.

R.W. Davenport, Program in Neuronal Growth and Development, Department of Anatomy and Neurobiology, Colorado State University, Ft. Collins, CO 80523, USA.

E. de Schutter, Division of Biology, California Institute of Technology, Pasadena, CA 91125, USA.

S.S. Easter Jr., Department of Biology, University of Michigan, Ann Arbor, MI 48109, USA.

R. Eckhorn, Department of Biophysics, Phillips-University, Renthof 7, Marburg 35032, Germany.

W.J. Freeman, Department of Molecular and Cell Biology, LSA 129, University of California at Berkeley, CA 94720, USA.

J.L. Gaiarsa, INSERM U-29, 123 Blvd de Port-Royal, Paris 75014, France.

P.S. Goldman-Rakic, Section of Neurobiology, Yale University School of Medicine, New Haven, CT 06510, USA.

T.M. Gomez, Department of Cell Biology and Neuroanatomy, 4-135 Jackson Hall, University of Minnesota, Minneapolis, MN 55455, USA.

G. Gutierrez-Ospina, Department of Neurobiology, Duke University Medical Center, Durham, NC 27710, USA.

P.B. Guthrie, Program in Neuronal Growth and Development, Department of Anatomy and Neurobiology, Colorado State University, Ft. Collins, CO 80523, USA.

M. Hofer, Department of Biology, Yale University, Kline Biology Tower, New Haven, CT 06511, USA.

L.J. Houenou, Neuroscience Program, Department of Neurobiology and Anatomy, Bowman Gray School of Medicine, Wake Forest University, Winston-Salem, NC 27157, USA.

J.G.R. Jefferys, Department of Physiology, St. Mary's Hospital Medical School, Imperial College, London W2 1PG, UK.

S.P. Kater, Program in Neuronal Growth and Development, Department of Anatomy and Neurobiology, Colorado State University, Ft. Collins, CO 80523, USA.

R. Khazipov, INSERM U29, Blvd de Port-Royal, Paris 75014, France.

S.E. Koester, Molecular Neurobiology Laboratory, The Salk Institute, 10010 N. Torrey Pines Rd., La Jolla, CA 92037, USA.

A. LaMantia, Department of Neurobiology, Duke University Medical Center, Durham, NC 27710, USA.

C.C. Law, Department of Physics and Neuroscience, Institute for Brain and Neural Systems, Brown University, Providence, RI 02912, USA.

P.C. Letourneau, Department of Cell Biology and Neuroanatomy, 4-135 Jackson Hall, University of Minnesota, Minneapolis, MN 55455, USA.

L. Li, Departments of Ophthalmology and of Neurobiology and Anatomy, Bowman Gray School of Medicine, Wake Forest University, Winston-Salem, NC 27157, USA.

A.C. Lo, Neuroscience Program, Department of Neurobiology and Anatomy, Bowman Gray School of Medicine, Wake Forest University, Winston-Salem, NC 27157, USA.

F.H. Lopes da Silva, Graduate School Neurosciences, and Institute of Neurobiology, Amsterdam, University of Amsterdam, Kruislaan 320, 1098 SM Amsterdam, The Netherlands.

R.C. Marcus, Department of Pathology, College of Physicians and Surgeons, Columbia University, NY 10032, USA.

K.D. Miller, Departments of Physiology and Otolaryngology, W.M. Keck Foundation, Center for Integrative Neuroscience, University of California at San Francisco, CA 94143, USA.

D.M. O'Leary, Molecular Neurobiology Laboratory, The Salk Institute, 10010 N. Torrey Pines Rd., La Jolla, CA 92037, USA.

R.W. Oppenheim, Neuroscience Program, Department of Neurobiology and Anatomy, Bowman Gray School of Medicine, Wake Forest University, Winston-Salem, NC 27157, USA.

H. Parnas, Department of Neurobiology, Life Science Institute and Center for Neural Computation, Hebrew University, Jerusalem, Israel.

J.G. Parnavelas, Department of Anatomy and Developmental Biology, University College, Gower Street, London WC1E 6BG, UK.

J.-P. Pijn, Institute for Epilepsy Prevention 'Meer en Bosch', De Cruqiushoeve, P.O. Box 21, Heemstede 2100 AA, The Netherlands.

Y. Prut, Department of Physiology, School of Medicine, the Hebrew University, P.O. Box 12272, Jerusalem 91120, Israel.

D. Purves, Department of Neurobiology, Duke University Medical Center, Durham, NC 27710, USA.

D. Raggozzino, INSERM U-29, 123 Blvd de Port-Royal, Paris 75014, France.

P. Rakic, Section of Neurobiology, Yale University School of Medicine, New Haven, CT 06510, USA.

D. Riddle, Department of Neurobiology, Duke University Medical Center, Durham, NC 27710, USA.

E.T. Rolls, Department of Experimental Psychology, South Parks Road, The University of Oxford, Oxford OX1 3UD, UK.

L.S. Ross, Department of Biological Sciences, Ohio University, Athens, OH 45701, USA.

A. Ruiz-Marcos, Department of Biophysics, Instituto Cajal, Doctor Arce 37, Madrid 28002, Spain.

A.K. Schierwagen, Institut für Informatik, FG Neuroinformatik, Universität Leipzig, Augustplatz 10/11, D-04109 Leipzig, Germany.

E. Sivan, Department of Neurobiology, Life Science Institute and Center for Neural Computation, Hebrew University, Jerusalem, Israel.

D.M. Snow, Department of Cell Biology and Neuroanatomy, 4-135 Jackson Hall, University of Minnesota, Minneapolis, MN 55455, USA.

N.C. Spitzer, Department of Biology and Center for Molecular Genetics, University of California at San Diego, La Jolla, CA 92093, USA.

J.S.H. Taylor, Department of Human Anatomy, South Parks Road, The University of Oxford, Oxford OX1 3QX, UK.

J.G. Taylor, Centre for Neural Networks, Department of Mathematics, King's College, Strand,

London WC2R 2LS, UK.

R.D. Traub, IBM T.J. Watson Research Center, Yorktown Heights, NY 10598, and Department of Neurology, College of Physicians and Surgeons, Columbia University, NY 10032, USA.

A. Treves, Department of Experimental Psychology, South Parks Road, The University of Oxford, Oxford, OX1 3UD, UK.

V. Tseeb, INSERM U-29, 123 Blvd de Port-Royal, Paris 75014, France.

H.B.M. Uylings, Graduate School Neurosciences, and Netherlands Institute for Brain Research, Meibergdreef 33, 1105 AZ Amsterdam, The Netherlands.

E. Vaadia, Department of Physiology, School of Medicine, the Hebrew University, P.O. Box 12272, Jerusalem 91120, Israel.

A. van Ooyen, Graduate School Neurosciences, and Netherlands Institute for Brain Research, Meibergdreef 33, 1105 AZ Amsterdam, The Netherlands.

J. van Pelt, Graduate School Neurosciences, and Netherlands Institute for Brain Research, Meibergdreef 33, 1105 AZ Amsterdam, The Netherlands.

M.P. van Veen, RIVM unit BFT, P.O. Box 1, Bilthoven 3720 BA, The Netherlands.

W.J. Wadman, Graduate School Neurosciences, and Institute of Neurobiology, University of Amsterdam, Kruislaan 320, 1098 SM Amsterdam, The Netherlands.

L.E. White, Department of Neurobiology, Duke University Medical Center, Durham, NC 27710, USA.

S.W. Wilson, Developmental Biology Research Centre, King's College, London WC2B 5RL, UK.

Q. Yan, Department of Neuroscience, Amgen Inc., Thousand Oaks, CA 91320, USA.

Preface

The nervous system is undoubtedly one of the most complex of all biological systems, and its understanding can be considered as a major challenge in biological research, with spin-off and implications far beyond the boundaries of its specific field of research. Its complex organization emerges over a wide range of scales in both space and time (molecular, cellular, network, system) and its understanding requires insight into its phenomenological manifestations on all these scales. This is an immense task, but present developments in, for instance, molecular neurobiology and neuronal networks, along with myriad new technological developments in neurobiology, stem hopeful.

The 18th International Summer School of Brain Research, held in 1993 in Amsterdam, and organized by The Netherlands Institute for Brain Research (NIBR) according to a long-standing biannual tradition, concentrated on the organizational level of neurons and neuronal networks under the unifying theme 'The Self-Organizing Brain: From Growth Cones to Functional Networks'. Such a theme is attractive because it incorporates all phases in the emergence of complexity and (adaptive) organization, as well as involving processes that remain operative in the mature state. Physiological activity is now known to have profound effects at all levels of structural organization, and this phenomenon has been given extensive treatment during the Summer School. This reflects not only the growing general interest in activity-dependent maturation, but also our conviction that this is one of the key mechanisms involved in the adaptive 'tuning' of structure and function to environmental demands.

Such reciprocal interaction between structure and function is characteristic for the process of self-organization, thus making developmental histories crucial for future operational characteristics. These histories include intrinsically generated bioelectric activity in the early organism, regulating details of initial network formation, as well as the later experience of patterned sensory stimulation in leaving lasting 'imprints' on the brain. This concept also motivates the need for understanding (nonlinear) dynamics in neurons and networks, as they may well play an important role in these developmental and adaptive processes. The keywords – *structure—function, development, dynamics and self-organization* – have therefore run through the program like a continuous thread.

Mathematical modelling and computer simulations form indispensible tools for understanding the behavior of complex dynamical systems, making integrative empirical and theoretical approaches especially pertinent in the case of brain research. Since it has been one of the aims of this Summer School to explore the fruitfulness of such an approach, those areas are emphasized where complementary theoretical and empirical studies are actively on the forefront of modern research. The order of the sections follows successive levels of organization from neuronal growth cones, neurite formation, neuronal morphology and signal processing to network development, network dynamics and, finally, to the formation of functional circuits.

We are most gratified at the enthusiasm with which a large number of scientists have contributed to the meeting and to this volume. In addition, we wish to acknowledge the generosity of both the Royal Netherlands Academy of Sciences and the Neurosciences Graduate School Amsterdam, under whose joint auspices the Summer School was held. Additionally, we are grateful to the other generous financial supporters mentioned on a separate page. We would also like to thank Drs. H. Parnas and A. Schierwagen for their advice as members of the Advisory Committee, and Arjen van Ooyen and Mark van Veen for their contribution to the conceptual planning. Finally, we thank, Tini Eikelboom, Aad Jansen, Wilma Top and Wilma Verwey for taking care of most of the organizational work, along with Henk Stoffels for inspired artwork.

Jaap van Pelt
Michael A. Corner
Harry B.M. Uylings
Fernando H. Lopes da Silva

Acknowledgements

The 18th International Summer School of Brain Research was organized under the auspices of:

Royal Netherlands Academy of Sciences
Graduate School Neurosciences Amsterdam

Financial support was also obtained from:

ABN-AMRO Bank
Academic Press Ltd.
Blackwell Science Publishers Ltd.
Cambridge Electronic Design Ltd.
The Company of Biologists Ltd.
Graduate School Neurosciences Amsterdam
Harry Fein, World Precision Instruments
Marion Merrell, Dow Research Institute
Organon International B.V.
Oxford University Press
Remmert Adriaan Laan Fund
Royal Netherlands Academy of Sciences
for Sigma-Tau
Smith Kline Beecham Pharmaceuticals
Van den Houten Foundation

Contents

SECTION I

Introduction

J. van Pelt, M.A. Corner H.B.M. Uylings and F.H. Lopes da Silva (Eds.)
Progress in Brain Research, Vol 102

CHAPTER 1

Reciprocity of structure-function relations in developing neural networks: the Odyssey of a self-organizing brain through research fads, fallacies and prospects

Michael Alan Corner

Graduate School of Neuroscience and Netherlands Institute for Brain Research, Amsterdam, The Netherlands

Introduction

In the editorial introduction we emphasized the need for empirical and theoretical studies to complement one another when analyzing complex, dynamic non-linear interactions in systems such as neural networks. The design of crucial experiments in order to test competing hypotheses (Platt, 1964) can only be accomplished by simulating the consequences of all plausible mechanisms by means of appropriate mathematical models, since only in exceptionally simple situations will deductive logic suffice to generate the required 'truth tables' (see below). Second of all, we felt justified in concentrating on the level of multi-cellular 'circuitry' to the exclusion of the molecular and behavioral dimensions below and above it, respectively. Finally, we emphasized not only the importance of network models for understanding how the nervous system operates, but also the reciprocal question (Gottlieb, 1973) of how neural function can feed back upon the structure generating it so as to modify its organization, often in a long-lasting way.

This last point provides me with a welcome opportunity for putting the principle of *spontaneous* physiological activity into the limelight as an exciting, but hitherto insufficiently appreciated (e.g. Purves and Lichtman, 1985; Edelman, 1987; Kandel et al., 1991), principle in the development and function of neural networks (but see Corner, 1963, 1964, 1967, 1987, 1990; Hamburger, 1973; Changeux and Mikoshiba, 1978; Oppenheim, 1982). Since these are all issues with which I've been involved personally from the start, I'm in a position to illustrate most points with material taken from direct experience. I'll be addressing, then, the following three issues in turn: (1) the need for theoretical guidance in empirical research, (2) multi-cellular 'circuitry' as an invaluable intermediate level construct, and (3) intrinsically generated electrical activity as an 'epigenetic' factor in structural and functional brain maturation.

On the need for theoretical rigor in experimental neurobiology

The genius of Sherlock Holmes lay primarily in his realization (most instructive even for professional scientists when applying 'the method'!; see Platt, 1964) that truth can be arrived at most reliably by working from *multiple* working hypotheses (as many of them as common sense dictates to be plausible) and most efficiently by looking specifically for those *crucial* bits of empirical evidence which enable some of the hypotheses to be rejected. Indeed, I once actually spared myself years of inherently inconclusive research in just this fashion, by subjecting a well-meaning suggestion for a technically high-powered study of the neurophysiological basis for 'misdirected' tactile reflexes (in skin-grafted frogs: e.g. Baker, 1978) to a simple but rigorous truth-table analysis, using nothing more than simple Aristotelian logic, before committing ourselves to the project. When it became clear that the three most serious hypotheses all predicted the same result from the proposed experiment (i.e. intracellular recordings from motoneurons, mapping the monosynaptic spinal reflex connections originating from different sensory locations; see Table I: 'central switch?') we were still in a position to drop it in favor of a design which *could*, in principle, decide among alternative growth mechanisms (Table I; see also Baker and Corner, 1978, 1981).

The approach we eventually came up with (intracellular recording from identified sensory ganglion cells: see Fig. 1), besides providing most of the answers we were looking for, proved to be far less time-consuming and even to require a much less complicated setup. An earlier proposed mechanism ('innervation selectivity': the re-establishment of the original connections following early skin rotation) could be refuted in a more thorough and convincing fashion, while a mechanism not previously considered seriously at all ('neuronal competition': the overproduction of neurons and elimination of the excess) was shown to be, in fact, the only plausible alternative which was consistent with the total data set (Fig. 2; Table II). The classical alternative to the selective-innervation hypothesis ('end-organ modulation': the switching of central sensory projections according to the precise nature of the peripheral innervation; e.g. Weiss, 1969) could thus, at long last, be shown to be at best highly unlikely.

Attempts have since been made (see Frank et al., 1988) to approach this same question with respect to the ontogeny of proprioceptive rather than cutaneous reflex selectivity but, falling into exactly the same design trap that we had luckily managed to avoid (see above), their results have necessarily remained inconclusive as far as developmental mechanisms are concerned. Therefore, despite some specious argumentation to the contrary (e.g., Frank et al., 1988), there are no more grounds for believing that the *end-organ specification* hypothesis is correct for the afferent (primary sensory) side of the motoneuron than there are for taking it seriously on the efferent side (i.e. the neuromuscular junction, where this putative mechanism has long been in disrepute — for review see Landmesser, 1980).

In a truly brilliant article, the physicist J. Platt (1964) reminded us a long time ago that what Baker and I had done was nothing more than to apply the method of 'strong inference', formalized by Francis Bacon already in the 17th century. Yet, how many of us are not, in practise, more like poor Inspector Lastrade (of Scotland Yard), who incorrigibly pounces upon the first logically sound explanation compatible with the clues available at that moment (so far so good) but then, under the delusion that he is thereby 'supporting his hypothesis', proceeds merely to accumulate additional data consistent with this initial candidate? Inefficient at best, and dangerously prone to 'miscarriages of justice', we can easily afford to ridicule the detective who goes about his business in this fashion .... as long as Holmes is around to remind us how it *should* be done: "elementary, my dear Watson!". Left to our own devices, however, 'horrible examples' abound of how easily we, too, fall into the trap of the single working hypothesis — or worse yet, of having none at all worthy of the name: the empirical *fallacy* in place of inductive *reasoning* (Popper, 1965)!

Even the simplest non-linear systems, I have learned, routinely display counter-intuitive behavior, and the differential equations describing such systems quickly become so complicated that they are incapable of solution using analytic methods, even after simplification into a 'lumped' network model (e.g. Lopes da Silva et al., 1976). To make matters worse, when the interacting neural ele-

TABLE I

'Truth-table' of predictions deduced from different proposed mechanisms for the ontogeny of sensory reflex connections in skin-grafted frogs

Hypothetical mechanism	Stimulus site?	Central switch?	Peripheral switch?	Cell death?	Selective outgrowth?	Altered survival?
Impulse patterning	NO	No	No	Maybe	Maybe	(NO)
Innervation selectivity	Yes	Yes	YES	Maybe	Yes	(NO)
End-organ modulation	Yes	Yes	No	Maybe	Maybe	NO
Neuronal competition	Yes	Yes	Maybe	Yes	Yes	Yes

The specific prediction which, in the light of discordant experimental findings, falsified the hypothetical mechanism in question is indicated in capitals (see, respectively, Corner and Baker, 1978; Baker et al., 1978, 1981a,b for details). The final listing (i.e. competition among 'redundant' neurons) is the only presently plausible hypothesis which makes sense of all the experimental findings, and is not in contradiction with any of them (but see Fig. 1 for further refinement).

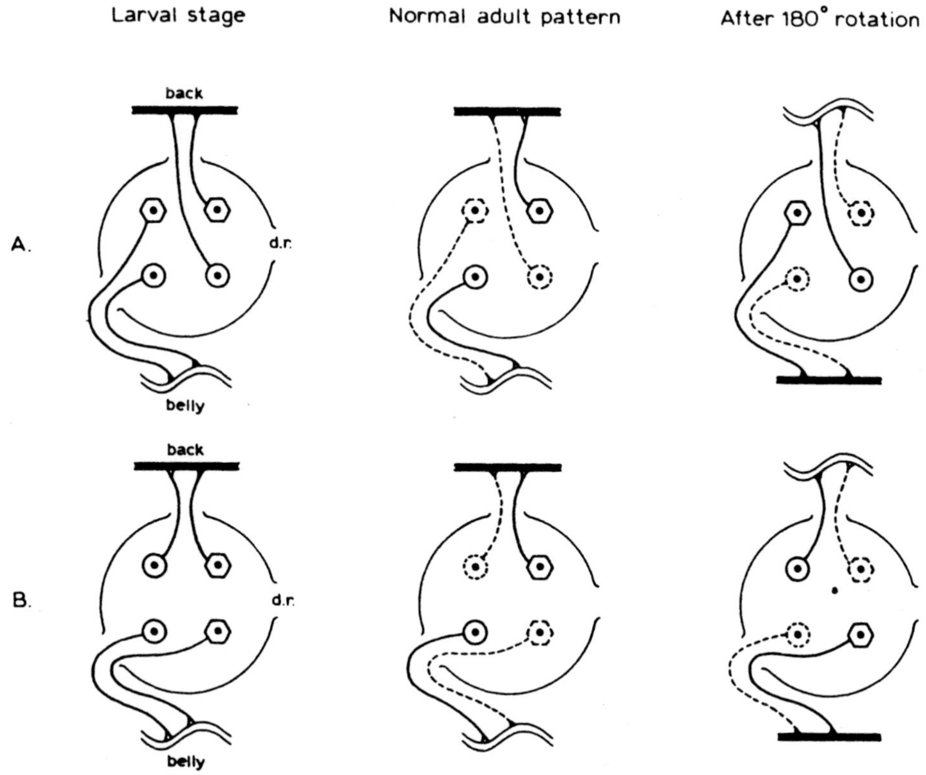

Fig. 1. Two possible mechanisms, in accordance with the redundant sensory ganglion cell hypothesis, for the selective replacement of one cutaneous neuronal population by another. (A) Topographic separation (left) of pre-specified back- and belly-skin neurons in, respectively, the dorsal and ventral parts of the ganglion. Stippled cells show the neuron types which would degenerate in normal as compared with skin-rotated animals: the predominant ganglionic projections would thereby become reversed. (B) Admixture of pre-specified 'back' and 'belly' cell types throughout the ganglion but with preferential projections of dorsally and ventrally situated neurons to, respectively, the animal's back and belly: with this model no detectable changes would take place, as indeed turned out to be the case (see Baker and Corner, 1978, 1981).

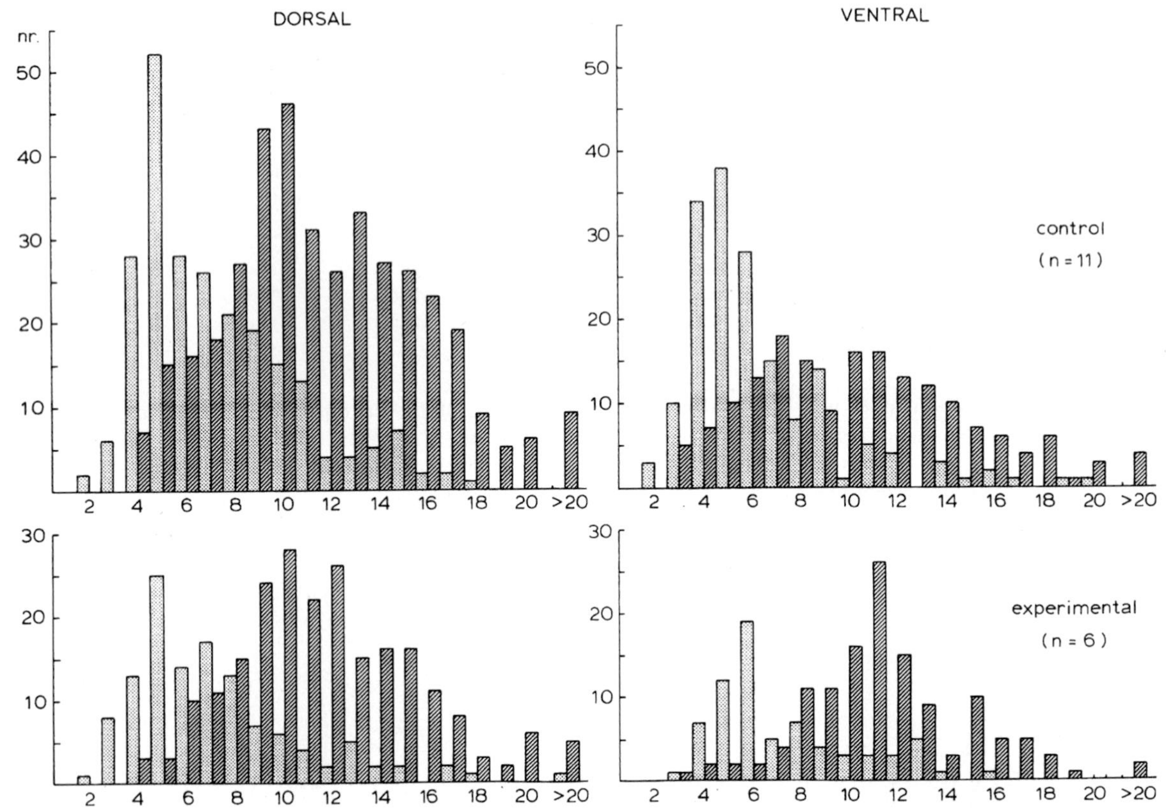

Fig. 2. Size distributions of neuronal soma profiles counted in 20-μm thick sections through the dorsal and ventral halves of silver stained thoracic dorsal root ganglia of normal and skin-rotated frogs (see Baker et al., 1981a,b). The number of cells falling in each size class (in microns on the abscissa) is indicated on the ordinate for both the interior (predominantly smaller neurons: see stippled bars) and the 'cortex' of the ganglia. The latter group deviates significantly (see Table II) from the normal size distribution, although the total number of neurons per ganglion was unaffected.

TABLE II

Statistically significant DRG neuronal size differences in frogs showing misdirected reflexes versus control frogs

Experimental group	(Mean over $N=$)	Coeff.-variation		Kurtosis	
		dorsal	ventral	dorsal	ventral
Skin-rotated	(6)	0.27	0.28	1.0	0.8
Un-operated	(11)	0.36	0.39	0.0	−0.1
Skin-rotated	(4)	0.17	0.21	1.5	0.4
Sham-operated	(5)	0.32	0.27	0.1	0.0

Misdirected skin-wiping responses are reproducibly associated with (i) a reduced variance in, and (ii) a non-Gaussian distribution of cell soma sizes (for details see Baker et al., 1981a,b).

TABLE III

A listing of some of the fundamental (Big!) questions in modern science

Origin of	The big	Key concept	Discipline
Energy	Secret!	Unified force field	Mathematics
Matter	Bang!	Fundamental particles	Physics
Order	Flow!	Self-organization	Thermodynamics
Earth	Drift!	Continental crust	Geology
Life	Flash!	Negative entropy	Biochemistry
Species	Brawl!	Evolutionary fitness	Biology
Death	Sleep!	Sexual reproduction	Botany
Behavior	*Burst!*	*Spontaneous motility*	*Zoology*
Arousal	Startle!	Sleep-wake rhythms	Physiology
Memory	Time!	Conditioned learning	Psychology
Mind	Mystery!	Conscious experience	Philosophy
(Wo)man	Trip!	Cultural progress	Anthropology
Self	Joke!	Subjective duality	Metaphysics

See Corner (1987, 1990) for a thorough on-going discussion of the 'zoology' entry: the phenomenology and putative mechanisms of endogenous behavioral rhythmicity.

ments need to be simulated more realistically, only a 'distributed' model (e.g. Kowalski et al., 1992; Van Ooyen et al., 1992) suffices, thus requiring the services of computers with still greater power and speed. Since the essence of 'scientific method' remains the reasonably accurate prediction of consequences to be expected from a given working hypothesis, thus enabling empirical validation, it is clear that only the most elementary questions about nervous function (see Table I) can be tackled without the use of numerical methods. Mathematical models also require us to make fully explicit all of the assumptions embodied in our hypotheses, for without such explication the risk of fallacious or tautological deductions will usually be very great indeed. The time may not be far off, as a matter of fact, when neurobiologists will wonder how their predecessors ever managed — especially amazing in an age when powerful computers are so readily available — to muddle through using, at most, old-fashioned deductive logic to guide decisions about how to optimally apply all their sophisticated (and expensive!) technology!

Structural bases for function in neuronal networks

The 'cell' as starting point in the hierarchical analysis of biological sub-systems

The *cell* will perhaps forever remain the pivotal concept (see Weiss, 1969; Moore, 1987) for defining the elements of any 'bottom-up' biological theory, analogous as it were to 'words' in which the language of living systems is written (see Van der Weele, 1993 for a discussion of 'top-down' developmental theories, using the concept of *field* as the elemental construct). We must not forget, however, that a dissection of individual words into smaller parts only explains how each one came to have its unique definition as a linguistic building-block, but provides little or no insight into the meaning of the messages made by linking them together to form speech or literature. Similarly, although molecular mechanisms must be held accountable for all the observable *properties* of particular nerve cells, they are not directly relevant for understanding the behavior of networks assembled from such elements (but see Black, 1991, arguing for causal the primacy of certain 'symbolic' molecules).

Above and beyond the cell membrane a higher-order analysis is required, and this need not wait until 'deeper' insights have been obtained at lower levels of material organization. Quite the contrary, a 'top-down' conceptual strategy appears best suited (see Boyd and Noble, 1993) for guaranteeing that our attempts to penetrate nature's secrets will unearth insights into matters which are truly of interest to us as biologists: ecology, evolution, animal behavior, the human mind..... (Table III). Otherwise, many of the details that we bring to light may wind up only adding to the 'noise' (information overload), rather than being the insightful 'signals' we would like them to be!

Wiener's (1961) 'cybernetic' vision was that the material realization of a given network design — whether in metal, silicon, vacuum tubes or protoplasm — is actually but a footnote as far as an understanding of its functional organization is concerned (cf. Black's attempt (1991) to debunk such a 'functionalist' perspective). What is needed for comprehension at the 'circuitry' level of abstraction — embodied in the form of a sufficiently complete mathematical representation to enable construction of the system out of *any* suitable material — is knowledge of (i) the response characteristics of the functional elements of which the network is composed, and (ii) how each such element is structurally interconnected with the rest of the system. From the foregoing discussion it will be clear that conceptually isolated neuronal sub-systems can yield biologically interesting information only insofar as their input-output characteristics ('transfer functions') are, in turn, integrated with other components of the nervous system to form a 'machine' — i.e. *mechanism(s)* in the broadest sense of the word: Craik, 1952 — capable of generating behavior (e.g. Fig. 3) which is well-adapted to a given species' ecological niche.

Morphology and physiology: an indissoluble multidisciplinary interpenetration — the cerebrum's 'basic bioelectric waveform'

The fundamental 'forces' which can — and must — be considered when analyzing a network's behavior reduce to our old friends, *structure* and *function*. We must, first of all, have precise knowledge of which neuronal elements are directly coupled to which others, and what is the spatial configuration of each such connection. In addition, we need precise knowledge of the way each neuron converts afferent into efferent signals (i.e. its 'transfer function'), and the kinetic aspects of impulse conduction and transmission to 'downstream' target cells. Anatomical elaboration certainly opens up ever new possibilities but these are realized to a widely varying extent at different stages of development and under different environmental conditions. Form does not generally turn out to be the rate-limiting factor in functional ontogeny: more often there is a considerable degree of 'forward reference' to later performance (see Oppenheim, 1982), a consideration which is perhaps the main factor responsible for the disappointing yield of interesting conclusions from many of the ambitious multi-disciplinary studies which were so popular 30 years or so ago (see e.g. Purpura and Schadé, 1964).

An illustrative case in point is provided by my own experience along these lines when working on the development of the cerebral hemispheres in chick embryos. Both ongoing 'spontaneous' electrical activity ('EEG') and direct evoked surface potentials show a rapid transition to mature looking patterns about 4 days prior to hatching (Corner et al., 1974a,b). Stereotyped isolated waveforms, resembling the 'K-complexes' seen during the onset of slow-wave sleep in adult mammals (see e.g. Kostopoulos and Gotman, 1989), appear initially at long intervals in late stage 42, but gradually merge by early stage 44 into an almost continuous sleep-like (see Corner et al., 1966) 'delta wave' pattern (Fig. 4) — which, incidentally, is largely generated within the forebrain itself (Corner and Bot, 1969; Corner et al., 1972): slow-wave sleep as functional decerebration (also see Fig. 5, and Steriade et al., 1990, 1993)! To my surprise, however, cerebral cytomorphological and ultrastructural development proved to be at a standstill at that point, and in a highly immature condition to boot (Fig. 6). Strik-

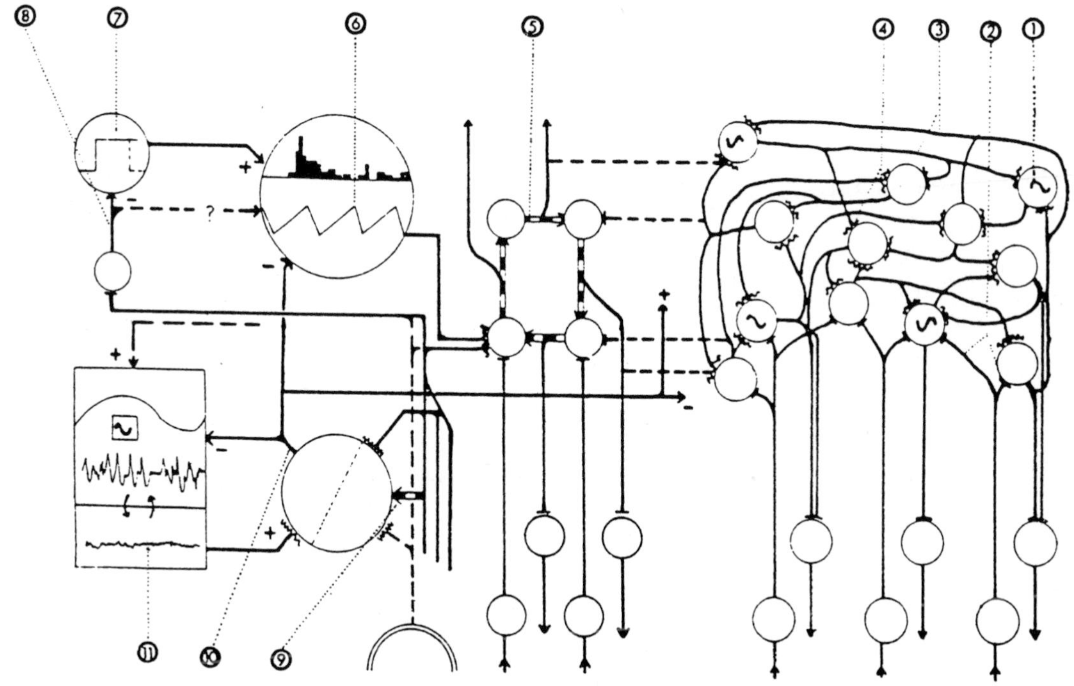

Fig. 3. Schematic summary of deductions from available behavioral and electrophysiological data concerning the functional organization of the central nervous sytem in chicken embryos (from Corner, 1967): (1) neurons discharging without any spike-driven presynaptic input (represented as A.C. generators); (2) incoming impulses via sensory ganglion neurons; (3) diffuse 'reticular' brainstem/spinal cord interneuronal connections; (4) fluctuating thresholds for 'local circuit' interactions; (5) pattern generating circuitry for stereotyped behaviors such as respiration, locomotion, vocalization and swallowing; (6) brainstem/cord rhythm-generating network, triggering movement bursts when motor output thresholds are sufficiently low; (7) hypothalamic 'command' network, triggering persistent movement bursts during hatching; (8) inhibitory effects of vestibular and other sensory inputs upon the 'hatch-generating system'; (9) extraordinary efficacy of neck proprioceptive inputs for triggering waking behavior, as putatively mediated by a 'central activating system' (CAS); (10) complex effects of CAS activity to inhibit EEG slow-wave (sleep) activity as well as stereotyped or chaotic sleep-like spontaneous movements; (11) activated 'desynchronized' state of the forebrain, associated with reverberating excitation whenever endogenous discharges are inhibited by the CAS.

ing as they indisputably are, the coherent 'brain waves' observed during deep sleep already close to a century ago (see Steriade et al., 1990) close to a century ago apparently require nowhere near the full synaptic potential available to a mature cerebrum.

The rate-limiting factor turns out in this case to almost certainly be neuro-pharmacological (Corner et al., 1974a) rather than structural (see above) or bio-energetic (Jongkind et al., 1972) in nature: the efficacy of excitatory amino acid neurotransmission (for review, see Corner and Romijn, 1983). Glutamic acid levels (Table IV) and metabolism (Rudnick and Waelsch, 1955), for instance, are closely correlated with the above-mentioned electrophysiological developments, while exogenous glutamate is capable of precipitating the premature appearance of virtually identical waves (Fig. 4). Comparable results were later obtained in neonatal rat pups, where the rapid appearance of mature slow-wave sleep EEG and neuronal firing patterns (Mirmiran and Corner, 1982; Corner et al., 1984) on about postnatal day 13 — which makes it analogous, functionally speaking, to stage 43 in the chick embryo — is followed within a day or two by a clearcut intensification of slow-wave suppression during 'paradoxical' sleep epochs (Fig. 7). This highly com-

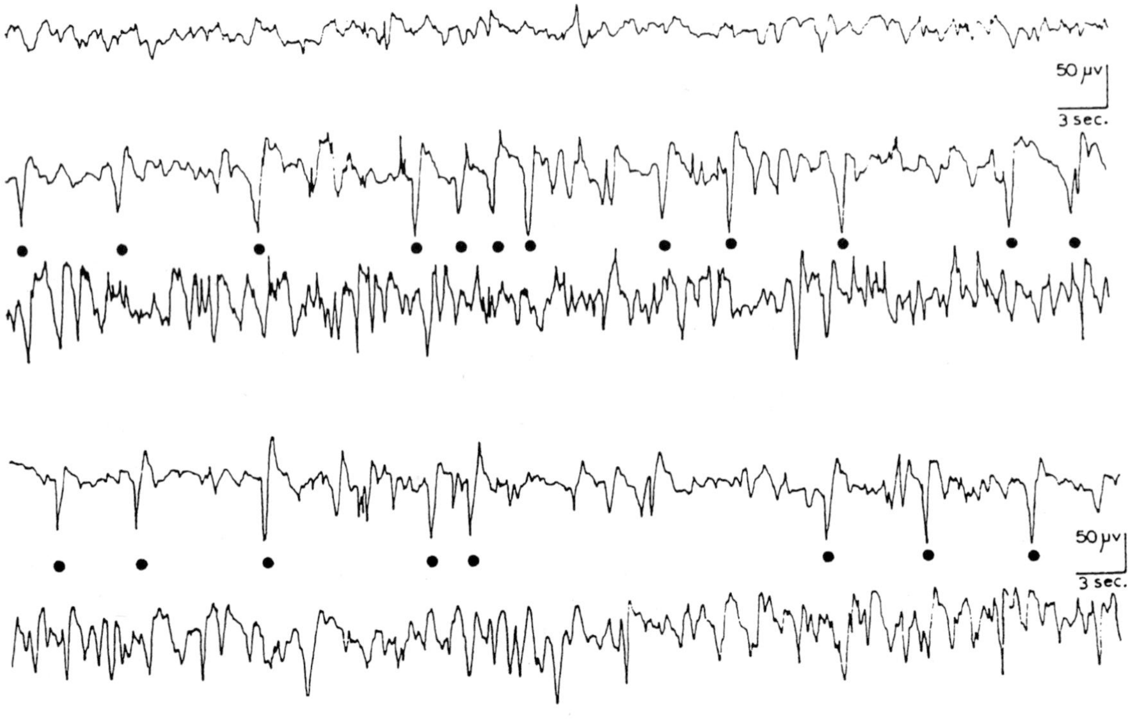

Fig. 4. Development of stereotyped high amplitude EEG slow waves in the embryonic chicken cerebral hemisphere (accessory hyperstriatum). Normal stage 42 pattern (top) followed by sample records taken 10 and 30 min following sodium glutamate administration, respectively, mimicking the stage 43 and 44 EEG patterns (bottom two records). Clearly identifiable bioelectric 'basic waveforms' are indicated by dots (see Corner et al., 1974a for details).

pressed physiological development is in good agreement with the relatively early maturation of excitatory cortical synapses in the rat (Aghajanian and Bloom, 1967) in comparison with the chick (see below).

A 'basic (spontaneous) waveform' lasting several hundred milliseconds also occurs in the sleeping mammalian cerebral cortex in situ, where it is associated with a burst of neuronal firing followed by a longer inhibitory 'pause' (Fig. 8)

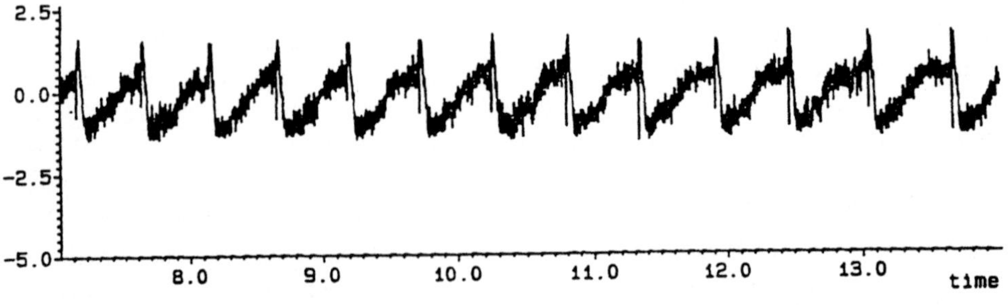

Fig. 5. Unusually regular 'EEG delta' wave-like field potentials recorded in a 2-week-old organotypic cerebral (neo)cortex explant (see Baker and Ruijter, 1991 for details).

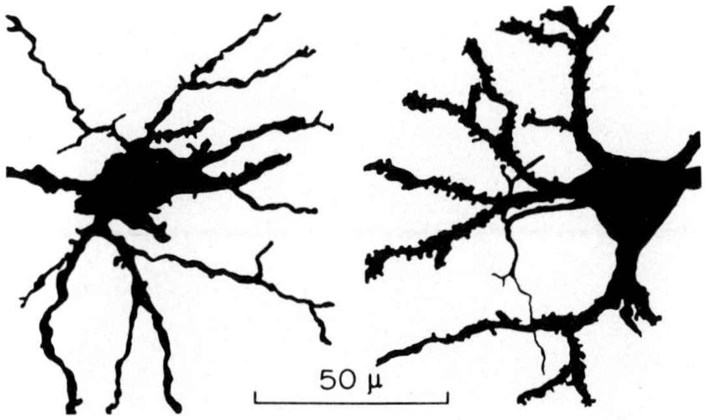

Fig. 6. Camera lucida drawing of representative neurons in the dorsal accessory hyperstriatum of the domestic chicken at embryonic stages 44 and 45. The former still shows long, but thin and varicose, dendrites such as are present already from stage 40 (but note the onset of spine formation close to the soma now), whereas the latter shows uniformly thickened dendrites with abundant spines (rapid Golgi preparation; see Corner and Romijn, 1983 for details).

TABLE IV

Maturation of cerebral glutamic acid levels and integrated electro-cephalic (EEG) waves in chicken embryos

Stage	[Glu]	[Delta]	[Theta]
41	3.1	12.9	5.1
42	3.3	32.1***	11.9***
43	3.7***	45.0**	19.5**
44	4.2*	63.0***	21.4*

Median values in μM/g and μV, respectively; *$P < 0.05$; **$P < 0.02$; ***$P < 0.005$ (Mann-Whitney U-test). Note that the abrupt rise in [Glu] during stage 43 is correlated with, specifically, the onset of slow-wave complexes (see Fig. 5 and text), not with the progressive increase in 'background' EEG activity from inception (early in stage 41) to mature amplitude (stage 44) (for details see Corner et al., 1974a).

(see Kostopoulos and Gotman, 1989). The resemblance to primitive brainstem and spinal cord spontaneous firing patterns discussed earlier is striking and, in all three cases, this basic activity pattern is beautifully preserved in neural tissues cultured in vitro (Corner and Crain, 1972; Crain, 1976). In accordance with Claude Bernard's well-known injunction to investigate the simplest

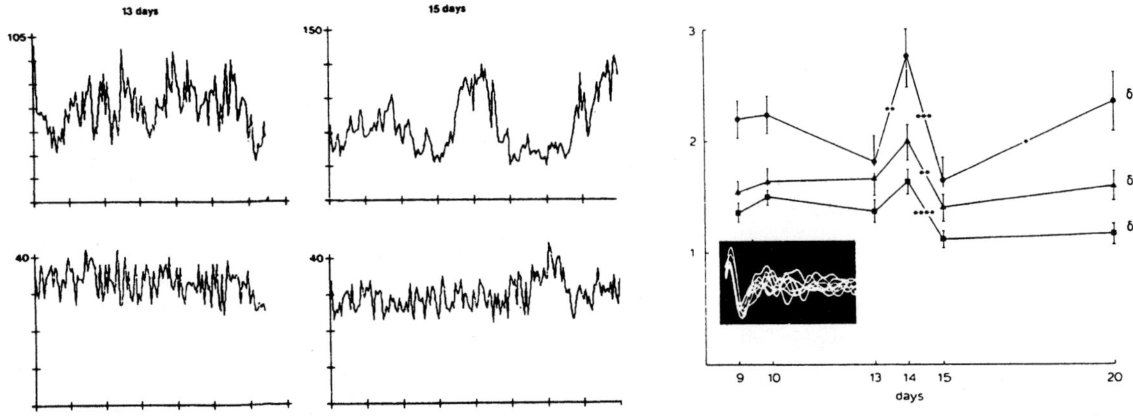

Fig. 7. Maturation of sleep-EEG patterns in the rat (abscissa: 3 min per division; the ordinate gives the mean integrated amplitude (in μV) for the delta (top) and beta frequency bands over consecutive 10-sec epochs), showing the onset of sleep-like neocortical slow waves around postnatal day 13 (thus analogous to the stage 43 chick embryo: see Fig. 4) followed on day 15 by more pronounced slow-wave suppression during episodes of 'paradoxical' sleep (see Fig. 10: stage 45 chick embryos). These events are manifested graphically on the right, the inset of which shows the 'basic waveform' as recorded inside the occipital cortex of an 11-day-old rat pup (superimposed 500-msec sweeps) about 15 min after an injection of sodium glutamate. Mean broad-band EEG ratios are plotted as a function of age, with asterisks indicating a significant difference between successive age groups; the final gradual increase in relative slow-wave activity is correlated with declining REM-sleep levels (see Corner et al., 1980, 1984 for details).

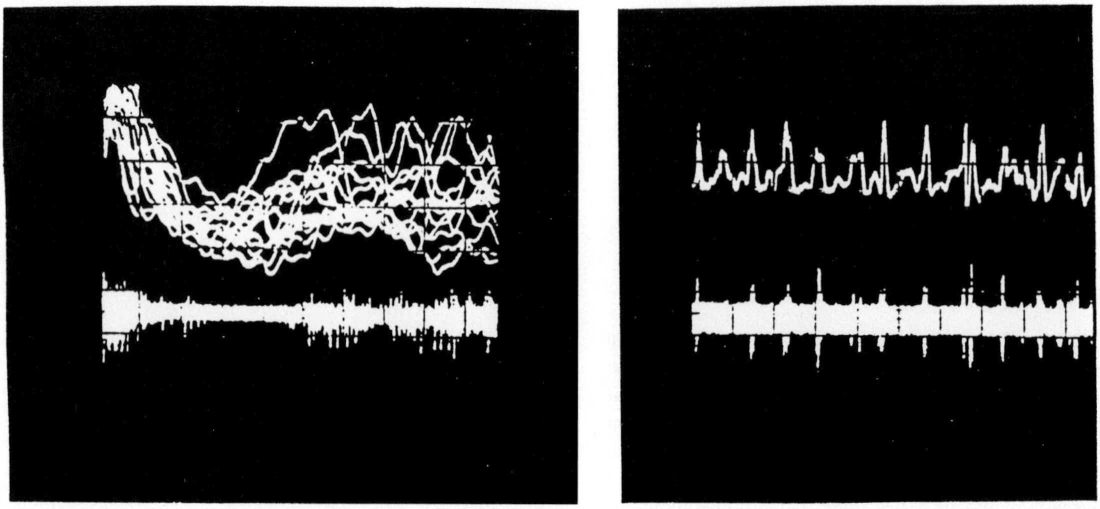

Fig. 8. Stereotyped EEG waveforms during slow-wave sleep in a 20-day-old rat showing: (left) ten superimposed 300-msec sweeps of waveform-triggered field (above) and action potentials, recorded with the same intracortical electrode and revealing a consistent cessation of neuronal firing during the local positive wave; (right) a single 5-sec sweep illustrating the association of spiking with the negative phase (upward deflection) of each field potential (see Mirmiran and Corner, 1982 for details).

preparation which still shows the phenomenon of interest (see Popper, 1965; Singer, 1959), this fortunate circumstance opens up possibilities for experimentally approaching the question of a possible developmental 'purpose' for naturally occurring spontaneous neuronal firing, a point which will be taken up extensively in the last section of my paper.

Maximum capacity versus effective activation: hypersynchronized and 'desynchronized' physiological states

Conversely, the explosion of dendritic spines and new synapses — now chock full of vesicles for the first time (Corner et al., 1977) — 1 or 2 days prior to hatching brings no further increase in either the frequency or amplitude of EEG potentials (Corner et al., 1974a,b). That is, unless we take as our starting point — since structural parameters are more a reflection of capacity than of momentary performance — not what the system manifests on its own initiative or in response to partial stimulation but, rather, what it is capable of doing when fully challenged. Therefore, since epileptiform potentials reflect the synchronous activation of all or most of the excitable elements within a defined 'focus' (see Steriade et al., 1990), the maximum amplitudes of paroxysmal EEG waveforms elicited by epileptogenic substances might be expected to be positively correlated with the extent of dendritic branching. This type of quantification indeed yields a developmental curve (Fig. 9) which, in contrast to the maturation of normal EEG activity (and despite some intriguing deviations) bears a striking similarity to the evolving structural picture in the same region (see Corner and Romijn, 1983). No comparable studies appear to have been made for mammalian brain development, although qualitative reports sometimes give the same impression of a positive correlation between maximum EEG waveform amplitudes and cyto-morphological development (see Aghajanian and Bloom, 1967; Staudacherova et al., 1978).

In addition to this expected large increase in the amplitude of paroxysmal potentials generated by the embryonic (domestic) chicken forebrain, the late maturation of dendritic spines and

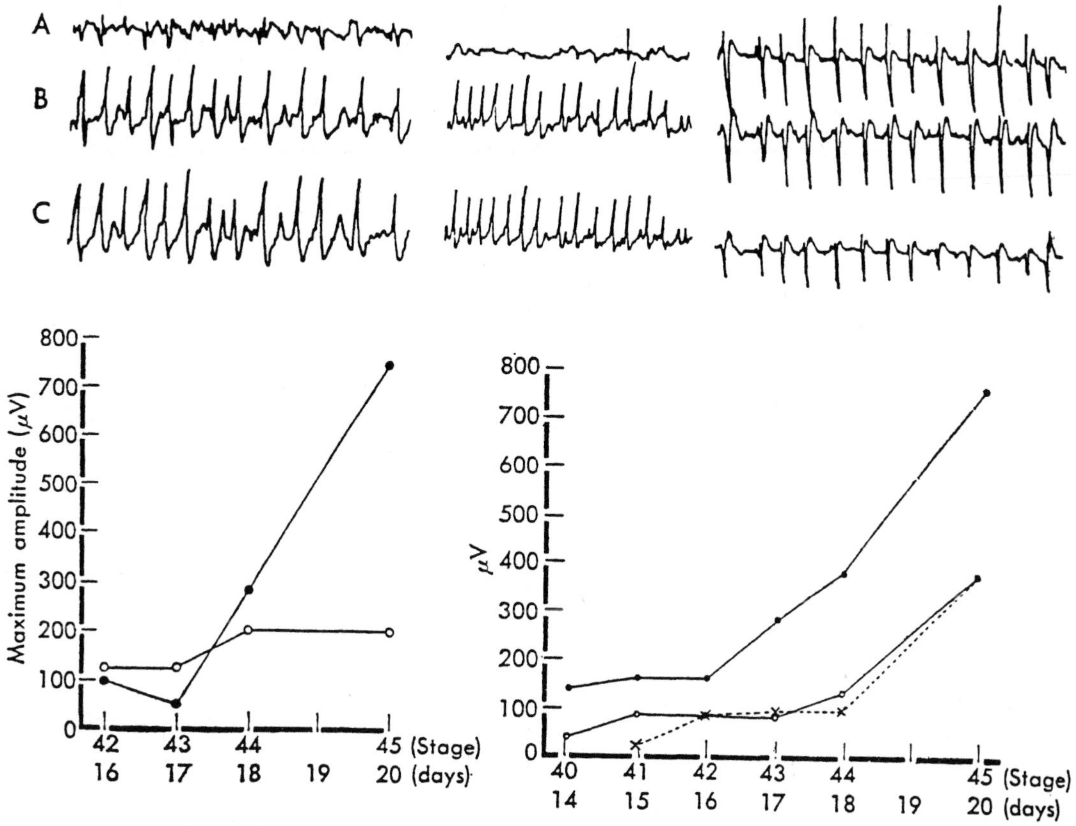

Fig. 9. Quantitative ontogeny of epileptiform waveforms in the embryonic chicken cerebral hemispheres. Illustrative discharges elicited by, respectively, *methionine sulfoximine* (MSO; left), *ouabain* (middle) and *metrazol* (right) are shown above: differential fronto-occipital surface EEGs (A) are compared with the monopolar records (B and C). The maximum amplitude attained during each type of seizure (grand-mean values per age group) is given on the left for surface negative (filled circles) and positive MSO-waves, and on the right for surface positive ouabain-waves (crosses) and for negative (filled circles) and positive metrazol-waves (for details see Bot and Corner, 1973).

synapses is paralleled in the non-epileptic cerebrum by the onset of EEG events of an entirely different kind. Thus, repetitive brief episodes of low-voltage electrical activity make their appearance, coming and going in a cycle of several minutes. In the space of a day, these epochs of presumably partially desynchronized neuronal firing (see Steriade et al., 1990, 1993) coalesce into prolonged periods of continuous activation, still alternating with comparable periods of uninterrupted slow-wave activity (Fig. 10). This most dramatic example of a classical 'REM' sleep-like EEG pattern ever recorded in any avian species has, by the time of hatching, already been supplanted once again by almost continuous high-voltage 'delta' waves, once again demonstrating the permissive, non-rate-limiting nature of structural constraints in the nervous system. Indeed, it is only several hours after hatching that an EEG 'arousal' response is observable in the chicken, despite clearcut behavioral wakefulness (Corner et al., 1972).

The most subtle putative structural determinant of physiological maturation described so far in the literature involves the 'overshoot' phenomenon recently described morphometrically for

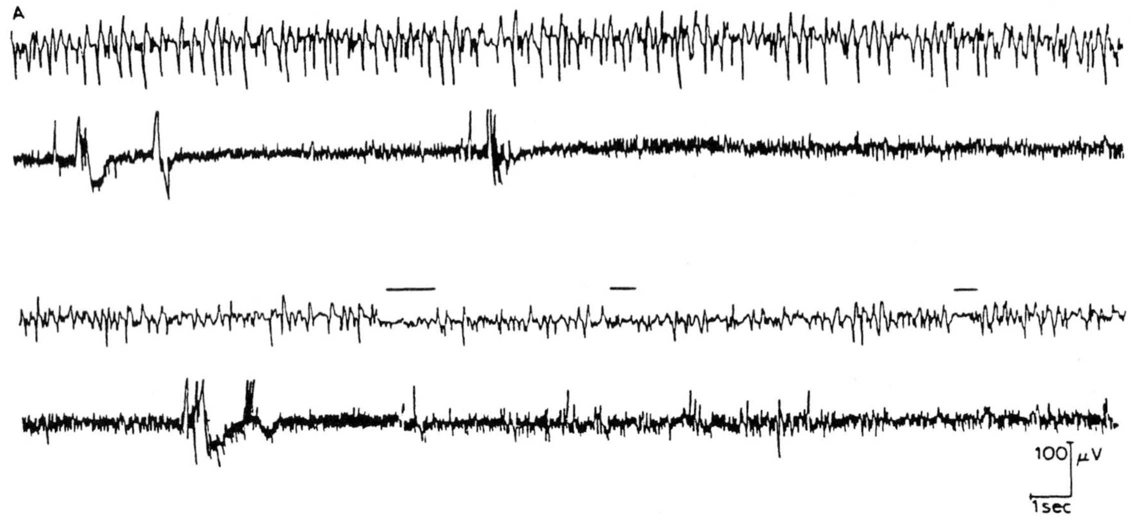

Fig. 10. Embryonic EEG cyclicity in the stage-45 domestic chicken. Continuous recording of cerebral electrical activity (upper traces) with concomitant body movements (lower traces) showing a transition from a high to a low amplitude phase of the cycle. Especially flattened stretches are indicated by horizontal bars (for details see Corner et al., 1972).

the rat prefrontal cortex (PFC) (Van Eden and Uylings, 1985). Since a quantitative analysis of neuronal spike trains recorded from the occipital cortex during slow-wave sleep (Corner and Mirmiran, 1990) had suggested a dramatic change in firing patterns during the fourth postnatal week, thus corresponding to the period of anatomical overshoot, the experiment was repeated for the PFC itself using more closely spaced measurements (Corner et al., 1992). From 14 to 24 days of age, single-unit firing rates and modal interspike intervals increase in parallel (Fig. 11), suggesting that inhibitory and excitatory connections are being elaborated simultaneously (see Corner and Ramakers, 1992) as the PFC increases in size. By postnatal day 28, however, not only has the PFC volume fallen to a considerably lower plateau level but, on average, modal interspike interval durations have shortened abruptly to a quarter of their peak value. This last development can be explained on the basis of a selective pruning of excitatory synapses (Corner and Ramakers, 1992), as has indeed been observed in cortical neuronal networks as they mature in vitro (Van Huizen et al., 1987; Van Ooyen and Van Pelt, 1994). The extremely long modal intervals which appear, along with higher mean firing rates, in the sleeping cortex during the third postnatal month (Fig. 11) (see also Corner and Mirmiran, 1990) are suggestive of a new wave of increasingly effective intracortical connections (presumably inhibitory as well as excitatory: Corner and Ramakers, 1992) — an unexpected development for which no particular mechanistic explanation can as yet be favored.

The 'take-home message' here seems to be that, apart from exceptional cases where clearcut non-trivial correlations are possible, a biomathematical approach will be needed both for pinpointing those structural features most likely to be critical for the specific behaviors observed in a given neural network, and for predicting the testable functional consequences of competing hypotheses (see Table I) about how a given system might be anatomically organized. The chick results make clear, first of all, that easily observable physiological events such as stereotyped and highly synchronized burst-pause firing during

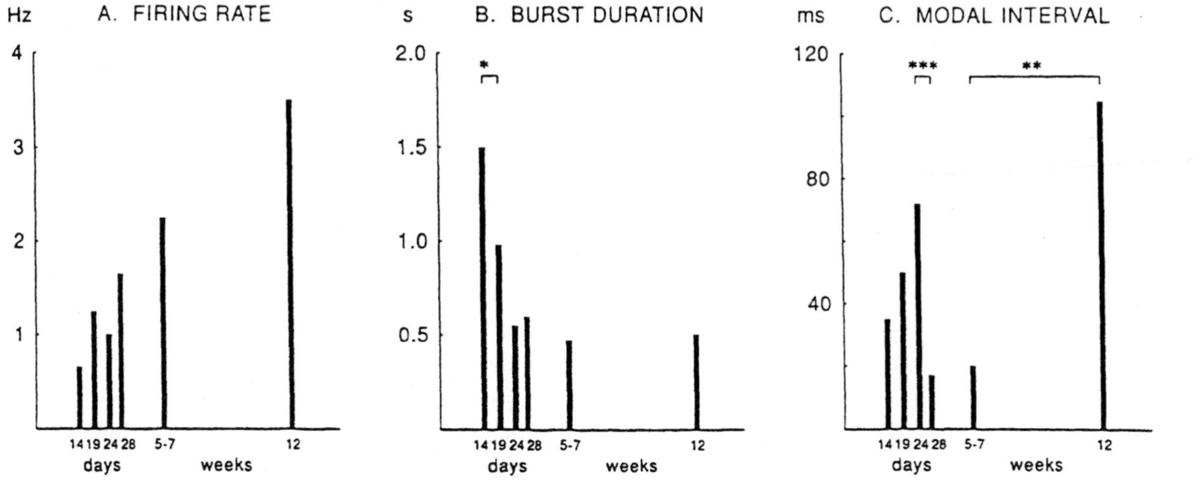

Fig. 11. Maturation of spike-train parameters in the prefrontal cortex of postnatal rats during urethane induced slow-wave sleep (Armstrong-Jones and Fox, 1984). Grand-mean values in successive age groups are given for the three parameters most clearly reflecting distinctive developmental time courses. (A) *Mean firing rates* (spikes per second) calculated per neuron; $P < 0.01$ for the overall age effect. (B) *Mean burst duration* (seconds) per neuron, defined as a run of shorter than average interspike intervals; $P < 0.02$ for the overall age effect. (C) *Modal interval* (milliseconds) derived from individual interval histograms; $P < 0.01$ for the overall age effect. $^*P < 0.05$, $^{**}P < 0.01$, $^{***}P < 0.005$ for differences between adjacent age groups (see Corner and Mirmiran, 1990; Corner et al., 1992 for full details).

slow-wave sleep require only a minimal structural basis. Conversely, cyto-morphological complexity (besides making possible the complicated 'calculations' normally performed on incoming signals by an alert brain) may be an essential condition for global state changes, such as occur within sleep (Jouvet 1967; Steriade et al., 1990, 1993) as well as wakefulness, the functional implications of which may be unpredictable without mathematical simulations as a guideline.

The 'epigenetic' character of neuronal self-organization

On the formal aspects of causal interactions during ontogeny

Once the fallacy had taken root that the 'genome' (i.e. the totality of DNA molecules) is the major repository for developmental information, i.e. that there exists a *genetic* 'program' or 'blueprint' (e.g. Black, 1991) theoretically capable of creating something analogous to a 'Jurassic Park' (see Paabo, 1993) even when operating within alien cytoplasm, several serious misinterpretations of classical experimental embryological terminology (see Weiss, 1969; Moore, 1987) also made their appearance in the literature. Thus, neglecting the essential difference between mechanical and organic design (i.e. construction from parts according to a blueprint vs. self-organization from an undifferentiated whole) concern began to be expressed that no organism's genetic information would suffice to specify, for instance, all the synaptic connections within its nervous system (e.g. Changeux and Mikoshiba, 1978; Edelman, 1987). The concept of 'epigenesis' then became subjected to a radical transformation from its original meaning so as to denote extra information from the (micro- or macro-)environment which was needed to alleviate this purported shortage. The traditional confusion between 'nature versus nurture' as, supposedly, the dual *additive* determinants of adult phenotypes thus contaminated the new discipline of 'molecular biology' from the outset. This unbiological approach overlooks, of course, the basic fact that far less information is needed to specify a process of 'becoming' (e.g. Monod, 1971; Prigogine and Senghers, 1984; Oyama, 1985; Edelman, 1987; Van Ooyen and Van Pelt, 1994) than to describe the resulting state of 'being'!

Linear summation of 'hereditary' and environmental determinants of development is in reality a formula appropriate only for the analysis of *individual differences* in phenotypic expression (see Van der Weele, 1993): $\Delta P_t = \Delta E_0..._t + \Delta H_0$, where the Δs signify intra-group variability or, more precisely, measurable differences among the individual phenotypes (e.g. Corner et al., 1966). In other words, part of the variance within a given population at a given time (t) will be due to the differential cumulative experience of its members, from an arbitrary starting point down to the present (0...t), while the rest will be due to protoplasmic differences among zygotes at, say, the time of fertilization ($t = 0$). (Note, too, that I prefer 'H', for *heredity*, above the more popular 'G' for *genome*, since it is by no means clear that *cytoplasmic* variations among egg cells can afford to be neglected in such a factor analysis: e.g. Gurdon, 1974; Chandebois, 1983; Schwabl, 1993).

For understanding how, throughout the life-cycle, interactions between an organism and its environment (or between a given cell and the 'milieu interior' — including adjacent cells — in which it is immersed) cooperate to direct developmental changes, a radically different equation is required: $P_t = H_0(E_0..._t)$. That is, an organismic or cellular phenotype (P) at a given point in time (t) can be completely described as a function of its cumulative 'experience' (E...), permissive as well as instructive, while its 'heredity' (H_0) — the sum total of maturational processes culminating in the fertilized egg (i.e. '$t = 0$') — *defines the mathematical nature* of that ongoing function.

Environmental factors thus play a decisive role in *all* ontogenetic processes (Corner and Schade, 1967) but it is the organism itself that, as an integrated system, dictates the nature of each and every developmental responses. More dynamically, $dP/dt = P_t(E_t)$ where E_t represents the totality of environmental factors impinging on the system at time t, while P_t (the equivalent of which in 'hereditary' notation would be $H_0..._t$, i.e. the 'heritage' of the organism in terms of its sequential life history) represents the instantaneous phenotypic state at time t. In other words, it is a series of contingent developmental *responses* which may be said to be 'programmed' within the egg cell (Chandebois, 1983), rather than phenotypic *endpoints* of any sort (Oyama, 1985): the living organism or cell self-organizes on the basis of its own internal structuring, in continuous interaction with the environment in which it finds itself (see Corner and Schade, 1967).

Rethinking 'genetic determination' — a modern guide for the perplexed in these confused times

Unrealistic demands upon the genetic *code* (no more, no less! — see, e.g., Boyd and Noble, 1993) need therefore to be abandoned in favor of the realization that any evolutionarily 'successful' organism possesses at least enough DNA to enable the synthesis of all the structural and enzymatic proteins which go into the composition of its myriad cell types — plus the proteins required for selectively activating different sets of genes inside each 'determined' cell (e.g. Wolpert 1991; Portin, 1993). No additional information of an equivalent nature is ever needed or supplied — except in the case of a viral infection or, perhaps, very early in the evolution of living systems (see Monod, 1971). Enough is known about the genetic machinery by now (see above) to appreciate, furthermore, that this is virtually the only kind of information which polynucleotide molecules are inherently capable of containing (e.g. Black, 1991; Portin, 1993): nothing there at all about *which* proteins will be expressed in which cells, at what time and in what quantities. These latter instructions can only come from the cytoplasm of the individual cells themselves, as they 'place their orders' (each from an identical little DNA workshop: Gurdon, 1974) for the mRNA needed, under a specified set of environmental contingencies, for the completion and maintainence of their distinctive paths of chemical differentiation (e.g. Chandebois, 1983). Using military terminology, the genome cannot (despite widespread partial misunderstanding of this point: see Changeux and Mikoshiba, 1978; Edelman, 1987; Black, 1991) by any stretch of the imagination be identified with the general staff, but rather with the quartermaster corps.

Using a musical analogy, the cellular 'orchestra' can be said to provide but a single type of instrument for all its members, all of whom play from different 'scores'. As in any well-prepared jam-session, the instructions possess enough degrees of freedom to allow for choices to be made along the way — depending on audience response, hall acoustics, outside traffic noise, and other external ('milieu') influences. The cloned intracellular 'conductors' (genome) in this metaphor are located off-stage and have no knowledge whatsoever about the actual performance, only a detailed set of instructions dictating which messengers (mRNA) should — upon arrival inside the nucleus of appropriate 'promotor' signals — be sent 'onstage' (i.e. into the cytoplasm) with instructions for particular proteins to start, or continue, to play for awhile (for details see Portin, 1993). The degree to which onstage 'players' do actually participate in the performance (e.g. RNA transduction, enzyme activation, second messenger systems) and what the nature of their contribution will be is determined solely by their own properties, as these unfold in the course of interactions with other molecular players.

What we are dealing with in a living organism, then, is a structured improvisation involving myriad contingent choices, rather than the predetermined unfolding of a blueprint or 'recipe' (for developmental details, see Weiss, 1969; Edelman, 1987). This cellular 'happening', finally takes place in partial view (only some of the action is 'up front') of an 'audience' — i.e. the external milieu — the presence of which, including adjacent other cells, also exerts an important influence (in most cases of a 'permissive' rather than an 'instructive' nature; see Oyama 1985; Moore, 1987). In digital computer parlance, the organized substance of the egg cell itself contains the complete 'software package' needed for *programming* (NOT 'predetermining'!) its own life-cycle. The genome can best be regarded for the most part as a portion of the 'hardware', albeit one that must be well-matched with the rest of the cellular machinery if the whole is to work (e.g. Oyama, 1985; Portin, 1993). The countless 'epigenetic' factors by which the environment (of each cell as well as of the whole organism; see Corner and Schade, 1967) affects the course of development are not making up for any incompleteness in the programming but, rather, providing answers to the seqential 'questions' generated once any such program is up and running (e.g. Chandebois, 1983; Wolpert, 1991).

Thus, the term 'epigenesis' (also see Edelman, 1987) does denote the addition of a kind of ontogenetic information, after all, but of a qualitatively very different sort than that which is embodied in the protoplasm. No amount of the former can ever make up for a deficiency in the latter; on the contrary, the more extensive the 'genetically determined' potentialities available to a cellular program, the *more* will be the (epigenetic) information about environmental contingencies which needs to be fed into it. Hopefully, the time has come for us to exorcise those pre-scientific metaphysical assumptions which have stood in the way of a clear formulation of 'the nature-nurture question' (see Corner and Schade, 1967; Oyama, 1985, and the above discussion): especially (i) what we may call the *Hebraic* fallacy, i.e. the conception of living matter in general (and the cytoplasm in particular) as a passive, clay-like, substance which needs to be 'organized' and/or 'animated' from the outside — if not by a *deus ex machina*, then by the natural environment — along with (ii) the *Grecian* fallacy, i.e. viewing the female life principle (the ovum or, in up-to-date terms, the cytoplasm in general) as being merely nutritive and supportive (see Singer, 1959; Moore, 1987), with 'order' and 'intelligence' thus being the exclusive province of the male principle, i.e. sperm or (in its modern generalization) the genome-containing nucleus.

It is only the phenomenal *self*-organizing capacity of especially the protein constituents of living cells (see Monod, 1971; Boyd and Noble, 1993) that makes possible the ontogenetic emergence of a viable organic structure out of an indescribable 'chaos' of highly complicated sequential chemical reactions (see Prigogine and Senghers, 1984; Edelman, 1987). These reaction chains are often potentially oscillatory, owing to feedback inhibition exerted by intermediate enzyme products which, in turn, may be common to more than one metabolic process (Goodwin, 1963; Monod, 1971). In effect, every living cell therefore constitutes a highly diversified network of coupled molecular oscillators, one which has been refined over the eons into an integrated system that works in an adaptive manner within its natural limits of tolerance. Such a labyrinth of intracellular regulatory mechanisms (see Oyama, 1985; Black, 1991) undoubtedly demands a non-linear systems theoretical approach to even begin to do justice to its complexity. When, added to this, all the extra complications are considered of having such biochemical networks outfitted with mechanisms for responding selectively to all manner of impinging stimuli, the limitations of straight-forward deductive logical methods (e.g. Table I) become painfully obvious. In consequence, mathematical studies involving the simulation of non-linear interactions may eventually come to play as essential a role in a 'general theory of biochemical (super) systems' as they are now beginning to do for *neuronal* network theory (e.g. this volume).

Spontaneous bioelectric activity and self-organization in developing neural networks

Having in the preceding section presented a richly metaphoric framework for understanding the concept of 'epigenesis' — which, as a card-carrying experimental embryologist (Corner, 1963, 1964), I gladly donate to the neuroscience community — I would like to devote the rest of my presentation to a discussion of the role of *spontaneous* bioelectric activity (SBA) and of, the other side of the same coin, excitatory and inhibitory synaptic transmission in neural network formation. This is, it goes without saying, a perfectly natural preoccupation for a developmental biologist interested in the mechanisms and significance of neuronal rhythmicity (see Corner, 1987, 1990; Corner et al., 1980) and it nicely complements the other 'plasticity' studies presented in this volume, where the emphasis is mainly upon afferent sensory stimulation. Indeed, one of SBA's most interesting putative ontogenetic effects is to 'gate' or modulate the potency of sensory mediated experience for inducing lasting changes in the functional organization of the brain during sensitive periods of development (Singer, 1985; Mirmiran and Uylings, 1983; Mirmiran et al., 1983). Chronically stimulated sensory ganglion cells, for example, make more and stronger monosynaptic contacts with developing spinal cord target neurons in vitro than do unstimulated cells growing in from the contralateral ganglion, an effect which appears to be dependent upon the network's being spontaneously active during the 'training' period (Fields and Nelson, 1991).

A probably closely related phenomenon involves the visual system in pre- and neonatal cats, where each retina acts as an independent source of synchronized activity that is transmitted to the developing central visual structures (Meister et al., 1991; Wong et al., 1993). Experimental suppression of this coherent spontaneous activity prevents the appearance of the ocular preference patterns in forebrain target neurons that would otherwise be present even in dark-reared kittens (Stryker and Harris, 1986). Such 'priming' of the functional organization of visual projections could be a permissive factor in the later maturation of ocular dominance columns under the influence of patterned visual stimulation.

Since this last process depends upon functioning NMDA receptors during the period of visual experience (Shatz, 1990), *target* as well as source networks may need to be spontaneously active in order for plasticity to occur. In any event, a critical level of background excitation needs to be supplied for a visual stimulus to be able to permanently modify brain response patterns (Singer, 1985). Since brainstem systems crucial for arousal

TABLE V

Calculated cell number (from DNA measurements) in spinal cord explants cultured under different conditions

Group	$n =$	Cell number ($\times 10^{-3}$)
Basal medium	4	109.9 (90.0 — 241.8)
plus:		
Galactose	3	106% (99 — 133)
Gangliosides	2	97% (86 — 108)
GM1-gangl.*	2	102% (85 — 118)
(plus tetrodotoxin):		
Basal medium	3	41% (38 — 57)
Galactose	2	64% (52 — 76)
Gangliosides	2	66% (46 — 85)
GM1-gangl.	2	61% (57 — 66)

$^*P < 0.01$ (Mann-Whitney U-test) for the difference between the pooled TTX-treated versus the untreated cultures ($n = 9$). The TTX groups are presented as percentages of the corresponding control values in a given series of measurements (median and range). The untreated chemical additive groups are given as a percentage of the concomitantly cultured basal-medium group (see Baker et al., 1992 for details).

and attention appear to be implicated in this 'gating' mechanism, endogenously active neurons regulating circadian and other biorhythmical state changes might be an indispensible feature of any reasonably complete model purporting to account for lasting 'imprinting' of sensory experiences into the brain, from early network formation ('primary process': Edelman, 1987) up to and including 'memory' traces (see Rolls and Treves, 1994).

The pioneer of developmental brain/behavior studies, G.E. Coghill, early on subjected his wonder about the possible significance of early movements (see Gottlieb, 1973) to an experimental test: amphibian larvae were reared in the presence of an anesthetic which suppressed all visible motility, evoked as well as spontaneous. His conclusions, repeatedly confirmed since then (see Oppenheim, 1982), were that such deprivation has little or no effect upon tadpole behavioral maturation. Modern studies employing chicken embryos, however, have revealed a scala of muscular and skeletal defects, including the sensational discovery that naturally occurring motoneuron cell death can be prevented by chronic neuromuscular blockade in ovo (e.g. Oppenheim, 1982). Also the 'pruning' of initially excessive synapses (Changeux and Mikoshiba, 1978) and, putatively, the definitive choice of transmitter to be synthesized by autonomic neuroblasts (Black and Patterson, 1980) and of whether early differentiating cerebellar granule cells will live or die (Balazs et al., 1992) all appear to be subject to regulation by spontaneous electrical activity originating in spinal interneurons (see Hamburger, 1973).

This question was pursued more rigorously by Crain and collaborators, using a tissue culture 'model system' to carry out a series of multi-disciplinary studies (for review see Crain, 1976). Despite a total suppression of spike firing throughout development in vitro, no obvious structural or functional abnormalities could be detected in such 'virginal' spinal cord networks, attesting to the considerable degree of self-organization possible even in the absence of any neurophysiological expression. Quantitative studies, however, later

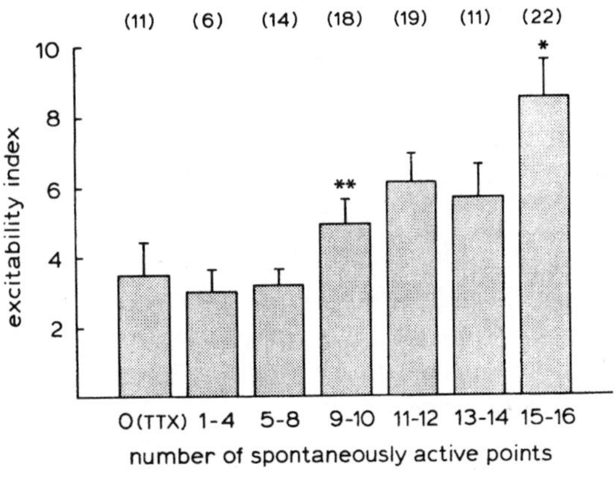

Fig. 12. Mean excitability scores (based on sensory ganglion evoked polysynaptic spike discharges: see Corner and Baker, 1987 for details) as a function of the incidence of spontaneous bioelectric activity in cultured spinal cord explants. $^*P < 0.05$ vis-a-vis cultures ($n = 30$) with 11–14 active sites; $^{**}P < 0.01$ vis-a-vis cultures ($n = 20$) with 1–8 active sites (Mann-Whitney U-test). TTX data are taken from Baker et al., 1986.

TABLE VI

Regional innervation preferences of sensory ganglion fibers projecting to 'organotypic' spinal cord explants (cultured for 4–6 weeks in vitro) having different spontaneous discharge levels

Group	$n =$	Dorsal	D/V	Ventral
TTX**	72	0.31	0.44	0.25
Silent*	42	0.48	0.26	0.26
Weakly***	18	0.50	0.39	0.11
Strongly	25	0.88	0.04	0.08

Indicated are the percentages of tetrodotoxin (TTX)-treated and control cultures displaying, respectively, a predominantly dorsal, indeterminate (D/V) or ventral fiber *entrance* (only the two spontaneously active groups deviated significantly from a 1:1 distribution). *$P < 0.05$ for active vs. silent control cultures (pooled); **$P < 0.025$ for TTX-treated cultures vs. the pooled controls; ***$P < 0.005$ for strongly vs. weakly active cultures: 'strong' means that spontaneous firing was noted at more than half of the 16 recording points monitored in each spinal cord explant. (Cultures younger than 4 weeks in vitro did not show preferential projection routes for sensory ganglion fibers into the cord no matter how active the latter was; see Corner et al., 1987a for details).

revealed a strong and selective reduction in the number of surviving neurons (Fields and Nelson, 1991) (Table V), in the physiological effectiveness of monosynaptic sensory projections (Fig. 12) (see also Fields and Nelson, 1991) and in the regional specificity for target selection within the cord (Fig. 13; Corner et al., 1987b; and Table VI). This last finding is of special interest here since it implicates *intrinsic* network activity as an epigenetic factor equivalent in its effects to *sensory* mediated stimulation, such as has been abundantly demonstrated for especially the developing (vertebrate) visual system (e.g. this volume).

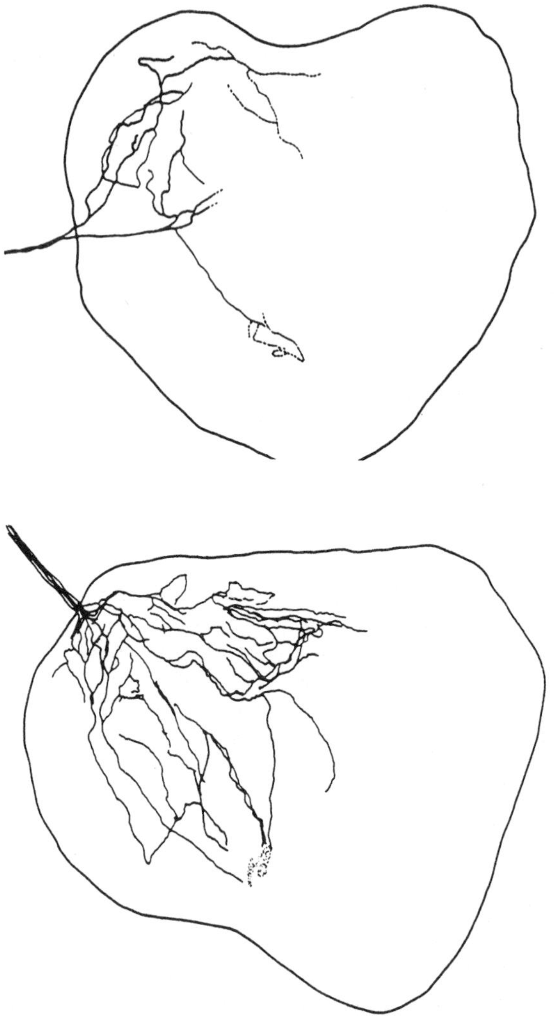

Fig. 13. Camera lucida drawings of sensory afferent projections (horseradish peroxidase-filled dorsal root ganglia) into cultured spinal cord explants. A representative control (top) is compared with a chronically tetrodotoxin treated specimen (see Baker et al., 1982, 1984 for quantification and further details).

The cerebral cortex as a 'model' system for studying spontaneous activity as an epigenetic factor

Again making use of the accessibility and control afforded by an in vitro culture approach, some of the same conclusions have recently been reached with respect to the cerebral (neo)cortex where, too, qualitative morpho-physiological studies in explanted tissue slices had earlier failed to reveal any effect of functional blockade throughout early maturation (see Crain, 1976). With quantification, in contrast, an approximately about 50% reduction could be demonstrated in the number of surviving neurons when spontaneous electrical interactions were totally suppressed either by tetrodotoxin (TTX) or by an appropriately elevated (12.5 mM) magnesium concentration (Baker and Ruijter, 1991; Baker et al., 1991b; Ruijter et al., 1991). The entire loss occurred cumulatively over the first 2 weeks in culture and was followed by compensatory neuropil formation, presumably by the remaining neurons, as had been observed in TTX-treated cortical cell cultures as well (Van Huizen et al., 1987).

Of considerable interest, furthermore, is the opposite effect which was observed when spontaneous firing (but not the *ability* to generate action potentials) was largely eliminated by means of high (25 mM) potassium-induced membrane depolarization: despite their electrical inactivity, no cortical neurons appeared to be lost under these conditions — even in combination with TTX treatment (also see Ling et al., 1991) — but they remained in a highly immature state, functionally as well as structurally (Fig. 14) (Baker et al., 1991a; Ruijter and Baker, 1990). It would appear that the membrane potentials of neuroblasts may need to reach a critical level (Messenger and Warner, 1979) in order for phenotypic differentiation (including a phase of dependency upon adequate functional stimulation in order to survive: Balazs et al., 1992) to be initiated.

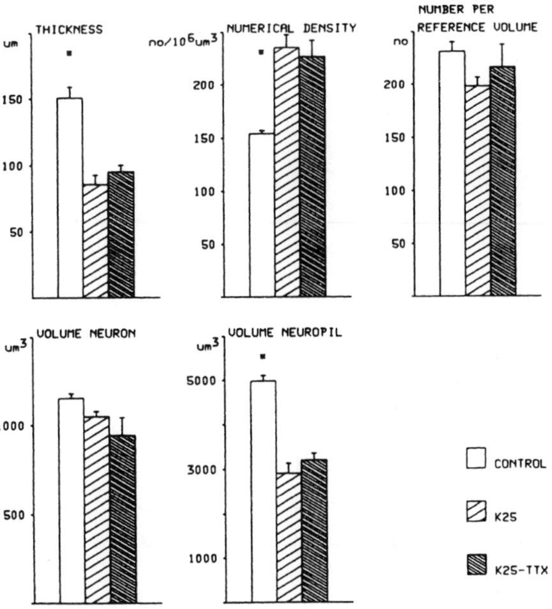

Fig. 14. Morphometric analysis of 2-week-old neocortical explants grown in normal ('control') or elevated ('K25') potassium concentrations, with or without added tetrodotoxin (TTX). The number of neurons per 'reference volume' was obtained by multiplying thickness and numerical density for each culture; mean neuronal volume and neuropil volume per neuron were calculated per culture from the overall volumetric and numerical density measurements (see Ruijter et al., 1991 for details).

Dissociated cortical cells taken from perinatal rat pups are capable of developing into a functional network which approximates many of the corresponding in vivo features, including the generation of spatio-temporally patterned neuronal discharges (Fig. 15). Whereas bioelectrically active networks show an 'overshoot' in the production of synapses — a sigmoidal growth curve followed by a rapid decline to a plateau level of about half the maximal value — chronically blocked cultures show an indefinite persistence of the peak numerical density attained during the growth phase, even long after normal activity has returned (Van Huizen et al., 1987; Van Ooyen and Van Pelt, 1994). A number of physiological abnormalities have also been documented following such treatment; for example, stereotyped bursts of synchronized firing — reminiscent of 'interictal' epileptic discharges and primitive firing patterns (Habets et al., 1987; also see Corner and Ramakers, 1992) — are routinely observed at developmental stages when control preparations have progressed to spatio-temporally less stereotyped, more 'chaotic' appearing spontaneous activity patterns (Fig. 15) (see also Ramakers et al., 1993).

A two-fold increase in the extent and the frequency of network bursting activity as a consequence of chronic functional blockade has been observed in organotypic *neocortical* explant cultures as well (Figs. 16, 17). The temporal ordering of neuronal action potential sequences, too, deviates from the normal pattern in developmentally silenced networks: histograms reveal a considerable shortening of the modal intervals, together with a large increase in the incidence of complex,

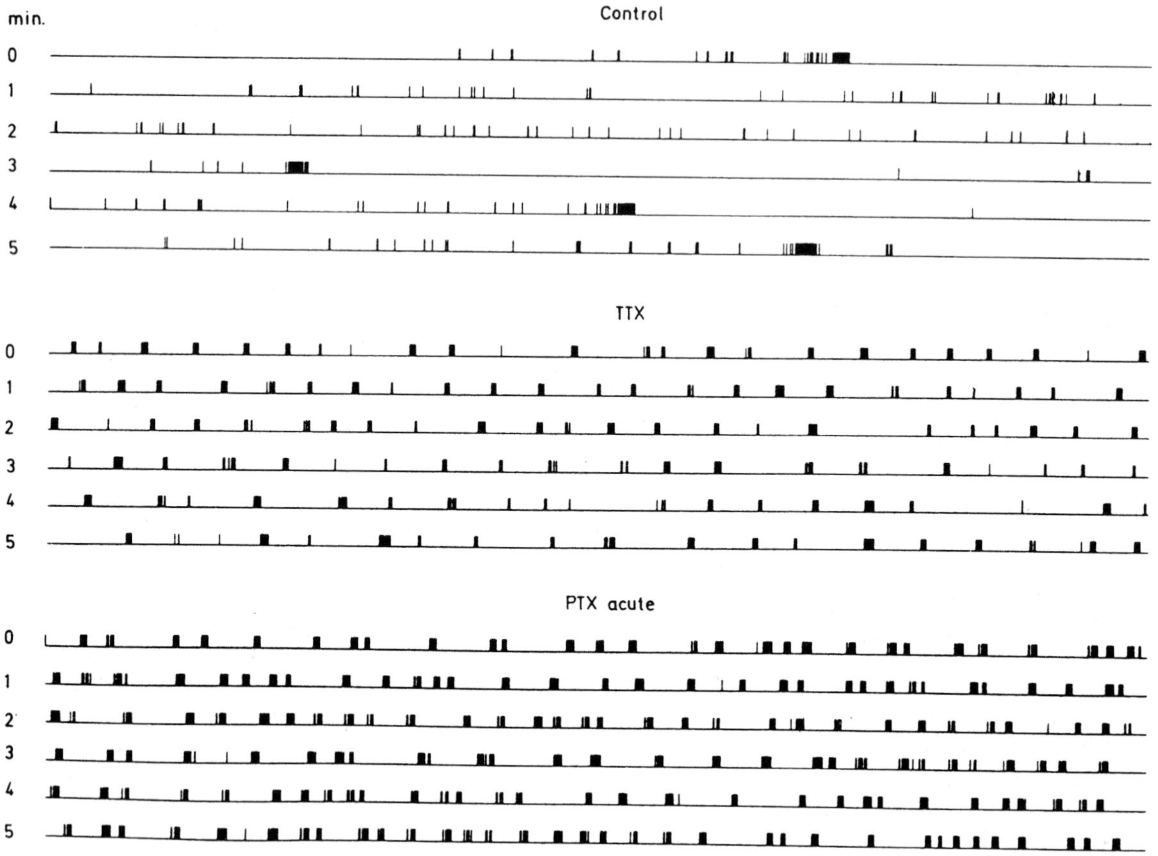

Fig. 15. Representative firing patterns after 3 weeks in vitro in untreated ('control') neocortical cell cultures, showing 'variable/non-burst' firing in normal medium (above) and 'non-variable/burst' firing in the presence of picrotoxin (below: 'PTX acute'), in comparison with chronically tetrodotoxin (TTX) treated cultures when recorded in control medium ('non-variable bursting': see Habets et al., 1987; Ramakers et al., 1990 for details of the quantitative spike-train analysis procedure).

i.e. bi- or tri-modal, interspike interval distributions P Nuijtinck and M.A. Corner, unpublished observations. This difference is strikingly reminiscent of the abrupt change in firing patterns observed when dissociated control (i.e. spontaneously active) cortical cell cultures are acutely disinhibited using picrotoxin (Ramakers et al., 1991; Corner and Ramakers, 1992). The implication here is that a relative deficiency in GABAergic synaptic drive could underlie abnormal neuronal discharges resulting from early interference with spontaneous bioelectric activity (see below for further discussion of this intriguing possibility).

Inhibitory maturation as a target for neurophysiological activation: a homeostatic mechanism?

The failure of excess axo-dendritic spine (thus presumably excitatory) synapses to be trimmed back to control levels is probably not the only, or even main, reason for the hyperactivity seen following chronic functional blockade in developing neocortical networks. Long before synapse density counts have attained their maximum value, significant differences in network physiology have appeared as a result of the persistence in chronically deprived networks (when placed in normal growth medium for monitoring purposes) of

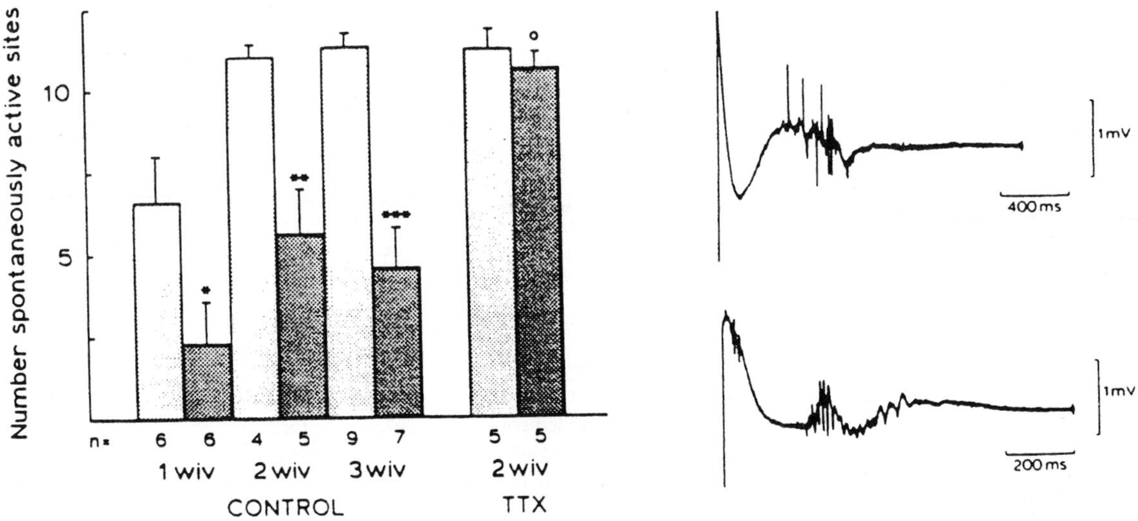

Fig. 16. Bar graphs indicating the mean number of sites (out of 12 per explant) from which spontaneous bioelectric activity — see inset on the right — could be observed in 'organotypic' rat cortical cultures; the light bars represent auditory cortex, the heavy bars visual cortex. Asterisks indicate a significant difference between the two regions, while the open circle refers to the difference ($P < 0.025$) between the corresponding TTX and control groups (see Baker and Ruijter, 1991 for details).

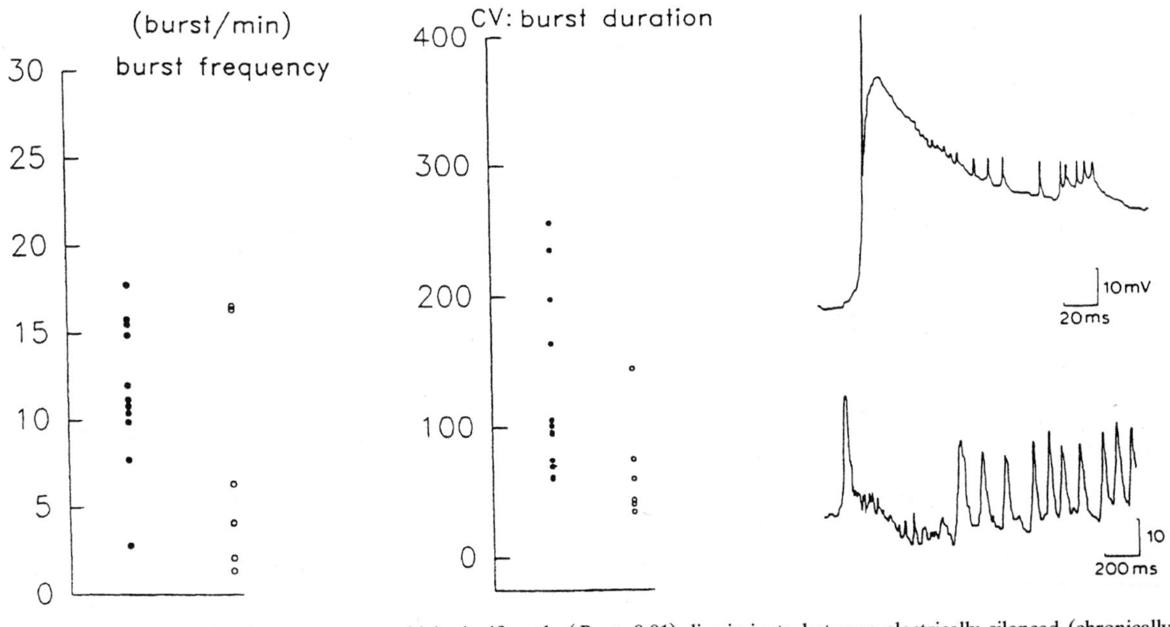

Fig. 17. Scattergrams showing parameters which significantly ($P < 0.01$) discriminate between electrically silenced (chronically high-magnesium treated: filled circles) and spontaneously active control (open circles) organotypic neocortex cultures. The *incidence* of polyneuronal 'bursts', defined on the basis of a 500-msec maximal interspike interval criterion, is twice as high in the experimental as in the control group ('burst frequency'). With 50 msec as the criterion, the coefficient of variation (C.V.) for burst *durations* is lower in control than in treated cultures. Inset on the right shows sample intracellular records illustrating the 'paroxysmal depolarizing shifts' (top) and repetitive afterdischarges which are relatively often seen in chronic high-Mg^{2+} treated cortex neurons (see Baker et al., 1991b for details).

TABLE VII

Ontogeny of amino acid levels in functionally silenced dissociated neonatal rat cerebral cortex cultures (in proportion to corresponding control values)

Weeks in vitro	1	2	3	4	5	(5:1)
Serine	0.90	0.75	0.65	0.50	0.40	(0.9)
Glutamate	1.00	0.50	0.40	0.35	0.30	(0.7)
Aspartate	0.95	0.35	0.35	0.40	0.25	(0.7)
GABA	1.00	0.20	0.15	0.05	0.05	(0.4)

Whereas serine levels faithfully reflect the reduction in calculated neuronal cell number in chronically silent (tetrodotoxin treated) vs. untreated cultures (see Ramakers et al., 1994), there is a trend in the excitatory amino acids *glutamate* and *aspartate* towards a disproportionately large reduction, with respect both to age-matched control cultures and to their own peak values (always at 1 week in vitro: see column '5:1'). These reductions are quite small, however, in comparison with the decline in *GABA* levels; all of the significant differences between TTX and untreated cultures (i.e. from 2 weeks on) reach the 1% confidence level: Student '*t*'-test, $n = 5$; (see Ramakers et al., 1994 for details).

primitive, often quite stereotyped, episodic spontaneous firing patterns (Ramakers et al., 1990). The chief underlying mechanism appears to be a failure of inhibitory synaptic drive to mature fully in the absence of bioelectric activity: preliminary pharmacological evidence reveals a much greater reduction in both the level (Table VII) and the release (Fig. 18) of GABA than of excitatory amino acids in chronically tetrodotoxin treated cell cultures (Ramakers et al., 1994).

Furthermore, the acute experimental reduction of GABAergic inhibition by means of picrotoxin (PTX) in mature control cultures precipitates a dramatic return to 'epileptiform' bursting (Figs. 15, 19), such as is characteristic for very young cortical networks (in which inhibitory mechanisms are disproportionately weak: e.g. Habets et al., 1987; Luhmann and Prince, 1990, 1991) (see also Fig. 18). The same treatment carried out in cortex cultures which had first been continuously *disinhibited* throughout development (chronic PTX treatment) has a significantly weaker epileptiform effect than in untreated control networks (Ramakers et al., 1991; Corner and Ramakers, 1992). Since, in addition, chronically PTX-treated cultures showed even less of a tendency for spontaneous spike discharges to cluster into 'phasic bursts' than did the control cultures, GABA blockade during development appears to be provoking a compensatory strengthening of synaptic inhibition relative to excitation (compare Fig. 20 with Fig. 21). Such a selective coupling of inhibitory neuron maturation to the level of ongoing excitatory activity (Corner and Ramakers, 1992; see also Seil and Drake-Baumann, 1994; Seil et al., 1994) implies the existence of homeostatic mechanisms for keeping the working balance between excitation and inhibition in the network within biologically tolerable limits (see Lopes da Silva et al., 1994).

Since mean single-unit firing levels in cultured neocortex neurons were not noticeably altered during chronic PTX treatment, despite the ongoing epileptiform discharges triggered by such disinhibition (see Figs. 15, 19), 'phasic' *burst* firing — whether by virtue of the spatial (widespread synchronization) or the temporal (stereotypy and regularity) aspects of such activity — is implicated as a critical factor in activity-dependent development. This is not completely unexpected since, owing to the voltage dependency of certain ion channels (see McCormick et al., 1993), calcium entrance into the cell — the putative intervening variable in most known forms of neuronal plasticity (e.g. Kater et al., 1994) — will be greater for a short but intense spike barrage than when the same number of action potentials is more evenly spread out over time (Spitzer, 1994). In a similar vein, the relative roles of NMDA receptor activation and of bioelectric interactions per se in mediating neuroplasticity has become a meaningful question to pose (Constantine-Paton et al., 1990; Hofer and Constantine-Paton, 1994): although the two variables overlap slightly (NMDA

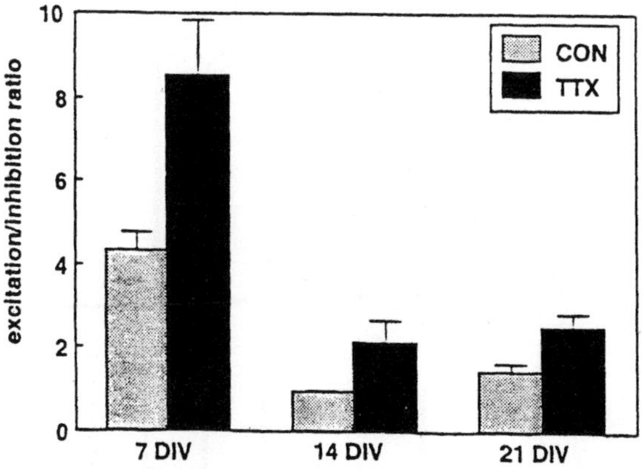

Fig. 18. 'Excitation/inhibition ratio' as a function of *age* and bioelectric *activity* in neocortical cell cultures. The E:I ratio was obtained by dividing the release (i.e. K^+-provoked minus basal levels) of the excitatory amino acids, glutamate plus aspartate, by that of the inhibitory amino acid GABA (see Ramakers et al., 1994 for details). A one-way analysis-of-variance indicated a statistically significant difference between combined control (CON) and tetrodotoxin (TTX) treated cultures ($P < 0.05$); in both groups E:I declines drastically between 7 and 14 days in vitro ($P < 0.01$).

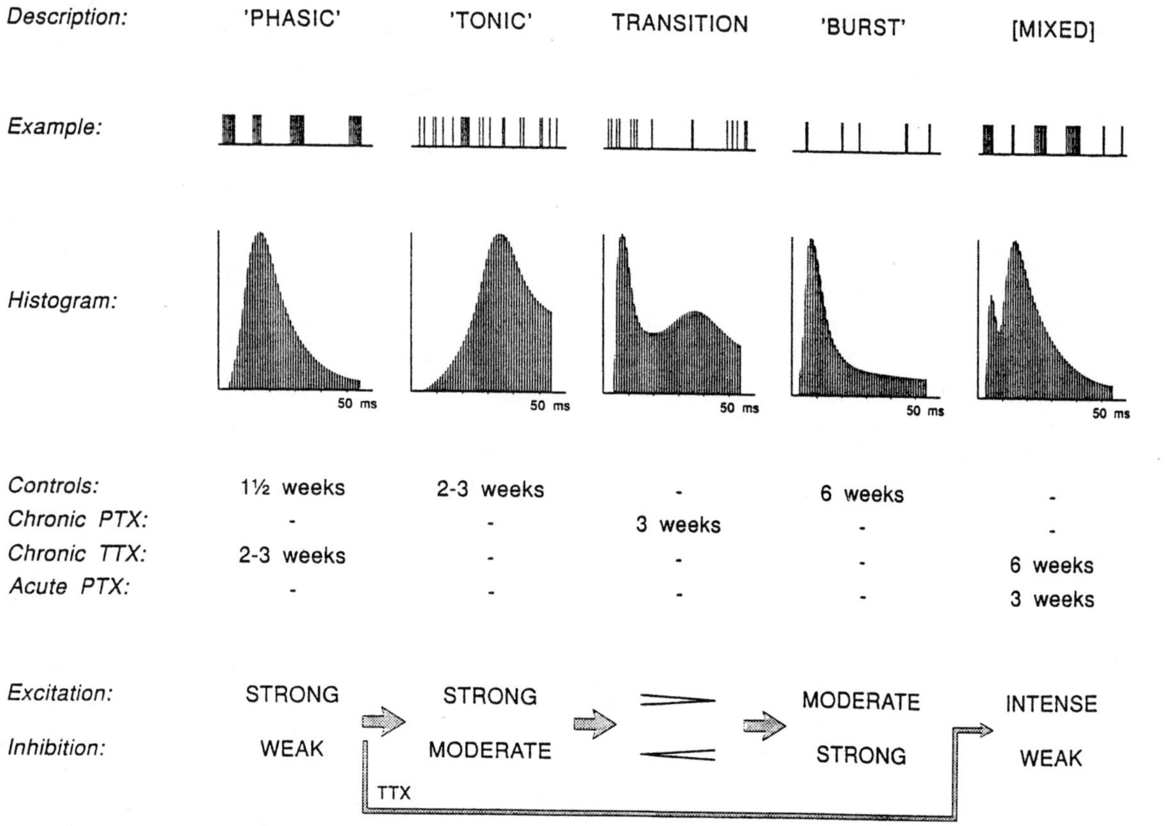

Fig. 19. Distinctive firing patterns encountered in developing neocortical cell cultures under different growth conditions: their designation, characteristic interspike interval histogram, predominant age for each pattern, and the postulated balance between overall excitatory and inhibitory synaptic drive within the network (see Corner and Ramakers, 1992 for details and discussion).

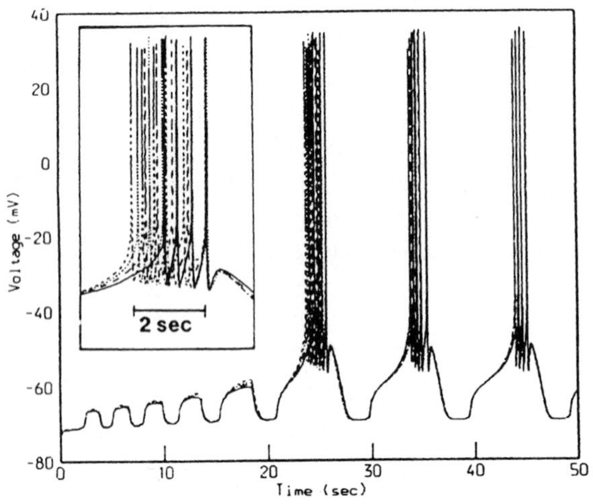

Fig. 20. Synchronous repetitive firing (stable over at least 500 sec) generated by a network of four identical, computer simulated, excitatory neurons with all-to-all connections. All traces are superimposed, and the initial burst is shown expanded in the inset (see Kowalski et al., 1992 for details).

receptors prolong glutamatergic synaptic potentials) the main avenue for NMDA-dependent calcium entrance will be via receptor linked ion channels.

In view of the fact that chronic NMDA blockade during cortical network formation in vitro has recently been discovered to cause only a transient reduction in spontaneous firing (M.A. Corner and R. Nuijtinck), the question of the respective roles of these two variables is now amenable to experimental evaluation. If the 'calcium hypothesis' is valid in its strongest formulation (i.e. all avenues of entry into the cell synergistically activate calcium-dependent processes), one would predict that selective NMDA antagonists will interfere less with normal development than will treatments which, in addition, completely suppress

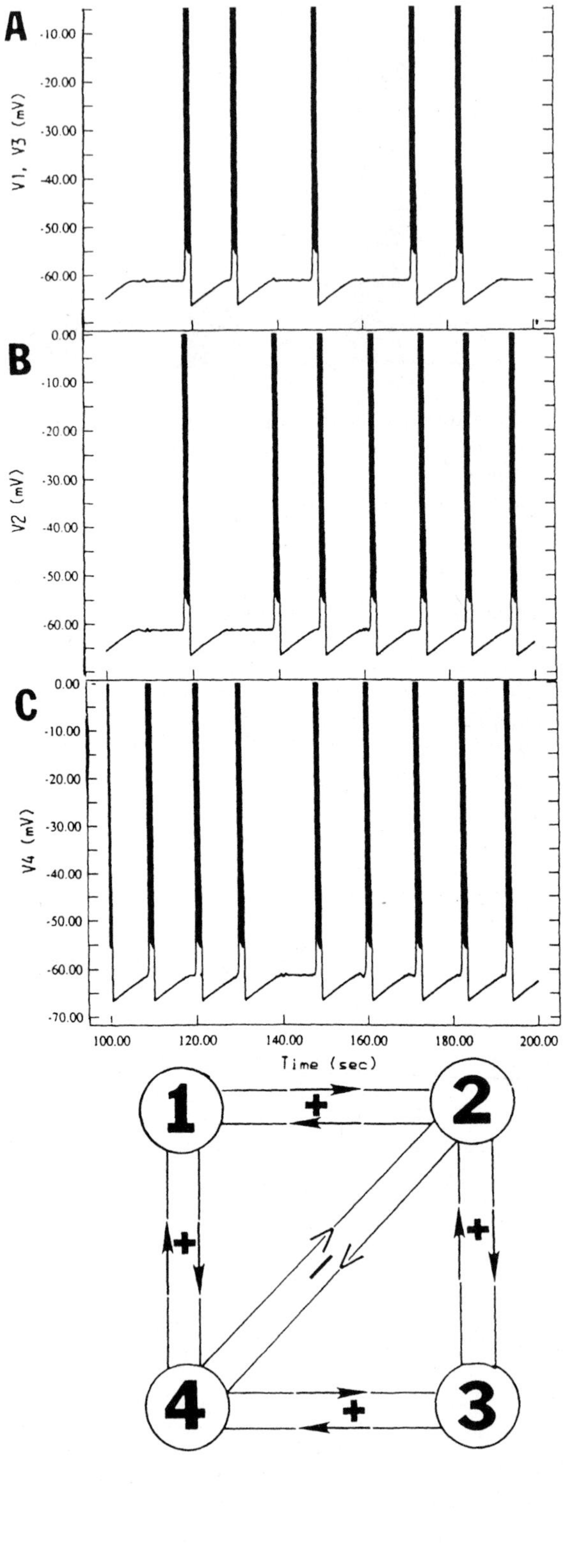

Fig. 21. Apparently chaotic, non-synchronized bursting (except for neurons 1 and 3: see connection diagram) in a computer simulated network into which synaptic inhibition has been introduced (see Fig. 20; and see Kowalski et al., 1992).

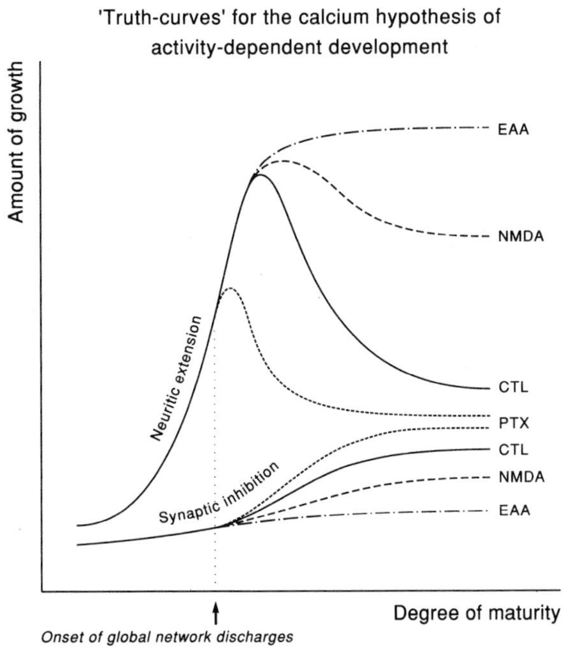

Fig. 22. 'Truth graphics' for the experimental analysis of the role of amino acid neurotransmitter mediated activity in the ontogeny of neocortical neuronal networks, based upon the 'calcium hypothesis' of structural and functional differentiation (see Kater et al., and Van Ooyen and Van Pelt, this volume). CON, physiologically active controls; PTX, chronic picrotoxin disinhibition; NMDA, chronic partial suppression of (excitatory) synaptic activity; EAA, chronic total blockade of excitatory amino-acid mediated neurotransmission. Based upon empirical findings (see text), both the 'pruning' of excess synapses and the full maturation of GABAergic inhibitory mechanisms are postulated as being function-dependent to the extent that calcium is enabled to enter the neurons.

neurotransmitter mediated electrical discharges (Fig. 22). Conversely, since synaptic disinhibition of the network should intensify calcium entry, GABA antagonists are expected to accelerate maturation (also see Van Ooyen, 1994). Spontaneous neurophysiological biorhythmicity ends up in our story, then, as a class of phenomena which not only mirror the primal sources of animal behavior and cortical function (see previous sections of this article), but one which could open up new avenues of approach towards answering basic questions about the mechanisms underlying neuroplasticity in general.

Concluding remarks

I've tried to make it clear, in a wide-ranging and sometimes light-hearted analytic and personal narrative, why I believe that the time may be ripe for making systematic interactions between empirical and theoretical neuroscientists into one of the highest research priorities. In addition, I hope to have amply justified our having concentrated pretty much exclusively on the level of cellular 'circuitry' in this Progress in Brain Research volume on the theme of 'Neural Networks'. I hope at the same time to have convinced you, albeit implicitly, that rigorous guidance by systems-analytical techniques would be an excellent precedent also for experimental studies at intra- and supra-cellular levels, i.e. the *molecular* and the *histological*, respectively.

Finally, I've presented a few of the arguments for regarding intrinsically generated bioelectric activity as a key variable in understanding many features of plasticity and self-organization in the developing nervous system. I've occasionally felt myself, over the years, to be cast somewhat in the role of a 'lone haranguer' on this last point (but see Oppenheim, 1982), so I'm gratified to note that interest in the phenomenon appears to be entering a rapid growth phase (e.g. Shatz, 1990; Spitzer, 1991; Fields and Nelson, 1991; Agston et al., 1991a,b; Ling et al., 1991; Murphy et al., 1992; Seil et al., 1994); this interest is also evident in several of the chapters in the present volume besides my own: e.g. Ben-Ari et al., Houenou et al., Spitzer, Van Ooyen and Van Pelt).

More generally, it's a special source of personal gratification to have been able to participate in a Summer School in Amsterdam on precisely this subject, thus reflecting the coming of age of concerns which have occupied me personally for some 30 odd years — concerns which also belong to the long-standing traditions of the Amsterdam

'Brain Research Institute' (Corner and Boer, 1979) and which, in an up-to-date fashion, are still being productively incorporated into our research program (e.g. Van Ooyen and Van Pelt, 1994).

References

Aghajanian, G.K. and Bloom. F.E. (1967) The formation of synaptic junctions in developing rat brain: a quantitative electron microscopic study. *Brain Res.,* 6: 716–727.

Agston, D.V., Eiden, L.E. and Brenneman, D.E. (1991a) Calcium-dependent regulation of the enkephalin phenotype by neuronal activity during early ontogeny. *J. Neurosci. Res.,* 28: 140–148.

Agston, D.V., Eiden, L.E., Brenneman, D.E. and Gozes, I. (1991b) Spontaneous electrical activity regulates vasoactive intestinal peptide expression in dissociated spinal cord cell cultures. *Mol. Brain Res.,* 10: 235–240.

Armstrong-James, M. and Fox, K. (1989) Similarities in unitary cortical activity between slow-wave sleep and light urethane anaesthesia. *J. Physiol.* (London), 346: 55–56.

Baker, R.E. (1978) Synapse selectivity in somatic afferent systems. In: M. Corner, R. Baker, N. van de Poll, D. Swaab and H. Uylings (Eds.), *Progress in Brain Research, Vol. 48, Maturation of the Nervous System,* Elsevier, Amsterdam, pp. 77–98.

Baker, R.E. and Corner, M.A. (1978) Possible mechanisms underlying the development of central and peripheral cutaneous afferent connections. In: M. Cuenod, G. Kreutzberg and F. Bloom (Eds.), *Progress in Brain Research, Vol. 51, Chemical Specificity of Neurons,* Elsevier, Amsterdam, pp. 117–121.

Baker, R.E. and Corner, M.A. (1981) Sensory end-organ modulation vs. nerve cell redundancy as possible mechanisms in the development of misdirected reflex responses in skin-grafted anurans. *Adv. Physiol. Sci.,* 36: 35–46.

Baker, R.E. and Ruijter, J.M. (1991) Chronic blockade of bioelectric activity in neonatal rat neocortex in vitro: physiological effects. *Int. J. Dev. Neurosci.,* 9: 321–329.

Baker, R.E., Corner, M.A. and Veltman, W. (1978) Topography of cutaneous mechanoceptive neurones in dorsal root ganglia of skin-grafted frogs. *J. Physiol. (London),* 284: 181–192.

Baker, R.E., Matesz, K. and Corner, M.A. (1981a) Shifts in dorsal root ganglion cell sizes following early skin rotation in the frog *D. pictus*. *IRCS Med. Sci.,* 9: 1073.

Baker, R.E., Matesz, K., Corner, M.A. and Szekely, G. (1981b) Peripheral reinnervation patterns and dorsal root ganglion topography in skin-grafted frogs: a behavioral and histological examination. *Dev. Neurosci.,* 4: 134–141.

Baker, R.E., Habets, A.M., Brenner, E. and Corner, M.A. (1982) Influence of growth medium, age in vitro and spontaneous bioelectric activity on the distribution of sensory ganglion evoked activity in spinal cord explants. *Dev. Brain Res.,* 5: 329–341.

Baker, R.E., Corner, M.A. and Habets, A.M. (1984) Effects of chronic suppression of bioelectric activity on the development of sensory ganglion evoked responses in spinal cord explants. *J. Neurosci.,* 4: 1187–1192.

Baker, R.E., Corner, M.A., Lammertse, T. and Furth, E. (1986) Some functional effects of suppressing bioelectric activity in fetal mouse spinal cord-dorsal root ganglion explants. *Exp. Neurol.,* 94: 426–430.

Baker, R.E., Ruijter, J.M. and Bingmann, D. (1991a) Elevated potassium prevents neuronal death but inhibits network formation in neocortical cultures. *Int. J. Dev. Neurosci.,* 9: 339–345.

Baker, R.E., Ruijter, J.M. and Bingmann, D. (1991b) Effect of chronic exposure to high magnesium on neuron survival in long-term neocortical explants of neonatal rats in vitro. *Int. J. Dev. Neurosci.,* 9: 597–606.

Baker, R.E., Ruijter, J.M., Corner, M.A. and Mud, M.T. (1992) Effects of spontaneous bioelectric activity and gangliosides on cell survival in vitro. *Brain Res. Bull.,* 28: 975–978.

Balazs, R., Hack, N. and Jorgenson, O.S. (1992) Neurobiology of excitatory amino acids. In: A. Schousboe, N. Diemer and H. Kofod (Eds.), *Drug Research Related to Neuroactive Amino Acids,* Munskgaard, Copenhagen, pp. 397–410.

Black I.B. (1991) *Information in the Brain,* Mass. Inst. Technol. Press, Cambridge, USA.

Black I.B. and Patterson, P.H. (1980) Developmental regulation of neurotransmitter phenotype. *Curr. Topics Dev. Biol.,* 15: 27–40.

Bot, A.P.C. and Corner, M.A. (1973) Cerebral seizure susceptibility and patterns of paroxysmal activity during embryonic development in the domestic chicken. *Israel J. Med. Sci.,* 9 (Suppl.): 43–54.

Boyd, C. and Noble, D. (1993) *The Logic of Life: the Challenge of Integrative Physiology,* Oxford University Press, New York.

Chandebois, R. (1983) Automation in animal development. *Monogr. Dev. Biol.,* 16: – .

Changeux, J.-P. and Mikoshiba, K. (1978) Genetic and 'epigenetic' factors regulating synapse formation in vertebrate cerebellum and neuromuscular junction. In: M. Corner, R. Baker, N. van de Poll, D. Swaab and H. Uylings (Eds.), *Progress in Brain Research, Vol. 48, Maturation of the Nervous System,* Elsevier, Amsterdam, pp. 43–68.

Constantine-Paton, M., Cline, H.T. and Debski, E. (1990) Patterned activity, synaptic convergence, and the NMDA receptor in developing visual pathways. *Annu. Rev. Neurosci.,* 13: 129–154.

Corner, M.A. (1963) Development of the brain of *Xenopus laevis* after removal of portions of the neural plate. *J. Exp. Zool.,* 153: 301–311.

Corner, M.A. (1964) Localization of capacities for functional development in the neural plate of *Xenopus laevis*. *J. Comp. Neurol.,* 123: 243–256.

Corner, M.A. (1967) Aspects of the functional organization of the central nervous system during embryonic development in the chick. In: L. Jilek and S. Trojan (Eds.), *Ontogenesis of the Brain,* Charles Univ. Press, Prague, pp. 85–96.

Corner, M.A. (1987) *Of Mouse and Mind — Prolegomena to a Neuro(psycho)physiology of Development.* Inaugural lecture, University of Amsterdam, privately printed (available upon request).

Corner, M.A. (1990) Brainstem control of behavior: ontogenetic aspects. In: W. Klemm and R. Vertes (Eds.), *Brainstem Mechanisms of Behavior,* Wiley, New York, pp. 239–266.

Corner, M.A. and Baker, R.E. (1978) Wiping reflexes and nerve impulse patterns evoked by electrical stimulation of the skin in frogs. *Physiol. Behav.,* 21: 789–792.

Corner, M.A. and Baker, R.E. (1987) Central neuronal responsiveness to sensory ganglion stimulation is correlated with the incidence of spontaneous bioelectric activity in developing spinal cord cultures. *Pfluegers Arch. / Eur. J. Physiol.,* 410: 563–565.

Corner, M.A. and Boer, K. (1979) Neuroscience centers of the world — the Netherlands Institute for Brain Research. *Trends Neurosci.,* 2: VIII–IX.

Corner, M.A. and Bot, A.P.C. (1969) Electrical activity in the isolated forebrain of the chick embryo. *Brain Res.,* 12: 473–476.

Corner, M.A. and Crain, S.M. (1972) Patterns of spontaneous bioelectric activity during maturation in culture of fetal rodent medulla and spinal cord tissues. *J. Neurobiol.,* 3: 25–45.

Corner, M.A. and Mirmiran, M. (1990) Spontaneous neuronal firing patterns in the occipital cortex of developing rats. *Int. J. Dev. Neurosci.,* 8: 309–316.

Corner, M.A. and Ramakers, G.J.A. (1992) Spontaneous firing as an epigenetic factor in brain development — physiological consequences of chronic tetrodotoxin and picrotoxin exposure on cultured rat neocortex neurons. *Dev. Brain Res.,* 65: 57–64.

Corner, M.A. and Romijn, H.J. (1983) Ontogeny of cerebral seizure characteristics and their morpho-physiological substrate in the chick embryo. In: T. Ookawa (Ed.), *The Brain and Behavior of the Fowl,* Japan Sci. Soc. Press, Tokyo, pp. 297–306.

Corner, M.A. and Schadé, J.P. (1967) Developmental patterns in the central nervous system of birds IV. Cellular and molecular basis of functional activity. In: C.E. Bernhard and J.P. Schadé (Eds.), *Progress in Brain Research, Vol. 26, Developmental Neurology*, Elsevier, Amsterdam, pp. 237–250.

Corner, M.A., Peters, J.J. Rutgers Van der Loeff, P. (1966) Electrical activity patterns in the cerebral hemisphere of the chick during maturation correlated with behaviour in a test situation. *Brain Res.*, 2: 274–292.

Corner, M.A., Bakhuis, W.L. and van Wingerden, C. (1972) Sleep and wakefulness during early life in the domestic chicken, and their relationship to hatching and embryonic motility. In: G. Gottlieb (Ed.), *Behavioral Embryology,* Academic Press, New York, pp. 245–279.

Corner, M.A., Roholl, P.J.M. and Bot, A.P.C. (1974a) Effects of glutamate on spontaneous electrical activity in the embryonic chick cerebrum. *Exp. Neurol.,* 44: 229–245.

Corner, M.A., Smith, J. and Romijn, H.J. (1974b) Maturation of cerebral bioelectric activity in the chick embryo in relation to morphological and biochemical factors. In: L. Jilek and S. Trojan (Eds.), *Ontogenesis of the Brain, Vol. 2,* Charles Univ. Press, Prague, pp. 21–32.

Corner, M.A., Romijn, H.J. and Richter, A. (1977) Synaptogenesis in the cerebral hemisphere (accessory hyperstriatum) of the chick embryo. *Neurosci. Lett.,* 4: 15–19.

Corner, M.A., Mirmiran, M., Bour, H., Boer, G., Van de Poll, N., Van Oyen, H. and Uylings, H. (1980) Does rapid-eye-movement sleep play a role in brain development? In: P. McConnell, G. Boer, H. Romign, N. Van de Poll and M. Corner (Eds.), *Adaptive capabilities of the Brain, Progress in Brain Research, Vol 53*, Elsevier, Amsterdam, pp. 347–356.

Corner, M.A., Scholte, R., Korf, J. and Mirmiran, M. (1984) Electrocortical activity patterns during chloral hydrate induced sleep in developing rats. *Brain Res. Bull.,* 12: 77–81.

Corner, M.A., Baker, R.E. and Habets, A.M. (1987a) Regional specificity of functional sensory connections in developing spinal cord cultures varies with the incidence of spontaneous bioelectric activity. *Roux' Arch. Dev. Biol.,* 196: 401–404.

Corner, M.A., Habets, A.M. and Baker, R.E. (1987b) Bioelectric activity is required for regional specificity of sensory ganglion projections to spinal cord explants cultured in vitro. *Roux' Arch Dev. Biol.,* 196: 133–136.

Corner, M.A., van Eden, C.G. and de Beaufort, A.J. (1992) Spike-train analysis reveals 'overshoot' in developing rat prefrontal cortex function. *Brain Res. Bull.* 28: 799–802.

Craik, K.J.W. (1952) *The Nature of Explanation,* University of Cambridge Press.

Crain, S.M. (1976) *Neurophysiologic Studies in Tissue Culture,* Raven Press, New York.

Edelman, G.M. (1987) *Neural Darwinism — The Theory of Neuronal Group Selection,* Basic Books, New York.

Fields, R.D. and Nelson, P.G. (1991) Activity-dependent development of the vertebrate nervous system. *Annu. Rev. Neurosci.,* 14: 133–214.

Frank, E., Smith, C. and Mendelson, B. (1988) Strategies for selective synapse formation between muscle sensory and motor neurons in the spinal cord. In: S. Easter, D. Barald and B. Carlson (Eds.), *From Message to Mind — Directions in Developmental Neurobiology,* Sinauer, Sunderland MA, pp. 180–202.

Goodwin, B. (1963) *Temporal Organization in Cells — A Dynamic Theory of Cellular Control Processes,* Academic Press, New York.

Gottlieb, G. (1973) Introduction of behavioral embryology. In: G. Gottlieb (Ed.), *Behavioral Embryology,* Academic Press, New York, pp. 3–45.

Gurdon, J.B. (1974) *The Control of Gene Expression in Animal Development,* Clarendon Press, Oxford.

Habets, A., van Dongen, A., van Huizen, F. and Corner, M. (1987) Spontaneous neuronal firing patterns in fetal rat cortical networks during development in vitro: a quantitative analysis. *Exp Brain Res.,* 69: 43–52.

Hamburger, V. (1973) Anatomical and physiological basis of embryonic motility in birds and mammals. In: G. Gottlieb (Ed.), *Behavioral Embryology,* Academic Press, New York, pp. 51–76.

Hofer, M. and Constantine-Paton, M. (1994) Regulation of *N*-methyl-D-aspartate (NMDA) receptor function during the rearrangement of developing neuronal connections. In: J. van Pelt, M.A. Corner, H.B.M. Uylings and F.H. Lopes da Silva (Eds.), *Progress in Brain Research, Vol. 102, The Self-organizing Brain — From Growth Cones to Functional Networks,* Elsevier Science Publishers, Amsterdam, this volume.

Jongkind, J.F., Corner, M.A. and Bruntink, R. (1972) Glycolytic substrate utilization and energy consumption in the cerebral hemispheres of the chick embryo during the period of EEG development. *J. Neurochem.,* 19: 389–394.

Jouvet, M. (1967) Neurophysiology of the states of sleep. *Physiol. Rev.,* 47: 117–177.

Kandel, E.R., Schwartz, J.H. and Jessell, T.M. (1991) *Principles of Neural Science,* 3rd ed., Elsevier, Amsterdam.

Kater, S.B., Davenport, R.W. and Guthrie, P.B. (1994) Filopodia as detectors of environmental cues: signal integration through changes in growth cone calcium levels. In: J. van Pelt, M.A. Corner, H.B.M. Uylings and F.H. Lopes da Silva (Eds.), *Progress in Brain Research, Vol. 102, The Self-organizing Brain — From Growth Cones to Functional Networks,* Elsevier Science Publishers, Amsterdam, this volume.

Kostopoulos, G. and Gotman, J. (1989) Computer assisted analysis of relations between single-unit activity and spontaneous EEG. *Electroenceph. Clin. Neurophysiol.,* 57: 69–82.

Kowalski, J.M., Albert, G.L., Rhoades, B.K. and Gross, G.W. (1992) Neuronal networks with spontaneous, correlated bursting activity: theory and simulations. *Neural Networks,* 5: 805–822.

Landmesser, L. (1980) The generation of neuromuscular specificity. *Annu. Rev. Neurosci.,* 3: 279–301.

Ling, D.S.F., Petroski, R.E. and Geller, H.M. (1991) Both survival and development of spontaneously active rat hypothalamic neurons in dissociated culture are dependent on membrane depolarization. *Dev. Brain Res.,* 59: 99–103.

Lopes da Silva, F.H., van Rotterdam, A., Barts, P., van Heusden, E. and Burr, W. (1976) Models of neuronal populations: the basic mechanisms of rhythmicity. In: M.A. Corner and D.F. Swaab (Eds.), *Progress in Brain Research, Vol. 45, Perspectives in Brain Research,* Elsevier, Amsterdam, pp. 281–308.

Lopes da Silva, F.H., Pijn, J.-P. and Wadman, W.J. (1994) Dynamics of local neuronal networks: control parameters and state bifurcations in epileptogenesis. In: J. van Pelt, M.A. Corner, H.B.M. Uylings and F.H. Lopes da Silva (Eds.), *Progress in Brain Research, Vol. 102, The Self-organizing Brain — From Growth Cones to Functional Networks,* Elsevier Science Publishers, Amsterdam, this volume.

Luhmann, H.J. and Prince, D.A. (1990) Control of NMDA receptor-mediated activity by GABAergic mechanisms in mature and developing rat neocortex. *Dev. Brain Res.* 54: 287–290.

Luhmann, H.J. and Prince, D.A. (1991) Postnatal maturation of the GABAergic system in rat neocortex. *J. Neurophysiol., 65: 247–263.*

McCormick, D.A., Wang, Z. and Huguenard, J. (1993) Neurotransmitter control of neocortical neuronal activity and excitability. Cerebral Cortex, 3: 387–398.

Meister, M., Wong, R., Baylor, D.A. and Shatz, C.J. (1991) Synchronous bursts of action potentials in ganglion cells of the developing mammalian retina. *Science,* 252: 939–943.

Messenger, E.A. and Warner, A.E. (1979) The function of the sodium pwm during differentiation of amphibian embryonic neurones. *J. Physiol* (London), 292: 85–105.

Mirmiran, M. and Corner, M.A. (1982) Neuronal discharge patterns in the occipital cortex of developing rats during active and grief sleep, *Dev. Brain Res.,* 3: 37–48.

Mirmiran, M., Uylings and Corner, M.A. (1983) Chronic REM-sleep deprivation prior to weaning in male rats counteracts the effectiveness of sub segment environmental enrichment on cortical growth. *Dev. Brain Res.*, 7: 102–105.

Mirmiran, M. and Uylings, H. (1983) The environmental enrichment effect upon cortical growth is neutralized by concomitant pharmacological suppression of active sleep in female rats. *Brain Res.*, 261: 331–334.

Monod, J. (1971) *Chance and Necessity,* Random House, New York.

Moore, J.A. (1987) Science as a way of knowing — developmental biology. *Am. Zool.,* 27: 415–573.

Murphy, T.H., Blatter, L.A., Wier, W.G. and Baraban, J.M. (1992) Spontaneous synchronous synaptic calcium transients in cultured cortical neurons. *J. Neurosci.,* 12: 4834–4845.

Oppenheim, R.W. (1982) The neuroembryological study of behavior: progress, problems and perspectives. *Curr. Topics Dev. Biol.,* 17: 257–309.

Oyama, S. (1985) *The Ontogeny of Information — Developmental Systems and Evolution,* Cambridge University Press.

Paabo, S. (1993) Ancient DNA. *Sci. Am.,* 269: 60–66.

Platt, J.R. (1964) Strong inference. *Science,* 146: 347–353.

Popper, K. (1965) *The Logic of Scientific Discovery,* Hutchinson, London.

Portin, P. (1993) The concept of the gene: short history and present status. *Q. Rev. Biol.,* 68: 173–223.

Prigogine, I. and Senghers, I. (1984) *Order out of Chaos — Man's New Dialogue with Nature,* Heinemann, London.

Purpura, D.P. and Schade, J.P. (1964) Growth and maturation of the brain. In: *Progress in Brain Research, Vol. 4,* Elsevier, Amsterdam.

Purves, D. and Lichtman, J.W. (1985) *Principles of Neural Development,* Sinauer, Sunderland (MA).

Ramakers, G.J., Corner, M.A. and Habets, A.M. (1990) Development in the absence of spontaneous bioelectric activity results in increased stereotyped burst firing in cultures of dissociated cerebral cortex. *Exp. Brain Res.,* 79: 157–166.

Ramakers, G.J., Corner, M.A. and Habets, A.M. (1991) Abnormalities in the spontaneous firing patterns of cultured rat neocortical neurons after chronic exposure to picrotoxin during development in vitro. *Brain Res. Bull.,* 26: 429–432.

Ramakers, G., de Wit, C., Wolters, P. and Corner, M. (1993) A developmental decrease in NMDA-mediated spontaneous firing in cultured rat cerebral cortex. *Int. J. Dev. Neurosci.,* 11: 25–32.

Ramakers, G, van Galen, H., Feenstra, M, Corner, M and Boer, G. (1994) Activity-dependent plasticity of inhibitory and excitatory amino acid transmitter systems in cultured rat cerebral cortex. Submitted for publication.

Rolls, E.T. and Treves, A. (1994) Neural networks in the brain involved in memory and recall. In: J. van Pelt, M.A. Corner, H.B.M. Uylings and F.H. Lopes da Silva (Eds.), *Progress in Brain Research, Vol. 102, The Self-organizing Brain — From Growth Cones to Functional Networks,* Elsevier Science Publishers, Amsterdam, this volume.

Rudnick, D. and Waelsch, H. (1955) Development of glutamotransferase and glutamine synthetase in the nervous system of the chick. *J. Exp. Zool.,* 129: 309–326.

Ruijter, J.M. and Baker, R.E. (1990) The effects of potassium-induced depolarization, glutamate receptor antagonists and *N*-methyl-D-aspartate on neuronal survival in cultured neocortex explants. *Int. J. Dev. Neurosci.,* 8: 361–370.

Ruijter, J.M., Baker, R.E., de Jong, B.M. and Romijn, H.J. (1991) Chronic blockade of bioelectric activity in neonatal rat cortex grown in vitro: morphological effects. *Int. J. Dev. Neurosci.,* 9: 331–338.

Schwabl, H. (1993) Yolk is a variable source of maternal testosterone for developing birds. *Proc. Natl. Acad. Sci.* USA, 90: 11446–11450.

Seil, F., Drake-Baumann, R., Leiman, A., Herndon, R. and Tiekotter, K. (1994) Morphological correlates of altered neuronal activity in organotypic cerebellar cultures chronically exposed to anti-GABA agents. *Dev. Brain Res.,* 77: 123–132.

Seil, F.J. and Drake-Baumann, R. (1984) Reduced cortical inhibitory synaptogenesis in organtypic cerebellar cultures developing in the absence of neuronal activity. *J. Comp. Neuronal.,* 342: 366–377.

Shatz, C.J. (1990) Impulse activity and the patterning of connections during CNS development. *Neuron,* 5: 745–756.

Singer, C. (1959) *A Short History of Scientific Ideas to 1900,* Oxford University Press, New York.

Singer, W. (1985) Activity-dependent self-organization of the mammalian visual cortex. In: D. Rose, and V. Dobson (Eds.), *Models of the Visual Cortex,* Wiley, Chichester, pp. 123–136.

Spitzer, N.C. (1991) A developmental handshake: neuronal control of ionic currents and their control of neuronal differentiation. *J. Neurobiol.,* 22: 659–673.

Spitzer, N.C. (1994) Development of voltage-dependent and ligand-gated channels in excitable membranes. In: J. van Pelt, M.A. Corner, H.B.M. Uylings and F.H. Lopes da Silva (Eds.), *Progress in Brain Research, Vol. 102, The Self-organizing Brain — From Growth Cones to Functional Networks,* Elsevier Science Publishers, Amsterdam, this volume.

Staudacherova, D., Mares, P., Kozakova, H. and Camutaliova, M. (1978) Ontogenetic development of acetylcholine and atropine epileptogenic cortical foci in rats. *Neuroscience,* 3: 749–753.

Steriade, M. Gloor, P., Llinas, R., Lopes da Silva, F. and Mesulam, M. (1990) Basic mechanisms of cerebral rhythmic activities. *Electroenceph. Clin. Neurophysiol.,* 76: 481–508.

Steriade, M., McCormick, D.A. and Sejnowski, T.J. (1993) Thalamocortical oscillations in the sleeping and aroused brain. *Science,* 262: 679–685.

Stryker, M. and Harris, W. (1986) Binocular impulse blockade prevents the formation of ocular dominance columns in cat visual cortex. *J. Neurosci.,* 6: 2117–2133.

Van der Weele (1993) Explaining embryological development: should integration be the goal? *Biol. Philos.,* 8: 385–397.

Van Eden, C.G. and Uylings, H.B.M. (1985) Postnatal volumetric development of the prefrontal cortex in the rat. *J. Comp. Neurol.,* 241: 268–274.

Van Huizen, F., Romijn, H.J. and Corner, M.A. (1987) Indications for a critical period for synapse elimination in developing rat cerebral cortex cultures. *Dev. Brain Res.,* 31: 1–6.

Van Ooyen, A. and van Pelt, J. (1994) Activity-dependent neurite outgrowth and neural network development. In: J. van Pelt, M.A. Corner, H.B.M. Uylings and F.H. Lopes da Silva (Eds.), *Progress in Brain Research, Vol. 102, The Self-organizing Brain — From Growth Cones to Functional Networks,* Elsevier Science Publishers, Amsterdam, this volume.

Van Ooyen, A., van Pelt, J., Corner, M.A. and Lopes da Silva, F.H. (1992) The emergence of long-lasting transients of activity in simple neural networks. *Biol. Cybern.,* 67: 269–277.

Weiss, P.A. (1969) *Principles of Development,* Revised ed., Harper, New York.

Wiener, N. (1961) *Cybernetics,* 2nd ed., Wiley, New York.

Wolpert, L. (1991) *The Triumph of the Embryo,* Oxford University Press, New York.

Wong, R., Meister, M and Shatz, C.J. (1993) Transient period of correlated bursting activity during development of the mammalian retina. *Neuron,* 11: 923–938

SECTION II

From Growth Cone to Neuron

A. Growth Cone Dynamics and Neuritic Outgrowth

J. van Pelt, M.A. Corner H.B.M. Uylings and F.H. Lopes da Silva (Eds.)
Progress in Brain Research, Vol 102

CHAPTER 2

Growth cone motility: substratum-bound molecules, cytoplasmic [Ca^{2+}] and Ca^{2+}-regulated proteins

Paul C. Letourneau, Diane M. Snow and Timothy M. Gomez

Department of Cell Biology and Neuroanatomy, 4–135 Jackson Hall, University of Minnesota, Minneapolis, MN 55455, U.S.A.

Introduction

Neuronal circuitry develops by way of highly stereotyped patterns of axonal elongation. For example, axons extend directly from motor neurons in the spinal cord to the developing muscles of a limb, where the axons establish topographically organized connections, in which neighboring muscle fibers are innervated by axons projected from neighboring neuronal somata in the spinal cord. (Tosney and Landmesser, 1985a,b; Goodman and Schatz, 1993). Growing neurites undergo characteristic activities of elongation, retraction, turning, branching, and finally synaptogenesis (Letourneau, 1989). These basic features of neurite growth are largely determined by the motile behavior of the tips of elongating neurites, called growth cones by their discoverer, Santiago Ramon y Cajal (1890). Growth cone motility involves the continuous extension and exploratory movements of sensory filopodial and lamellipodial protrusions, which are endowed with a variety of cell surface receptors (Fig. 1). Filopodial protusions can extend up to 100 μm from the neurite tip, and, thereby, significantly expand the opportunities for direct interaction of a growth cone with its environment.

Embryonic growth cones, especially those pioneers that first establish axonal pathways, actively navigate as they penetrate intervening tissues on their way to their target regions. By saying navigate, we propose that the nerve growth cone is a sensory-effector system, involving the surface expression of cell surface receptors for environmental molecules that serve as cues or signals. Binding of such cues to their receptors prompts transmembrane signals that regulate the growth cone machinery to produce appropriate motility, including advance, retreat, turning, and branching (Fig. 1). It is becoming apparent that growth cones are subject to a complex variety of positive and negative signals as they migrate towards their targets. A number of recent papers have discussed issues related to growth cone navigation or guidance (Bray and Hollenbeck, 1988; Lankford et al., 1990; Bixby and Harris, 1991; Letourneau et al., 1991; Baier and Bonhoeffer, 1992; Kapfhammer and Schwab, 1992).

In this chapter we present some of our investigations of this navigational hypothesis for growth cone behavior. We are analyzing the behaviors of growth cones at boundaries between combinations of permissive and inhibitory substratum-bound molecules. In addition, we are investigating the involvement of one cytoplasmic regulatory factor, namely cytoplasmic [Ca^{2+}], in controlling growth cone behaviors. Our results suggest that (1) growth cone behavior changes at boundaries

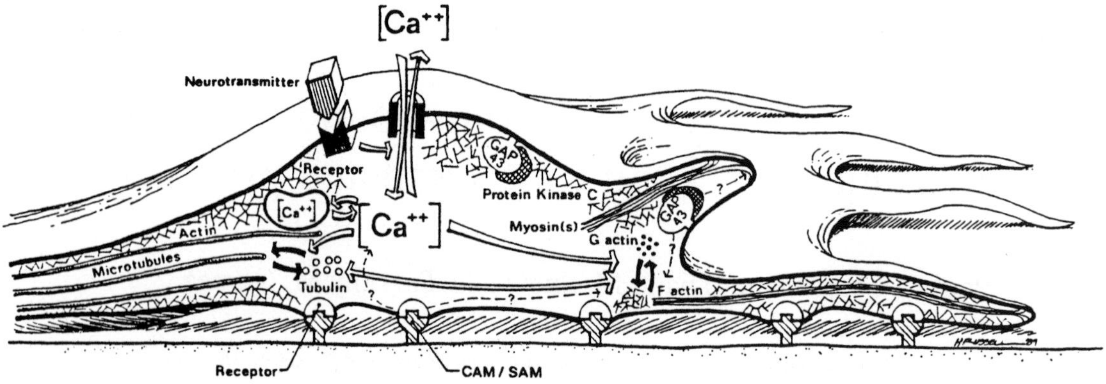

Fig. 1. A model of the sensory-effector machinery of growth cone navigation. To simplify the drawing, only the calcium second messenger system is shown. Open arrows are regulatory relationships between calcium extracellular signals, and cytoskeletal assembly. Changes in internal calcium ion levels regulate the assembly/disassembly of actin and tubulin. The movements of calcium ions between compartments are reversible and regulated. Possible relationships between binding of adhesion receptors to cell adhesion molecules (CAMs) and substratum adhesion molecules (SAMs) and intracellular calcium or the cytoskeleton are indicated with question marks.

between substratum-adsorbed molecules, (2) interactions with surface bound molecules can trigger changes in growth cone $[Ca^{2+}]_i$, (3) recurrent spikes of cytoplasmic $[Ca^{2+}]$ can be generated in growth cones, and (4) growth cones contain Ca^{2+}-regulated proteins that control the organization and functions of actin filaments.

Machinery of growth cone movement

The machinery that drives elongation of neurites involves distinct activities of the two major cytoskeletal systems, microtubules and actin filaments. These two systems generate distinct forces of ‘push’ and ‘pull’ that contribute to neurite elongation (Bray, 1987; Letourneau et al., 1987; Mitchison and Kirschner, 1988; Goldberg and Burmeister, 1989; Heidemann et al., 1990).

The ‘push’ for the neurite to advance is produced by microtubules, which have two roles: (1) as the major supportive components of neurites, and (2) as the tracks along which many organelles and other components are transported. For neurites to grow, microtubules must be advanced, and this occurs both via microtubule translocation and by tubulin polymerization onto the distally oriented plus ends of microtubules (see Mitchison and Kirschner, 1988; Hollenbeck, 1989; Reinsch et al., 1991; Okabe and Hirokawa, 1992, 1993; Joshi and Baas, 1993; Black 1994). Since growth cones contain many plus-oriented microtubule ends, which are the sites of tubulin polymerization in neurites (Letourneau, 1983; Mitchison and Kirschner, 1988; Okabe and Hirokawa, 1988; Robson and Burgoyne, 1988; Joshi and Baas, 1993), they are an important site for regulating microtubule advance in growing neurites.

The ‘pull’ for directing the growing neurite is produced by actin filaments (Letourneau, 1983; Smith, 1988; Bridgman and Dailey, 1989). Bundles and networks of actin filaments fill the transient filopodial and lamellipodial protrusions of the growth cone. These actin filaments are required for the protrusive motility of growth cones, as demonstrated by the inhibition of protrusive activity in the presence of the drug cytochalasin B (Marsh and Letourneau, 1984; Letourneau et al., 1987). The interactions of actin filaments are diverse, dynamic, and central to growth cone motility. These include associations of actin filaments (1) with the plasma membrane, where they are linked to adhesive molecules and transmit mechanical forces to points of adhesive contact, (2) with microtubules, which may influence the position of microtubules and, consequently, other neuritic components, and (3) with mechanochemical enzymes such as myosin(s), which produce the forces that move growth cone protrusions and pull the neurite forward (Letourneau, 1983; Bridgman and Dailey, 1989; Letourneau and Shattuck, 1989).

On the basis of these 'push' and 'pull' activities, microtubules and actin filaments comprise the major components that drive neurite elongation. Our proposal of growth cone navigation is based on the hypothesis that regulation of the organization and physiology of these two filament systems is the primary focus of transmembrane signals triggered by environmental cues.

The 'push' of microtubule advance, accompanied by organelle transport and membrane assembly is necessary and sufficient for neurite elongation. 'Push' alone can sustain neurite elongation even without growth cone motility (Marsh and Letourneau, 1984; Bentley and Torian-Raymond, 1987; Letourneau et al., 1987), whereas treatments that inhibit the 'push' of microtubule advance effectively block neurite elongation (Letourneau and Ressler, 1984; Bamburg et al., 1986). Figure 3 shows frames from a video record of a neurite elongating in the presence of cy-

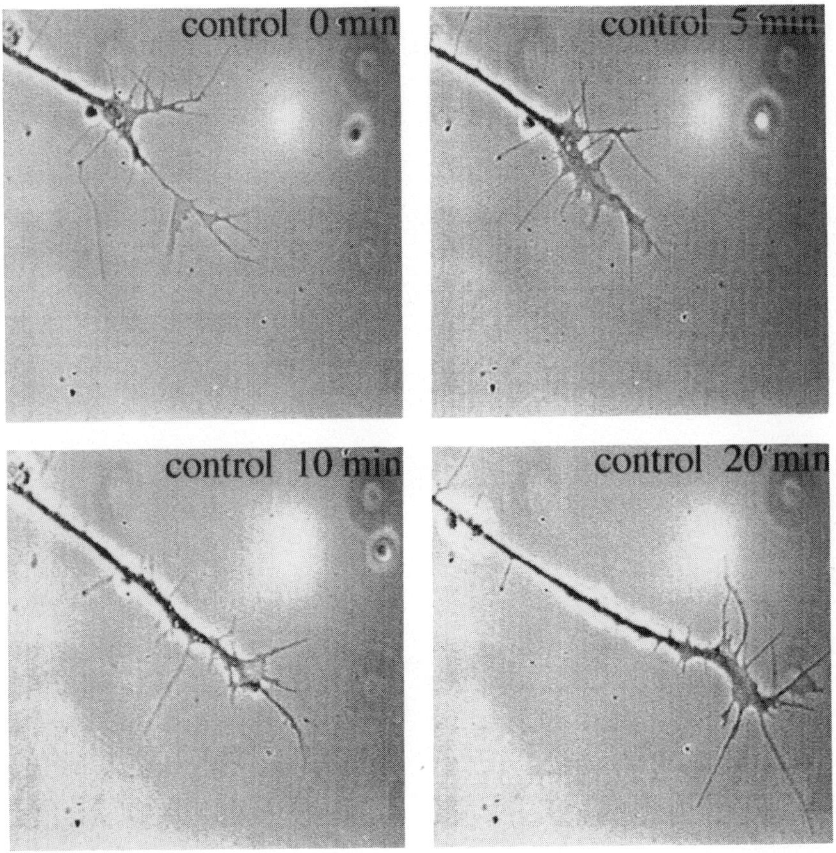

Fig. 2. Motility of a normal growth cone. This growth cone is migrating across an unpatterned laminin-treated substratum. Protrusion of filopodia and lamellipodia create a constantly changing growth cone shape, and greatly expand the area of growth cone exploration.

tochalasin B on a laminin (LM)-treated substratum. The neurite advances slowly without filopodial protrusions. Its path is relatively straight, with few of the side-to-side movements that normally accompany the advance of the growth cone leading margin (see Fig. 2) (Katz, 1985; Van Veen et al., personal communication).

Growth cone behavior at boundaries

Although neurite elongation can occur without actin filament-driven protrusion and motility, growth cone navigation usually involves changes in growth cone position, movement, or direction that require the participation of filopodia (Raper et al., 1983; O'Connor et al., 1990). We investigated this role of filopodia previously by analyses of growth cone movements between LM-coated islands separated by a non-adhesive agar surface (Hammarback and Letourneau, 1986). The upper curve in Fig. 4 presents the frequency of growth cones crossing a non-adhesive gap as a function of the width of the gap. The lower curve presents the average distribution of filopodial lengths protruded from sensory growth cones. Note that only about 5% of filopodia are greater than 30 μm long, yet 65% of the growth cones can cross a 39 μm wide non-adhesive gap. This suggests that the ability of growth cones to cross between islands of LM depends on the infrequent extension

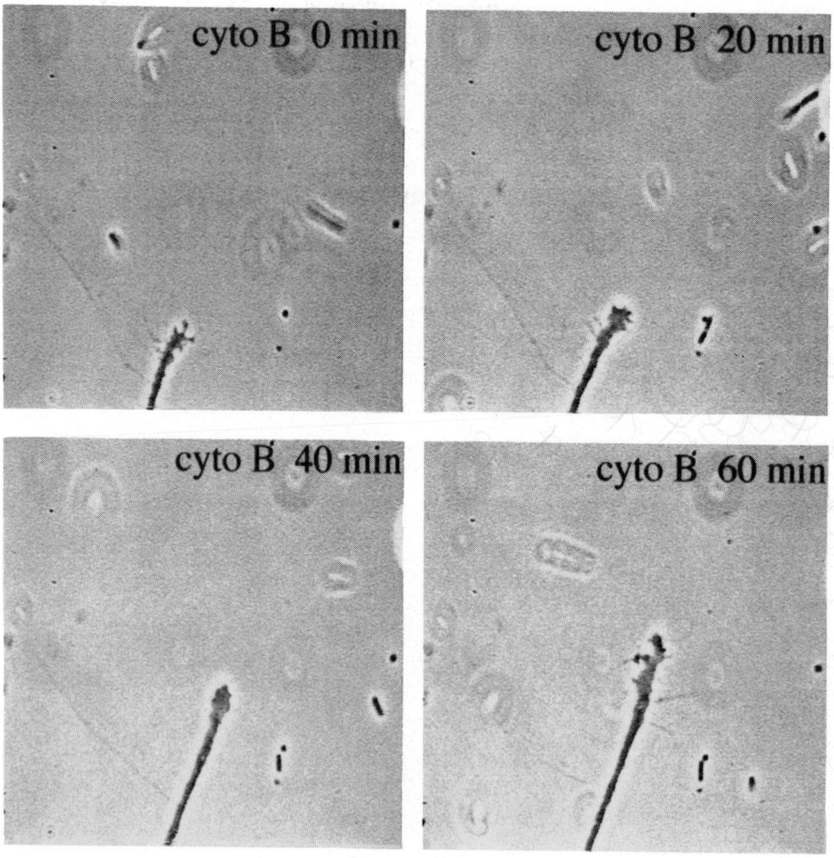

Fig. 3. Motility of a cytochalasin B-treated growth cone. This growth cone is migrating across a laminin-treated substratum in the presence of 0.05 μg/ml cytochalasin B, which inhibits polymerization of actin, and, thereby, protrusion of filopodia. The growth cone migrates forward at a rate less than one-third the normal migration rate.

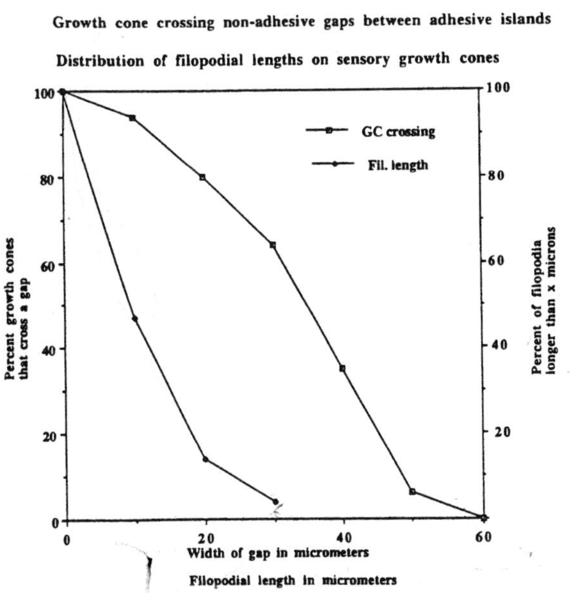

Fig. 4. In the upper curve this graph presents the frequency of growth cones crossing a non-adhesive gap as a function of the width of the gap. The lower curve presents the average distribution of filopodial lengths protruded from sensory growth cones. Note that only about 5% of filopodia are longer than 30 μm long, yet 65% growth cones can cross a 30 μm non-adhesive gap.

of long filopodia that can span the non-adhesive gulf to another LM surface. The surfaces of filopodia contain many molecules that are involved in adhesive interactions (Fig. 5) (Tsui et al., 1985; Letourneau and Shattuck, 1989), and once a filopodium makes contact across the gulf, the growth cone moves rapidly from one island to the next. It is unknown whether growth cone advance is triggered solely by mechanical 'pull', exerted by filopodia that have crossed the gap and adhered to LM, or whether the stimulus to advance is prompted by transmembrane signals that regulate motile activities in response to activation of LM receptors on the filopodial membrane.

We have continued to investigate the navigational behavior of growth cones by analyses of growth cone behavior at boundaries between substrata coated with molecules that either promote or inhibit growth cone migration. We have been especially interested in the behaviors of growth cones that are in contact with one substratum, and are making filopodial contacts ahead of the body of the growth cone at the boundary with another substratum. Although the underlying

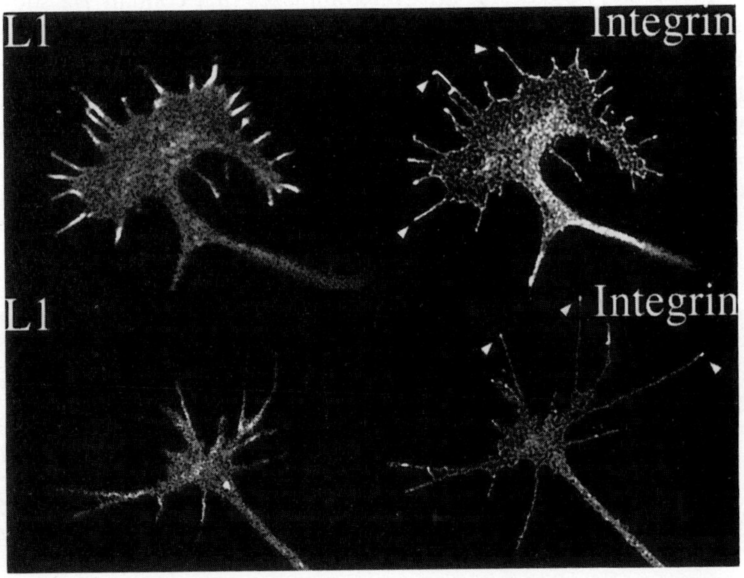

Fig. 5. Immunofluorescence localization using antibodies against two adhesive molecules, L1 and the $\beta 1$ subunit of integrin, on two double-labeled growth cones. Note that filopodia are strongly labeled by both antibodies, especially the anti-L1. Also, note that the tips of filopodia (arrowheads) are more strongly labeled by anti-integrin than by anti-L1.

mechanisms of filopodial signaling remain unclear, the results so far confirm our ideas about the regulatory roles of filopodia.

One experimental design that we have investigated involves growth cones that encounter proteoglycans (PGs), many of which inhibit migration. It has been recently recognized that inhibitory cues are also very important in regulating growth cone migration (Kapfhammer and Raper, 1987; Patterson, 1988; Davies and Cook, 1991; Fishman and Strittmatter, 1993). In several regions of developing neural tissues, migrating growth cones do not enter spaces that are rich in extracellular PGs (Tosney and Oakley, 1990; Snow et al., 1990, 1991), such as chondroitin sulfate proteoglycan (CSPG). We, therefore, analyzed the behavior of sensory and retinal neuronal growth cones that migrate on the adhesive glycoprotein LM and encounter a stripe of chondroitin sulfate, deposited on LM. Growth cones stop or turn aside upon contact with CSPG, mixed with LM, and then migrate along the LM/CSPG border (Fig. 6) (Snow et al., 1991).

Although the arrest of growth cone migration at a boundary with CSPG depends on the concentration of CSPG deposited on the substratum we also found an interesting modulation of the response to CSPG by growth cones on a step gradient of CSPG deposition (Snow and Letourneau, 1992). Retinal neuronal growth cones can cross onto a surface containing a higher concentration of CSPG, if they first encounter a lower concentration of deposited CSPG (Fig. 7). That is to say, they can adapt to substratum-bound CSPG by prior exposure to low concentrations. It is still unclear how this adaptation occurs, whether a decrease in an inhibitory signal produced by CSPG, or whether by an increase in the ability of growth cones to respond to the growth-promoting LM substratum, e.g. by an up regulation of integrin expression on the growth cone surface. Another possible mechanism of interaction with CSPG may involve regulation of cytoplasmic $[Ca^{2+}]$. In a later section we will present evidence for the modulation of growth cone $[Ca^{2+}]_i$ by CSPG.

Another experimental situation that we have investigated involves growth cones moving along substrata containing alternating stripes of LM and fibronectin (FN), two adhesive glycoproteins of extracellular matrices (ECM) that mediate substratum adhesion of chick sensory neurons and promote neurite elongation (Rogers et al., 1983). Both these molecules have multiple domains that mediate interactions with neurons, while neurons, in turn, have multiple distinct surface components that recognize both FN and LM, including several integrin complexes that contain $\beta 1$ subunits (Bixby and Harris, 1991; Reichardt and Tomaselli, 1991; Letourneau et al., 1992). Figure 5 shows the immunofluorescence distribution of integrin complexes containing the $\beta 1$ integrin subunit on filopodia of sensory growth cones. It can be seen (Fig. 5) that integrins are present all

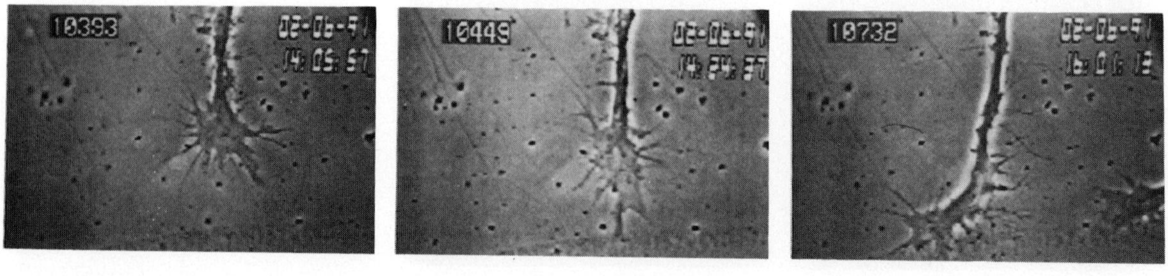

Fig. 6. Arrest of growth cone migration at a boundary with CSPG. A growth cone advances on a LM substratum towards a boundary where CSPG is deposited over LM. In the middle figure, 20 min later, filopodial and lamellipodial extensions are seen contacting the CSPG, and in the right figure the growth cone has turned and is migrating along the boundary with CSPG.

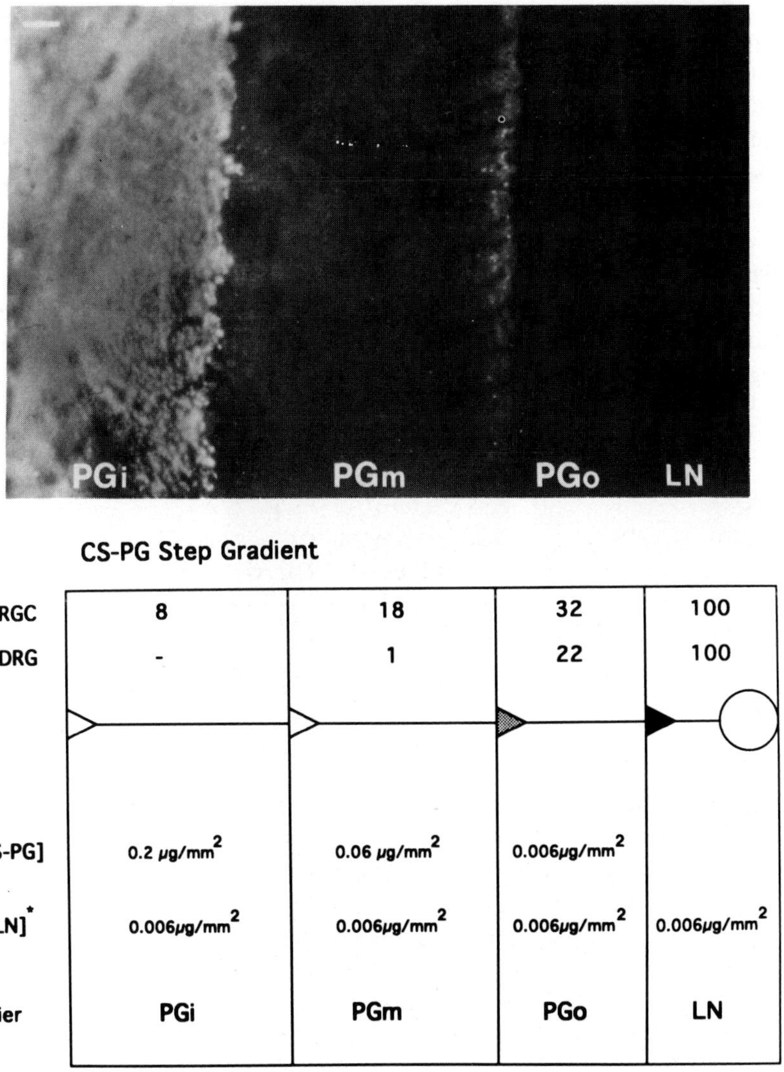

PGi
PGm
PGo
LN
CS-PG Step Gradient
%RGC
%DRG
8
-
18
1
32
22
100
100
[CS-PG]
[LN]*
0.2 μg/mm²
0.06 μg/mm²
0.006μg/mm²
0.006μg/mm²
0.006μg/mm²
0.006μg/mm²
0.006μg/mm²
Tier
PGi
PGm
PGo
LN

Fig. 7. Migration of DRG and retinal growth cones on a step gradient of CSPG deposition. Immunofluorescence staining for CSPG in the photograph at the top shows the step gradient of CSPG deposition on LM. The figure at the bottom indicates the approximate concentrations of LM and CSPG deposition, as well as the percent of DRG and retinal growth cones that can cross from LM to successively higher concentrations of deposited CSPG. Growth cones are totally blocked by a single step of CSPG deposition of 0.02 $\mu g/mm^2$, but some of them can cross onto a substratum of 0.06 $\mu g/mm^2$ after first encountering a low level of CSPG.

along filopodia but that they are more concentrated at filopodial tips. Sensory neurons initiate neurite growth on either FN or LM, so that their growth cones are observed on both FN and LM as they approach a boundary with the other ECM component.

The behaviors we have observed do not suggest a clear preference for one ECM molecule over another, since the majority of growth cones continue in the direction that they are headed and cross the boundary from one substratum to the other. Importantly, however, growth cones usually change their behavior in response to the component on the other side of the boundary, and, not surprisingly, these responses are initiated by filopodia that extend across the boundary and contact the alternate substratum. Some growth cones seem to choose to cross the boundary, since they accelerate and/or turn towards the boundary, whereas other growth cones choose to remain on their original substratum, turning and/or decelerating at the boundary. Both kinds of behavior are initiated by filopodial interactions, since these behavioral changes begin before the bodies of the growth cones have touched either a boundary or the alternative substratum (Fig. 8). We do not know whether these different responses reflect the presence of different distributions of receptors for ECM components on distinct sub-populations of DRG neurons.

As with the behavior of growth cones at a boundary with CSPG deposited on LM, the mechanisms of this modulation of growth cone behavior are unclear. We have examined growth cone-substratum contact, using the technique of interference reflection microscopy, and the re-

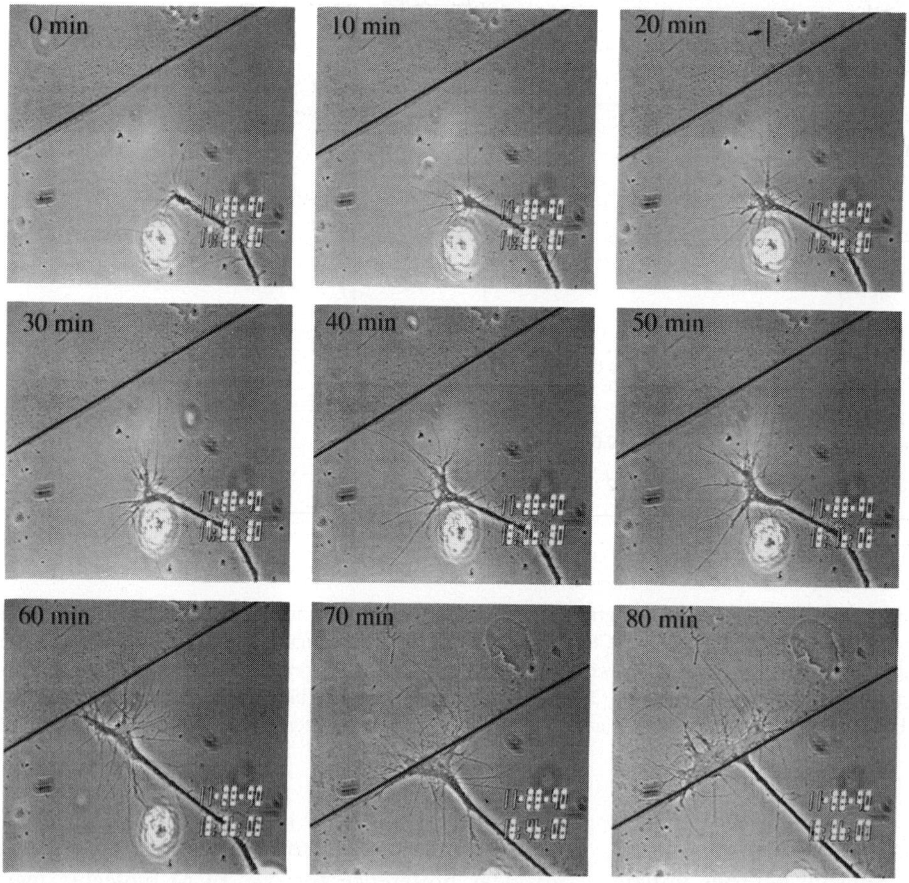

Fig. 8. Growth cone behavior at a boundary between FN and LM. The growth cone is initially on FN and approaches the boundary with LM (indicated by a black line). At 20 min the tip of single long filopodium has contacted the LM (arrow and line). In subsequent images the growth cone is seen to orient towards the LM surface. It accelerates and flattens dramatically after the body of the growth cone contacts the LM.

sults indicate that the choice to cross or turn at the boundary is not correlated with either maintaining or achieving the closest contact with the substratum. Given the multiplicity of receptors for FN and LM on sensory neurons, plus the potential for involvement of several cytoplasmic second messenger systems, it will be a formidable task unravelling the mechanisms of these changes in growth cone behavior, even in a simple in vitro system (Bixby and Harris, 1991; Reichardt and Tomaselli, 1991; Hynes, 1992; Damsky and Werb, 1992; Doherty and Walsh, 1992; Hynes and Lander, 1992; Schweighoffer and Shaw, 1992; Letourneau et al., 1992; Rivas et al., 1992; Fishman and Strittmatter). A factor that has been implicated in regulation of cell motility is cytoplasmic [Ca^{2+}], and in the next section we will describe studies of cytoplasmic [Ca^{2+}] as a possible regulator of growth cone motility, as well as in response to substratum-bound molecules.

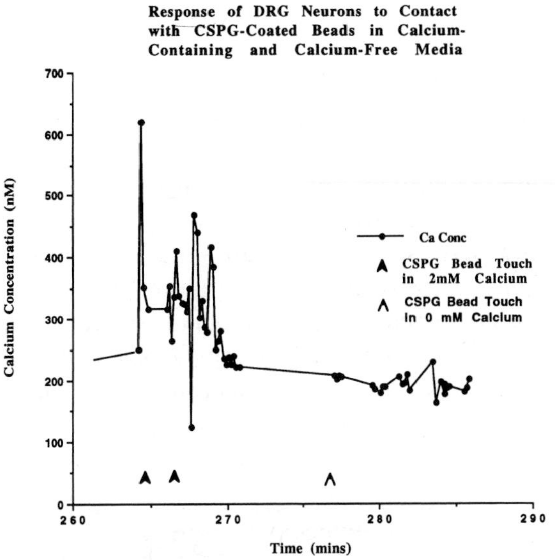

Fig. 9. Contact with a CSPG-coated bead induces [Ca^{2+}] transients in a DRG neuron. A DRG neuron was touched with a CSPG-coated polystyrene bead at the first arrow head, inducing a large rise in $[Ca^{2+}]_i$. Another touch induces a similar response, but when the neuron was touched again after changing the bath to a Ca^{2+} free medium, no rise in $[Ca^{2+}]_i$ occurred.

Cytoplasmic [Ca^{2+}] as a regulator of growth cone motility

Cytoplasmic [Ca^{2+}] is one of the most important second messengers involved in the immediate regulation of cell behaviors, including growth cone migration (Kater et al., 1988; Lankford et al., 1990; Kater and Mills, 1991; Letourneau and Cypher, 1991; Berridge, 1993). Changes both in growth cone behavior and cytoskeletal organization are associated with changes of growth cone $[Ca^{2+}]_i$ (Goldberg, 1988; Lankford and Letourneau, 1989, 1991; Davenport and Kater, 1992; Rehder and Keter, 1992; Bandtlow et al., 1993; Davenport et al., 1993). Experimental manipulations that greatly elevate growth cone $[Ca^{2+}]_i$ lead to breakdown of actin filaments, while manipulations that reduce growth cone $[Ca^{2+}]_i$ reduce the dynamic reorganization of actin filaments (Lankford and Letourneau, 1989, 1991; Lankford et al., 1990). We have, therefore, begun to investigate the involvement of growth cone $[Ca^{2+}]_i$ in the navigational behaviors of growth cones at boundaries between substratum-associated ECM molecules.

Recall that growth cones do not cross a boundary onto LM coated with a sufficient level of CSPG (Fig. 6) (Snow et al., 1991). We have found that contact of a growth cone with CSPG, bound to either a microbead or to the substratum in a striped pattern, leads to rapid and significant elevation of growth cone $[Ca^{2+}]_i$ (Fig. 9) (Snow et al., 1993). This elevation occurs by influx of extracellular Ca^{2+}, not by release from intracellular stores. Enzymatic removal of the chondroitin sulfate glycosaminoglycan, which abolishes CSPG inhibition of neurite elongation, also abolishes the elevated [Ca^{2+}] response, indicating that the glycosaminoglycan of the CSPG is the active component. The magnitude of the $[Ca^{2+}]_i$ elevation is consistent with a $[Ca^{2+}]_i$ increase that leads to breakdown of actin filaments in the leading margin of the growth cone (Lankford and Letourneau, 1989, 1991; Lankford et al., 1990). Such a disruption of the leading margin would explain the arrest of growth cone migration at a boundary with CSPG.

In other studies, we are investigating the mechanisms underlying spontaneous rapid changes in cytoplasmic [Ca^{2+}] in sensory growth cones. Elevation of extracellular [Ca^{2+}] from the normal 2 mM level to 20 mM often triggers repetitive spikes of [Ca^{2+}]$_i$ (Fig. 10). These spikes have been proposed as an important regulatory mechanism for cellular behaviors in many cell types (Tsien and Tsien, 1990). They are initiated within a minute of elevating external Ca^{2+}, and spike duration is about 15 sec. Multiple spikes are common, and the intervals between spikes can be highly irregular. The mean spike height is about 200–300 nM, with [Ca^{2+}]$_i$ increasing from about 100 nM to 400 nM. Initiation and maintenance of these [Ca^{2+}]$_i$ transients is not inhibited by drugs that block voltage-gated Ca^{2+} channels, but 1 mM La^{3+} blocks spikes or terminates spikes in the presence of 20 mM external Ca^{2+}. These [Ca^{2+}]$_i$ transients seem to be locally regulated, since there is no correlation between the spiking behavior of different growth cones at the ends of several branches of a parent neurite. In addition, when a growth cone exhibits [Ca^{2+}]$_i$ spikes, the adjacent neurite and neuronal perikaryon do not show [Ca^{2+}]$_i$ transients.

We are continuing to characterize this dynamic regulation of growth cone [Ca^{2+}]$_i$. Clearly, these [Ca^{2+}]$_i$ transients in grwoth cones involve Ca^{2+} influx from extracellular spaces. It is unclear what are the channels that admit the Ca^{2+} ions, since our data indicate that voltage-gated channels are not involved. It is not always necessary to elevate external Ca^{2+} to 20 mM in order to initiate spikes, since we have observed growth cones spontaneously exhibiting [Ca^{2+}]$_i$ transients in 2 mM Ca^{2+}. We are also unsure of the mechanisms responsible for the rapid decrease of cytoplasmic [Ca^{2+}] at the end of a spike. This may result from both Ca^{2+} pumps in the plasma membrane and in cytoplasmic organelles, as well as inactivation of influx from the outside. Our initial findings on responses to two inhibitors of the Ca^{2+} pump in

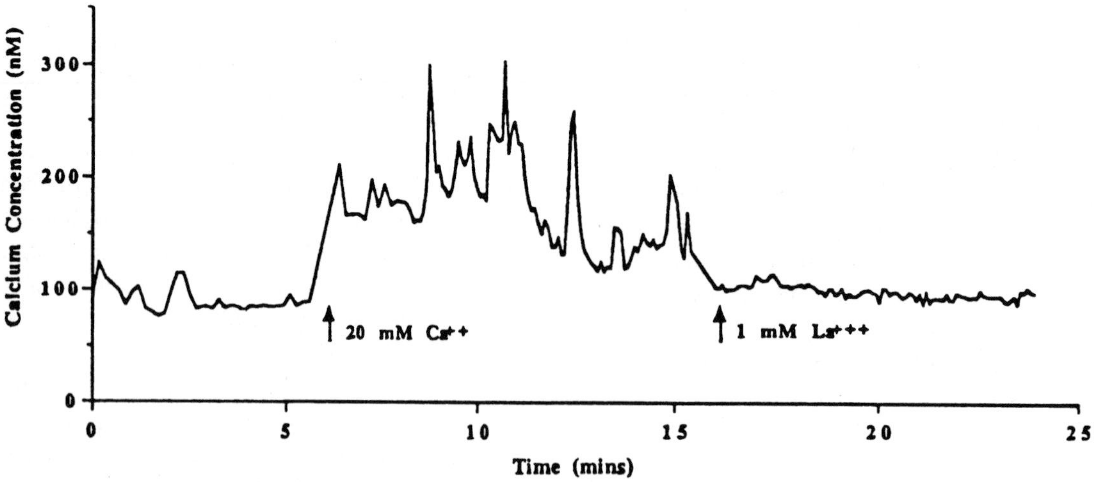

Fig. 10. Induction of Ca^{2+} spiking by exposure to 20 mM external [Ca^{2+}] and inhibition of spiking by La^{3+}. At the indicated arrow the external [Ca^{2+}] was shifted from 2 mM to 20 mM, and spiking cytoplasmic [Ca^{2+}] began. This behavior was immediately stopped by the addition of 1 mM La^{3+} in the continued presence of 20 mM external [Ca^{2+}].

the endoplasmic reticulum, thapsigargin and cyclopiazonic acid (Takemura et al., 1989), indicate that growth cones indeed contain membranous organelles that can remove Ca^{2+} from the cytoplasmic compartment.

Thus, growth cones contain mechanisms for rapidly changing cytoplasmic [Ca^{2+}] and producing complex signals consisting of $[Ca^{2+}]_i$ spikes of varying height, duration and frequencies. There are many implications of these events for regulation of growth cone behavior. The regulatory actions of calcium ions are mediated and modulated by a rich diversity of kinases, actin-associated proteins, and regulatory proteins in both a direct and an indirect manner (Tsien and Tsien, 1990). We have previously reported that sensory growth cones contain several cytoskeletal-associated proteins that are regulated by [Ca^{2+}] (Letourneau and Shattuck, 1989), and, recently, we have examined growth cones for the presence of several other Ca^{2+}-regulated proteins that may regulate growth cone motility.

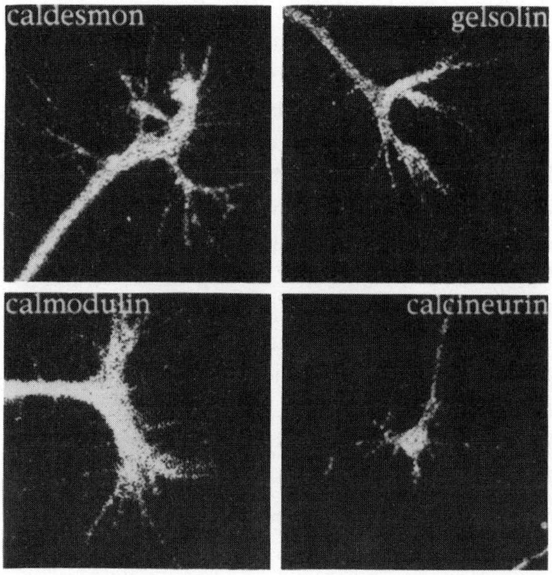

Fig. 11. Immunofluorescence localization using antibodies against four calcium-regulated proteins in sensory neuronal growth cones. Caldesmon and gelsolin are actin-binding proteins, calmodulin is a ubiquitous regulator of enzymes and other proteins, and calcineurin is a protein phosphatase.

Ca^{2+}-regulated proteins in growth cones

The organization of actin filaments is regulated by a variety of proteins that control actin polymerization, interactions of actin filaments with other filaments and organelles, and the integrity of filaments. Caldesmon, which is regulated by Ca^{2+}/calmodulin, may regulate interactions of actin filaments with other cytoskeletal components (Hartwig and Kwiatowski, 1991; Matsumura and Yamashiro, 1993). Figure 11 shows the immunofluorescence localization of caldesmon in growth cones. Gelsolin binds to actin filaments, and in the presence of $[Ca^{2+}]_i > 100$ nM, it severs actin filaments. This causes the breakdown of filament networks, but also creates new filament ends for actin re-polymerization (Hartwig et al., 1989; Hartwig and Kwiatowski, 1991). Figure 11 shows an immunofluorescence demonstration of enriched localization of gelsolin in the leading margin and filopodia of growth cones. We have also demonstrated the presence fo gelsolin and caldesmon by Western blot analysis of growth cone preparations from chick embryo brain.

The actions of $[Ca^{2+}]_i$ are modulated by many Ca^{2+}-binding regulatory proteins (Baimbridge et al., 1992). Figure 11 shows additional immunocytochemical evidence for the presence of two such proteins in growth cones, calmodulin, a ubiquitous regulator of protein kinases and other enzymes, and calcineurin, a Ca^{2+}-and calmodulin dependent serine/threonine phosphatase.

Conclusions

The model for the navigational machinery of growth cones proposed at the beginning of this article depends on cell surface receptors that respond to the presence of environmental cues by transmembrane signaling to cytoplasmic regulatory systems. The identification of these putative receptors and the mode of their transmembrane signaling is the focus of research in an increasing number of laboratories. We have focused our efforts on growth cone interactions with surface-bound environmental components, and we find that the inhibitory effects of PGs and glycosaminoglycans may be mediated in part by rapid elevation of growth cone cytoplasmic [Ca^{2+}]. We also are investigating how growth cones produce their calcium signals, perhaps via distinct plasma membrane channels and Ca^{2+} pumps that are not yet identified or located. Growth cones contain a diversity of Ca^{2+}-regulated proteins that could participate in the regulation of growth cone motility, and elucidation of the roles of all these molecules and metabolic events in growth cone navigation will continue to be a challenging and fertile ground for research.

Acknowledgments

This research was supported by NIH research grants HD19950 and NS28807 to PCL, NIH NRSA EY06331 to DMS, and NIH grant EY07133 predoctoral traineeship to TMG. We thank Jerry Sedgewick for excellent assistance in producing the figures.

References

Baier, H. and Bonhoeffer, F. (1992) Axon guidance by gradients of a target-derived component. *Science*, 255: 472–475.

Baimbridge, K.G., Celio, M.R. and Rogers, J.H. (1992) Calcium-binding proteins in the nervous system. *Trends Neurosci.*, 8: 303–308.

Bamburg, J.R., Bray, D. and Chapman, K. (1986) Assembly of microtubules at the tip of growing axons. *Nature*, 321: 788–790.

Bandtlow, C.E., Schmidt, M.F., Hassinger, T.D., Schwab, M.E. and Kater, S.B. (1993) Role of intracellular calcium in NI-35 evoked collapse of neuronal growth cones. *Science*, 259: 80–83.

Bentley, D. and Torian-Raymond, A. (1986) Disoriented pathfinding by pioneer neuron growth cones deprived of filopodia by cytochalasin treatment. *Nature*, 323: 712–715.

Berridge, M.J. (1993) Inositol triphosphate and calcium signaling. *Nature*, 36: 315–325.

Bixby, J.L. and Harris, W.A. (1991) Molecular mechanisms of axon growth and guidance. *Ann. Rev. Cell Biol.*, 7: 117–159.

Black, M.M. (1994) Microtubule transport and assembly cooperate to generate the microtubule array of growing axons. Chapter 4, this volume.

Bray, D. (1987) Growth cones: do they pull or are they pushed? *Trends Neurosci.*, 10: 431–434.

Bray, D. and Hollenbeck, P.J. (1988) Growth cone motility and guidance. *Annu. Rev. Cell Biol.*, 4: 43–61.

Bridgman, P.C. and Dailey, M.E. (1989) The organization of myosin and actin in rapidly frozen nerve growth cones. *J. Cell Biol.*, 108: 95–109.

Damsky, C.H. and Werb, Z. (1992) Signal transduction by integrin receptors for extracellular matrix: cooperative processing of extracellular information. *Curr. Opin. Cell Biol.*, 4: 772–781.

Davenport, R.W. and Kater, S.B. (1992) Local increases in intracellular calcium elicit local filopodial responses in Helisoma neuronal growth cones. *Neuron*, 9: 405–416.

Davenport, R.W., Dou, P., Rehder, V. and Kater, S.B. (1993) A sensory role for neuronal growth cone filopodia. *Nature*, 361: 721–724.

Davies, J.A. and Cook, G.M.W. (1991) Growth cone inhibition —an important mechanism in neural development? *BioEssays*, 13(1): 11–15.

Doherty, P. and Walsh, F.S. (1992) Cell adhesion molecules, second messengers and axonal growth. *Curr. Opin. Neurobiol.*, 2: 595–601.

Fishman, M.C. and Strittmatter, S.M. (1993) Detection and amplification of inhibitory signals at the neuronal growth cone. *Neurosci. Facts*, 4(6): 21–22.

Goldberg, D.J. (1988) Local role of Ca^{2+} in formation of veils in growth cones. *J. Neurosci.*, 8: 2596–2605.

Goldberg, D.J. and Burmeister, D.W. (1989) Looking into growth cones. *Trends Neurosci.*, 12: 503–506.

Goodman, C.S. and Schatz, C. (1993) Developmental mechanisms that generate precise patterns of neuronal connectivity. *Cell/Neuron*, 72/10: 77–98.

Hammarback, J.A. and Letourneau, P.C. (1986) Neurite extension across regions of low cell-substratum adhesivity: implications for the guidepost hypothesis of axonal pathfinding. *Dev. Biol.*, 177: 655–662.

Hartwig, J.H. and Kwiatowski, D.J. (1991) Actin-binding proteins. *Curr. Opin. Cell Biol.*, 3: 87–97.

Hartwig, J.H., Chambers, K.A. and Stossel, T.P. (1989) Association of gelsolin with actin filaments and cell memebranes of macrophages and platelets. *J. Cell Biol.*, 108: 467–479.

Heidemann, S.R., Lamoureux, P. and Buxbaum, R.E. (1990) Growth cone behavior and production of traction force. *J. Cell Biol.*, 111: 1949–1957.

Hollenbeck, P.J. (1989) The transport and assembly of the axonal cytoskeleton. *J. Cell Biol.*, 108: 223–227.

Hynes, R.O. (1992) Integrins: versatility, modulation, and signaling in cell adhesion. *Cell*, 69: 11–25.

Hynes, R.O. and Lander, A.D. (1992) Contact and adhesive

specificities in the associations, migrations, and targeting of cells and axons. *Cell*, 68: 302–322.

Joshi, H.C. and Baas, P.W. (1993) A new perspective on microtubules and axon growth. *J. Cell Biol.*, 121: 1192–1196.

Kapfhammer, J. and Raper, J. (1987) Collapse of growth cone structure on contact with specific neurites in culture. *J. Neurosci.*, 7(1): 201–212.

Kapfhammer, J.P. and Schwab, M.E. (1992) Modulators of neuronal migration and neurite growth. *Curr. Opin. Cell Biol.*, 4: 863–868.

Kater, S.B. and Mills, L.R. (1991) Regulation of growth cone behavior by calcium. *J. Neurosci.*, 11: 891–899.

Kater, S.B., Mattson, M.P., Cohan, C. and Connor, J. (1988) Calcium regulation of the neuronal growth cones. *Trends Neurosci.*, 11: 315–321.

Katz, M.J. (1985) How straight do axons grow. *J. Neurosci.*, 5: 589–595.

Lankford, K.L. and Letourneau, P.C. (1989) Evidence that calcium may control neurite outgrowth by regulating the stability of actin filaments. *J. Cell Biol.*, 109: 1229–1243.

Lankford, K.L. and Letourneau, P.C. (1991) Roles of actin filaments and three second-messenger systems in short-term regulation of chick dorsal root ganglion neurite outgrowth. *Cell Motil. Cytoskel.*, 20: 7–29.

Lankford, K., Cypher, C. and Letourneau, P.C. (1990) Nerve growth cone motility. *Curr. Opin. Cell Biol.*, 2: 80–85.

Letourneau, P.C. (1983) Differences in the organization of actin in the growth cones compared with the neurite of cultured neurons from chick embryos. *J. Cell Biol.*, 97: 963–973.

Letourneau, P.C. (1989). Nerve cell shape. In: *Cell Shape, Determinants, Regulation and Regulatory Role*, Academic Press, New York, pp. 247–289.

Letourneau, P.C. and Cypher, C. (1991) Regulation of growth cone motility. *Cell Motil. Cytoskel.*, 20: 267–271.

Letourneau, P.C. and Ressler, A.H. (1984) Inhibition of neurite initiation and growth by taxol. *J. Cell Biol.*, 98: 1355–1362.

Letourneau, P.C. and Shattuck, T.A. (1989) Distribution and possible interactions of actin-associated proteins and cell adhesion molecules of nerve grwoth cones. *Development*, 105: 505–519.

Letourneau, P.C., Shattuck, T.A. and Ressler, A.H. (1987) 'Pull' and 'push' in neurite elongation: observations on the effects of different concentrations of cytochalasin B and taxol. *Cell Motil. Cytoskel.*, 8: 193–209.

Letourneau, P.C., Kater, S.B. and Macagno, E.R. (Eds.) (1991) *The Nerve Growth Cone*, Raven Press, New York, 535 pp.

Letourneau, P.C., Condic, M.L. and Snow, D.M. (1992) Extracellular matric and neurite outgrowth. *Curr. Opin. Gen. Dev.*, 2: 625–634.

Marsh, L. and Letourneau, P.C. (1984) Growth of neurites without filopodial or lamellipodial activity in the presence of cytochalasin B. *J. Cell Biol.*, 99: 2041–2047.

Matsumura, F. and Yamashiro, S. (1993) Caldesmon. *Curr. Opin. Cell Biol.*, 5: 70–76.

Mitchison, T. and Kirschner, M. (1988) Cytoskeletal dynamics and nerve growth. *Neuron*, 1: 761–772.

O'Connor, T.P., Duerr, J.S. and Bentley, D. (1990) Pioneer growth cone steering decisions mediated by single filopodial contact in situ. *J. Neurosci.*, 10: 3935–3946.

Okabe, S. and Hirokawa, N. (1988) Microtubule dynamics in nerve cells: analysis using microinjection of biotinylated tubulin into PC12 cells. *J. Cell Biol.*, 107: 651–664.

Okabe, A. and Hirokawa, N. (1992) Differential behavior of photoactivated microtubules in growing axons of mouse and frog neurons. *J. Cell Biol.*, 117: 105–120.

Okabe, S. and Hirokawa, N. (1993) Do photobleached fluorescent microtubules move? Re-evaluation of fluorescent laser photobleaching both in vitro and in growing *Xenopus* axons. *J. Cell Biol.*, 120: 1177–1186.

Patterson, P.H. (1988) On the importance of being inhibited, or saying no to growth cones. *Neuron*, 1: 263–267.

Ramon y Cajal, S. (1890) Sur l'origine et les ramifications des fibres nerveuses de la moelle embryonalre. *Anat. Anz.*, 5: 609–613, 631–639.

Raper, J.A., Bastiani, M. and Goodman, C.S. (1983) Pathfinding by neuronal growth cones in grasshopper embryos. I. Divergent choices made by the growth cones of sibling neurons. *J. Neurosci.*, 3: 20–30.

Rehder, V. and Kater, S.B. (1992) Regulation of neuronal growth cone filopodia by intracellular calcium. *J. Neurosci.*, 12: 3175–3186.

Reichardt, L.F. and Tomaselli, T.J. (1991) Extracellular matrix molecules and their receptors. *Annu. Rev. Neurosci.*, 14: 531–570.

Reinsch, S.S., Mitchison, T.J. and Kirschner, M.W. (1991) Microtubule polymer assembly and transport during axonal elongation. *J. Cell Biol.*, 115: 365–380.

Rivas, R.J., Burmeister, D.W. and Goldbeg, D.J. (1992) Rapid effects of laminin on the growth cone. *Neuron*, 8: 107–115.

Robson, S.J. and Burgoyne, S.D. (1988) Differential levles of tyrosinated, detyrosinated, and acetylated alpha-tubulin in neurites and growth cones of dorsal root ganglion neurons. *Cell Motil. Cytoskel.*, 12: 273–282.

Rogers, S.L., Letourneau, P.C., Palm, S.L., McCarthy, J.B., Furcht, L.T. (1983) Neurite extension by peripheral and central nervous system neurons in response to substratum-bound fibronectin and laminin. *Dev. Biol.*, 96:212–220.

Schweighoffer, T. and Shaw, S. (1992) Adhesion cascades: diversity through combinatorial strategies. *Curr. Opin. Cell Biol.*, 4: 824–829.

Smith, S.J. (1988) Neuronal cytomechanics: the actin-based motility of growth cones. *Science*, 242: 708–715.

Snow, D.M. and Letourneau, P.C. (1992) Neurite outgrowth on a step gradient of chondroitin sulfate proteoglycan (CS-PG). *J. Neurobiol.*, 23(3): 322–336.

Snow, D.M., Steindler, D.A. and Silver, J. (1990) Molecular and cellular characterization of the glial roof plate of the spinal cord and optic tectum: a possible role for a proteoglycan in the development of an axon barrier. *Dev. Biol.*, 138: 359–376.

Snow, D., Watanabe, M., Letourneau, P.C. and Silver, J. (1991) A chondroitin sulfate proteoglycan may influence the direction of retinal ganglion cell outgrowth. *Development*, 113: 1473–1485.

Snow, D.M., Atkinson, P., Hassinger, T., Kater, S.B. and Letourneau, P.C. (1993) Growth cone intracellular calcium levels are elevated upon contact with sulfated proteoglycans. *Soc. Neurosci. Abst.*, 19: 876

Takemura, H., Hughes, A.R., Thastrup, O. and Putney, J.W. (1989) Activation of calcium entry by the tumor promoter thapsigargin in parotid acinar cells. *J. Biol. Chem.*, 264: 12266–12271.

Tosney, K.W. and Landmesser, L.T. (1985a) Development of the major pathways for neurite outgrowth in the chick hind limb. *Dev. Biol.*, 109: 193–214.

Tosney, K.W. and Landmesser, L.T. (1985b) Growth cone morphology and trajectory in the lumbrosacral region of the chick embryo. *J. Neurosci.*, 5: 2345–2358.

Tosney, K. and Oakley, R. (1990) The perinotochordal mesenchyme acts as a barrier to axon advance in the chick embryo: Implications for a general mechanism of axonal guidance. *Exp. Neurol.*, 109: 75–89.

Tsien, R.W. and Tsien, R.Y. (1990) Calcium channels, stores, and oscillations. *Annu. Rev. Cell Biol.*, 6: 715–760.

Tsui, H.C.T., Lankford, K.L. and Klein, W.L. (1985) Differentiation of neuronal growth cones: specialization of filopodial tips for adhesive interactions. *Proc. Natl. Acad. Sci. USA*, 82: 8256–8260.

J. van Pelt, M.A. Corner H.B.M. Uylings and F.H. Lopes da Silva (Eds.)
Progress in Brain Research, Vol 102

CHAPTER 3

Filopodia as detectors of environmental cues: signal integration through changes in growth cone calcium levels

S.B. Kater, R.W. Davenport and P.B. Guthrie

Program in Neuronal Growth and Development, Department of Anatomy and Neurobiology, Colorado State University, Ft. Collins, CO 80523, U.S.A.

Introduction

Many answers to questions about how specific neuronal circuity is formed can be found at the level of the neuronal growth cone (Kater and Letourneau, 1985; Letourneau et al., 1992). This structure, found at the tips of elongating neurites, is also primarily responsible for regrowth and guidance during regeneration and repair. Even more important is the presence of growth cones in mature undamaged nervous systems undergoing remodeling. The repeated appearance of the neuronal growth cone clearly indicates that this structure is one of the primary decision-making elements throughout the life of the organism, whose actions result in a vast spectrum of dynamic morphological events. Given the importance of this structure, it is not surprising that it has received a great deal of attention ranging from morphological to biochemical studies. Neuronal growth cones, with their characteristic, long filopodia and expanded lamellipodium, appear to contain all of the machinery for the generation of the neuritic processes laid down behind them. The growth cone also contains machinery capable of directed navigation (decision making) along the tortuous course of the various molecular terrains encountered during pathfinding. Finally the growth cone appears to be the primary structure for the initial events in the formation of a synaptic terminal.

The net result of the events governed by the neuronal growth cone generates a mature neuronal structure. This neuronal structure determines how the neuron will function within the circuitry within which it participates. It is important to note that the final established morphology is going to be a product of two important sets of developmental forces: the genotype of the neuron and the environmental influences that impinge upon that neuron during its development. Clearly, different neurons respond differently to the same environmental cues because of their genetic makeup. This will be discussed in detail below. Equally important is the fact that different neurons will respond differently to the same environmental cues as a result of their developmental history; the previously encountered cues can alter the response of a growth cone to new environmental signal. Finally, the environmental terrain over which the developing growth cone must travel can be highly varied. Thus, only by understanding a combination of these two factors can one precisely define why any given neuron takes the form it does and participates within the circuitry in the precise manner in which we find it.

In this article we examine a selected series of environmental stimuli that are known to interact with growth cones. We examine how these stimuli are translated by the second messenger systems intrinsic to the growth cone proper. In particular, this article emphasizes the role of intracellular calcium ions as the messenger translating extracellular environmental cues into changes in neuronal growth cone behavior. The first part of this article examines the effects of two distinctive classes of external cues: neurotransmitters and membrane protein NI-35, derived from central nervous system oligodendrocytes. We discuss how these two sets of stimuli interact with intracellular systems within the cell. In the next section, we provide a description of what is now termed 'the calcium hypothesis for the control of growth cone behavior' (Kater et al., 1988). Finally, in the last sections of this paper, we examine an extended set of roles for intracellular calcium and how individual components of the growth cone interact. In doing so, we will present the antennae-amplifier model of filopodia/growth cone interaction and how this model incorporates changes in intracellular calcium from distant filopodia into a graded set of actions within the growth cone proper.

Stimuli from the external environment alter growth cone behavior

The behaviors exhibited by the neuronal growth cone can be simply classified as elongating, branching, turning, or stopping. For example, a neurite elongating in essentially a straight line might turn in response to an environmental cue, might cease motility in response to another cue or might reach its target where motility must also stop. Each of these conditions appears possible as a result of specific interactions between the neuronal growth cone and specific elements of the molecular terrain over which it traverses. For a long time, the focus of most investigations examining environmental cues was on factors which enhance neurite outgrowth. Molecules such as laminin or other extracellular matrix proteins certainly can enhance the activity of the growth cone (Patterson, 1988).

It is now clear that stop signals play an equally important role in development. Figure 1 demonstrates one of the most striking transitions that a neuronal growth cone can make: namely, from a highly dynamic, motile structure with numerous filopodia that is actively elongating, to one that has reached a stable state. Such growth cones often have no filopodia and generate no additional neuritic length. We now know that a variety of external cues (S_1, S_2, S_3, ...S_n in Fig. 1) can bring about such a transition (Kater and Mills, 1991). Below we describe two very different classes of molecule that we have investigated and then examine briefly similarities in their regulation of growth cone behavior.

The role of neurotransmitters in regulating growth cone behavior

By examining the very large growth cones from the snail, *Helisoma*, it has been possible to demonstrate a highly selective effect of particular neurotransmitters (Haydon et al., 1984). Individual identified neurons of *Helisoma* will respond to some transmitters while others will not. For example growth cones from neuron B19, and several other identified neurons from the buccal ganglia, respond to the presence of low concentrations of the neurotransmitter, serotonin, with complete inhibition of their outgrowth: filopodia widens, the lamellipodium shrinks and motility ceases. However, outgrowth from a number of other buccal neurons, for example, neurons B5, do not respond to serotonin. There is, indeed, a very highly selective effect of serotonin (Haydon et al., 1984). In addition, other neurotransmitters can affect these neurons in different ways. For example, the neurotransmitter acetylcholine has, by itself, no obvious effects on the growth cones of neuron B19. When acetylcholine is present, however, the subsequent application of serotonin fails to alter behavior of B19 growth cones (McCobb et al., 1988). Thus, the growth cone is integrating information from the two different neurotransmitters (Fig. 2). These results in vitro are paralleled by findings in vivo which demonstrate that the presence of serotonin does, indeed, alter neuronal morphology during embryonic development (Goldberg and Kater, 1989).

The initial discovery of a role for neurotransmitters in growth cone guidance in *Helisoma* has now been confirmed in a variety of nerve cell types. For instance, cultured rat hippocampal pyramidal neurons from day 18 embryos also display a selective responsiveness to an excitatory neurotransmitter. Glutamate applied to these neurons results in a complete inhibition of outgrowth of dendritic growth cones (Mattson et al., 1988). Interestingly, the selectivity here is for dendritic versus axonal growth cones; there is no obvious effect of glutamate at moderate concentrations on axonal growth cones. In continued parallel with the findings in *Helisoma*, the inhibitory transmitter γ-aminobutyric acid (GABA) has, by itself, no obvious effect on any of the growth cones of hippocampal pyramidal neurons. However, when GABA is present, the subsequent application of glutamate fails to inhibit outgrowth. Taken together, these findings strongly suggest that, early in development, neurotransmitters have a role which proceeds their role in neural transmission. The ability of neurotransmit-

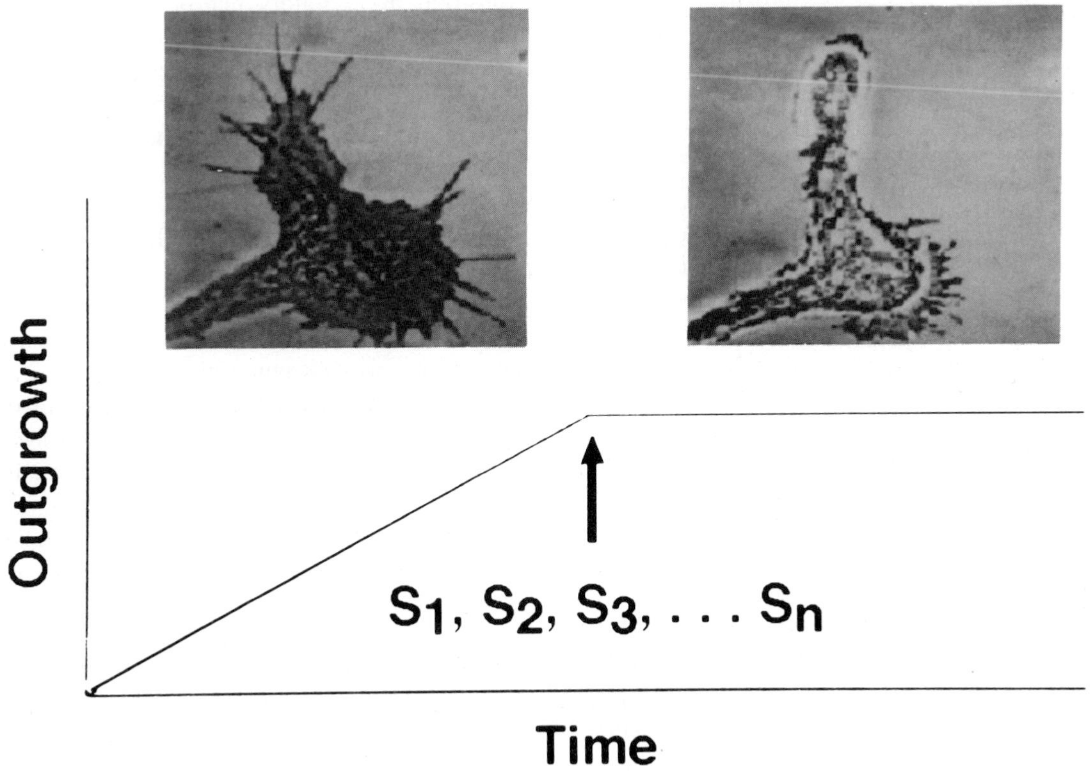

Fig. 1. A semi-diagrammatic representation of the effect of neuronal growth cone 'stop signals'. A variety of signals (S_1, S_2, S_3, ...S_n) in the environments encountered by growth cones are now known to inhibit neurite outgrowth. Stimuli such as the generation of action potentials, the application of specific neurotransmitters and contact with certain cell surface molecules can result in the abrupt inhibition of motility (the graphic line) and a dramatic change in the morphology of the growth cone from one containing multiple filopodia and a broad lamellipodium, to a much more simple and inanimate morphology.

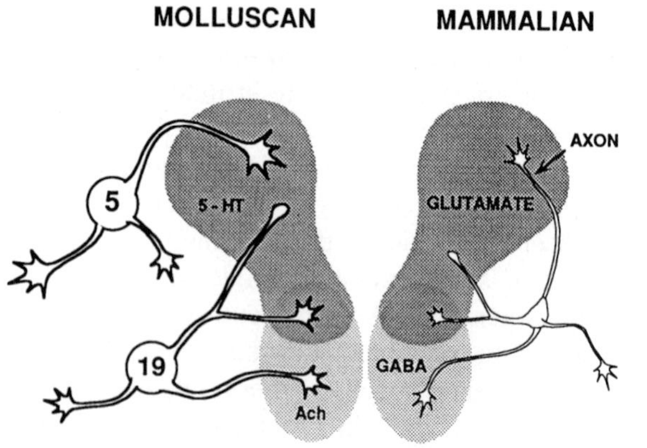

Fig. 2. Neurotransmitters can have highly specific effects on growth cones of a broad range of neuronal cell types. Both molluscan and mammalian cell types have been investigated in this laboratory. The specificity of the effect in molluscan systems is demonstrated from work on *Helisoma* where neuron B5 is totally unaffected by the presence of serotonin (5HT). However, the growth cones of neuron B19 are inhibited by the presence of this neurotransmitter. None of the *Helisoma* neurons are significantly affected by acetylcholine alone. However, when a neuron B19 growth cone encounters acetylcholine, it is protected against the inhibitory effects of serotonin. Rat hippocampal pyramidal neurons show a similar specificity. Axonal growth cones are relatively unaffected by glutamate, whereas dendritic growth cones are inhibited. The presence of GABA can negate the glutamate-induced inhibition of dendritic growth cones. Neurotransmitters have been studied now in a wide variety of laboratories and have shown to have prominent effects on neuronal development.

ters to regulate the behavior of neuronal growth cones sparks the imagination about the kind of interactions that neurons might have with one another during the complex development that occurs in vivo (Lipton and Kater, 1989).

The second messenger systems mediating these neurotransmitter effects on growth cone behavior have been investigated in detail. Intracellular calcium levels have been found to play a prominent role in transducing the effects of neurotransmitters. Fura-2 imaging of intracellular calcium changes in living growth cones has demonstrated that excitatory neurotransmitters evoke large rises in intracellular calcium in the growth cone proper (Connor, 1986; Cohan et al., 1987). Furthermore, these calcium increases can be negated by the appropriate inhibitory neurotransmitter (acetylcholine in the case of *Helisoma* and GABA in the case of pyramidal neurons of the hippocampus (McCobb et al., 1988; Mattson and Kater, 1989). The effects of these neurotransmitters on growth cone behavior are blocked when calcium influx is prevented (Mattson and Kater, 1987; Mattson et al., 1988). Thus, calcium influx is necessary to bring about inhibition. At sufficient concentrations, growth cones from many different animal types are found to respond to addition of the ionophore, A23187, which carries calcium from the outside to the inside of the cell, by an inhibition of outgrowth. Thus, calcium is *sufficient* to bring about inhibition. These results have strongly implicated calcium as the primary second messenger system mediating inhibition of active elongation by excitatory neurotransmitters. The finding that a calcium-mediated change in growth cone behavior can be evoked by an environmental stimulus has laid the groundwork for investigations on other classes of environmental cues that inhibit growth cone motility.

The CNS myelin-associated growth inhibitory protein, NI-35

Largely as a result of the work of Dr. Martin Schwab and his colleagues, we now know of the existence of an important molecule derived from oligodendrocytes in the CNS (Caroni and Schwab, 1988). The molecule NI-35 appears to have the capability of inhibiting neurite outgrowth. The potential importance of this molecule in regulating regeneration within the CNS cannot be understated since antibody neutralization of NI-35 significantly enhances regeneration within the CNS. We have undertaken an investigation of the second messenger systems underlying the NI-35 induced transformation of growing growth cones to inhibited growth cones. An important foundation for this study was laid by the work of (Bandtlow et al., 1990a) who demonstrated that the activity of dorsal root ganglion neuron growth cones is abruptly altered upon encountering an oligodendrocyte (Fig. 3). Contact of even a single

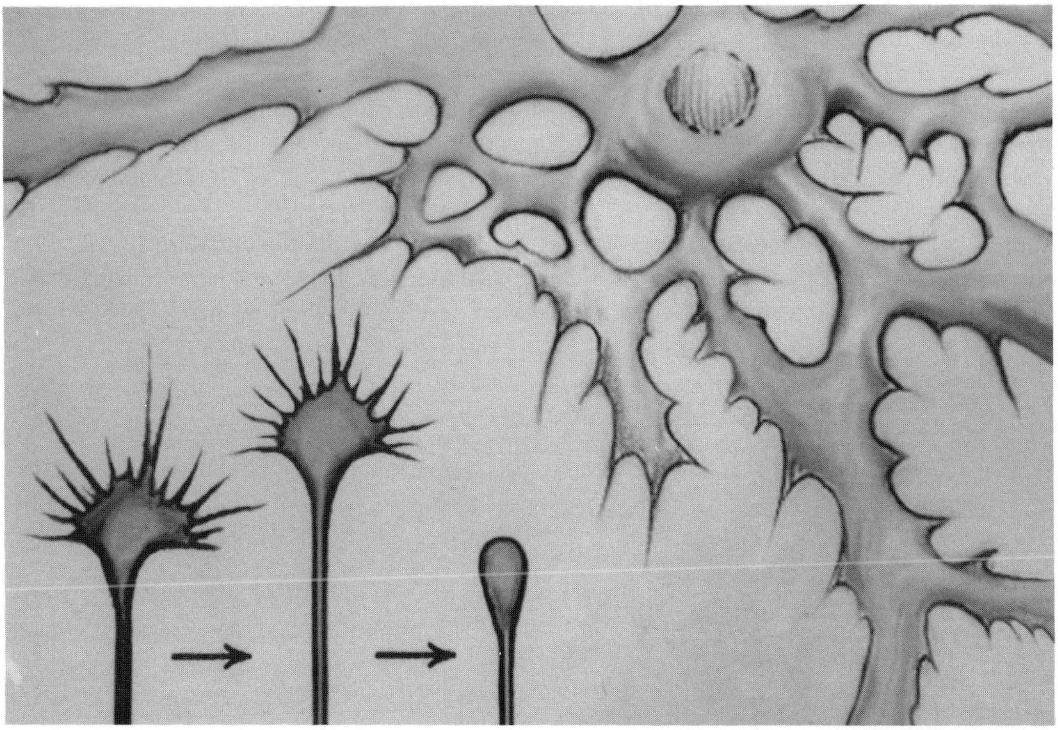

Fig. 3. A schematic representation (adapted from the work of Bandtlow et al., 1990a) of the interaction of a dorsal root ganglion neuron growth cone with an oligodendrocyte. Contact by even a single filopodium from the neuron can result in an oligodendrocyte induction of growth cone collapse. This collapse is characterized by paralysis of the growth cone for a considerable length of time.

filopodium with an oligodendrocyte surface results in an abrupt collapse; while the initial filopodium may remain extended, the growth cone proper shrinks and elongation is halted. These collapsed growth cones remain immobile for considerable periods of time. In order to investigate the nature of the second messenger(s) underlying this effect, we incorporated NI-35 into liposomes, as a vehicle for controlled stimulation. Application of such liposomes to growing dorsal root ganglions (DRG) growth cones resulted in growth cone collapse which is essentially indistinguishable from that evoked by contact with oligodendrocytes (Fig. 4). This collapse did not occur with empty vesicles or vesicles containing control proteins. Furthermore, when NI-35 liposomes were neutralized by the antibody IN-1, addition of the liposomes failed to alter growth cone behavior. The striking nature of this transformation is shown in Fig. 5. Thus, this is a highly specific protein which, upon contact with a growth cone, evokes a highly specific behavior (Bandtlow et al., 1990b).

NI-35 has been shown to exert its effects by evoking a release of calcium from intracellular stores (Bandtlow et al., 1993). Figure 6 illustrates the time-course of growth cone collapse as well as the time course in the rise in intracellular calcium. Blocking this rise in intracellular calcium using the agent dantrolene, which prevents calcium release from intracellular stores, also blocks growth cone collapse. Furthermore, the evoked rise in intracellular calcium, as well as the growth cone collapse itself, is prevented by the blocking antibody IN-1. Taken together, these results demonstrate that for NI-35, as for neurotransmitters, calcium is the primary second messenger conveying information which results in this stark change in behavior of the growth cone.

Intracellular calcium is a primary regulator of growth cone behavior

A large volume of data (see Kater and Mills, 1991) demonstrates a prominent quantitative association between intracellular calcium levels and changes in growth cone behavior. In brief, it appears that there is an optimum calcium concentration for the various behaviors that growth cones can exhibit. For instance, for elongation, a typical growth cone might extend at the fastest rate at, perhaps, 100 nM free intracellular calcium concentration. Either decreasing intracellular calcium concentrations to 50 nM or increasing concentration to 800 nM would result in a diminution of this peak rate. This is shown in the model depicted in Fig. 7.

This model was developed (Kater et al., 1988) to explain various apparently contradictory experimental results. On the one hand were observa-

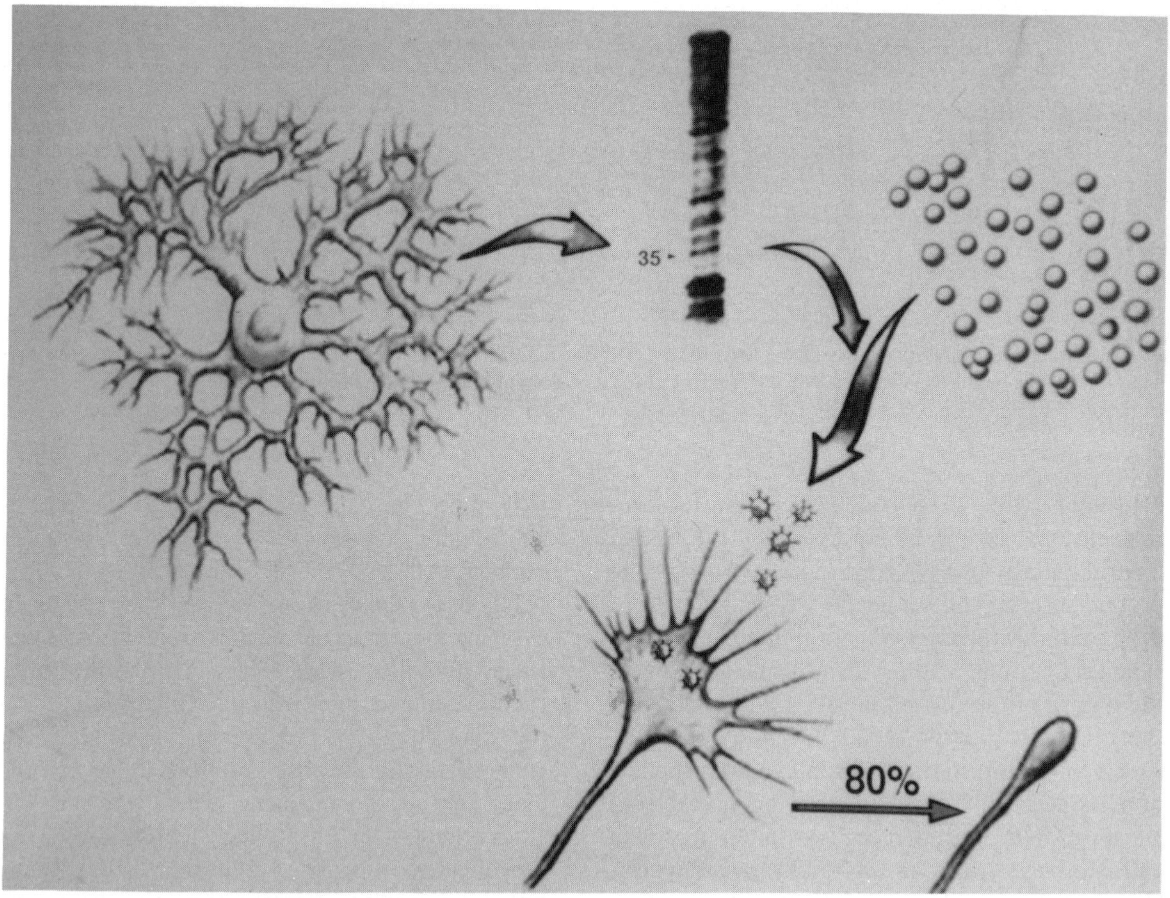

Fig. 4. A diagrammatic representation of the experimental paradigm used to investigate the role of intracellular calcium levels in mediating the effects of NI-35, the CNS myelin-associated growth inhibiting protein. The gel depicted above (adapted from Caroni and Schwab, 1988) indicates the presence of NI-35 which can be incorporated into liposomes. These liposomes can be used effectively as an experimental means of delivering NI-35 to growth cones. Contact of these liposomes with the growth cone result in a collapse, essentially indistinguishable from that observed with oligodendrocyte contact. This induction of collapse is highly specific to liposomes to incorporating NI-35 and can be neutralized by the blocking antibody IN-1 (see Bandtlow et al., 1990b).

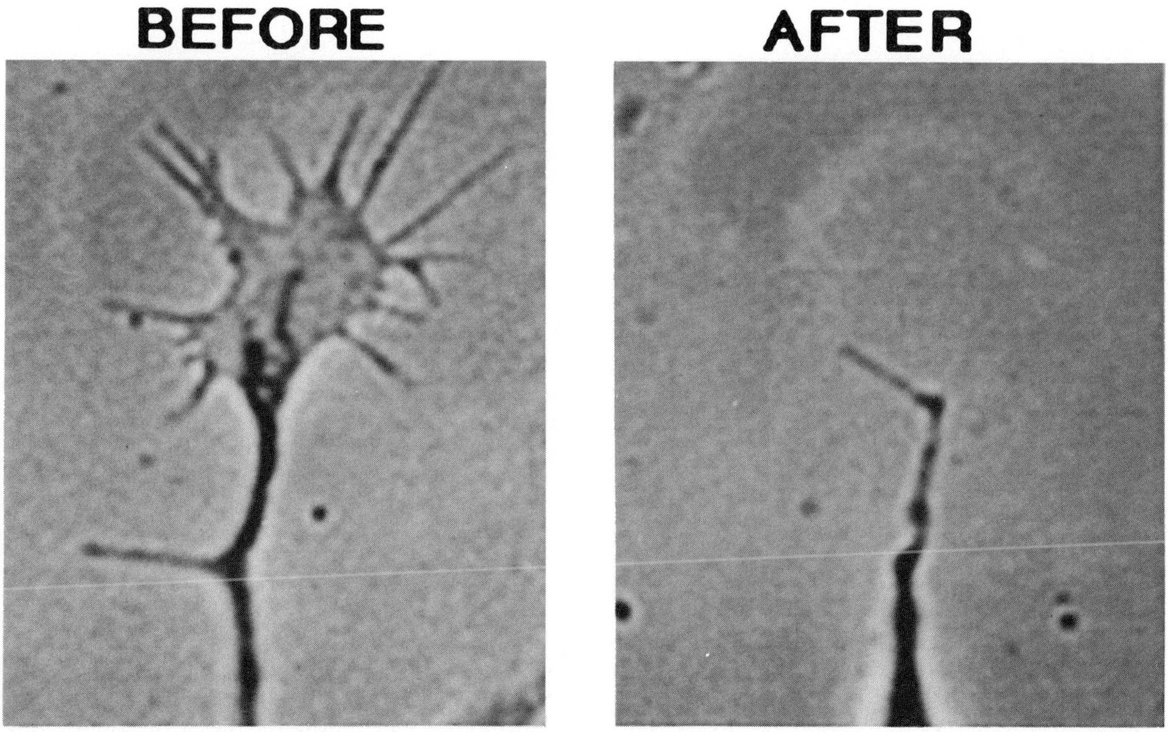

Fig. 5. Computer-derived images of IN-35 induced collapse of a DRG growth cone. Before application of NI-35 loaded vesicles, growth cones displayed characteristic filopodial and lamellipodial structure along with a high level of motility. After contact with NI-35, the growth cone has collapsed, displaying no filopodia, no lamellipodia, and paralysis (modified from Bandtlow et al., 1990b).

tions that actions which would be expected to decrease levels of intracellular (addition inorganic calcium channel blockers or removal of extracellular calcium) inhibited outgrowth (Nishi and Berg, 1981; Suarez-Isla et al., 1984), or had no effect (Letourneau and Wessells, 1974; Bixby and Spitzer, 1984). On the other hand, treatments expected to increase levels of intracellular calcium were reported to either inhibit (Haydon et al., 1984; Suarez-Isla et al., 1984; Hantaz-Ambroise and Trautmann, 1989; Robson and Burgoyne, 1989) or accelerate (Hinnen and Monard, 1980; Nishi and Berg, 1981) outgrowth. With the development of the calcium indicator fura-2 (Grynkiewicz et al., 1985), it became possible to quantitatively determine the relationship between intracellular calcium and growth cone behavior. In fact, a careful quantitative study of intracellular calcium levels and neurite outgrowth rates has confirmed the bell-shaped relationship in dorsal root ganglion neurons (Al-Mohanna et al., 1992). The model is further supported by the observation that, in mammalian dorsal root ganglion neurons, the growth cone actin cytoskeleton (and, therefore, lamellipodial extent) shows a bell-shaped dependence on intracellular calcium (Lankford and Letourneau, 1989).

Today, after nearly a decade of experiments utilizing calcium indicators to investigate the role calcium serves in controlling neuronal outgrowth, numerous data from different cell types indicate there is an optimal calcium concentration for the various behaviors that growth cones can exhibit. Different aspects of growth cone behavior, such

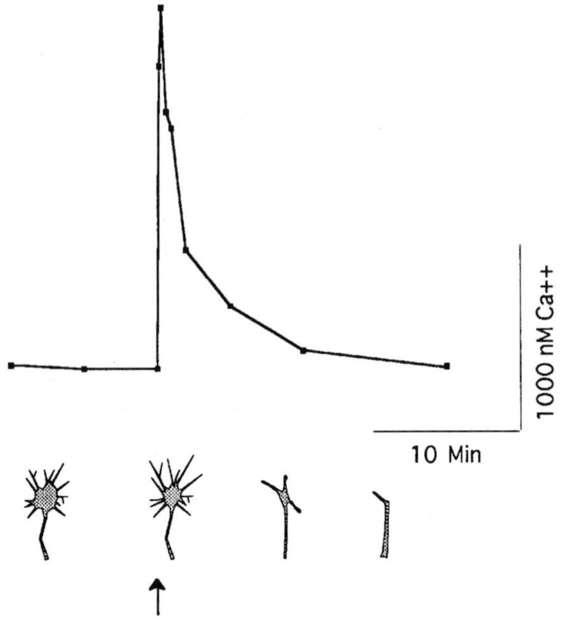

Fig. 6. A diagrammatic representation of the calcium rise induced by application of NI-35 and the subsequent collapse of a DRG growth cone. Application of the blocking antibody IN-1, prevents both the calcium rise and all effects on growth cone morphology and motility (modified from Bandtlow et al., 1990b).

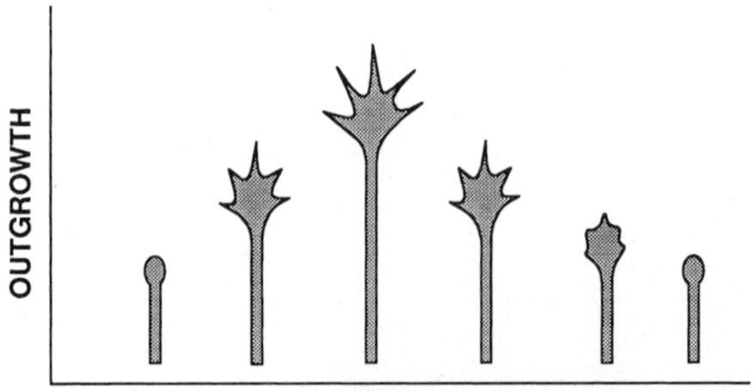

Fig. 7. A schematic representation of the calcium hypothesis of growth cone regulation. Growth cones and their neurites are displayed in graphic form to indicate the degree of outgrowth (as an indication of rate) a neuron would generate when its growth cone had a particular intracellular calcium level. A clear optimum is observed with a particular intracellular calcium giving the highest rate of outgrowth. A complete arrest of growth cone motility can occur either at very low or very high intracellular calcium concentrations.

as filopodial activity, might in fact have slightly different optimal calcium ranges, allowing, for example, for somewhat independent regulation of elongation rates, filopodial number, filopodial length and lamellipodial extent. Though there is a permissive range of 100 or 200 nM intracellular calcium concentration, it, appears that an absolute minimum and maximum calcium level does exist for several of the behaviors that growth cones express.

Filopodia as antennae probing distant environmental terrains

One of the most striking features of growth cones is the characteristic formation and regression of filopodia along the growth cone surface. These filopodia can reach out and contact specific environments spanning a wide radius around the growth cone. Furthermore, they can contact distant points up to 2 h before the advancing growth cone (Rehder and Kater, 1992). In our recent investigations, we have been able to demonstrate that filopodia are endowed with sensory capabilities. This was possible by virtue of the use of the large growth cones of identified neurons of *Helisoma* and our ability to isolate filopodia from the rest of the growth cone (Rehder and Kater, 1992). It has been difficult in past studies to determine the precise role of filopodia because of their intimate association with the growth cone. In our studies, neurons were filled with the dye fura-2, which diffused into the growth cone and into the associated filopodia. Figure 8 demonstrates the method employed and how, using a microelectrode as a surgical tool, it is possible to isolate individual filopodia filled with fura-2 (Davenport et al., 1993). These filopodia showed calcium responses that were indistinguishable from that which had been observed in the growth cone proper (Fig. 9). For instance, application of serotonin caused a distinct rise in intracellular cal-

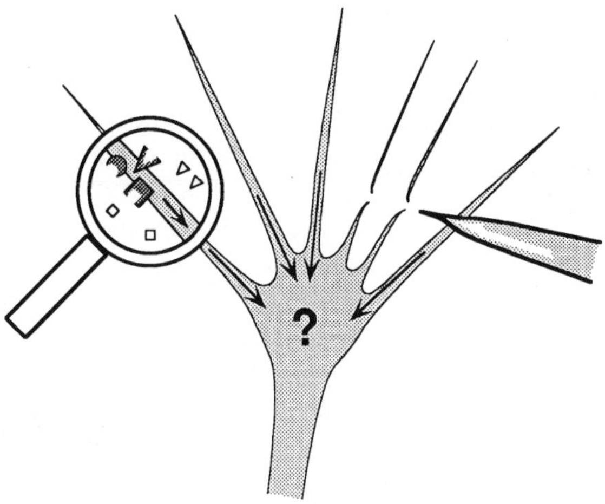

Fig. 8. Experiments were performed to test the hypothesis that filopodia contained all of the machinery to act as receptive structures. An antennae-like role for filopodia could be determined precisely only by isolating these entities from the growth cone itself. Fura-2 filled neurons were employed and a glass microelectrode used to surgically isolate individual filopodia from the growth cone proper. These fura-2-filled isolated filopodia were then exposed to agents which these structrues might detect in their local environment.

cium; subsequent application of acetylcholine would restore calcium to original levels. These findings indicate clearly that filopodia contain receptive systems to respond to environmental cues, channels for allowing calcium influx and also mechanisms for clearing calcium once it has entered the filopodium.

Clearly, the filopodium is a basic unit of organization which can obtain information considerably in advance of the growth cone itself. In addition, and quite remarkably, isolated filopodia can, themselves contract in response to environmental cues. Excitatory neurotransmitters specifically inhibit growth cone motility—causing filopodia and lamellipodia to retract leaving the growth cone in a non-motile form. Application of the excitatory neurotransmitters dopamine or serotonin to isolated filopodia also caused significant and dose-dependent reduction in filopodial lengths. Thus, in addition to their receptive systems, filopodia appear to contain the mechanisms necessary for contraction. Taken together, these data suggest the filopodium may represent the minimal sensory-motor unit of the developing neuron.

The antennae-amplifier model of growth cone function

A large amount of data has now accrued that allows us to generate a working model that includes an extended role for filopodia and a prominent role for intracellular calcium. This model (Fig. 10), suggests that actively elongating filopodia act as sensors in advance of the growth cone. While this model was developed to explain observations employing calcium indicators, other second messenger systems will need to be included into the amplifier model. Individual filopodia may, in fact, contact agents which cause calcium to rise. For the most part, however, such an isolated event would be lost due to the large sink provided by the growth cone proper. On the other hand, if several filopodia were to contact environmental information which causes rises in calcium within these filopodia, a local rise in calcium could occur within the growth cone. Such a local rise might, in fact, be responsible for inhibiting or, given the bell shaped curve of Fig. 7, accelerating the local events in that side of the growth cone. When even more filopodia contact environmental cues that elevate filopodial calcium, calcium may increase sufficiently within the growth cone to result in a calcium-induced calcium release from intracellular stores. This type of interaction would result in small calcium rises being *amplified* to a much larger one, bringing total inhibition of the growth cone itself. Such an inhibition would surely produce collapse and cessation of growth cone activities. Taken together, this model, which is presently under detailed investigation, suggests a prominent (though not exclusive) role for intracellular calcium as a key regulator that ultimately can determine central aspects of the formation, repair, and modification of functional neuronal circuitry.

Conclusion

Taken together, the work we have described illustrates the importance in considering not only the differences between particular neurons (i.e. their genetic predispositions) but also the specific environments that they encounter. Clearly, the environment of the developing nervous system, as well as the environment encountered by the mature nervous system during repair and remodelling, contains a wide diversity of environmental cues which range from having subtle to extremely powerful effects on the behavior of the neuronal growth cone. The work briefly described here demonstrates that changes in intracellular calcium concentration can account for both of these kinds of behavioral responses. Global changes in intracellular calcium can evoke a complete paralysis of the growth cone through growth cone collapse. On the other hand, highly local changes in intracellular calcium can result in navigation or pathfinding through more subtle changes in growth cone morphology and motility. It is also recognized that a good model for filopodial-growth cone interactions can be found in one that considers this to be an *antennae-amplifier* system. The long evanescent filopodia contact distant environmental terrains significantly in advance of the more slowly moving growth cone proper. Environmental stimuli can produce changes in intracellular calcium within filopodia; when these changes are large enough they produce local changes in the growth cone. Finally, when such local changes reach a high enough value, they can result in amplification through calcium-induced calcium release. Taken together, these data illustrate the powerful role of this specific second messenger system in governing the final morpho-

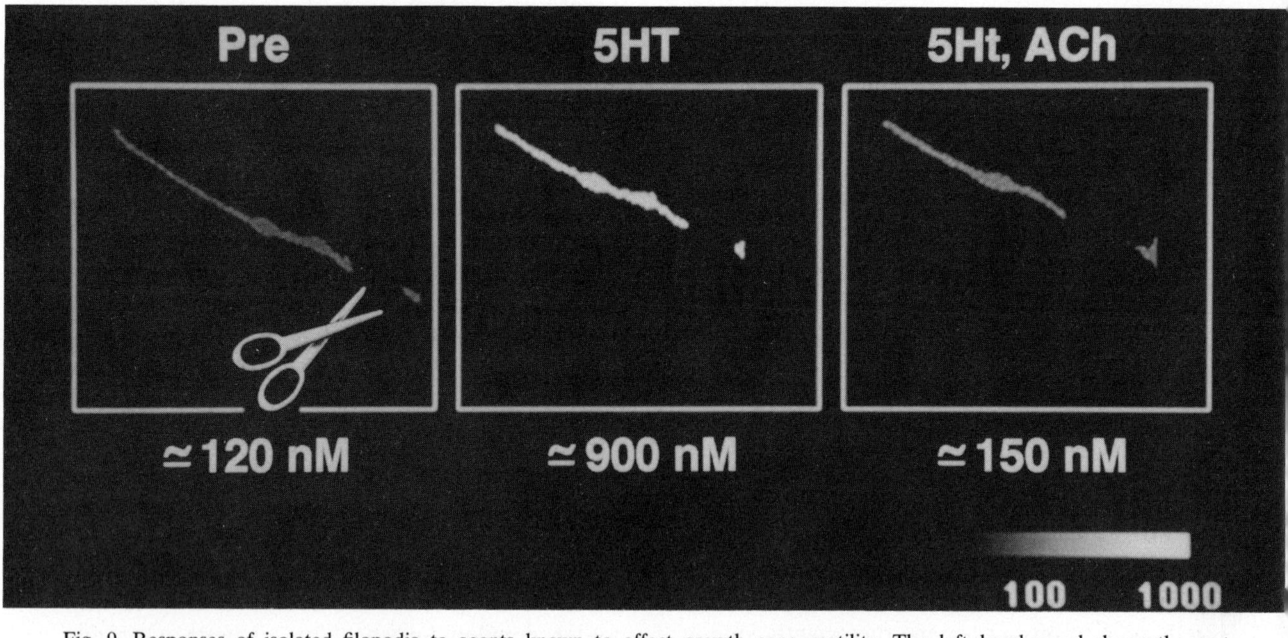

Fig. 9. Responses of isolated filopodia to agents known to affect growth cone motility. The left hand panel shows the rest intracellular calcium levels of a surgically isolated individual filopodium from a *Helisoma* neuron B19 growth cone. The middle panel shows a very large rise in intracellular calcium induced by the presence of serotonin. Rest levels are restored in the last panel by the presence of acetylcholine in combination with serotonin. Note also the shortening of the filopodium on exposure to serotonin. These results demonstrate that the isolated filopodium can, indeed, serve an antennae-like function in the context of its role in the growth cone. Filopodia, undoubtedly, contain receptive structures for both serotonin and acetylcholine, systems such as channels for an influx of intracellular calcium and efflux mechanisms which allow calcium in the last panel to be restored to near rest levels.

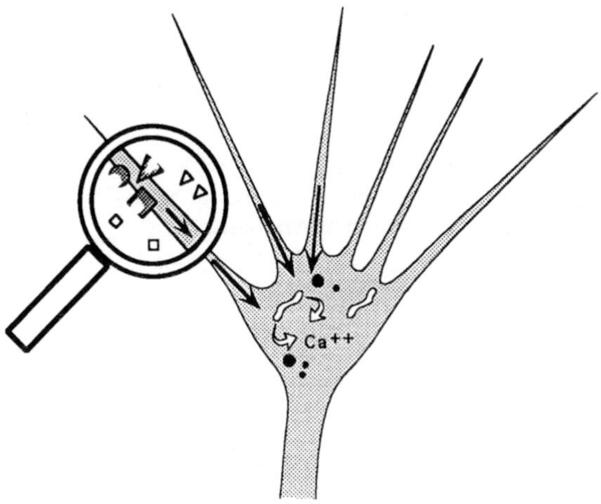

Fig. 10. The antennae-amplifier model of growth cone function. According to this model, filopodia serve as primary antennae for sensing distant information. Appropriate external cues activate filopodia with a rise in intracellular calcium which can diffuse along the length of the filopodium to the growth cone proper (blank arrows indicate calcium movement). According to this model, an individual filipodium could contribute only a very small rise in intracellular calcium to the growth cone. A second filopodium contacting a similar cue might also result in a small rise again due to the large sink capacity of the growth cone proper. However, when a sufficient number of filopodia contact on environmental cue (or contact of a single filopodium creates a sufficiently large source of calcium influx), a rise in intracellular calcium can be initiated in the growth cone. Rises in calcium within the growth cone proper can become sufficient to trigger a calcium-induced calcium release from intracellular stores. This results in a massive intracellular calcium release which (1), if it is located in a specific side of the growth cone, could alter the direction of navigation of the growth cone, or (2) if it is general to the overall growth cone, could result in collapse of the growth cone structure and the inhibition of growth cone motility.

logical structure for neurons and, thereby, the final dynamics of neuronal circuitry within the brain.

References

Al-Mohanna, F.A., Cave, J. and Bolsover, S.R. (1992) A narrow window of intracellular calcium concentration is optimal for neurite outgrowth in rat sensory neurones. *Dev. Brain Res.*, 70: 287–290.

Bandtlow, C., Zachleder, T. and Schwab, M. (1990a) Oligodendrocytes arrest neurite growth by contact inhibition. *J. Neurosci.*, 10: 3837–3848.

Bandtlow, C.E., Schmidt, M.F., Kater, S.B. and Schwab, M.E. (1990b) Inhibition of neurite growth by a CNS myelin-specific protein (NI-35) is correlated to changes of [Ca^{2+}]. *Cell Biol. Int. Rep.*, 14: S138.

Bandtlow, C.E., Schmidt, M.F., Hassinger, T.D., Schwab, M.E. and Kater, S.B. (1993) Role of intracellular calcium in NI-35-evoked collapse of neuronal growth cones. *Science*, 259: 80–83.

Bixby, J.L. and Spitzer, N.C. (1984) Early differentiation of vertebrate spinal neurons in the absence of voltage-dependent Ca^{2+} and Na^+ influx. *Dev. Biol.*, 106: 89–96.

Caroni, P. and Schwab, M.E. (1988) Two membrane protein fractions from rat central myelin with inhibitory properties for neurite growth and fibroblast spreading. *J. Cell Biol.*, 106: 1281–1288.

Cohan, C.S., Connor, J.A. and Kater, S.B. (1987) Electrically and chemically mediated increases in intracellular calcium in neuronal growth cones. *J. Neurosci.*, 7: 3588–3599.

Connor, J.A. (1986) Digital imaging of free calcium changes and of spatial gradients in growing processes in single, mammalian central nervous system cells. *Proc. Natl. Acad. Sci. U.S.A.*, 83: 6179–6183.

Davenport, R.W., Dou, P., Rehder, V. and Kater, S.B. (1993) A sensory role for neuronal growth cone filopodia. *Nature* 361: 721–724.

Goldberg, J.I. and Kater, S.B. (1989) Expression and function of the neurotransmitter serotonin during development of the *Helisoma* nervous system. *Dev. Biol.*, 131: 483–495.

Grynkiewicz, G., Poenie, M. and Tsien, R. (1985) A new generation of Ca^{2+} indicators with greatly improved fluorescence properties. *J. Biol. Chem.*, 260: 3440–3450.

Hantaz-Ambroise, D. and Trautmann, A. (1989) Effects of calcium ion on neurite outgrowth of rat spinal cord neurons in vitro: the role of non-neuronal cells in regulating neurite sprouting. *Int. J. Dev. Neurosci.*, 7: 591–602.

Haydon, P.G., McCobb, D.P. and Kater, S.B. (1984) Serotonin selectively inhibits growth cone dynamics and synaptogenesis of specific identified neurons of *Helisoma*. *Science*, 226: 561–564.

Hinnen, R. and Monard, D. (1980) Involvement of calcium ions in neuroblastoma neurite extension.In: L. Jimener de Ausa (Ed.), *Control Mechanisms in Animal Cells*, Raven Press, New York, pp. 315–323.

Kater, S.B. and Letourneau, P. (1985) *The Biology of the Neuronal Growth Cone*, Alan R. Liss, New York.

Kater, S. and Mills, L. (1991) Regulation of growth cone behavior by calcium. *J. Neurosci.*, 11: 891–899.

Kater, S.B., Mattson, M.P., Cohan, C.S. and Connor, J.A. (1988) Calcium regulation of the neuronal growth cone. *Tends Neurosci.*, 11: 315–321.

Lankford, K.L. and Letourneau, P.C. (1989) Evidence that

calcium may control neurite outgrowth by regulating the stability of actin filaments. *J. Cell Biol.*, 109: 1229–1243.

Letourneau, P.C. and Wessells, N.K. (1974) Migratory cell locomotion versus nerve axon elongation. *J. Cell Biol.*, 61: 56–69.

Letourneau, P.C., Kater, S.B. and Macagno, E.R. (1992). *The Nerve Growth Cone*, Raven Press, New York.

Litpon, S.A. and Kater, S.B. (1989) Neurotransmitter regulation of neuronal outgrowth, plasticity, and survival. *Trends Neurosci.*, 12 (7): 265–270.

Mattson, M.P. and Kater, S.B. (1987) Calcium regulation of neurite elongation and growth cone motility. *J. Neurosci.*, 7: 4034–4043.

Mattson, M.P. and Kater, S.B. (1989) Excitatory and inhibitory neurotransmitters in the generation and degeneration of hippocampal neuroarchitecture. *Brain Res.*, 478: 337–348.

Mattson, M.P., Dou, P. and Kater, S.B. (1988) Outgrowth-regulating actions of glutamate in isolated hippocampal pyramidal neurons. *J. Neurosci.*, 8: 2087–2100.

McCobb, D.P., Cohan, C.S., Connor, J.A. and Kater, S.B. (1988) Interactive effects of serotonin and acetylcholine on neurite elongation. *Neuron*, 1: 377–385.

Nishi, R. and Berg, D.K. (1981) Effects of high K^+ concentrations on the growth and development of ciliary ganglion neurons in cell culture. *Dev. Biol.*, 87: 301–307.

Patterson, P. (1988) On the importance of being inhibited, or saying no to growth cones. *Neuron*, 1: 263–267.

Rehder, V. and Kater, S.B. (1992) Regulation of neuronal growth cone filopodia by intracellular calcium. *J. Neurosci.*, 12: 3175–3186.

Robson, S.J. and Burgoyne, R.D. (1989) L-type calcium channels in the regulation of neurite outgrowth from rat dorsal root gangion neurons in culture. *Neurosci. Lett.*, 104: 110–114.

Suarez-Isla, B.A., Pelto, D.J., Thompson, J.M. and Rapoport, S.I. (1984) Blockers of calcium permeability inhibit neurite extension and formation of neuromuscular synapses in cell culture. *Dev. Brain Res.*, 14: 263–270.

J. van Pelt, M.A. Corner H.B.M. Uylings and F.H. Lopes da Silva (Eds.)
Progress in Brain Research, Vol 102

CHAPTER 4

Microtubule transport and assembly cooperate to generate the microtubule array of growing axons

Mark M. Black

Department of Anatomy and Cell Biology, Temple University School of Medicine, 3400 N. Broad Street, Philadelphia, PA 19140, U.S.A.

Introduction

Nervous systems collect, process, store and transmit information. The way in which nervous systems perform these tasks is defined by the pattern of interconnections of their neurons. The structural basis of these interconnections is the elongate neurite, of which there are two kinds: the axon and the dendrite. To develop a nervous system, neurons must extend their axons and dendrites often over great distances in order to establish contact with appropriate targets cells. The mechanisms of axonal and dendritic growth are complex, involving the coordination of many intracellular events. These events in turn are regulated by interactions between the neuron and its environment which represent the pathfinding mechanisms that guide axons and dendrites to their appropriate destinations.

This article focuses on one of the essential events in axon growth, namely the elaboration of the microtubule (MT) array of these neurites. In neurons, as in many other cell types, MTs provide an architectural framework for establishing and maintaining cellular asymmetries. They also organzie cytoplasm to carry out the variety of motile and metabolic processes that are essential to life. In this article, I discuss the current understanding of the dynamic processes that shape the MT array of growing neurites.

Specialized MT arrays are generated by growing neurons that are essential for the elaboration of axonal structure. This was first indicated by electron microscopic studies which revealed organized arrays of parallel MTs that appeared coincident with the initiation of axon outgrowth (Luckenbill-Edds et al., 1979; Stevens et al., 1988). The amount of polymer comprising these arrays increases coordinately with increases in axon length, due to increases in both the number and length of MTs in the axon (Stevens et al., 1988). This steady addition of polymer to growing axons maintains continuity of the MT array from the cell body to the tips of these neurites. These observations suggest that MTs play a role in establishing the elongate shape of the axon, a view that has been confirmed by the observation that treatment of growing neurons with agents that block MT assembly prevents further axon outgrowth and can cause axons to retract (Yamada et al., 1970; Daniels, 1973; Stevens et al., 1988). Thus, the steady addition of MT polymer to elongating axons is required to sustain their growth.

MT polymer for the axon is formed by the assembly of tubulin (and other MT proteins) into MTs. A variety of different approaches have been used to explore where this assembly occurs. While it is now clear that MT assembly activity varies dramatically along the length of the axon, the relative contribution of assembly at different sites to the net addition of MT polymer required for axon growth remains a matter of speculation. Also relevant to this issue is the delivery of MT proteins to the axon. These proteins are synthesized in the cell body, and then they are conveyed into the axon by active transport mechanisms (reviewed in Grafstein and Forman, 1980; Lasek, 1988). These transport processes play a crucial role in the elaboration of the MT array of developing axons by supplying the MT proteins required for growth. In addition, the available evidence indicates that MTs themselves are conveyed by the transport machinery and that the transport mechanisms directly determine aspects of MT organization in axons. In this article, I review current information on MT assembly and transport in growing neurons, focusing on how these processes cooperate to generate the MT array of the axon.

MT assembly / disassembly dynamics varies dramatically in the cell body and along the axon

To understand how MT assembly within the axon contributes to the elaboration of the axonal MT array, it is necessary to know where this assembly occurs and how it contributes to the addition of new MT polymer required to sustain axon elongation. MT assembly/disassembly dynamics occurs throughout the axon. This has been demonstrated by microinjection (Okabe and Hirokawa, 1988; Keith and Blane, 1990), photobleaching (Lim et al., 1989, 1990; Okabe and Hirokawa, 1990, 1993), photoactivation (Reinsch et al., 1991; Okabe and Hirokawa, 1992), and immunohistochemical approaches (Baas and Black, 1990; Arregui et al., 1991). Many authors have focused attention on the growth cone as the principle site of MT assembly for the axon (reviewed in Hollenbeck, 1989; Nixon, 1992). However, more recent studies indicate that the cell body and proximal part of the axon are also active sites of MT assembly in growing neurons, and that more newly assembled polymer for the axon may be generated at these proximal sites than at the growth cone.

Studies using tyrosinated α-tubulin (tyr-tubulin) as a marker of newly assembled polymer have provided several insights into sites of MT assembly in growing neurons (Baas and Black, 1990; Arregui et al., 1991; Brown et al., 1992). Most α-tubulins are synthesized with a C-terminal tyrosine that can be removed by tubulin carboxypeptidase and then replaced by tubulin tyrosine ligase (reviewed in Barra et al., 1988; Bulinski and Gundersen, 1991). The substrate specificities of these enzymes are distinct: the tubulin carboxypeptidase acts on assembled tubulin, whereas the tubulin tyrosine ligase acts on unassembled tubulin, and rapidly retyrosinates any detyrosinated α-tubulin released from MTs. Tyr-tubulin is assembly competent and, once assembled into MTs, its detyrosination proceeds in a time-dependent manner. As a result, the relative amount of detyrosinated subunits in MTs increases progressively with their age and, because this increase occurs at the expense of tyrosinated subunits, the proportion of tyrosinated subunits in MTs decreases as they age. Thus, newly assembled MTs contain higher proportions of tyr-tubulin than older MTs (Kreis, 1987; Wehland and Weber, 1987; Bre et al., 1987; Gundersen et al., 1987; Schulze et al., 1987; Sherwin et al., 1987; Webster et al., 1987; Sherwin and Gull, 1989). Because the tubulin carboxypeptidase is active throughout growing neurons (Arregui et al., 1991; our own unpublished data), the relative tyr-tubulin content of MTs in the cell body and along the axon represents an indirect measure of the relative amount of newly assembled MT polymer in these regions.

Studies with rat sympathetic neurons have revealed three relatively discrete regions based on the relative tyr-tubulin content of their MT polymer; a proximal region which includes the cell body and the initial approximately 40 μm of the axon, a distal region, which includes the distal approximately 100 μm of the axon and the growth cone, and the axon shaft, which includes the axon in between the proximal and distal regions (Brown et al., 1992). The relative tyr-tubulin content of MTs was much higher in the proximal and distal regions compared with the axon shaft. Comparable results have also been obtained in studies on cerebellar neurons (Arregui et al., 1991). Analyses at the level of individual MTs indicate that most of the polymer in the proximal and distal regions of the axon is rich in tyr-tubulin. By contrast, in the axon shaft, only discrete MT segments are rich in tyr-tubulin, and these represent a minority of the total polymer in this region (Baas and Black, 1990; Ahmad et al., 1993; Baas et al., 1993; Brown et al., 1993). These observations indicate that while newly assembled polymer is present all along the axon, the youngest (most recently assembled) MT polymer is concentrated in the proximal and distal regions of growing neurons. This in turn suggests that these proximal and distal regions are especially active sites of MT assembly dynamics in growing neurons. These interpretations are fully supported by observations on the recovery of fluorescence after photobleaching in the cell bodies and neurites of PC12 cells (Lim et al., 1989) and by studies that have microinjected biotin-tubulin into neurons and then quantified its incorporation into MTs of the cell body and along the axon (Keith and Blane, 1990; Black and Brown, 1993).

The lengths of these proximal and distal axonal regions are relatively independent of axon length, at least in rat sympathetic neurons (Brown et al., 1992; Brown and Black, unpublished data). These studies examined axons that ranged in length from 161 μm to 1270 μm. The length of the proximal and distal axon regions were 41 $\pm$ 24 μm (mean $\pm$ S.D, n = 87 axons) and 116 $\pm$ 45 μm, respectively. Thus, the factors responsible for the enrichment of newly assembled polymer proximally and distally within growing axons are apparently not influenced by axon length.

These observations focus attention on two relatively discrete regions of growing neurons, one situated proximally and one situated distally, as especially active sites of MT assembly and, thereby, potential sites for the formation of new MT polymer for the axon during axon growth. The available data do not directly reveal the extent to which these proximal and distal regions contribute to the net addition of polymer to the axon during axon growth. However, a sense of this can be obtained by using the amount of tyr-tubulin in polymer in the cell body and along the axon as a rough approximation of the amount of newly assembled polymer in these regions. On this basis, images of growing neurons stained to reveal total tyr-tubulin in polymer indicate that the amount of newly assembled polymer in the cell body exceeds by many-fold that anywhere else in the neuron, and that the amount of newly assembled polymer in the proximal 10–20 μm of the axon is at least equivalent to that in the distal 10–20 μm of the axon. Thus, more polymer for growing axons may be generated in the cell body and proximal axon than in the distal axon and growth cone.

Organizing microtubule assembly in the axon

MTs are intrinsically polar in their kinetics of assembly; one end is favored for assembly over the other, and these two ends are referred to as plus and minus, respectively (Binder et al., 1975). Under appropriate in vitro conditions, MT proteins can self-assemble into new MTs. However, in cells, such self-assembly occurs to a very limited extent, if at all (Kirschner and Mitchison, 1986). Instead, discrete MT templates spatially organize MT formation in situ by directing the assembly of tubulin subunits into MTs (Brinkley, 1985). In this section, I consider the nature of the templates that organise MT assembly in growing axons.

The identity of MT templates in non-neuronal cells such as fibroblasts has been revealed in part by analyses of the sites of assembly of exogenous tubulin that has been microinjected into these cells (Soltys and Borisy, 1985; Schulze and Kirschner, 1986). These analyses showed that the injected tubulin assembled from the ends of existing MTs and from the centrosome. The centrosome consists of a pair of centrioles surrounded by a cloud of relatively dense staining material

(Brinkley, 1985). This cloud contains elements that nucleate entirely new MTs from tubulin subunits (Brinkley, 1985; Joshi et al., 1992). In many cell types, MTs nucleated by the centrosome remain attached to it via their minus ends, and as they elongate, their plus ends radiate away from the centrosome. In this way, the centrosome generates an array of MTs that has a uniform polarity orientation (Brinkley, 1985). The centrosome also influences the number of MTs formed in cells (Brinkley, 1985) and the lattice structure of these MTs (Evans et al., 1985).

It is generally assumed that MT assembly in axons is organized by discrete MT templates, in part because axonal MTs are uniform in polarity orientation: their plus ends point away from the cell body toward the axon tip (Heidemann et al., 1981; Burton and Paige, 1981). This pattern is established early in development (Baas et al., 1989), and is recapitulated during recovery after treatments that promote MT depolymerization (Heidemann et al., 1984: Baas and Ahmad, 1992). Thus, mechanisms exist within the axon to spatially organize MT dynamics so that uniform polarity orientation is established and maintained. In this regard, neurons contain a centrosome; it is situated in the cell body and it nucleates MTs (Yu et al., 1993). However, MTs are not continuous from the centrosome into the axon. Rather, axonal MTs start and stop within the axon itself. Thus, the centrosome does not organize MT assembly in growing axons.

An important clue to understanding how MT assembly is organized in axons is the observation that axonal MTs are composite structures, consisting of two distinct domains that differ in their stability properties and composition (Baas and Heidemann, 1986; Sahenk and Brady, 1987; Baas and Black, 1990; Brown et al., 1993). This was unambiguously demonstrated by staining axonal MTs for tyr-tubulin, which is a relatively specific marker for newly assembled polymer (see above). Individual MTs stained poorly over a portion of their length, while they stained relatively strongly over the remainder, and the transition between these regions was abrupt. The tyr-tubulin-poor domain was situated at the minus end of the MT, whereas the tyr-tubulin-rich domain extended from the plus-end of the tyr-tubulin-poor domain to the end of the MT. The tyr-tubulin-rich domain and tyr-tubulin-poor domain of composite MTs also differ in their sensitivity to nocodazole, a potent MT depolymerizing agent (Baas and Black, 1990; Baas et al., 1991). Specifically, the tyr-tubulin-rich domain depolymerizes rapidly in the presence of 2 μg/ml nocodazole ($t_{1/2}$ approximately 3.5 min), whereas the tyr-tubulin-poor domain depolymerizes much more slowly ($t_{1/2}$ approximately 130 min).

These differences in drug sensitivity and composition suggest that the two domains of composite MTs differ in their stability properties. Specifically, the rate of loss of MTs in the presence of depolymerizing drugs such as nocodazole is dependent, in part, on their normal turnover dynamics, with MTs that turnover rapidly depolymerizing faster than MTs that turn over slowly (Cassimeris et al., 1986; Kreis, 1987; Wadsworth and McGrail, 1990; Prescott et al., 1992). If this is true for axonal MTs, then the drug-resistant domains of axonal MTs are more stable than the drub-labile domains (Baas and Black, 1990). This interpretation is reinforced by the observed differences in the tyr-tubulin content of these domains (see above and Brown et al., 1993). These considerations suggest that axonal MTs consist of a relatively stable domain from which extends a more recently assembled domain. Direct evidence for this has been obtained by microinjecting biotin-tubulin (Bt-tubulin) into neurons, and then analyzing the incorporation of Bt-tubulin into the tyr-tubulin-rich and tyr-tubulin-poor polymer of the axon. These experiments showed that most of the tyr-tubulin-rich polymer contained Bt-tubulin by 2 h after injection, whereas it was necessary to wait 8 h after injection to observe comparable labeling of the tyr-tubulin-poor polymer (Li and Black, unpublished data). Thus, the tyr-tubulin-poor domains are more long-lived than the tyr-tubulin-rich domains.

These observations indicate that axonal MTs are composite, consisting of a relatively stable domain in direct continuity with a more dynamic, newly assembled domain. This continuity suggests that the newly assembled domain elongated directly from the stable domain. If this is correct, then stable MTs are assembly-competent, and thus represent relatively long-lived templates that direct the assembly of tubulin subunits into MTs locally in the axon. The assembly competence of the stable polymer in axons has been demonstrated by MT regrowth studies (Baas and Ahmad, 1992). Briefly, neurons were treated with nocodazole to deplete the axons of all of the drug-labile, tyr-tubulin-rich polymer. The drug was then washed out of the cultures, the neurons were allowed to recover from the drug for varying periods of time, after which they were prepared for immunoelectron microscopic localization of tyr-tubulin in polymer. Under these conditions, any tyr-tubulin-staining polymer detected in the axon represents newly assembled polymer. Tyr-tubulin-staining polymer initially appeared as discrete MT segments all along the axon, indicating that newly assembled polymer formed at multiple sites in the axon. Serial reconstruction of consecutive thin sections indicated that, in all cases, the newly formed polymer assembled in direct continuity with the plus end of the stable polymer remaining in the axon after drug treatment. No evidence was obtained either for self-assembly of tubulin subunits into new MTs or for MT assembly from structures other than the plus ends of stable MTs. Furthermore, when MTs were completely depleted from the axon by treatment with nocodazole, MT regrowth did not occur (Baas and Heidemann, 1986). Collectively, these findings indicate that the stable polymer of axons is assembly-competent at its plus ends and that MT assembly in the axon occurs exclusively by elongation of this MT polymer. This interpretation is further supported by studies that have microinjected biotin-labeled tubulin (Bt-tubulin) into neurons and then identified where the Bt-tubulin assembled into MTs within the axon; assembly occurred exclusively from existing MTs (Okabe and Hirokawa, 1988; Li and Black, unpublished data).

These and other features of MT organization in axons are shown schematically in Fig. 1. Axonal MTs do not radiate from a single, centrally situated nucleating structure (Lyser, 1968; Hinds and Hinds, 1974). Rather, MTs start and stop within the axon itself, and are staggered such that MT ends are distributed all along the axon (Bray and Bunge, 1981; Tsukita and Ishikawa, 1981; Stevens et al., 1988). Most if not all MTs in the axon are composite with regard to their dynamic properties, consisting of a relatively stable domain which is situated at the minus end of the MT and which appears to persist for many hours (Edson et al., 1993 and see above). The plus ends of these stable domains are assembly-competent, and they nucleate MT assembly within the axon. Thus, extending from the plus end of this stable polymer is a more recently assembled domain. This latter domain is generated locally in the axon, and turns over relatively rapidly (Edson et al., 1993; Okabe and Hirokawa, 1990; and see above). In this perspective, stable MTs represent the only long-lived template in the axon for nucleating MT assembly, and all assembly in the axon is associated with these MTs (see above).

An obvious question that arises from this model concerns where stable MTs for the axon are generated. The model shows that while newly stabilized polymer is present throughout the axon, it is much more concentrated in the proximal region of the axon than anywhere else along its length. This is based on immunologic studies that examined the staining of stable polymer for tubulin variants that identify newly assembled polymer (Baas et al., 1993). These studies showed that most of the stable polymer of the axon that stained in this way was located in its most proximal 10 μm. Because stable polymer that stains in this manner should be more recently formed than stable polymer that does not stain, this observation suggests that the proximal axon is the principal site for the addition of new stable polymer to the axon. At first glance, this conclusion appears to conflict with the fact that stable polymer is not restricted to the proximal part of the axon, but is found throughout the axon shaft, where it comprises approximately 50% of the total MT polymer (Baas and Black, 1990; Bass et al., 1991, 1993). One feasible explanation for this apparent paradox is that much of the stable polymer in the axon shaft is not stabilized therein, but arrives there after its stabilization elsewhere. In this regard, the proximal axon, with its high levels of newly stabilized polymer, is a reasonable source for some of the stable polymer in the axon shaft. If this is correct, then mechanisms must exist in the axon to transport stable polymer generated proximally to more distal sites. As I discuss below, there is compelling evidence that MTs for the axon in fact originate proximally in the cell body

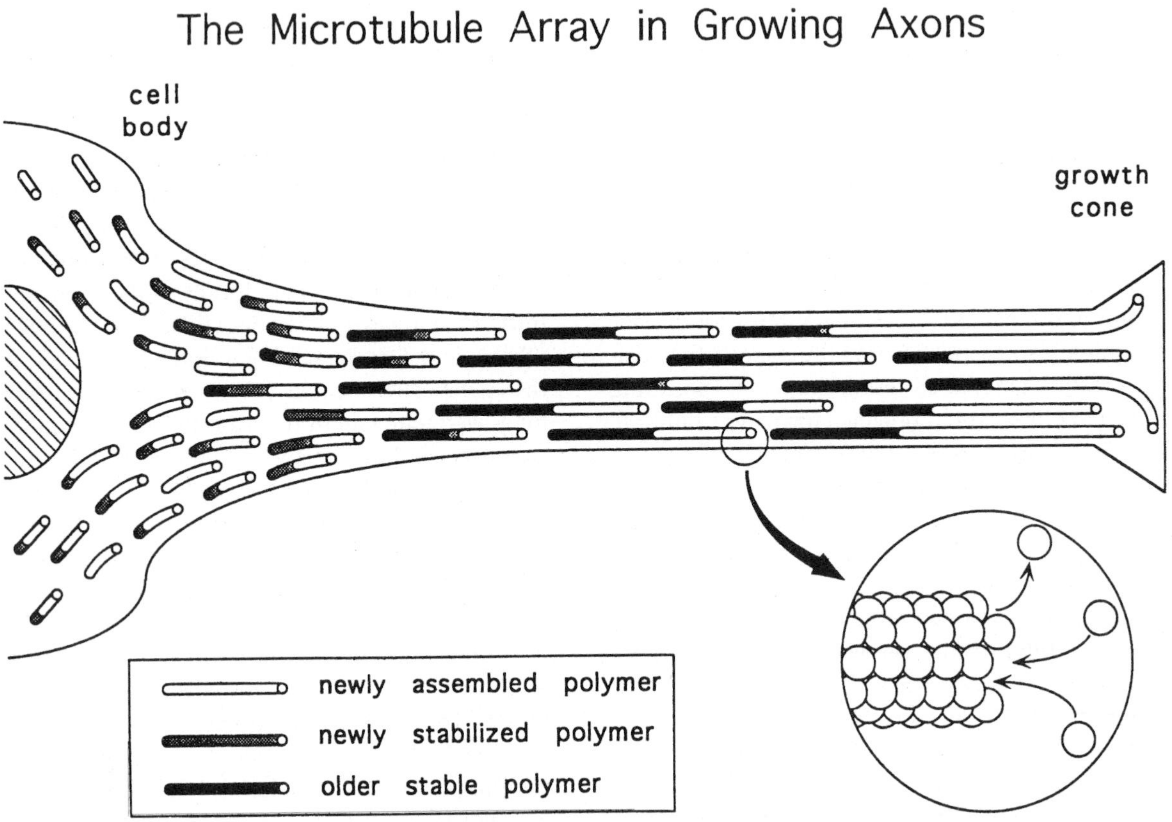

Fig. 1. A model of the MT array in growing axons. MTs form a continuous array that extends from the cell body to the growth cone. The MTs comprising this array do not radiate from a discrete nucleating structure, but are staggered all along the axon. Thus, the continuity within this array results from the overlap of MTs that are shorter than the length of the axon. MTs of the axon are composite structures, consisting of a stable domain situated at the minus end of the polymer and a newly assembled domain that extends from the plus end of the stable domain to the end of the MT (Baas and Black, 1990; Brown et al., 1993). The newly assembled polymer is generated locally within the axon (see inset), and appears to turnover relatively rapidly (Li and Black, unpublished observations). The model also suggests that the stable polymer of the axon is generated primarily proximally, in the cell body and the initial part of the axon (stippled polymer) (Baas et al., 1993). The proportions of total stable, newly stabilized, and newly assembled polymer at different sites along the axon are based on the data of Baas et al., 1991 and 1993, and are accurately represented in the schematic. The diagram also depicts the proximal-to-distal decrease in MT number that occurs in the proximal region of the axon (Brown et al., 1992; Black et al., 1994), the transport of MTs from the cell body into the axon toward the axon tip, and a progressive lengthening of MTs as they are transported through the proximal part of the axon (see text for additional details).

and that they are delivered to the axon by active transport mechanisms.

Mechanisms in addition to MT assembly are essential to the generation of the MT array of the axon

The finding that MT assembly in the axon is restricted to elongation of stable MTs (see Fig. 1 and above) has important implications as to how this assembly contributes to the generation of the axonal MT array. During axon growth, the amount of MT polymer in the axon increases coordinately with increases in axon length, and this results from increases in both the number of MTs in the axon and in the length of these MTs (Stevens et al., 1988). Because MT assembly in the axon occurs only by elongation of existing MTs, this assembly can only lengthen existing axonal MTs. It cannot, in and of itself, cause the increases in MT number that accompany axon growth. Other mechanisms must be invoked to account for the addition of new MTs to the axon during axon growth. Two mechanisms can be considered which are not mutually exclusive. One possibility is that a MT severing activity exists in the axon which generates short MT segments from longer MTs (Joshi and Baas, 1993). These short MT segments can then elongate by assembly mechanisms that operate locally within the axon. MT severing activity has been identified in a number of cell types (Sanders and Salisbury, 1987; Vale, 1991; Shiina et al., 1992). This activity has been demonstrated most frequently in the cytoplasm of mitotic cells, in which it is thought to contribute to the very rapid turnover of MTs that comprise the mitotic spindle. An important question for future research is to determine whether such an activity exists in neurons, and if so, whether it contributes to the dynamic properties of MTs in axons.

A second possible mechanism involves the assembly of MTs in the cell body, followed by their translocation into the axon. In this view, new MTs for the axon are assembled in the cell body, and then they are conveyed into the axon toward the axon tip by active transport mechanisms (Lasek, 1986, 1988). The length of these MTs is then modulated locally along the axon by MT assembly mechanisms. In this perspective, both MT transport and assembly contribute to the generation of the MT array in growing axons. This perspective has a long history, and numerous observations have provided compelling support for the hypothesis that MTs are actively transported in axons (Lasek, 1988). In the following section, I discuss the development of this model from a historical perspective, and then review some of the more recent studies that have revealed the importance of the polymer transport mechanisms in axon growth.

The structural hypothesis of axonal transport: a unified perspective

Much of the current understanding of axonal transport has derived from studies using radioisotopic labeling procedures to study the translocation of proteins in axons. Radiolabeled amino acids are injected in the immediate vicinity of the neuron cell bodies, where they are rapidly incorporated into proteins, thereby producing a pulse of labeling in vivo. The transport of these pulse-labeled proteins into the axon is then studied by removing the nerve containing the labeled axons, dividing it into consecutive pieces of a few millimeters in length, and then quantifying the total radioactivity in each segment. Various biochemical methods have also been used to detect specific labeled proteins in the nerve segments. These analyses have provided a picture of the distribution of total as well as individual pulse-labeled axonally transported proteins in the axon. By determining such distributions at multiple post-injection times, a fairly complete picture of protein transport in axons has been obtained.

One of the more remarkable findings from these studies is that of the several hundred proteins conveyed within the axon, most are transported in only one of three rate classes: slow component *a* (SCa or group V), slow component *b* (SCb or group IV), and fast component (FC or group I) (reviewed in Grafstein and Forman,

1980). The transport rates of SCa, SCb, and FC vary somewhat from one type of neuron to another and also as a function of development (Grafstein and Forman, 1980; Baitinger et al., 1983). However, within a particular system, the rates are relatively discrete. For example, in the adult rat, the modal transport rates of SCa, SCb, and FC are 0.2 mm/day, 2.0 mm/day and 250 mm/day, respectively, in retinal ganglion cell axons; in dorsal root ganglion cell axons, the rates are 1.5 mm/day, 3 mm/day and 400 mm/day respectively.

The discrete nature of axonal transport is further illustrated by the observation that each rate class has a distinct protein composition (Willard et al., 1974; Tytell et al., 1981). Nerve segments containing labeled proteins of SCa, SCb, or FC were anlyzed by both one- and two-dimensional gel electrophoresis and fluorography in order to reveal the profile of proteins conveyed by these rate classes. Both FC and SCb consist of hundreds of distinct proteins, whereas SCa consists of only five major proteins plus a similar number of minor proteins. With few exceptions, the proteins comprising a given rate class are present at only trace levels in the other rate classes. To account for such specificity, the *structural hypothesis of axonal transport* was put forth; this proposes that each rate class represents one or more macromolecular assemblies or structures that are moved by the transport machinery (Lasek, 1980; Tytell et al., 1981). Thus, the proteins comprising each rate class will be those comprising the specific *structures* transported at that rate.

Direct support for the structural hypothesis has been provided by the demonstration that the FC represents the movement of vesicular structures in the axon. Biochemical studies showed that the FC contained macromolecules typically associated with membranes, such as lipids and glycoproteins, and that FC as a whole co-purified with membranes during differential centrifugation and sucrose-gradient analyses (reviewed in Grafstein and Forman, 1980). Intrinsic as well as peripheral membrane components, such as GAP-43, Na-K ATPase, and N-CAM, were also identified in FC (Skene and Willard, 1981; Garner et al., 1986; Spect and Sweadner, 1984). Electron microscopic autoradiographic studies showed that FC proteins were associated with vesicular structures in situ (reviewed in Grafstein and Forman, 1980) and experimental studies showed that focal blockade of fast transport resulted in the rapid accumulation of vesicular structures just proximal to the site of blockade (Tsukita and Ishikawa, 1980; Fahim et al., 1985). The final proof that FC represents the movement of membranous elements came in the early-to-mid 1980s with the direct visualization of vesicular structures moving at the rates of fast transport in living axons (reviewed in Vallee and Bloom, 1991).

Circumstantial support for the hypothesis that SCa and SCb represent the transport of macromolecular assemblies stems from fractionation studies which have shown that the proteins of SCa and SCb form discrete non-membranous complexes which can be easily sedimented (Lorenz and Willard, 1978). With regard to SCb, it has proven difficult to identify its structural correlates in part because of its enormous compositional complexity. However, many specific proteins of SCb form multi-protein assemblies in vitro and/or in vivo. For example, actin, spectrin, synapsin, clathrin, and hsc70 are transported in SCb (Black and Lasek, 1979; Willard et al., 1979; Levine and Willard, 1980; Garner and Lasek, 1981; Baitinger and Willard, 1987; Black et al., 1991), and all of these proteins form oligomeric complexes under appropriate in vitro conditions and in intact cells. In addition, many enzymes of intermediary metabolism are transported in SCb (Brady and Lasek, 1981; Oblinger et al., 1988) and these can form particulate complexes in vitro (Knull, 1978). These biochemical findings support the hypothesis that the proteins of SCb are organized into one or more structural assemblies that are transported in the axon.

The relatively simple composition of SCa has greatly facilitated consideration of its structural correlates. SCa consists principally of tubulin and neurofilament proteins (reviewed in Lasek et al., 1984), which are the principal subunits of MTs and neurofilaments (NFs), respectively. The structural hypothesis proposes that tubulin and NF proteins are transported as MT and NF, resepctively. The following observations indicate that NF proteins are transported as NF. First, the only structure formed by NF proteins so far identified in intact axons is the NF. Second, NFs are remarkably stable polymers and, at any moment in time, all but a trace amount of the NF protein in axons is polymerized into NFs (Morris and Lasek, 1982; Black et al., 1986). Because NF proteins are actively transported in axons, and these proteins also appear to be stably polymerized, it is logical to conclude that these proteins are transported in the form of NF. This view is fully supported by experimental studies in which manipulations that alter the transport of NF proteins have concomitant effects on the number of NFs in the axon (Clarke et al., 1980; Hoffman et al., 1988; Monaco et al., 1989). Although a strong case can be made for NF transport in axons, direct observation of NF translocation in axons has not been achieved. Nonetheless, we are further encouraged to believe that NF are the transport form for NF proteins because of the proven utility of the structural hypothesis of axonal transport (see above). This hypothesis provides a unified framework for understanding the organization of proteins during their transit in the axon. It proposes that proteins are transported as component parts of discrete structures.

Based on the structural hypothesis of axonal transport, it has been proposed that tubulin is actively transported in the form of MTs (Tytell et al., 1981). The possibility that MTs are actively translocated in cells is not without precedent. For example, MT translocation has been directly visualized in the cytoplasm of the fresh water amoeba *Reticulomyxa* (Koonce et al., 1987), and the light-induced length changes of teleost retinal photoreceptors are based in part on the active sliding of MTs (Burnside, 1989). Furthermore, MT translocation occurs under a variety of in vitro conditions in which MTs are mixed with MT-based translocators (Vale, 1987), and there is growing evidence that similar types of movements contribute to the behavior of MTs in the mitotic spindle in vivo (Hyman and Mitchison, 1991; Nislow et al., 1992; Sawin et al., 1992). Attempts to directly study the transport of MTs in axons have led to conflicting interpretations, with some authors arguing for MT transport in axons, while others argue that MTs are stationary and that tubulin is transported in some other form. The issue of MT transport has been further clouded by several authors arguing that the occurrence of MT assembly dynamics in the axon is inconsistent with polymer transport (Gordon-Weeks, 1991). These latter arguments are weak because there is no a priori reason to asume that polymer transport and polymer dynamics are inconsistent with each other. In fact, in vitro studies have clearly revealed that MTs can simultaneously undergo active translocation and either elongation or shortening (Belmont et al., 1990). In my opinion, the original proposal based on the structural hypothesis of axonal transport is strongly supported by the available data, and in the next section I will review these observations.

Active transport mechanisms deliver MTs from the cell body into the axon

The most direct experiments on the transport form of tubulin have used photobleaching or photoactivation approaches on cultured neurons. In theses studies, neurons injected with fluorescent or caged-fluorescent tubulin are incubated for a time sufficient to allow the injected tubulin to equilibrate with the endogenous tubulin pool in the cell. After this incubation, the modified tubulin is either photobleached or photoactivated at a discrete site in the axon so that it can be distinguished from the rest of the modified tubulin. This marked tubulin is then observed over time and inferences are made about its transport based on the extent to which it undergoes vectorial movement. Using these approaches, the directed movement of MTs at the rates of slow axonal transport has been unambiguously demonstrated in the axons of *Xenopus* neurons (Reinsch et al., 1991; Okabe and Hirokawa, 1992, 1993). However, similar studies on chick and mouse neurons have not detected any directed movement of tubulin at all (Okabe and Hirokawa, 1990, 1992; Lim et al., 1990), and these latter studies have been used to argue that MTs are not transported in axons.

In principle, the techniques of photobleaching and photoactivation are capable of detecting tubulin movement regardless of the form in which it moves. Thus, the significance of studies that have not detected directed movement of tubulin is unclear because tubulin is actively transported in some form; these studies certainly do not effectively address the question of the transport form of tubulin in axons. Several possible explanations can account for the detection of movement in some experiments but not in others. For example, photobleaching and photoactivation have a number of side-effects that might interfere with tubulin transport (Vigers et al., 1988; Reinsch et al., 1991), with different types of neurons varying in their sensitivity to these side-effects. It has been argued that these side-effects are not a serious concern because the axons continue to grow during such experiments (Okabe and Hirokawa, 1990; Lim et al., 1990). However, the side-effects may be local, and thus alter tubulin transport in the vicinity of the photobleached or photoactivated area, without noticeably affecting axon growth during the time course of the experiment. It is also possible that only a portion of the total photobleached or photoactivated material moves during the time course of these experiments, and that under some experimental conditions this amount is too small to be detected. In this regard, Lim et al. (1990) did not find any tubulin movement in their photobleaching studies. While they concluded that MTs are not transported, they also indicated that approximately 20% of the total photobleached material would have had to move in order for them to have detected it. Whatever the explanation, axonal transport of tubulin is a fundamental and constitutive process in neurons, and those studies that have succeeded in detecting this movement indicate that tubulin is axonally transported in the form of MTs (Reinsch et al., 1991; Okabe and Hirokawa, 1992, 1993).

The view that MTs are actively transported in axons is also supported by a number of indirect observations. For example, real-time observation of MT configurations in living growth cones are consistent with active translocation mechanisms driving MTs from the axon shaft into the growth cone (Tanaka and Kirschner, 1991; Sabry et al., 1991; Lin and Forscher, 1993). Implicit in the polymer transport model is that new MTs are added to the axon proximally, by translocation from their sites of assembly in the cell body into the axon. In this regard, several observations implicate the cell body as a principle source of MT polymer for the axon. First, newly assembled polymer is concentrated in the initial segment of the axon, and there appears to be more newly assembled polymer in the first few microns of the axon than anywhere else along its length (Brown et al., 1992). Second, this proximal region also appears to be especially active in generating the long-lived polymer that is found throughout the axon (Baas et al., 1993). Third, the fact that new MTs are not nucleated locally in the axon implicates the cell body as a source of the new MTs for the axon as it grows (see above).

Finally, evidence for MT transport in axons has also been provided by pharmacological experiments (George et al., 1988; Baas and Ahmad, 1993). I shall discuss the latter because these authors make a particularly compelling case. Their studies used the drug vinblastine to examine MT transport from the cell body into the axon. When administered in millimolar amounts, vinblastine causes MT depolymerization and the formation of tubulin paracrystals. In nanomolar amounts, however, vinblastine can block both MT assembly and disassembly (Jordan and Wilson, 1990; Jordan et al., 1992). Baas and Ahmad (1993) showed that sympathetic neurons plated in the presence of vinblastine at concentrations ranging from 4 nM to 100 nM extended axons, and although these axons were considerably shorter than those of control neurons, they nonetheless contained MTs. If MT assembly is effectively suppressed at such vinblastine concentrations, then the presence of MTs in the axon must reflect their transport from the cell body. Morphometric analyses showed that MT levels in control neurons increased by more than two-fold during the first 20 h in culture. By contrast, the amount of MT polymer in neurons cultured in 16 nM vinblastine remained unchanged during the same period, while at 50 nM vinblastine there was an actual decline in MT polymer levels. Furthermore, at both of these vinblastine concentrations, there was a dramatic redistribution of polymer from the cell body into the axon, such that by 20 h after plating approximately 80% of the polymer in drug-treated cells was located in their axons. The most straightforward interpretation of these data is that MT assembly was effectively suppressed and, therefore, that the observed redistribution of MTs from the cell body into the axon reflects their translocation from the cell body. Alternative explanations that do not involve MT transport must invoke levels of both MT assembly and disassembly in the presence of vinblastine that are inconsistent with the current knowledge of the actions of this drug.

The issue of MT transport in axons has been controversial largely because of the results from those photobleaching and photoactivation studies where no tubulin transport was detected. However, given that tubulin is actively transported in axons, the value of such negative observations is questionable. Furthermore, the interpretation of such findings to mean that MTs are not transported in axons is contradiced by other studies that clearly demonstrate vectorial movement of MTs in axons. These latter observations, taken together with the studies discussed above indicate that MTs are conveyed from the cell body into the axon.

MT transport and assembly cooperate to generate the MT array of growing axons

Available information on the relative contributions of polymer transport and MT assembly within the axon to the elaboration of the MT array during axonal growth suggests a balanced perspective in which each contributes in a distinct, but complementary manner (see Lasek, 1988; Cleveland and Hoffman, 1992; Black and Brown, 1993; Joshi and Baas, 1993). The MT array of the axon is continuous from the cell body to the tip of the axon, and the amount of polymer comprising this array increases coordinately with axon growth due to increases in both the number and length of MTs. The cell body is a principal site for the generation of new MTs for the axon. Once formed these MTs are conveyed into the axon by the polymer transport mechanisms. Thus, polymer transport contributes to the expansion of the MT array during axon growth by adding new MTs to the axon. MT assembly within the axon further shapes the MT array of growing axons. MT assembly occurs throughout growing axons, and it is especially active in the proximal and distal regions. This assembly modulates the length of MTs conveyed by the polymer transport mechanisms. However, the extent to which local assembly contributes to the expansion of the MT array in growing axons is not clear. In many cell types, MTs can be highly dynamic structures, undergoing alternating phases of assembly and disassembly (Kirschner and Mitchison, 1986), and this is also true for a portion of the MT polymer in growing axons (Baas and Black, 1990; Okabe and Hirokawa, 1990; Edson et al., 1993). Thus, the occurrence of MT assembly per se is not automatically indicative of net polymer addition. For local MT assembly to increase the amount of MT polymer in the axon, assembly must exceed disassembly. As discussed below, the nature of tubulin delivery to the axon limits the extent to which this can occur.

Tubulin is supplied from the cell body to the axon by diffusion and by MT transport. The extent to which diffusion can deliver tubulin subunits to the axon declines with the square of the distance from the cell body (Popov and Poo, 1992). Thus, diffusion will more effectively deliver tubulin subunits to the proximal part of the axon than to the distal part. Over the relatively short distance between the cell body and the proximal part of the axon, diffusion is relatively rapid and can deliver tubulin for assembly onto the ends of transported MTs (Okabe and Hirokawa, 1988; Popov and Poo, 1992). This can potentially increase the total amount of polymer in the axon. However, diffusion is limited in its ability to deliver tubulin to more distal sites in the axon during time periods relevant to axon elongation, and this limitation will increase progressivly as the axon grows (Reinsch et al., 1991). Thus, as axons lengthen, the tubulin for local assembly at more distal sites increasingly is supplied by transported MT that give up subunits which then reassemble locally onto other MTs. This is an exchange process, and it can change the length of axonal MTs. It cannot, however, result in a net increase in MT polymer because the tubulin for this assembly is supplied by the disassembly of MTs that were transported into the axon. Thus, MT assembly within the axon can contribute to the expansion of the MT array during axon growth only to the extent that tubulin diffuses from the cell body into the axon and then assembles onto existing MTs. The potential for this to occur is limited spatially within the axon; it is greatest in the proximal region of the axon, and declines progressively with distance from the cell body, and as the axon continues to grow, it becomes negligible at more distal sites along the axon.

These considerations suggest the following perspective for the elaboration of the MT array of the axon during axon growth (see Fig. 1). Tubulin is continually synthesized in the cell body, where it assembles into MTs. Some of these MTs are then translocated into the axon by the polymer transport mechanisms, thereby adding new polymer to the axon. Because axonal MTs are much longer than those in the cell body (Stevens et al., 1988), these MTs grow many-fold in length as they are transported through the proximal part of the axon. Some of this increase in MT length is due to tubulin subunits that diffuse from the cell body and then assemble onto the ends of transported MTs. This assembly contributes to the net addition of MT polymer to the growing axon. In addition, the proximal-to-distal decrease in MT number that occurs in the region of the proximal axon (Brown et al., 1992; Black et al., 1994) suggests that some subunits also come from transported MTs that give up their subunits for the elongation of other MTs. This exchange process facilitates the overall lengthening of MTs that occurs as they are transported through the proximal part of the axon, but it does not result in the net addition of MT polymer to the axon. More distally within the axon, subunits for assembly also are supplied by transported MTs that give up subunits to the unassembled pool. These released subunits can then assemble onto other MTs. This exchange process also does not increase the amount of MT polymer in the axon as a whole. Rather, it provides a mechanism for shaping the MT array locally along the axon by changing MT length. This occurs to a particularly great extent in the region of the growth cone, where MT assembly/disassembly dynamics are especially active (Lim et al., 1989; Brown et al., 1992). These dynamics presumably help coordinate growth cone motility with axon extension. Other mechanisms may also modulate MT arrangements locally in the axon. For example, MT number and length may be modulated within the axon by MT severing, in which individual MTs are cut into multiple segments that then change their length through subunit exchange mechanisms (Joshi and Baas, 1993). Although the contribution of such a mechanism to the dynamics of the axonal MT system is a matter of speculation, MT severing appears to contribute substantially to MT dynamics in other cell types (Vale, 1991).

In addition to adding new MTs to the axon, the polymer transport mechanisms also establish aspects of MT organization in the axon. In particular, axonal MTs have a uniform polarity orientation; their plus-ends point away from the cell body toward the axon tip (Burton and Paige, 1981; Heidemann et al., 1981; Baas et al., 1988). The MT array of many non-neuronal cells also exhibits uniform polarity orientation, and in such cells this arises from the continuity of the MTs with a discrete nucleating structure (Brinkely, 1985). Such continuity cannot explain the uniform polarity orientation of axonal MTs because these MTs do not radiate from a single, discrete site in the cell but, rather, are staggered along the axon (Bray and Bunge, 1981; Stevens et al., 1988). Recent studies indicate that the transport mechanisms convey MTs from the cell body into the axon specifically with their plus ends leading (Baas and Ahmad, 1993). Thus, the establishment of uniform polarity orientation apparently is a direct consequence of the polymer transport mechanisms.

Finally, the motility that underlies MT transport can also be linked to axonal extension. Several studies have pointed out the importance of tension within the axon as a determinant of its growth behavior (Bray, 1984; Dennerll et al., 1989, Zheng et al., 1991). In particular, experimental studies have shown that mechanical tension applied along the axon stimulates extension. While the explanation for this phenomenon remains uncertain, it illuminates the potential role of the internal tension generated within a growing axon in the control of axon elongation. The origin of tension within the axon is a matter of speculation. Some have proposed that is is generated solely at the growth cone (Bray, 1984). However, studies on cut axons in vitro indicate that longitudinal tension is generated all along the axon (George et al., 1988; Dennerll et al., 1989). This tension apparently results in part from the polymer transport machinery that slides MTs and other cytoskeletal polymers within the axon (Lasek, 1988; George et al., 1988). This tension generates both a backward pull on the growth cone and a forward pull on the cytoskeleton. Depending on the adhesivity of the axon and growth cone to the substrate, this tension can lead to elongation or retraction of the axon. In this view, the polymer transport mechanisms generate forces that directly impact on the process of axonal growth.

Summary

MTs are major architectural elements in growing axons. MTs overlap with each other along the axon, forming an array that is continuous from the cell body to the tip of the axon. The MT array constitutes a scaffolding that mechanically supports the elongate shape of the axon and also contributes directly to its shape. MTs also direct the transport of vesicular organelles between the cell body and the axon, and thereby determine, in part, the composition of the axon. In this article, I have discussed mechanisms involved in the elaboration of the MT array in growing axons, and I have emphasized the distinct but complementary roles of polymer transport mechanisms and local assembly dynamics. MTs for the axon originate in the cell body, and they are delivered to the axon by the polymer transport mechanisms. These mechanisms thus contribute directly to the shape of the axon by supplying it with essential architectural elements. The shape of the axon is further modulated by dynamic processes that alter cytoskeletal structure locally along its length. These dynamic processes include the assembly/disassembly mechanisms which influence polymer length and possibly number locally along the axon by subunit exchange between the monomer and polymer pools. In addition, the polymer transport mechanisms themselves are subject to modulation along the axon, as demonstrated by the observation that transport rate of MTs varies along the length of individual axons (Reinsch et al., 1991). Such local variations can, in and of themselves, change the number of MTs along the axon, and thereby focally affect axon shape. Thus, the dynamic processes of polymer transport and local assembly act cooperatively to shape the MT array of the axon, and thereby contribute directly to the elaboration of axonal morphology.

Acknowledgements

The author is supported by a grant from the N.I.H. (NS17681).

References

Ahmad, F.J., Pienkowski, T.P. and Baas, P.W. (1993) Regional differences in microtubule dynamics in the axon. *J. Neurosci.*, 13: 856–866.

Arregui, C., Busciglio, J., Caceres, A. and Barra, H.S. (1991) Tyrosinated and detryosinated microtubules in axonal processes of cerebellar macro-neurons grown in culture. *J. Neurosci. Res.*, 28: 171–181.

Baas, P.W. and Ahmad, F.J. (1992) The plus ends on stable microtubules are the exclusive nucleating structures for microtubules in the axon. *J. Cell Biol.*, 116: 1231–1241.

Baas, P.W. and Ahmad, F.J. (1993) The transport properties of axonal microtubules establish their polarity orientation. *J. Cell Biol.*, 120: 1427–1437.

Baas, P.W. and Black, M.M. (1990) Individual microtubules in the axon consist of domains that differ in both composition and stability. *J. Cell biol.*, 111: 495–509.

Baas, P.W. and Heidemann, S.R. (1986) Microtubule reassembly from nucleating fragments during the regrowth of ammputated neurites. *J. Cell Biol.*, 103: 917–927.

Baas, P.W., Deitch, J.S., Black, M.M. and Banker, G.A. (1988) Polarity orientation of microtubules inhippocampal neurons: uniformity in the axon and nonuniformity in the dendrite. *Proc. Natl. Acad. Sci. USA*, 85: 8335–8339.

Baas, P.W., Black, M.M. and Banker, G.A. (1989) Changes in microtubule polarity orientation during the development of hippocampal neurons in culture. *J. Cell Biol.*, 109: 3085–3094.

Baas, P.W., Slaughter, T., Brown, A. and Black, M.M. (1991) Microtubule dynamics in axons and dendrites. *J. Neurosci. Res.*, 30: 134–153.

Baas, P.W., Ahmad, F.J., Pienkowski, T.P., Brown, A. and Black, M.M. (1993) Sites of stabilization for axonal microtubules. *J. Neurosci.*, 13: 2177–2185.

Baitinger, C. and Willard, M. (1987) Axonal transport of synapsinI-like proteins in rabbit retinal ganglion cells. *J. Neurosci.*, 7: 3723–3735.

Baitinger, C., Cheney, R., Clements, D., Glicksman, M., Hirokawa, N., Levine, J., Meiri, K., Simon, C., Skene, P. and Willard, M. (1983) Axonally transported proteins in axon development, maintenance, and regeneration. *Cold Spring Harbor Symp.*, 48: 791–802.

Barra, H.S., Arce, C.A. and Argaraña, C.E. (1988) Post-translational tyrosination/detyrosination of tubulin. *Mol. Neurobiol.*, 2: 133–153.

Belmont, L.D., Hyman, A.A., Sawin, K.E. and Mitchison, T.J. (1990) Realtime visualization of cell cycle-dependent changes in microtubule dynamics in cytoplasmic extracts. *Cell*, 62: 579–589.

Binder, L.I., Dentler, W.L. and Rosenbaum, J.L. (1975) Assembly of chich brain tubulin onto flagellar microtubules from *Chlamydomonas* and sea urchin sperm. *Proc. Natl. Acad. Sci. USA*, 72: 1122–1126.

Black, M.M. and Brown, A. (1993) Sites of microtubule assembly in growing axons. In: N. Hirokawa (Ed.) *Neuronal Cytoskeleton: Morphogenesis and synaptic transmission*, Japan Scientific Societies Press, CRC Press, pp. 171–182.

Black, M.M. and Lasek, R.J. (1979) Axonal transport of actin: slow component b is the principal source of actin for the axon. *Brain Res.*, 171: 401–413.

Black, M.M., Keyser, P. and Sobel, E. (1986) Interval between the synthesis and assembly of cytoskeletal proteins in cultured neurons. *J. Neurosci.*, 6: 1004–1012.

Black, M.M. Chestnut, M.H., Pleasure, I.T., and Keen, J.H. (1991) Stable clathrin: uncoating protein (hsc70) complexes in intact neurons and their axonal transport. *J. Neurosci.*, 11: 1163–1172.

Black, M.M., Slaughter, T. and Fischer, I. (1994) Microtubule-associated protein 1b (MAP1b) is concentrated in the distal region of growing axons. *J. Neurosci.*, 14: 857–870.

Brady, S.T. and Lasek, R.J. (1981) Nerve specific enolase and creatine phosphokinase in axonal transport: soluble proteins and the axoplasmic matrix. *Cell*, 23: 515–523.

Bray, D. (1984) Axonal growth inresponse to experimentally applied mechanical tension. *Dev. Biol.*, 101: 379–389.

Bray, D. and Bunge, M.B. (1981) Serial analysis of microtubules of cultured rat sensory neurons. *J. Neurocytol.*, 10: 589–605.

Bré, M-H., Kreis, T. and Karsenti, E. (1987) Control of microtubule nucleation and stability in Madin-Darby canine kidney cells: the occurrence of non-centrosomal, stable detyrosinated microtubules. *J. Cell Biol.*, 105: 1283–1296.

Brinkley, B.R. (1985) Microtubule organizing centers. *Annu. Rev. Cell Biol.*, 1: 145–172.

Brown, A., Slaughter, T. and Black, M.M. (1992) Newly assembled microtubules are concentrated in the proximal and distal regions of growing axons. *J. Cell Biol.*, 119: 867–882.

Brown, A., Slaughter, T. and Black, M.M. (1993) Composite microtubules of the axon: quantitative analysis of tyrosinated and acetylated α-tubulin along axonal microtubules. *J. Cell Sci.*, 104: 339–352.

Bulinski, J.C. and Gundersen, G.G. (1991) Stabilization and post-translational modification of microtubules during cellular morphogenesis. *BioEssays*, 13: 285–293.

Burnside, B. (1989) Microtubule sliding and the generation of force for cell shape change. In: F.D. Warner and J.R. Macintosh (Eds), *Cell Movement, Vol 2. Kinesin, Dynein, and Microtubules Dynamics*, Alan R. Liss, New York, pp. 169–189.

Burton, P.R. and Paige, J.L. (1981) Polarity of axoplasmic microtubules in the olfactory nerve of the frog. *Proc. Natl. Acad. Sci. USA*, 78: 3269–3273.

Cassimeris, L.U., Wadsworth, P. and Salmon, E.D. (1986) Dynamics of microtubule depolymerization in monocytes. *J. Cell. Biol.*, 102: 2023–2032.

Clarke, A.W., Griffin, J.W. and Price, D.L. (1980) The axonal pathology in chronic IDPN intoxication. *J. Neuropathol. Exp. Neurol.*, 39: 42–55.

Cleveland, D.W. and Hoffman, P.N. (1992) Slow axonal transport models come full circle: evidence that microtubule sliding mediates axonal elongation and tubulin transport. *Cell*, 87: 453–456.

Daniels, M.P. (1973) Fine structural changes in neurons and nerve fibers associated with colchicine inhibition of nerve fiber formation in vitro. *J. Cell Biol.*, 58: 463–470.

Dennerll, T.J., Lamoureux, P., Buxbaum, R.E. and Heidemann, S.R. (1989) The cytomechanics of axonal elongation and retraction. *J. Cell Biol.*, 109: 3073–3083.

Edson, K.J., Lim, S-S., Borisy, G.G. and Letourneau, P.C. (1993) FRAP analysis of the stability of the microtubule population along the neurites of chick sensory neurons. *Cell Motil. Cytoskel.*, 25: 59–72.

Evans, L., Mitchison, T. and Kirschner, M. (1985) Influence of the centrosome on the structure of nucleated microtubules. *J. Cell Biol.*, 100: 1185–1191.

Fahim, M.A., Lasek, R.J., Brady, S.T. and Hodge, A.J. (1985) Identification of the membranous organelles moving in fast axonal transport in squid giant axons: a correlative electron microscopy and video microscopy study. *J. Neurocytol.*, 14: 689–704.

Garner, J.A. and Lasek, R.J. (1981) Clathrin is axonally transported as part of slow component b: the microfilament complex. *J. Cell Biol.*, 88: 172–178.

Garner, J.A., Watanabe, M. and Rutishauser, U. (1986) Rapid axonal transport of the neural cell adhesion molecule. *J. Neurosci.*, 6: 3242–3249.

George, E.B., Schneider, B.F., Lasek, R.J. and Katz, M.J. (1988) Axonal shortening and the mechanisms of axonal motility. *Cell Motil. Cytoskel.*, 9: 48–59.

Gordon-Weeks, P.R. (1991) Control of microtubule assembly in growth cones. *J. Cell Sci. Suppl.*, 15: 45–49.

Grafstein, B. and Forman, D.S. (1980) Intracellular transport in neurons. *Physiol. Rev.*, 60: 1167–1282.

Gundersen, G.G., Khawaja, S. and Bulinski, J.C. (1987) Postpolymerization detyrosination of α-tubulin: a mechanism for subcellular differentiation of microtubules. *J. Cell Biol.*, 105: 251–264.

Heidemann, S.R., Landers, J.M. and Hamborg, M.A. (1981) Polarity orientation of axonal microtubules. *J. Cell Biol.*, 91: 661–665.

Heidemann, S.R., Hamborg, M.A., Thomas, S.J., Song, B., Lindley, S. and Chu, D. (1984) Spatial organization of axonal microtubules. *J. Cell Biol.*, 99: 1289–1295.

Hinds, J.W. and Hinds, P.L. (1974) Early ganglion cell differentiation in the mouse retina: an electron microscopic analyses using serial sections. *Dev. Biol.*, 37: 381–416.

Hoffman, P.N., Koo, E.H., Muma, N.A., Griffin, J.W. and Price, D.L. (1988) Role of neurofilaments in the control of axonal caliber in myelinated nerve fibers. In: R.J. Lasek and M.M. Black (Eds.), *Intrinsic Determinants of Neuronal Form and Function*, Alan R. Liss, New York, pp. 389–402.

Hollenbeck, P.J. (1989) The transport and assembly of the axonal cytoskeleton. *J. Cell Biol.*, 108: 223–227.

Hyman, A.A. and Mitchison, T.J. (1991) Two different microtubule-based motor activities with opposite polarities in kinetochores. *Nature (London)*, 351: 206–211.

Jordan, M.A. and Wilson, L. (1990) Kinetic analysis of tubulin exchange at microtubule ends at low vinblastine concentrations. *Biochemistry*, 29: 2730–2739.

Jordan, M.A., Thrower, D. and Wilson, L. (1992) Effects of vinblastine, podophyllotoxin, and nocodazole on mitotic spindles: implications for the role of microtubule dynamics in mitosis. *J. Cell Sci.*, 102: 402–416.

Joshi, H.C. and Baas, P.W. (1993) A new perspective on microtubules and axon growth. *J. Cell Biol.*, 121: 1191–1196.

Joshi, H.C., Palacois, M.J., McNamara, L. and Cleveland, D.W. (1992) Gamma-tubulin is a centrosomal protein required for cell cycle-dependent microtubule nucleation. *Nature (London)*, 356: 80–83.

Keith, C.H. and Blane, K. (1990) Sites of tubulin polymerization in PC12 cells. *J. Neurochem.*, 54: 1258–1268.

Kirschner, M. and Mitchison, T. (1986) Beyond self assembly: from microtubules to morphogenesis. *Cell*, 45: 329–342.

Knull, H.R. (1978) Associations of glycolytic enzymes with particulate fractions from nerve endings. *Biochim. Biophys. Acta*, 522: 1–9.

Koonce, M.P., Tong, J., Euteneur, U. and Schliwa, M. (1987) Active sliding between cytoplasmic microtubules. *Nature (London)*, 328: 737–739.

Kreis, T. (1987) Microtubules containing detyrosinated tubulin are less dynamic. *EMBO J.*, 6: 2597–2606.

Lasek, R.J. (1980) The dynamics of neuronal structures. TINS, 3: 87–91.

Lasek, R.J. (1986) Polymer sliding in axons. In: *The Cytoskeleton: Cell Function and Organization. J. Cell Sci. Suppl.*, 5: 161–179.

Lasek, R.J. (1988) Studying the intrinsic determinants of neuronal form and function. In: (R.J. Lasek and M.M. Black (Eds.), *Intrinsic Determinants of Neuronal Form and Function*, Alan R. Liss, New York, pp. 3–58.

Lasek, R.J., Garner, J.A. and Brady, S.T. (1984) Axonal transport of the cytoplasmic matrix. *J. Cell Biol.*, 99: 212s–221s.

Levine, J. and Willard, M. (1980) Fodrin: axonally transported polypeptides associated with the internal periphery of many cells. *J. Cell Biol.*, 90: 631–643.

Lim, S-S., Sammak, P.J. and Borisy, G.G. (1989) Progressive and spatially differentiated stability of microtubules in developing neuronal cells. *J. Cell Biol.*, 109: 253–263.

Lim, S-S., Edson, K.J., Letourneau, P.C. and Borisy, G.G. (1990) A test of microtubule translocation during neurite elongation. *J. Cell Biol.*, 111: 123–130.

Lin, C-H. and Forscher, P. (1993) Cytoskeletal remodeling during growth cone-target interactions. *J. Cell Biol.*, 121: 1369–1383.

Lorenz, T. and Willard, M. (1978) Subcellular fractionation of intraaxonally transported polypeptides in the rabbit visual system. *Proc. Natl. Acad. Sci. USA*, 75: 505–509.

Luckenbill-Edds, L., Van Horn, C. and Greene, L.A. (1979) Fine structure of initial outgrowth of processes induced in a pheochromocytoma cell line (PC12) by nerve growth factor. *J. Neurocytol.*, 8: 493–511.

Lyser, K.M. (1968) An electron microscopic study of centrioles of differentiating neuroblasts. *J. Embryol. Exp. Morphol.*, 20: 343–354.

Monaco, S., Autilio-Gambetti, L., Lasek, R.J., Katz, M.J. and Gambetti, P. (1989) Experimental increase of neurofilament transport rate: decrease in neurofilament number and in axon diameter. *J. Neuropathol. Exp. Neurol.*, 48: 23–32.

Morris, J.R. and Lasek, R.J. (1982) Stable polymers of the axonal cytoskeleton: the axoplasmic ghost. *J. Cell Biol.*, 92: 192–198.

Nixon, R.A. (1992) Slow axonal transport. *Curr. Opin. Cell Biol.*, 4: 8–14.

Nislow, C., Lombillo, V.A., Kuriyama, R. and McIntosh, J.R. (1992) A plus-end directed motor enzyme that moves antiparallel microtubules in vitro localizes to the interzone of mitotic spindles. *Nature (London)*, 359: 543–547.

Oblinger, M.M., Foe, L.G., Kwiatkowska, D. and Kemp, R.G. (1988) Phosphofructokinase in the rat nervous system: regional differences in activity and characteristics of axonal transport. *J. Neurosci. Res.*, 21: 25–34.

Okabe, S. and Hirokawa, N. (1988) Microtubule dynamics in nerve cells: analysis using microinjection of biotinylated tubulin into PC12 cells. *J. Cell Biol.*, 107: 651–664.

Okabe, S. and Hirokawa, N. (1990) Turnover of fluorescently labelled tubulin and actin in the axon. *Nature (London)*, 343: 479–482.

Okabe, S. and Hirokawa, N. (1992) Differential behavior of photoactivated microtubules in growing axons of mouse and frog neurons. *J. Cell Biol.*, 117: 105–120.

Okabe, S. and Hirokawa, N. (1993) Do photobleached fluorescent microtubules move? Re-evaluation of fluorescence laser photobleaching both in vitro and in growing *Xenopus* axons. *J. Cell Biol.*, 120: 1177–1187.

Popov, S. and Poo, M. (1992) Diffusional transport of macromolecules in developing nerve processes. *J. Neurosci.*, 12: 77–85.

Prescott, A.R., Dowrick, P.G. and Warn, R.M. (1992) Stable and slow-turning-over microtubules characterize the processes of motile epithelial cells treated with scatter factor. *J. Cell Sci.*, 102: 103–112.

Reinsch, S.S., Mitchison, T.J. and Kirschner, M.W. (1991) Microtubule polymer assembly and transport during axonal elongation. *J. Cell Biol.*, 115: 365–379.

Sabry, J.H., O'Connor, T.P., Evans, L., Toroian-Raymond, A., Kirschner, M. and Bently, D. (1991). Microtubule behavior during guidance of pioneer neuron growth cones in situ. *J. Cell Biol.*, 115: 381–395.

Sahenk, Z. and Brady, S.T. (1987) Axonal tubulin and microtubules: morphological evidence for stable regions on axonal microtubules. Cell Motil. Cytoskel., 8: 155–164.

Sanders, M.A. and Salisbury, J.L. (1987) Centrin-mediated microtubule severing during flagellar excision in *Chlamydomonas reinhardtii*. *J. Cell Biol.*, 108: 1751–1760.

Sawin, K.E., LeGuellec, K., Philippe, M. and Mitchison, T.J. (1992) Mitotic spindle organization by a plus-end-directed microtubule motor. *Nature (London)*, 359: 540–543.

Schulze, E. and Kirschner, M. (1986) Microtubule dynamics in interphase cells. *J. Cell Biol.*, 102: 1020–1031.

Schulze, E., Asai, D.J., Bulinski, J.C. and Kirschner, M. (1987) Post-translational modification and microtubule stability. *J. Cell Biol.*, 105: 2167–2177.

Sherwin, T. and Gull, K. (1989) Visualization of detyrosination along single microtubules reveals novel mechanisms of assembly during cytoskeletal duplication in trypanosomes. *Cell*, 57: 211–221.

Sherwin, T., Schneider, A., Sasse, R., Seebeck, T. and Gull, K. (1987) Distinct localization and cell cycle dependence of COOH terminally tyrosinolated α-tubulin in microtubules of *Trypanosoma brucei*. *J. Cell Biol.*, 104: 439–446.

Shiina, N., Gotoh, Y. and Nishida, E. (1992) A novel homooligomeric protein responsible for an MPF-dependent microtubule-severing activity. *EMBO J.*, 11: 4723–4731.

Skene, J.H.P. and Willard, M. (1981) axonally transported proteins associated with axon growth in rabbit central and peripheral nervous systems. *J. Cell Biol.*, 89: 96–103.

Soltys, B.J. and Borisy, G.G. (1985) Polymerization of tubulin in vitro: direct evidence for assembly onto microtubule ends and from centrosomes. *J. Cell Biol.*, 100: 1682–1689.

Spect, S. and Sweadner, K.J. (1984) Two different Na, K-ATPases in the optic nerve: cells of origin and axonal transport. *Proc. Natl. Acad. Sci. USA*, 81: 1234–1238.

Stevens, J.K., Trogadis, J. and Jacobs, J.R. (1988) Development and control of axial neruite form: a serial electron microscopic analysis. In: (R.J. Lasek and M.M. Black (Eds.), *Intrinsic Determinants of Neuronal Form and Function*, Alan R. Liss, New York, pp. 115–146.

Tanaka, E.M. and Kirschner, M.W. (1991) Microtubule behavior in the growth cones of living neurons during axon elongation. *J. Cell Biol.*, 115: 345–363.

Tsukita, S. and Ishikawa, H. (1980) The movement of membranous organelles in axons: electron microscopic identification of anterogradely and retrogradely transported organelles. *J. Cell Biol.*, 84: 513–530.

Tsukita, S. and Ishikawa, H. (1981) The cytoskeleton in myelinated axons: serial section study. *Biomed. Res.*, 2: 424–437.

Tytell, M., Black, M.M., Garner, J.A. and Lasek, R.J. (1981) Axonal transport: Each major rate component reflects the movement of distinct macromolecular complexes. *Science*, 214: 179–181.

Vale, R.D. (1987) Intracellular transport using microtubule-based motors. *Annu. Rev. Cell Biol.*, 3: 347–378.

Vale, R.D. (1991) Severing of stable microtubules by a mitotically activated protein in *Xenopus* egg extracts. *Cell*, 64: 827–839.

Vallee, R.B. and Bloom, G.S. (1991) Mechanisms of fast and slow axonal transport. *Annu. Rev. Neurosci.*, 14: 59–92.

Vigers, G.P.A., Coue, M. and McIntosh, J.R. (1988) Fluorescent microtubules break up under illumination. *J. Cell Biol.*, 107: 1011–1024.

Wadsworth, P. and McGrail, M. (1990) Interphase microtubule dynamics are cell type-specific. *J. Cell Sci.*, 95: 23–32.

Webster, D.R., Gundersen, G.G., Bulinski, J.C. and Borisy, G.G. (1987) Differential turnover of tyrosinated and detyrosinated microtubules. *Proc. Natl. Acad. Sci.*, 84: 9040–9044.

Wehland, J. and Weber, K. (1987) Turnover of the carboxy-terminal tyrosine of α-tubulin and means of reaching elevated levels of detyrosination in living cells. *J. Cell Sci.*, 88: 185–203.

Willard, M., Wiseman, M., Levine, J. and Skene, P. (1979) Axonal transport of actin in rabbit retinal ganglion cells. *J. Cell Biol.*, 81: 581–591.

Willard, M., Cowan, W.M. and Vagelos, P.R. (1974) The polypeptide composition of intra-axonally transported polypeptides: evidence for four transport velocities. *Proc. Natl. Acad. Sci. USA*, 71: 2183–2187.

Yamada, K.M., Spooner, B.S. and Wessells, N.K. (1970) Axon growth: roles of microfilaments and microtubules. *Proc. Natl. Acad. Sci. USA*, 66: 1206–1212.

Yu, W., Centonze, V.E., Ahmad, F.J. and Baas, P.W. (1993) Microtubule nucleation and release from the neuronal centrosome. *J. Cell Biol.*, 349–360.

Zheng, J., Lamoureux, P., Santiago, V., Dennerll, T., Buxbaum, R.E. and Heidemann, S.R. (1991) Tensile regulation of axonal elongation and initiation. *J. Neurosci.*, 11: 1117–1125.

J. van Pelt, M.A. Corner H.B.M. Uylings and F.H. Lopes da Silva (Eds.)
Progress in Brain Research, Vol 102

CHAPTER 5

Initial tract formation in the vertebrate brain

Stephen S. Easter, Jr.[1], John Burrill[1], Riva C. Marcus[2], Linda S. Ross[3], Jeremy S.H. Taylor[4] and Stephen W. Wilson[5]

[1]*Department of Biology, U. Michigan, Ann Arbor, MI 48109-1048, U.S.A.,* [2]*Department of Pathology, Columbia University, New York, NY, U.S.A.,* [3]*Department of Biological Sciences, Ohio University, Athens, OH, U.S.A.,* [4]*Department of Human Anatomy, University of Oxford, Oxford, U.K. and* [5]*Developmental Biology Research Centre, Kings College, London, U.K.*

Introduction

One of the very first steps in the formation of a nervous system is the creation of tracts. Many of the founders of neuroanatomy, such as Ramon y Cajal and Herrick, were concerned with this event, and they used the best techniques of their days (reduced silver stains and the Golgi method) to label axons in early embryonic brains. The information that they could obtain was limited by the techniques, and by about the 1950s, the systematic study of tract formation had ceased, probably because the methods had shown all that they could. The consensus was that the medial longitudinal fasciculus was the first tract to appear, pioneered by axons from somata that would become the interstitial nucleus of Cajal in the adult (e.g. Rhines and Windle, 1941). Other tracts appeared quickly thereafter, but their nomenclature and their presumed relation to adult tracts were problematic.

With the introduction of degeneration methods by Nauta and his colleagues, the field of neuroanatomy was reborn in the 1960s, and pathways that could only be guessed at prior to that time were revealed with great clarity. A tract or a group of somata were lesioned, Wallerian degeneration ensued, and days or weeks later, the specialized staining protocols revealed selectively the degenerating axons and/or terminals on an unstained background. These very powerful techniques were never of much use to development, however, for several reasons. First, the delay needed for degeneration to occur was incompatible with the rapidly developing embryos; one hopes to see the degenerating tract in the brain where one made the lesion, but by the time it can be stained, the brain has changed. Second, most of the staining was of neurofilaments, and they are slow to appear in early axons. Third, embryonic nervous systems have abundant degenerating axons in them naturally, as we now know, so whatever staining did appear was not restricted to the ones intentionally damaged.

The brute force approach of serially sectioned tissue examined electron microscopically and reconstructed with computer graphics was briefly successful. The work of Levinthal and his colleagues revealed beautifully the formation of the retinofugal projection in the tiny crustacean, *Daphnia magna* (LoPresti et al., 1973), and in the zebrafish, *Brachydanio rerio* (Bodick and Levinthal, 1980). This same approach was used successfully in the study of the nematode, *Caenorhabditis elegans*, and it has been used with great success by Goodman and his colleagues, especially Bastiani, in conjunction with other techniques in locust and *Drosophila* (e.g. Raper et al., 1983), but the method is so labor intensive, and the amount of labor is so sensitive to the size of the structures being examined, that it has not been widely adopted as a first order approach to describing the nervous system of larger animals.

The biggest step forward in the study of early axonogenesis came with the introduction of fluorescent tracers: the procion dyes, Lucifer Yellow, fluorescent dextrans, and most recently the carbocyanines. The embryonic CNS of the locust was explanted; it was thin and transparent enough that it could be cultured and the cells in it visualized. Individual cells were injected with fluorescent tracers directly and the outgrowth of the axons and dendrites were monitored over time (Shankland and Goodman, 1982). This is perhaps the ideal way to study tract development, particularly since these fluorescent tracers can often be made electron-dense, allowing electron microscopic examination of identified processes. Some investigators have been successful in injecting individual cells in the vertebrate CNS (Kuwada, 1986; Holt, 1989) but most of the recent reports on axonogenesis have employed markers applied extracellularly, either horseradish peroxidase (HRP) or carbocyanine dyes applied locally or antibodies against molecules associated with early axons applied more broadly.

In this article, we summarize some of this recent work, with the emphasis on our own.

Scaffolds

Jacobs and Goodman (1989) have identified a set of glial cells in the *Drosophila* CNS on which the earliest set of axons grow out, and called them the 'glial scaffold'. These cells prefigure the location of the first tracts, and the earliest axonal growth cones contact them and are wrapped by them, suggesting that they provide some support for the tracts. We have used the term 'axonal scaffold' (Easter and Taylor, 1989; Wilson et al., 1990) to refer to an early set of tracts, the first ones to appear, on which a great many of the later axons seem to grow. Both uses of the term, scaffold, imply an early structure that both prefigures and facilitates the formation of a later one. Neither the glial nor the axonal scaffolds are demonstrably temporary, and for that reason they may not be in accord with the dictionary's definition of 'scaffold', which is: a temporary wooden or metal frame-work for supporting workmen and materials during the erecting, repairing, or painting of a building, etc. (*Webster's New Twentieth Century Dictionary*). As we shall see below, the idea that the axonal scaffold provides support for the axons that apparently grow along it is also in question. We may conclude that 'substrate' or some other term with fewer connotations might have been preferable to 'scaffold', but for the moment we shall use that term nonetheless. We find the structure interesting on two counts. The first is functional — the possibility that it may actually serve the support role implied by the term, scaffold. The second is comparative and evolutionary — the hypothesis that this early network of tracts may be conserved across vertebrates, and thus represent the earliest solution to the problem of how to begin an axonal network.

Axonal scaffold of anamniotes

The zebrafish has recently become a popular embryological preparation, owing to the easy availability of fertilized eggs, the extreme transparency of the embryo, and the fact that the embryo develops well outside its shell. We have investigated the development of the first tracts in the presumptive fore- and mid-brains. Unlike other vertebrates, fish do not form a neural plate that folds up to form a neural tube. Instead, presumptive nerve cells gather together on the dorsal side of the yolk to form a solid mass of cells (the neural keel), which then hollows out (cavitation) to form a central canal throughout the length of the neuraxis, including the optic stalks and the optic primordia. Despite this early morphogenetic difference, the development of the fish CNS is presumed to have much in common with other vertebrates, and this presumption is supported by the results reported below. Figure 1 shows a few scanning electron micrographs of various stages of the zebrafish, and some of these are illustrated

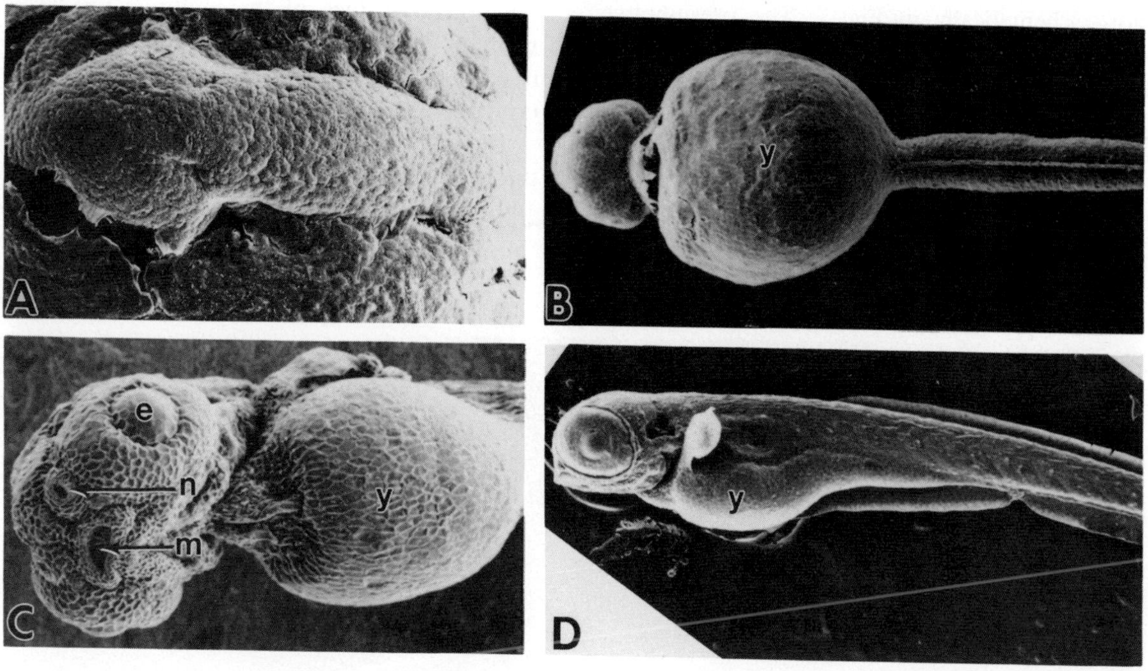

Fig. 1. Scanning electron micrographs of zebrafish embryos. In all cases, anterior is to the left. (a) 12 h. Dorsal view. The broad left end is the forebrain with the wing-shaped eye primordia flanking it. The tubular embryo rests on the massive yolk cell. (b) 24 h. Ventral view. The yolk cell (y) is smaller, and the head is foreshortened. (c) 48 h. Ventrolateral view. The yolk cell (y) is further reduced, the mouth (m) is open, and the external nares (n) and the eye (e) are evident. (d) 96 h. Left view. The yolk cell (y) is nearly gone, and the larva looks like a fish.

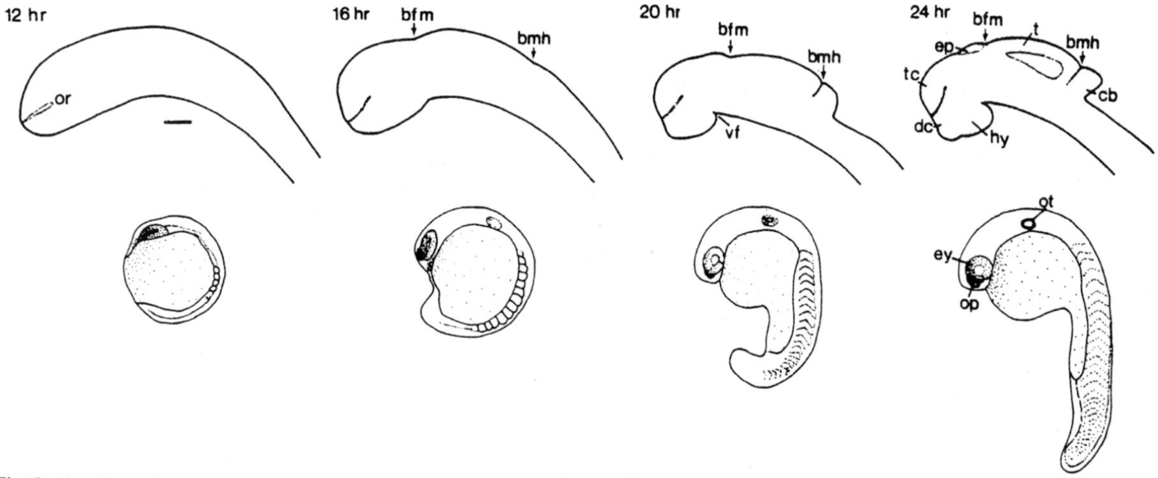

Fig. 2. A schematic view of the development of the brain of zebrafish. All lateral views, from the left. The entire embryo is figured below, and just the presumptive brain above, at higher magnification. bfm: border between fore- and mid-brains; bmh: boundary between mid- and hind-brains; cb: cerebellum; dc: diencephalon; ep: epiphysis; ey: eye; hy: hypothalamus; op: olfactory placode; or: optic recess; ot: otocyst; t: tectum; tc: telencephalon; vf: cephalic flexure. Scale bar: 50 μm. (From Ross et al., 1992.)

more schematically in Fig. 2. The presumptive CNS, apart from the eyes, is initially a uniform tube (12 h), the front end of which shortens to create the boundaries between presumptive fore-, mid-, and hind-brains (16 h). The presumptive hypothalamus enlarges and the neuraxis bends, forming a prominent ventral (cephalic) flexure on the ventral surface of the midbrain (20 h), and many of the structures in the adult become evident, including the epiphysis or pineal body, the optic tectum, and the cerebellum (24 h).

The first neurons appear in the basal plate near the presumptive cephalic flexure, between 14 and 16 h. They are identified first by acetyl-

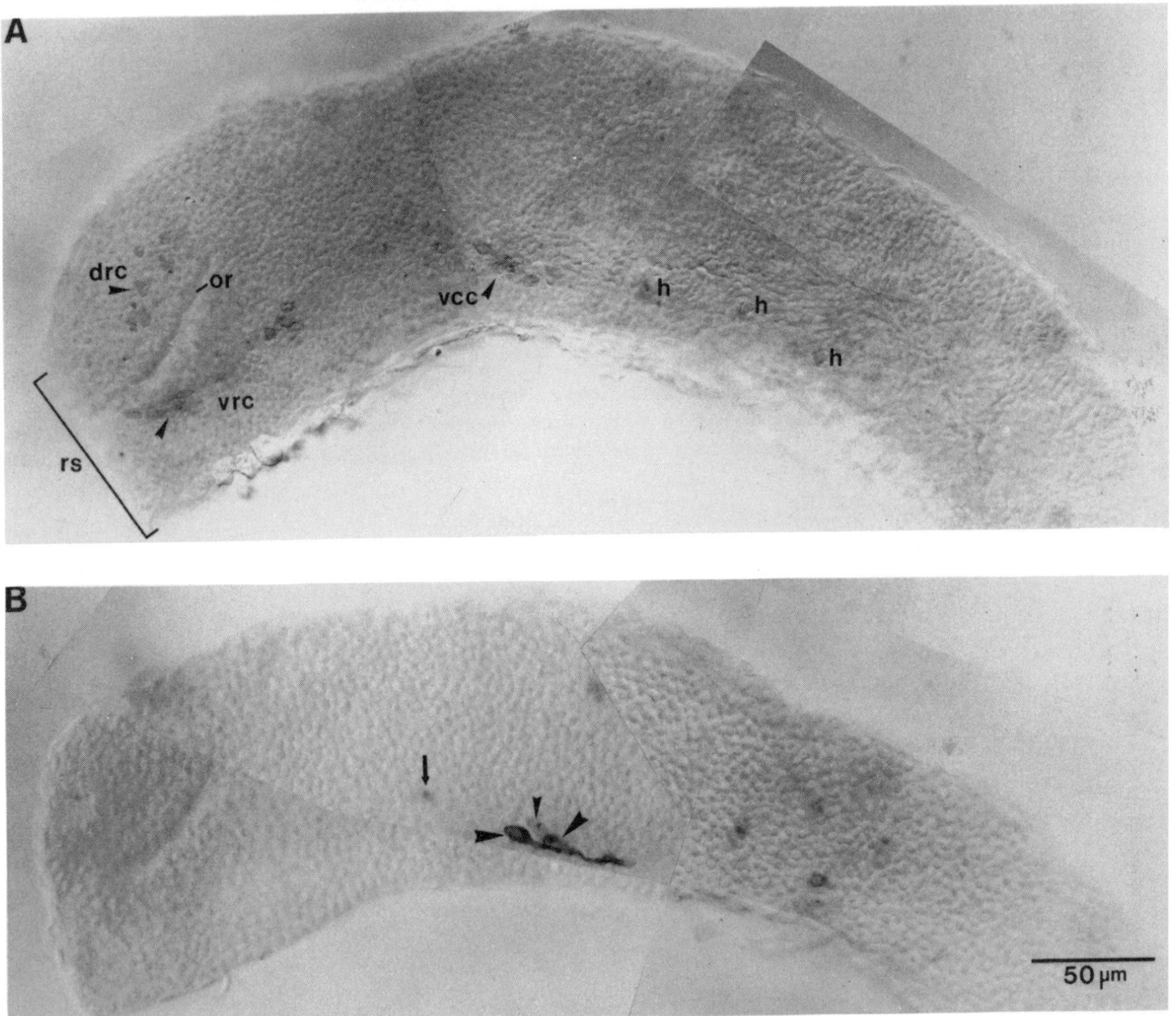

Fig. 3. Lateral views of two whole-mounted 16-h brains reacted for acetylcholinesterase (A) and HNK-1 antibody (B). Dorsal is up and rostral is to the left. (A) Labeled cells (arrowheads) are in three clusters — dorsorostral, ventrorostral, and ventrocaudal — and are separated from one another and from isolated labeled cells in the hindbrain by gaps of unlabeled cells. (B) Only cells in the ventrocaudal cluster are labeled with HNK-1 (arrowheads) and they send out growth cones caudally. drc: dorsorostral cluster; h: labeled hindbrain cells; or: optic recess; rs: rostral surface of the neural tube; vcc: ventrocaudal cluster; vrc: ventrorostral cluster. (From Ross et al., 1992.)

cholinesterase activity (Layer, 1983), as Fig. 3a shows. They number two or three on both sides of the midline, and they send growth cones caudally to pioneer the medial longitudinal fasciculus (MLF), as shown in Fig. 3b. At this stage, the cells and their processes are labeled by the HNK-1 antibody, which labels an epitope common to many cell adhesion molecules (Kruse et al., 1984). A similar pattern of labeling is obtained with an antibody against acetylated alpha tubulin (Piperno and Fuller, 1985). The formation of tracts proceeds as shown in Fig. 4. By 18 h the number of axons has increased, the presumptive MLF has elongated, and a new cluster of neurons near the base of the optic stalk has sent out caudally directed growth cones to pioneer the second major longitudinal tract, the tract of the postoptic commissure (TPOC). The next 4 h see the appearance of one more longitudinal tract (the tract of the anterior commissure), three dorsoventral tracts (the supraoptic tract, the dorsoventral diencephalic tract, and the tract of the posterior commissure), and three commissures (the anterior, postoptic, and posterior), thus producing the 'axonal scaffold' that we referred to in a preceding paragraph (Ross et al., 1992). The reason for calling it a scaffold, with the implication that it supports something that appears later, is evident from the examination of the sketches of the 30 and 36 h brains. Note that only one new tract has appeared (the tract of the habenular commissure) but all the pre-existing tracts are considerably larger than at 24 h, reflecting the addition of new axons.

These initial tracts are striking in several respects. First, the small number (six or seven tracts per side and three commissures) occupy very little of the surface of the brain. One wonders why they don't spread more widely — what keeps them together? Second, the axons are all superficial, lying immediately below the glial end-feet that underlie the basal lamina that surrounds the brain. Third, they are orthogonal. Allowing for the contortion of the neuraxis associated with the cephalic flexure, the longitudinal tracts are clearly distinct from the dorsoventral ones, and at this stage, there are no oblique ones. Fourth, the basal plate clearly matures ahead of the alar plate; the clusters of cells that give rise to the MLF and the TPOC develop first, and the dorsal ones then produce axons that grow down to the basal tracts and join them, turning either caudally or rostrally, and making scarcely any mistakes (Wilson et al., 1990; Chitnis and Kuwada, 1990; Wilson and Easter, 1991a,b).

How certain are we that the axons revealed

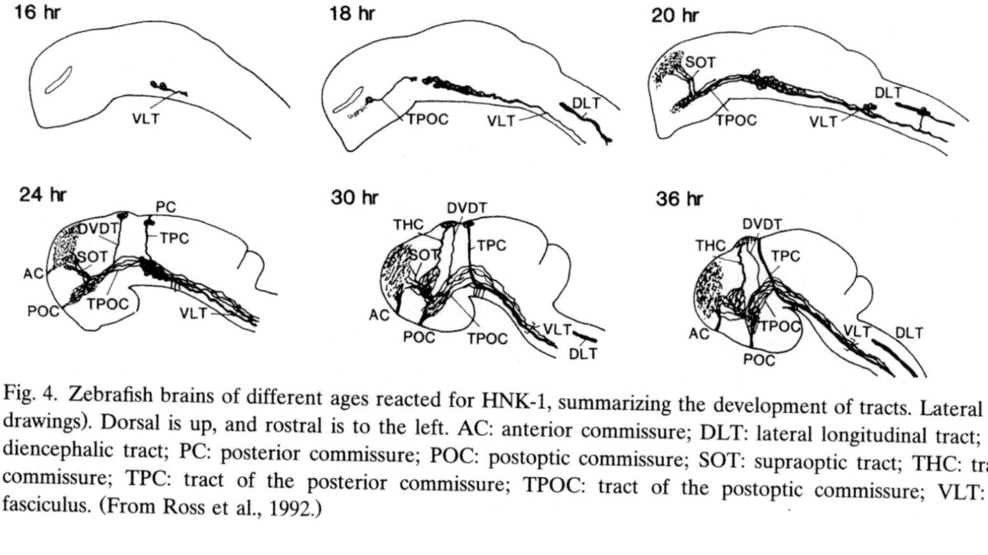

Fig. 4. Zebrafish brains of different ages reacted for HNK-1, summarizing the development of tracts. Lateral views (camera lucida drawings). Dorsal is up, and rostral is to the left. AC: anterior commissure; DLT: lateral longitudinal tract; DVDT: dorsoventral diencephalic tract; PC: posterior commissure; POC: postoptic commissure; SOT: supraoptic tract; THC: tract of the habenular commissure; TPC: tract of the posterior commissure; TPOC: tract of the postoptic commissure; VLT: medial longitudinal fasciculus. (From Ross et al., 1992.)

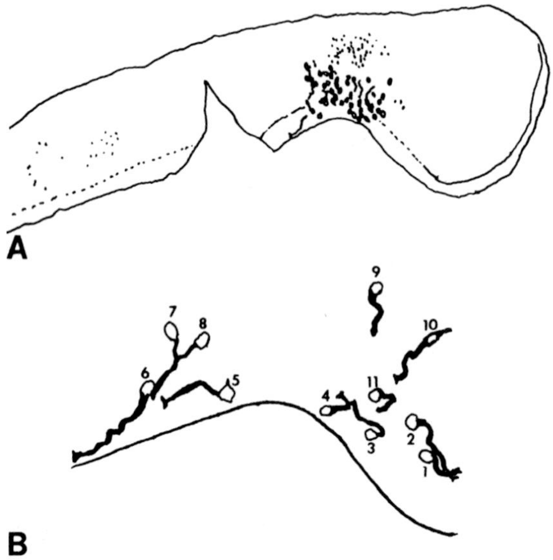

Fig. 5. Chick embryo brain, Hamburger-Hamilton Stage 12, lateral view of the right half, labeled with the TuJ1 antibody. Dorsal is up, and rostral is to the right. Camera lucida drawings. (A) Low magnification view to show the location of a large cluster of labeled cells above the cephalic flexure. (B) More highly magnified view of ten selected cells in this region, showing the diverse trajectories of their initial axons.

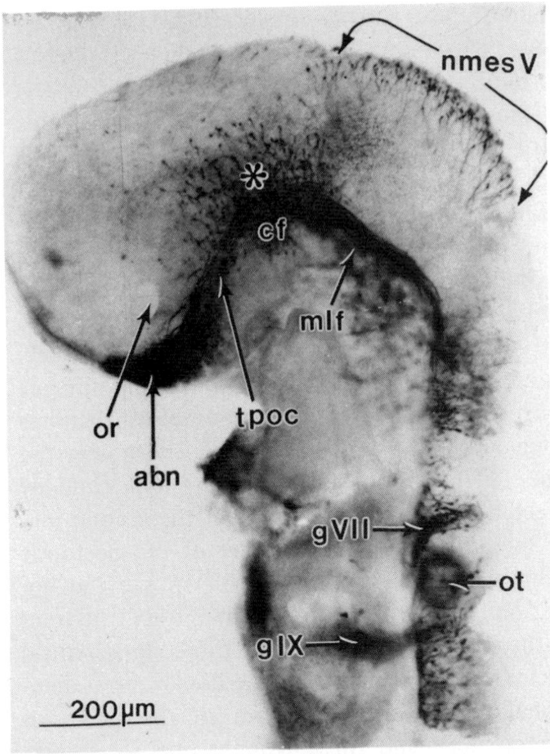

Fig. 6. Chick embryo brain, Hamilton-Hamburger Stage 14, lateral view of the left half, labeled with the TuJ1 antibody. Dorsal is up, and rostral is to the left. abn: anterobasal nucleus; cf: cephalic flexure; gVII: ganglion of the VIIth cranial nerve; gIX: ganglion of the IXth cranial nerve; mlf: medial longitudinal fasciculus; nmesV: mesencephalic nucleus of the trigeminal nerve; or: optic recess; ot: otocyst; tpoc: tract of the postoptic commissure.

immunocytochemically are all the axons in the brain? This was checked in several ways. First, the fact that two antibodies gave essentially the same pattern argues in favor. Second, ultrathin sections viewed electron microscopically revealed axons only in those tracts that we had identified light microscopically. Third, and this is the most convincing, micro-applications of HRP all over the embryonic nervous system revealed axons only in the previously identified tracts. When HRP was applied elsewhere, it produced reaction product at the site of application, but never revealed any axons of passage. If others were there, the HRP should have labeled them (Wilson et al., 1990).

Using the same methods that had worked well with zebrafish, we examined the development of early tracts in the anuran amphibian, *Xenopus laevis* (Taylor and Easter, unpublished). The same antibodies (HNK-1 and anti-acetylated alpha tubulin) label axons, and the pattern of tracts in the Nieuwkoop-Faber Stages 24-30 embryos was essentially identical to that of zebrafish 24-h embryos. The only difference lay in the anterior end, where axons were found to cross between the tracts of the anterior and postoptic commissures; no such crossing was seen in zebrafish.

Evolution is believed to proceed conservatively, in the sense that a successful solution to a developmental problem will be conserved even as further changes are made in the developmental program. This conservatism is the explanation usually given for why early vertebrate embryos resemble one another more closely than the adults. Applying this to the evolution of the nervous system, one would anticipate that the early brains of all vertebrates, including the amniotes, would have much in common, including perhaps the early network of tracts. As we have seen, the two representatives of the classes of fish and amphibia fulfill this expectation. What about the amniotes?

When we first asked that question, we anticipated that the scientific literature would have an answer. But it did not, as the techniques that had been applied to the study of early axonogenesis in birds and mammals were inadequate. As noted in the Introduction, silver staining and the Golgi method gave only partial answers. But the limitation was not just in the staining, but in the visualization of stained neurons; the need to section material and then reconstruct it is very disadvantageous when one is looking for a relatively few

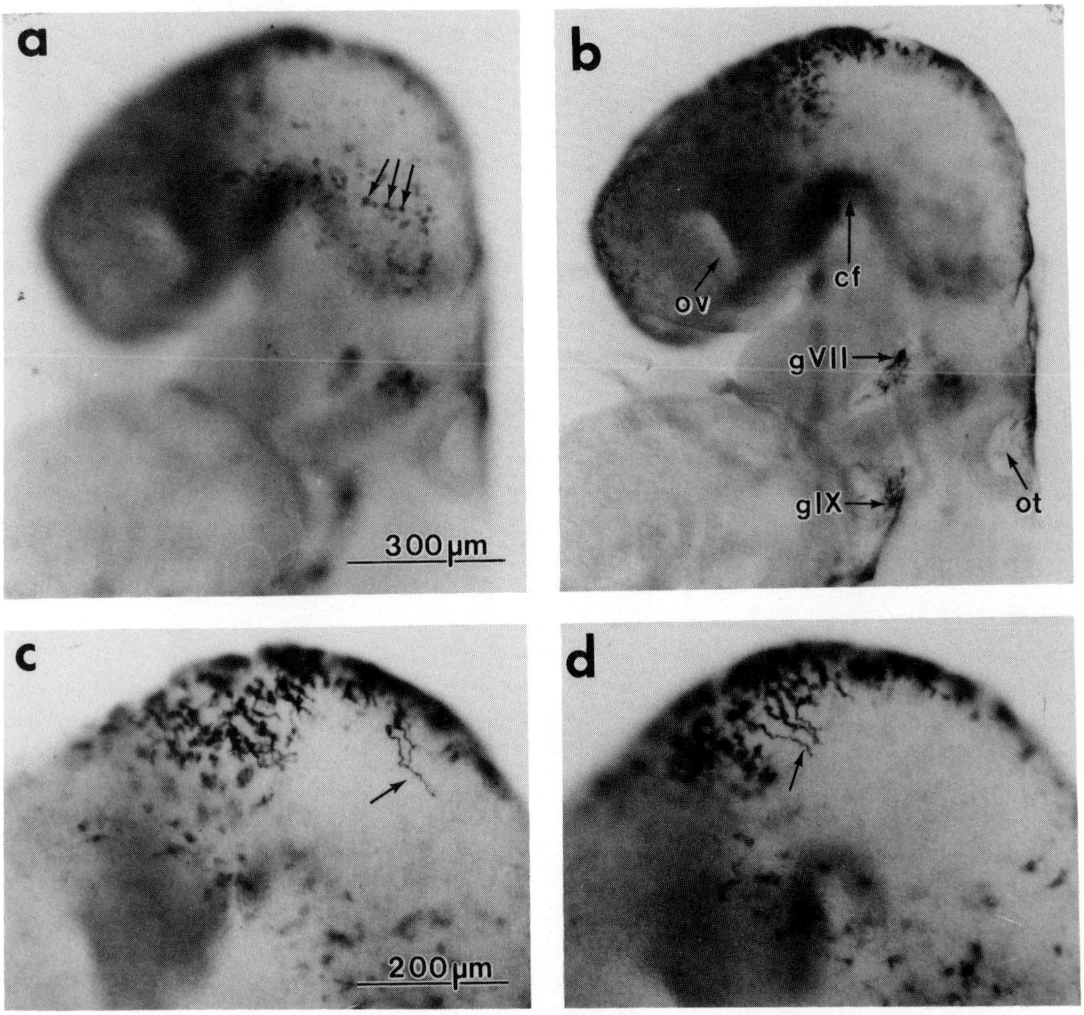

Fig. 7. ←
Mouse embryo brain, E10.0, lateral view of the left half, reacted with the TuJ1 antibody. The numbers of labeled cells are markedly increased relative to E9.0. (a) Dorsal is up and rostral is to the left. The biggest central tract is in the alar plate, from the mesencephalic nucleus of the trigeminal nerve; the mlf and the tpoc are evident. The dashed lines labeled a, c, and d refer to Fig. 10, panels a, c, and d, respectively. (b) Detail of the box labeled, b, in panel a. (c) The cerebral vesicle. Detail of the box labeled, c,

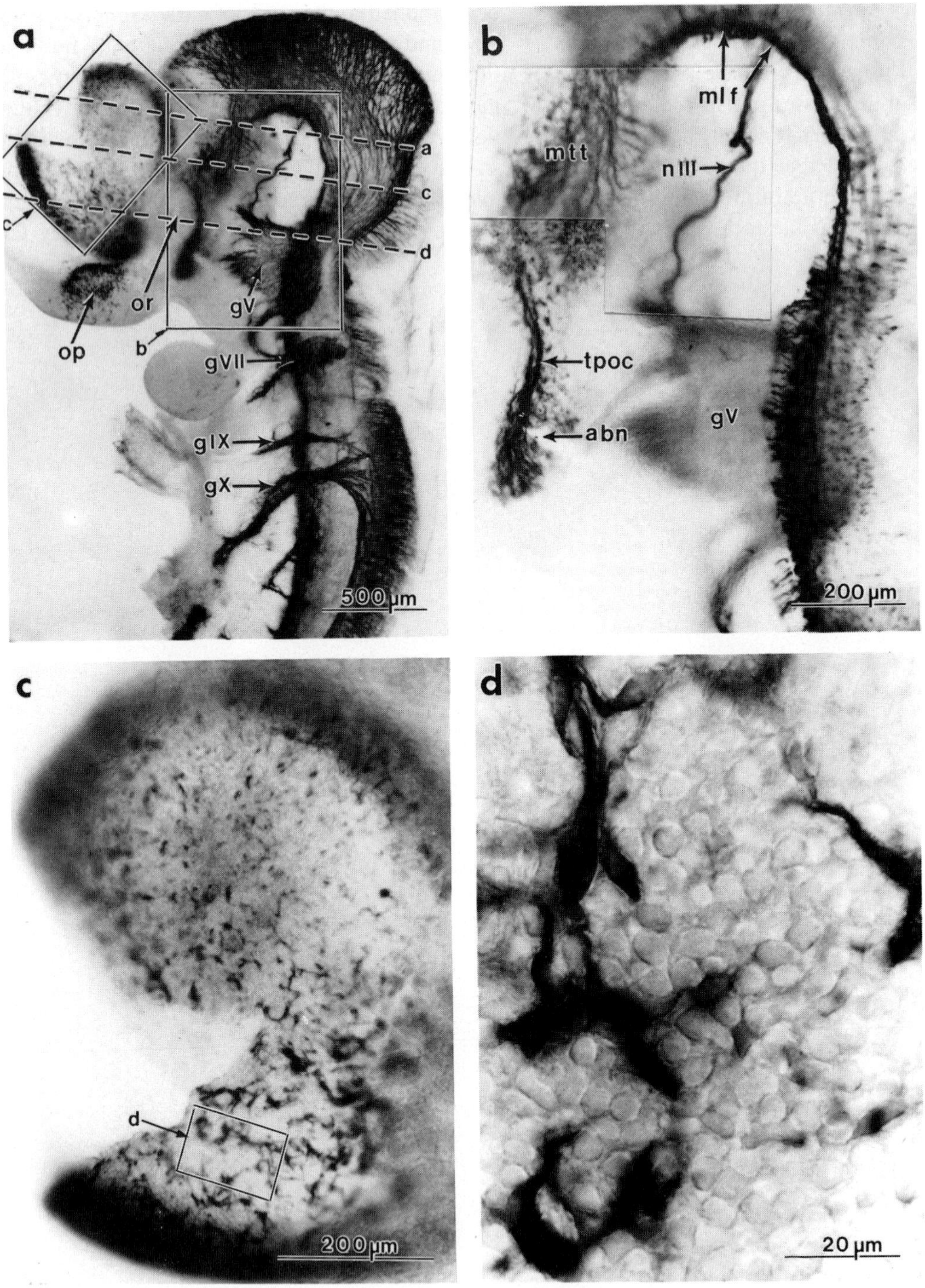

Fig. 8. Mouse embryo brain, E9.0, lateral view of the left half, reacted with the TuJ1 antibody. Dorsal is up and rostral is to the left. (a) Superficial focal plane; arrows indicate labeled cells in the skin and/or mesenchyme. (b) Deeper focal plane showing most of the labeled cells in the CNS are along the dorsal midline of the mesencephalon and in a broad wedge-shaped band covering part of the diencephalon. (c and d) More highly magnified views of the dorsal mesencephalic cells, showing (arrows) the pioneering axons. cf: cephalic flexure; gVII: ganglion of the VIIth cranial nerve; gIX: ganglion of the IXth cranial nerve; ot: otocyst; ov: optic vesicle. (From Easter et al., 1993.)

labeled neurons. Whole-mounts are much better, as they show single neurons in their entirety, in high contrast (see Fig. 3b), thus permitting one to identify a tiny number of labeled cells among thousands of unlabeled ones.

We examined embryonic chick and mouse brains with methods similar to those used on zebrafish and frog; i.e. we fixed lightly in formaldehyde and reacted with neuron-specific antibodies in whole-mounts and sections. By far the best antibody was TuJ1, raised by Frankfurter and his associates against a neuron-specific class III beta-tubulin (Lee et al., 1990).

In the chick, the first axons in the brain were the pioneers of the MLF, as in both zebrafish and frog. Figure 5 shows a camera lucida drawing of a TuJ1-reacted embryo (Hamburger-Hamilton Stage 12: Hamburger and Hamilton, 1951). Note that the number of cells is substantially larger than in the zebrafish, and the unanimity of direction that was evident in the fish is not so evident here. The initial group of ten axons shown in the detail of Fig. 5b have a caudal bias, but in several cases, the caudally directed axons have apparently made a corrective turn away from the rostral direction. Figure 6 shows a slightly older embryo, Stage 14-, and the precocity of the basal plate that was so evident in zebrafish is once again evident. The two longitudinal tracts, the TPOC and the MLF, are quite well developed. The cluster of cells giving rise to the MLF has expanded, but the anterobasal nucleus, the origin of the TPOC, is still quite compact. Several of the rhombomeres are immunoreactive basally, but the only differentiated cells in the alar plate are adjacent to the midline in the mesencephalon. Their axons descend ventrally and turn caudally at the boundary of alar and basal plates to enter the hindbrain. These cells have been identified as the mesencephalic nucleus of the trigeminal nerve on the basis of two observations. First, the birth-dating study of Rogers and Cowan (1973) showed that they were the only cells in the dorsal midbrain born this early. Second, we have labeled these cells retrogradely with diI applied to the trigeminal ganglion in older embryos (Stage 20: Easter, unpublished). We have not examined the development of the chick's brain beyond about Stage 20, by which time the alar plate is still mostly unlabeled, so we can not comment on alar plate homologies between anamniotes and chick. The two basal tracts seem clearly homologous with the MLF and the TPOC of the anamniotes.

In the mouse, our work (Easter et al., 1993) has shown that the first TuJ1-positive cells appear in the neural plate on E8.5, prior to neural tube closure, and are almost certainly the precursors of the mesencephalic nucleus of the trigeminal nerve (Taber and Pierce, 1973). By E9.0, these cells have begun to send out their first axons (Fig. 7), which grow away from the dorsal midline in a generally ventral and caudal direction. Each seems to advance independently of the others, although when they meet, they often fasciculate together, so that by E9.5, the dorsal wall of the mesencephalon is thick with their fascicles, which descend through a narrow opening into the hindbrain. This precocity of development in the alar plate is quite unexpected, and none of the earlier workers on mammalian embryos noticed it. Instead, the MLF was generally reported to be the first tract. We have no explanation for this oversight; perhaps the alar cells did not label well with the silver stains that were used in the earlier studies. The MLF and TPOC are once again recognizable early, on E9.5, and they are quite long by E10.0 (Fig. 8). Figure 8 also illustrates the first TuJ1-positive cells in the cerebral cortex, the

in panel a. (d) Cajal-Retzius cells in the cerebral cortex. Detail of the box labeled, d, in panel c. abn: anterobasal nucleus; gV: ganglion of the Vth cranial nerve; gVII: ganglion of the VIIth cranial nerve; gIX: ganglion of the IXth cranial nerve; gX: ganglion of the Xth cranial nerve; mlf: medial longitudinal fasciculus; mtt: mammillotegmental tract; nIII: oculomotor nerve; op: olfactory placode; or: optic recess; tpoc: tract of the postoptic commissure. (From Easter et al., 1993.)

Cajal-Retzius cells of the primordial plexiform layer (or preplate) (Marin-Padilla, 1971).

A summary diagram of the development of the tracts in the mouse brain is shown in Fig. 9. It should be compared with Fig. 4, an analogous diagram for the zebrafish. The question that motivated the study — is the set of early tracts the same in all vertebrates? — can now be answered, and the answer is 'no.' The well isolated dorsoventral tracts in the alar plate of the zebrafish do not apparently have their counterparts in the mouse, or if they do, they appear later than we examined. Likewise, the precocious alar plate in the mouse, particularly the mesencephalic nucleus of the trigeminal nerve, does not have a counterpart in the zebrafish. The most striking similarities are in the basal plate, where the MLF and TPOC are immediately recognizable as early tracts in all four vertebrates thusfar examined. Differences between animals in the schedule of development (signs of heterochrony) were once thought to be important phenomena that had to be accounted for in evolutionary terms (see Gould, 1977). The more current interpretation of developmental schedules emphasizes the onset and offset of expression of different genes, and we are not surprised to see that the order of appearance of certain structures differs across species. However, the striking similarity shared by all four of the vertebrates examined here is in the precocity of the two longitudinal basal tracts, the MLF and the TPOC. The fact that they appear so early in all animals suggests that they play an important role in subsequent development, but in the absence of any experimental evidence, we can only speculate on what that role is. The most likely seems to be as an avenue along which the alar tracts can distribute their endings to other sites along the neuraxis. This interpretation could be tested by removing, either surgically or genetically, one or both of these two longitudinal tracts, and determining whether the brain could organize along the rostrocaudal axis without them.

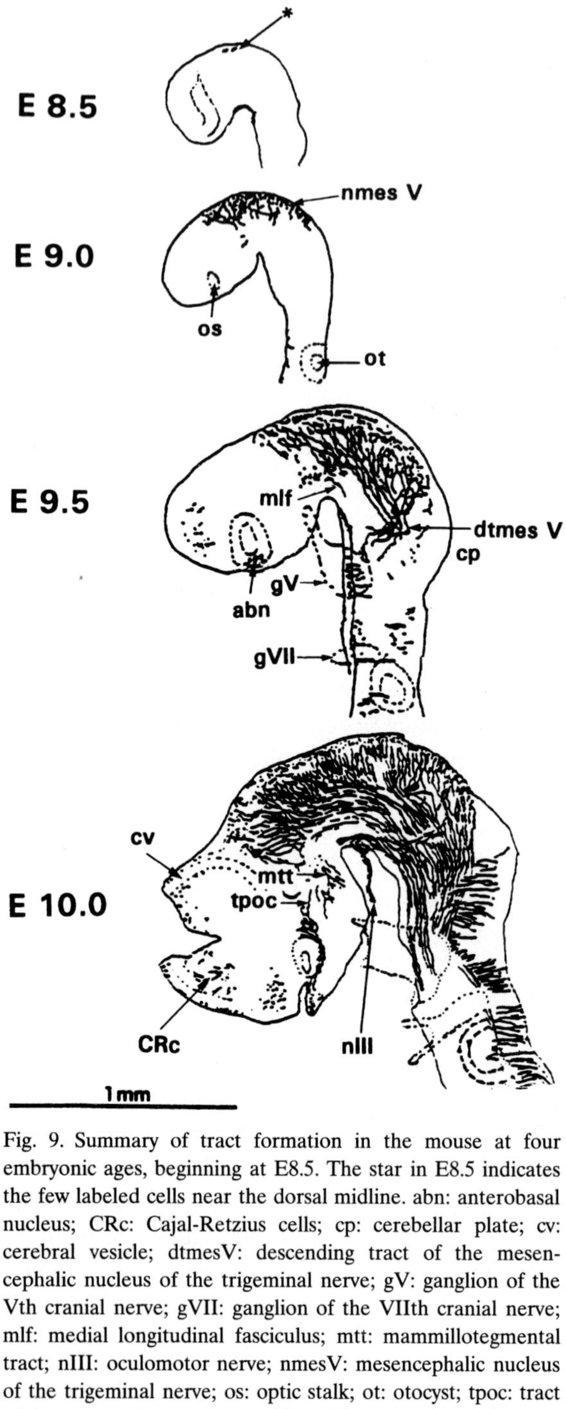

Fig. 9. Summary of tract formation in the mouse at four embryonic ages, beginning at E8.5. The star in E8.5 indicates the few labeled cells near the dorsal midline. abn: anterobasal nucleus; CRc: Cajal-Retzius cells; cp: cerebellar plate; cv: cerebral vesicle; dtmesV: descending tract of the mesencephalic nucleus of the trigeminal nerve; gV: ganglion of the Vth cranial nerve; gVII: ganglion of the VIIth cranial nerve; mlf: medial longitudinal fasciculus; mtt: mammillotegmental tract; nIII: oculomotor nerve; nmesV: mesencephalic nucleus of the trigeminal nerve; os: optic stalk; ot: otocyst; tpoc: tract of the postoptic commissure. (From Easter et al. 1993.)

Location of the tracts

We commented above on the fact that all of the early axons course in very restricted regions, leaving most of the surface of the brain axon-free. In the other dimension, perpendicular to the surface, the axons are similarly selective. All of the early axons are superficial, occupying a very thin lamina subjacent to the pia, separated from the basal lamina that surrounds the neural tube by thin neuroepithelial end-feet (Easter and Taylor,

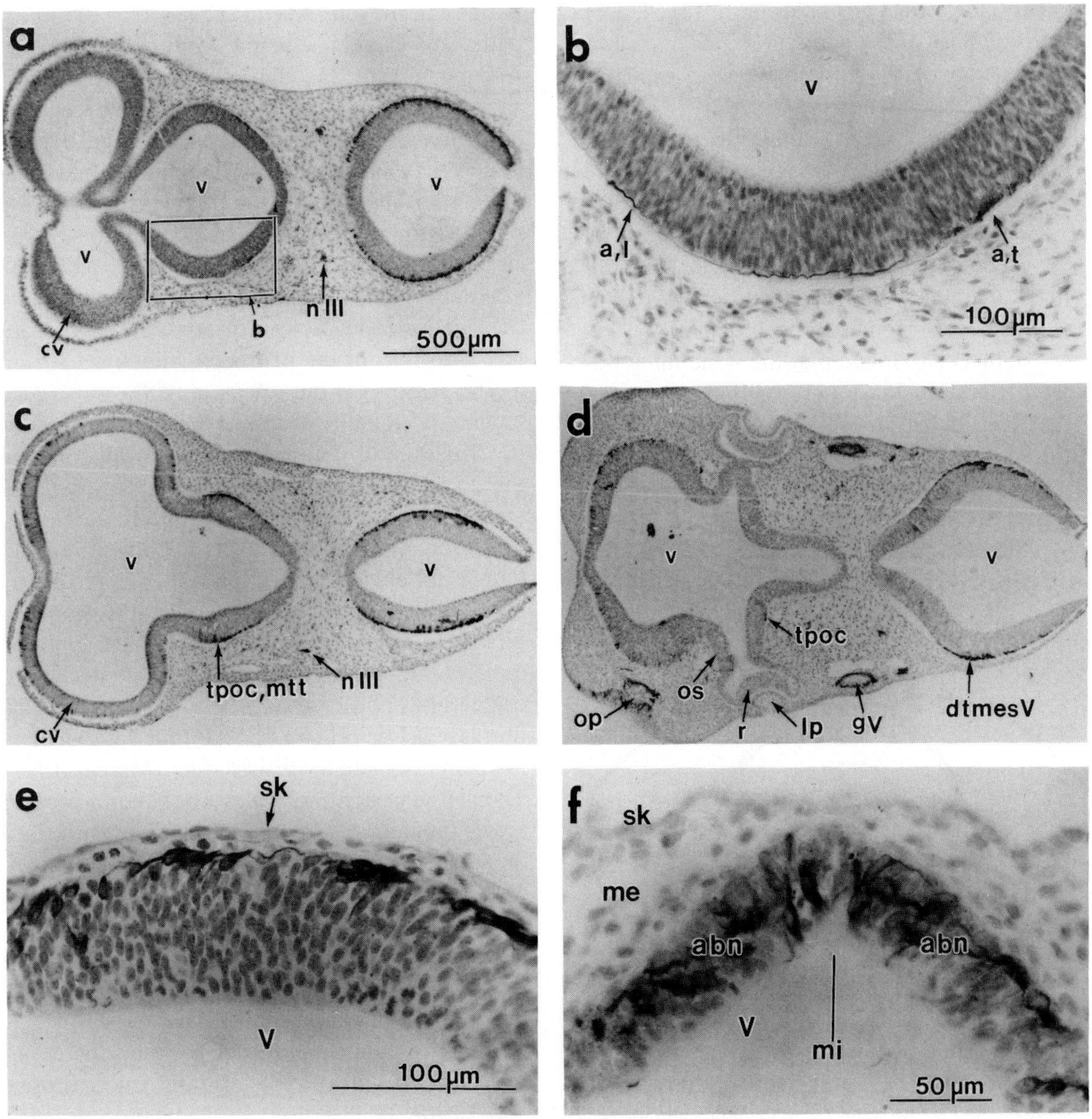

Fig. 10. Sections through E10 mouse embryonic brains to show the superficial location of axons. Panels a–d are all from the same embryo, cut along the planes indicated by the dashed lines in Fig. 8a, rostral to the left. Panel b is a highly magnified view of the box labeled, b, in panel a. Panels e and f are from different embryos, cut transversely through the mesencephalic roof (e) and the anterobasal nucleus (f). abn: anterobasal nucleus; a,l: axon cut longitudinally; a,t: axon cut transversely; cv: cerebral vesicle; dtmesV: descending tract of the mesencephalic nucleus of the trigeminal nerve; gV: ganglion of the Vth cranial nerve; lp: lens placode; m: mesenchyme; mi: midline; mtt: mammillotegmental tract; op: olfactory placode; os: optic stalk; nIII: oculomotor nerve; r: retina; tpoc: tract of the postoptic commissure; sk: skin; v: ventricle. (From Easter et al., 1993.)

1989; Wilson et al., 1990). What features lead the axons to grow in such well- defined locations rather than spreading across the breadth and depth of the neuroepithelial wall? This question is most generally approached in terms of whether the fibers are attracted to the path, repelled by the surroundings of the path, or are channeled by a combination of the two influences. Available data are not very instructive.

Several workers have noted that the tracts seem to mark the boundaries of regions of selective gene expression. Lumsden and Keynes (1989) were the first to suggest this in the hindbrain of the chick. They remarked that the axons emerging from neurons within the rhombomeres gathered in the interrhombomeric boundaries. Trevarrow et al. (1990) found that the same thing was true for zebrafish. Krauss et al. (1991) examined the regions of expression of *pax* genes in zebrafish embryonic brain, and noted that part of the axonal scaffold outlined one of those regions. Others have adopted this same assumption in their interpretations of early tracts, and the subject has been reviewed by Wilson et al. (1993).

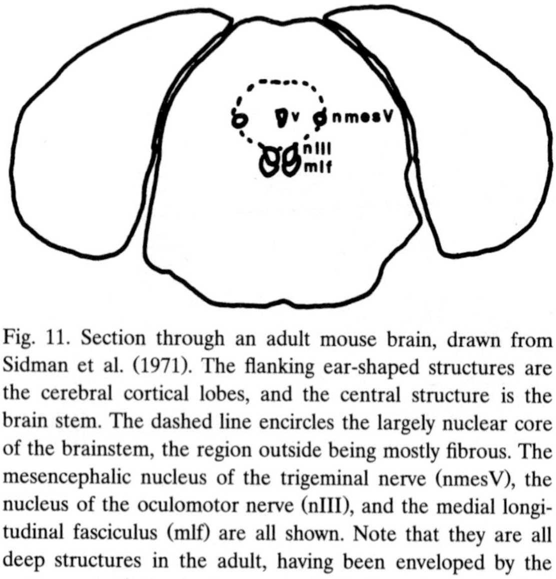

Fig. 11. Section through an adult mouse brain, drawn from Sidman et al. (1971). The flanking ear-shaped structures are the cerebral cortical lobes, and the central structure is the brain stem. The dashed line encircles the largely nuclear core of the brainstem, the region outside being mostly fibrous. The mesencephalic nucleus of the trigeminal nerve (nmesV), the nucleus of the oculomotor nerve (nIII), and the medial longitudinal fasciculus (mlf) are all shown. Note that they are all deep structures in the adult, having been enveloped by the outer part of the brain stem added since they were first formed.

The only proximate cause that has been proposed is greater growth cone mobility in the interrhombomeric region owing to the increased intercellular space there (Lumsden and Keynes, 1989; Guthrie et al., 1991). Studies of pathfinding in the peripheral nervous system suggest that the initial paths are largely a result of repulsion by the surrounding media. Keynes and Stern (1984) showed that the posterior sclerotome repelled motor axons from embryonic chick spinal cord, and Tosney and Landmesser (1985) extended that finding to show that several other tissues also repelled the motor axons, and in doing so, constrained them to follow the 'normal' paths. As yet, there have been no experiments to evaluate the relative repulsiveness or attractiveness of different regions of the neuroepithelium.

The reason for the selection of the superficial lamina is equally obscure. Wilson and Easter (1991b) showed that pioneering growth cones in the zebrafish extended processes deeply, but the trailing axon was always left in the most superficial layer. This shows that the deep neuroepithelium is not able to prevent processes from advancing into it, but the deep is apparently less hospitable to axons than the shallow. Cell adhesion molecules are preferentially distributed superficially (Silver and Rutishauser, 1984), as is the glial fibrillary acidic protein (GFAP), the intermediate filament associated with astrocytes (Marcus, 1992). The cell adhesion molecules would be of obvious utility (assuming that growth cones advance best on adhesive surfaces). GFAP is initially expressed widely, surrounding the neural tube on all sides except the dorsal and ventral midlines, but later, as axons appear, GFAP disappears from most of those places and becomes restricted to the glial wrappings around the tracts.

We have commented extensively on the exclusively superficial location of early tracts (Easter and Taylor, 1989; Wilson et al., 1990; Wilson and Easter, 1991a,b; Easter et al., 1993), and Fig. 10 shows one example of this in mice. To our knowledge, there is only one exception to this general rule, and that is the thalamocortical and corticothalamic projection in mammals, which has been shown to be pioneered in a lamina deep to the cortical plate (McConnell et al., 1989). We know that most tracts in the adult CNS are not superficial, however; if most begin superficial, how do they become deep? We suggest that this occurs by secondary envelopment of the superficial tracts by new neurons that are produced at the ventricular layer and then migrate out beyond the fiber layer to form a new outer surface of the brain. Figure 11 shows a drawing of a cross-section through the adult mouse brain. Note that the structures that were superficial in the embryos are quite deep in the adult, presumably submerged by ventriculofugal migration of new neurons and glia.

This suggests a 'chronotopic' stratification of the brain, in which the deeper structures are the older. Such a stratification is consistent with the 'inside-out' development of the cerebral cortex (Rakic', 1974), but it is only approximately true even for the cerebral cortex, as the oldest cells of the primordial plexiform layer (pre-plate) are divided by the cortical plate, so that some of them remain superficial while others (the subplate cells) are deep. A more fundamental exception to the 'inside-out' chronotopic model is found in the cerebellar cortex, which can better be described as 'outside-in,' because adult granule cells originally entered from the outside (Rakic', 1971). Thus, the generalization of chronotopic stratification is only an approximation, useful in the formulation of a first guess about which structures are older than others, but not a certain basis for prediction.

Development of the retinofugal projections

Herrick (1938) described a commissure behind the optic stalk, linked to a stirrup-shaped pair of tracts that coursed up the wall of the presumptive diencephalon. This commissure and its associated tracts were later joined by optic fibers. We (Easter and Taylor, 1989) rediscovered this and proposed that the tract of the postoptic commissure (TPOC) provided a substrate for the optic fibers as they exited from the optic stalk and entered the brain. Wilson et al. (1990) noted a similar arrangement in zebrafish, and argued for a similar function.

The hypothesis that the TPOC plays a crucial role in the guidance of optic fibers from the optic stalk to the brain has been examined in two different ways, one on *Xenopus*, the other on zebrafish, and has been found wanting.

Burrill (1993) examined the leading optic growth cones in zebrafish in a combined light- and electron-microscopic study, and showed that the optic fibers were always separate from the TPOC. Their filopodia occasionally touched axons in the TPOC, but they never fasciculated with them. Indeed, the optic growth cones were restricted to a preformed ribbon-shaped strip adjacent to the TPOC on its anterior side. The ultrastructure in this region resembled neither the axon-filled space of the TPOC and nor the tight array of basal end-feet that prevailed over most of the neural tube. Instead, it was intermediate between the two, a thin (< 2 μm deep) lamina filled with small processes and a lot of virtual space through which the growth cones advanced. This 'intermediate zone' extended out in front of the leading optic growth cones, and was for that reason probably not caused by the growth cones. It is reminiscent of the 'preformed channels' associated with the 'blueprint model' of axonal growth (Singer et al., 1979). The intermediate zone warrants further investigation: does an intermediate zone appear in the absence of the TPOC? Does it appear if the optic axons are absent? Is an intermediate zone found in association with other early tracts?

Cornel and Holt (1993) evaluated the importance of the TPOC in an experimental study on *Xenopus*. They used various tricks to present an eye capable of axonal outgrowth with a diencephalon lacking a TPOC, and they found that the optic fibers could find the tectum anyway. Thus we must conclude that the TPOC is not an essential structure for optic axon outgrowth and navigation. The most likely conclusion to draw from this experiment is that the optic growth cones may follow the same cues that the TPOC axons did, and the tendency of late axons to join pre-existing tracts is attributable to the hospitality of the same region to both sets of axons, early and late.

Acknowledgments

The work described in this paper was supported by research grants to SSE from NIH (R01-EY-00168) and from the Office of the Vice President for Research of the University of Michigan.

References

Bodick, N. and Levinthal, C. (1980) Growing optic nerve fibers follow neighbors during embryogenesis. *Proc. Natl. Acad. Sci. USA,* 77: 4374–4378.

Burrill, J. D. (1993) The early development of retinal ganglion cell projections in the zebrafish, *Brachydanio rerio*. Ph. D. Dissertation, U. Michigan, Ann Arbor., 149pp.

Chitnis, A.B. and Kuwada, J.Y. (1990) Axonogenesis in the brain of zebrafish embryos. *J. Neurosci.,* 10: 1892–1905.

Cornel, E., and Holt, C. (1993) Precocious pathfinding: retinal axons can navigate in an axonless brain. *Neuron* 9: 1001–1011.

Easter, Jr., S.S. and Taylor, J.S.H. (1989) The development of the *Xenopus* retinofugal pathway: optic fibres join a pre-existing tract. *Development,* 107: 553–573.

Easter, Jr., S.S., Ross, L.S. and Frankfurter A. (1993) Initial tract formation in the mouse brain. *J. Neurosci.,* 13: 285–299.

Gould, S.J. (1977) *Ontogeny and Phylogeny,* Harvard U. Press, Cambridge, 501 pp.

Guthrie, S., Butcher, M. and Lumsden, A. (1991) Patterns of cell division and interkinetic nuclear migration in the chick embryo hindbrain. *J. Neurobiol.,* 22: 742–754.

Hamburger, V. and Hamilton, H.L. (1951) A series of normal stages in the development of the chick embryo. *J. Morphol.,* 88: 49–92.

Herrick, C.J. (1938) Development of the cerebrum of *Amblystoma* during the early swimming stages. *J. Comp. Neurol.,* 68: 203–241.

Holt, C.E. (1989) A single-cell analysis of early retinal ganglion cell differentiation in *Xenopus*: from soma to axon tip. *J. Neurosci.,* 9: 3123–3145.

Jacobs, J.R. and Goodman, C.S. (1989) Embryonic development of axon pathways in the *Drosophila* CNS. I. A glial scaffold appears before the first growth cones. *J. Neurosci.,* 9: 2402–2411.

Keynes, R.J. and Stern, C.D. (1984) Segmentation in the vertebrate nervous system. *Nature,* 310: 786– 789.

Krauss, S., Johansen, T., Korzh, V. and Fjose, A. (1991) Expression pattern of zebrafish *pax* genes suggests a role in early brain regionalisation. *Nature,* 353: 267–270.

Kruse, J., Mailhammer, R., Wernecke, H., Faissner, A., Sommer, I., Gondis, C. and Schachner, M. (1984) Neural cell adhesion molecules and myelin-associated glycoproteins share a common carbohydrate moiety recognised by monoclonal antibodies L2 and HNK-1. *Nature,* 311: 153–155.

Kuwada, J.Y. (1986) Cell recognition by neuronal growth cones in a simple vertebrate embryo. *Science,* 233: 740–746.

Layer, P.G. (1983) Comparative localization of acetylcholinesterase and pseudocholinesterase during morphogenesis of the chick brain. *Proc. Natl. Acad. Sci. USA,* 8: 6413–6417.

Lee, M.K., Tuttle, J.B., Rebhun, L.I., Cleveland, D.W. and Frankfurter, A. (1990) The expression and posttranslational modification of a neuron-specific beta tubulin isotype during chick embryogenesis. *Cell Motil. Cytoskel.,* 17: 118–132.

LoPresti, V., Macagno, E.R. and Levinthal, C. (1973) Structure and development of neuronal connections in isogenic organisms: cellular interactions in the development of the optic lamina of *Daphnia*. *Proc. Natl. Acad. Sci. USA,* 70: 433–437.

Lumsden, A. and Keynes, R. (1989) Segmental patterns of neuronal development in the chick hindbrain. *Nature,* 337: 424–428.

Marcus, R.C. (1992) Intermediate filaments in the nervous system of the embryonic zebrafish. Ph.D. Dissertation, U. Michigan, Ann Arbor., 162 pp.

Marin-Padilla, M. (1971) Early prenatal ontogenesis of the cerebral cortex (neocortex) of the cat (*Felis domestica*). A Golgi study. I. The primordial neocortical organization. *Z. Anat. Entwicklungsgesch.,* 134: 117–145.

McConnell, S.K, Ghosh, A. and Shatz, C.J. (1989) Subplate neurons pioneer the first axon pathway from the cerebral cortex. *Science,* 245: 978–982.

Piperno, G. and Fuller, M.T. (1985) Monoclonal antibodies specific for an acetylated form of alpha-tubulin recognise the antigen in cilia and flagella from a variety of organisms. *J. Cell Biol.,* 101: 2085–2094.

Rakic', P. (1971) Neuron-glia relationship during granule cell migration in developing cerebellar cortex. A Golgi and electromicroscopic study in *Macacus rhesus*. *J. Comp. Neurol.,* 141: 283–312.

Rakic', P. (1974) Neurons in rhesus monkey visual cortex: systematic relation between time of origin and eventual disposition. *Science,* 183: 425–427.

Raper, J.A., Bastiani, M.J. and Goodman, C.S. (1983) Pathfinding by neuronal growth cones in grasshopper embryos II. Selective fasciculation onto specific axonal pathways. *J. Neurosci.,* 3: 31–41.

Rhines, R. and Windle, W.F. (1941) The early development of the fasciculus longitudinalis medialis and associated secondary neurons in the rat, cat and man. *J. Comp. Neurol.,* 75: 165–183.

Rogers, L.A. and Cowan, W.M. (1973) The development of the mesencephalic nucleus of the trigeminal nerve in the chick. *J. Comp. Neurol.,* 147: 291–320.

Ross, L.S., Parrett, T. and Easter, Jr., S.S. (1992) Axonogenesis and morphogenesis in the embryonic zebrafish brain. *J. Neurosci.,* 12: 467–482.

Shankland, M. and Goodman, C.S. (1982) Development of the dendritic branching pattern of the Medial Giant Interneuron in the grasshopper embryo. *Dev. Biol.,* 92: 483–500.

Sidman, R.L., Angevine, Jr., J.B. and Taber Pierce, E. (1971) *Atlas of the Mouse Brain and Spinal Cord,* Harvard University Press, Cambridge, MA., 180 pp.

Silver, J. and Rutishauser, U. (1984) Guidance of optic axons in vivo by a preformed adhesive pathway on neuroepithelial endfeet. *Dev. Biol.,* 106: 485–499.

Singer, M., Nordlander, R.H. and Egar, M. (1979) Axonal guidance during embryogenesis and regeneration in the spinal cord of the newt: the blueprint hypothesis of neuronal pathway patterning. *J. Comp. Neurol.,* 185: 1–22.

Taber Pierce, E. (1973) Time of origin of neurons in the brain stem of the mouse. *Prog. Brain Res.,* 40: 53– 65.

Tosney, K.W. and Landmesser, L.T. (1985) Development of the major pathways for neurite outgrowth in the chick hindlimb. *Dev. Biol.,* 109: 193–214.

Trevarrow, B., Marks, D.L. and Kimmel, C.B. (1990) Organization of hindbrain segments in the zebrafish embryo. *Neuron,* 4: 669–679.

Wilson, S. and Easter, Jr., S.S. (1991a) A pioneering growth cone in the embryonic zebrafish brain. *Proc. Natl. Acad. Sci. USA,* 88: 2293–2296.

Wilson, S.W. and Easter, Jr., S.S. (1991b) Stereotyped pathway selection by growth cones of early epiphysial neurons in the embryonic zebrafish. *Development,* 112: 723–746.

Wilson, S., Ross, L., Parrett, T. and Easter, Jr., S.S. (1990) The development of a simple scaffold of axon tracts in the brain of the embryonic zebrafish, *Brachydanio rerio. Development,* 108: 121–147.

Wilson, S.W., Placzek, M. and Furley, A.J. (1993) Border disputes: do boundaries play a role in growth-cone guidance? *Trends NeuroSci.,* 16: 316–322.

J. van Pelt, M.A. Corner H.B.M. Uylings and F.H. Lopes da Silva (Eds.)
Progress in Brain Research, Vol 102

CHAPTER 6

Dynamic mechanisms of neuronal outgrowth

M.P. van Veen and J. van Pelt

Graduate School of Neuroscience and Netherlands Institute for Brain Research, Amsterdam, The Netherlands

Introduction

Mature neurons are characterized by dendrites and axons with a complex and widely varying shape. These structures often exist as branched trees, which range from simple, as in pyramidal cells, to elaborate, as in Purkinje cells. The cells attain their shape during early development, after migration to their specific positions in the brain. Many studies have been and are concerned with the description of dendritic and axonal morphology, ranging from the famous studies of Ramon y Cajal (1911) to quantitative metric (Uylings, 1977; Uylings et al., 1989; Burke et al., 1992) and topological descriptions (Kliemann, 1987, Horsfield et al., 1989; Van Pelt et al., 1992).

In the formation of dendritic and axonal branching patterns the nerve growth cone and the cytoskeleton play a key role. The growth cone is a motile structure at the tips of outgrowing axons and dendrites, and is thought to act as the 'eyes and ears' of the neurite (see Letourneau et al., 1991, 1994; Kater et al., 1994). It is the structure which directly contacts the immediate environment of the neuron and which is assumed to sense other cells and gradients of soluble and membrane-bound chemical compounds. The growth cone adheres to the surrounding substrate, thereby establishing a close contact between neurite and environment. The importance of the growth cone for the development of neuritic shape lies in its involvement in guidance, elongation and branching of neurites.

The cytoskeleton is a second key factor crucial in neuritic outgrowth and maintenance of form (see Black, 1994). The cytoskeleton is a collective term for three types of molecules: intermediate filaments, actin filaments and microtubules. Intermediate filaments are the most and actin filaments the least stable molecules (Amos and Amos, 1991). Disrupting actin results in a much slower elongation rate and a lower branching probability (Marsh and Letourneau, 1984). Disrupting the microtubules can result in retraction of neurites (Yamada et al., 1970) and preventing their polymerization stops outgrowth (Bamburg et al., 1986). Among the elements of the cytoskeleton, the microtubules form a continuous core within neurites and sustain the complex neuronal shape.

What is the effect of changes in growth cone behavior or microtubule elongation on neuronal morphogenesis? And, how do these two key factors interact in neuronal outgrowth? These questions will be the subject of this chapter. Questions regarding the role of the growth cone in neuronal outgrowth will be treated in the first part of this chapter. Questions regarding the elongation of the cytoskeleton, as a limiting factor in neuronal outgrowth, are treated in the second part of this chapter. The third part aims at integrating and discussing the genesis of neuronal shape, focussing on the elongation of the neurite and cytoskeleton, the branched structure of neurites and the role of cellular contacts.

The growth cone

During axonal and dendritic outgrowth, the growth cone is the structure where much of the growth activity appears (see Letourneau et al., 1991, for reviews). It is the site where three kinds of actions are performed: elongation, steering and branching. The first of these depends on elongation of the cytoskeleton as well, and its discussion will be deferred to part 2 of the chapter. The latter two are mediated by the growth cone itself, in interaction with its immediate environment.

Growth cone morphology

There are two, light microscopically distinguishable domains in the growth cone, the central and the peripheral domains (Fig. 1). The central domain (or C-domain) is directly connected to the neuritic shaft, the microtubules end here like a fan and cell organelles are present. The outer part of the growth cone, the peripheral domain (or P-domain), is flattened and lacks cell organelles, and microtubules only occasionally protrude into this region (Gordon–Weeks, 1991).

The P-domain can be subdivided into two parts: the filopodia and, proximal to the filopodia, a basal part containing lamellae and veils. The lamellae and veils have a high actin content and show protrusive movements, comparable with the movements of pseudopods of amoebae (Lewis and Murray, 1992). Filopodia have an internal core of aligned actin bundles (Letourneau and Ressler, 1984; Mitchinson and Kirschner, 1988) and are very motile structures. They can be extended within minutes, have a lifespan of tens of minutes, and can be retracted again within minutes (Bray and Chapman, 1985). They adhere to the environment by forming bonds with cell adhesion molecules (CAMs), among which NCAM, N-cadherin, L1, and the integrin receptor family rank as the most important (Dodd and Jessell, 1988).

Steering of the growth cone

The filopodia are the parts of the growth cone which have to contact a substrate for guidance. Caudy and Bentley (1986) and Bentley and

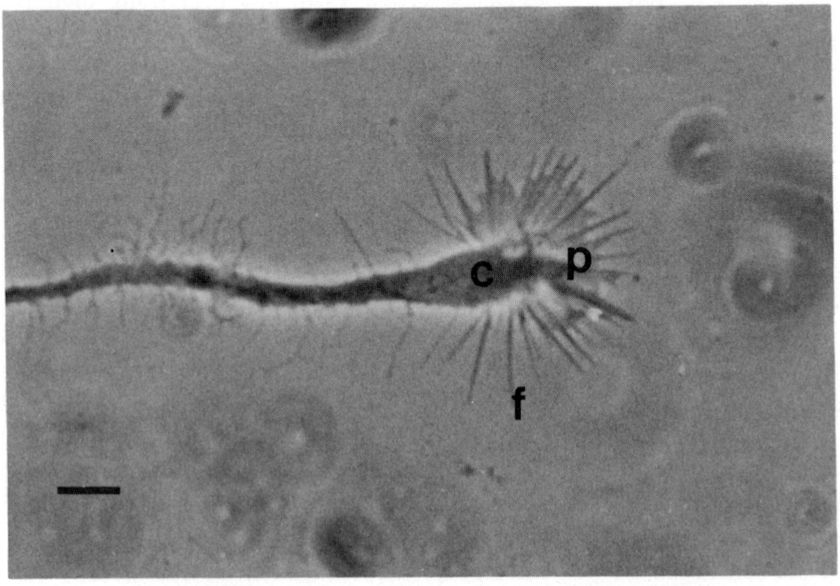

Fig. 1. Morphology of the growth cone. c: central domain, p: peripheral domain, f: filopodium. Measure bar is 5 μm.

O'Connor (1991) show in the grasshopper that a filopodium has to touch a guidepost cell in order to redirect the growth cone. The distance between these guidepost cells is approximately the length of a filopodium (Ho and Goodman, 1982), allowing for a 'stepping stone' mode of guidance. Removing the filopodia by the addition of cytochalasin B in situ (Bentley and Toroian-Raymond, 1986) results in undirected growth by axons which would otherwise use the guidepost cells to orient themselves. Hammarback and Letourneau (1986) provided a tissue culture model of this mode of outgrowth by seeding neurons in an environment where adhesive squares were positioned within nonadhesive lanes. At small nonadhesive lane width, the neurites were able to grow over the lanes onto other adhesive squares. At a width larger than the longest filopodial length, the neurites could not cross the lanes and remained confined within their square. Filopodia and growth cones will not only adhere to substrates, but they can also be repelled by contact mediated repulsion (Kapfhammer and Raper, 1987; Schwab, 1990; Bantlow et al., 1993).

The model favored by these observations is one in which filopodia are probing the environment and transduce a signal to the growth cone, which will be received by the basal part of the P-domain. The translation from filopodial signals to growth cone displacement could be based on interference with the formation of lamellae and lamellipodia. The protrusion of lamellipodia is determined by the state of the actin cortex inside a cell or growth cone (Oster, 1988; Alt, 1990; Vailiev, 1991; Lewis and Murray, 1992; Stossel, 1993). If signals from the filopodia are able to effect the state of the actin cortex, they will be able to interfere with lamellipodial protrusions. A strong filopodial signal could cause either actin disassembly or cause a change in proteins affecting the state of the actin cortex, thereby causing lamellipodial protrusions (Alt, 1990; Lewis and Murray, 1992; Stossel, 1993). The result is that the growth cone will move in the direction of the protruding lamellipod. On the other hand, inhibitive signals (Kapfhammer and Raper, 1987; Schwab, 1990; Bantlow et al., 1993) would have an opposite effect on the actin cortex, thus causing retraction of lamellipodia.

Branching of the growth cone

One of the first hypotheses regarding the role of filopodia in branching was put forward by Vaughan et al. (1974), who conjectured that a branch would be formed when two filopodia simultaneously contacted synapses. This hypothesis was later generalized (Berry et al., 1980), stating that branching would occur when two filopodia simultaneously made strong adhesive bounds to the environment. Wessells and Nuttall (1978) showed that growth cones could be induced to branch when their front filopodia were lifted from the substrate, leaving only adhered filopodia at the sides. Not merely the adhesion of the filopodia, but their angular distribution was shown to be a main determinant of branching by these experiments. This idea was also put forward by Bray (1979), who attributed the increased branching of growth cones after dissecting their trailing neurite to the increased forces of sidewards directed filopodia. Observations on spontaneously branching growth cones show an increased branching probability when the filopodia happen to be directed to the sides of the growth cone (Van Veen et al., 1993).

Branching is thought to be dependent not only on the angular distribution of the filopodia, but also on an interplay between filopodia and the microtubules more centrally in the neuritic shaft (Fig. 2) (Letourneau et al., 1986). Branching would occur when the filopodia orient the microtubules into distinct clusters. This configuration is attained most easily when the filopodia are oriented in distinct, sidewards directed clusters. Interestingly, the addition of low doses of taxol completely prevents branching, while the filopodia are less widely spread around the central axis than in the control situation (Letourneau et al., 1987).

A growth cone model

To investigate the importance of filopodia in growth cone behavior, we have modeled the

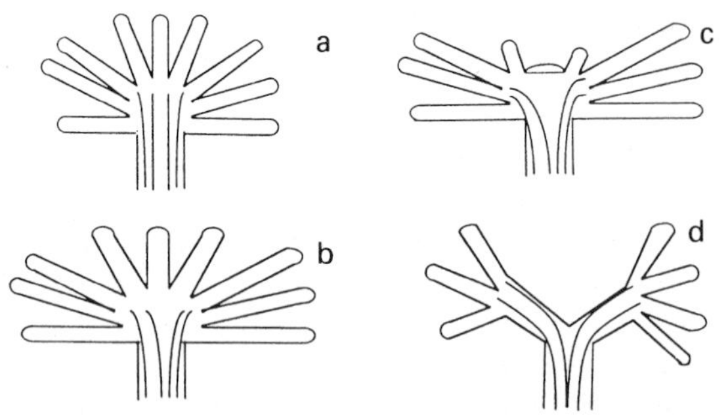

Fig. 2. Interplay between the filopodia and the microtubules in the neuritic shaft as proposed by Letourneau et al. (1986), leading to branching (reproduced with permission).

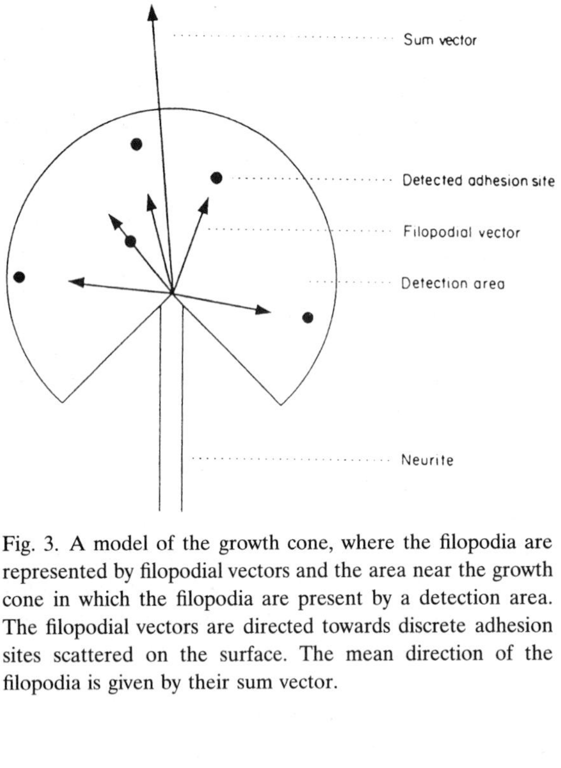

Fig. 3. A model of the growth cone, where the filopodia are represented by filopodial vectors and the area near the growth cone in which the filopodia are present by a detection area. The filopodial vectors are directed towards discrete adhesion sites scattered on the surface. The mean direction of the filopodia is given by their sum vector.

growth cone as a structure existing solely of filopodia, which are present in a limited area near the basis of growth cone (Fig. 3). Both the number and directions of the filopodia are assumed to follow a stochastic distribution. These distributions are generated in the model by scattering discrete 'adhesion sites' on the substrate on which the modeled growth cone is positioned. Filopodial vectors are projected towards all adhesion sites within the detection area of the modeled growth cone (Fig. 3). The filopodial vectors represent the filopodia, while the detection area represents the zone around the growth cone which can be reached by filopodia. It is not proposed that in reality the number and direction of filopodia are determined by discrete adhesion sites on the substrate, but the adhesion site distribution is used to generate the stochastic distributions for the number and direction of filopodia.

The direction of growth cone displacement can be defined by the sum vector of the filopodial vectors (Fig. 3), with components:

$$x_s = \sum_{i=1}^{n} a_i x_i$$

$$y_s = \sum_{i=1}^{n} a_i y_i$$

where x_i, y_i are the coordinates of filopodial vector i and normalized such that $\sqrt{x_i^2 + y_i^2} = 1$, a_i is the weight factor of filopodial vector i, n is the total number of filopodial vectors and x_s, y_s are the coordinates of the sum vector. By definition, all vectors have their origin in the basis of the growth cone and the vector $x = 1$, $y = 0$ is drawn in the direction opposite to the trailing neurite. In this approach the direction of each filopodial vector accounts for the direction, while the weight factor accounts for the strength of the filopodial action on the growth cone. A problem is that the weight factors must be determined, and presently it is difficult, if not impossible, to measure filopodial weights, either in terms of force or intracellular signals. As a first approximation, we will take each weight factor to be equal to 1, thereby focussing on the number of filopodia around the growth cone rather than the strength of the signals. In the model, the growth cone will turn left if there are more filopodia pointing to the left than to the right of the growth cone.

Experimental evidence (Wessells and Nutall, 1978; Van Veen et al., 1993) suggests that the growth cone will branch when the filopodia are mainly directed towards the sides of the growth cone. This distribution can be quantified by the perpendicular variance B of the filopodial vectors:

$$B = \frac{1}{\sum_{i=1}^{n} a_i} \sum_{i=1}^{n} a_i \sin^2(\phi_i - \phi_s).$$

where a_i, i and n are as defined before, ϕ_i is the angle of filopodial vector i and ϕ_s is the angle of the sum vector. These angles can be calculated from the coordinates of the vectors: $\phi_i = \arcsin(y_i)$ and $\phi_s = \arcsin(y_s/\sqrt{x_s^2 + y_s^2})$. Again, the weights of the filopodial vectors are assumed to be equal to each other. B has a value of 0 when all filopodial tips have the same direction and a value of 1 when exactly half the number of filopodia point to each side of the growth cone.

In a homogeneous environment, the path of a tissue-cultured growth cone is a restricted random walk (Katz et al., 1984; Katz and George, 1985; Van Veen et al., 1993). This random walk of tissue cultured growth cones is mimicked to a great extent by the random walk of modeled growth cones (Van Veen et al., 1993). Van Veen and Van Pelt (1992) showed that in simulations of modeled growth cone paths the tortuosity of the path is caused by the random positions of the filopodia, and restrictions on the direction of movement imposed by the trailing neurite. In tissue-cultured growth cones, the randomness in the path is probably not caused by the random placement of the filopodia, given the lack of correlation between the angular distribution of filopodial numbers and the direction of growth cone displacement (Van Veen et al., 1993). Nevertheless, the similarity between both random walks (Van Veen et al., 1993) suggests that the random movement of tissue-cultured growth cones is also caused by random, uniformly distributed directional cues.

The microtubule cytoskeleton

The growth cone may be one of the principal structures in outgrowth, but the neurite formed by the growth cone must later be maintained. The main cellular components responsible for this maintenance are the cytoskeleton proteins. Of these cytoskeleton proteins, the microtubule cytoskeleton forms a rigid, continuous core within the whole neuritic tree. The microtubules are shown to be important in the maintenance of neuritic shape: disrupting the microtubules can result in retraction of neurites (Yamada et al., 1970) and preventing the polymerization of microtubules stops outgrowth (Bamburg et al., 1986). Although the microtubles cytoskeleton as such spans the whole neurite, the separate microtubules can stop and start everywhere within the neurite (Bray and Bartlett Bunge, 1981).

Microtubules have a chemical polarity which is reflected in the assembly/diassembly rates at the ends of a microtubule, these being the sites where most assembly activity takes place (Bergen and Borisy, 1980; Soifer and Mack, 1984; Amos and Amos, 1991). One end is known as the fast assembly (or +-end) the other as the slow assembly (or −-end). On a time scale of hours, the assembly dynamics of microtubules have been shown to be linear, with the assembly rate depending on the concentration of soluble tubulin and a constant disassembly rate (Bergen and Borisy, 1980; Soifer and Mack, 1984):

$$\frac{1}{e}\frac{\mathrm{d}l_\mu}{\mathrm{d}t} = aQ - b \tag{1}$$

where l_μ is the length of the microtubule, e is the length of a tubulin dimer, Q is the concentration of tubulin, a is the assembly constant and b is the disassembly rate.

The assembly dynamics defined in this way suggest a steady elongation, but on a time scale of seconds, microtubules may change from an elongating to a shrinking state, a process known as dynamic instability (Mitchinson and Kirschner, 1988; Amos and Amos, 1991; Pryer et al., 1992). Dynamic instability can be fully described by the frequencies of the transitions between the elongation and the shrinkage states, and the rate constants of the elongation and shrinkage phases (Verde et al., 1992; Dogterom and Leibler, 1993).

The interplay between these four parameters determines if the average elongation rate of microtubules, as defined by Eqn. (1), will be positive, negative or zero (Dogterom and Leibler, 1993). Microtubule-associated proteins appear to increase the transition from shrinkage to elongation, thereby promoting a positive average elongation rate (Pryer et al., 1992).

There are two regions in the neurite with a high level of newly assembled parts of microtubules, namely the proximal 40 μm and the distal 100 μm (Black, 1994). It is still an open question which of these regions contributes significantly to the elongation of the neurite. Two kinds of studies suggest that assembly in the distal region is important in neuritic elongation. Bamburg et al. (1986) showed that inhibition of the tubulin polymerization at the proximal part of a neurite does not affect the elongation rate, while inhibition at the distal part immediately stopped the elongation. Shaw and Bray (1977) measured the migration rate of isolated growth cones. They showed that these growth cones elongated the trailing piece of neurite with a speed leveling off in time, suggesting that assembly takes place from a pool of unassembled material near the growth cone.

The production of tubulin takes place in the soma. The half life of tubulin mRNA is only 1–2 h (Cleveland et al., 1981), ensuring that most of it will be contained in the soma because it is degraded before it is able to reach the dendrites. The spatial difference in the production and polymerization site of the tubulin protein poses a transport problem: tubulin has to be transported from the soma to the neuritic tips in order to be available for assembly. In favor of the hypothesis that tubulin is transported passively through the neurite in the form of tubulin dimers are studies which show that the concentrations profiles of a single pulse of labeled tubulin dimers (Okabe and Hirokawa, 1988; Lim et al., 1990) look very much like those of diffusing macromolecules (Popov and Poo, 1992) and also like the profiles expected from theoretical studies (Okubo, 1980; Murray, 1989). Further, experiments by Tashiro and Komiya (1992) and Hoffman et al. (1992) suggest that a large pool of soluble tubulin exists inside neurites, which is susceptible to diffusion.

It has been argued that diffusive transport is not able to carry tubulin dimers to be tips of axons over a distance in the order of millimeters (Black and Smith, 1988; Ahmad et al., 1993). However, this view is based on a solution of the diffusion equation without sources and sinks of soluble tubulin. Then, a single pulse of tubulin diffuses out forming a bell-shaped profile in space. Because no tubulin is produced, the concentration gradient levels off, and the flux of tubulin is proportional to the square root of time (Okubo, 1980). In a nerve cell, the soma forms a source of tubulin, while the assembly sites form sinks. The source and sinks keep up the gradient of soluble tubulin, which attains a constant slope and the flux of tubulin becomes constant (Van Veen and Van Pelt, 1994). In this case, diffusion is able to account for a constant transport rate of tubulin over long distances. When the drive behind tubulin transport is taken to be diffusion, the flux J of material at each point between soma and tip is given by (Okubo, 1980; Murray, 1989):

$$J = -D\frac{dQ}{dx} \tag{2}$$

where Q is the concentration of soluble tubulin at position x and time t and D is the diffusion constant.

The combined result of tubulin production, diffusive transport and polymerization, determining the elongation rate of the microtubules, has been studied by means of a mathematical model by Van Veen (Van Veen, 1993; Van Veen and Van Pelt, 1994). The model, assuming a constant tubulin production at the soma, predicts a constant elongation rate of the microtubules. This constant elongation rate is similar to observations on microtubules in tissue culture (Bergen and Borisy, 1980; Verde et al., 1992).

In the model, the tips of the neurites, being the main polymerization sites of tubulin, compete for soluble tubulin present at a common branch point (Fig. 4). Tips with a high assembly rate a or a low

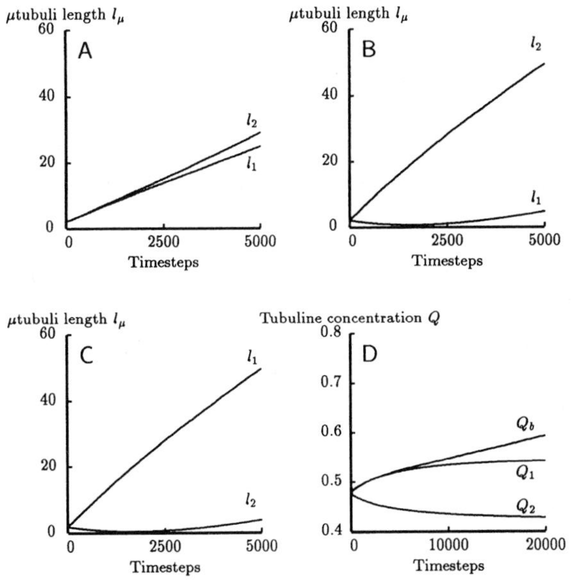

Fig. 4. Evolution of unstretched length in time when daughter segments 1 and 2 differ in value of the parameters a, b or D. (A) Difference in D: $D_1 = 5$, $D_2 = 20$. (B) Difference in a: $a_1 = 0.1$, $a_2 = 0.13$. (C) Difference in b: $b_1 = 0.05$, $b_2 = 0.064$. (D) Evolution in time of the tubulin concentration in the tips of both daughter branches (Q_1 and Q_2) and in the branch point (Q_b) with parameter a differing for the daughter segments: $a_1 = 0.1$, $a_2 = 0.13$.

disassembly rate b compete successfully for soluble tubulin with other tips (Figs. 4b, 4c) by creating a low tubulin concentration in the tip, and establishing a steep tubulin gradient. Differences in the tubulin diffusion constants o, determining the transport rate towards the tip, are of minor importance in the competition between tips (Fig. 4a).

Less successful tips not necessarily disappear because the tubulin concentrations in the soma and branch points are increasing with time (Fig. 4d). At a certain moment, less successful competitiors can experience tubulin concentrations which allow them to grow. In Figs. 4b and 4c, this sequence of events is displayed, while Fig. 4d gives the tubulin concentrations accompanying Fig. 4b. Thus, competition for soluble tubulin can result in delayed outgrowth, or even retraction, of the tip with the lowest assembly or the highest disassembly rate.

The stability of neuronal microtubules increases as maturation proceeds. In highly branched dendrites, such as those of Purkinje cells, the fraction of cold stable microtubules changes from almost zero to almost one during maturation (Faivre et al., 1985). The cold stability measures the stability of microtubules by measuring their resistance to sudden cooling, which normally causes disassembly. Important factors in stabilisation are the microtubuli associated proteins (MAPs) (Matus, 1988; Amos and Amos, 1991). The microtubule assembly model predicts that unstable microtubules will retract when the production of tubulin in the soma is diminished. It is proposed here that stabilization, for instance by MAPs, is essential to sustain the neuritic tree formed during outgrowth.

The genesis of neuronal shape

The growth cone and cytoskeleton determine the branching probability, the direction of elongation and the elongation rate of an emerging neuritic tree. To investigate how these actions interact in morphogenesis, the following section will focus on the elongation of segments and the formation of the branched structure. Because nerve cells seldom grow in isolation, in influence of other nerve cells on the morphogenesis will also be discussed.

Elongation

A complicating factor in relating the polymerization of the microtubule cytoskeleton to the elongation rate of the neurite is the elasticity of the neurite. Dennerl et al. (1989) and Lamoureux et al. (1992) have shown that neurites behave like simple elastics, and that they follow Hook's law relating force to stretching:

$$F_e = \frac{EA(l - l_0)}{l_0}$$

where A is the transsectional area of the neurite, F_e is the elastic force of a segment with stretched length l and unstretched length l_0, and E is a proportionality constant. Because the growth cone exerts force on its neurite, the neurite is always in a stretched state (Lamoureux et al., 1989). As shown by Van Veen and Van Pelt (1994), the stretched length is proportional to the unstretched length, given that changes in force exerted on the neurite lead quickly to a new state of stretching. It follows also that the change in stretched length is proportional to the change in unstretched length:

$$\frac{d(l)}{dt} = k\frac{d(l_0)}{dt} \tag{3}$$

where $k = 1 + F_g/(EA)$ is a proportionality constant which is expressed in terms of the force F_g that the growth cone exerts, E and A. If the force exerted at the growth cone increases, for example by 'towing' the growth cone, the stretched length l will increase faster in time, solely because of a larger value for k. This stretching effect on the elongation rate might partially explain towed growth, as demonstrated by Bray (1984) and Zheng et al. (1990). By towing, the force at the growth cone is increased, resulting in an increase in elongation rate of the stretched length.

A stretched rubber band serves as a good analogy for this type of outgrowth. Take a rubber band in both hands (grab the band somewhat from the ends, leaving these free for some distance), and stretch the band. Now, while trying to exert a constant pulling force at one hand, slip a bit of rubber band through your other hand. This results in an increase of the unstretched length of the rubber band in between your hands. Because you are trying to exert a constant force at one hand, your hands will move apart. If you slip the rubber band with a constant speed, you will notice that your hands will separate with constant speed, and the slipping speed and the separating speed are related by Eqn. (3). On top of this passive stretching effect, Buxbaum and Heidemann (1988, 1992) argue for a dependence of the assembly rate of microtubules on the amount of stretching. In their view, microtubules are compressed structures and stretching the neurite translates into a release of compression at the microtubules, giving rise to a larger assembly rate.

The interaction between tension generated by the growth cone and polymerization of the microtubules links the function of the peripheral domain, which is presumed to generate tension, to that of the central domain of the growth cone, which is presumed to generate tension, to that of the central domain of the growth cone, which is the site of tubulin assembly. This approach combines the 'push' and the 'pull' hypotheses for neuritic elongation, respectively, put forward by Bray (1987) and by Goldberg and Burmeister (1989). Bray argued that growth cones pull their neurite forward while Goldberg and Burmeister argued that the growth cone is pushed forward by material deposited in the growth cone. The mechanism put forward here postulates both a tension generating pull of the growth cone and a microtubule elongating input of tubulin into the growth cone.

Branching

The branched appearance of dendritic and axonal trees is highly characteristic for neurons and the name 'tree' refers intrinsically to a branched structure. While describing the actions of the growth cone, branching was correlated with the angular spread of the filopodia. However, this correlation gives neither information on the branch probability nor on the dependence of neuritic morphology on the branch probability.

A simplified description of the formation of dendrites and axons can be given by topological means. In this approach, the neuronal trees are treated as a collection of segments, disregarding their length, connected by branch points. At each branch point three or more segments are connected. Van Pelt and colleagues (Van Pelt and Verwer, 1986; Van Pelt et al., 1986; Verwer et al., 1992), Horsfield et al. (1987) and Kliemann (1987) developed and analyzed outgrowth models based solely on these features. Van Pelt et al. (1986) derived a model which accounts for the probability of occurrence of neuronal branching patterns, and which is attractive because of its simplicity. It is based on only two parameters, the branch probability and an exponential function defining the dependence of the branch probability on the order of a segment. 'Order' is the centrifugal order, which is defined as the number of branch points between the segment and the basis of the tree. Fitting their model, Van Pelt and colleagues concluded that in rat pyramidal cells (Van Pelt et al., 1986) and rat Purkinje cells (Van Pelt et al., 1986, 1992) the branch probability decreases with increasing centrifugal order, that is towards the top of the tree, whereas multipolar non-pyramidal cells have a constant branch probability along the tree.

Van Veen and Van Pelt (1993) developed a stochastic model to study the impact of the branch probability on segment lengths within the neurite, assuming that the branch probability is constant over the whole length of the neurite. This implies that the branching probability of a piece of neurite is proportional to its length. One important constraint was incorporated: a neurite has some finite length, restricting the maximal length of a segment. With these assumptions, Van Veen and Van Pelt (1993) show that terminal segments are expected to be longer than intermediate segments, and that both types of segments decrease in length with centrifugal order. These predictions agree with measurements of rat pyramidal cells in the neocortex (Uylings et al., 1978).

Van Veen and Van Pelt (1993) show that the in vitro neurons depicted by Bray (1973) agreed nicely with the predicted results. However, the terminal segments of the in vivo basal dendrites of rat pyramidal cells were much longer than expected. This difference might well be attributed to a decreasing branch probability per unit length during development, causing terminal segments of higher order to experience a progressively lower branching probability. Burke et al. (1992) also found increasing segment lengths at smaller segment diameter in dendrites of in vivo motoneurons. Because of tapering towards the tips, thin segments generally have a high order. The branching rate is found to decrease with decreasing segment diameter and thus increasing order. When using the measured taper and branching rates in a discrete time outgrowth model, Burke et al. (1992) reproduced their measurements, indicating that a branching probability which decreases with increasing order is an important determinant of segment lengths.

Van Veen and Van Pelt (1992) and Van Veen (1993) present simulation studies of neuritic outgrowth, where branching is determined by the angular distribution of the filopodia on the growth cone. In a homogenous environment this results essentially in a binomial branch probability per unit of length. The branch probability was shown to depend on the number of filopodia, such that the branching probability is small at very low or very high filopodial number and larger at intermediate numbers.

Forcing a *constant* branch probability per unit of length, the simulations yield longer teminal than intermediate segments, and decreasing segments lengths with higher order. This is exactly the pattern predicted from the stochastic model of Van Veen and Van Pelt (1993), also based on a constant branch probability per unit length. However, the simulations yield a larger topological asymmetry (expresses, on average, how much the left and right of a tree look alike at each branch point: see Van Pelt et al., 1992) and too many terminal segments in comparison with cultured and in vivo neurons. Changing the branch probability to other constant values or changing the mean number of filopodia did not change these patterns much. Instead, a decreasing branch probability with increasing centrifugal order generated more realistic values for the topological asymmetry and the number of terminal segments. These modeling results suggest that the branching pattern of neurons depends sensitively on changes in branch probability during outgrowth and that the branching probability is expected to decrease with increasing centrifugal order in real neurons.

This conclusion agrees with results from topological growth models fitted to rat pyramidal cells, where decreasing branch probabilities with increasing centrifugal order have also been found

(Van Pelt et al., 1986). Also, these topological growth models show that the expected topological asymmetry depend sensitively on an alteration of the branch probability with order.

Activity dependent outgrowth

A number of experiments using electrical stimulation (Cohan and Kater, 1986; Fields et al., 1990) or application of neurotransmitters (Haydon et al., 1987; Mattson and Kater, 1989) to an outgrowing neuron show that these treatments are able to stop or reverse the outgrowth. In tissue culture, Van Huizen and Romijn (1987) gained an increase in branching frequency after chronic blockade of bioelectric activity by tetrodotoxin in rat cerebral cells. The effect of both electrical stimulation and neurotransmitters is reversible; cessation results in a restart of outgrowth (Cohan and Kater, 1986; Fields et al., 1990). In *Helisoma*, a sea snail, the growth cone structure changed markedly after evoking action potentials on its trailing neurite. The number of filopodia decreased and lamellipodia retracted (Cohan and Kater, 1986; Fields et al., 1990), and, eventually, the growth cone disappeared. The effects of serotonin on neuron 19 (Haydon et al., 1987) and experimentally rising the internal calcium concentration (Rehder and Kater, 1992) are comparable with the effects of evoked action potentials.

The first steps in the cascade induced by neuro-transmission are a depolarization of the membrane potential and the elevation of the internal calcium concentration (Parnas and Segel, 1980; Wadman and Connor, 1992). This elevation of the internal calcium has been made visible by calcium sensitive dyes and is observed in reaction to electrical stimulation (Kater et al., 1988), during excitatory neurotransmitter application (Wadman and Connor, 1992) and during activation by other neurons (Schilling et al., 1991). By blocking the elevation of the calcium concentration during iontophoretic application of glutamate, pyramidal cell dendrites could be prevented from retraction (Mattson and Kater, 1989), suggesting that neuritic retraction in response to neurotransmitters or electrical activity is mediated by an elevated calcium concentration.

The cascade following the elevation of calcium concentrations affects, amongst others, neuritic elongation and branching. Elevated internal calcium concentrations interfere with filopodial and veil protrusions (Goldberg et al., 1991; Rehder and Kater, 1992), probably by interfering with actin polymerization (Amos and Amos, 1991; Lewis and Murray, 1992). Calcium is also known to induce microtubule disassembly (Dustin, 1978; Gordon-Weeks, 1991; Diaz-Nido et al., 1991) at concentrations in the range of 10^{-6} M to 10^{-9} M. These concentrations are well within the physiological ranges measured in neuronal growth cones. Bantlow et al. (1993) measured a calcium concentration of 157 nM in control growth cones, while the concentration rose to 2132 nM in growth cones touching an inhibitive substratum. Al-Mohanna et al. (1992) have reported optimal neurite outgrowth at 35 nM cytosolic free calcium, while already at a calcium concentration of 60 nM a reduced neuritic outgrowth was found.

The observations indicate that calcium concentrations become elevated as soon as functional neuronal contacts are established. These elevated calcium concentrations interfere with microtubule assembly, and thus with the cytoskeleton which supports neuronal shape, explaining how neuritic outgrowth can be reduced by calcium influx resulting from neuronal contacts. In this view, the size of a neuron will be the result of a dynamic interaction with its neuronal environment. If the intensity of functional contacts changes, the morphology of the neurite will change appropriately. This principle is used by Van Ooyen and Van Pelt (1992, 1994a,b) to explain how networks of stable size are formed after an intial overshoot in neuronal connections. Neurons do change their morphology during the whole lifespan of an animal by retracting and extending branches. These morphological changes can be monitored by observing the positions of synapses (Purves et al., 1987) or by directly observing dendritic or axonal branches (Speidel, 1933; O'Rourke and Fraser, 1990).

Conclusion

During neuronal morphogenesis, branching and steering of the growth cone and microtubule elongation have been shown to be cooperating factors. While the growth cone accounts for the initiation of branch points and the stretching of neuritic segments, the tubulin dynamics determine whether branch points will survive and the elongation of the microtubule cytoskeleton supports ongoing neuritic elongation. Calcium is an important regulator of this interaction, because it changes both the functioning of the growth cone and the rate of microtubule elongation. Because calcium concentrations and electric or neurotransmitter stimulation are correlated, functional contacts are expected to change the interaction between growth cone and microtubule dynamics and, thus, to change neuronal morphology.

In this chapter, much of the interaction between the growth cone and the cytoskeleton has been explored by mathematical modeling of neuronal outgrowth. These models show that realistic neuronal outgrowth can be simulated by solely incorporating branching, steering and elongation of the growth cone and the microtubules. Of course, these models implement an hypothesis about the functioning of the growth cone and the microtubules, but the present chapter shows that there is much experimental evidence in support of the models.

References

Ahmad, F.J., Pienkowski, Th. P. and Baas, P.W. (1993) Regional differences in microtubule dynamics in the axon. *J. Neurosci.* 13: 856–866.

Al-Mohanna, F.A., Cave, J. and Bolsover, S.R. (1992) A narrow window of intracellular calcium concentration is optimal for neurite outgrowth in rat sensory neurons. *Dev. Brain Res.*, 70: 287–290.

Alt, W. (1990) Mathematical models and analysing methods for the lamellipodial activity of leucocytes. In: N. Akkas (Ed), *Biomechanics of Active Movement and Deformation of Cells*, NATO ASI series, 42, Springer Verlag, Heidelberg.

Amos, L.A. and Amos, W.B. (1991) *Molecules of the Cytoskeleton*, MacMillan Molecular Biology Series, MacMillan Education Ltd., London.

Bamburg, J.R., Bray, D. and Chapman, K. (1986) Assembly of microtubules at the tip of growing axons. *Nature*, 321: 788–790.

Bantlow, C.E., Schmidt, M.F., Hassinger, T.D., Schwab, M.E. and Kater, S.B. (1993) Role of intracellular calcium in NI-35-evoked collapse of neuronal growth cones. *Science*, 259: 80–83.

Bentley, D. and O'Connor, T.P. (1991) Guidance and steering of peripheral pioneer growth cones in grasshopper embryos. In: P.C. Letourneau, S.B. Kater and E.R. Macagno (Eds.) *The Nerve Growth Cone*, Raven Press, New York, pp. 265–282.

Bentley, D. and Toroian-Raymond, A. (1986) Disoriented pathfinding by pioneer growth cones deprived of filopodia by cytochalasin treatment. *Nature*, 323: 712–715.

Bergen, L.G. and Borisy, G.G. (1980) Head-to-tail polymerization of Microtubules in vitro. *J. Cell Biol*, 84: 141–150.

Berry, M., McConnell, P. and Sievers, J. (1980) Dendritic growth and the control of neuronal form. In: R.K. Hunt (Ed.), *Current Topics in Developmental Biology*, Academic Press, London.

Black, M.M. (1994) Microtubule transport and assembly cooperate to generate the microtubule array of growing axons. Chapter 4, this volume.

Black, M.M. and Smith, W. (1988) Regional differentiation of the neuronal cytoskeleton with an appendix: diffusion of proteins in the neuron cell body—mathematical approximations and computer simulations. In: R.J. Lasek and M.M. Black (Eds.), *Intrinsic Determinants of Neuronal Form and Function*, A.R. Liss Inc., New York.

Bray, D. (1973) Branching patterns of individual sympathetic neurons in culture. *J. Cell Biol.*, 56: 702–712.

Bray, D. (1979) Mechanical tension produced by nerve cells in tissue culture. *J. Cell Sci.*, 37: 391–410.

Bray, D. (1984) Axonal growth in response to experimentally applied mechanical tension. *Dev. Biol.*, 102: 379–389.

Bray, D. (1987) Growth cones: do they pull or are they pushed? *Trends Neurosci.*, 10(10): 431–434.

Bray, D. and Bartlett Bunge, M. (1981) Serial analysis of microtubules in cultured rat sensory axons. *J. Neurocytol.*, 10: 589–605.

Bray, D. and Chapman, K. (1985) Analysis of microspike movement on the neutral growth cone. *J. Neurosci.*, 12: 3204–3213.

Burke, R.E., Marks, W.B. and Ulfhake, B. (1992) A parsimonious description of motoneuron dendritic morphology using computer simulation. *J. Neurosci.*, 12: 2403–2416.

Buxbaum, R.E. and Heidemann, S.R. (1988) A thermodynamic model for force integration and microtubule assembly during axonal elongation. *J. Theor. Biol.*, 134: 379–390.

Buxbaum, R.E. and Heidemann, S.R. (1992) An absolute rate

theory model for tension control of axonal elongation. *J. Theor. Biol.*, 155: 409–426.

Caudy, M. and Bentley, D. (1986) Pioneer growth cone steering along a series of neuronal and non-neuronal cues of different affinities. *J. Neurosci.*, : 1781–1795.

Cleveland, D.W., Lopata, M.A., Sherline, P. and Kirschner, M.W. (1981) Unpolymerized tubulin modulates the level of tubulin mRNAs. *Cell*, 25: 537–546.

Cohan, C.S. and Kater, S.B. (1986) Suppression of neurite elongation and growth cone motility by electrical activity. *Science*, 232: 1638–1640.

Dennerl, T.J., Lamoureux, P., Buxbaum, R.E. and Heidemann, S.R. (1989) The cytomechanics of axonal elongation and retraction. *J. Cell Biol.*, 109: 3073–3083.

Diaz-Nido, J., Armas-Portela, R., Martinez, A., Rocha, M. and Avila, J. (1991) Role of microtubules in neurite outgrowth. In: P.C. Letourneau, S.B. Kater and E.R. Macagno (Eds.), *The Nerve Growth Cone, Raven Press, New York, pp.* 39–53.

Dodd, J. and Jessell, T.M. (1988) Axon guidance and the patterning of neuronal projections in vertebrates. *Science*, 242: 692–699.

Dogterom, M. and Leibler, S. (1993) Physical aspects of the growth and regulations of microtubule dynamics. *Phys. Rev. Lett.*, 70: 1347–1350.

Dustin, P. (1978) *Microtubules*, Springer Verlag, Berlin.

Faivre, C., Legrand, Ch. and Rabié, A. (1985) The microtubular apparatus of cerebellar Purkinje cell dendrites during post-natal development of the rat: the density and cold-stability of microtubules increase with age and are sensitive to thyroid hormone deficiency. *Int. J. Dev. Neurosci.*, 3: 559–565.

Fields, R.D., Neale, E.A. and Nelson, P.G. (1990) Effects of patterned electrical activity on neurite outgrowth from mouse sensory neurons. *J. Neurosci.*, 10: 2950–2964.

Goldberg, D.J. and Burmeister, D.W. (1989) Looking into growth cones. *Trends Neurosci.*, 12: 503–506.

Goldberg, D.J., Burmeister, D.W. and Rivas, R.J. (1991) Video microscopic analysis of events in the growth cone underlying axon growth and the regulation of these events by substrate bound proteins. In: P.C. Letourneau, S.B. Kater and E.R. Macagno (Eds.), *The Nerve Growth Cone*, Raven Press, New York, pp. 39–53.

Gordon-Weeks, P.R. (1991) Growth cones: the mechanism of neurite advance. *BioEssays*, 13: 235–239.

Hammarback, J.A. and Letourneau, P.C. (1986) Neurite extension across regions of low cellsubstratum adhesivity: implications for the guidepost hypothesis of axonal pathfinding. *Dev. Biol.*, 117: 655–662.

Haydon, P.G., McCobb, D.P. and Kater, S.B. (1987) The regulation of neurite outgrowth, growth cone motility and electrical synaptogenesis by serotonin. *J. Neurobiol.*, 18: 197–215.

Ho, R.K. and Goodman, C.S. (1982) Peripheral pathways are pioneered by the array of central and peripheral neurones in grasshopper embryos. *Nature*, 297: 404–406.

Hoffman, P.N., Lopata, M.A., Watson, D.F. and Luduena, R.F. (1992) Axonal transport of class II and III β-tubulin: evidence that the slow component wave represents the movement of only a small fraction of the tubulin in mature axons. *J. Cell Biol.*, 119: 595–604.

Horsfield, K., Woldenberg, M.J. and Bowes, C.L. (1987) Sequential and synchronous growth models related to vertex analysis and branching ratios. *Bull. Math. Biol.*, 49: 413–429.

Kapfhammer, J.P. and Raper, J.A. (1987) Collapse of growth cone structure on contact with specific neurites in culture. *J. Neurosci.*, 7: 201–212.

Kater, S.B., Mattson, M.P., Cohan, C. and Connor, J. (1988) Calcium regulation of the neuronal growth cone. *Trends Neurosci.*, 11: 315–321.

Kater, S.B., Davenport, R.W. and Guthrie, P.B. (1994) Filopodia as detectors of environmental cues: signal integration through changes in growth cone calcium levels. In: J. van Pelt, M.A. Corner, H.B.M. Uylings and F.H. Lopes da Silva (Eds.), *Progress in Brain Research*, *Vol.* 102, *The Self-organizing Brain—from Growth Cones to Functional Networks*, Elsevier Science Publishers, Amsterdam, this volume.

Katz, M.J. and George, E.B. (1985) Fractals and the analysis of growth path. *Bull. Math. Biol.*, 47: 273–286.

Katz, M.J., George, E.B. and Gilbert, L.J. (1984) Axonal elongation as a stochastic walk. *Cell Motil.*, 4: 351–370.

Kliemann, W. (1987) A stochastic dynamical model for the characterization of the geometrical structure of dendritic processes. *Bull. Math. Biol.*, 49: 135–152.

Lamoureux, P., Buxbaum, R.E. and Heidemann, S.R. (1989) Direct evidence that growth cones pull. *Nature*, 340: 159–162.

Lamoureux, P., Zheng, J., Buxbaum, R.E. and Heidemann, S.R. (1992) A cytomechanical investigation of neurite growth on different culture surfaces. *J. Cell Biol.*, 118: 655–661.

Letourneau, P.C. and Ressler, A.H. (1984) Inhibiition of neurite initiation and growth by taxol. *J. Cell Biol.*, 98: 1355–1362.

Letourneau, P.C., Shattuck, T.A. and Resslear, A.H. (1986) Branching of sensory and sympathetic neurites *in vitro* is inhibited by treatment with taxol. *J. Neurosci.*, 6(7): 1912–1917.

Letourneau, P.C., Shattuck, T.A. and Ressler, A.H. (1987) ‘Pull’ and ‘push’ in neuritic elongation: observations on the effects of different concentrations of cytochalasin B and taxol. *Cell Motil. Cytoskel.*, 8: 193–209.

Letourneau, P.C., Snow, D.M. and Gomez, T.M. (1994) Growth cone motility: substratum-bound molecules, cytoplasmic [Ca^{2+}] and Ca^{2+}-regulated proteins. In: J. van Pelt, M.A. Corner, H.B.M. Uylings and F.H. Lopes da Silva (Eds.),

Progress in Brain Research, Vol. 102, *The Self-organizing Brain—from Growth Cones to Functional Networks*, Elsevier Science Publishers, Amsterdam, this volume.

Letourneau, P.C., Kater, S.B. and Macagno, E.R. (1991) *The Nerve Growth Cone*, Raven Press, New York.

Lewis, M.A. and Murray, J.D. (1992) Analysis of dynamic and stationary pattern formation in the cell cortex. *J. Math. Biol.*, 31: 25–71.

Lim, S., Edson, K.J., Letourneau, P.C. and Borisy, G.G. (1990) A test of microtubule translocation during neurite elongation. *J. Cell Biol.*, 111: 123–130.

Marsh, L. and Letourneau, P.C. (1984) Growth of neurites without filopodial or lamellipodial activity in the presence of cytochalasin B. *J. Cell Biol.*, 99: 2041–2047.

Mattson, M.P. and Kater, S.B. (1989) Excitatory and inhibitory neurotransmitter in the generation and degeneration of hippocampal neuroarchitecture. *Brain Res.*, 478: 337–348.

Matus, A. (1988) Microtubule-associated proteins: their potential role in determining neuronal morphology. *Annu. Rev. Neurosci.*, 11: 29–44.

Mitchinson, T. and Kirschner, M. (1988) Cytoskeletal dynamics and nerve growth. *Neuron*, 1: 761–772.

Murray, J.D. (1989) *Mathematical Biology*. Springer Verlag, Berlin, pp: 1–767.

Okabe, S. and Hirokawa, N. (1988) Microtubule dynamics in nerve cells: analysis using microinjection of biotinylated tubulin into PC12 cells. *J. Cell Biol.*, 107: 651–664.

Okubo, A. 1980. *Diffusion and Ecological Problems: Mathematical Models*. Springer Verlag, Berlin, pp. 1–254.

O'Rourke, N.A. and Fraser, S.E. (1990) Dynamic changes in optic fiber terminal arbors lead to retinotopic map formation: an in vivo confocal microscopic study. *Neuron*, 5: 159–171.

Oster, G.F. (1988) Cell motility and tissue morphogenesis. In: W. Stein and F. Bronner (Eds), *Cell Shape: determinants, regulation and regulatory control*, Academic Press, New York.

Parnas, H. and Segel, L. (1980) A theoretical explanation for some effects of calcium on the facilitation of neurotransmitter release. *J. Theor. Biol.*, 84: 3–29.

Popov, S. and Poo, M.M. (1992) Diffusional transport of macromolecules in developing nerve processes. *J. Neurosci.*, 12: 77–85.

Pryer, N.K., Walker, R.A., Skeen, V.P., Bourns, B.D., Soboeiro, M.F. and Salmon, E.D. (1992) Brain microtubule-associated proteins modulate microtubule dynamic instability in vitro. Real time observations using video microscopy. *J. Cell Sci.*, 103: 965–976.

Purves, D., Voyvodic, J.T., Magrassi, L. and Yawo, H. (1987) Nerve terminal remodeling visualized in living mice by repeated examination of the same neuron. *Science*, 238: 1122–1126.

Ramon y Cajal, S. (1911) *Histologie du système nerveux de l'homme et des vertértebrés*. Vols. I and II, A. Maloine, Paris.

Rehder, V. and Kater, S.B. (1992) Regulation of neuronal growth cone filopodia by intracellular calcium. *J. Neurosci.*, 12: 3175–3186.

Schilling, K., Dickinson, M.H., Connor, J.A. and Morgan, J.I. (1991) Electrical activity in cerebellar cultures determines Purkinje cell dentritic growth patterns. *Neuron*, 7: 891–902.

Schwab, M.E. (1990) Myelin-associated inhibitors of neurite growth and regeneration in the CNS. *Trends Neurosci.*, 13: 452–456.

Shaw, G. and Bray, D. (1977) Movement and extension of isolated growth cones. *Exp. Cell Res.*, 104: 55–62.

Soifer, D. and Mack, K. (1984) *Microtubules in the nervous system*. In: Lajtha A. (Ed.), Handbook of Neurochemistry, Volume 7, Plenum Press, New York, pp. 245–280.

Speidel, C.C. (1933) Studies of living nerves. II. Activities of ameboid growth cones, sheath cells, and myelin segments, as revealed by prolonged observation of individual nerve fibers in frog tadpoles. *Am. J. Anat.*, 52: 1–79.

Stossel, T.P. (1993) On the crawling of animal cells. *Science*, 260: 1086–1094.

Tashiro, T. and Komiya, Y. (1992) Slow axonal transport and the dynamics of the axonal cytoskeleton. In: M. Satake (Ed.), *Molecular Basis of Neuronal Connectivity*, Kohko-do, Tokio, pp. 187–190.

Uylings, H.B.M. (1977) A study on morphometry and functional morphology of branching structures, with applications to dendrites in visual cortex of adult rats under different environmental conditions. Ph.D. Thesis UvA, Amsterdam.

Uylings, H.B.M., Kuypers, K., Diamond, M.C. and Veltman, W.A.M. (1978) Effects of differential environments on plasticity of dendrites of cortical pyramidal neurons in adult rats. *Exp. Neurol.*, 62: 658–677.

Uylings, H.B.M., van Pelt, J., Verwer, R.W.H. and McConnell, P. (1989) Statistical analysis of neuronal populations In: J.J. Capowski (Ed.) *Computer Techniques in Neuroanatomy*, Plenum, New York.

Van Huizen, F. and Romijn, H.J. (1987) Tetrodoxin enhances initial neurite outgrowth from fetal rat cerebral cortex cells in vitro. *Brain Res.*, 408: 271–274.

Van Ooyen, A. and van Pelt, J. (1992) Phase transitions, hysteresis and overshoot in developing neural networks. In: I. Aleksander and J. Taylor (Eds.) *Artificial Neural Networks* 2: 907–910.

Van Ooyen, A. and van Pelt, J. (1994a) Activity dependent outgrowth of neurons and overshoot phenomena in developing neural networks. *J. Theor. Biol.*, in press.

Van Ooyen, A. and van Pelt, J. (1994b) Activity-dependent neurite outgrowth and neural network development. In: J.

van Pelt, M.A. Corner, H.B.M. Uylings and F.H. Lopes da Silva (Eds.), *Progress in Brain Research, Vol.* 102, *The Self-organizing Brain—from Growth Cones to Functional Networks*, Elsevier Science Publishers, Amsterdam, this volume.

Van Pelt, J. and Verwer, R.W.H. (1986) Topological properties of binairy trees grown with order-dependent branching probabilities. *Bull. Math. Biol.*, 48: 197–211.

Van Pelt, J., Verwer, R.W.H. and Uylings, H.B.M. (1986) Application of growth models to the topology of neuronal branching patterns. *J. Neurosci. Methods*, 18: 153–165.

Van Pelt, J., Uylings, H.B.M., Verwer, R.W.H., Pentney, R.J. and Woldenberg, M.J. (1992) Tree asymmetry—a sensitive and practical measure for binairy topological trees. *Bull. Math. Biol.*, 54: 759–784.

Van Veen, M.P. (1993) *Dynamic mechanism of neuronal outgrowth*. Ph.D. thesis, V.U,. Amsterdam.

Van Veen, M.P. and van Pelt, J. (1992) A Model for outgrowth of branching neurites. *J. Theor. Biol.*, 159: 1–24.

Van Veen, M.P. and van Pelt, J. (1993) Terminal and intermediate segment lengths in neuronal trees with finite length. *Bull. Math. Biol.*, 55: 277–294.

Van Veen, M.P. and van Pelt, J. (1994) Neuritic growth rate described by modeling microtubule dynamics. *Bull. Math. Biol.*, 56: 249–273.

Van Veen, M.P., Letourneau, P.C. and van Pelt, J. (1993) Behaviour of growth cones in a homogeneous environment: comparison between model and experiment. In: *Dynamic mechanisms of Neuronal Outgrowth*, Ph.D. Thesis, Free University, Amsterdam.

Vaughn, J.E., Hendrikson, C.K., Grieshaber, J.A. (1974) A quantitative study of synapses on motor neuron dendritic growth cones in developing mouse spinal chord. *J. Cell Biol.*, 60: 664–672.

Vasiliev, J.M. (1991) Polarization of pseudopodial activities: cytoskeletal mechanisms. *J. Cell Sci.*, 98: 1–4.

Verde, F., Dogterom, M., Stelzer, E., Karsenti, E. and Leibler, S. (1992) Control of microtubule cynamics and length by cyclin A- and cylclin B-dependent kinases in *Xenopus* egg extracts. *J. Cell Biol.*, 118: 1097–1108.

Verwer, R.W.H., van Pelt, J. and Uylings, H.B.M. (1992) An introduction to topological analysis of neurones. In: Steward, F. (Ed.) *Quantitative Methods in Neuroanatomy*, Wiley, London.

Wadman, W.J. and Connor, J.A. (1992) Persisting modification of dendritic calcium influx by excitatory amino acid stimulation in isolated CA1 neurons. *Neuroscience*, 48: 293–305.

Wessells, N.K. and Nuttall, R.P. (1978) Normal branching, induced branching, and steering of cultured parasympathetic motor neurons. *Exp. Cell Res.*, 115: 111–122.

Yamada, K.M., Spooner, B.S. and Wessells, N.K. (1970) Axon growth: roles of microfilaments and microtubules. *Proc. Natl. Acad. Sci. USA*, 66: 1206–1212.

Zheng, J., Lamoureux, P., Santiago, V., Dennerl, T., Buxbaum, R.E. and Heidemann, S.R. (1990) Tensile regulation of axonal elongation and initiation. *J. Neurosci.*, 11: 1117–1125.

J. van Pelt, M.A. Corner H.B.M. Uylings and F.H. Lopes da Silva (Eds.)
Progress in Brain Research, Vol 102

CHAPTER 7

Geometrical and topological characteristics in the dendritic development of cortical pyramidal and non-pyramidal neurons

Harry B.M. Uylings[1], Jaap van Pelt[1], John G. Parnavelas[2] and Antonio Ruiz-Marcos[3]

[1]*Graduate School of Neuroscience and Netherlands Institute for Brain Research, Amsterdam, The Netherlands,* [2]*Department of Anatomy and Developmental Biology, University College London, UK and* [3]*Instituto Cajal, Madrid, Spain*

Introduction

The literature indicates that in the cerebral cortex neuronal development in general and dendritic growth in particular, do not simply involve a progression in size. For instance, cortical neurons are generated at different times, for the most part according to an inside-out pattern, i.e. the majority of neurons in the upper layers are generated later than those in the lower layers (see Miller, 1988). In rat cerebrum, nearly all neocortical neurons are generated between embryological day 11 (E11) and E21 in the ventricular and subventricular zones (Fig. 1) (Uylings et al., 1990). They begin to proliferate dendrites mainly after they have reached the plexiform primordium (preplate) and cortical plate (Fig. 1). However, during normal development some neuronal death occurs in the neocortex at different stages (for review see Finlay, 1992; Ferrer et al., 1992). In addition, we know that several surviving neurons show an overgrowth, with a clear regression in the dendritic field, which even leads to a different morphological appearance in some cell types, e.g. Cajal-Retzius neurons, subplate neurons, and the small layer V pyramidal neurons.

In this chapter we will first deal with regression or reduction in the size of dendritic fields in some cell types during normal development, while the general developmental characteristics of cortical neurons will be discussed in greater detail. This part of the chapter will be illustrated using data from layer II/III pyramidal neurons, layer IV multipolar non-pyramidal neurons, small layer V pyramidal neurons and large layer V pyramidal neurons in rat visual cortex. These neurons show some different developmental characteristics.

Regression of dendrites during normal development

The outgrowth of individual, living cortical neurons in vivo (i.e. a longitudinal study) cannot be followed throughout their development. Therefore such phenomena as the occurrence of regression and spatial reorientation of dendritic fields can only be derived from tissues taken from different animals and at different ages. Thus, due to interneuronal and interanimal variability only relatively large-scale regression phenomena can be reliably detected.

For neocortical neurons such regression has so far been reported for a number of neuronal types, i.e. Cajal-Retzius cells, subplate neurons, and

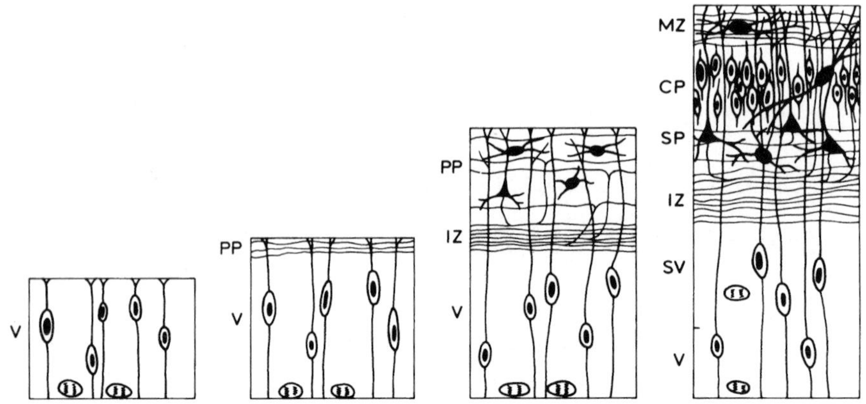

Fig. 1. A scheme of four stages in fetal development of cortical lamination in the cerebral wall. V, ventricular zone; PP, plexiform primordium (also called preplate); IZ, intermediate zone; SV, subventricular zone; SP, subplate; CP, cortical plate; MZ, marginal zone (fetal layer I). The horizontal Cajal-Retzius cells in MZ and the subplate neurons in SP are co-generated in a similarly restricted period. The few more differentiated neurons, which traverse the CP in early stages, are probably also co-generated with SP neurons and Cajal-Retzius neurons (modified after Uylings et al., 1990).

layer V callosal pyramidal neurons. Marin-Padilla (1990) has shown that the dendrites and the axonal field of Cajal-Retzius cells transform the Cajal-Retzius cell in the marginal zone (Fig. 1) into a large horizontal multipolar neuron during ontogenesis in human neocortex. In the rat, Parnavelas and Edmunds (1983) have found also cellular transformation of the Cajal-Retzius cells, while Derer and Derer (1990), in their EM study have observed that some Cajal-Retzius cells die during normal development in mouse neocortex. In this respect it is important to note that the cell density measures mentioned in some studies will not be sufficient to detect the occurrence of cell death. They can even be misleading, especially during early development, when the tissue volume increases considerably (Swaab and Uylings, 1987).

Another clear example of neuronal shape transformation is seen in the group of subplate neurons. The subplate is a prominent zone situated below the cortical plate in the human brain (e.g. Kostović and Rakic, 1990; Mrzljak et al., 1990) (see Fig. 1). This zone is also clearly present in cat cortex (Shatz et al., 1988) and in rat cortex (Uylings et al., 1990). In the human brain the subplate is prominent during the third quarter of the gestation, when it reaches its peak (± 26–30 weeks of gestation), and is then about five times as thick as the cortical plate (Mrzljak et al., 1990). After that age the subplate diminishes or stretches to a thin layer which is difficult to discern 1 year after birth in Nissl-stained sections. The studies of Kostović and Rakic (1980), Luskin and Shatz (1985) and Wahle and Meyer (1987) indicate that many subplate cells die, but that a significant number remains, albeit in a more diluted fashion, in the white matter and in layer VI. With immunocytochemical staining, using neuropeptide Y (NPY) antibodies, it appears that the persisting NPY-positive subplate neurons in the human cortex become smaller during the first year and reorient their dendritic fields tangentially, i.e. parallel, to the direction of the axonal fibers of the white matter (Uylings and Delalle, 1994). The period of reshaping and reorientation is also the period in which degeneration features of other subplate neurons are observed (Uylings and Delalle, 1994). There is some suggestion that this also takes place in other mammalian species.

A third example of an obvious dendritic regression and reshaping is the apical dendritic field in the callosal, small layer V pyramidal cell (Koester and O'Leary, 1992). Parnavelas et al. (1977) and Wise and Jones (1977) described the presence of different pyramidal cell forms in layer V and other lower layers in rat visual cortex. Later studies in rat and cat cortex (Hübener and Bolz, 1988; Hallman et al., 1988; Mason and Larkman, 1990; Chagnac-Amitai et al., 1990; Hübener et al., 1990; Vercelli and Innocenti, 1993) showed that the subdivision of layer V pyramidal neurons, based upon dendritic criteria, corresponds to subdivisions made using either their axonal connections or their electrophysiological properties. Small layer V pyramidal neurons have an apical dendrite which does not reach the superficial layers I and II, and these neurons appear to give rise to callosal or to ipsilateral intracortical projections. Electrophysiologically, the regular-spiking neurons are included among the small layer V pyramidal cells (Chagnac-Amitai et al., 1990; Mason and Larkman, 1990). Large layer V pyramidal neurons have a large apical dendrite which reaches the superficial layers I and II. These neurons appeared to project to subcortical structures. Electrophysiologically, the intrinsically bursting cells of layer V are included among the large layer V pyramidal neurons (Chagnac-Amitai et al., 1990; Mason and Larkman, 1990; Kawaguchi, 1993).

Koester and O'Leary (1992) and Vercelli et al. (1992) indicate that during normal development the callosal layer V neurons lose a superficial part or even whole apical dendrites during the first week of postnatal life. Taking literature data into consideration (Finlay, 1992; Ferrer et al., 1992), this period of regression coincides with the period of reported cortical cell death in the rodent cortical layers I–VI. The partial regression of the callosal apical dendrite is nicely demonstrated for rat cortex in the study of Koester and O'Leary (1992), which showed that all apical dendrites of all layer V pyramidal neurons reach layer I until postnatal day 4. After day 4, however, the apical dendrites of callosal pyramidal neurons regress so that they lose their contact with the superficial layer I (see Fig. 2). From this age on, dendritic differences in forms of apical dendrites distinguish the cortical projection pyramidal neurons from the subcortical projection neurons. Regression in their basal dendrites has so far not been noted, but was nevertheless also examined in our data on postnatal development (see section below).

Extensive dendritic regression during normal development is typical only for some neocortical neuronal types. Pronounced dendritic regression appears to be the exception rather than the rule for the neocortical neurons. This will be obvious from our data on pyramidal and non-pyramidal cortical neuronal development, given in the next two sections.

Geometrical development

Several hypotheses on the maturation sequence in dendritic development have been proposed in the past: (a) earlier generated neurons located in ontogenetically older layers mature earlier; (b) projection neurons mature earlier than local circuit neurons; (c) larger cells differentiate earlier than smaller cell types (e.g. Jacobson, 1978; Lund, 1978). Several qualitative and quantitative studies on cortical development, mostly carried out in rat visual cortex (Parnavelas et al., 1978; Juraska and Fifkova, 1979; Parnavelas and Uylings, 1980; Juraska, 1982; Hedlich and Winkelmann, 1982; Uylings et al., 1983, 1990; Miller, 1988; Petit et al., 1988), have led to considerable modification of these hypotheses (e.g. Uylings et al., 1990).

In 120 μm thick Golgi-Cox stained sections of the visual cortex in female Sprague–Dawley rats we studied the large and the small layer V pyramidal neurons separately, since they make up distinct subpopulations. On postnatal day 6 (the earliest age at which measurements were done) these two different groups of neurons were easily discernible (Fig. 2). In addition, layer IV multipolar non-pyramidal and layer II/III pyramidal neurons were examined. From this study we noted a different time course for the branching of basal dendrites (Fig. 3), the increase in somatic size (Fig. 4) and the total dendritic length per neuron (Fig. 5). In her rapid-Golgi study of 100-μm thick sections of the visual cortex of male and female hooded rats, Juraska (1982) indicated that no further increase in branching occurs after day 15 for layers III and V pyramidal neurons. We found no significant increase in branching of the two layer V pyramidal groups after day 10, but a decrease between days 10 and 14 in the large layer V pyramidal neurons and a decrease after day 18 in the small layer V pyramidal neurons (Fig. 3). After day 10, the number of basal dendrites reached a stable figure (i.e. mean values of 5.2 for layer II/III neurons, 5.0 for small layer V neurons, and 6.2 for the large layer V pyramidal neurons). After day 14, the number of dendrites per layer IV multipolar neurons reached the stable mean value of 6.0. These data on dendrite

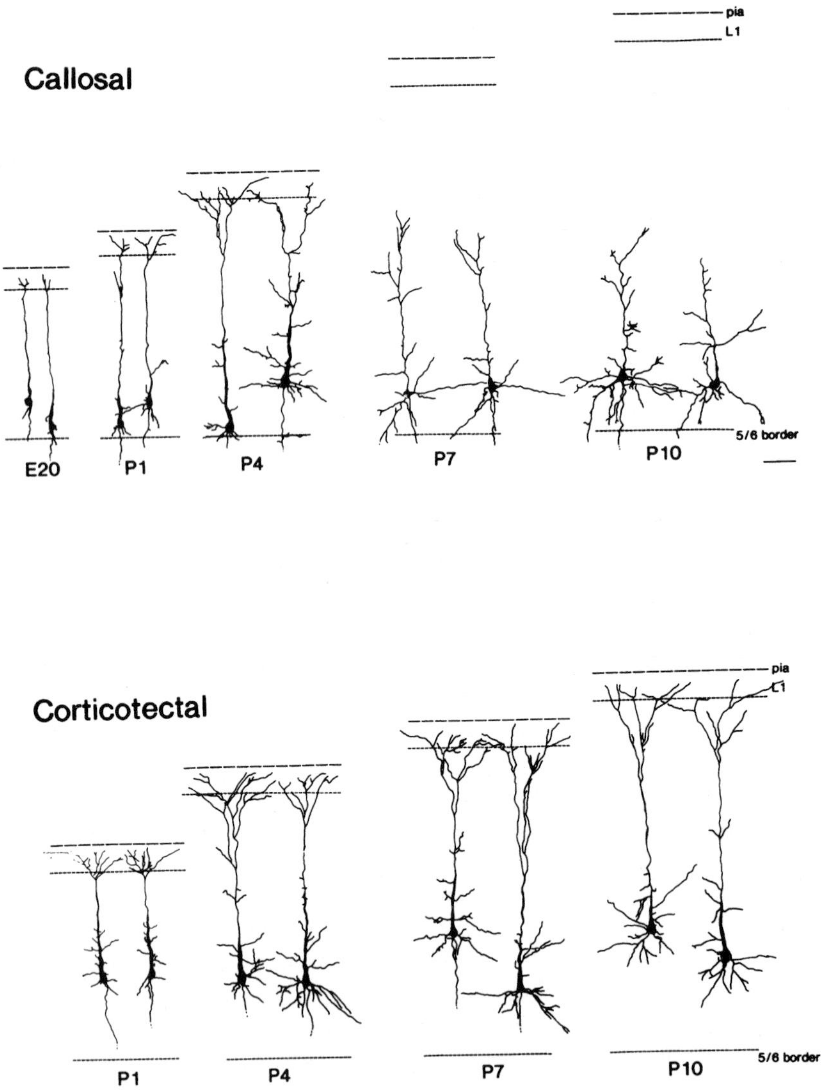

Fig. 2. An example of large dendritic regression during normal postnatal development in rat. Apical dendrites in callosal, small layer 5 pyramidal neurons regress in size between postnatal days (P) 4 and 7, whereas this does not occur in the large layer 5, long distance projection, pyramidal neurons. The borders of pial surface, layer I and layer 5/6 are indicated. Scale bar is 100 μm (reproduced with permission from Koester and O'Leary, 1992).

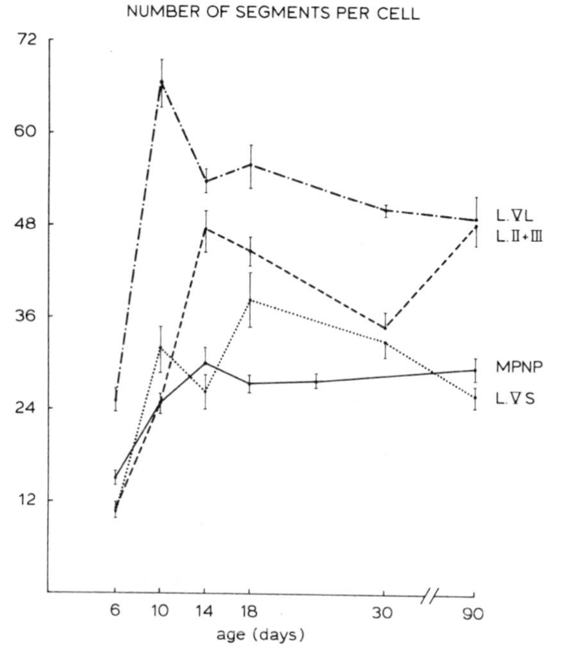

Fig. 3. Number of basal dendrite segments per pyramidal neuron: L.II + III: layer II/III neurons (total number of cells studied 135); L.Vs: small layer V neurons (total number of cells studied 143); L.VL: large layer V neurons (total number of cells studied 124). The number of dendritic segments per multipolar non-pyramidal neuron (MPNP) is indicated by the uninterrupted line (total number of cells studied is 220).

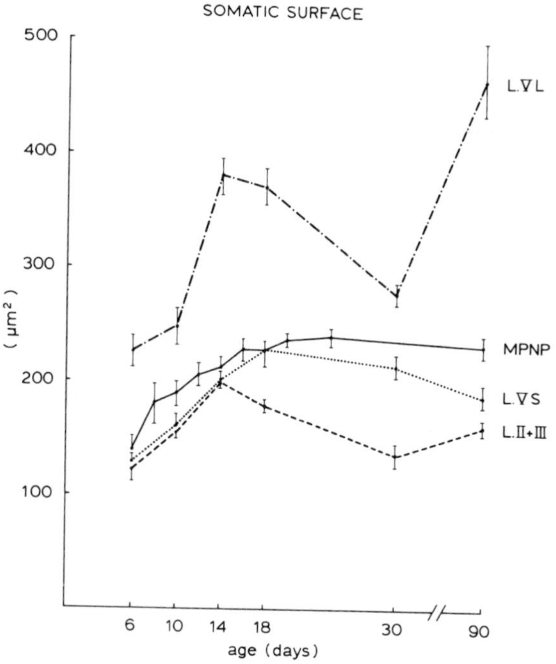

Fig. 4. The surface of the Golgi-Cox stained somata, projected on to the section-plane is indicative of somatic size, given the 3-D shape. Abbreviations as given in legends of Fig. 3.

number are comparable with those of Miller (1988) who also studied Sprague–Dawley rats.

The lengthening of dendrites, however, continues until a later age, day 18 (Fig. 5), at which time the maximal values are reached for both types of layer V pyramidal cells. This maximum is reflected in the individual segment length values (Fig. 6) as well as in the developmental curve of the radial distance from terminal tips (Fig. 7), so that we can assume that cutting did not interfere with our interpretation (Uylings et al., 1989b). The same assumption, i.e. that cutting does not complicate the interpretation of Fig. 5, holds for the increase in length through 90 days in multipolar non-pyramidal neurons and layer II/III pyramidal neurons (see Figs. 5, 6, 7). Our data on total dendritic lengths are comparable with, and an extension of, the data of Juraska (1982), who did not distinguish small from large layer V pyramidal neurons. Our data, however, differ from the developmental curves given for the large ('giant') layer V pyramidal neurons by Petit et al. (1988). In their study of 90-μm thick rapid Golgi stained sections of the visual cortex of male hooded rats they observed no clear maximal value on day 18. They reported that after day 15 the layer V 'total length of basal dendrites' values increase, but not significantly so. Their values for the 'giant' layer V pyramidal neurons are comparable only with our much lower values for the small layer V neurons. This holds also for their values of the individual terminal segments.

As reported earlier (Uylings et al., 1989b), no strong positive correlation exists in adulthood between somatic size (or projected somatic surface) and total dendritic length of basal dendrites. This is even weaker during development, probably

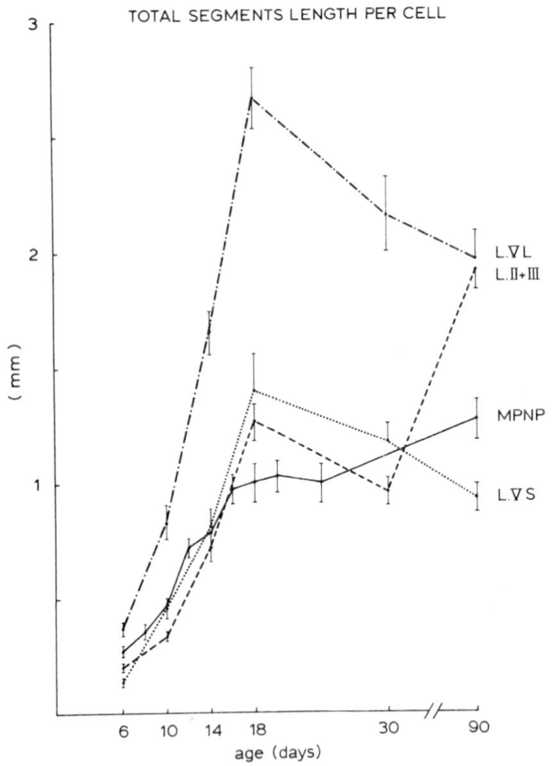

Fig. 5. The total length of all dendritic segments (including cut ones) per neuron (per pyramidal neuron only basal dendrites are taken). See text for interpretation of curves. The length values are based on 3-D measurements in the 120-μm thick sections (e.g. Overdijk et al., 1978). Reproduced with permission from Uylings et al., 1990.

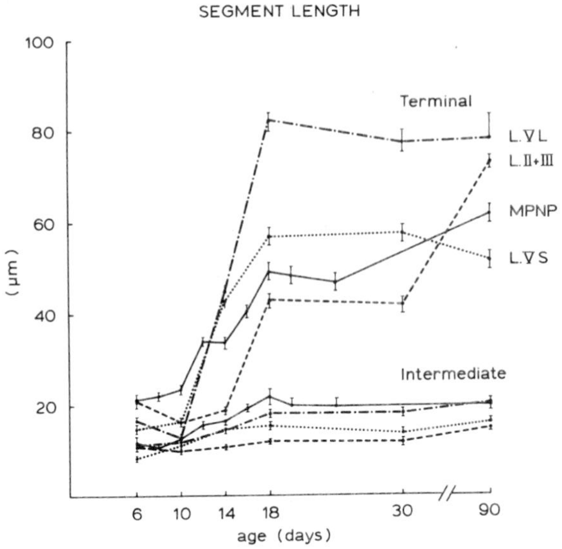

Fig. 6. The length of individual terminal segments (excluding cut ones) and individual intermediate segments of basal dendrites of pyramidal neurons and of multipolar non-pyramidal dendrites. For abbreviations see legend to Fig. 3.

due to an unequal pattern in the development of axonal, apical and basal dendritic processes) (see Figs. 4 and 5). Our developmental curves for somatic sizes are compatible with those shown in Miller (1988) in rat and Ramirez and Kalil (1985) with the exception of the single very low value for large layer V pyramidal neurons on day 30.

In our data from Golgi-Cox stained sections the length of terminal segments accounts for 80–90% of the total length of (basal) dendrites per neuron. This is similar to the data on HRP-stained, fully reconstructed pyramidal neurons in young adult rats, as obtained by Larkman (1991). The large increase in the length of individual terminal segments of the basal dendrites in layer II/III pyramidals is noticeable after day 30, in contrast to those of the two layer V pyramidal cell groups (Fig. 6). This increase is thus also present in the total dendritic length per neuron and the radial distance of terminal tips of basal dendrites (Figs. 5, 7). On day 90 the length of the individual terminal segments of layer II/III does not differ significantly from those of the large layer V pyramidal neurons. Larkman (1991) reported that in young adult rats the values for both layer II/III and the large layer V pyramidal neurons are comparable, but that those for the small layer V were even larger. We did not find the latter result for the small layer V pyramidal neurons in our own data (Fig. 6; Uylings et al., 1990). Our data show a noticeable decline after day 18 in the curve for total length of basal dendrites per layer II/III neuron, whereas further growth has been detected even after day 30 (Figs. 5, 6, 7). This is further illustrated in the orientation-density analysis with the method of Ruiz-Marcos (1983) (see also Ruiz-Marcos and Ipiña, 1986 and Uylings et al., 1989b).

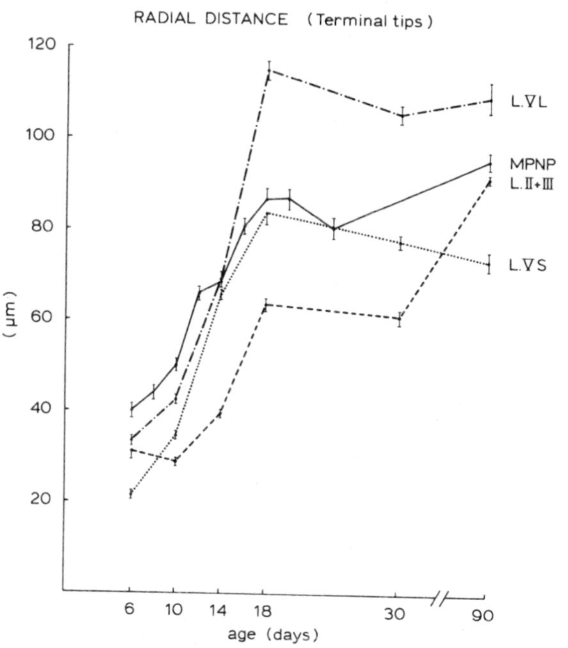

Fig. 7. The radial distance of terminal tips to center of soma in basal dendrites and in multipolar non-pyramidal dendrites. The number of terminal tips varies with cell type and age, but is larger than 200, frequently over 400 and twice over 700, per age per cell type. Reproduced with permission from Uylings et al., 1990.

The developmental alterations in extension and relative density of all of the pyramidal neurons examined is shown in Fig. 8. Statistical comparison of the homologous matrix elements (see Fig. 9) shows a slight decline in density for the two layer V and layer II/III pyramidal cell types after day 18, but also a clear increase in the extension and density of the layer II/III pyramidal neurons between days 30 and 90. The pictorial 2-D orientation-density analysis method of Ruiz-Marcos deepens our insight into the total length data presented above. It shows that minor dendritic regression in basal dendritic fields of pyramidals

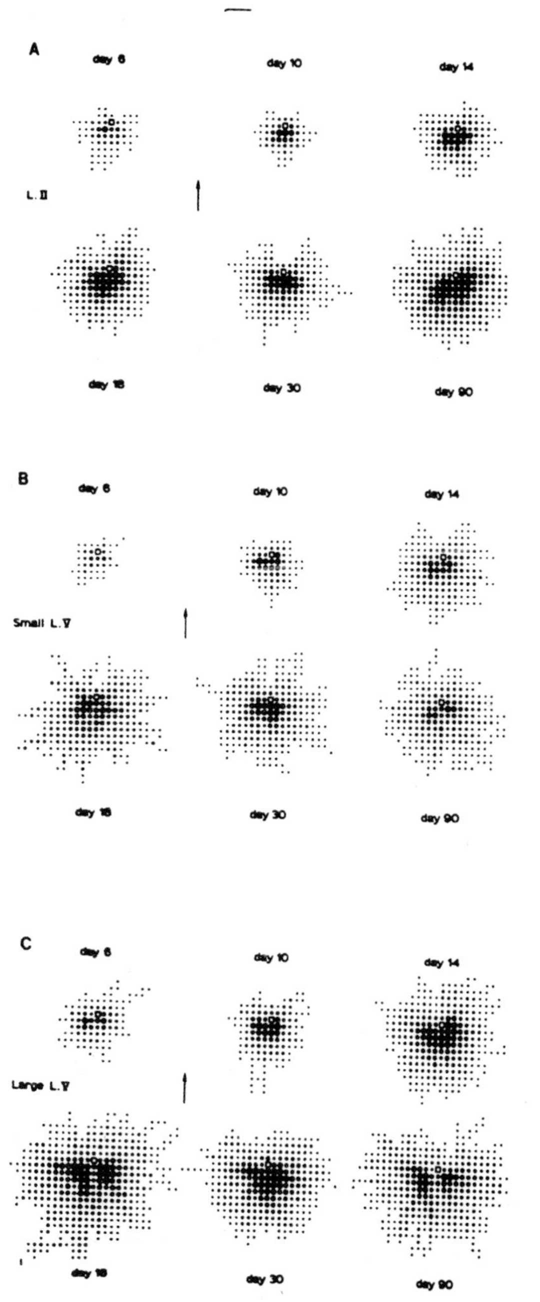

Fig. 8. The density/orientation of basal dendritic fields of the group of layer II/III (A), small layer V (B) and large layer V (C) pyramidal neurons displayed at various developmental ages. The basal dendritic fields have been projected on to a grid or matrix with 15 × 15 μm square elements. This matrix is parallel to the plane of sectioning. Per cell type and per age, the mean dendritic length value per 15 × 15 μm is indicated by six rank numbers visualized by six different sizes of dots, respectively. The arrow points to the pial surface (see Ruiz-Marcos, 1983 for a detailed description of the method).

occurs after day 18, and that this is distributed over different parts of the basal dendritic field. However, in the layer II/III pyramidal neurons the regression is apparently reversed after day 30.

Dendritic regression of cortical pyramidal neurons is not generally reported in the rat studies (Wise et al., 1979; Petit et al., 1988) or else the reported trend did not reach statistically significant values (Juraska, 1982). Some regression is also indicated in the developmental study of layer V pyramidal neurons in rabbit visual cortex of Murphy and Magness (1984). In our human studies (Koenderink and Uylings, 1994; Koenderink et al., 1994), however, we do not observe a clear-cut maximal peak-decline pattern during development in the basal dendritic field of the layer IIIc and V pyramidal cells.

As mentioned above, this kind of study is hampered by the fact that a longitudinal study of particular neurons is presently impossible. In contrast to human studies, animal studies have the advantage that they can be designed so that the genetic and environmental conditions are under control, so that inter-individual variations can be minimized. In addition, it appears to be of importance in animal studies to examine a rather extensive and sufficiently fine-graded age series in order to avoid missing periods of rapid growth or decline. In addition, the possibility is not excluded (indeed for particular brain regions even likely: Greenough et al., 1977; Goldstein et al., 1990; Sasaki and Arnold, 1991; Cherry et al., 1992) that the developmental pattern differs for male and female animals (Muñoz-Cueto et al., 1991). Hemispheric asymmetry can also influence the developmental pattern in particular cortical areas but, viewing the possible alterations in asymmetry

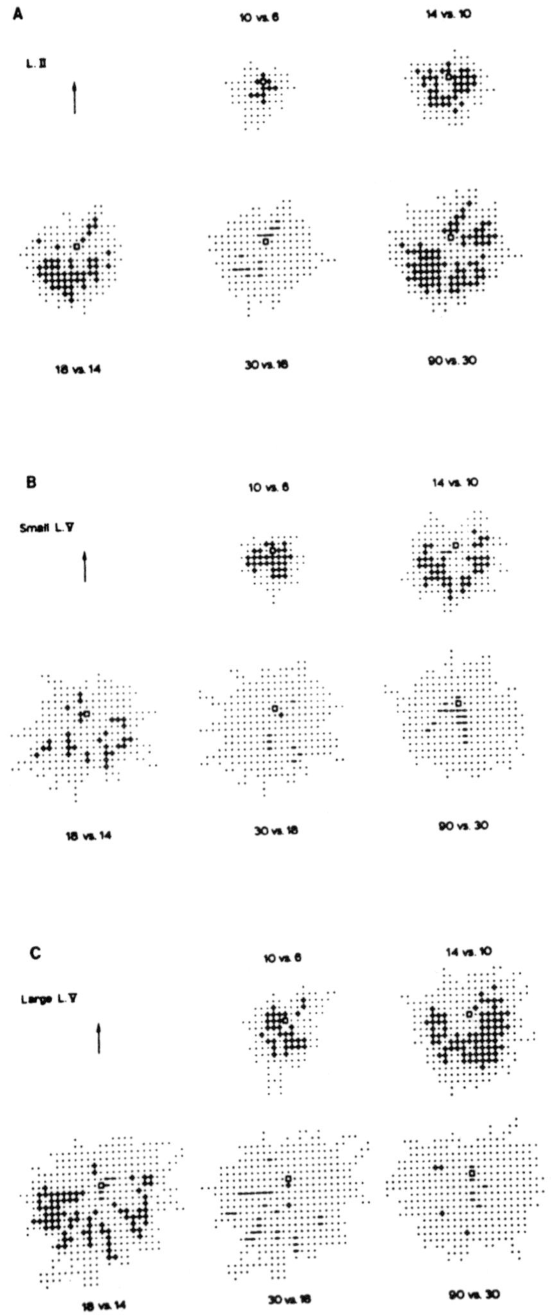

Fig. 9. The comparison of the density of the basal dendritic fields of the set of layer II/III (A), of small layer V (B) and of large layer V (C) pyramidal neurons at successive ages studied. The size of each square element is 15 × 15 μm. The plane of projection is the coronal plane, and the arrow points to the pial surface. A + and a − indicate a significantly higher and lower dendritic length value, respectively, per square element according to a two-tailed *t*-test ($P < 0.05$; see Uylings et al., 1986 for interpretation of statistical differences). In this pictorial test a single significant matrix element is not meaningful; only clusters of significantly differing matrix elements are relevant and indicative of a significant development in a particular direction.

during volumetric development of cortical areas, a study of this kind will require that many more animals and neurons be examined (Van Eden et al., 1984; Uylings et al., 1990).

So far, the data indicate that in the rat visual cortex the dendritic outgrowth of cortical neurons reach a plateau at the same age, i.e. 18 days, irrespective of layer, birth date of neuron, type of neuron or initial differences in maturation. After this age, dendritic fields either stay the same size or decline, without or with subsequent regrowth. Examples of these categories in the rat visual cortex are, respectively, (1) the multipolar layer IV neurons, (2) the two layer V pyramidal types, and (3) the layer II/III pyramidal neurons. A pattern of growth, decline and regrowth has also been detected in cerebellar Purkinje cells (Sadler and Berry, 1984; Pentney, 1986; Quakenbush et al., 1990; Woldenberg et al., 1993).

Mode of growth derived from segment length data

When no longitudinal data on growing trees are available, the analysis of growth pattern can best be performed with tree topology methods (e.g. Van Pelt and Verwer, 1986; Uylings et al., 1989a; Van Pelt et al., 1989, 1992; Verwer et al., 1992). The distribution of length values for individual segments, however, can also indicate to some extent how dendrites grow and bifurcate. The large differences between lengths of terminal segments and intermediate segments during development (Fig. 6) and for stimulated growth in adulthood (Uylings et al., 1978) indicate that branching occurs mainly in terminal segments, but not necessarily at their tips, due to 'dormant' or laterally induced growth cones; see for this aspect also the outgrowth model based on microtubule dynamics in the chapter by Van Veen and Van Pelt in this volume (Van Veen and Van Pelt, 1994).

The frequency distributions of terminal segment lengths are fairly symmetrical and show only a few values lower than 40 μm, whereas the frequency distributions of intermediate segments are skewed with a modal value around 10 μm and have relatively few values over 40 μm (see figures of these distributions in Uylings et al., 1978; Larkman, 1991). Segment length distributions can be simulated on the basis of growth algorithms (Nowakowski et al., 1992; Van Veen and Van Pelt, 1993). From the analysis of Van Veen and Van Pelt (1993) it appears that the difference in length between terminal and intermediate segments can be understood by (a) assuming a Poisson process for the occurrence of branch points along dendrites, i.e. assuming a constant branching probability per unit length of a neurite, (b) elongation at the terminal tip, and (c) branching at terminal tips or 'dormant' growth 'cones' along the terminal segments. In our previous review (Uylings et al., 1986) we described that, in nearly all neuronal types studied, the terminal segments are (considerably) longer than those of intermediate segments. In addition, we noted that the mean values for different types of intermediate segments vary widely among multipolar non-pyramidal neurons (Uylings et al., 1986), neocortical pyramidal neurons and hippocampal granule cells (Uylings et al., 1989b). The frequency distributions for different types of intermediate segments of cortical pyramidal and non-pyramidal neurons, respectively, show that these groups mainly differ in the tails towards higher values (Figs. 10 and 11). The longest tail is present in the segments of degree-2, i.e. segments which branch into two terminal segments. The modal value for all the different types is around 7–12 μm. This indicates the possibility of branching at long segments of degree-2. This is suggested by the following: (a) the mode and mean values for the group of intermediate segments with four or more terminal tips ahead (i.e. degree $\geq$ 4) remains about the same for all of the pyramidal cell types as well as for the multipolar non-pyramidal type after day 6; (b) the developmental increase in intermediate segments displayed in Fig. 6 for some types up to about day 18 is mainly due to the increase in degree-2 segment values (especially due to a longer and thicker tail of higher values) whereas the modal values for each cell type remain the

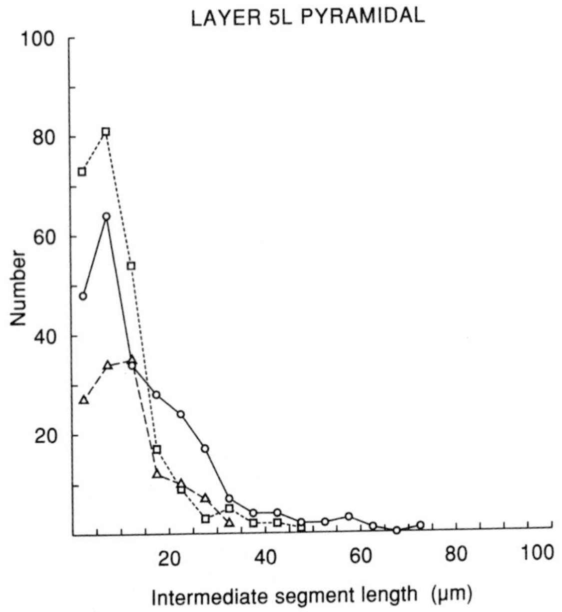

Fig. 10. The frequency distribution of the length of individual intermediate segments of basal dendrites in large layer V pyramidal cells (age: 10 days). The class width is 5 μm, the points in the graph are the class-midpoints. The open circles represent the degree-2 intermediate segments, i.e. segments branching into two terminal segments. Their number is 239 and mean length value is 15.0 μm. The open triangles display the degree-3 intermediate segments; $n = 127$ and the mean value is 11.3 μm. The open squares represent the degree equal to and larger than four segments; $n = 247$ and the mean value is 9.7 μm.

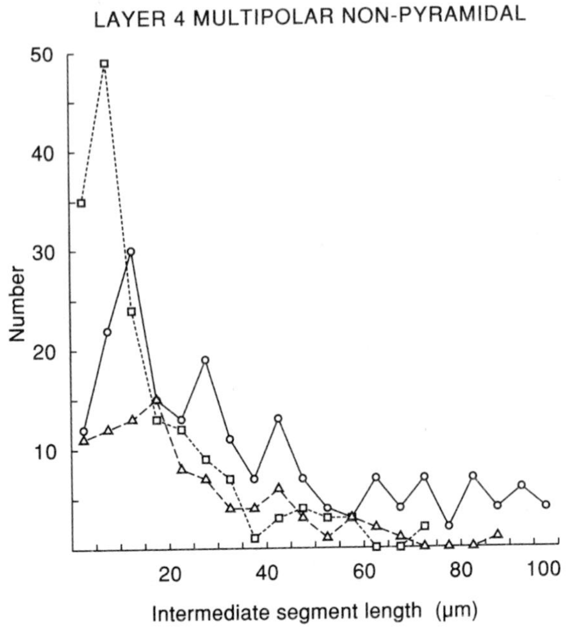

Fig. 11. The frequency distribution of the length of individual intermediate segments of layer IV multipolar non-pyramidal dendrites (age: 16 days). Class width is 5 μm and class values are displayed by class midpoints. The open circles display the values for degree-2 segments; $n = 225$ and the mean length is 28.8 μm. The open triangle values represent the degree-3 segments; $n = 97$ and the mean value is 20.0 μm. The open square values display the degree $\geq$ 4 segments; $n = 165$, and the mean value is 9.7 μm. Note the large variability in length values of the non-pyramidal neurons as compared with pyramidal neurons.

same. Another possible hypothesis is that especially degree-2 segments have a length frequency distribution with a long tail of high value due to incidental delayed branching, especially when the branching probability becomes reduced.

A characteristic difference between the frequency distributions of all the pyramidal cell types and the non-pyramidal cell types (e.g. multipolar, bitufted) is the low variability and the more clearly peaked modal values in the degree-2 and degree-3 segments in pyramidal neurons (see e.g. Figs. 10 and 11). The larger variability in length values for the intermediate segments of multipolar non-pyramidal neurons (Fig. 11), in comparison with those of pyramidal neurons, can be explained (using the model of Van Veen and Van Pelt, 1993) by a lower branching probability per unit length for the multipolar non-pyramidal neurons. This explanation is also compatible with the significantly lower mean number of terminal segments (which is equivalent to the number of bifurcations) per dendrite of multipolar non-pyramidal cells. These values make up nearly half of the values found in the basal dendrites of layer II/III pyramidal neurons after day 10.

Topological analysis of dendritic growth

When no longitudinal data are available for individual growing trees, topological analysis is the method of choice for examining the mode of branching. A topological description of a tree defines how its different segments are interconnected, irrespective of metrical dimensions; each unique description is called a topological tree type (Van Pelt and Verwer, 1984a; Uylings et al., 1989a; Verwer et al., 1992). The topological size of a rooted tree, i.e. the number of segments, is directly related to the number of terminal segments per tree (e.g. Uylings et al., 1989a) or number of branching points per tree (Verwer et al., 1992). The topological analysis cannot be applied to one tree alone, since a given topological tree type can occur after different ways of branching. The frequencies of different topological tree types of a group of neurons, however, are indicative of the mode of branching (Van Pelt and Verwer, 1984a, 1986), even when some trees are cut due to sectioning (Van Pelt and Verwer, 1984b; Van Pelt, in preparation). A major difficulty in topological analysis is that, with increasing size of a tree, the number of topological tree types becomes unmanageably large: e.g. for trees with 14 terminal tips 2179 topological tree types exist, and for trees with 20 terminal tips (degree = 20) 293,547 topological tree types exist (Van Pelt et al., 1989; Uylings et al., 1989a). For large trees (degree $n \geq 100$) there are roughly 2.4^n different topological tree types (Van Pelt et al., 1992).

For a long time we have been looking for topological measures which make a statistical and topological analysis of groups of observed trees

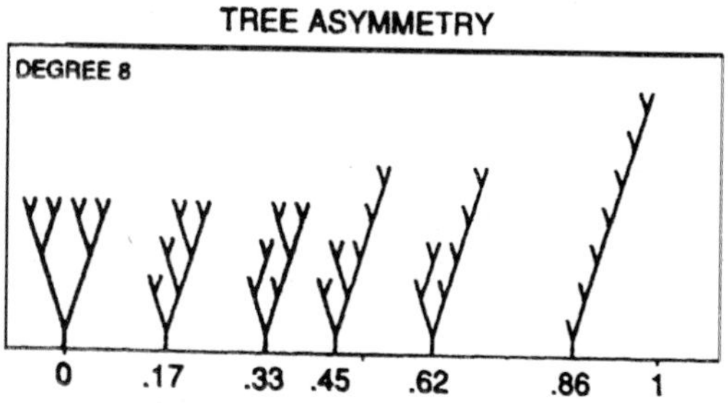

Fig. 12. Topological 'tree asymmetry' values are shown for six of the 23 different topological tree types of degree 8 trees (i.e. trees with 8 terminal tips) in order to illustrate the practical meaning of this variable.

manageable (Uylings et al., 1989a; Van Pelt et al., 1989; Verwer et al., 1992). We think we have managed to develop a practical measure, 'tree asymmetry' (Verwer and Van Pelt, 1986; Van Pelt et al., 1992), which has a strong discriminative power for topological differences (Van Pelt et al., 1989) and is very suitable for comparing sets of branching patterns (almost independent of topological size) as well as for indicating the mode of branching, since the mean and variance of tree asymmetry of a group of neurons depend on the mode of growth (Van Pelt et al., 1992). At each bifurcation, the number of terminal tips is partitioned. When they are partitioned into two equal sets the bifurcation is symmetrical and the asymmetry value is then zero. The 'tree asymmetry' is defined as the mean value of asymmetry of all bifurcations in a tree. This 'tree-asymmetry' measure has been normalized so that it ranges between 0 and the limit 1 (see for further details Van Pelt et al., 1992). To obtain a qualitative impression, Fig. 12 illustrates different 'tree-asymmetry' values for some trees with eight terminal tips.

Analysis of growth models shows that the mean 'tree asymmetry' values differ significantly for sets of trees grown under different branching rules (Van Pelt et al., 1992). For a set of trees with a number of terminals between 4 and 15 per tree, grown according to the random segmental branching model, mean tree asymmetry is equal to 0.59, whereas according to the random terminal branching model it is equal to 0.45 (see Table 3 in Van Pelt et al., 1992). These and other branching models predict a much larger variability in 'tree asymmetry' values for trees with less than 15 terminal tips (i.e. degree ≤ 15) in comparison with large trees (e.g. Purkinje cell trees with a degree between 150 and 800). This is observed for cortical pyramidal and non-pyramidal neurons in which almost all the dendrites have less than 15 terminal tips. Therefore, in groups of these small trees, testing of a slightly differing branching mode is relatively difficult (see Van Pelt et al., 1992). Still, the available data on

TABLE I
Mean tree asymmetry ($\bar{A}_t$) during development (for trees with degree ≥ 4)

Age (days)	Layer II/III pyr.			Small layer V pyr.			Large layer V pyr.			Layer IV multipolar nonpyramidal		
	$\bar{A}_t$	S.E.M.	(no.)	$\bar{A}_t$	S.E.M.	(no.)	$\bar{A}_t$	S.E.M.	(no.)	$\bar{A}_t$	S.E.M.	(no.)
6	—	—	(5)	—	—	(7)	0.43	0.04	(46)	0.46	0.06	(17)
10	0.45	0.04	(45)	0.47	0.03	(59)	0.48	0.02	(91)	0.43	0.04	(41)
14	0.44	0.02	(97)	0.44	0.04	(41)	0.47	0.02	(92)	0.43	0.04	(41)
18	0.44	0.03	(74)	0.43	0.03	(33)	0.37	0.02	(59)	0.44	0.03	(52)
30	0.44	0.03	(52)	0.38	0.03	(56)	0.45	0.03	(66)	—	—	—
90	0.42	0.02	(89)	0.48	0.04	(40)	0.37	0.03	(46)	0.45	0.03	(64)

pyr., pyramidal neurons.

different types of cortical pyramidal and non-pyramidal neurons (see Table I) point clearly to a topological mode of branching which is (1) similar for the layer II/III pyramidal and layer IV non-pyramidal multipolar neurons during the various developmental stages in rat visual cortex, and (2) consistent with the random terminal branching model. This model predicts a mean 'tree asymmetry' of 0.45 and a variance which is similar to the observed variances. The values of the two layer V pyramidal cell types are more variable during the different ages, although in general these values are also consistent with a model which assumes branching randomly at terminal segments. The lower values at some ages for the layer V pyramidal neurons are below the expected values according to the random terminal branching model and reflect a non-random regression in such a way that the trees become topologically more symmetrical. This needs to be tested in further research.

Conclusions

We may conclude from the data available concerning cortical dendrite development that, irrespective of birth date or an earlier start of dendritic maturation, the phase of rapid dendritic growth ends around the same age for all types of cortical pyramidal and non-pyramidal neurons. In rat cortex this is around postnatal day 18; at that age the electrophysiological development of rat cortex, though quite advanced, is not yet finished (e.g. Corner and Mirmiran, 1990; Hamill et al., 1991; Corner et al., 1992; Connors, 1994). After day 18 this is true also for dendritic development: the basal dendrites of layer II/III pyramidals continue to grow at later ages, after a small decline, while a persisting minor regression can be noted in the basal dendrites of both types of layer V pyramidal neurons. No further large alteration has been found in the layer IV multipolar non-pyramidal neurons.

In general, the mode of branching appears to be the same for all the types of cortical neurons examined at the various ages studied, i.e. mainly random branching at the terminal segments. However, the branching probability per dendritic unit of length is lower in non-pyramidal neurons, resulting in fewer bifurcations per dendrite.

Dendritic regression during the period of normally occurring cell death (apoptosis) is large in only a few neuronal types. In addition, our results indicate the occurrence of some regression in the basal dendrites of pyramidal cell types during normal development and puberty. The major part of dendritic development appears to follow an intrinsic 'program' which can be modified, however, by environmental influences without altering the mode of branching (e.g. early malnutrition, and exposure to an enriched environment, even in adulthood) (Uylings et al., 1978; Juraska et al., 1980). In cases of major pathology, on the other hand, such intrinsic 'programs' of dendritic development may be drastically changed.

Acknowledgements

Henk Stoffels and Gerben van der Meulen are gratefully acknowledged for their help with the figures, and we would like to thank Wilma Verweij for her secretarial help.

References

Chagnac-Amitai, Y., Luhmann, H.J. and Prince, D.A. (1990) Burst generating and regular spiking layer 5 pyramidal neurons of rat neocortex have different morphological features. *J. Comp. Neurol.,* 296: 598–613.

Cherry, J.A., Tobet, S.A., DeVoogd, T.J. and Baum, M.J. (1992) Effects of sex and androgen treatment on dendritic dimensions of neurons in the sexually dimorphic preoptic/anterior hypothalamic area of male and female ferrets. *J. Comp. Neurol.,* 323: 577–585.

Connors, B.W. (1994) Intrinsic neuronal physiology and the functions, dysfunctions and development of neocortex. In: J. Van Pelt, M.A. Corner, H.B.M. Uylings and F.H. Lopes da Silva (Eds.), *Progress in Brain Research, Vol. 102, The Self-Organizing Brain — from Growth Cones to Functional Networks,* Elsevier, Amsterdam, this volume.

Corner, M.A. and Mirmiran, M. (1990) Spontaneous neuronal firing patterns in the occipital cortex of developing rats. *Int. J. Dev. Neurosci.,* 8 (3): 309–316.

Corner, M.A., van Eden, C.G. and De Beaufort, A.J. (1992) Spike-train analysis reveals 'overshoot' in developing rat prefrontal cortex function. *Brain Res. Bull.,* 28: 799–802.

Derer, P. and Derer, M. (1990) Cajal-Retzius cell ontogenesis and death in mouse brain visualized with horseradish peroxidase and electron microscopy. *Neuroscience,* 36: 839–856.

Ferrer, I., Soriano, E., Del Rio, J.A., Alcántara, S. and Auladell, C. (1992) Cell death and removal in the cerebral cortex during development. *Prog. Neurobiol.,* 39: 1–43.

Finlay, B.L. (1992) Cell death and the creation of regional differences in neuronal numbers. *J. Neurobiol.,* 23: 1159–1171.

Goldstein, L.A., Kurz, E.M. and Sengelaub, D.R. (1990) Androgen regulation of dendritic growth and retraction in the development of a sexually dimorphic spinal nucleus. *J. Neurosci.,* 10: 935–946.

Greenough, W.T., Carter, C.S., Steerman, C. and De Voogd, T.J. (1977) Sex differences in dendritic patterns in hamster preoptic area. *Brain Res.,* 126: 63–72.

Hallman, L.E., Schofield, B.R. and Lin, C-S. (1988) Dendritic morphology and axon collaterals of corticotectal, corticopontine, and callosal neurons in layer V of primary visual cortex of the hooded rat. *J. Comp. Neurol.,* 272: 149–160.

Hamill, O.P., Huguenard, J.R. and Prince, D.A. (1991) Patch-clamp studies of voltage-gated currents in identified neurons of the rat cerebral cortex. *Cerebral Cortex,* 1: 48–61.

Hedlich, A. and Winkelmann, E. (1982) Neuronentypen des visuellen Cortex der adulten and juvenilen Ratte. *J. Hirnforsch.,* 23: 353–373.

Hübener, M. and Bolz, J. (1988) Morphology of identified projection neurons in layer 5 of rat visual cortex. *Neurosci. Lett.,* 94: 76–81.

Hübener, M., Schwarz, C. and Bolz, J. (1990) Morphological types of projection neurons in layer 5 of the cat visual cortex. *J. Comp. Neurol.,* 301: 655–674.

Jacobson, M. (1978). *Developmental Neurobiology,* 2nd ed., Plenum Press, New York, 562 pp.

Juraska, J.M. (1982) The development of pyramidal neurons after eye opening in the visual cortex of hooded rats: a quantitative study. *J. Comp. Neurol.,* 212: 208–213.

Juraska, J.M. and Fifkova, E. (1979) A Golgi study of the early postnatal development of the visual cortex of the hooded rat. *J. Comp. Neurol.,* 183: 247–256.

Juraska, J.M., Greenough, W.T., Elliot, C., Mack, K.J. and Berkowitz, R. (1980) Plasticity in adult rat visual cortex: An examination of several cell populations after differential rearing. *Behav. Neur. Biol.,* 29: 157–167.

Kawaguchi, Y. (1993) Groupings of non-pyramidal and pyramidal cells with specific physiological and morphological characteristics in rat frontal cortex. *J. Neurophysiol.,* 69: 416–431.

Koenderink, M.J.Th. and Uylings, H.B.M. (1994) Postnatal maturation of the layer V pyramidal neurons in the human prefrontal cortex: a quantitative Golgi analysis. Submitted.

Koenderink, M.J.Th., Uylings, H.B.M. and Mrzljak, L. (1994) Postnatal maturation of the layer III pyramidal neurons in the human prefrontal cortex: a quantitative Golgi analysis. *Brain Res.*, in press.

Koester, S.E. and O'Leary, D.D.M. (1992) Functional classes of cortical projection neurons develop dendritic distinctions by class-specific sculpting of an early common pattern. *J. Neurosci.,* 12: 1382–1393.

Kostović, I. and Rakic, P. (1980) Cytology and time of origin of interstitial neurons in the white matter in infant and adult human and monkey telencephalon. *J. Neurocytol.,* 9: 219–242.

Kostović, I. and Rakic, P. (1990) Developmental history of the transient subplate zone in the visual and somatosensory cortex of the Macaque monkey and human brain. *J. Comp. Neurol.,* 297: 441–470.

Larkman, A.U. (1991) Dendritic morphology of pyramidal neurones of the visual cortex of the rat: I. branching patterns. *J. Comp. Neurol.,* 306: 307–319.

Lund, R.D. (1978) *Development and Plasticity of the Brain,* Oxford Press, New York, 370 pp.

Luskin, M.B. and Shatz, C.J. (1985) Studies of the earliest generated cells of the cat's visual cortex: cogeneration of subplate and marginal zones. *J. Neurosci.,* 5: 1062–1075.

Marin-Padilla, M. (1990) Three-dimensional structural organization of layer I of the human cerebral cortex: a Golgi study. *J. Comp. Neurol.,* 299: 89–105.

Mason, A. and Larkman, A. (1990) Correlations between morphology and electrophysiology of pyramidal neurons in slices of rat visual cortex. *J. Neurosci.,* 10: 1415–1428.

Miller, M.W. (1988) Development of projection and local circuit neurons in neocortex. In: A. Peters and E.G. Jones (Eds.), *Development and Maturation of Cerebral Cortex, Cerebral Cortex,* Vol. 7, Plenum Press, New York, pp. 133–175.

Mrzljak, L., Uylings, H.B.M., van Eden, C.G. and Judáš, M. (1990) Neuronal development in human prefrontal cortex in prenatal and postnatal stages. In: H.B.M. Uylings, C.G. Van Eden, J.P.C. de Bruin, M.A. Corner and M.G.P. Feenstra (Eds.), *Progress in Brain Research, Vol. 85, The Prefrontal Cortex: its Structure, Function and Pathology,* Elsevier, Amsterdam, pp. 185–222.

Muñoz-Cueto, J.A., Garcia-Segura, L.M. and Ruiz-Marcos, A. (1991) Regional sex differences in spine density along the apical shaft of visual cortex pyramids during postnatal development. *Brain Res.,* 540: 41–47.

Murphy, E.H. and Magness, R. (1984) Development of the rabbit visual cortex: a quantitative Golgi analysis. *Exp. Brain Res.,* 53: 304–314.

Nowakowski, R.S., Hayes, N.L. and Egger, M.D. (1992) Competitive interactions during dendritic growth: a simple stochastic growth algorhithm. *Brain Res.,* 576: 152–156.

Overdijk, J., Uylings, H.B.M., Kuypers, K. and Kamstra, A.W. (1978) An economical, semi-automatic system for measuring cellular tree structures in three dimensions, with special emphasis on Golgi-impregnated neurons. *J. Microsci. (Oxford),* 114: 271–284.

Parnavelas, J.G. and Uylings, H.B.M. (1980) The growth of non-pyramidal neurons in the visual cortex of the rat: a morphometric study. *Brain Res.,* 193: 373–382.

Parnavelas, J.G. and Edmunds, S.M. (1983) Further evidence that Retzius-Cajal cells transform to non-pyramidal neurons in the developing rat visual cortex. *J. Neurocytol.,* 12: 863–871.

Parnavelas, J.G., Lieberman, A.R. and Webster, K.E. (1977) Organization of neurons in the visual cortex, area 17, of the rat. *J. Anat.,* 124: 305–322.

Parnavelas, J.G., Bradford, R., Mounty, E.J. and Lieberman, A.R. (1978) The development of non-pyramidal neurons in the visual cortex of the rat. *Anat. Embryol.,* 155: 1–14.

Pentney, R.J. (1986) Quantitative analysis of dendritic networks of Purkinje neurons during aging. *Neurobiol. Aging,* 7: 241–248.

Quakenbush, L.J., Ngo, H. and Pentney, R.J. (1990) Evidence for non-random regression of dendrites of Purkinje neurons during aging. Neurobiol. Aging 11: 111–115.

Petit, T.L., LeBoutillier, J.C., Gregario, A. and Libstug, H. (1988) The pattern of dendritic development in the cerebral cortex of the rat. Dev. Brain Res., 41: 209–219.

Ramirez, L.F. and Kalil, K. (1985) Critical stages for growth in the development of cortical neurons. J. Comp. Neurol., 237: 506–518.

Ruiz-Marcos, A. (1983) Mathematical models of cortical structures and their application to the study of pathological situations. In: S. Grisolía, C. Guerri, F. Samson, S. Norton and F. Reinoso-Suárez (Eds.), *Ramá y Cajal's Contribution to the Neurosciences,* Elsevier, Amsterdam, pp. 209–222.

Ruiz-Marcos, A. and Ipiña, S.L. (1986) Hypothyroidism affects preferentially the dendritic densities on the more superficial region of pyramidal neurons of the rat cerebral cortex. *Dev. Brain Res.,* 28: 259–262.

Sadler, M. and Berry, M. (1984) Remodelling during development of the Purkinje cell dendritic tree in the mouse. *Proc. R. Soc. London,* 221: 349–368.

Sasaki, M. and Arnold, A.P. (1991) Androgenic regulation of dendritic trees of motoneurons in the spinal nucleus of the bulbocavernosus: reconstruction after intracellular iontophoresis of horseradish peroxidase. *J. Comp. Neurol.,* 308: 11–27.

Shatz, C.J., Shun, J.J.M. and Luskin, M.B. (1988) The role of the subplate in the development of the mammalian telencephalon. In: A. Peters and E.G. Jones (Eds.), *Development and maturation of cerebral cortex. Cerebral Cortex,* Vol. 7 Plenum Press, New York, pp. 35–58.

Swaab, D.F. and Uylings, H.B.M. (1987) Comments on review by Coleman and Flood 'Neuron numbers and dendritic extent in normal aging and Alzheimer's disease. Density measures: parameters to avoid. *Neurobiol. Aging,* 8: 574–576.

Uylings, H.B.M. and Delalle, I. (1994) Morphology of NPY-ir neurons and fibers in human prefrontal cortex during development. Submitted.

Uylings, H.B.M., Kuypers, K., Diamond, M.C. and Veltman, W.A.M. (1978) Effects of differential environments on plasticity of dendrites of cortical pyramidal neurons in adult rats. *Exp. Neurol.,* 62: 658–677.

Uylings, H.B.M., Verwer, R.W.H., Van Pelt, J. and Parnavelas, J.G. (1983) Topological analysis of dendritic growth at various stages of cerebral development. *Acta Stereol.,* 2: 55–62.

Uylings, H.B.M., Ruiz-Marcos, A. and Van Pelt, J. (1986) The metric analysis of three-dimensional dendritic tree patterns: a methodological review. *J. Neurosci. Methods,* 18: 127–151.

Uylings, H.B.M., Van Pelt, J. and Verwer, R.W.H. (1989a)

Topological analysis of individual neurons. In: J.J. Capowski (Ed.), *Computer Techniques in Neuroanatomy,* Plenum Publishing Corporation, New York, pp. 215–239.

Uylings, H.B.M., Van Pelt, J., Verwer, R.W.H. and McConnell, P. (1989b) Statistical analysis of neuronal populations. In: J.J. Capowski (Ed.), *Computer Techniques in Neuroanatomy,* Plenum Publishing Corporation, New York, pp. 241–264.

Uylings, H.B.M., Van Eden, C.G., Parnavelas, J.G. and Kalsbeek, A. (1990) The prenatal and postnatal development of rat cerebral cortex. In: B. Kolb and R.C. Tees (Eds.), *The Cerebral Cortex of the Rat,* MIT Press, Cambridge, pp. 35–76.

Van Eden, C.G., Uylings, H.B.M. and Van Pelt, J. (1984) Sex-difference and left-right asymmetries in the prefrontal cortex during postnatal development in the rat. *Dev. Brain Res.,* 12: 146–153.

Van Pelt, J. and Verwer, R.W.H. (1984a) New classification methods of branching patterns. *J. Microsci.,* 136: 23–34.

Van Pelt, J. and Verwer, R.W.H. (1984b) Cut trees in the topological analysis of branching patterns. *Bull. Math. Biol.,* 46: 283–294.

Van Pelt, J. and Verwer, R.W.H. (1986) Topological properties of binary trees grown with order-dependent branching probabilities. *Bull. Math. Biol.,* 47: 323–336.

Van Pelt, J., Uylings, H.B.M. and Verwer, R.W.B. (1989) Distributional properties of measures of tree topology. *Acta Stereol.,* 8: 465–470.

Van Pelt, J., Uylings, H.B.M., Verwer, R.W.H., Pentney, R.J. and Woldenberg, M.J. (1992) Tree asymmetry a sensitive and practical measure for binary topological trees. *Bull. Math. Biol.,* 54: 759–784.

Van Veen, M.P. and Van Pelt, J. (1993) Terminal and intermediate segment lengths in neuronal trees with finite length. *Bull. Math. Biol.,* 55: 277–294.

Van Veen, M.P. and Van Pelt, J. (1994) Dynamic mechanisms of neuronal outgrowth. In: J. Van Pelt, M.A. Corner, H.B.M. Uylings and F.H. Lopes da Silva (Eds.), *Progress in Brain Research, Vol. 102, The Self-Organizing Brain — from Growth Cones to Functional Networks,* Elsevier, Amsterdam, this volume.

Vercelli, A. and Innocenti, G.M. (1993) Morphology of visual callosal neurons with different locations, contralateral targets or patterns of development. *Exp. Brain Res.,* 94: 393–404.

Vercelli, A., Assal, F. and Innocenti, G.M. (1992) Emergence of callosally projecting neurons with stellate morphology in the visual cortex of the kitten. *Exp. Brain Res.,* 90: 346–358.

Verwer, R.W.H. and Van Pelt, J. (1986) Descriptive and comparative analysis of geometrical properties of neuronal tree structures. *J. Neurosci. Methods,* 18: 179–206.

Verwer, R.W.H., Van Pelt, J. and Uylings, H.B.M. (1992) An introduction to topological analysis of neurones. In: M.G. Stewart (Ed.), *Quantitative Methods in Neuroanatomy,* Wiley, Chichester, pp. 295–323.

Wahle, P. and Meyer, G. (1987) Morphology and quantitative changes of transient NPY-ir neuronal populations during early postnatal development of the cat visual cortex. *J. Comp. Neurol.,* 261: 165–192.

Wise, S.P. and Jones, E.G. (1977) Cells of origin and terminal distribution of descending projections of the rat somatic sensory cortex. *J. Comp. Neurol.,* 175: 129–158.

Wise, S.P., Fleshman, J.W. and Jones, E.G. (1979) Maturation of pyramidal cell form in relation to developing afferent and efferent connections of rat somatic sensory cortex. *Neuroscience,* 4: 1257–1297.

Woldenberg, M.J., O'Neill, M.P., Quackenbush, L.J. and Pentney, R.J. (1993) Models for growth, decline and regrowth of the dendrites of rat Purkinje cells induced from magnitude and link-length analysis. *J. Theor. Biol.,* 162: 403–429.

SECTION II

From Growth Cone to Neuron

B. Morphology, Excitability and Signal Processing

J. van Pelt, M.A. Corner H.B.M. Uylings and F.H. Lopes da Silva (Eds.)
Progress in Brain Research, Vol 102

CHAPTER 8

Electrotonic properties of passive dendritic trees—effect of dendritic topology

Jaap van Pelt[1] and Andreas Schierwagen[2]

[1]*Graduate School of Neuroscience and Netherlands Institute for Brain Research, Amsterdam, The Netherlands and*
[2]*Department of Informatics, University of Leipzig, Germany*

Introduction

Neuronal information processing is thought to be primarily based on the transformation of spatio-temporal patterns of synaptic inputs into particular time sequences of outgoing action potentials. Many mechanisms are involved in this transformation process, among which synaptic actions and the transduction of postsynaptic signals towards the neuron's spike initiating site play a major role. This transduction process is based on the flow of postsynaptic currents through the dendrites and their depolarizing effects at the axon hillock. This electrical process takes place in a branching structure with a conductive intracellular medium, a conductive and capacitive membrane and driven ionic channels in the membrane with voltage- and time-dependent conductances. The effectiveness of the transduction process is therefore not easily expressed, because of the dependence on the momentary state of the dendrite.

In this chapter the role of dendritic geometry in the transduction process will be emphasized under the simplifying assumption that the dendrite is in a stable state, i.e. with fixed conductances, capacitances and geometry. A branching pattern consists of segments, connected to each other in a particular way. In expressing dendritic geometry it is therefore not sufficient to specify the metrics (lengths and diameters), but also the connectivity pattern of the segments, referred to as the topological structure or tree type. Dendritic geometrical variability has consequently both a metrical and a topological origin (Verwer and van Pelt, 1986; Verwer et al., 1992). Recent studies of dendritic topology have shown that topological variability is specific and that only a fraction of all possible tree types have a high enough probability of occurrence to actually be observed. It has further been shown that such probability patterns originate from, and are dependent upon, the mode of dendritic outgrowth. These morphological findings have raised the question of whether topology per se is an important determinant in dendritic signal processing, and to what extent such limited topological variance has significant consequences for functional operations. In the model approach used to answer this question, dendritic geometries will be parameterized and the effect of topological variance will be studied by keeping the metrical and electrical parameters constant. The topological structure is also involved in the assumptions underlying the equivalent cylinder (EC) concept. One of these assumptions concerns the equal electrotonic distance from soma to all terminal tips, a condition which can be fulfilled in a fully symmetric tree. In view of topological variability such tree types appear to have a very low probability to occur, when the trees are not too small. The question of how topology interferes with the EC concept will therefore also be addressed in this study.

Dendritic structure

Dendritic morphology is complex and shows a large natural variability. Being interested in the effect of topology on electrotonic properties, we may profit from the distinction between metrics and topology and reduce the metrical complexity and variability of segments by assuming simple cylinders with fixed lengths and diameters. The topological variability will be maintained including the effect of extreme (symmetric and asymmetric) tree types.

Dendritic topology

Branching patterns consist of segments connected to each other in a particular way. For a given number of segments only a finite number of different connectivity patterns (tree types) is possible (Van Pelt and Verwer, 1983), although this number rapidly increases with the number of segments (tree size)(Van Pelt et al., 1989). An efficient numerical label for topology is the tree asymmetry A_t, defined as the mean value of the partition asymmetries in the tree

$$A_t = \frac{1}{n-1} \sum_{i=1}^{n-1} A_p(r_i, s_i) \qquad (1)$$

The summation runs over all $n - 1$ branch points in the tree of degree n, i.e. with n terminal segments, while the partition (r_i,s_i) at branch point i denotes the number of terminal segments in both subtrees. A_p denotes the partition asymmetry, defined as

$$A_p(r, s) = \frac{|r-s|}{(r+s-2)} \text{ if } r+s > 2 \text{ and}$$

$$A_p(1, 1) = 0 \qquad (2)$$

(Van Pelt et al., 1992). The tree-asymmetry values range from 0 for fully symmetric trees to 1 for fully asymmetric and infinitely large trees. The tree types of degree range 4–8 are displayed in Fig. 1 versus their tree-asymmetry value. Among the measures that have been used in the literature to express topology in a single measure, the tree-asymmetry measure is the most discriminative one (Van Pelt et al., 1989). Analyzing the frequencies of occurrence in observed sets of dendritic trees, it is apparent that the different types do not occur with equal likelihood. Model studies of the growth of branching patterns have shown that the probabilities of occurrence are strongly dependent on the mode of growth (e.g. Berry et al., 1975; Van Pelt and Verwer, 1983; Horsfield et al., 1987; Van Pelt et al., 1992). In the used QS growth model, so-called after its two parameters Q and S (Van Pelt and Verwer, 1986), the growth of trees is described by a series of branching events. During each event a new (terminal) segment is attached to an existing one. Each segment has a particular probability of branching and the model is defined by the branch probabilities of all segments. The parameter Q defines the ratio for the branching probabilities of intermediate and terminal segments, and the parameter S defines how the branching probabilities depend on the centrifugal order of the segments, i.e, the number of branch points between the root and the segment.

Two particular growth modes have received much attention, i.e. the random terminal growth mode in which only terminal segments are allowed to branch, each with the same probability (i.e. $Q = 0$, $S = 0$), and the random segmental growth mode in which all segments have an equal probability of branching (i.e. $Q = 0.5$, $S = 0$) (e.g. Hollingworth and Berry, 1975; Van Pelt and Verwer, 1983). An example of the probabilities of occurrence of tree types of degree 8 for three modes of growth is given in Fig. 2. How the probability distributions differ between growth modes is clearly shown. For instance, symmetrical trees have a probability that is high for the $(Q,S) = (0,3)$ mode, low for the $(Q,S) = (0,0)$ mode and

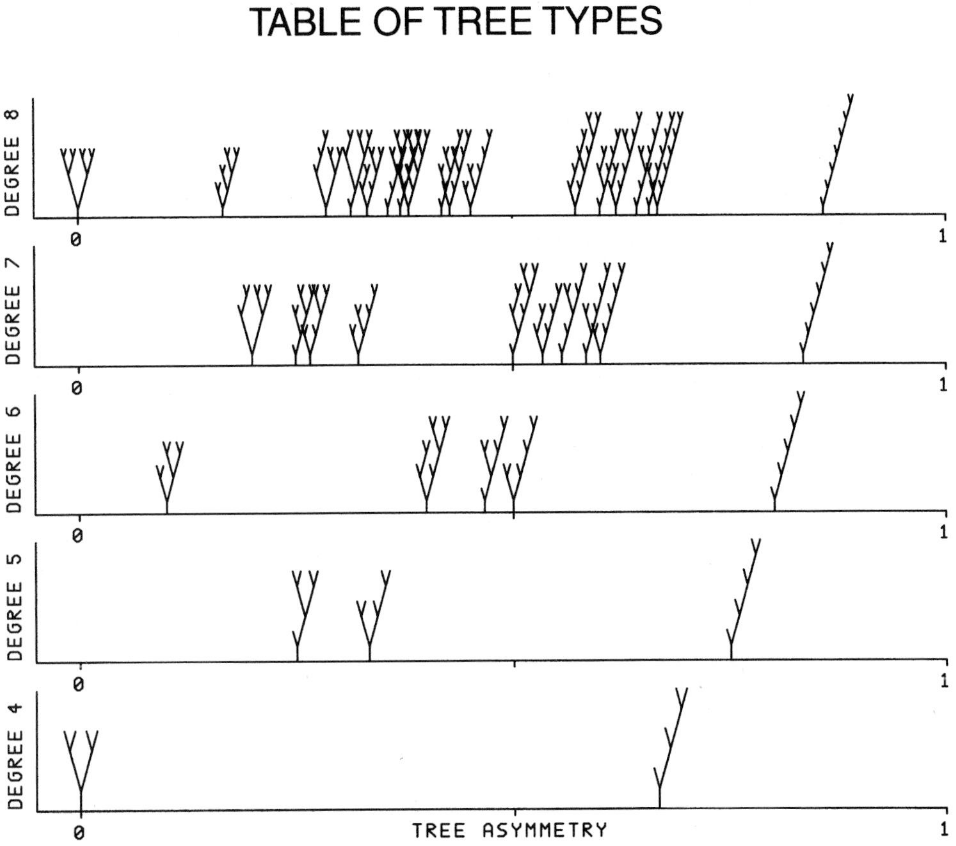

Fig. 1. Display of tree types plotted versus their tree-asymmetry value. The figure shows all tree types in the degree range 4–8. The asymmetry values range from zero for fully symmetric trees (only possible when the degree is a power of two) to a maximum value for fully asymmetric trees, which value approaches one when the asymmetric tree becomes large.

very small for the $(Q,S) = (0.5,0)$ mode of growth. Consequently, also the expectation of the tree-asymmetry measure differs for different modes of growth, e.g. having a value of about 0.46 for the random terminal growth mode, and a value of about 0.59 for the random segmental growth mode (for small trees). It also appears that in contrast with other measures for branching pattern topology, the tree-asymmetry expectations do not depend much on the size of the tree (Van Pelt et al., 1992).

The observed mean (S.D.) values for tree asymmetry are 0.49 (0.02) for Purkinje cell dendritic trees, 0.38 (0.22) for pyramidal cell basal dendrites and 0.43 (0.26) for multipolar non-pyramidal dendrites (Van Pelt et al., 1992). By appropriate choices for the two parameters, the model has been shown to be able to account for all the topological variance in observed dendritic tree sets (Van Pelt et al., 1992). Using only the axes in the QS parameter plane, the basal dendrites of pyramidal cells were best described by $(Q,S) = (0,0.87)$, the multipolar non-pyramidal cell dendrites by $(Q,S) = (0,0)$ and the Purkinje cells by $(Q,S) = (0.11,0)$. The random trees produced by the QS growth model are therefore assumed to be realistic with respect to their topological variability. The procedure for obtaining

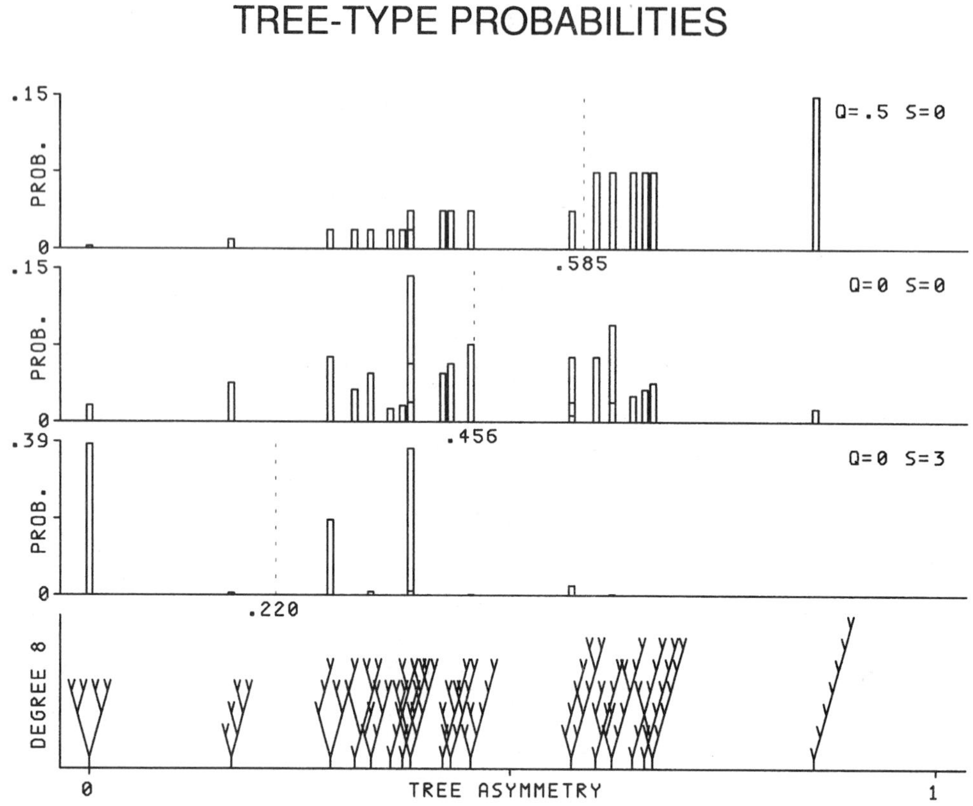

Fig. 2. Plot of probabilities of occurrence of the tree types of degree 8. The trees are plotted versus the tree-asymmetry value in the bottom row. The probabilities have been calculated for three modes of dendritic growth, i.e. $(Q,S) = (0.3)$ (second row); the random terminal growth mode $(Q,S) = (0,0)$ (third row) and the random segmental growth mode $(Q,S) = (0.5,0)$ in the fourth row. The figure shows how the probabilities of occurrence of the tree types depend on the mode of growth. Also indicated in the figure is that the tree-asymmetry expectation depends on the mode of growth.

such random trees is decribed in Van Pelt et al. (1992, pp. 773–774).

Dendritic metrical properties

Segment length. Distributions of dendritic segment lengths are generally very broad, spanning the full range from almost zero to maximal values. When the segments are grouped according to their type, position in the tree, type of dendrite or the cell type from which they originate, this large variation appears to have a certain structure. A general observation is that terminal segments in almost all neuron types have a greater mean length than intermediate segments (e.g. Schierwagen and Grantyn, 1986; Uylings et al., 1986; Larkman, 1991; Burke et al., 1992, Woldenberg et al., 1993). Terminal segments may also have a decreasing length with increasing centrifugal order and may depend on the degree of the remote subtree (Uylings et al., 1986). Woldenberg et al. (1993) distinguished the short and long terminal segments in a pair (i.e originating from the same branch point) and single (not paired) ones and found for these groups different mean lengths in Purkinje cell dendritic trees. Intermediate segments from apical terminal arbors are longer than those of basal dendrites (Larkman, 1991). Nevertheless, even the segment lengths of a particular subgroup show considerable variation and both symmetric and highly skewed distributions may be observed.

In a theoretical study on segment léngth, Van Veen and Van Pelt (1993) showed that some of these length characteristics can be understood on the basis of a simple probabilistic scheme, i.e. by assuming that the branch points along a path are distributed according to a Poisson process. Typical values for the means of, respectively, terminal and intermediate segment length distributions in different cells are: 70.5 μm and about 19 μm in rat pyramidal cell basal dendrites (Uylings et al., 1989), 117 μm and 11 μm in rat visual cortex layer 5 pyramidal cell basal dendrites (Larkman, 1991), 132 μm and 59.5 μm in human dentate granule cells (Uylings et al., 1989), 69.7 μm and 22.1 μm in rat multipolar non-pyramidal neurons (Uylings et al., 1986), 104.7 μm and 78.2 μm in superior colliculus neurons (Schierwagen and Grantyn, 1986), and 5 μm and 13 μm or 8 μm in rat Purkinje cells (Woldenberg et al., 1993).

Segment diameter. Hillman (1979) found that diameters of terminal segments are consistently constrained by minimal and maximal values, with a mean value of 1.1 μm for Purkinje and pyramidal neurons and 0.7 μm for granule and stellate neurons. This difference is possibly associated with the spiny nature of the former group. Schierwagen and Grantyn (1986) found terminal segment diameters to fall off at about 1 μm for three superior colliculus neurons. Larkman (1991) also noted the narrow range of terminal segment diameters and reported mean values of about 0.7 μm for the rat visual cortex pyramidal cell basal and oblique dendrites. Intermediate segment diameters are also subject to large variation. Possible correlations between parent-daughter segment diameters may be studied by looking for those values of the branch power e for which the ratio $(d_1^e + d_2^e)/d_p^e$ has, on average, a value of unity. Hillman (1979) reported a value of about $e = 2$ for rat Purkinje and neocortical pyramidal neurons. Schierwagen and Grantyn (1986) found mean values in the range $e \in [1.4 - 1.53]$ (S.D. $\approx$ 0.3) for dendrites of cat superior colliculus neurons, concluding that these dendrites basically conform to the 3/2 power relationship. Culheim et al. (1987) reported a value 1.13 $\pm$ 0.34 for the ratio $(d_1^{1.5} + d_2^{1.5})/d_p^{1.5}$ in cat motoneurons. Larkman et al. (1992) found for rat visual cortex pyramidal neurons a value for the branch power between $e = 1.5$ and $e = 2$.

Morphological parametrization

A parametrization of dendritic geometry may now be given by (i) the number of segments n, (ii) the length of terminal segments l_t, (iii) the length ratio of intermediate and terminal segments r, (iv) the terminal segment diameter d_t, (v) the branch power e and (vi) the topological tree type, randomly produced by a growth model defined by the parameters Q and S. Although both the segment lengths and the diameters show considerable variation, within as well as between trees, it suffices in the context of the present study to take only fixed (but characteristic) values for these variables. For the terminal and intermediate segment lengths we have chosen for the values from human dentate granule cells (Uylings et al., 1989), i.e. 132 μm and 59.5 μm ($r = 0.45$), respectively. A branch power relation is used for the calculation of the segment diameters. Because of the variations found in this parameter, and in order to estimate how sensitively it determines electrotonic properties, the calculations are done for three different values: $e = 1$, 1.5 and 2. Because of the restricted range for terminal segment diameters a fixed value of 0.7 μm has been chosen for all electrotonic calculations. Applying the branch power rule, the root diameter d_r is given by $d_r = d_t n^{1/e}$, with n denoting the number of terminal tips. For $e = 1$ we have $d_r = nd_t$ while for $e = 2$ it results in $d_r = \sqrt{n}\, d_t$. (Note that the root diameter does not depend on the topological structure of the tree.) The tree types are produced by the random terminal growth mode $(Q,S) = (0,0)$ or by the random segmental growth mode $(Q,S) = (0.5,0)$, while the most symmetrical and asymmetical trees are included so as to span the whole topological range.

Electrical properties of dendrites

There is general agreement on the value for the membrane capacitance (1 $\mu F/\text{cm}^2$) and on the range of possible values for the intracellular resistance (i.e. $R_i = 70\text{–}300\ \Omega\text{cm}$), (e.g. Holmes, 1992; Larkman et al., 1992). We have selected the frequently used value of 100 Ωcm. For the membrane resistance, on the other hand, many different values have been reported for different types of neurons, ranging from values of 1500 Ωcm^2 (see overview in Schierwagen, 1986) to about 195 $\text{k}\Omega\text{cm}^2$ (Larkman et al., 1992). For the purpose of this paper, however, we need only a representative value and will choose a value of 3 $\text{k}\Omega\text{cm}^2$, which lies well within the range given for many types of neurons.

Transient responses in a branched structure

Vector cable equations

The electrical behavior in a cylinder with a leaky capacitive membrane and conductive intracellular medium is governed by the well-known cable equation

$$\lambda^2 \frac{\partial^2 v}{\partial x^2} - v - \tau \frac{\partial v}{\partial t} = 0 \qquad (3)$$

(e.g. Rall, 1959, 1977; Jack et al., 1975), with $v(x,t)$ the transmembrane potential, $\lambda = \sqrt{dR_m/4R_i}$ the space constant, $\tau = R_\text{m}C_\text{m}$ the membrane time constant and d the diameter of the cylinder. The electrical properties R_m, C_m and R_i denote the specific membrane resistance and capacitance and the specific axial resistance, respectively. Several mathematical/numerical procedures exist for finding solutions of these equations for given sets of boundary conditions, such as branching-cable and compartmental procedures (e.g. Holmes, 1986; DeSchutter, 1992; Segev, 1992). One of the branching-cable approaches makes use of the Laplace transform of the cable equation, thus transforming the set of differential equations into a set of linear equations (e.g. Holmes, 1986). This set of linear equations gets the very simple structure

$$\mathbf{I} = \mathbf{GV} \qquad (4)$$

when a vectorial representation is used for the currents $\mathbf{I}$ and the voltages $\mathbf{V}$ at the nodes in the branched structure (Van Pelt, 1992). Especially the conductance matrix $\mathbf{G}$ has a structure that forms a one-to-one mapping of the connectivity pattern of the segments in the tree and thus of the topological structure. The conductance matrix contains only non-zero matrix elements for all pairs of nodes that are directly connected. Its general structure can be summarized by the following rules:

- node i has a diagonal term, $G_{ii} = \sum_j \tau_{ij} + G_{io}$, with j running over all nodes directly connected to node i with G_{io} indicating the additional conductance to the outside at node i.
- node i has an off-diagonal term $G_{ij} = -\sigma_{ij}$ for any node $j (j \neq i)$ directly connected to node i. All other off-diagonal matrix elements in row i (i.e. from all nodes not directly connected to node i) are zero.

The σs and τs are functions of the cylinder geometry, electrical properties and Laplace variable s, and are defined as

$$\sigma_{ij} = \frac{\gamma_{ij}}{r_{ij}} \frac{1}{\sinh \gamma_{ij} l_{ij}} \text{ and } \tau_{ij} = \frac{\gamma_{ij}}{r_{ij}} \frac{1}{\tanh \gamma_{ij} l_{ij}} \qquad (5)$$

with $\gamma_{ij} = \sqrt{1 + \tau s}/\lambda_{ij}$, l_{ij} and d_{ij} denoting the length and diameter, respectively, of the segment connecting node i and j, and $r_{ij} = 4R_i/\pi d_{ij}^2$ denoting the axial resistance per unit length in the segment ij. The diagonal terms in $\mathbf{G}$ (i.e. the τs) relate the current at a node with the local potential. The off-diagnonal terms (i.e. the σs) relate the current at a node with the potential at a directly connected node. Because the σs depend only upon the cylinder, this relation is symmetric such that matrix $\mathbf{G}$ is a symmetric one. An example of the structure of the Laplace-transformed vector cable equation is given in Fig. 3. This

Structure of matrix G is a direct mapping of the connectivity pattern of the segments in the tree !

Example: Neuron with three dendrites and in total 10 nodes:

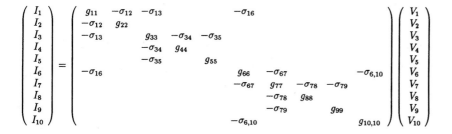

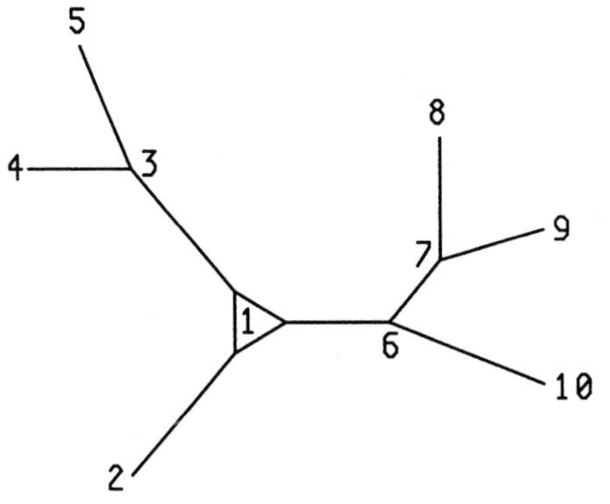

The current-voltage equation I = G V is given by:

$$\begin{pmatrix} I_1 \\ I_2 \\ I_3 \\ I_4 \\ I_5 \\ I_6 \\ I_7 \\ I_8 \\ I_9 \\ I_{10} \end{pmatrix} = \begin{pmatrix} g_{11} & -\sigma_{12} & -\sigma_{13} & & & -\sigma_{16} & & & & \\ -\sigma_{12} & g_{22} & & & & & & & & \\ -\sigma_{13} & & g_{33} & -\sigma_{34} & -\sigma_{35} & & & & & \\ & & -\sigma_{34} & g_{44} & & & & & & \\ & & -\sigma_{35} & & g_{55} & & & & & \\ -\sigma_{16} & & & & & g_{66} & -\sigma_{67} & & & -\sigma_{6,10} \\ & & & & & -\sigma_{67} & g_{77} & -\sigma_{78} & -\sigma_{79} & \\ & & & & & & -\sigma_{78} & g_{88} & & \\ & & & & & & -\sigma_{79} & & g_{99} & \\ & & & & & -\sigma_{6,10} & & & & g_{10,10} \end{pmatrix} \begin{pmatrix} V_1 \\ V_2 \\ V_3 \\ V_4 \\ V_5 \\ V_6 \\ V_7 \\ V_8 \\ V_9 \\ V_{10} \end{pmatrix}$$

$$\begin{aligned} g_{11} &= \tau_{12} + \tau_{13} + \tau_{16} + G_s & g_{22} &= \tau_{12} \\ g_{33} &= \tau_{13} + \tau_{34} + \tau_{35} & g_{44} &= \tau_{34} \\ g_{55} &= \tau_{35} & g_{66} &= \tau_{16} + \tau_{67} + \tau_{6,10} \\ g_{77} &= \tau_{67} + \tau_{78} + \tau_{79} & g_{88} &= \tau_{78} \\ g_{99} &= \tau_{79} & g_{10,10} &= \tau_{6,10} \end{aligned}$$

Fig. 3. Example of the structure of the current-voltage matrix equation **I** = **GV** for a neuron with three dendrites. The vectors **I** and **V** have ten entries describing the Laplace-transformed currents and voltages at the ten nodes in the branching system. The structure of the conductance matrix **G** is defined in the text. Only directly connected nodes give rise to non zero matrix elements in the conductance matrix **G**. The empty spaces in the matrix denote the zero matrix elements. The variables τ and σ are defined in Eqn. (7).

system with ten nodes (terminal tips, branch points, soma) is described by a 10×1 current and voltage vector, and by a 10×10 conductance matrix. Any point along a cylinder can be regarded as an additional node which would give new entries to the vectors.

Steady-state properties of a single cylinder.

For a single cylinder the Laplace-transformed current voltage (cv) equation becomes

$$\begin{pmatrix} I_1 \\ I_2 \end{pmatrix} = \begin{pmatrix} \tau & -\sigma \\ -\sigma & \tau \end{pmatrix} \begin{pmatrix} V_1 \\ V_2 \end{pmatrix} \qquad (6)$$

with I_1 and I_2 the currents and V_1 and V_2 the voltages at both ends of the cylinder while σ_{12} and τ_{12}, or in short σ and τ, are defined as

$$\sigma = \frac{\gamma}{r} \frac{1}{\sinh \gamma l} \text{ and } \tau = \frac{\gamma}{r} \frac{1}{\tanh \gamma l} \qquad (7)$$

with $\gamma = \sqrt{1 + \tau s}/\lambda$ and $\lambda = \sqrt{dR_m/4R_i}$. Note, that $\tau^2 - \sigma^2 = (\gamma/r)^2$ and $\sigma/\tau = 1/\cosh L$. For an infinitely long cylinder ($l = \infty$), $\tanh \gamma l = 1$ and $\sinh \gamma l = \infty$, such that $\tau_\infty = \gamma/r$ and $\sigma_\infty = 0$.

Under steady-state conditions we may set the Laplace variable s to zero such that $\gamma = 1/\lambda$ and interpret I and V as the steady-state currents and voltages. Defining additionally the electrotonic length L of the cylinder as $L = l/\lambda$ the variables τ and σ become

$$\sigma = \frac{1}{\lambda r} \frac{1}{\sinh L} \text{ and } \tau = \frac{1}{\lambda r} \frac{1}{\tanh L} \qquad (8)$$

Well-known expressions (e.g. Rall, 1977; Rall et al., 1992) for the steady-state input conductance and voltage and current attenuations can easily be derived from the cv-equation (6). For instance, we obtain for the input conductance G_1 and output conductance G_2

$$G_1 = \frac{I_1}{V_1} = \frac{\tau V_1 - \sigma V_2}{V_1} \text{ and}$$

$$G_2 = \frac{-I_2}{V_2} = \frac{-\sigma V_1 + \tau V_2}{V_2} \qquad (9)$$

G_1 can be expressed in terms of G_2 via

$$G_1 = \frac{G_2 + (\tau^2 - \sigma^2)/\tau}{G_2/\tau + 1} \qquad (10)$$

For a sealed-end condition ($G_2 = 0$), we obtain

$$G_1^s = \frac{\tau^2 - \sigma^2}{\tau} = \frac{1}{\lambda r} \tanh L \qquad (11)$$

For a killed-end condition ($G_2 = \infty$), we obtain

$$G_1^k = \tau = \frac{1}{\lambda r} \frac{1}{\tanh L} \qquad (12)$$

For an infinitely long cylinder ($L = \infty$), $\tanh L = 1$, and thus

$$G_\infty = \frac{1}{\lambda r} = \frac{\pi}{2} \frac{d^{3/2}}{\sqrt{R_m R_i}} \qquad (13)$$

such that σ, τ and G_1 can also be expressed as

$$\sigma = \frac{G_\infty}{\sinh L}, \tau = \frac{G_\infty}{\tanh L} \text{ and}$$

$$G_1 = \frac{G_2 + G_\infty \tanh L}{\frac{G_2}{G_\infty} \tanh L + 1} \qquad (14)$$

The ratio G_2/G_∞ is equal to the variable B used by Rall (1959). The course of G_1 versus L for the three boundary conditions of G_2 is displayed in Fig. 4.

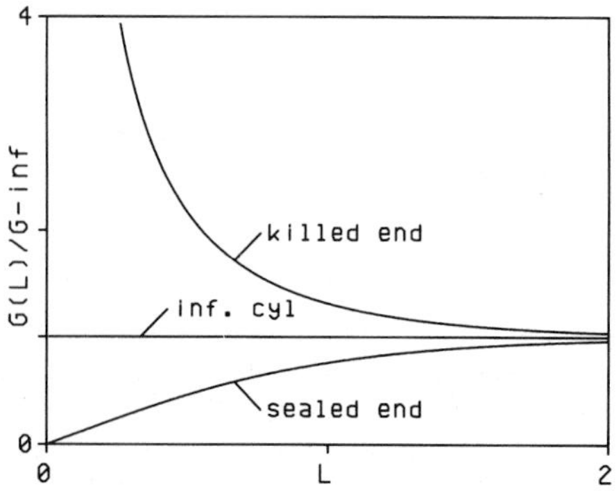

Fig. 4. Ratio of input conductance $G(L)$ of a cylinder with electrotonic length L and of the infinite cylinder as function of L and of sealed-end and killed-end boundary conditions.

From the cv-equation (6), we may also derive an expression for the voltage attenuation along the cylinder as

$$\frac{V_2}{V_1} = \frac{\sigma}{G_2 + \tau} = \frac{1}{\frac{G_2}{G_\infty}\sinh L + \cosh L} \quad (15)$$

This expression shows how the voltage attenuation in a cylinder depends on the cylinder properties (i.e. its electrotonic length L and the input conductance G_∞ of the infinite cylinder) and on boundary conditions (i.e. the output conductance G_2 at the distal node).

For a sealed-end condition, $G_2 = 0$, and thus $V_2/V_1 = 1/\cosh\ L$. For a killed-end condition, we have by definition $V_2 = 0$. For the infinite cylinder, $G_2 = G_\infty$, and $V_2/V_1 = 1/[\sinh\ L + \cosh\ L] = e^{-L}$.

The potential profile along the cylinder can be calculated on the basis of the potentials at both ends, V_1 and V_2, by means of

$$v(x) = \frac{1}{\sinh L}\left[v_1 \sinh\frac{x_2 - x}{\lambda} + v_2 \sinh\frac{x - x_1}{\lambda}\right] \quad (16)$$

Examples of the potential profile for different electrotonic lengths are given in Fig. 5.

Measures for electrotonic extent

Just as the geometric extent of a dendrite is important from a morphological point of view, so is its electrotonic extent assumed to play a critical role in electrical properties such as the steady-state input conductance and transient signal transfer properties. For that reason, it is important to have measures for the electrotonic extent of dendrites that can be appropriately related to these properties.

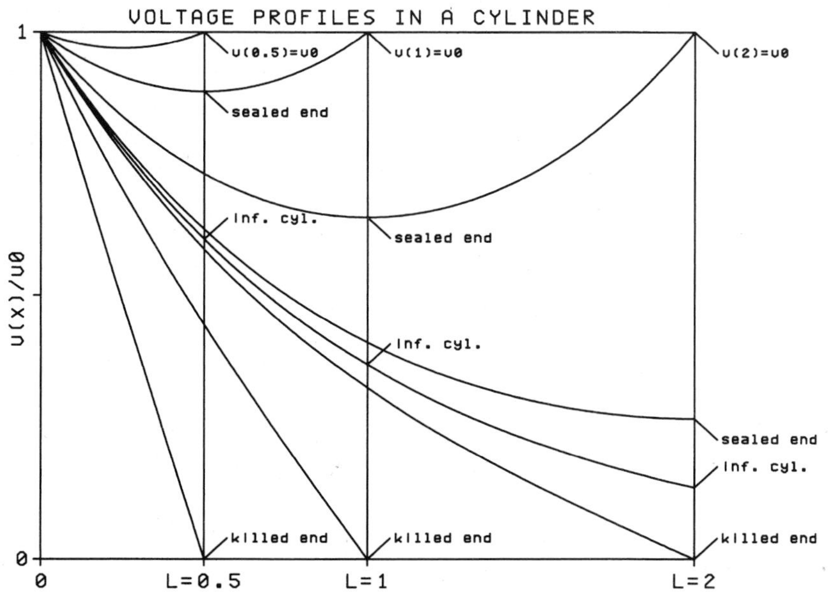

Fig. 5. Composite figure with voltage profiles in three cylinders with electrotonic length $L = 0.5$, $L = 1$ and $L = 2$. For each cylinder are the profiles calculated for different boundary conditions at the distal end, i.e. sealed ($G_L = 0$), killed ($G_L = \infty$ and $v(L) = 0$), or voltage clamped ($v(L) = v0$). Note, that the voltage profile for a sealed-end condition coincides with the voltage profile in a double-length cylinder with a symmetric voltage clamp.

Mean terminal-tip electrotonic length $\overline{L_{tip}}$

The electrotonic extent of a dendrite may be defined in terms of the electrotonic distances of the terminal tips from the soma. The electrotonic length of a path from the soma to a terminal tip is equal to the sum of the electrotonic lengths of all the segments in the path. Generally, the terminal tips are located at different electrotonic distances from the root. We will therefore use the mean value $\overline{L_{tip}}$ of the electrotonic distance distribution.

Electrotonic length estimates on the basis of input conductance

An alternative way to express the electrotonic extent is by the length of the single (equivalent) cylinder that matches particular properties of the complex dendritic branching pattern. For instance, a cylinder may be defined having the same specific resistance and capacitance as the dendrite, with equal boundary conditions, a diameter equal to the stem diameter of the dendrite, and a length such that its input conductance is equal to that of the dendritic structure. This relation is expressed by Eqn. (14) which also can be written as

$$L = \operatorname{arc\,tanh} \frac{G_1 - G_2}{G_\infty - G_1 G_2/G_\infty} \tag{17}$$

Because the argument in this inverse hyperbolic function should be less than or equal to 1, a value of L exists only when

$$G_1 \leq \frac{G_\infty + G_2}{1 + G_2/G_\infty} \quad \text{or} \quad G_1 \leq G_\infty \quad \text{if} \quad G_2 = 0 \tag{18}$$

(A discussion on the equivalent cylinder concept and compatible dendritic geometries is given by Schierwagen (1994).)

Electrotonic length estimates on the basis of time constants

An equivalent cylinder may also be defined so as to match the time constants in the voltage response on a current step stimulus. Rall (1969) has provided the theoretical basis for estimating the electrotonic length L_{tc} of a cylinder on the basis of the time constants of passive membrane potential decay from an initially non-uniform distribution over the cylinder. Because of the boundary conditions (sealed ends), such a decay is described by a series of exponentials

$$V(t) = \sum_{n=0}^{\infty} C_n e^{-t/\tau_n} \tag{19}$$

with the membrane time constant τ_0 reflecting the leak of current through the membrane and with time constants $\tau_1, \ldots, \tau_\infty$ reflecting the rapid equalization of the membrane potential over the cylinder. Rall (1969) showed that these time constants τ_n, $(n = 1, \infty)$ are related to the electrotonic length of the cylinder via

$$\tau_n = \frac{\tau_0}{1 + (n\pi/L)^2} \quad \text{with } L = \frac{n\pi}{\sqrt{\tau_0/\tau_n - 1}} \quad \text{or}$$

$$L = \frac{\pi}{\sqrt{\tau_0/\tau_1 - 1}} \tag{20}$$

On the basis of this equation, the measurement of τ_0 and τ_1 suffices to obtain a reliable estimate for L_{tc}. In electrophysiological practise this relationship has also been used in the case of real neurons with branched dendritic structures, by applying a short current step pulse at the soma and recording the voltage response at the same location. By plotting the voltage response in a semi-logarithmic way, time constants could be recovered by the so-called 'peeling' technique (see example in Schierwagen, 1994). Because of noise and stimulus artifacts, generally only τ_0 and τ_1 could be recovered by this procedure. The electrotonic length obtained via Eqn. (20), has subsequently been used as a measure for dendritic electrotonic extent.

Complications in estimating electrotonic extent on the basis of time constants

Branched structure. The applicability of this procedure to branched structures has recently been studied by Glenn (1988), Rose and Dagum (1988) and, in greater detail, by Holmes et al.

(1992). These authors stress the point that, in branched structures, any path in the structure between points where boundary conditions are imposed gives rise to a series of exponentials in the passive decay of membrane potential. This means that in the exponential expansion of a transient voltage response to a current stimulus as many series of terms arise as there are paths in the dendrite. Holmes et al. (1992) also noted that the coefficients in the expansion depend upon the stimulus location and the point of observation.

For a single cylinder the largest equalizing time constant describes the spread of current from one end to the other, while the second time constant describes the current spread from the midpoint to the tips. Current injection and voltage recording at one end, therefore, will result in a large coefficient for the first term which will be dominant in the voltage response. Current injection and voltage recording at the midpoint, however, will make the second term dominant. A branched structure may have many tips and, consequently, many paths between pairs of tips. The site of stimulation/recording on these paths therefore determines the actual contribution of the exponential terms in the transient response. Without knowledge of the electrotonic structure it is no longer clear, therefore, how the sequence of terms in the expansion relates to the electrotonic length of the paths in the structure (e.g. Holmes et al., 1992; Rall et al., 1992).

Number of time constants obtained in an exponential decomposition. An additional complicating factor is that, in extracting the exponentials from a transient voltage response by means of curve-fitting procedures, only a limited number of exponential terms can be obtained. These exponentials may arise from dominant terms but also from clusters of closely spaced terms. In a detailed discussion of this problem, Holmes et al. (1992) showed that the first equalizing time constant for voltage transients in a motoneuron, as obtained by means of the peeling technique, corresponds to a weighted average over possibly more than 100 theoretical time constants. Experimentally obtained responses additionally suffer from noise and artifacts in the first channels. Therefore, only the data above a certain time threshold are used in the exponential expansion, and, in practice, no more than two or three terms can be extracted. Since calculated responses do not suffer from experimental noise or artifacts, and the full response can be used, more than three terms can generally be extracted.

Completeness of the function set. For an accurate fit, all time constants should be included that could contribute to the response function. This means that, assuming than an exponential function has a negligible amplitude after ten time constants, and using the response above a time threshold th, all time constants greater than $th/10$ have to included in the series of exponentials. Vice versa, only time constants greater than $th/10$ are recoverable from the measured response. In conclusion, it is (1) not possible to extract all time constants in the response function by non-linear regression techniques, and (2) the outcome for a limited number of terms in the expansion depends heavily on the number itself and on the part of the response that is actually used.

To indicate the conditions under which the time constants are derived, we will denote them by $\tau_i^j\,(th)$, with $i = 0,\ldots, j-1$, showing that this time constant is from the ith term in a j-term expansion with a time threshold th. The value τ_0^j is consistently used for the membrane time constant, and $i = 1,\ldots, j-1$ refer to the equalizing time constants (or clusters). In the following, we will omit the threshold indication it if equals zero.

Voltage attenuations

For a single cylinder, the steady-state voltage attenuation can directly be related to its electrotonic length L by means of Eqn. (15), $V_L/V_0 = 1/(\cosh L + G_L \lambda r \sinh L)$. Here, V_0 denotes the applied voltage at one side of the cylinder, and V_L the voltage at the remote side while the boundary condition is set by G_L, denoting the input conductance of the remote continuation of the cylinder. Because the input conductance of an infinitely long cylinder with diameter d equals $G_\infty = 1/\lambda r$, the term $G_L \lambda r$ can also be written as the ratio $B = G_L/G_\infty$. Then we obtain $V_L/V_0 = 1/(\cosh L + B \sinh L)$. For a sealed end condition (no current flow), the conductance G_L is zero ($B = 0$) and the voltage attenuation at the remote end becomes a function of the electrotonic length L only, via the simple relation $V_L/V_0 = 1/\cosh L$.

Assuming sealed tips in a dendrite, this formula then applies for terminal segments with a centrifugal current flow. Intermediate segments, however, have conductive continuations and show voltage attenuations depending on the input conductances of the remote subtrees. The total voltage attenuation for a path from the root to a terminal tip equals the product of the attenuations for all the segments in the path. Therefore, the centrifugal attenuation along a particular path can only simply be expressed in terms of the electrotonic length L_{tip} of the total path under particular conditions, as the ones formulated by Rall (1977) (i.e. a branch power exponent of 3/2 throughout the dendrite, and equal L_{tip} values and boundary conditions for all terminal tips).

When a stimulus is applied at a terminal tip, we have a centripetal signal flow. In the centripetal direction, the quantity G_L denotes the input conductance of the part of the dendrite at the proximal site of the particular segment. The proximal and distal parts of the dendrite, as seen from the particular segment, will have different input conductances and the voltage attenuation in centripetal direction will consequently differ from the centrifugal attenuation, as well as the current flow (Wolf et al., 1992). Nitzan et al. (1990) used a particular measure $\psi = AF_{\mathrm{pet}}/AF_{\mathrm{fug}}$ for the ratio between centripetal and centrifugal voltage attenuation factors between soma and terminal tips under steady-state conditions.

Results

Electrotonic extent measures $\overline{L_{tip}}$ and L_G

A qualitative example of the effect of topology on $\overline{L_{tip}}$ and L_G is given in Fig. 6. For each dendrite of degree 16 in the figure, the terminal-tip electrotonic lengths are calculated and profiles in the dendrite are drawn for the minimal, mean and maximal value found. Below each dendrite, the equivalent cylinder is drawn matching the dendritic input conductance (black bar). Both the dendrites and the equivalent cylinders are plotted in metrical scale. The electrotonic length profiles are mapped onto the equivalent cylinders (or their extensions) and labeled with the value for the electrotonic length. The calculations are done for the most symmetrical tree (column 1), a random tree grown according to the random terminal growth mode (column 2), and the most asymmetric tree (column 3). The value for the tree asymmetry is given at the top of each column. The branch power has given the values $e = 1$ (first row), $e = 1.5$ (second row) and $e = 2$ (third row). All terminal tips in the most symmetrical tree are at equal distance from the root. Note that such a tree is possible only when the degree is a power of two. In the random and the asymmetric tree the terminal tips are at variable distances from the root with the largest variation in the asymmetric tree. The mean value $\overline{L_{tip}}$ appears to increase with the topological asymmetry and to increase with the value for the branch power. From the mapping of $\overline{L_{tip}}$ onto the equivalent cylinder it appears that $\overline{L_{tip}}$ equals L_G only in the symmetric tree with $e = 1.5$, although the difference for the random tree is very small. In all other cases, however, the difference is substantial. For $e = 1$, the value $\overline{L_{tip}}$ is more than twice the value L_G. No comparison could be made for $e = 2$ because the equivalent cylinder and, thus, L_G are undefined. Such a situation occurs when the input conductance of the dendrite is larger than that of the infinite cylinder, as is demonstrated in Eqn. (18).

Example of the calculation and regression of transient responses.

The transient voltage response at the root has been calculated using a current stimulus of delta-impulse shape. The responses have been calculated for 200 time steps covering a total time span of 6 μsec and a time interval between two time

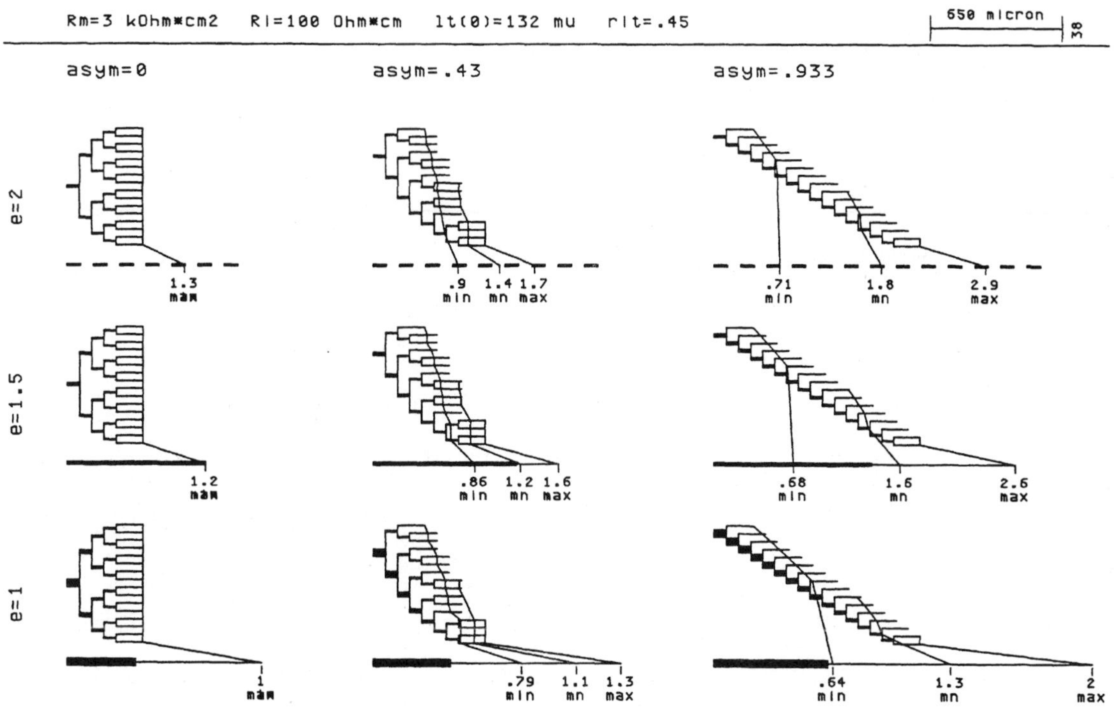

Fig. 6. Display of iso-electrotonic distance curves in dendrites of different topological structure and different values for the branch power. In each dendrite of degree 16, the profiles are drawn for the minimal, mean and maximal value of the terminal-tip electrotonic lengths. Below each dendrite is drawn the equivalent cylinder matching the dendritic input conductance (black bar). Both the dendrites and the equivalent cylinders are plotted in metrical scale. The electrotonic length profiles are mapped onto the equivalent cylinders (or their extensions) and labeled with the value for the electrotonic length. The calculations are done for three different topological structures, i.e. the most symmetrical tree (column 1), a random tree grown according to the random terminal growth mode $(Q,S) = (0,0)$ (column 2) and the most asymmetrical tree (column 3). The value for the tree asymmetry is given at the top of each column. Three different values for the branch power have been used in the diameter calculations, i.e. $e = 1$ in the first row, $e = 1.5$ in the second row and $e = 2$ in the third row. Because the degree is a power of two, all terminal tips in the most symmetric tree are on equal distance from the root.

steps of 30 μsec. For the decomposition of the response into a series of exponentials we used the unconstrained formulation of the optimization problem (D'Aguanno et al., 1986) in combination with the least-squares fitting routine MRQMIN from Numerical Recipes (Press et al., 1986) which implements the Marquardt algorithm. A typical example of the voltage response to a current delta pulse at the root is shown in Fig. 7 for both a symmetric (Fig. 7a) and an asymmetric tree (Fig. 7b). The fast decay in the response reflects the equalizing phenomenon, while the slow decay is due to the membrane time constant. The exponential decomposition has been illustrated for a number of terms. The quality of the fit is shown in the difference curves (bottom row) and demonstrates the improvement of the fit when more terms are used. For the symmetric tree, a stable expansion was obtained with five terms. The sixth term did not noticeably improve the fit because its time constant is much smaller than the bin width ($\tau_5^6 = 3$ μsec) with a relatively moderate coefficient. Consequently, this term contributed only moderately to this bin.

The five-term expansion for the symetric tree resulted in a very good fit, which differed only in

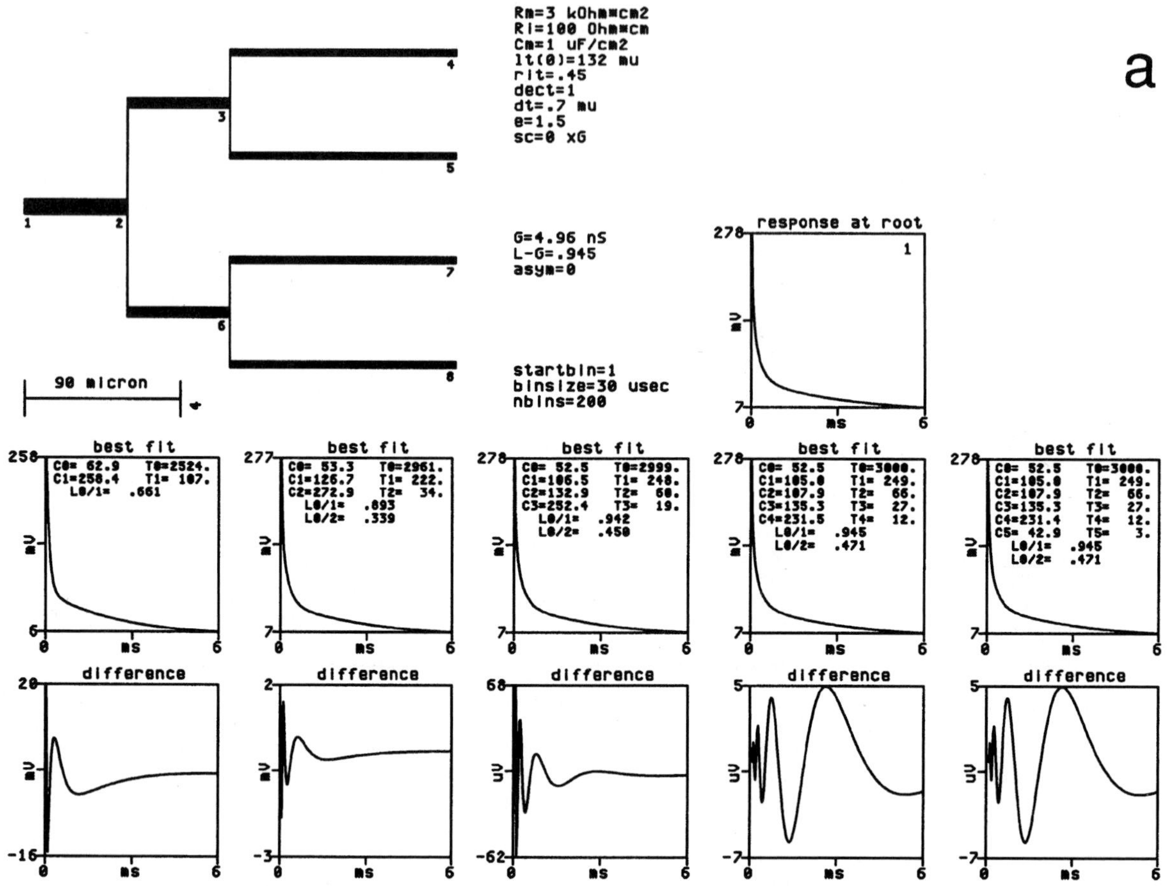

Fig. 7. Example of the voltage response at the root (upper right panel) of a symmetric (a) and an asymmetric (b) tree of degree 4 on a delta-function shaped current pulse at the root. Examples of the exponential decomposition of the voltage response are given for a number of exponential terms, ranging from 2 to 7. The difference between actual response and the best fit from the optimization

the μV range with the response curve. The first time constant, $\tau_0 = 3000$ μsec, is in perfect agreement with the membrane's $R_m C_m = 3$ msec time constant. In addition, the estimate of $L_{0/1} = 0.945$ is in perfect agreement with the estimate of $L_G = 0.945$ on basis of the input conductance. For the asymmetric tree, a stable expansion was obtained with six terms because the seventh time constant became much smaller than the first bin width. A good fit was obtained also in this case, with differences from the response curve only in the μV range. This resulted in a correct estimation of the membrane time constant, $\tau_0^6 = 2999$ μsec. However, the estimate $L_{0/1} = 1.451$ disagrees considerably with the expected value $L_G = 0.961$. If the second equalizing time constant, τ_2^6, is used for the esimation of L, a value of $L_{0/2} = 0.901$ is obtained; this agrees better, but still poorly, with L_G. These examples show that:

- the response function can accurately be described by a series of exponential terms;
- a correct estimation of the membrane time constant is obtained only when all time constants contributing to the response function are represented in the series expansion;
- for the symmetrical tree, the estimation of the electrotonic extent $L_{0/1}$, using the first equaliz-

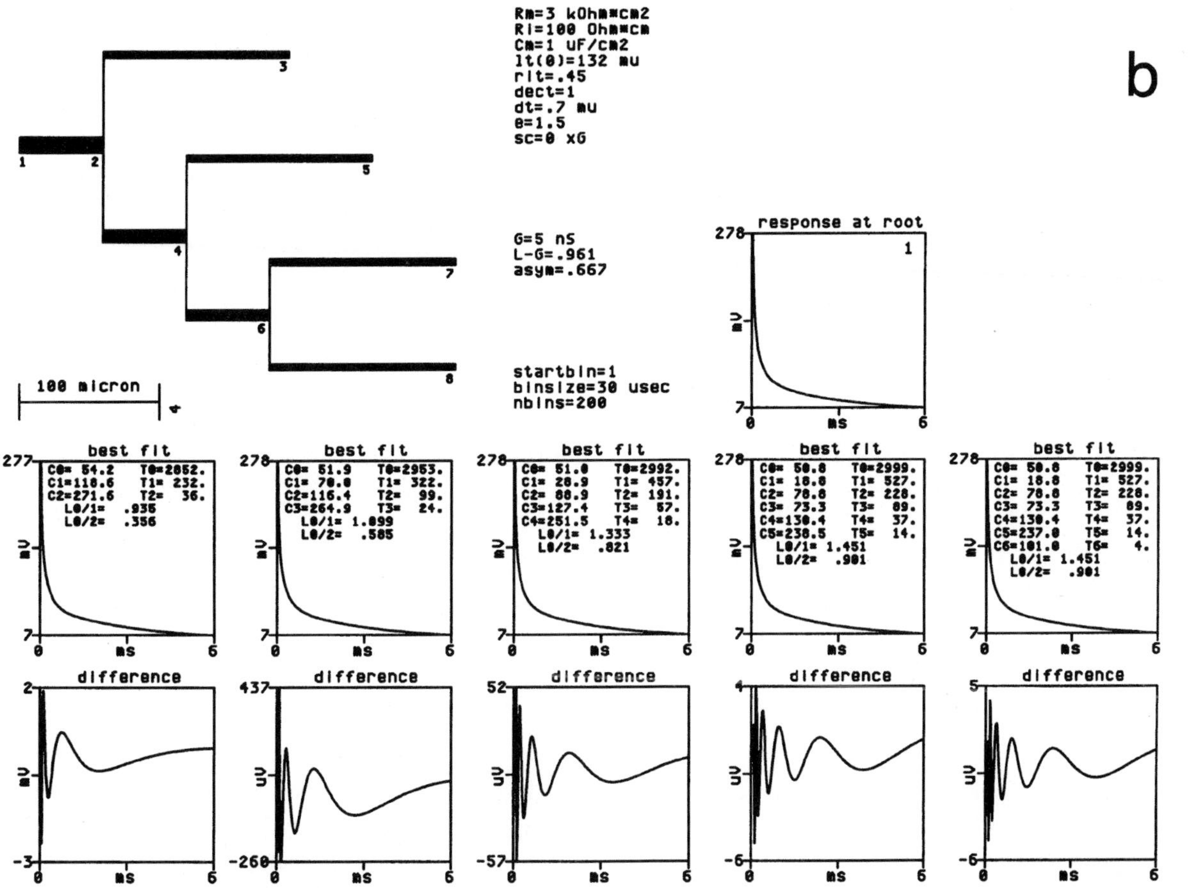

procedure is displayed in the bottom panels. Note the difference in ordinate scale for the best fit curves (mV) and the difference curves (mV or μV). The best-fit panels contain the coefficients and time constants of these terms. Estimates for the electrotonic length of the equivalent cylinder are calculated on the basis of the first ($L_{0/1}$) and the second ($L_{0/2}$) equalizing time constants.

ing time constant τ_1, agrees perfectly with the calculated L_G value;

- for the asymmetrical tree $L_{0/1}$ is not in good agreement with L_G. $L_{0/2}$ is in better (but still poor) agreement with L_G. The second equalizing time constant, τ_2^6, probably corresponds better with the root-tip equalization than does τ_1^6.

Estimation of L_{tc} and comparison with $\overline{L_{tip}}$ and L_G

Estimations of the electrotonic extent measure, L, have also been made for the dendrites displayed in Fig. 6. The time constants in the voltage response on a delta-function shaped current stimulus have been obtained from six- or seven-term exponential expansions of a total time period of 6 msec. An accurate fit was obtained in all cases, with differences between response function and best fit better than 1 to 1000. Also the membrane time constant was correctly estimated by τ_0, with an accuracy better than 2%. The results are given in Table 1, together with the outcomes for $\overline{L_{tip}}$ and L_G. Both the first and the second equalizing time constants are used for the estimation of L (see Eqn. (20)). A consistent estimation of L by the three measures $\overline{L_{tip}}$, L_G

TABLE I

Electrotonic extent of dendrites of degree 16, calculated as function of dendritic topology and of branch power e, using three different measures for dendritic electrotonic extent L_G, $\overline{L_{tip}}$, and $L_{0/1}$. The calculations are done for a symmetric tree (asym = 0), a random tree produced by the random terminal growth mode (asym = 0.43), and an asymmetric tree (asym = 0.93). Three values of the branch power e are used, e = 1, 1.5 and 2.

	asym = 0	asym = 0.43	asym = 0.93
$e = 1$	$L_G = 0.36$ $\overline{L_{tip}} = 1.05$ (0) $L_{0/1} = 1.18$ $L_{0/2} = 0.53$	$L_G = 0.41$ $\overline{L_{tip}} = 1.08$ (0.17) $L_{0/1} = 1.34$ $L_{0/2} = 1.00$	$L_G = 0.61$ $\overline{L_{tip}} = 1.28$ (0.47) $L_{0/1} = 1.79$ $L_{0/2} = 1.04$
$e = 1.5$	$L_G = 1.18$ $\overline{L_{tip}} = 1.18$ (0) $L_{0/1} = 1.18$ $L_{0/2} = 0.59$	$L_G = 1.21$ $\overline{L_{tip}} = 1.25$ (0.22) $L_{0/1} = 1.67$ $L_{0/2} = 1.09$	$L_G = 1.35$ $\overline{L_{tip}} = 1.60$ (0.63) $L_{0/1} = 2.42$ $L_{0/2} = 1.24$
$e = 2$	$L_G = \infty$ $\overline{L_{tip}} = 1.26$ (0) $L_{0/1} = 1.18$ $L_{0/2} = 0.61$	$L_G = \infty$ $\overline{L_{tip}} = 1.35$ (0.26) $L_{0/1} = 1.96$ $L_{0/2} = 1.14$	$L_G = \infty$ $\overline{L_{tip}} = 1.81$ (0.74) $L_{0/1} = 2.58$ $L_{0/2} = 1.30$

and L_{tc} is obtained only for the symmetric tree with e = 1.5. In all other cases they give different outcomes, caused by the different dependencies on the tree asymmetry, A_t, and on the branch power e. All the measures show an increase with A_t, with L_G most slowly, and $L_{0/1}$ most prominently. All measures also show an increase with branch power, but the increase of L_G is most prominent (such that, for e = 2, L_G is even undefined).

Voltage attenuations

An example of the transient response in a dendrite on an alpha-function shaped voltage stimulus is given in Fig. 8. This figure shows the voltage response at all nodes in a random (Q,S) = (0,0) dendritic tree with tree asymmetry A_t = 0.52, when a voltage stimulus is given at (a) the root, (b) a terminal tip and (c) simultaneously at all the tips. The shapes of the voltage responses depend on the point of observation, on the point of stimulation, and on the type of stimulation. Shape characteristics are the amplitude, time-to-peak and the width at half-maximum. It is clear how the amplitude decreases, the peak delays and the width increases with distance from the stimulus site. There is a large difference in the attenuation of the peak value between a centrifugal and a centripetal signal flow (Rall and Rinzel, 1973). The response at the root is very small when only one of the terminal tips is stimulated. When all the tips are stimulated simultaneously, the centripetal attenuation becomes of the same order of magnitude as the centripetal attenuation. The effect of boundary conditions on the attenuation in a single cylinder is illustrated by the potential decay over the terminal segment between nodes 30 and 32. In the centrifugal direction the signal encounters a sealed-end tip (node 32) and the attenuation is given by the formula V_{32}/V_{30} = $1/\cosh L$, with L being the electrotonic length of the terminal segment. In the centripetal direction the signal encounters the input conductance of the centripetal part of the dendrite, as seen from node 30. The attenuation is now given by $V_{30}/V_{32} = 1/(\cosh L + G_{30}^{\text{pet}}\ \lambda r \sinh L)$ which, because of the second positive term in the denominator, is considerably smaller than the centrifugal attenuation.

The different electrotonic distances of the terminal tips from the root also result in different signal attenuations for the individual tips. This is best illustrated in the histograms of Fig. 9a, showing the distribution of transfer ratios in the displayed dendrite for both a centrifugal (upper histograms) and a centripetal stimulation (lower histograms). The attenuations are given for both steady-state (left part) and transient conditions (right part). The centrifugal attenuation distribution is obtained by stimulating the root, and then calculating the attenuations at all terminal tips. The centripetal attenuation distribution is obtained by separately stimulating all terminal tips, and calculating for each case the attenuation at the root. Under transient conditions, the signal undergoes a stronger attenuation than under steady-state conditions, because the signal's current not only leaks away through the conductive membrane to the outside but also smears out in time because of the membrane capacitance (e.g. Rinzel and Rall, 1974).

In all cases the attenuations show a considerable variation. The centripetal transfer ratios are very small, and they concentrate in the first bin(s) of the distribution. The ψ value appears to be larger under transient conditions than under steady-state conditions, and also to depend sensitively on the value for the branch power e. For the three dendrites with e = 2, 1.5 and 1, respectively, the steady-state ψ values are 5.8, 8.4 and 15.4, and the transient ψ values are 8.28, 14 and 34.8. The steady-state values are well within the order of magnitude of the value of 10, estimated

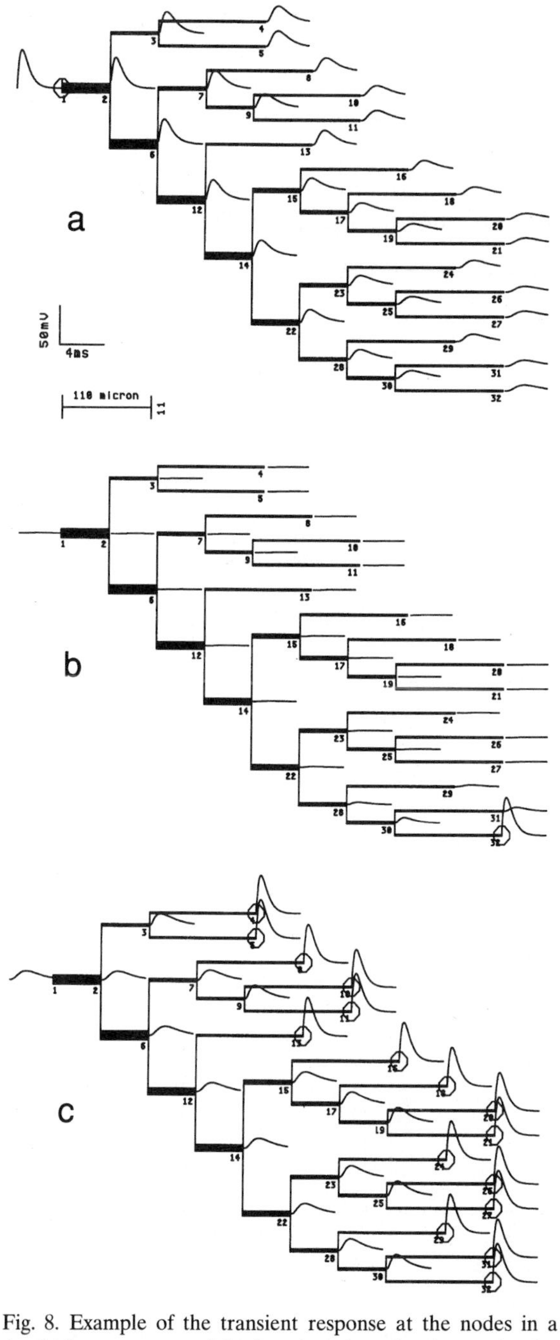

Fig. 8. Example of the transient response at the nodes in a dendritic tree on an alpha-function shaped voltage stimulus given (a) at the root, (b) at a terminal tip and (c) simultaneously at all the tips. The figures show how the shape of the voltage response depend on the point of observation, on the point of stimulation and the pattern of stimulation. Shape characteristics are the amplitude, time-to-peak and the width at half maximum. The dendritic structure is defined by the parameters: $R_{\text{m}} = 3\ \text{k}\Omega\text{cm}^2$, $R_i = 100\ \Omega\text{cm}$, $C_{\text{m}} = 1\ \mu F/\text{cm}^2$, $l_{\text{t}} = 132\ \mu\text{m}$, $r = 0.45$, $d_{\text{t}} = 0.7\ \mu\text{m}$ and $e = 1.5$ (see also text).

by Nitzan et al. (1990) for guinea pig vagal motoneurons. The simultaneous stimulation of all tips results in a centripetal attenuation (dashed line) that closely corresponds to the mean value of the centrifugal attenuation distribution. There is also good agreement with the attenuations, calculated for the equivalent cylinders with electrotonic length $\overline{L_{tip}}$ and L_G. These results apply for the dendrite, calculated with a branch power value of $e = 1.5$.

Effect of branch power

The effect of branch power is illustrated in Figs. 9b and 9c, calculated for dendrites with

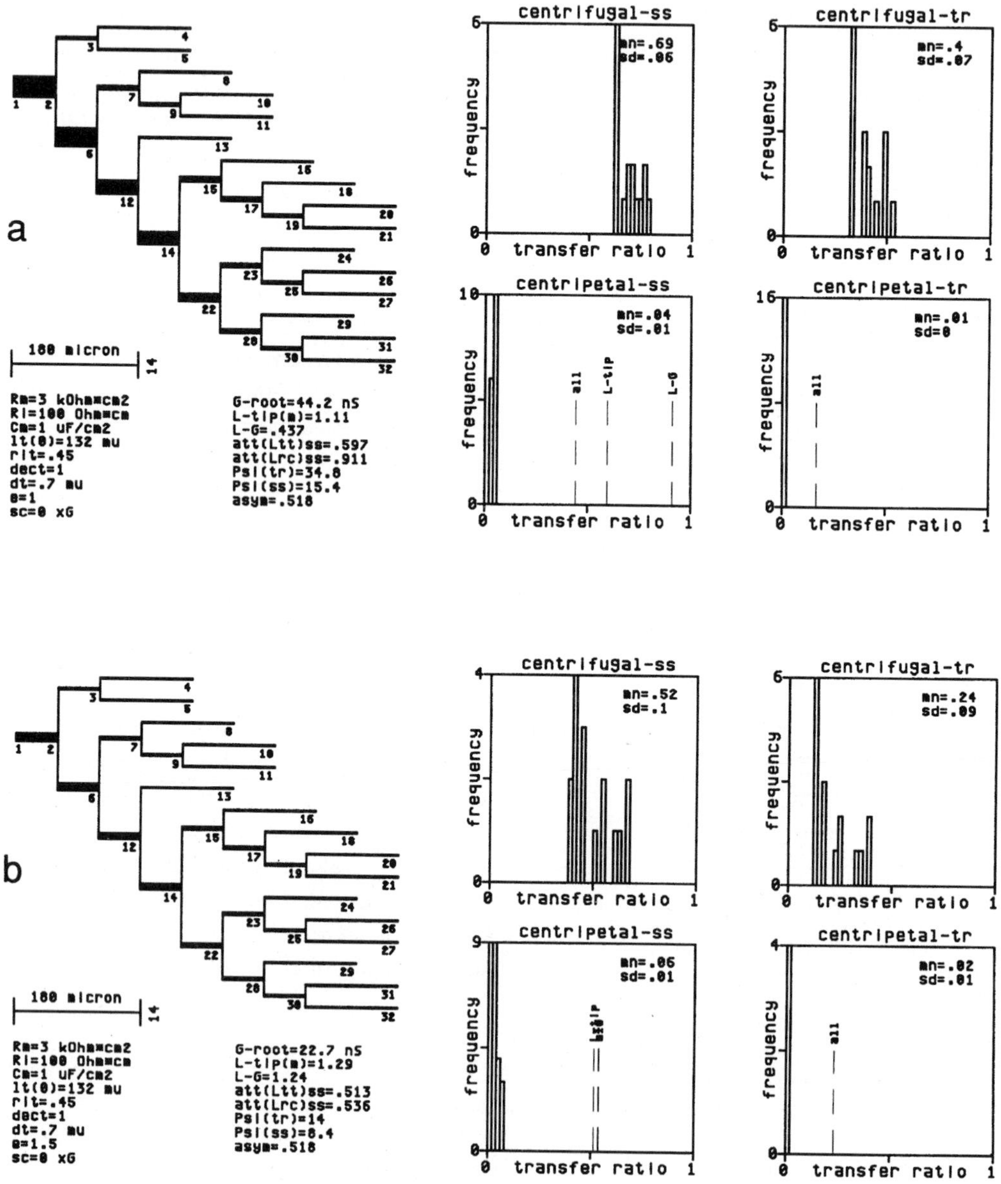

branch power valued $e = 1$ and $e = 2$. The transfer ratios in the tree with $e = 1$ appear to be larger, and in the tree with $e = 2$ smaller than in the case of $e = 1.5$. These differences arise because of the changed diameter values for the intermediate segments, resulting in a changed balance between internal and membrane resistance and changed boundary conditions at the distal ends of the intermediate segments. An important finding is that, both in trees with $e = 1$ and with $e = 2$, the correspondence between the mean centrifugal attenuation and the combined centripetal attenuation and the predictions for the equivalent cylinders $\overline{L_{tip}}$ and L_G is lost. In the tree with $e = 2$, L_G is even undefined.

Effect of topology

The effect of topology is illustrated by comparing the transfer ratios for the most symmetric, most asymmetric and a random $(Q,S) = (0,0)$ tree of a particular degree. The outcomes under transient conditions are plotted in Fig. 10 for both centrifugal (left panels) and centripetal (right panels) attenuations and for the branch-power values $e = 1$ (upper), 1.5 (middle) and 2 (lower panels). The mean transfer ratios in symmetric trees appear in all cases to be greater than in asymmetric trees with a difference, increasing (i) decreasing branch power, and (ii) increasing size of the trees. All transfer ratios decrease with increasing degree of the trees. The rate of decrease, however, becomes very small in symmetric trees when $e = 1$. The mean attenuation in random $(Q,S) = (0,0)$ trees differs only slightly from the mean attenuation in symmetric trees with a strong overlap of the standard deviations. Larger values for the branch power result in lower values for the centrifugal transfer ratio, and in slightly larger values for the centripetal transfer ratio, (and thus in a smaller ψ value (Nitzan et al., 1990) for the ratio between centripetal and centrifugal attenuation factors).

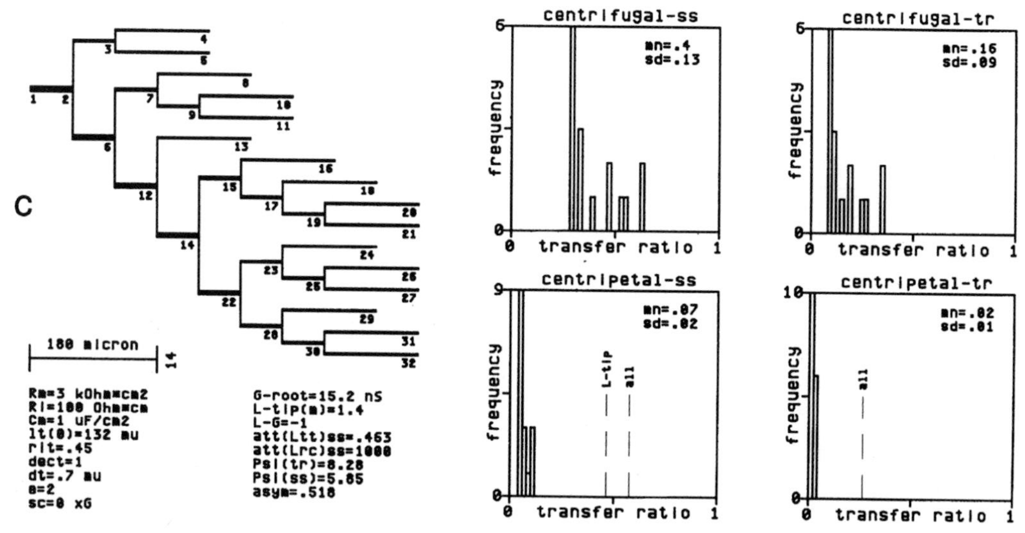

Fig. 9. Example of voltage attenuations in a random $(Q,S) = (0,0)$ dendritic tree calculated with branch power (a) $e = 1$, (b) $e = 1.5$ and (c) $e = 2$. The attenuations are calculated in both centrifugal (upper panels) and centripetal direction (lower panels) under both steady-state (left panels) and transient conditions (right panels). The frequency distribution of centrifugal transfer ratios is obtained by stimulating the root and calculating the attenuations at all terminal tips. The frequency distribution of centripetal transfer ratios is obtained by separately stimulating all terminal tips and calculating for each case the attenuation at the root. The dashed lines in the lower panels indicate the centripetal transfer ratio under simultaneous stimulation of all tips ('all'), and the transfer ratios calculated for the equivalent cylinders with lengths $\overline{L_{tip}}$ and L_G. Note that for $e = 2$ (Fig. 9c), L_G is undefined.

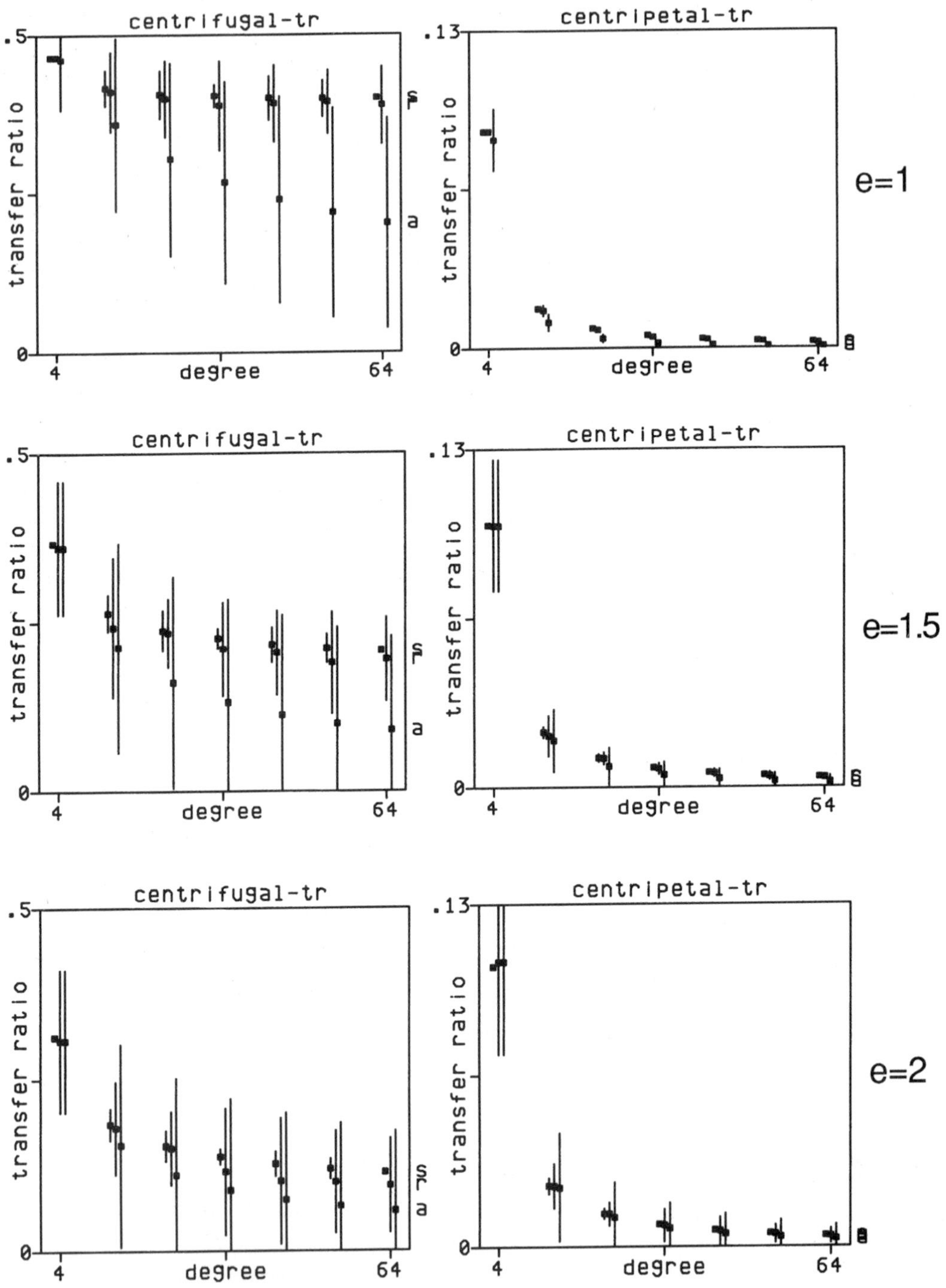

Fig. 10. Transfer ratios, calculated for symmetric ('s'), asymmetric ('a') and random $(Q,S) = (0,0)$ trees. Displayed are the mean and S.D. of transfer ratios of transient signals between root and terminal tips in centrifugal (left panels) and centripetal (right panels) direction and for branch power values $e = 1$ (upper panels), $e = 1.5$ (middle panels) and $e = 2$ (lower panels).

Discussion

In the present paper it is shown how dendritic topology interferes with electrotonic extent and voltage attenuations. The motivation for the study is based on the observation that dendritic morphological variation finds its origin to a large extent in variations in topological structure. The topological variation can be expressed by a measure for the topological asymmetry, called tree asymmetry, with a value of zero for fully symmetric trees and a value approaching unity for fully asymmetric trees. Model studies of dendritic growth have demonstrated that the topological variability is highly specific and determined by the mode of dendritic growth. That means that each tree type has a particular probability of occurrence, which depends strongly on the mode of growth. The topological variability in real dendrites appears to correspond closely with branching at terminal segments, with a small decrease in branching probability with increasing centrifugal order. Such realistic modes of growth result in tree-type distributions in which fully symmetric trees have a very low probability of occurrence.

Our studies of the impact of dendritic topology on electrotonic properties have been done using modelled dendrites with fixed choices for the metrical and electrical parametrization, but with varying topological structure. It is shown that dendritic topology is directly related to the variations in electrotonic distances from the root to the terminal tips and with the distribution of segment diameters. Consequently, the electrotonic extent of a dendrite is sensitively dependent on its topological structure. Three different measures for electrotonic extent have been used i.e. the mean terminal-tip electrotonic distance $\overline{L_{tip}}$, the electrotonic length L_G of a cylinder with the same input conductance and L_{tc}, estimated from time constants in transient voltage responses. These measures appear to give different outcomes in dendrites with realistic topological structures. Consistent results are obtained only for fully symmetric trees with a value of $3/2$ for the branch power, i.e. when Rall's equivalent cylinder conditions are met. The same findings are obtained for signal attenuations, estimated from the electrotonic measures. Additionally, these estimates do not correspond either with the mean of the actual attenuations when $e \neq 1.5$.

The correspondence between mean centrifugal attenuation and centripetal attenuation with simultaneous stimulation of all the terminal tips is lost when the above mentioned conditions are not met. The mean centripetal and centrifugal attenuations depend on the topological structure, with the largest differences being between symmetric and asymmetric trees. These differences increase with decreasing branch power and with increasing tree size. The attenuations of transient signals in random $(Q,S) = (0,0)$ trees, however, differ only slightly from those in symmetric trees of the same degree for all three values of the branch power $e = 1$, 1.5 and 2. The variance in root-tip attenuations increases with increasing asymmetry and is already substantially larger in the random trees than in the symmetrical ones. In conclusion, dendritic topology plays a crucial role (in interference with the branch power) when measures for dendritic electrotonic extent are used in relation to dendritic input conductance and signal attenuations. It also has a substantial influence on both the mean and the variance of signal attenuations between root and terminal tips. However, the mean attenuation in the random trees with realistic topologies and metrical parametrization as used in this study, appears to differ only slightly from those in symmetric trees.

Acknowledgement

This work was supported by NATO grant CRG 930426.

References

Berry, M., Hollingworth, T., Anderson, E.M. and Flinn, R.M. (1975) Application of network analysis to the study of the branching patterns of dendritic fields. *Adv. Neurol.*, 12: 217–245.

Burke, R.E., Marks, W.B. and Ulfhake, B. (1992) A parsimonious description of motoneuron dendritic morphology using computer simulation. *J. Neurosci.*, 12: 2403–2416.

Culheim, S., Fleshman, J.W., Glenn, L.L. and Burke, R.E. (1987) Membrane area and dendritic structure in type-identified triceps surae alpha-motoneurons. *J. Comp. Neurol.*, 25: 68–81.

D'Aguanno, A., Bardakijan, B.L. and Carlen, P.L. (1986) Passive neuronal membrane parameters: comparison of optimization and peeling methods. *IEEE Trans. Biomed. Eng. MBE*33, 12: 1188–1195.

DeSchutter, E. (1992) A consumer guide to neuronal modelling software. *Trends Neurosci.*, 15: 462–464.

Glenn, L.L. (1988) Overestimation of the electrotonic length of neuron dendrites and synaptic electrotonic attenuation. *Neurosci. Lett.*, 91: 112–119.

Hillman, D.E. (1979) Neuronal shape parameters and substructures as a basis of neuronal form. In: F.O. Schmitt, F.G. Worden (Eds.), *The Neurosciences, 4th Study Program*, MIT Press Cambridge, pp. 477–498.

Hollingworth, E.M. and Berry, M. (1975) Network analysis of dendritic fields of pyramidal cells in neocortex and Purkinje cells in the cerebellum of the rat. *Phil. Trans. R. Soc. London, B. Biol. Sci.*, 270: 227–264.

Holmes, W.R. (1986) A continuous cable method for determining the transient potential in passive dendritic trees of known geometry. *Biol. Cybern.*, 55: 115–124.

Holmes, W.R. (1992) Computer simulations of individual neurons. In: *Methods in Neurosciences, Vol.* 10, *Computers and Computations in the Neurosciences*, Academic Press, New York, pp. 503–525.

Holmes, W.R., Segev, I. and Rall, W. (1992) Interpretation of time constant and electrotonic length estimates in multicylinder or branched neuronal structures. *J. Neurophys.*, 68: 1401–1420.

Horsfield, K., Woldenberg, M.J. and Bowes, C.L. (1987) Sequential and synchronous growth models related to vertex analysis and branching ratios. *Bull. Math. Biol.*, 49: 413–430.

Jack, J.J.B., Noble, E. and Tsien, R.W. (1975) *Electric Current Flow in Excitable Cells*. Oxford University Press, Oxford.

Larkman, A.K. (1991) Dendritic morphology of pyramidal neurones of the visual cortex of the rat: I. Branching patterns. *J. Comp. Neurol.*, 306: 307–319.

Larkman, A.K., Major, G., Stratford, K.J. and Jack, J.J.B. (1992) Dendritic morphology of pyramidal neurones of the visual cortex of the rat. IV: Electrical geometry. *J. Comp. Neurol.*, 323: 137–152.

Nitzan, R., Segev, I. and Yarom, Y. (1990) Voltage behavior along the irregular dendritic structure of morphologically and physiologically characterized vagal motoneurons in the guinea pig. *J. Neurophys.*, 63: 333–346.

Press, W.H., Flannery, B.P., Teukolsky, S.A. and Vetterling, W.T. (1986) *Numerical Recipes: The Art of Scientific Computing*. Cambridge University Press, Cambridge, pp. 526–527.

Rall, W. (1959) Branching dendritic trees and motoneuron membrane resistivity. *Exp. Neurol.*, 1: 491–527.

Rall, W. (1969) Time constants and electrotonic length of membrane cylinders and neurons. *Biophys. J.*, 9: 1483–1508.

Rall, W. (1977) Core conductor theory and cable properties of neurons. In: R. Kandel (Ed.), *The Nervous System: Cellular Biology of Neurons, Section* 1, *Handbook of Physiology, Part* 1, *Vol.* 1, American Physiological Society, Bethesda, pp. 39–97.

Rall, W. and Rinzel, J. (1973) Branch input resistence and steady-state attenuation for input to one branch of a dendritic neuron model. *Biophys. J.*, 13: 648–688.

Rall, W., Burke, R.E., Holmes, W.R. Jack, J.J.B., Redman, S.J. and Segev, I. (1992) Matching dendritic neuron models to experimental data. *Physiol. Rev.*, 72: 159–186.

Rinzel, J. and Rall, W. (1974) Transient response in a dendritic neuron model for current injected at one branch. *Biophys. J.*, 14: 759–790.

Rose, P.K. and Dagum, A. (1988) Nonequivalent cylinder models of neurons: interpretation of voltage transients generated by somatic current injection. *J. Neurophys.*, 60: 125–148.

Schierwagen, A. (1986) Segmental cable modelling of electrotonic transfer properties of deep superior colliculus neurons in the cat. *J. Hirnforsch*, 27: 679–690.

Schierwagen, A.K. (1994) Exploring the computational capabilities of single neurons by continuous cable modelling. In: J. van Pelt, M.A. Corner, H.B.M. Uylings and F.H. Lopes da Silva (Eds.), *Progress in Brain Research, Vol.* 102, *The Self-Organizing Brain—from Growth Cones to Functional Networks*. Elsevier Science Publishers, Amsterdam, this volume.

Schierwagen, A. and Grantyn, R. (1986) Quantitative morphological analysis of deep superior colliculus neurons stained intracellularly with HRP in the cat. *J. Hirnforsch.*, 27: 611–623.

Segev, I. (1992) Single neurone models: oversimple, complex and reduced. *Trends Neurosci.*, 15: 414–421.

Uylings, H.B.M., Ruiz-Marcos, A. and van Pelt, J. (1986) The metric analysis of three-dimensional dendritic tree patterns: a methodological review. *J. Neurosci. Methods*, 18: 127–151.

Uylings. H.B.M., van Pelt, J., Verwer, R.W.H. and McConnell, P.M. (1989) Statistical analysis of neuronal populations. In: J.J. Capowski (Ed.), *Computer Techniques in Neuroanatomy*, Plenum, New York, pp. 241–263.

Van Pelt, J. (1992) A simple vector implementation of the Laplace-transformed cable equations in passive dendritic trees. *Biol. Cybern.*, 68: 15–21.

Van Pelt, J. and Verwer, R.W.H. (1983) The exact probabilities of branching patterns under segmental and terminal growth hypotheses. *Bull. Math. Biol.*, 45: 269–285.

Van Pelt J. and Verwer R.W.H. (1986) Topological properties of binary trees grown with order-dependent branching probabilities. *Bull Math. Biol.*, 48: 197–211.

Van Pelt, J., Uylings, H.B.M. and Verwer, R.W.H. (1989) Distributional properties of measures of tree topology. *Acta Stereol.*, 8: 465–470.

Van Pelt, J., Uylings, H.B.M., Verwer, R.W.H., Pentney, R.J. and Woldenberg, M.J. (1992) Tree asymmetry—a sensitive and practical measure for binary topological trees. *Bull. Math. Biol.*, 54: 759–784.

Van Veen, M.P. and van Pelt, J. (1993) Terminal and intermediate segment lengths in neuronal trees with finite length. *Bull Math. Biol.*, 55: 277–294.

Verwer, R.W.H. and van Pelt. (1986) Descriptive and comparative analysis of geometrical properties of neuronal tree structures. *Neurosci. Neth.* 10: 179–206.

Verwer, R.W.H., van Pelt and H.B.M. Uylings (1992). An introduction to topological analysis of neurones. In: M.G. Stewart (Ed.), Quantitative Methods in Neuroanatomy, Wiley, Chicester, pp. 295–323.

Wolf, E., Dirinyi, A. and Székely, G. (1992) Simulation of the effect of synapses: the significance of the dendrite diameter in impulse propagation. *Eur J. Neurosci.*, 4: 1013–1021.

Woldenberg, M.J., O'Neill, M.P., Quackenbush, L.J. and Pentney, R.J. (1993) Models for growth, decline and regrowth of the dendrites of rat Purkinje cells induced from magnitude and link-length analysis. *J. Theor. Biol.*, 162: 403–429.

J. van Pelt, M.A. Corner H.B.M. Uylings and F.H. Lopes da Silva (Eds.)
Progress in Brain Research, Vol 102

CHAPTER 9

Exploring the computational capabilities of single neurons by continuous cable modelling

A.K. Schierwagen

Universität Leipzig, Institut für Informatik, FG Neuroinformatik, Augustusplatz 10 / 11, D-04109 Leipzig, Germany

Introduction

Current models of artificial neural networks use large numbers of very simple, homogeneous but highly interconnected processing units, 'artificial neurons'. In comparison with most known neurons these units are poor caricatures, and the need for more realistic neuron models has been stressed (Sejnowski et al., 1988; Schierwagen, 1989a). Such models must consider that very many factors controlling neuronal input-output relationship can be listed, among them electrical membrane and cell properties, dendritic geometry and patterns of input connectivity (Graubard and Calvin, 1979; Burke, 1987).

One of the most striking features of vertebrate neurons is the immense diversity of their dendritic branching patterns being indicative of the variety of operations possibly subserved by dendritic membranes. The role of dendritic geometry in single neuron computation has been difficult to assess (see Koch and Segev, 1989; Durbin et al., 1989; McKenna et al., 1992). Recent advances in single cell staining techniques allow the derivation of correlated electro-physiological and morphological data from the same cell, which are then used for model construction. In general, these models are based on the cable equations, i.e. linear partial differential equations, and aim at reconstructing electrical phenomena recorded in the cell body, often far from their dendritic sites of origin.

In the following, the basic principles underlying electrotonic modelling will be outlined. Then the main types of neuronal cable models will be reviewed, and their use as tools in evaluating various aspects of neuronal computing capabilities demonstrated. We then discuss some basic mathematical problems arising in neuronal electrotonic modelling and how these might best be approached. Special emphasis is laid on exactly solvable, reduced models that retain important features of dendritic branching patterns. Such models of intermediate anatomical complexity are necessary if we are ever to succeed understanding information processing in neuronal networks.

Basic concepts in neuronal modelling

Neuron form and regional specification of function

One of the main forms of communication in the nervous system is by means of electrical signals. Throughout this paper, the term 'neuronal function' will refer to this aspect of neuronal activity, i.e. the ability of neurons to receive, transmit and generate electrical signals. The signalling function of a nerve cell depends on both its structure, i.e. geometrical arrangement, and its membrane properties. The main anatomical features of neurons are the dendrites and the axon.

Generally, the former are regarded to provide receptive surfaces for input signals to the neuron, which are mainly mediated by synapses distributed primarily over the widely branched dendritic trees. These graded signals are conducted with decrement to the soma and the axon hillock, where they usually are converted into sequences of nerve impulses (spikes) which are propagated without attenuation along the axon to target cells, i.e. other neurons, muscle cells, etc.

We now know of many neurons where this classical identification of the processing steps within a neuron must be supplemented with additional processes, such as dendritic spikes, intermittent conduction or spikeless transmission (Graubard and Calvin, 1979; Shepherd, 1992). Even if, therefore, neurons may deviate in various ways from the above concepts which collectively comprise an idealized 'standard neuron', the basic principles appear to be common to almost all nerve cells, and probably provide the basis of their operation.

The idealized neuron exhibits regionally different electrical characteristics. The soma and the dendrites have fixed ionic permeabilities, so that a change of polarization produced somewhere on the dendritic tree will spread and decay as it is conducted 'electrotonically', just as in a passive leaky cable. In the axon hillock, on the other hand, ionic permeabilities depend upon the membrane potential, and the integration of the electrotonic potentials may result in the initiation of spike trains.

These two kinds of membranes are referred to as passive and active, respectively. It has been shown, that in the former the electrotonic spread of current can be studied with the aid of linear cable theory (Rall, 1977).

Linear cable theory

Figure 1 shows a diagram of a passive cable and the way it can be represented as an equivalent electrical circuit. The major assumptions underlying this approximation are the following (Scott, 1975; Jack et al., 1975; Rall, 1977; Tuckwell, 1988): (a) the spatial dependence of membrane voltage V can be reduced to one dimension, i.e. the distance along the cable; radial components are negligible; (b) the intracellular medium provides a purely ohmic resistance R to electric current flow; (c) the extracellular space has negligible resistance.

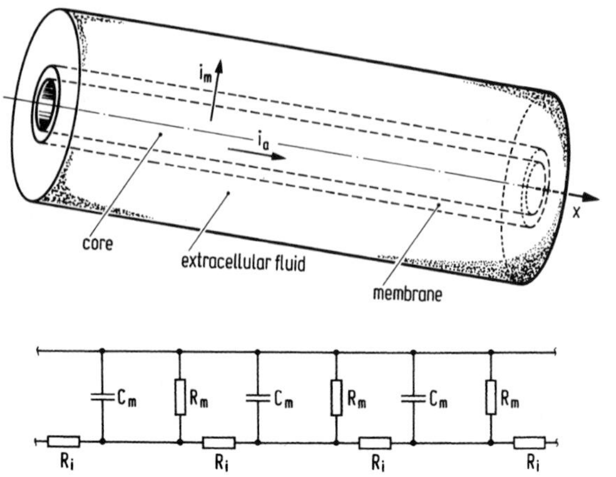

Fig. 1. Scheme of a membrane cylinder (above) and its equivalent electrical circuit (below).

Given this electrical circuit, the cable equations for membrane voltage $V = V(x,t)$ and axial current $i_a = i_a(x,t)$ in a cylindrical segment of length l and diameter d are as follows (x is distance, $0 \le x \le l$, and t is time)

$$\frac{\partial V}{\partial x} = -R\, i_a, \quad \frac{\partial i_a}{\partial x} = -i_m \tag{1}$$

with

$$i_m = C\,\frac{\partial V}{\partial x} + G \cdot V \tag{2}$$

where $i_m = i_m(x,t)$ denotes membrane current consisting of a capacitive and a resistive component. $R = 4R_i/(\pi d^2)$ is the axial resistance, $C = C_m \pi d$ is the membrane capacitance and $G = G_m \pi d$ is the membrane conductance (all quantities per unit length). Combining these equations we obtain

$$C\frac{\partial V}{\partial t} - \frac{1}{R}\,\frac{\partial^2 V}{\partial x^2} + G \cdot V = 0 \tag{3}$$

or, equivalently

$$\tau_m \frac{\partial V}{\partial t} - \lambda^2 \frac{\partial^2 V}{\partial x^2} + V = 0 \qquad (4)$$

where $\lambda = (GR)^{-0.5}$ is the length constant, while $\tau_m = C/G$ denotes the membrane time constant (R_τ, C_m and G_m are the specific parameters axial resistivity, membrane capacitance and membrane conductance). The constants τ_m and λ have been used to express time and distance, respectively, in terms of dimensionless variables $T = t/\tau_m$ and $X = x/\lambda$.

After Laplace transformation, the cable equation (1) reads

$$\frac{d^2\tilde{V}(x,\omega)}{dx^2} = \gamma^2(\omega) \cdot \tilde{V}(x,\omega), \qquad (5)$$

where Laplace transformed variables are denoted by a tilde.

In the frequency domain, the membrane conductance G is to be replaced by the corresponding admittance $\tilde{Y}(\omega)$,

$$\tilde{Y}(\omega) = G \cdot (1 + j\omega)^{-1}. \qquad (6)$$

The propagation constant $\gamma(\omega)$ in Eqn. (5) is given by $\gamma^2(\omega) = R \cdot \tilde{Y}(\omega)$, which leads in the DC case ($\omega = 0$) to $\gamma^2 = RG = 1/\lambda^2$.

The solution to Eqn. (5) in terms of hyperbolic functions is

$$\tilde{V}(x,\omega) = A(\omega)\cosh(\gamma(\omega)x) + B(\omega)\sinh(\gamma(\omega)x), \qquad (7)$$

where $A(\omega)$ and $B(\omega)$ depend on the specific initial and boundary conditions of the problem (Rall, 1977). For a cable of finite length, extending from $x = 0$ to $x = l$, two boundary conditions must be specified. We assume at $x = 0$ a current source with zero internal admittance, but at $x = 1$ no electrical sources. Then, a particularly useful form of solution to Eqn. (4) is (see Rall, 1977, Eqn. (3.23)):

$$\tilde{V}(x,\omega)/\tilde{V}(l,\omega) = \cosh(\gamma(\omega)(l-x)) + \tilde{Y}_t^*(\omega)\sinh(\gamma(\omega)(l-x)). \qquad (8)$$

The values of $\tilde{V}(l,\omega)$ and $\tilde{Y}_t^*(\omega)$ still need to be determined from the pair of boundary conditions. Let $\tilde{V}(0,\omega) = \tilde{V}_0(\omega)$ be one boundary condition, and $\tilde{V}(l,\omega) = \tilde{V}_l(\omega)$ the resulting voltage at $x = l$. It follows from Eqn. (8) that

$$\tilde{V}_l(\omega)/\tilde{V}_0(\omega) = \left[\cosh(\gamma(\omega)l) + \tilde{Y}_t^*(\omega)\sinh(\gamma(\omega)l)\right]^{-1}. \qquad (9)$$

This quotient defines the attenuation factor for a steady AC voltage from $x = 0$ to $x = l$. Equation (9) may be used to eliminate $\tilde{V}(l,\omega)$ from Eqn. (8):

$$\begin{aligned}\tilde{V}(x,\omega)/\tilde{V}_0(\omega) &= \left[\cosh(\gamma(\omega)(l-x)) + \tilde{Y}_t^*(\omega)\sinh(\gamma(\omega)(l-x))\right] \\ &\div \left[\cosh(\gamma(\omega)l) + \tilde{Y}_t^*(\omega) \times \sinh(\gamma(\omega)l)\right] \qquad (10)\end{aligned}$$

In this solution $\tilde{Y}_t^*(\omega)$ provides for different boundary conditions at $x = l$. It can be interpreted as the terminal admittance of the cable at $x = l$, normalized by the characteristic admittance $\tilde{Y}_c(\omega) = \gamma(\omega)/R$, i.e.

$$\tilde{Y}_t^*(\omega) = \tilde{Y}_t(\omega)/\tilde{Y}_c(\omega). \qquad (11)$$

$\tilde{Y}_t(\omega)$ is a formal admittance relating axial current $\tilde{i}_a(x,\omega)$ and voltage $\tilde{V}(x,\omega)$ at $x = l$,

$$\tilde{i}_a(l,\omega) = \tilde{Y}_t(\omega)\,\tilde{V}(l,\omega). \qquad (12)$$

Two extreme cases are often considered: the cable termination is assumed to be either of the 'sealed end' (i.e. $\tilde{Y}_t(\omega) = 0$) or of the 'open end' type (i.e. $\tilde{Y}_t(\omega) = Y_c(\omega)$), implying $\tilde{Y}_t^*(\omega) = 0$ or 1. From Eqns. (8) and (12), the input admittance $\tilde{Y}_{in}(x,\omega) = \tilde{i}_a(x,\omega)/\tilde{V}(x,\omega)$ can be obtained. In particular, the estimation of model parameters (see below) requires the determination of the input admittance $\tilde{Y}_{in}(0,\omega)$. We find

$$\begin{aligned}\tilde{Y}_{in}(0,\omega) &= \tilde{Y}_c(\omega)\left[\sinh(\gamma(\omega)l) + Y_t^*(\omega)\cosh(\gamma(\omega)l)\right] \\ &\div \left[\cosh(\gamma(\omega)l) + Y_t^*(\omega)\sinh(\gamma(\omega)l)\right] \qquad (13)\end{aligned}$$

Parameter identifiability in cable models

The cable Eqn. (3) and its solutions Eqns. (7)–(10) refer to the assumed simple geometry of a cylindric process. In the case of branching structures such as dendrites, the situation becomes much more complex. One has to include boundary conditions at each branch point and terminal tip, resulting in complicated expressions for the corresponding solutions. To cope with these problems, models of different degrees of complexity have been developed. Among them are compartmental models, which have become very popular because of the widely accessible computer power (see e.g. De Schutter (1992) for a recent review of available software packages based on the compartmental approach). However, since all these models are finite-difference approximations of the continuous cable models, only the latter will be considered here.

It is worthwhile remembering that any modelling attempt serves two complementary purposes (see Schierwagen, 1988, 1990 for full discussion): (1) given the model structure known and all parameter values specified, the behavior of the model is calculated (the forward problem); (2) some of the parameters and/or system structures are unknown, and are to be determined by working 'backwards' from the observed behavior of the system being modelled (inverse problem).

The inverse problem consists of several stages, among them model specification, identifiability and parameter estimation. The identifiability problem deals with the practical question of whether or not it is possible to determine one or more distinct sets of solutions for the unknown parameters of a model, with specified structure, from an input-output experiment.

The application of intracellular recording and staining techniques makes it possible to obtain data on electrophysiological and anatomical properties of single neurons. Due to the difficulties of impaling a neuron in vivo with two separate microelectrodes — one for stimulus application, the other for response recording — in general only a single electrode, penetrating the soma, has been employed. Dynamic response properties of single neurons are traditionally investigated by applying step and impulse stimuli as well as sinusoidal stimuli. These response properties, together with anatomical information, provide for the model specification and parameter estimation.

The corresponding inverse and forward problems for dendritic cable models have systematically been considered in Schierwagen (1988, 1990); other authors have adopted this approach so far (Holmes and Rall, 1992; White et al., 1992). These studies have shown that the problem of identifying model parameters for any cable model is ill-posed, in that a complete and unique determination of these parameters is not possible for the typical experimental situation. In particular, applying results on parameter identifiability in linear, one-dimensional, parabolic partial differential equations, it was shown that for the usual case of one-point recordings at the soma of the nerve cell, the parameters of the cable models can be estimated only under the assumption of spatial constancy (Schierwagen, 1990). For example, to identify two constant parameters it is necessary that at least two modes of neuron behavior (e.g. steady-state and transient responses) can be observed. Accordingly, in many cable models the passive electrical parameters — R_i, C_m, G_m in Eqns. (1) and (2) — are assumed to be constant over the neuron. None of these parameters, however, is directly accessible in the course of most electrophysiological experiments. Instead, if only steady-state recordings are possible, values for two of them (usually C_m and R_i) must be assumed from the literature. The remaining parameter G_m can then be calculated from a unique relationship between G_m, on the one hand, and the experimentally measured value of the input resistance R_N, together with anatomical information about neuron dimensions, on the other hand (see next section).

Model classes for dendritic neurons

The large body of results derived in cable theory and briefly reviewed in the preceding sections has been used to model the electrotonic structure of a neuron with branched dendrites. Since Rall's (1959) seminal study, two classes of passive dendritic cable models have been developed: equivalent-cylinder (EC) models and branching-cable (BC) models (Fig. 2).

Equivalent cylinder models

The EC models come in two main variants. In Rall's (1962) original model of the nerve cell, R_i, G_m and C_m are assumed to be uniform all over the neuron, and the soma to be an isopotential sphere. Each dendritic tree is then treated as an equivalent cylinder. Provided certain symmetry conditions are fulfilled, a further simplification can be made: the cylinders can be treated as a single equivalent cylinder.

The symmetry requirements for the equivalent-cylinder transformation are: (i) all dendritic terminals are at the same electrotonic distance from the soma; (ii) the boundary conditions at all terminals are the same; (iii) at branch points the 3/2 power relation holds, i.e. $d_0^{3/2} = d_1^{3/2} + d_2^{3/2}$ where d_0 and d_1, d_2 are the diameters of the parent and the daughter branches, respectively; (iv) the input current densities at all points on the dendritic tree, which are electrotonically equidistant from the soma, are the same.

Obviously, the reduction of an originally complicated dendritic tree into a single cable segment (Fig. 2D) represents a tremendous simplification. The identification problem for constant membrane and cell parameters (see section *Parameter identifiability in cable models*) can then be solved as follows. The voltage response of a neuron (as a passive system) to a current pulse applied at the soma can be represented as

$$V(t) = V_{ss} + \sum_{n=0}^{\infty} C_n e^{-t/\tau_n} \qquad (14)$$

where V_{ss} is the final steady state, τ_0 the membrane time constant and τ_n $(n \geq 1)$ are shorter equalizing time constants describing the redistribution of charge in the dendrites. Several methods are available for the extraction of time constants and coefficients from experimental transients, including graphical 'peeling', non-linear regression analysis and optimization methods (see Rall et al. (1992) for discussion). The peeling method is illustrated in Fig. 3 which is based on recordings from cat superior colliculus neurons (Grantyn et al., 1983).

In the reference case of a single cylinder with both ends sealed, the relationship

$$L = n\pi/\sqrt{\tau_0/\tau_n - 1} \qquad (15)$$

can be used to estimate the dimensionless, electrotonic length of the cylinder from ratios of time constants such as τ_0/τ_1 (see Van Pelt and Schierwagen, 1994). Based on these L-estimates, the specific electrical parameters R_m, C_m and R_i are calculated (for details see Grantyn et al., 1983, Schierwagen 1986). The procedure presumes knowledge of the input resistance R_N of the whole neuron, i.e. the DC value of the input impedance, $R_N = 1/\tilde{Y}_N(0)$, and of the anatomical parameters A_S (soma surface area) and D (diameter of the equivalent cylinder). Unfortunately, there is increasing evidence that some of the symmetry requirements necessary for the EC transformation are violated in many real dendrites (see Schierwagen, 1989b,1990; Rall et al., 1992 for discussion). For example, though some nerve cells, e.g. spinal motoneurons (Rall, 1977) and superior colliculus output neurons (Schierwagen and Grantyn, 1986) have been found to obey assumption (iii), many others do not (Hillman, 1979). The assumption (i) that all electrotonic path lengths in a dendritic tree are equal has been questioned recently (Schierwagen, 1986; Glenn, 1988; Fleshman et al., 1988). Condition (ii) is expected to mirror the real situation well, and if one is interested in the case of current injection at the soma or of equal synaptic inputs on regions of the dendritic tree which are at the same electrotonic distance from the soma, condition (iv) will be satisfied. In their recent review, Rall and co-workers have given some ideas on how to deal with these complications while preserving the basic concept (Rall et al., 1992; see also Van Pelt and Schierwagen, 1994).

Branching cable models

The morphology-based BC models (Rall, 1959; Barrett and Crill, 1974; Koch et al., 1982; Turner, 1984; Schierwagen, 1986) assume no constraints on the dendritic branching structure, i.e. only condition (ii) is supposed to hold. Representing the soma as an isopotential lumped element, the dendrites are modelled by a number of individual cable segments, the dimensions of which are based on anatomical reconstructions of the nerve cell. In this way, each dendritic tree is represented as a network of cylindrical cable segments (Fig. 2B),

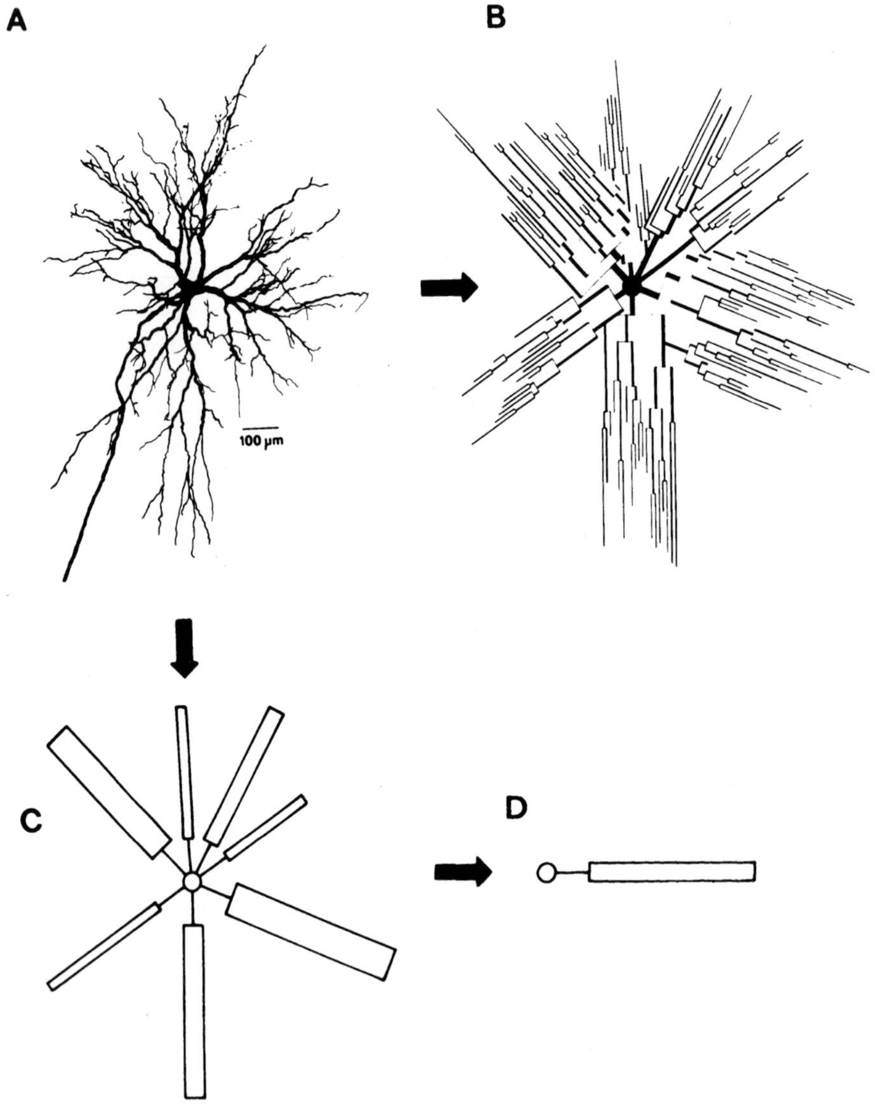

Fig. 2. Approximation levels for continuous cable models. The sequence of transformations A → C → D illustrates the steps which are made in the case of idealized branching, whereas A → B represents the derivation of a detailed segmental cable model as used in the general case. The neuron displayed in A belongs to a class of superior colliculus neurons (for full details, see text).

and either steady-state or transient calculations can be performed with specially developed computer programs employing the analytical solutions of the cable equations (1)–(3) (e.g. Van Pelt, 1992).

The parameter identification problem for the BC model can be solved by applying Eqn. (13) sequentially to each segment. At each branch point, the input admittances of the daughter segments are summed to give the terminating admittance of the parent segment; this yields, after normalization, the $\tilde{Y}_t^*$ of the parent segment for use of Eqn. (13) in the next iteration. Starting with the most distal segments of a tree for which the $\tilde{Y}_t^*$ may be set to zero or unity as discussed in the section *Linear cable theory*, the $\tilde{Y}_t^*$ calculations proceed towards the soma, until the input admittances $\tilde{Y}_{Dj}$ of the k dendritic stems are determined. These are summed, together with the soma admittance $\tilde{Y}_S$, to give the total neuron admittance,

$$\tilde{Y}_N = \tilde{Y}_N(\omega) = \tilde{Y}_S + \sum_{j=1}^{k} Y_{Dj} \qquad (16)$$

Exemplified by two neurons of different type from the superior colliculus (SC) (Grantyn et al., 1983; Schierwagen, 1986; Schierwagen and Grantyn, 1986), the method of parameter estimation is illustrated in Fig. 4: comparing the experimental with the calculated value of the total input resistance $R_N = 1/\tilde{Y}_N(0)$ of the neuron, a region of parameter combinations (R_m, R_i, $\tilde{Y}_t^*$) can be derived which is consistent with the experimental data. In these calculations values of R_i between 50 and 150 Ωcm centering around 100 Ωcm were used. This region is known from special measurements as the physiological range of variation in most neuron types (Rall, 1977; Rall et al., 1992). The assumption of perfectly insulated (sealed ends, $\tilde{Y}_t^* = 0$) and totally uninsulated (open ends, $\tilde{Y}_t^* = 1$) terminal segment corresponds to the extreme hypotheses bracketing the real situation, as discussed above. Recent experimental evidence, however, suggests $\tilde{Y}_t^*$ to be almost zero (see Rall et al., 1992).

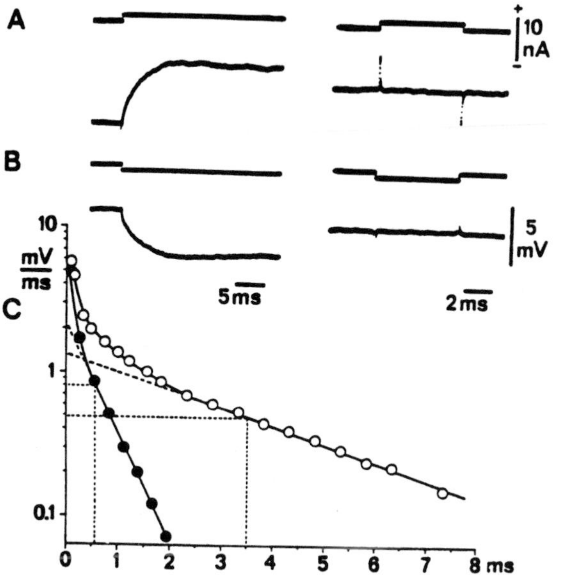

Fig. 3. Extraction of time constants by the 'peeling' method from transients recorded in SC neurons (see text). (A, B) Left column: specimen records of electrotonic potentials (lower traces) produced by depolarizing (A) and hyperpolarizing (B) current steps. Right column: extracellular control of bridge balance and of rectifying properties of the microelectrode. (C) Semilogarithmic plot of dV/dt against time t (○) for depolarizing current as shown in A. The curve was fitted by eye. Interrupted line, linear extrapolation to $t = 0$ of the late portion of the curve. Dotted line, determination of τ_0 (3.5 msec) at $dV/dt = 1/e \cdot dV(0)/dt$. ●, semilogarithmic plot of the deviation of the experimental curve from exponential course at $t < 2$ msec. The resulting exponential function decays with time constant $\tau_1 = 0.58$ msec. Reprinted from Grantyn et al. (1983).

The parameters thus estimated differ considerably in many cases among the various neuron classes, as illustrated in Fig. 4. Distinct differences are revealed when they are used in calculations of the neuronal behavior ('forward problem'). For example, iterative application of Eqn. (9) gives rise to the results of Fig. 5. Displayed are the variant courses of voltage and current attenuation for single synaptic inputs to the neurons used in the calculations of Fig. 4.

To indicate the degree of the electrotonic complexity of a dendritic tree, an asymmetry factor,

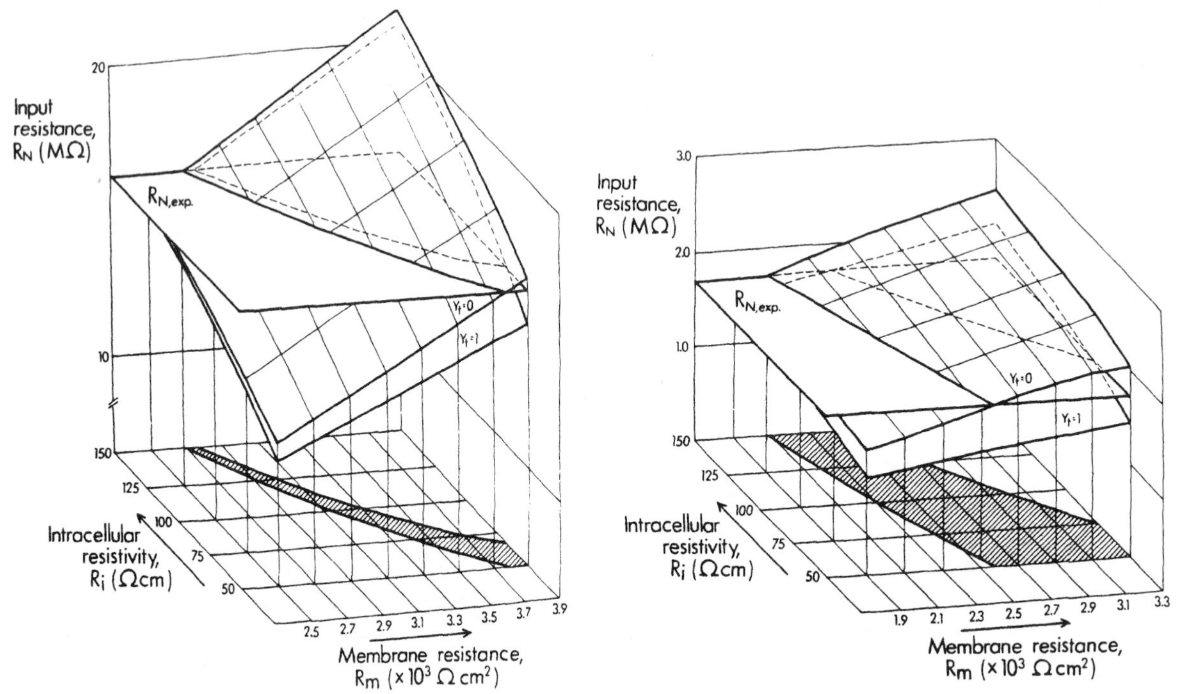

Fig. 4. Parameter estimation in branching cable models. Depicted are 3-D plots of calculated R_N vs. assumed R_m and R_i for sealed ($Y_t^* = 0$) and open ($Y_t^* = 1$) dendritic endings. The intersections of the two corresponding R_N surfaces with the plane $R_N = R_{N,\mathrm{exp}}$ denote the parameter region (hatched) which is consistent with the experimental R_N. The plots belong to the neurons shown in Fig. 5.

Φ, can be defined

$$\Phi = \frac{AF_{\mathrm{T}\rightarrow\mathrm{S}}}{AF_{\mathrm{S}\rightarrow\mathrm{T}}} \qquad (17)$$

$\mathrm{AF}_{\mathrm{T}\rightarrow\mathrm{S}}$ and $\mathrm{AF}_{\mathrm{S}\rightarrow\mathrm{T}}$ are the average voltage attenuation in the centripetal (from dendritic terminals to the soma) and the centrifugal (from soma to the dendritic terminals) direction. (Note that the definition (17) of Φ yields the reciprocal of the factor Ψ defined in Nitzan et al. (1990)). $\Phi = 1$ in the case of a single cylinder with both ends sealed but, in general, Φ will be either greater or less than unity, depending on whether centrifugal surpasses centripetal attenuation, or vice versa. Below, this relationship is worked out analytically for non-uniform cable models.

In a wide-field neuron of the rat SC, we estimated Φ to be 0.35, and in a narrow-field SC neuron to be 0.21, using $R_i = 100\ \Omega\mathrm{cm}$; R_m values were 10.25 $\mathrm{k}\Omega\mathrm{cm}^2$ and 10 $\mathrm{k}\Omega\mathrm{cm}^2$, calculated from input resistances R_N of 80 $\mathrm{M}\Omega$ and 70 $\mathrm{M}\Omega$, respectively (Schierwagen et al., 1993). Thus, the differences in electrotonic asymmetry as measured by Φ suggest a close relationship between dendritic shape and neuronal function.

Reduced but realistic: equivalent cable models

With respect to dendritic morphology, the model classes described above can be considered as two extreme cases within the spectrum of models ranging from most simplified (EC model) to almost realistic (BC model). However, due to the violation in many real neurons of some of the assumptions of the EC model, on the one hand, and the often considerable amount of computer power which is necessary when complex BC models are used, on the other hand, the need for reduced models which are nevertheless suffi-

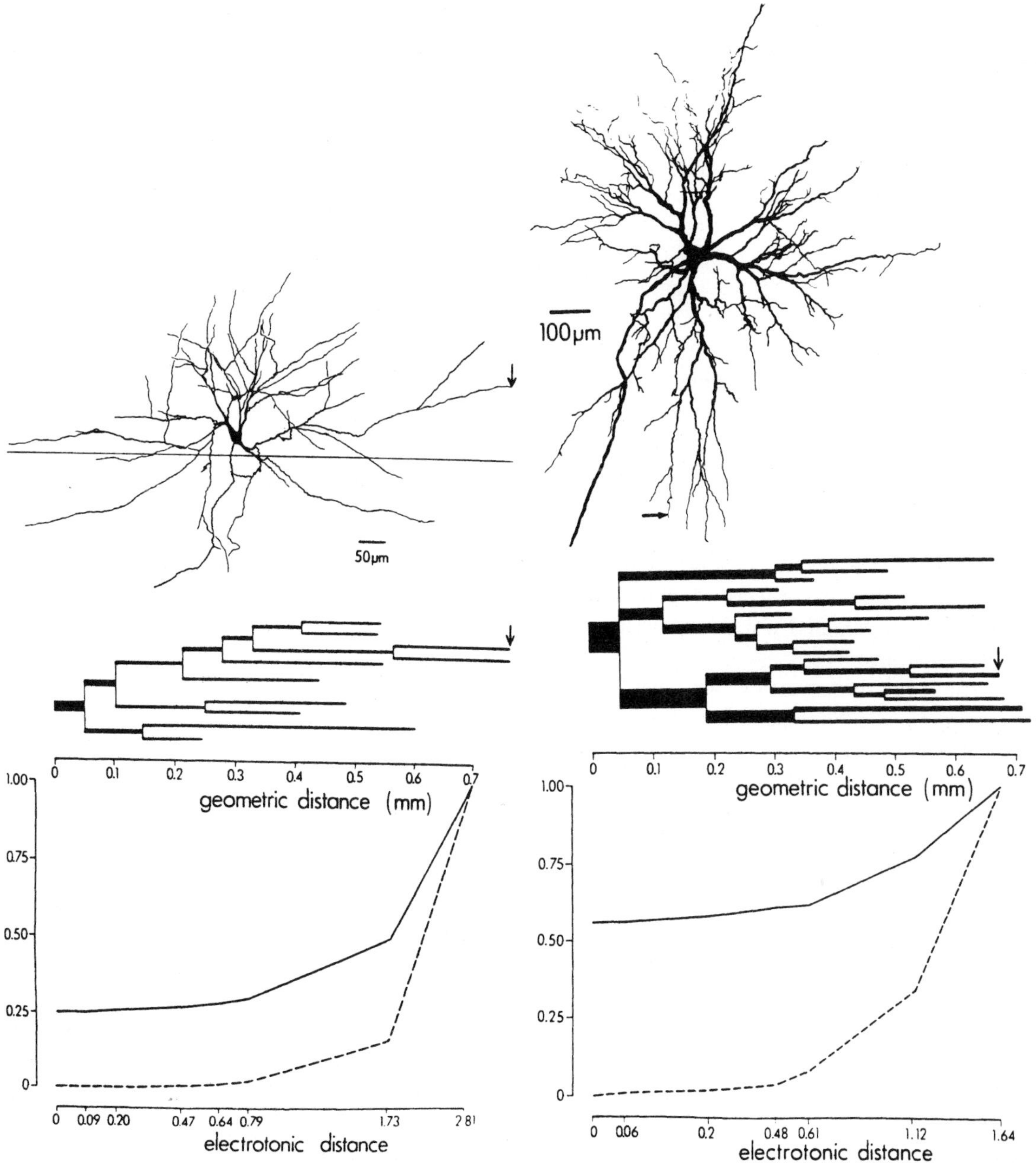

Fig. 5. Dendritic ramification patterns and transfer properties of two types of SC neurons. For each neuron are displayed the plane reconstruction (top), the Sholl diagram of one dendrite (middle) and the attenuation curves of current (continuous line) and of voltage (dashed line) produced by a steady synaptic input (arrow) on that dendritic tree.

ciently realistic has grown. Several authors have therefore employed a reduced approximation to branching dendritic trees (Rose and Dagum, 1988; Fleshman et al., 1988; Clements and Redman, 1989; Stratford et al., 1989), using the following ad hoc method. All segments of the tree located at the same electrotonic distance from the soma are combined into an unbranched equivalent cable (ECa) with variable diameter D:

$$\Sigma D_i^{3/2} = D^{3/2} \tag{18}$$

where D_i are the diameters of the segments. It turns out that such empirically derived ECa models are useful for solving inverse and forward problems in dendritic cable modelling (see section *Branching condition for dendritic trees* below). Common to those studies, however, is the employment of exclusively numerical methods for determining the solutions of the non-uniform cable equations defined by Eqn. (18), since no analytical solutions were available. This is unsatisfactory in the sense that the effects of parameter changes can only be studied using long series of computer simulations.

To overcome this imperfection, we have studied the non-uniform cable equation with spatially varying diameter. Analysis of this equation reveals that analytical solutions can be derived, provided certain conditions hold for the dendritic tree which is being modelled (see Schierwagen (1989b) for details).

The non-uniform cable equation

The non-uniform analogue to Eqns. (1), (2) or (3) is

$$C(x)\frac{\partial V(x,t)}{\partial t} - \frac{\partial}{\partial x}\left(\frac{1}{R(x)}\frac{V(x,t)}{\partial x}\right) + G(x)\cdot V(x,t) = 0 \tag{19}$$

i.e. in contrast to Eqn. (3), it has space-dependent electrical parameters (R, C and G). We suppose the cable to have a variable circular cross-section of diameter $d(x)$, while the specific electrical parameters R_m, C_m and G_m are assumed to be fixed along the cable length. Applying the variable transformations $T = t/\tau$, $\tau = C(x)/G(x)$ and

$$Z = \int_0^x \lambda(s)^{-1} ds,\ \lambda(x) = [R(x)G(x)]^{-1/2},$$

Eqn. (19) reads in dimensionless form

$$\frac{\partial^2 V}{\partial Z^2} + q\frac{\partial V}{\partial Z} - \frac{\partial V}{\partial T} - V = 0 \tag{20}$$

where

$$q = (G'R - GR')/(2GR) = 3/2\,(\ln D(Z))'. \tag{21}$$

$D(Z)$ is the function governing the diameter of the cable in Z-space, and the prime denotes $\partial/\partial Z$.

Using the relations

$$V(Z,T) = F(Z)\cdot W(Z,T)$$

and (22)

$$F(Z) = \exp\left(-\tfrac{1}{2}\int q\,dZ\right) = D(Z)^{-3/4},$$

Eqn. (20) can be written in normal form

$$\frac{\partial^2 W}{\partial Z^2} - \frac{\partial W}{\partial T} - (p+1)\,W = 0, \tag{23}$$

where the coefficient $p = p(Z)$ is defined by the simple Riccati equation

$$2q' = 4p - q^2 \tag{24}$$

which follows from Eqns. (20) and (22). Various classes of non-uniform cable geometries may be obtained by specifying the function $p = p(Z)$ in condition (24). The simplest class to consider are those cables for which p is a constant. This class can be determined by solving the Riccati differential Eqn. (24). The complete solution set and the corresponding diameter functions determined from Eqn. (21) are given in Table I.

Since the solutions to the Riccati Eqn. (24) allow one to adjust the two free parameters p and C shown in Table I, a rather extensive class of non-uniform cables geometries is obtained, for which analytical results can be given, as shown below.

The diameter functions $d(x)$ in the original

x-space can be obtained from the relation

$$\partial x/\partial Z = \lambda(x) = \lambda_0\sqrt{d(x)/d_0} = \lambda_0\sqrt{D(Z)/d_0},$$

from which follows

$$x(Z) = \lambda_0/\sqrt{d(0)} \int \sqrt{D(Z)}\, dZ. \quad (25)$$

λ_0 denotes the length constant belonging to the starting diameter d_0 of the cable. We note that only for the uniform, the power and the exponential geometry types, the integral in Eqn. (25) can be evaluated in closed form. The corresponding diameter functions $d(x)$ in x-space are of uniform, power (exponent $4/5$) and quadratic type, respectively.

Basic solutions to the non-uniform cable equation

The above results enable us to derive some basic solutions to Eqn. (19) in terms of the solutions to the uniform cable Eqn. (3). For this, the transformations (22) are used to calculate the voltage $V(Z,T)$ solving Eqn. (19) as

$$V(Z,T) = \exp(-pT)\cdot F(Z)\cdot U(Z,T) \quad (26)$$

with $U(Z,T)$ being any solution to the uniform cable Eqn. (3) in normalized form,

$$\frac{\partial^2 U}{\partial Z^2} - \frac{\partial U}{\partial T} - U = 0. \quad (27)$$

The relationship (26) shows that deviations from the case of a uniform cable described by Eqn.

TABLE I

Cable geometries as defined by the solutions of the special Riccati equation $2q' = 4p - q^2$

Geometry	p	$q(Z)$	diameter $D(Z)$
Uniform	$p=0$ $q=0$	0	d_0
Power	$p=0$ $q\neq 0$	$\frac{2}{Z-C}$	$d_0\cdot\left(1-\frac{Z}{C}\right)^{4/3}$
Exponential	$p>0$	q_1	$d_0\cdot\exp(\frac{2}{3}q_1 Z)$
Hyperbolic sine	$p>0$ $\lvert q\rvert > q_1$	$q_1\cosh\left(q_1\frac{Z-C}{2}\right)$	$d_0\cdot\left(\frac{\sinh\left(q_1\frac{Z-C}{2}\right)}{\sinh(-q_1\frac{C}{2})}\right)^{4/3}$
Hyperbolic cosine	$p>0$ $\lvert q\rvert < q_1$	$q_1\tanh\left(q_q\frac{Z-C}{2}\right)$	$d_0\cdot\left(\frac{\cosh\left(q_1\frac{Z-C}{2}\right)}{\cosh(q_1\frac{C}{2})}\right)^{4/3}$
Trigonometric	$p<0$	$-\lvert q_1\rvert\tan\left(\lvert q_1\rvert\frac{Z-C}{2}\right)$	$d_0\cdot\left(\frac{\cos\left(\lvert q_1\rvert\frac{Z-C}{2}\right)}{\cos(\lvert q_1\rvert\frac{C}{2})}\right)^{4/3}$

The function $q(Z)$ is related to diameter $D(Z)$ through $q(Z) = \frac{3}{2}$ (In $D(Z))'$, and $q = q_1 = \pm\sqrt{4p}$ is the stationary solution to the Riccati equation.

(27) are mirrored by two factors, one depending only on time, the other only on space. Corresponding to the particular cable geometry defined by the values of the free parameters p and C (see Table I), various situations of amplification or attenuation are possible. Two examples are given for illustration. The general steady-state solution of the normal form (23) is

$$W(Z) = \begin{cases} A_1 Z + A_2, & p = -1 \\ A_1 \exp\left(\sqrt{p+1}\, Z\right) + \\ A_2 \exp\left(-\sqrt{p+1}\, Z\right), & p > -1 \\ A_1 \sin\left(\sqrt{-p-1}\, Z\right) + \\ A_2 \cos\left(-\sqrt{-p-1}\, Z\right), & p < -1 \end{cases} \qquad (28)$$

where the two constants A_1, A_2 must be determined from the boundary conditions belonging to any specific problem (see section *Steady voltage attenuation in a non-uniform cable*). Using $T \to \infty$ in Eqn. (26), the general steady-state solution to Eqn. (20) is obtained from Eqn. (28) via Eqn. (22), i.e. $V(Z) = F(Z) \cdot W(Z)$.

In a similar manner, the basic transient solution to the dimensionless form (20) of the non-uniform cable Eqn. (19) may be calculated from that of Eqn. (27):

$$V(Z,T) = \exp\left[(-p - 1 - \alpha^2)T\right] \cdot F(Z,T) \times \left[B_1 \sin(\alpha Z) + B_2 \cos(\alpha Z)\right] \qquad (29)$$

where B_1, B_2 and α depend on the specific initial and boundary conditions.

Steady voltage attenuation in a non-uniform cable

From the preceding derivations it should be clear that expressions for input impedances, voltage attenuation, etc. may be deduced from the basic solutions (28) and (29), analogous to the case of a uniform cable or membrane cylinder (see sections *Linear cable theory* and *Equivalent cylinder models*). This is illustrated here for the asymmetry factor Φ defined in Eqn. (17).

Equation (28) describes the steady voltage distribution along a non-uniform cable, the diameter of which is governed by one of the functions $D(Z)$ in Table I. To obtain explicit expressions, e.g. for voltage attenuation, the corresponding boundary value problem must be solved for Eqn. (20).

If we consider a non-uniform cable extending from $Z = 0$ to $Z = Z_t$, with both ends sealed, the centrifugal attenuation (constant voltage V_a applied at $Z = 0$) can be determined as follows. The boundary conditions are

$$V(0) = V_a, \quad V'(Z_t) = 0 \qquad (30)$$

which transform via Eqn. (22) into

$$W(0) = V_a / F(0)$$

and (31)

$$W'(Z_t) = -V(Z_t) \cdot F'(Z_t) / F(Z_t)^2 = -W(Z_t) \cdot F'(Z_t) / F(Z_t).$$

After transformation, the problem in W,Z space is now described by the solutions to the uniform case. Using Eqn. (10), the course of the (transformed) voltage W with distance Z is given by

$$(W(Z)/W(0)) = \left[\cosh(a(Z_t - Z)) + Y_t^* \sinh(a(Z_t - Z))\right] \div \left[\cosh(aZ_t) + Y_t^* \sinh(aZ_t)\right] \qquad (32)$$

where, for short, we set $a^2 = p + 1$.

For illustration, the case of a cable with the cosine diameter function (in Z-domain) is considered in Fig. 6. The course of the diameter in both Z-space and in physical x-space is presented in Fig. 6A. For this geometry type, the height of the maximum of $D(Z)$ is determined by the value of $p < 0$, whereas its location on the Z-axis depends on the value of C (see Table I). The values arbitrarily chosen in Fig. 6 are $p = -4.71$ and $C = 0.34$.

The voltage attenuation in such a cable is shown below (Fig. 6B). In an uniform cable with diameter d_0, attenuation is the same in both the centrifugal (from $Z = 0$ to $Z = 1$) and the centripetal (from $Z = 1$ to $Z = 0$) direction (dotted curves), and amounts to 65% of the applied volt-

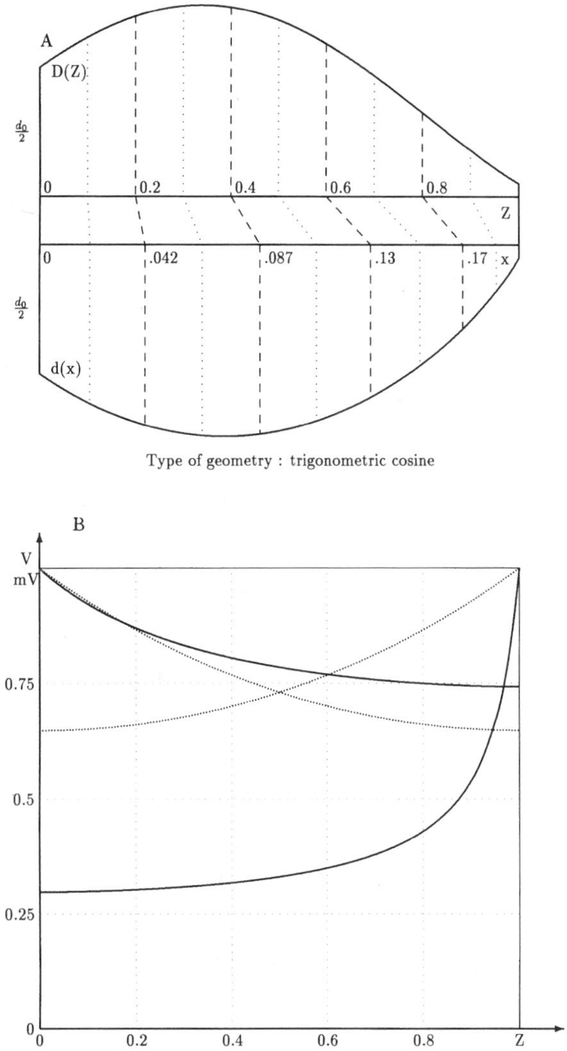

Fig. 6. Voltage attenuation in a non-uniform cable. (A) Displayed is the half-profile of a cable of cosine type both in Z and original x space. The parameter values determining this specific cable profile are $p = -4.71$ and $C = 0.34$. (B) Shown is the course of centrifugal and centripetal voltage attenuation both in an uniform cable of diameter d_0 (dotted curves) and in a cosine cable of corresponding initial diameter d_0 (see text).

age. Attenuation in the cable with cosine diameter depends on direction (continuous curves) for the chosen parameters. Centrifugal attenuation is lower (to 74% of the applied voltage) than attenuation in the uniform cable. In contrast, centripetal attenuation (to 30%) is much higher for this non-uniform geometry. The asymmetry factor Φ then amounts to $\Phi = 0.41$ which is somewhat larger than the values calculated from BC models of rat SC neurons (see section *Branching cable models*).

Equation (9) can be used to derive an expression for calculating the asymmetry factor Φ for any geometry defined in Table I. First, the centrifugal attenuation factor is given by

$$W(Z_t)/W(0) = \left[\cosh(aZ_t) + Y_t^* \sinh(aZ_t)\right]^{-1} \tag{33}$$

where $Y_t^* = -F'(Z_t)/F(Z_t)$.

Voltage decrement in the centripetal direction can be determined in the same way, using (in V,Z-space) the boundary conditions $V(Z_t) = V_t$, $V'(0) = 0$.

Since $V = F \cdot W$ (see Eqn. (22)), the attenuation factor in the original V,Z-space is given by

$$\frac{V_t}{V_0} = \frac{F(0)}{F(Z_t) \cdot \cosh(aZ_t) - F'(Z_t) \cdot \sinh(aZ_t)} \tag{34}$$

The asymmetry factor Φ then is readily calculated to be

$$\begin{aligned}\Phi = &\left[\cosh(aZ_t)\right. \\ &\left.+3/4 \cdot D(Z_t) \cdot D'(Z_t) \cdot \sinh(aZ_t)\right] \\ &\div \left[\cosh(aZ_t)\right. \\ &\left.+3/4 \cdot D(0) \cdot D'(0) \cdot \sinh(aZ_t)\right]\end{aligned} \tag{35}$$

where the relationship (22) was used to replace $F(Z)$ by the diameter function $D(Z)$.

The functional relationship between Φ and the two free parameters p and C (defining any particular $D(Z)$, see Table I) is illustrated in Fig. 7. Displayed is Φ versus p for certain values of C, $0 < C < 1$ (Fig. 7A). Symmetrical attenuation ($\Phi = 1$) in both directions is obtained with $C = 0.5$, independent of the value of p (Fig. 7A). All curves start from the point $p = 0$, which characterizes the case of a cable with constant diameter. It can readily be seen from the definition of Φ that a simple relationship holds for mirrored ca-

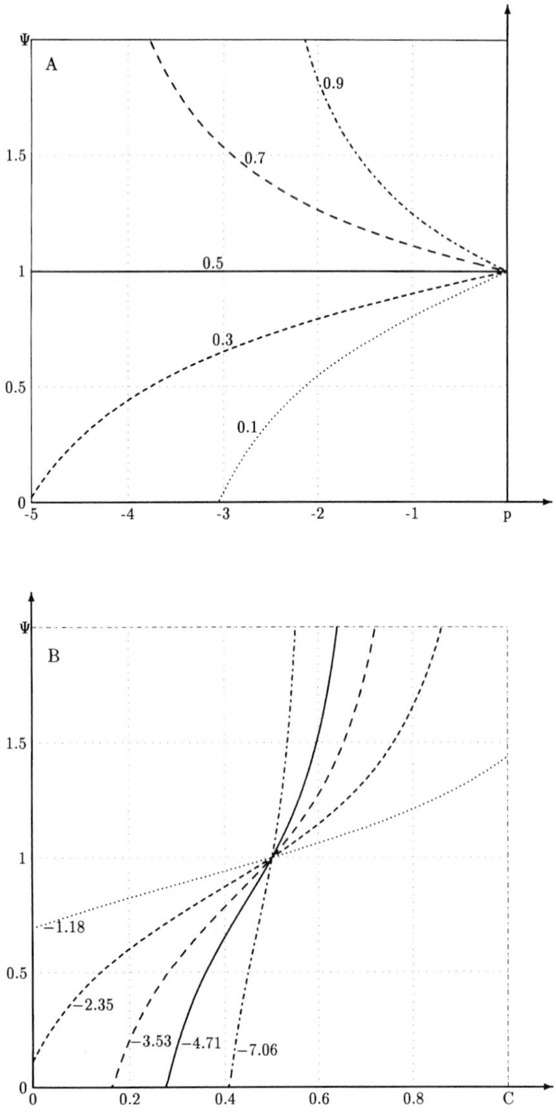

Fig. 7. Relationship between asymmetry factor Φ and cable geometry as defined by the parameters p and C (see Table I). (A) Displayed is Φ versus p for the values of C indicated on the curves. For any value of p, symmetrical attenuation ($\Phi = 1$) is obtained with $C = 0.5$. The point $p = 0$ characterizes uniform cables of any diameter. For new cable profiles, produced by mirroring a given profile at the perpendicular in $Z = 0.5$ to the Z axis, the corresponding Φ is obtained simply as the reciprocal of the original. (B) Depicted is Φ vs. C for the values of p indicated on the curves. Due to the symmetry property of the cosine geometry chosen, the point (0.5;1) is the common intersection point to all curves, irrespective of the value of p (see text).

ble geometry. If, for a given cable, its origin and termination are exchanged (i.e. the cable is mirrored at the perpendicular in $Z = 0.5$ to the Z axis), the new Φ is obtained simply by the reciprocal of the original (see Fig. 7A).

In Fig. 7B, Φ vs. C is depicted for several values of p; at the point (0.5;1) all curves intersect, irrespective of the value of p. This is so because, for the chosen cosine geometry, any cable of length 1 with $C = 0.5$ is symmetrical with respect to $Z = 0.5$, and thus $\Phi = 1$.

Branching condition for dendritic trees

To answer the question of what classes of dendritic trees allow a reduction to a non-uniform cable according to Eqns. (20) or (23), the relationship between macroscopic cable parameters and the microscopic quantities of the underlying dendritic tree must be analyzed. This relationship is expressed by Eqn. (21), assuming constant p in Eqn. (24). It allows the formulation of a general branching condition for trees that can be transformed to the non-uniform cable geometries in Table 1. After some lengthy calculations we obtain

$$\left\{\left(\Sigma\, D_i(Z)^2\right)\left(\Sigma\, D_i(Z)\right)\right\}^{1/2} = D(Z)^{3/2}$$

i.e.

$$n(Z)\,\overline{D}(Z)^{1/2}\cdot\hat{D}(Z) = D(Z)^{3/2} \quad (36)$$

where $n(Z)$, $\overline{D}(Z)$, $\hat{D}(Z)$ are the number, the arithmetical mean and the quadratic mean, respectively, of the diameters of dendritic branches, as a function of distance Z.

An important result is that Table I (last column) gives six different cable geometries which — if inserted into Eqn. (36) — determine the types of dendritic branching amenable to exact solution. It might be worth noting that Eqn. (36) provides a generalization to Rall's '3/2 power relationship' (Rall, 1959, 1962). To illustrate this, we remember that the conditions (i)–(iv) necessary for the EC transformation (see section *Equivalent cylinder models*) imply that the sum of the 3/2 powers of the diameters, for all branches at any given electrotonic distance from the soma, must remain constant out to the terminal points. In an ideal symmetrically branching dendritic tree, all branches of a given generation have equal diameters as well as equal lengths in the physical x-space, i.e. in Z space for $Z_j \leq Z \leq Z_j + 1$

$$D_i(Z) = D_j \ (i = 1,...,n_j) \qquad (37)$$

where branching occurs at distances Z_j ($j = 1,....,k$) with n_j branches of length $L_j = Z_{j+1} - Z_j$ between branch points at Z_j and Z_{j+1}. Obviously, Eqn. (36) simplifies in this case to Eqn. (18)

$$\Sigma D_i^{3/2} = n_j \cdot D_j^{3/2} = D^{3/2},$$

since the arithmetical and the quadratic mean of segment diameters are identical for any generation j, $\bar{D}(Z) = \hat{D}(Z) = D_j$.

A first conclusion to be drawn is that the acceptance of Eqn. (18) in cases where the above branching symmetry does not hold may yield erroneous results, since the non-uniform cable derived from Eqn. (18) reflects only imperfectly the electrotonic structure of the dendritic tree. The physical cause for this mismatch can be found if the relationships between electrotonic parameters and segment diameters are considered (see also the section *Linear cable theory*). In the present model, as well as in Rall's EC model, all dendritic branches at any distance Z are considered to be electrically in parallel. Therefore, combined axial resistance $R(Z)$ to current flow is inversely related to the total cross-sectional area $A(Z)$ of all branches at any distance Z. $A(Z)$ itself depends directly on the quadratic mean $\hat{D}(Z)$. Combined capacitance $C(Z)$ and conductance $G(Z)$ are directly related to summed membrane area, i.e. in the limit to perimeter $P(Z)$, which is directly related to the arithmetical mean $\bar{D}(Z)$.

Since the reduced ECa models employed so far are based on preserving $\Sigma D_i(Z)^{3/2}$, neither total membrane area (perimeter) $P(Z)$, nor cross-sectional area $A(Z)$ is conserved. Stratford et al. (1989) used therefore a 'cartoon model' for preserving the membrane area of pyramidal neurons in the rat visual cortex. Axial resistances, however, are different in the 'cartoon' and the corresponding BC model, causing differences in the synaptic attenuation. An alternative approach was suggested by Bush and Sejnowski (1993) whose reduced ECa model had the same axial resistance, but lower surface area than the full BC model. In the latter case, membrane conductance and capacitance must be accordingly scaled.

In contrast to these ad hoc models, the ECa model (19) and its normal form (23) preserve all the electrical parameters $R(Z)$, $G(Z)$ and $C(Z)$. This presupposes, however, that in the general, non-symmetrical case of dendritic branching Eqn. (18) is replaced by Eqn. (36)

$$\left\{\left(\Sigma D_i(Z)^2\right)\left(\Sigma D_i(Z)\right)\right\}^{1/2} = D(Z)^{3/2}$$

(see Schierwagen, 1989b). Thus, it seems quite possible that the differences between transient responses in a BC model and the corresponding ECa model reported in Rall et al. (1992) will decrease, if the profile of the ECa model is calculated according to Eqn. (36).

Conclusions

Although studies on artificial neural networks (ANN) as 'simplifying brain models' and 'prototypes for new computer architectures' (Sejnowski et al., 1988) are essential in elucidating general operation principles of 'intelligent' systems, they must be complemented by model studies which try to incorporate as much of the relevant data of a biological network as available. This review has addressed questions related to the computational organization of neuronal dendrites. Considering the fact that the role of dendritic geometry in neuron function has been difficult to study directly by experimental means, the advantage in this respect of neuron models was demonstrated. As usual, the primary function of the cable models discussed is to solve inverse and forward problems. The type of model chosen for this purpose depends on the available a priori information about neuronal structure and function as well as on the type of synaptic inputs and their locations.

Both the uniform EC models and their generalization, the non-uniform ECa models, were shown to possess an inherent non-uniqueness concerning parameter identifiability, even if spatially constant (specific) electrical parameters are assumed. However, regions in the parameter space can be bounded by imposing known physiological restrictions.

Another main purpose of this review has been to demonstrate how reduced, non-uniform ECa models of neuronal dendritic trees can be constructed, thus expanding the range of application for analytical models. They can be used in the same way as the EC models to solve inverse and forward problems. An important advantage of the ECa models, however, is the greater class of dendritic branching patterns to which it can be applied. Exemplified by the electrotonic asymmetry factor, Φ, we have shown that expressions can be derived relating characteristics of the neuron such as input resistance, electrotonic length, voltage and current attenuation to parameters defining any particular geometry of the non-uniform ECa model.

Elsewhere we have demonstrated that our results can be also applied for describing analytically the propagation of action potentials in non-uniform axons (Schierwagen, 1991; Schierwagen and Ohme, 1992), giving rise to new insights into the complex filtering operation that axons can perform.

Such results are obviously important in any modeling attempts where essential features of neuronal information processing such as graded, electrotonic potentials and dendritic and axonal time delays are included. These analog aspects of the computations performed by neurons point to a much greater variety of operations available when compared with the traditional view of spikes as information carriers.

Genuine insights into the computational tasks performed by single neurons, however, can only be gained from studies of networks composed of these neurons. The cable models discussed may be useful in this respect, since essential features of neuronal information processing such as synaptic background activity, synaptic attenuation and dendritic time delays can be included. By using such models to represent different types of neurons in structured neural networks, the gap between the overly simplified models currently employed in studies of artificial neural networks, on the one hand, and analyses of biological nervous systems, on the other hand, might be bridged.

A final remark concerns the role of analytical neuron models. Despite all the computer power readily available nowadays, we believe that fundamental significance is due to exactly solvable mathematical models in general, and to models of electrical nerve activity in particular. A computer model can be made increasingly realistic by increasing the number of variables and parameters, but this approach does not necessarily lead to a deeper understanding of the process studied. In contrast, exactly solvable models such as the ECa model force the researcher from the very beginning to concentrate on the main phenomena, thus reducing the high-dimensional problem space.

Acknowledgements

Thanks are due to Michael Corner and Jaap Van Pelt for critical reading of the manuscript. Michael Ohme produced Figures 6 and 7.

References

Barrett, J.N. and Crill, W.E. (1974) Specific membrane properties of cat motoneurones. *J. Physiol. (Lond.)*, 239: 301–324.

Burke, R.E. (1987) Synaptic efficacy and the control of neuronal input-output relations. *Trends Neurosci.*, 10: 42–45.

Bush, P.C. and Sejnowski, T.J. (1993) Reduced compartmental models of neocortical pyramidal cells. *J. Neurosci. Methods*, 46: 159–166.

Clements, J.D. and Redman, S.J. (1989) Cable properties of cat spinal motoneurones measured by combining voltage clamp, current clamp and intracellular staining. *J. Physiol. (Lond.)*, 409: 63–87.

De Schutter, E. (1992) A consumer guide to neuronal modeling software. *Trends Neurosci.*, 15: 462–464.

Durbin, R., Miall, C. and Mitchison, G. (1989) *The Computing Neuron*, Addison Wesley, Wokingham, England.

Fleshman, J.W., Segev, I. and Burke, R.E. (1988) Electrotonic architecture of type-identified α-motoneurons in the cat spinal cord. *J. Neurophysiol.*, 60: 60–85.

Glenn, L.L. (1988) Overestimation of the electrical length of neuron dendrites and synaptic electrotonic attenuation. *Neurosci. Lett.*, 91: 112–119.

Grantyn, R., Grantyn, A. and Schierwagen, A. (1983) Passive membrane properties, afterpotentials and repetitive firing of superior colliculus neurons studied in the anesthetized cat. *Brain Res.,* 50: 377–391.

Graubard, K. and Calvin, W.H. (1979) Presynaptic dendrites: implications of spikeless synaptic transmission and dendritic geometry. In: F.O. Schmitt and F.G. Worden (Eds.), *The Neurosciences,* Fourth Study Program, MIT Press, Cambridge, MA, pp. 317–331.

Hillman, D.E. (1979) Neuronal shape parameters and structures as a basis of neuronal form. In: F.O. Schmitt and F.G. Worden (Eds.), *The Neurosciences,* Fourth Study Program, MIT Press, Cambridge, MA, pp. 477–498.

Holmes, W.R. and Rall, W. (1992) Estimating the electrotonic structure of neurons with compartmental models. *J. Neurophysiol.,* 68: 1438–1452.

Jack, J.J.B., Noble, D. and Tsien, R.W. (1975) *Electric Current Flow in Excitable Cells,* Clarendon Press, Oxford.

Koch, C. and Segev, I. (1989) *Methods in Neuronal Modeling,* MIT Press, Cambridge, MA.

Koch, C., Poggio, T. and Torre, V. (1982) Retinal ganglion cells: a functional interpretation of dendritic morphology. *Phil. Trans. R. Soc. (Ser. B),* 298: 227–264.

McKenna, T., Davis, J. and Zornetzer, S.F. (1992) *Single Neuron Computation,* Academic Press, Boston.

Nitzan, R., Segev, I. and Yarom, Y. (1990) Voltage behavior along the irregular dendritic structure of morphologically and physiologically characterized vagal motoneurons in the guinea pig. *J. Neurophysiol.,* 63: 333–346.

Rall, W. (1959) Branching dendritic trees and motoneuron membrane resistivity. *Exp. Neurol.,* 1: 491–527.

Rall, W. (1962) Theory of physiological properties of dendrites. *Ann. NY Acad. Sci.,* 96: 1071–1092.

Rall, W. (1977) Core conductor theory and cable properties of neurons. In: E.R. Kandel (Ed.), *Cellular Biology of Neurons,* American Physiological Society, Bethesda, pp. 39–97.

Rall, W., Burke, R.E., Holmes, W.R., Jack, J.J.B., Redman, S.J. and Segev, I. (1992) Matching dendritic neuron models to experimental data. *Physiol. Rev.,* 72 (Suppl. 4): S159–S186.

Rose, P.K. and Dagum, A. (1988) Nonequivalent cylinder models of neurons: interpretation of voltage transients generated by somatic current injection. *J. Neurophysiol.,* 60: 125–148.

Schierwagen, A. (1986) Segmental cable modelling of electrotonic transfer properties of deep superior colliculus neurons in the cat. *J. Hirnforsch.,* 27: 679–690.

Schierwagen, A. (1988) Distributed parameter systems as models of current flow in nerve cells with branched dendritic trees: inverse and forward problems. *Syst. Anal. Model. Simul.,* 5: 455–473.

Schierwagen, A.K. (1989a) Real neurons and their circuitry : Implications for brain theory. In: L. Budach (Ed.), *Neural Informatics,* Informatik Informationen Reporte 12, Akademie der Wissenschaften, Berlin, pp. 17–20.

Schierwagen, A.K. (1989b) A non-uniform equivalent cable model of membrane voltage changes in a passive dendritic tree. *J. Theor. Biol.,* 141: 159–179.

Schierwagen, A.K. (1990) Identification problems in distributed parameter neuron models. *Automatica,* 26: 739–755.

Schierwagen, A.K. (1991) Travelling wave solutions of a simple nerve conduction equation for inhomogeneous axons. In: A.V. Holden, M. Markus and H. Othmer (Eds.), *Nonlinear Waves in Excitable Media,* Plenum, New York and London, pp. 107–114.

Schierwagen, A. and Grantyn, R. (1986) Quantitative morphological analysis of deep superior colliculus neurons stained intracellularly with HRP in the cat. *J. Hirnforsch.,* 27: 611–623.

Schierwagen, A. and Ohme, M. (1992) Excitation propagation in nonuniform axons: analytical solution. In: N. Elsner and D.W. Richter (Eds.), *Rhythmogenesis in Neurons and Networks,* Thieme, Stuttgart, P741.

Schierwagen, A., Gärtner, U. and Hilbig, H. (1993) Morphometrical and electrotonic characteristics of superior colliculus neurons in the rat. In: N. Elsner and H. Heisenberg (Eds.), Gene-Brain- Behaviour, Thieme, Stuttgart, P878.

Scott, A.C. (1975) The electrophysics of a nerve fiber. *Rev. Modern Phys.,* 47: 487–533.

Sejnowski, T.J., Koch, C. and Churchland, P.S. (1988) Computational neuroscience. *Science,* 241: 1299–1306.

Shepherd, G.M. (1992) Canonical neurons and their computational organization. In: T. McKenna, J. Davis and S.F. Zornetzer (Eds.), *Single Neuron Computation,* Academic Press, Boston, pp. 27–60.

Stratford, K., Mason, A., Larkman, A., Major, G. and Jack, J.J.B. (1989) The modeling of pyramidal neurones in the visual cortex. In: R. Durbin, C. Miall and G. Mitchison (Eds.), *The Computing Neuron,* Addison Wesley, Wokingham, England, pp. 296–321.

Tuckwell, H.C. (1988) *Introduction to Theoretical Neurobiology, Vol. 1: Linear Cable Theory and Dendritic Structure,* Cambridge University Press, Cambridge.

Turner, D.A. (1984) Segmental cable evaluation of somatic transients in hippocampal neurons. *Biophys. J.,* 46: 73–84.

White, J.A., Manis, P.B. and Young, E.D. (1992) The parameter identification problem for the somatic shunt model. *Biol. Cybernet.,* 66: 307–318.

Van Pelt, J. (1992) A simple vector implementation of the Laplace-transformed cable equations in passive dendritic trees. *Biol. Cybern.,* 68: 15–21.

Van Pelt, J. and Schierwagen, A. (1994) Electrotonic properties of passive dendritic trees — effect of dendritic topology. In: J. van Pelt, M.A. Corner, H.B.M. Uylings and F.H. Lopes da Silva (Eds.), *Progress in Brain Research, Vol. 102, The Self-organizing Brain — from Growth Cones to Functional Networks,* Elsevier Science Publishers, Amsterdam, this volume.

J. van Pelt, M.A. Corner H.B.M. Uylings and F.H. Lopes da Silva (Eds.)
Progress in Brain Research, Vol 102

CHAPTER 10

Development of voltage-dependent and ligand-gated channels in excitable membranes

Nicholas C. Spitzer

Department of Biology and Center for Molecular Genetics, University of California, San Diego, La Jolla, CA 92093, U.S.A.

Overview

Excitability is an essential feature of the nervous system, and the mechanisms and functions of rapid signaling in the millisecond time domain have been examined and understood in elegant detail since the first appreciation of its existence over 200 years ago. The basis of differentiation of excitability has been the subject only of more recent investigation. Voltage-gated and ligand-sensitive channels have been found to appear at very early stages in embryogenesis, often in a stereotyped manner, that may vary from one system to another. The initial expression of these properties is frequently different from that observed in the mature nervous system. These features raise the possibility of a developmental function for the forms of excitability characteristically displayed at these early stages of differentiation.

Strikingly, the early phases of electrical and chemical excitability have been shown to participate in spontaneous activity of the nervous system. This includes both signaling of the rapid form, involving impulses and synaptic-like potentials (Hamburger et al., 1966; Corner and Crain, 1972; Bergey et al., 1981; Baker et al., 1984) as well as much slower events involving elevations of intracellular calcium that occur on the time scale of seconds to minutes — three orders of magnitude slower than the rapid signaling events (Holliday and Spitzer, 1990; Yuste et al., 1992; Gu and Spitzer, 1993a,b; Gu et al., 1994). These fast and slow signals both appear to exert an important influence on later aspects of differentiation. This review focuses on the set of slow signals. In what forms are they expressed? In what patterns are they produced? How does their expression vary during development? How are they generated? What are their effects?

The *Xenopus* embryo is an attractive focus for investigation of these features of neuronal development. Indeed spontaneous activity and regionally specific patterning of rapid signaling have already been studied during the maturation of the nervous system of this species (Corner, 1964). Differentiation proceeds extraordinarily rapidly; the transition from the neurula stage, immediately prior to the closure of the neural tube, through the young swimming larva occurs in a period of a single day. This rapid development facilitates examination through the entire period. Moreover, the *Xenopus* embryo is especially accessible for experimental investigation as well as for various forms of perturbation experiments. Patch clamp recordings from the intact spinal cord and imaging of intercellular calcium in this developing preparation, as well as immunocytochemical and in situ hybridization analysis have been added to the repertoire of approaches successfully applied to this preparation. Moreover cells from the neural plate stage, prior to primary differentiation, can be explanted and grown in either organotypic (Corner, 1964) or dissociated cell culture (Spitzer and Lamborghini, 1976).

The latter cultures have been particularly useful in recent work since differentiation in vitro parallels that described in vivo in all respects so far examined. Cultures contain sensory, motor, and interneurons as well as a variety of other cell types. Isolated neurons attach to the tissue culture plastic substrate and begin to elaborate processes equipped with growth cones. The maturation of electrical excitability and sensitivity to neural transmitters proceeds as it does in the intact spinal cord. Neurotransmitter expression also begins just as it does in neurons developing in vivo. Finally, cells are grown in a simple, wholly defined culture medium that facilitates manipulation of the extracellular environment so as to permit examination of the functional role of various forms of excitability in implementing subsequent aspects of differentiation.

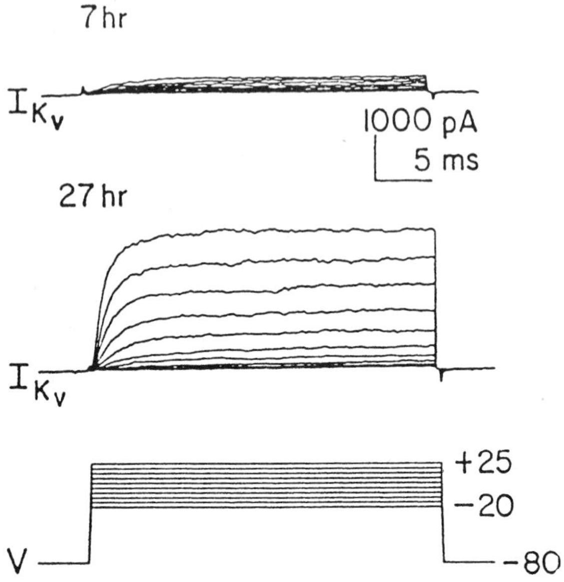

Fig. 1. Development of delayed rectifier potassium current (I_{Kv}) in amphibian spinal neurons cultured from neural plate-stage embryos. Records from two cells, at times indicated. Currents were recorded in the presence of TTX and the absence of calcium (from O'Dowd et al., 1988).

Action potentials and voltage-dependent currents

Inward calcium and sodium currents are expressed at high levels prior to the maturation of outward voltage- and calcium-gated potassium currents in these embryonic spinal neurons differentiating in vitro (Fig. 1) (O'Dowd et al., 1988; Ribera and Spitzer, 1990). Whole cell voltage-clamp recordings from neurons in the intact embryonic spinal cord indicate that differentiation proceeds similarily in vivo (Desarmenien et al., 1993). These observations are consistent with previous recordings of action potentials, which are long in duration and calcium dependent in early stages and subsequently become brief and sodium dependent with further maturation both in the spinal cord and in culture (Spitzer and Lamborghini, 1976; Baccaglini and Spitzer, 1977). Computer reconstruction of action potentials from whole cell currents, using a Hodgkin-Huxley formulation, reveals that the four voltage-dependent currents recorded from embryonic neurons (O'Dowd et al., 1988; Lockery and Spitzer, 1992) can fully account for the waveforms of action potentials at young and mature stages (see also Barish, 1986). Furthermore, the increase in amplitude of the voltage-dependent delayed rectifier potassium current is crucial for the maturation of the impulse (Fig. 2). The change in this current alone is sufficient to achieve 94% of the shortening of the action potential that suppresses calcium influx. Developmental changes in the other three currents make a much smaller contribution to the shift in ionic dependence of the nerve impulse from a largely calcium-dependent to a chiefly sodium-dependent event. This process requires transcription since it is blocked by α-amanitin or dichlororibobenzimidazole during an early critical period between 6 and 15 h in vitro (Fig. 3). *Xenopus* delayed rectifier potassium channels have now been cloned (Ribera, 1990; Ribera and Nguyen, 1993) in order to examine the molecular basis by which potassium currents are regulated.

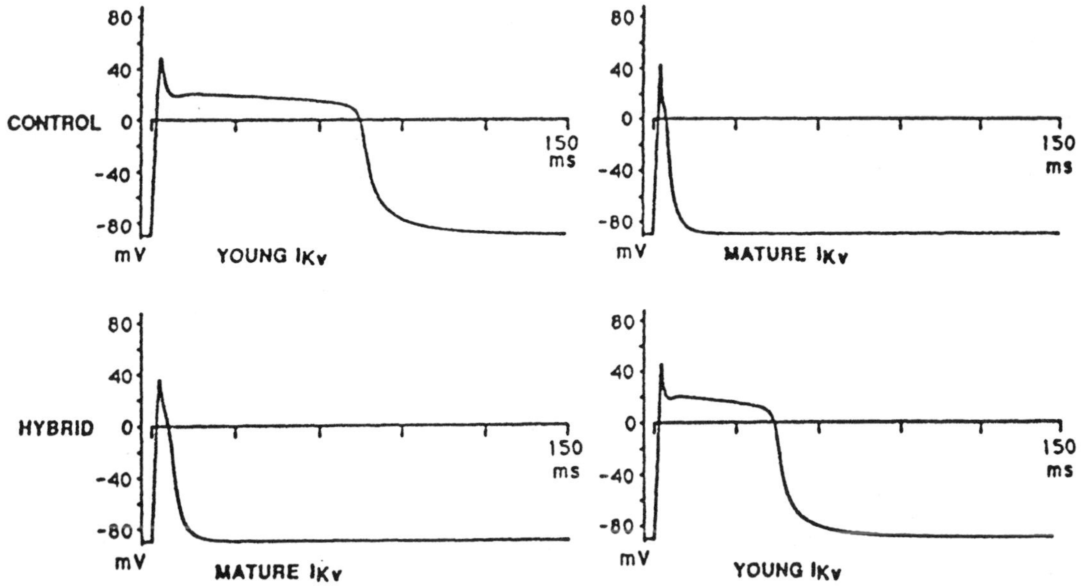

Fig. 2. The delayed rectifier potassium current plays the major role in the developmental shortening of the action potential of amphibian spinal neurons. Records of action potentials produced by modeled whole cell currents (I_{Na}, I_{Ca}, I_{Kc}, I_{Kv}) at two stages of development. Control young and mature action potentials closely resemble the mean recorded durations (top two panels). Selective introduction of the mature delayed rectifier alone causes a major decrease in duration of the young action potential (bottom left panel). On the other hand, retention of the young delayed rectifier while all other currents are in their mature form prevents the expression of the brief mature action potential (bottom right panel) (from Lockery and Spitzer, 1992).

These probes will be useful for determining the mechanisms by which the differentiation of action potentials is controlled, since they are likely to facilitate analysis of promoter sequences regulating transcription of genes encoding the delayed rectifier current.

The pattern of action potential differentiation illustrated by the spinal neurons of *Xenopus* embryos is observed for a wide variety of other differentiating neurons as well. Moreover the maturation of the action potential at least in *Xenopus* is a cell autonomous process once neural induction has occurred, since it can take place even when a neuron is cultured in the absence of other cells in a fully defined medium devoid of specific growth factors. The primary induction process has been investigated in embryos of the ascidian, *Halocynthia*. Neuronal differentiation requires intercellular interactions, since the anterior animal blastomere of a cleavage-arrested 8-cell embryo normally first expresses a long duration calcium-dependent action potential and then develops a brief sodium-dependent impulse. This process fails to occur when the cell is isolated, unless it is contacted by an anterior-vegetal blastomere (Fig. 4) (Okado and Takahashi, 1990a,b). Induction in this 2-cell model system may thus be homologous to neural induction from competent ectoderm by the chordamesoderm in the amphibian (Nieuwkoop, 1985), in regard to its topology and timing.

In other systems, however, a second pattern of differentiation is observed, in which the ionic basis of the action potential is relatively constant throughout development. Neural crest cells of quail embryos, like neurons in other systems, exhibit short duration sodium-dependent impulses at the earliest stages of excitability; long duration, calcium-dependent action potentials are observed only when potassium currents are sup-

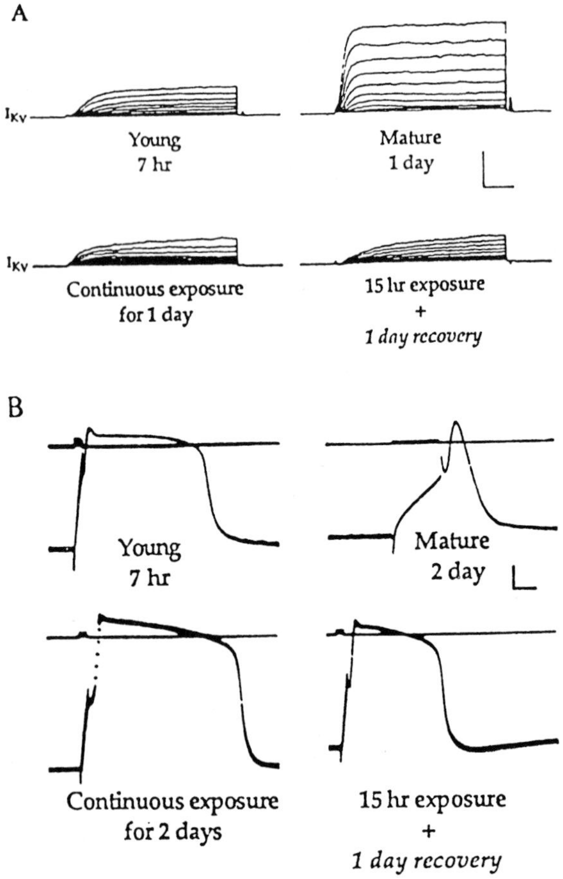

Fig. 3. Transcriptional blockade arrests the development of electrical excitability when applied during a critical period. Cultured amphibian spinal neurons were treated with dichlororibobenzimidazole (DRB) to block mRNA synthesis, and examined after varying times. (A) Currents from control young and 1-day neurons, and from neurons subjected to continuous exposure or a brief treatment. The current does not reappear following removal of the drug, although potassium A current is expressed (not shown). Scale bar is 1000 pA and 5 msec. (B) Action potentials from control young and 2-day neurons, and those subjected to DRB. The shortening of the action potential is prevented by the inhibitor. Scale bar is 20 mV and 10 msec or 1 msec (mature). Application of DRB at later times does not affect the maturation of excitability (from Ribera and Spitzer, 1989).

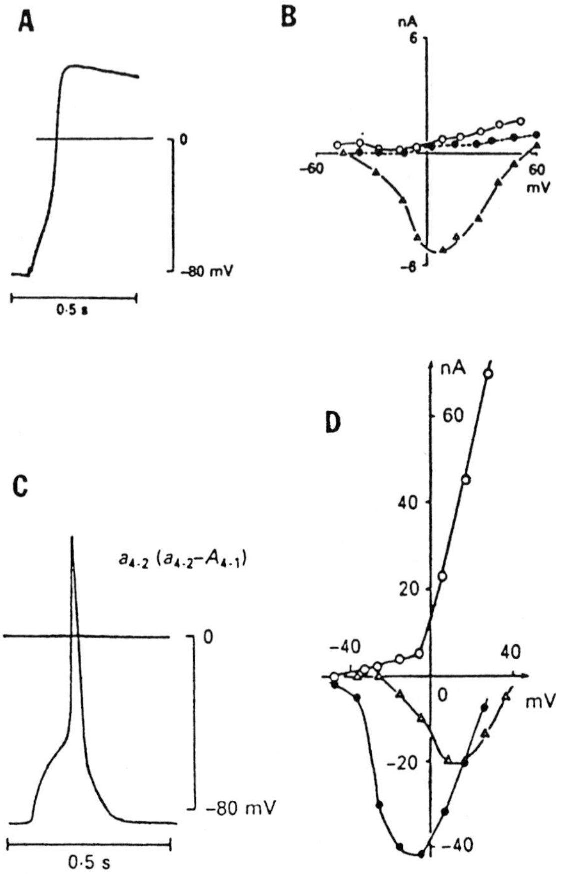

Fig. 4. Induction of electrical excitability in cleavage-arrested blastomeres isolated from 8-cell embryos of the ascidian *Halocynthia*. (A,B) Epidermal form of excitability in the $a_{4\text{-}2}$ blastomere. The action potential is long in duration and dependent on strontium (used in lieu of calcium); the current-voltage relation indicates that this current is larger than both the outward and leak currents. (C,D) Neural form of excitability in the cleavage-arrested $a_{4\text{-}2}$ blastomere isolated and cultured in contact with similarly treated $A_{4\text{-}1}$ blastomeres. The action potential is brief and sodium dependent; the current voltage relation indicates a substantial outward current in addition to inward sodium and calcium currents. △, ▲, calcium current; ○, potassium current; ●, leak current (B) and sodium current (D) (from Okado and Takahashi, 1990a,b).

pressed (Bader et al., 1985). The principal difference between these two classes of differentiation of electrical excitability therefore resides in the time course of maturation of outward potassium currents. In the first instance, potassium currents increase gradually, allowing the expression of long duration calcium-dependent events at early stages. The second class of maturation involves the initial expression of both inward sodium and calcium

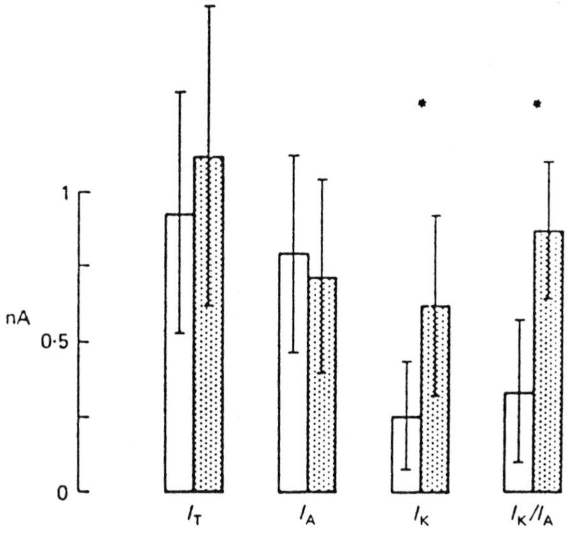

Fig. 5. Development of total outward potassium current (I_T), and inactivating (I_A) and delayed rectifier (I_K) current components in neurons from quail mesencephalic neural crest. Means ± S.D. for values at 1 day (open; $n = 21$) and more than 4 days (stippled; $n = 12$); asterisk indicates values significantly different ($P < 0.001$) (from Bader et al., 1985).

currents in conjunction with outward potassium currents, thereby restricting the expression of long duration calcium-dependent events (Fig. 5).

Spontaneous calcium transients

Initial observations

Embryonic *Xenopus* spinal neurons produce spontaneous transient elevations of intracellular calcium both in culture and in the intact spinal cord, evaluated with calcium indicators fura-2 and fluo-3. These events were first detected with a subtraction protocol whereby images of fluorescence from cultured cells were accumulated for two sequential periods of several minutes each, following which the first image was subtracted from the second (Holliday and Spitzer, 1990). This approach allowed detection of changes in fluorescence, and thus the levels of intracellular calcium, that occur during the period of time over which the two images are accumulated. The results show that the relative number of cells demonstrating these transient elevations of intracellular calcium is developmentally regulated as a function of time in culture. The percentage of cells is initially small, rises to roughly 20% around the time of primary neurite outgrowth, and then declines to 5% by the end of the first day in vitro (Fig. 6). Most of these transient elevations of intracellular calcium in the cell body are abolished either by removal of extracellular calcium or by agents that block its influx through high-voltage activated (HVA) channels (Holliday and Spitzer, 1990; Gu and Spitzer, 1993a,b). Furthermore, calcium influx triggers calcium-induced calcium-release from stores within the developing neuron. Depletion of these stores, using either ditertbenzohydroquinone or caffeine, suppresses the elevation of intracellular calcium achieved with depolarizations that stimulate calcium influx (Holliday et al., 1991). In contrast, there appear to be no large changes in the steady-state levels of intracellular calcium during the course of early neuronal differentiation (Holliday and Spitzer, 1990).

Low voltage-activated (LVA) T-type calcium

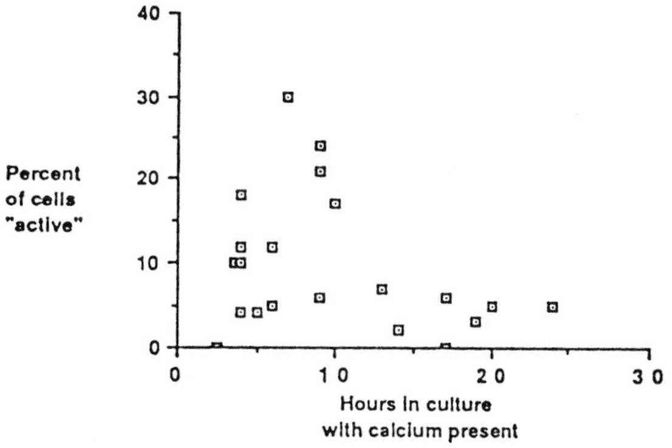

Fig. 6. Spontaneous calcium influx occurs early during differentiation of amphibian spinal neurons in vitro, when small potassium currents enable production of long calcium-dependent action potentials. Neurons were loaded with fura 2AM. Subtraction of images of cell bodies derived from sequential 2-min examination periods revealed increased fluorescence in some cells, termed 'active'. The spontaneously increased intracellular calcium in many 'active' cells is mediated by influx through voltage-dependent channels (not shown) (from Holliday and Spitzer, 1990).

currents are the immediate stimulus for most of these spontaneous elevations of intracellular calcium. These events are also blocked by low concentrations of nickel or amiloride that specifically block T current without significantly affecting HVA currents. In addition, the T current has the lowest threshhold of all known inward currents in these neurons and computer simulations establish that the T current can lower the voltage threshhold for initiation of action potentials at early stages (Gu and Spitzer, 1993a). The role of intracellular calcium stores appears to be developmentally regulated, since calcium release is especially sensitive to calcium influx at early stages of differentiation and becomes less sensitive as cells mature. The extent of elevation of intracellular calcium that can be provoked by depolarizing stimuli appears also to be greatest at early times.

Calcium spikes and waves in cultured neurons

Examination of spontaneous calcium transients has now been extended to encompass more substantial periods of the early differentiation of *Xenopus* spinal neurons. Initial analysis entailed observation over brief periods of several minutes. Confocal microscopy has made it possible to evaluate these spontaneous elevations of intracellular calcium for periods of 1 h in duration both in vitro and in vivo (see below), capturing images once every 5 sec (720 per hour).

This approach has allowed detection of two classes of spontaneous calcium transients occurring in the cell bodies of these neurons: rapid events, termed calcium spikes, and slow events, termed calcium waves (Gu et al., 1992; Gu and Spitzer, 1993b). Neurons can exhibit up to ten spikes in the course of a 1 h imaging period, although the mean frequency is lower. Spikes rise rapidly and have a double exponential decay with time constants of 10 sec and 3 min. In contrast, neurons exhibit calcium waves that occur at approximately the same mean frequency as spikes, which rise and decay more slowly. Over a period of 1 h, about one-third of the cells exhibit spikes, one-third produce waves and the rest remain silent. Neurons begin to extend neurites at 5–6 h in vitro, at the same time that neurite extension begins in the spinal cord (Hayes and Roberts, 1973; Spitzer and Lamborghini, 1976; Taylor and Roberts, 1983). Cultured neurons also exhibit these characteristic signatures of calcium elevation prior to the time of morphological differentiation. In contrast, calcium elevations in other cell types in these cultures exhibit slower rates of rise and longer durations.

Calcium spikes in the spinal cord

It has been useful to examine spontaneous elevations of intracellular calcium occurring in the intact spinal cord. For this purpose the spinal cord is dissected free of the embryo and mounted in a small well (Desarmenien et al., 1993) so as to allow stable imaging of intracellular calcium over periods of 1 h. A mean of 50% of neurons exhibit spikes during a 1 h period at an average frequency of 10 per hour. The incidence of spikes decreases with maturation of the neurons; by 1 day in culture, equivalent to a tailbud stage embryo, only a fifth of the cells display calcium spikes.

Cells also exhibit coactivity in the spinal cord; up to four cells show simultaneous elevation of intracellular calcium, producing spikes that are coincident within the resolution of the time-lapse analysis. Such cells may be contiguous or remote from each other by a distance of a 100 μm or even more. The latter observation is compatible with the observation of neurite extension in the spinal cord at about this time, providing the opportunity for these small ensembles of cells to coordinate their activity by anatomical interconnections. The probability of spontaneous occurrence of these events is low enough that some form of intercellular communication mechanism must be presumed. Whether it is electrical coupling (gap junctions), chemical synapses, or yet other forms of interaction that are causally involved remains to be explored.

Roles of spikes and waves

The functions of spikes and waves in influencing neuronal differentiation are now emerging as well. The opportunity to block calcium spikes specifically with Ni^{2+} (Gu and Spitzer, 1993a) allows determination of the requirement of spontaneous spikes for various aspects of later neuronal development. Thus, a population of interneurons acquires the ability to synthesize and store the neurotransmitter GABA at early stages of development, both in vivo (Dale et al., 1987; Roberts et al., 1987) and in culture (Spitzer et al., 1993). The time course of this process is parallel in the two cases, and is suppressed by removal of extracellular calcium from the medium in which dissociated neurons are growing in vitro. Interestingly, this aspect of differentiation is also sensitive to blockade by Ni^{2+}, and similar suppression of the appearance of GABA by blockers of messenger RNA synthesis raises the possibility that these events are controlled by regulation of transcripts for glutamic acid decarboxylase, the synthetic enzyme. The normal developmental increase in activation kinetics of the delayed rectifier potassium current is also suppressed in the same way by these agents, again raising the likelihood of transcriptional regulation (Ribera and Spitzer, 1989; Desarmenien and Spitzer, 1991).

In contrast, calcium waves appear to be important in regulating neurite lengths. Waves are generated in growth cones at a higher frequency than those seen in the soma. The role of extracellular calcium is demonstrated by its removal, leading to more rapid outgrowth of neuronal processes and the attainment of lengths that are two to three times greater than those grown in its presence (Bixby and Spitzer, 1984a; Holliday and Spitzer, 1990, 1993). Waves are thus in a position to control the extension of neurites. The developmental persistence of waves, in contrast with the developmental downregulation of spike incidence, is consistent with prolonged sensitivity of growth cones to extracellular calcium.

Calcium spikes thus appear to be important both for neurotransmitter accumulation and for the maturation of potassium channel kinetics, while calcium waves may be required for the regulation of neurite extension (see also Kater et al., 1994). Calcium is also important for target recognition and synapse formation by neurons in these cultures (Henderson et al., 1984) as well as for the regulation of neurotransmitter receptor function in these cells (Bixby and Spitzer, 1984a). Whether or not these processes, too, are governed by spontaneous calcium spikes and/or waves is currently under investigation. Spontaneous elevations of intracellular calcium generated by action potentials are also observed in endocrine cells, where they promote secretion of hormones (Schlegel et al., 1987; Ashcroft, 1988; Chen et al., 1990; Malgaroli et al., 1990; Petersen, 1992). Astrocytic glia exhibit spontaneous calcium transients (Fatatis and Russell, 1992), the function of which is still to be elucidated.

Signal transduction cascade

These results demonstrate a developmental signal transduction cascade in amphibian spinal neurons involving slow signaling by calcium transients, some of which (calcium spikes) may be generated by rapid action potentials. Calcium-dependent nerve impulses appear to produce calcium spikes by triggering intracellular calcium release and elevation of intracellular calcium. These elevations of intracellular calcium are necessary for further aspects of neural differentiation, and play a role in the feedback inhibition of calcium-dependent action potentials by enhancing repolarizing potassium currents. The mechanisms by which calcium T currents are activated and the details of the machinery by which calcium transients control later phenotypic differentiation remain to be established. Calcium waves provide a local, regionally autonomous mechanism that appears to be involved in the regulation of neurite extension. This process may be achieved by the disassembly and assembly of cytoskeletal components that results from the increase and decrease of intracellular calcium (Lankford and Letourneau, 1989).

Ligand gated channels

Changes in the responses to neurotransmitter agents also occur during neuronal differentiation, both through the appearance and in some cases the subsequent disappearance of receptors as a function of developmental age. The expression of a developmentally transient response to a neurotransmitter that disappears later has now been observed in many systems (Bixby and Spitzer, 1982; for review see Spitzer, 1991). Maturation of this kind parallels that of developmental changes in action potentials, which involve sequential rather than simultaneous activation of different classes of channels that vary proportion depending on the state of maturation. The serial acquisition of sensitivities to more than one transmitter leads to a change in the cellular response in other cases. This process has been shown to be regulated postsynaptically via the appearance of receptors for different ligands, and can be influenced by the presence of other cells (Clendening and Hume, 1990a,b). Receptors for different transmitters sometimes appear at essentially the same stage of development, however (Goodman and Spitzer, 1979; Bixby and Spitzer, 1984b). In some instances these processes entail changes in the properties of ligand-gated channels, such as the developmental switch in subunit expression of hetero-oligomeric channels of nicotinic acetylcholine receptors of muscle (Mishina et al., 1986; Gu and Hall, 1988).

Calcium influx

The early appearance of neurotransmitter receptors is particularly intriguing in cases in which NMDA receptors are involved. Influx of calcium and elevation of intracellular calcium have been demonstrated for NMDA receptors (Cline and Tsien, 1991), and the elevation of intracellular calcium that they may produce can be crucial in regulating further aspects of differentiation (Neve and Bear, 1989; Fox et al., 1989). NMDA receptors in rat hippocampus are less voltage- and magnesium-sensitive at early stages of development and may allow greater influx of calcium during this period (Ben-Ari et al., 1988; Bowe and Nadler, 1990; Brady et al., 1991). In addition, kainate/AMPA-gated glutamate receptors allow calcium influx and elevation of intracellular calcium (Linn and Christensen, 1992; Burnashev et al., 1992; Muller et al., 1992). The calcium permeability of glutamate receptors is governed by subunit composition (Hollmann et al., 1991), as well as by single amino acid substitutions generated by RNA editing (Hume et al., 1991; Keller et al., 1992; Kohler et al., 1993), and may be developmentally regulated. Excitatory transmitters are released spontaneously at early stages of development (Blanton and Kriegstein, 1991). Moreover the classically inhibitory transmitter GABA has been demonstrated to produce increases in intracellular calcium at early stages of neuronal differentiation (Connor et al., 1987; Yuste and Katz, 1991), probably via chloride conductances that depolarize cells sufficiently to activate LVA calcium currents (Mueller et al., 1984; Ben-Ari et al., 1988).

There are multiple potential roles for transmitter receptors expressed at early stages of neuronal development. The blockade of calcium influx via either voltage-gated or ligand-sensitive channels suppresses early neuronal migration neccessary for the formation of the cerebellar cortex (Komuro and Rakic, 1992, 1993). Release of transmitters by growth cones is consistent with a role in shaping patterns of neurite outgrowth, as suggested by the effects of application of transmitters to neurons in a variety of systems (Mattson and Kater, 1987; Cohan et al., 1987). Furthermore, calcium transients are features of pattern formation in the optic tectum as well as the visual and somatosensory cortex (Shatz, 1990; Constantine-Paton et al., 1990; Yuste et al., 1992). Localization of receptors in specific spatial regions of the neuron implicate a role in the process of synaptogenesis. Early sensitivities to neural transmitters are involved in the process of natural neuronal death as well (Oppenheim, 1991). Stimulation by neurotransmitter can lead to selective upregulation of its receptors (Balazs et al., 1992). Calcium entry by these routes may affect gene expression differentially via selective modes of calcium handling (Lerea and McNamara, 1993; Bading et al., 1993).

Conclusions

The importance of voltage-gated and transmitter-activated channels was first appreciated in the context of the mature nervous system where they function to promote rapid signaling and information processing important for generation of behaviors. The recent work on developmental expression of these components identifies additional roles in regulating the differentiation and assembly of the embryonic nervous system. Regulatory functions, expressed on this much slower time scale, appear also to be important for plasticity in the adult nervous system.

Acknowledgements

I thank my colleagues in my laboratory for stimulating discussions. My work is supported by NIH grants NS 15918 and NS 25916.

References

Ashcroft, F.M. (1988) Adenosine 5′-triphosphate-sensitive potassium channels. *Annu. Rev. Neurosci.*, 11: 97–118.

Baccaglini, P.I. and Spitzer, N.C. (1977) Developmental changes in the inward current of the action potential of Rohon-Beard neurones. *J. Physiol.*, 271:93–117.

Bader, C.R., Bertrand, D. and Dupin, E. (1985) Voltage-dependent potassium currents in developing neurones from quail mesencephalic neural crest. *J. Physiol.*, 366: 129–151.

Bading, H., Ginty, D.D. and Greenberg, M.E. (1993) Regulation of gene expression in hippocampal neurons by distinct calcium signaling pathways. *Science*, 260: 181–186.

Baker, R.E., Corner, M.A. and Habets, A.M.M.C. (1984) Effects of chronic suppression of bioelectric activity on the development of sensory ganglion evoked responses in spinal cord explants. *J. Neurosci.*, 4: 1187–1192.

Balazs, R., Resink, A., Hack, N., van der Valk, J.B., Kumar, K.N. and Michaelis, E. (1992) NMDA treatment and K + -induced depolarization selectively promote the expression of an NMDA-preferring class of the ionotropic glutamate receptors in cerebellar granule neurones. *Neurosci. Lett.*, 137: 109–113.

Barish, M.E. (1986) Differentiation of voltage-gated potassium current and modulation of excitability in cultured amphibian spinal neurones. *J. Physiol.*, 375: 229–250.

Ben-Ari, Y., Cherubini, E. and Krnjevic, K. (1988) Changes in voltage dependence of NMDA currents during development. *Neurosci. Lett.*, 94: 88–92.

Ben-Ari, Y., Cherubini, E., Corradetti, R., and Gaiarsa, J.L. (1989) Giant synaptic potentials in immature CA3 hippocampal neurones. *J. Physiol.*, 416: 303–325.

Bergey, G.K., Fitzgerald, S.C., Schrier, B.K. and Nelson, P.G. (1981) Neuronal maturation in mammalian cell cultures is dependent on spontaneous bioelectric activity. *Brain Res.*, 207: 49–58.

Bixby, J.L. and Spitzer, N.C. (1982) The appearaance and development of chemosensitivity in Rohon-Beard neurones of the *Xenopus* spinal cord. *J. Physiol.* 330: 513–536.

Bixby, J.L. and Spitzer, N.C. (1984a) Early differentiation of vertebrate spinal neurons in the absence of voltage-dependent Ca^{2+} and Na^+ influx. *Dev. Biol.* 106: 89–96.

Bixby, J.L. and Spitzer, N.C. (1984b) The appearance and development of neurotransmitter sensitivity in *Xenopus* embryonic spinal neurones in vitro. *J. Physiol.* 353: 143–155.

Blanton, M.G. and Kriegstein, A.R. (1991) Spontaneous action potential activity and synaptic currents in the embryonic turtle cerebral cortex. *J. Neurosci.* 11: 3907–3923.

Bowe, M.A. and Nadler, J.V. (1990) Developmental increase in the sensitivity to magnesium of NMDA receptors on CA hippocampal pyramidal cells. *Dev. Br. Res.* 56: 55–61.

Brady, R.J., Smith, K.L. and Swan, J.W. (1991) Calcium modulation of the N-methyl-D-asparatate (NMDA) response and electrographic seizures in immature hippocampus. *Neurosci. Lett.* 123: 92–96.

Burnashev, N., Monyer, H., Seeburg, P.H. and Sakmann, B. (1992) Divalent ion permeability of AMPA receptor channels is dominated by the edited form of a single subunit. *Neuron* 8: 189–98.

Chen, C., Zhang, J., Vincent, J.D. and Israel, J.M. (1990) Sodium and calcium currents in action potentials of rat somatotrophs: their possible functions in growth hormone secretion. *Life Sci.*, 46: 983–989.

Clendening, B. and Hume, R.I. (1990a) Expression of multiple neurotransmitter receptors by sympathetic preganglionic neurons in vitro. *J. Neurosci.*, 10: 3977–3991.

Clendening, B. and Hume, R.I. (1990b) Cell interactions regulate dendritic morphology and responses to neurotransmitters in embryonic chick sympathetic preganglionic neurons in vitro. *J. Neurosci.*, 10: 43992–4005.

Cline, H.T. and Tsien, R.W. (1991) Glutamate-induced increases in intracellular Ca^{2+} in cultured frog tectal cells mediated by direct activation of NMDA receptor channels. *Neuron*, 6: 259–67.

Cohan, C.S., Connor, J.A. and Kater, S.B. (1987) Electrically and chemically mediated increases in intracellular calcium in neuronal growth cones. *J. Neurosci.*, 7: 3599–3599.

Connor, J.A., Tseng, H.Y. and Hockberger, P.E. (1987) Depolarization- and transmitter-induced changes in intracellular Ca^{2+} of rat cerebellar granule cells in explant cultures. *J. Neurosci.*, 7: 1384–1400.

Constantine-Paton, M., Cline, H.T. and Debski, E. (1990) Patterned activity, synaptic convergence, and the NMDA receptor in developing visual pathways. *Annu. Rev. Neurosci.,* 13: 129–154.

Corner, M.A. (1964) Localization of the capacities for functional development in the neural plate of *Xenopus. J. Comp. Neurol.,* 123: 243–255.

Corner, M.A. and Crain, S.M. (1972) Patterns of spontaneous bioelectric activity during maturation in culture of fetal rodent medulla and spinal cord tissues. *J. Neurobiol.,* 3: 25–45.

Dale, N., Roberts, A., Ottersen, O.P. and Storm-Mathisen, J. (1987) The development of a population of spinal cord neurons and their axonal projections revealed by GABA immunocytochemistry in frog embryos. *Proc. R. Soc. London B,* 232: 205–215.

Desarmenien, M. G. and Spitzer, N.C. (1991) Determinant role of calcium and protein kinase C in development of the delayed rectifier potassium current in *Xenopus* spinal neurons. *Neuron,* 7: 797–805.

Desarmenien, M.G., Clendening, B. and Spitzer, N.C. (1993) In vivo development of voltage-dependent ionic currents in embryonic *Xenopus* spinal neurons. *J. Neurosci.,* 13: 2575–2581.

Fatatis, A. and Russell, J.T. (1992) Spontaneous changes in intracellular calcium concentration in Type I astrocytes from rat cerebral cortex in primary cultures. *Glia,* 5: 95–104.

Fox, K., Sato, H. and Daw, N. (1989) The location and function of NMDA receptors in cat and kitten visual cortex. *J. Neurosci.,* 9: 2443–2454.

Goodman, C.S. and Spitzer, N.C. (1979) Embryonic development of identified neurones: Differentiation from neuroblast to neurone. *Nature,* 280: 208–214.

Gu, Y. and Hall, Z.W. (1988) Immunological evidence for a change in subunits of the acetylcholine receptor in developing and denervated muscle. *Neuron,* 1: 117–125.

Gu, X., Olson, E.C. and Spitzer, N.C. (1992) Patterns of spontaneous transient elevations of intracellular calcium in embryonic spinal neurons prior to neurite extension. *Soc. Neurosci.,* 18: 1287.

Gu, X., Olson, E.C. and Spitzer, N.C. (1994) Spontaneous neuronal calcium spikes and waves during early differentiation. *J. Neurosci.* In press.

Gu, X. and Spitzer, N.C. (1993a) Low threshold Ca^{2+} current and its role in spontaneous elevations of intracellular Ca^{2+} in developing *Xenopus* neurons. *J. Neurosci.,* 13: 4936–4948..

Gu, X. and Spitzer, N.C. (1993b) Spontaneous Ca^{2+} spikes and waves in embryonic *Xenopus* spinal neurons in vitro and in vivo. *Soc. Neurosci.,* 19: 1111.

Hamburger, V., Wenger, E. and Oppenheim, R. (1966) Motility in the chick embryo in the absence of sensory input. *J. Exp. Zool.,* 162: 133–160.

Hayes, B.P. and Roberts, A. (1973) Synaptic junction development in the spinal cord of an amphibian embryo: an electron microscope study. *Z. Zellforsch.,* 137: 251–269.

Henderson, L.P., Smith, M.A. and Spitzer, N.C. (1984) The absence of calcium blocks impulse-evoked release of acetylcholine but not de novo formation of functional neuromuscular synaptic contacts in culture. *J. Neurosci.,* 4: 3140–3150.

Holliday, J. and Spitzer, N.C. (1990) Spontaneous calcium influx and its roles in differentiation of spinal neurons in culture. *Dev. Biol.,* 141: 13–23.

Holliday, J. and Spitzer, N.C. (1993) Calcium regulates neuronal differentiation both directly and via co-cultured myocytes. *J. Neurobiol.,* 24: 506–514.

Holliday, J., Adams, R.J., Sejnowski, T.J. and Spitzer, N.C. (1991) Calcium-induced release of calcium regulates differentiation of spinal neurons. *Neuron,* 7: 787–796.

Hollmann, M., Hartley, M. and Heinemann, S. (1991) Ca^{2+} permeability of KA-AMPA-gated glutamate receptor channels depends on subunit composition. *Science,* 252: 851–853.

Hume, R.I., Dingledine, R. and Heinemann, S.F. (1991) Identification of a site in glutamate receptor subunits that controls calcium permeability. *Science,* 253: 1028–1031.

Kater, S.B., Davenport, R.W. and Guthrie, P.B. (1994) Filopodia as detectors of environmental cues: signal integration through changes in growth cone calcium levels. In: J. van Pelt, M.A. Corner, H.B.M. Uylings and F.H. Lopes da Silva (Eds.), *Progress in Brain Research, Vol. 102, The Self-organizing Brain — from Growth Cones to Functional Networks,* Elsevier Science Publishers, Amsterdam, this volume.

Keller, B.U., Hollman, M., Heinemann, S. and Konnerth, A. (1992) Calcium influx through subunits GluR1/GluR3 of kainate/AMPA receptor channels is regulated by cAMP dependent protein kinase. *EMBO J.,* 11: 891–896.

Kohler, M., Burnashev, N., Sakmann, B. and Seeburg, P.H. (1993) Determinants of Ca^{2+} permeability in both TM1 and TM2 of high affinity kainate receptor channels: diversity by RNA editing. *Neuron,* 10: 491–500.

Komuro, H. and Rakic, P. (1992) Selective role of N-type calcium channels in neuronal migration. *Science,* 257: 806–809.

Komuro, H. and Rakic, P. (1993) Modulation of neuronal migration by NMDA receptors. *Science,* 260: 95–97.

Lankford, K.L. and Letourneau, P.C. (1989) Evidence that calcium may control neurite outgrowth by regulating stability of actin filaments *J. Cell Biol.,* 109: 1229–1243.

Lerea, L.S. and McNamara, J.O. (1993) Ionotropic glutamate receptor subtypes activate *c-fos* transcription by distinct calcium-requiring intracellular signaling pathways. *Neuron,* 10: 31–41.

Linn, C.P. and Christensen, B.N. (1992) Excitatory amino acid regulation of intracellular Ca^{2+} in isolated catfish cone horizontal cells measured under voltage- and concentration-clamp conditions. *J. Neurosci.,* 12: 2156–2164.

Lockery, S.R. and Spitzer, N.C. (1992) Reconstruction of action potential development from whole-cell currents of differentiating spinal neurons. *J. Neurosci.,* 12: 2268–2287.

Malgaroli, A., Fesce, R. and Meldolesi, J. (1990) Spontaneous

$[Ca^{2+}]_i$ fluctuations in rat chromaffin cells do not require inositol 1,4,5- trisphosphate elevations but are generated by a caffeine- and ryanodine-sensitive intracellular Ca^{2+} store. *J. Biol. Chem.* 265: 3005–3008.

Mattson, M.P. and Kater, S.B. (1987) Calcium regulation of neurite elongation and growth cone motility. *J. Neurosci.,* 7: 4034–4043.

Mishina, M., Takai, T., Imotu, K., Noda, M., Takahashi, T., Numa, S., Methfessel, C. and Sakmann, B. (1986) Molecular distinction between fetal and adult forms of muscle acetylcholine receptor. *Nature,* 321: 406–411.

Mueller, A.L., Taube, J.S. and Schwartzkroin, P.A. (1984) Development of hyperpolarizing inhibitory postsynaptic potentials and hyperpolarizing response to γ-aminobutyric acid in rabbit hippocampus studies *in vitro*. *J. Neurosci.*, 4: 860–867.

Muller, T., Moller, T., Berger, T., Schnitzer, J. and Kettenmann, H. (1992) Calcium entry through kainate receptors and resulting potassium-channel blockade in Bergmann glial cells. *Science,* 256: 1563–1566.

Neve, R.L. and Bear, M.F. (1989) Visual experience regulates gene expression in the developing striate cortex. *Proc. Natl. Acad. Sci.,* 86: 4781–4784.

Nieuwkoop, P.D. (1985) Inductive interactions in early amphibian development and their general nature. *J. Embryol. Exp. Morphol.,* 89 (Suppl.): 333–347.

O'Dowd, D.K., Ribera, A.B. and Spitzer, N.C. (1988) Development of voltage-dependent calcium, sodium and potassium currents in *Xenopus* spinal neurons. *J. Neurosci.,* 8: 792–805.

Okado, H. and Takahashi, K. (1990a) Differentiation of membrane excitability in isolated cleavage-arrested blastomeres from early ascidian embryos. *J. Physiol.,* 427: 583–602.

Okado, H. and Takahashi, K. (1990b) Induced neural-type differentiation in the cleavage-arrested blastomere isolated from early ascidian embryos. *J. Physiol.,* 427: 603–623.

Oppenheim, R.W. (1991) Cell death during development of the nervous system. *Annu. Rev. Neurosci.,* 14: 453–501.

Petersen, O.H. (1992) Stimulus-secretion coupling: cytoplasmic calcium signals and the control of ion channels in exocrine acinar cells. *J. Physiol.* 448: 1–51.

Ribera, A.B. (1990) A potassium channel gene is expressed at neural induction. *Neuron,* 5: 691–701.

Ribera, A.B. and Nguyen, D.-A. (1993) Primary sensory neurons express a Shaker like potassium channel gene. *J. Neurosci.,* 13: 4988–4996.

Ribera, A.B. and Spitzer, N.C. (1989) A critical period of transcription required for differentiation of the action potential of spinal neurons. *Neuron,* 2: 1055–1062.

Ribera, A.B. and Spitzer, N.C. (1990) Differentiation of I_{KA} in amphibian spinal neurons. *J. Neurosci.,* 10: 1886–1991.

Roberts, A., Dale, N., Ottersen, O.P. and Storm-Mathisen, J. (1987) The early development of neurons with GABA immunoreactivity in the CNS of *Xenopus* laevis embryos. *J. Comp. Neurol.,* 261: 435–449.

Schlegel, W., Winiger, B.P., Mollard, P., Vacher, P., Wuarin, F., Zahnd, G.R., Wollheim, C.B. and Dufy, B. (1987) Oscillations of cytosolic Ca^{2+} in pituitary cells due to action potentials. *Nature,* 329: 719–721.

Shatz, C.J. (1990) Impulse activity and the patterning of connections during CNS development. *Neuron,* 5: 745–756.

Spitzer, N.C. (1991) A developmental handshake: neuronal control of ionic currents and their control of neuronal differentiation. *J. Neurobiol.,* 22: 659–673.

Spitzer, N.C. and Lamborghini, J.E. (1976)The development of the action potential mechanism of amphibian neurons isolated in cell culture. *Proc. Natl. Acad. Sci.,* 73: 1641–1645.

Spitzer, N. C., deBaca, R. C., Allen, K. and Holliday, J. (1993) Calcium dependence of differentiation of GABA immunoreactivity in spinal neurons. *J. Comp. Neurol.,* 337: 168–175.

Taylor, J.S.H. and Roberts, A. (1983) The early development of the primary sensory neurones in an amphibian embryo: a scanning electron microscope study. *J. Embryol. Exp. Morphol.,* 75: 49–66.

Yuste, R. and Katz, L.C. (1991) Control of postsynaptic Ca^{2+} influx in developing neocortex by excitatory and inhibitory neurotransmitters. *Neuron,* 6: 333–344.

Yuste, R., Peinado, A. and Katz, L.C. (1992) Neuronal domains in developing neocortex. *Science,* 257: 665–669.

J. van Pelt, M.A. Corner H.B.M. Uylings and F.H. Lopes da Silva (Eds.)
Progress in Brain Research, Vol 102

CHAPTER 11

Theoretical models for describing neural signal transduction

Hanna Parnas and Ehud Sivan

Department of Neurobiology, Life Science Institute, and the Center for Neural Computation Hebrew University, Jerusalem, Israel

Introduction

Neural signal transduction is conducted by the synapse, which is comprised of pre and postsynaptic regions. An important question when dealing with a network of neurons, even a small one, is how to represent the synapses in a network. Some of the various alternatives are as follows. (1) The usual approach when dealing with very large neural networks is to consider the synapse just as a simple $+$, $-$ switch (denoted here as a sign synapse): an excitable synapse which, when active, will instantaneously increase the level of depolarization in the postsynaptic cell, and vice versa for an inhibitory synapse. Such synapses not only lack any connection to biology but are also devoid of important features such as synaptic delays. (2) A synapse is described more realistically than in (1), for example, the description may include a time course, but it is represented by means of a purely arbitrary function (usually an α function). Such a function might mimic the shape of the biological synaptic time course but, lacking an underlying mechanism, it has no predictive ability whatsoever. (3) A synapse is taken to include all biological details known to exist in the nerve terminals as well as in the postsynaptic cell. That is, all of the release machinery is accurately described, and the various states of the postsynaptic receptors (i.e. activation, desensitization, block, and so forth) are accurately depicted. In terms of formal presentation, such a synapse will involve many differential equations. (4) A synapse will be described by one or by a few analytical equations, describing succinctly only the key features of the underlying mechanisms, while omitting details which are involved only in the fine tuning of synaptic action.

Of the approaches mentioned, (3) can be immediately rejected. Not only is such a representation non-economical from a computational point of view but, even more important, understanding is masked when too many details are included, so that it becomes nearly impossible to link results to the underlying mechanism. The present study shows that, for many purposes, approaches (1) and (2) must also be rejected as being insufficient, leaving (4) as the optimal approach. Selecting (4) means that previously developed analytical equations can be implemented.

We have chosen the auditory communication mechanism of the cricket to demonstrate features of synaptic action that are important to capture in a synaptic representation. Many such features are derived from either the nerve terminal or the postsynaptic cell. In this paper, however, we confine our treatment to the presynaptic nerve terminal. This means that only models associated with the release of neurotransmitter (and not activation of the postsynaptic receptor) will be described. In conclusion, we will briefly describe how a succinct analytical expression has been derived from the many differential equations associated with the detailed model that we believe best account for the process of release of neurotransmitter.

Cricket auditory communication

We have chosen this example as it addresses a biologically critical process, i.e. mating. It is initiated by the female cricket's recognition of the song of the male. The female must recognize this song, distinguishing it from other sounds, and assess the location of her suitor. Here we will discuss only the question of distinguishing between the courtship song and other sounds. The neuronal system responsible for this recognition in the female is depicted in Fig. 1. Figure 1A shows how the male produces his courtship song (bottom) by rubbing his two front wings together (Fig. 1A middle): when the wings are closed, they are set briefly into oscillation at a frequency of about 5 kHz. As a result, each closing of the wings produces a 5 kHz sound, called a *syllable*, that lasts for 15–20 msec. The male produces these syllables at a rate of 25–30 syllables/sec. A group of syllables is denoted as a *chirp* (Fig. 1A bottom). The usual kind of chirp emitted by the

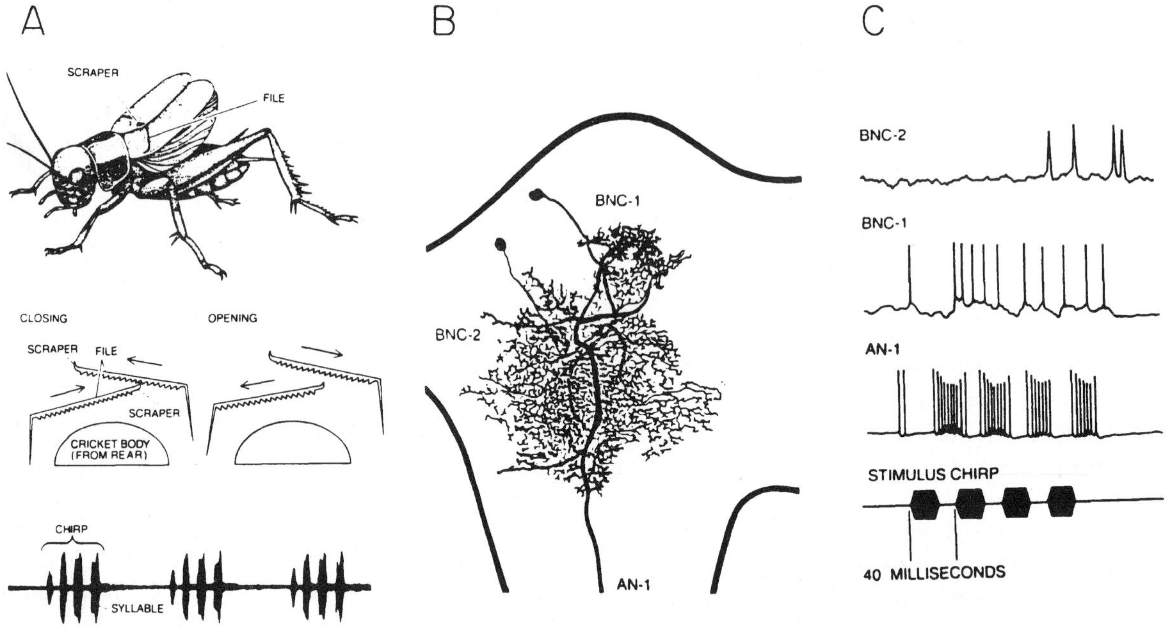

Fig. 1. The cricket auditory communication system. (A) The cricket song is produced by the male when it rubs its wings together. The highly schematic diagrams in the middle show the sound-making mechanism: each wing incorporates a scraper and a file, so that every time the wings close they are set in a brief vibration at a frequency of about 5 kHz. The song is compromised of 'chirps' in which 'syllables' of 5 kHz occur at a certain frequency (bottom). (B) A schematic diagram of the brain of the female cricket. It includes neurons that appear to participate in recognizing the temporal pattern of the male calling song. Here the left side of the brain and two of the neurons within it are drawn. The axon of the AN-1 cells in the prothoracic ganglion arrives from below. Its most forward synaptic terminals overlap the forward branchings of a population of neurons designated BNC-1. In turn, branchings of the BNC-1 cells overlap those of another neuron type, designated BNC-2. (C) Recordings from the neurons revealed that the AN-1 simply copy the temporal pattern of the male song. BNC-1 and BNC-2 on the other hand serve as low and high-pass filters (redrawn from Huber and Thorson, 1985).

courting cricket is a train of four syllables followed by a brief pause. The chirps are usually repeated at a rate of 2—4 chirps per second.

Huber and Thorson (1985) found that the most important factor in the male's song is the syllable rate within each chirp: the female stops responding to the male's song if the syllables are produced at frequencies lower or higher than the desired 25–30 Hz. The female's song recognition system that is illustrated in Figs. 1B, and 1C, is based on the work of Huber and Thorson (1986). The female cricket hears the chirps using a sound-transduction system including ear-drums that excites the auditory neurons through auditory receptor cells. The axons emerging from these cells are destined for the auditory neuropil in the prothoracic ganglion. Some of the neurons in the auditory neuropil have axons that travel upward to the brain. In particular, some neurons called AN-1 (ascending neurons class 1) were found to be highly sensitive to the 5 kHz calling song carrier frequency, such that they produce a burst for each syllable irrespective of the rate that the syllables arrive. In the brain, the AN-1 produce dense terminal arborizations which overlap the arborizations of a population of brain interneurons known as BNC-1 (brain neurons, class 1). In turn, the arborizations of the BNC-1 terminals overlap the arborizations of other brain neurons, called BNC-2 (Fig. 1B).

It was found that in the BNC-1 population some of the neurons serve as *low-pass* filters by firing only when the syllable's frequency is below 30 Hz. In BNC-2, on the other hand, *high-pass* filter cells exist which fire only when the syllable frequency is above 25 Hz. In addition, *band-pass* filters exist in BNC-2, which fire at frequencies between 25 and 30 Hz (Fig. 1C). These findings corroborate the suggestion that the song recognition is accomplished in two steps: first, low-pass and high-pass filters are activated, a set of band-pass filters becomes active at a second stage, provided both the low and the high-pass filters have been synchronously activated.

Model for discerning the courtship song

The model presented below aims at describing one plausible mechanism for recognizing the courtship song described above, not necessarily the biologically correct one. While other models are possible, this particular one has been selected because it relies on plausible physiological properties of synapses and, as such, demonstrates possible roles of various synaptic features in an example of network behavior. Simulations were carried out by means of SONN (for details see H. Parnas and Rudolph, 1991). Briefly, each neuron in SONN is represented by several computation units corresponding to the key processes taking place in various parts of a neuron (the units are termed: SOMA, AXON, TERMINAL, and so forth). Each unit, excluding SOMA, includes one or two analytical expressions derived from the many involved differential equations associated with the particular unit. The SOMA is the only unit which still includes differential equations. The various units are linked to each other by traansfer functions, translating the output of one unit to the proper input of the unit linked to it. Finally, the simulations are based on an event-driven method.

The particular model network associated with discerning the courtship song is depicted in Fig. 2. In our model, a population of neurons, (namely the population of AN-1, BNC-1 (low-pass), BNC-2 (high-pass) and the BNC-2 (band-pass) is represented by a single neuron. The model was further simplified by connecting the BNC-2 (high-pass) neuron directly to the AN-1 neuron and not through cells of the BNC-1 assembly. The response of an AN-1 cell to a syllable is fixed, and is always a burst of five action potentials lasting 16 msec. We did not speculate as to how the AN-1 cells are tuned to the 5 kHz carrier frequency of the song, and assume that the voice is always at the correct frequency. The neurons of all groups (AN-1 and others) were taken to exhibit identical behavior to that of follower cells (i.e. responding with an action potential to any input action potential). Consequently, the behavior of the network is determined by the synaptic connections between the cells. The characteristic time course of the various synapses in the model is seen on the upper part of Fig. 2. It can be seen that the synapses onto the low and high-pass neurons are long in duration. The process of release there lasts close to 30 msec. These are therefore slow synapses. In contrast, the synapse onto the band-pass neuron is a brief one, release lasting only a few milliseconds. In order that the quantal content (m, the area below the curves) in all these synapses be comparable, the amplitude of the brief synapse was greatly increased.

The characteristic features of the synapse onto the low-pass neuron as well as its resulting behavior are seen in Fig. 3. Recall that the low-pass neuron responds to frequencies of 30 Hz and

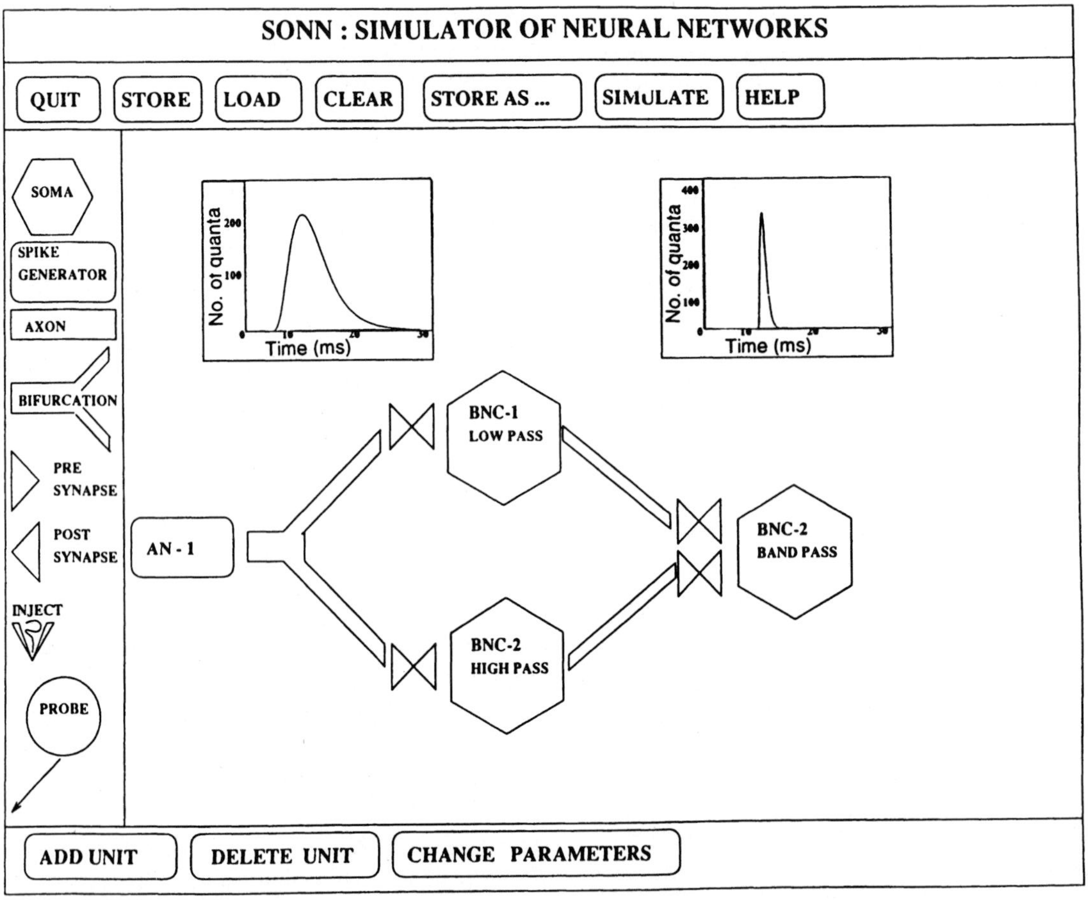

Fig. 2. The network model for the cricket song recognition. The network as modeled by SONN. Each of the cell groups (AN-1, BNC-1 and others) is represented by one neuron. AN-1 is represented by a spike generator unit. This unit can be programmed to fire any kind of firing frequency. AN-1 is connected to both BNC-1 (low-pass) and to BNC-2 (high-pass) by means of the unit AXON followed by a unit BIFURCATION and finally the presynaptic unit TERMINAL. In the current model the axon only serves as delay line (3 msec). TERMINAL is connected to a postsynaptic unit POSTSYNAPSE that translates the release into membrane conductance. POSTSYNAPSE is connected to a SOMA unit where, by a simplified version of the Hodgkin-Huxley model (Av-Ron et al., 1991), the membrane potential is calculated. The release duration of the terminal onto BNC-2 (band-pass) is seen on the upper right while the duration of release of the synapse onto both the BNC-1 (low-pass) and BNC-2 (high-pass) is seen on the upper left.

below, and does so by menas of a special characteristic of its synaptic input: strong depression and only moderate facilitation. The outcome is exemplified by its response to two frequencies (i.e. 33 Hz: left column; 20 Hz: right column). At 33 Hz, the moderate facilitation does not overcome the strong depression persisting in between syllables (see middle left). Consequently, only the first syllable (of a chirp) is able to produce action potentials in BNC-1 (low-pass) (Fig. 3, bottom left). On the other hand, at 20 Hz the synapse fully recovers from a previous syllable depression (Fig. 3, middle right) and, consequently, action potentials are generated in BNC-1 (low-pass) (Fig. 3, bottom right).

The behavior of the high-pass neurons as well as the characteristics of its synaptic input are depicted in Fig. 4. This synapse is characterized by high facilitation and only moderate depression. As before, its behavior is exemplified by its response to 20 Hz and 33 Hz. At 20 Hz, facilitation that has been accumulated in the course of one syllable is over when the subsequent one starts (Fig. 4, middle right). Moreover, during a single syllable, facilitation is not able to overcome the moderate yet existing depression. Consequently, BNC-2 (high-pass) does not respond at all (Fig. 4, bottom right). At 33 Hz, on the other hand, facilitation between syllables still exists (due to the high frequency) and depression is already over. As a result, BNC-2 (high-pass) responds from the second syllable on (Fig. 4, bottom left). Notice that this model guarantees that the two neurons will fire concurrently only at the frequencies of the BNC-2 (band-pass), i.e. 25–30 Hz. At lower frequencies BNC-2 (high-pass) will not fire at all, while at higher frequencies BNC-1 will fire only as a response to the first syllable while BNC-2 only to the subsequent ones.

Based on the above descriptions, the behavior of the band-pass, the discerning neuron, becomes apparent. At frequencies between 25 and 30 Hz both the low and the high-pass neurons respond. Consequently, both secrete neurotransmitter and the sum of the secretions of both (and only the sum) onto the band neuron is sufficient to generate an action potential in the latter.

The final pattern of behavior of the entire network, as well as the synapse onto the band-pass neuron, is seen in Fig. 5. In C and E, release corresponding to BNC-1 (low-pass) and BNC-2 (high-pass) are seen. Due to the short duration of the synapses onto the band-pass only a rigid synchronization between BNC-1 and BNC-2 releases

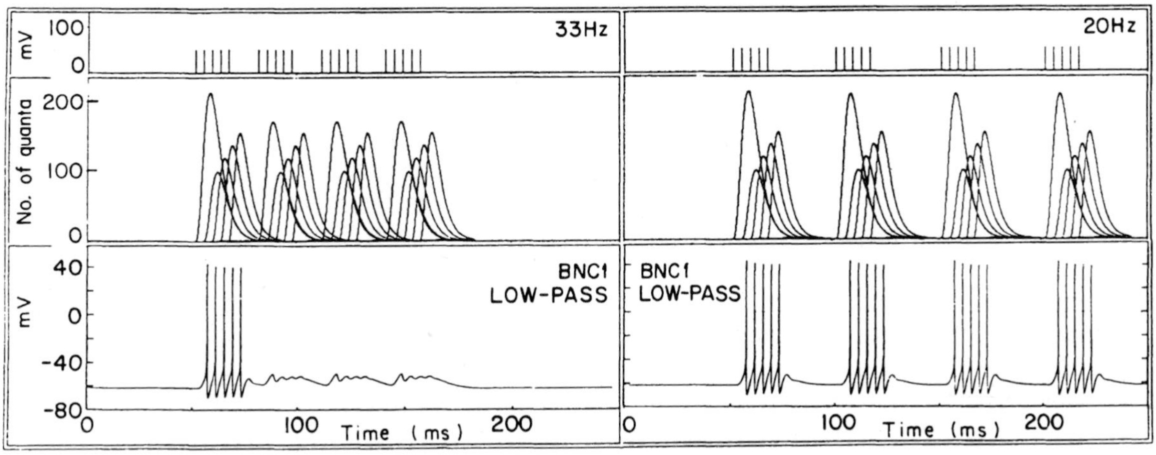

Fig. 3. BNC-1 (low-pass). The response of BNC-1 (low-pass) to syllables at two different rates: 20 Hz (right) and 33 Hz (left). In the upper strip syllables arriving from the AN-1 neuron are seen. The middle strip depicts the time course of release of the synapse onto the BNC-1 (low-pass) neuron. The last strip depicts the membrane potential in the unit SOMA.

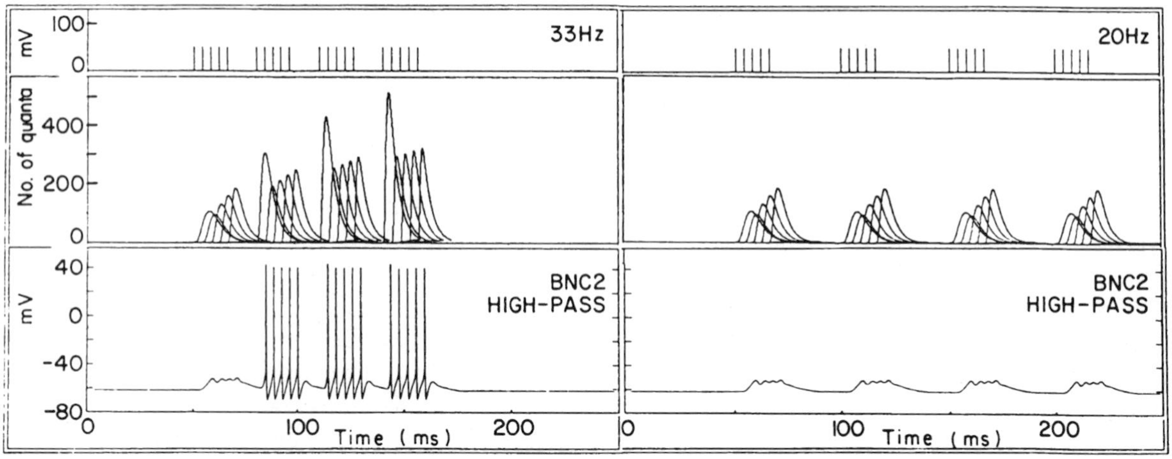

Fig. 4. BNC-2 (high-pass). The repsonse of BNC-2 (high-pass) to syllables at two different rates: 20 Hz (right) and 33 Hz (left). In the upper strip the syllables arriving from the AN-1 neuron are seen. The middle strip depicts the time course of release of the synapse onto the BNC-1 (low-pass) neuron. The last strip depicts the membrane potential in the unit SOMA.

will generate a response in the band-pass neuron (Fig. 5F). Although we do not claim that our model necessarily corresponds in all details to the real biological mechanism, it is interesting to note that it agrees with the observations of Huber and Thorson (1985) that the low-pass neuron always fires before the high-pass neuron. The model presented can be used as a case study to assess the important feature of synapses in the overall behavior of a network. It is already clear at this stage that synaptic amplitude, while important, is not sufficient to characterize the performance of synapses. To assess the significance of other synaptic features we will eliminate, in the next paragraph, one by one some of the key features of synapses and examine the outcome.

Synapses devoid of key features

δ-like synapses

The importance of a long duration synapse was demonstrated in Fig. 3. The first action potential in a syllable evoked a large and long release onto BNC-1 (low-pass). This release was sufficient to generate an action potential in the low-pass neuron. The next action potential evoked a much reduced release, due to a strong depression and insufficient facilitation to overcome it. This reduced release, however, still managed to produce an action potential in the low-pass neuron as it overlapped with the still unterminated previous release.

Figure 6 (left side) demonstrates the behavior of the network when this long synapse is replaced by a δ-like synapse. It can be seen that even at the proper frequency of 25 Hz only the first action potential of a syllable is able to generate a response in the low-pass neuron (bottom left). Under such conditions the band-pass neuron will not fire at any frequency.

Synapses with no synaptic delay

The importance of a rather strict synchronization between the synapses onto the band-pass is demonstrated in Fig. 5. Such a synchronization is achieved, in our model, by coordinated synaptic delays of the synapses onto the band-pass cell. Indeed, without synaptic delays, synchronization is lost and the band-pass does not fire at the desired frequency (Fig. 6, right). Notice that, other than the synaptic delay, the synaptic behavior looks identical to that observed earlier (compare with Figs. 5C and 5E). Nevertheless, the band-pass nueron fires a single action potential only.

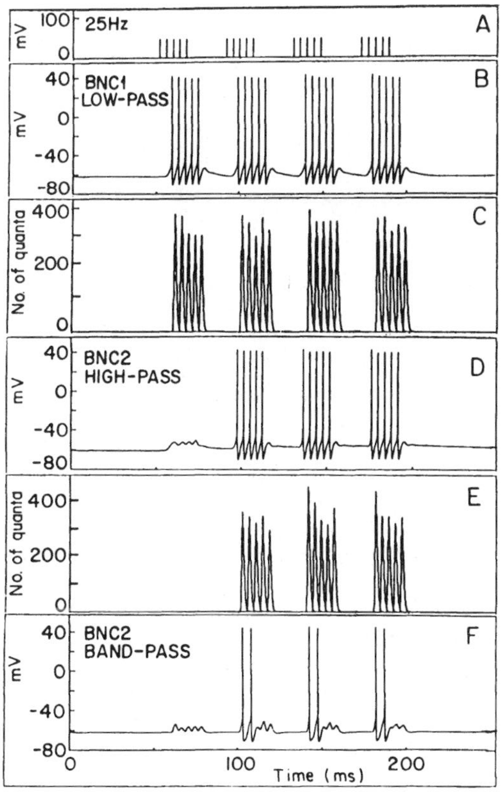

Fig. 5. BNC-2 (band-pass). The response of BNC-2 (band-pass) to syllables arriving at the correct 25 Hz. Both the BNC-1 (low-pass) and the BNC-2 (high-pass) fire (B and D) such that starting from the second syllable both cells activate concurrently their synapse onto the BNC-2 (band-pass) neuron.

Synapses devoid of facilitation

The role of facilitation in achieving high-pass filters was clearly demonstrated before. As expected, when facilitation is eliminated, release onto the high-pass neuron is too low (compare Fig. 7 upper left with Fig. 4 upper left). Consequently, the high-pass neuron does not fire at any frequency (Fig. 7 left bottom).

Synapses devoid of depression

Depression, as shown before, is important to produce the desired behavior of the low-pass neuron. As expected, without depression, the latter is unable to discern among different frequenceis, and fires at any input rate. In particular it fires at 33 Hz, where it should not fire (compare Fig. 7 right with Fig. 3 left). The analysis given above clearly reveals the importance of using biologically realistic synapses. Therefore, synapses represented by arbitrary function such as the α function, or the $+/-$ synapses, must be rejected as being unsuitable.

Models for release of neurotransmitter

Recall that representation of a synapse by a model which includes all the existing known details have been rejected as being inefficient. In spite of this, the adequate model to describe synaptic behavior must be assessed in order that a succinct analytical expression may be derived therefrom. As we are dealing in the present work with the presynaptic nerve terminal only, we provide below a brief account of how the proper model for release of neurotransmitter could be discerned.

To date two hypotheses exist to explain what controls the time course of neurotransmitter release. The Ca hypothesis, well established, implies that it is the rise of intracellular Ca^{2+} concentrations in the vicinity of the release sites which initiates release, while the rapid decline of intracellular Ca^{2+} terminates release (see review H. Parnas et al., 1990). The alternative Ca-Voltage hypothesis asserts that it is the depolarization-dependent activation of a different membranous component of the release machinery (i.e. other than the opening of Ca^{2+} channels) which initiates transmitter release, which is terminated by the subsequent inactivation of this mechanism upon membrane repolarization.

There is an inherent qualitative difference between the two hypotheses which enables them to be empirically distinguished. According to the Ca hypothesis, the time course of release must be tightly linked to the temporal distribution of intracellular Ca^{2+} concentration, whereas no such linkage is expected according to the Ca-Voltage hypothesis. Therefore, experiments in which conditions are set to significantly alter the temporal δ - LIKE SYNAPSE

NO SYNAPTIC DELAY

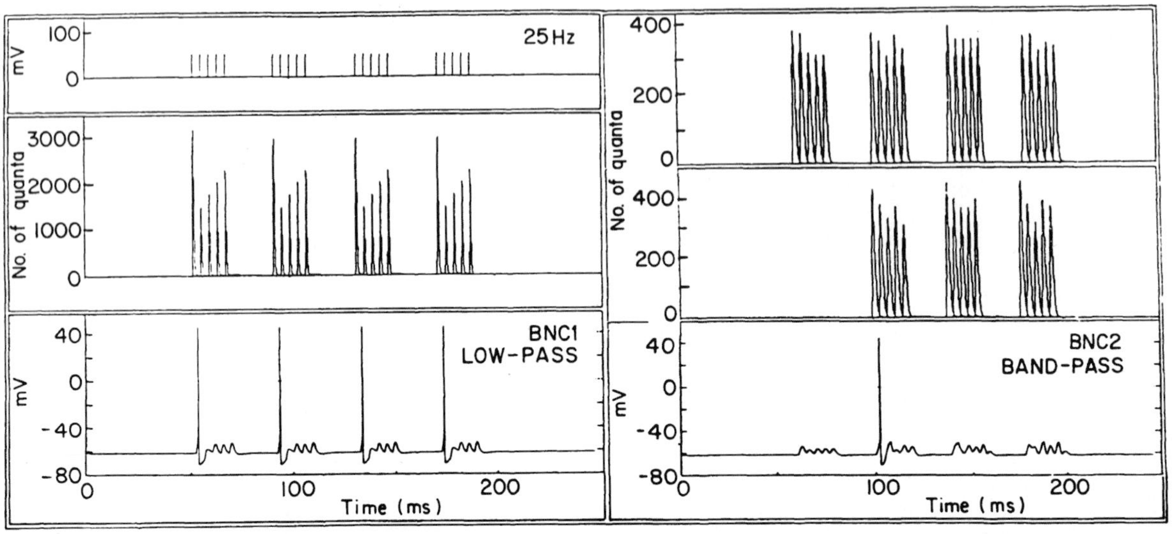

Fig. 6. Synapses devoid of key features: no synaptic delay and short duration. Left side: the synapse onto BNC-1 (low-pass) is changed such that its duration is only 2 msec (middle). As a result the neuron fires only one action potential at each burst (bottom). Right side: the synapse from BNC-2 (high-pass) onto BNC-2 (band-pass) exhibits no synaptic delay. Release from BNC-1 (low-pass) is seen in the upper strip, release from BNC-2 (high-pass), in the middle strip and the resulting membrane potential in the SOMA of the BNC-2 (band-pass) in the bottom.

NO FACILITATION

NO DEPRESSION

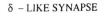

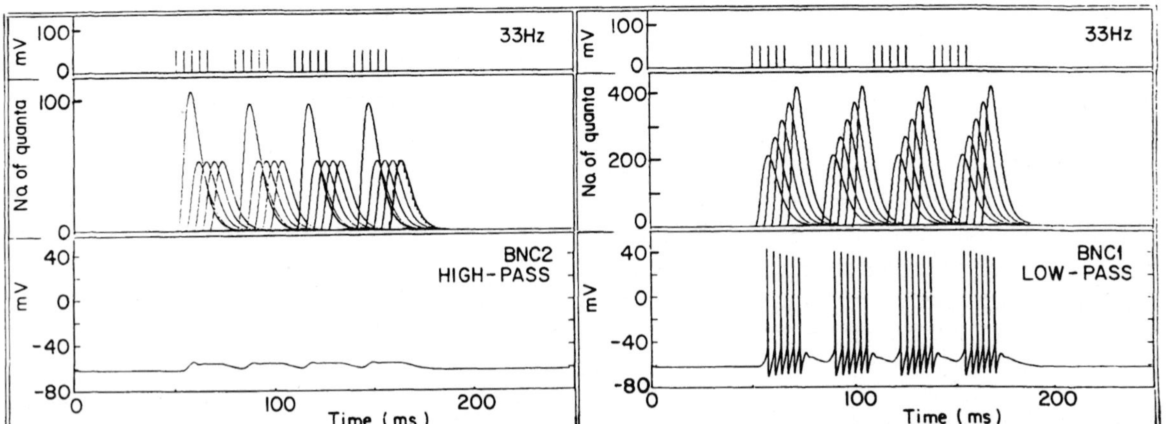

Fig. 7. Synapses devoid of key features: facilitation and depression. Left side: the synapse onto BNC-2 (high-pass) is devoid of facilitation. In the middle strip the pattern of release is seen and the resulting membrane potential is seen at the bottom. Right side: the synapse onto BNC-1 (low-pass) is devoid of depression. The consequent pattern of release is seen in the middle and the resulting membrane potential at the bottom.

distribution of intracellular Ca^{2+} should enable the rejection of one or the other of these hypotheses.

Figure 8 provides two examples of such experiments. It can be seen that if Ca^{2+} entering the cell is buffered faster than under normal conditions, by intracellularly injected Ca^{2+} buffer, the quantal content indeed declines, while the time course of release in not altered (Fig. 8a). The opposite effect on quantal content is seen in Fig. 8B, where a Ca^{2+} ionophore was added to the bathing solution: Ca^{2+} enters more or less continuously, independent of the depolarizing pulse, until a steady-state concentration is reached. Indeed, the quantal content was increased 5-fold in the presence of the Ca^{2+} ionophore. Nevertheless, in this case too, the time course of release remained the same as under control conditions.

These results add to a long list of experiments demonstrating a full independence between the temporal distribution of intracellular Ca^{2+} concentrations and the time course of release (e.g. Andreu and Barrett, 1980; Datyner and Gage, 1980; H. Parnas et al., 1989b. Arechiga et al., 1990). Based on the above, it seems safe to conclude that it is the Ca-Voltage hypothesis that should be pursued and used as a basis for developing analytical expressions for representing the presynaptic nerve terminal.

The Ca-Voltage hypothesis for release of neurotransmitter

The kinetic scheme of the Ca-Voltage hypothesis is as follows:

$$Ch_c \underset{k_c^{(\varphi)}}{\overset{k_o^{(\varphi)}}{\rightleftharpoons}} Ch_o \rightarrow \text{rise in } C \qquad (1)$$

$$T \underset{k_{-1}^{(\varphi)}}{\overset{k_1^{(\varphi)}}{\rightleftharpoons}} S;\ S + C \underset{k_{-2}}{\overset{k_2}{\rightleftharpoons}} Q;\ nQ + V_e \underset{k_{-3}}{\overset{k_3}{\rightleftharpoons}} V_e^*;$$

$$V_e^* \overset{k_4}{\longrightarrow} L$$

According to this scheme, during the depolarizing pulse (as assumed already by the Ca hypothesis) Ca channels (Ch_c) open in a voltage-dpeendent manner governed by the rate constant $k_o^{(\varphi)}$, which increases with depolarization. As a result, Ca^{2+} flows in and the concentration of intracellular Ca^{2+}, C, rises.

Concurrently and independently, according to the Ca-Voltage hypothesis, an inactive molecule T (probably membranous) is transformed during the depolarizing pulse into an active form, S, with a rate constant $k_1^{(\varphi)}$ which is voltage dependent (increases with rise in depolarization). Next, S binds to C so as to form a complex Q, a minimum number n of which is needed to render the vesicle, V_e, ready for release (this state denoted

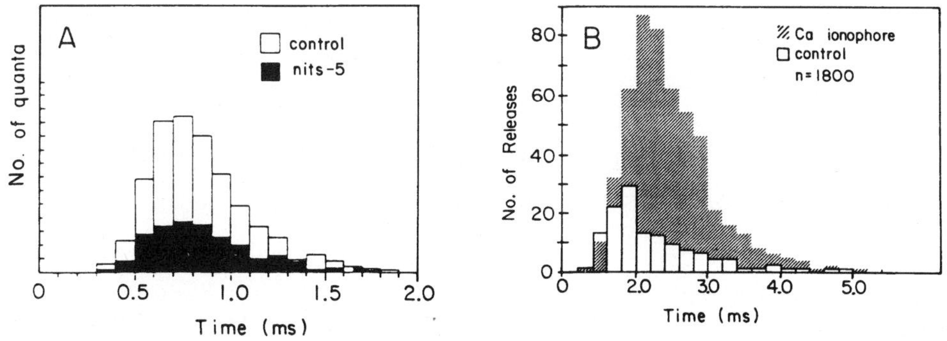

Fig. 8. Experiments in which the time course of release was measured under conditions that change the temporal distribution of Ca^{2+}. (A) Synaptic delay histograms in control and after injection of nitr-5 (redrawn from Hochner et al., 1991). (B) Synaptic delay histograms before and after application of a Ca^{2+} ionophore (redrawn from Parnas and Parnas, 1989).

here as V_e^*). Finally, V_e^* releases its content and, as a result, the release process, L, can finally commence. In (1) we assumed explicitly that S binds Ca^{2+}, but other variations of the Ca-Voltage hypothesis are possible without violating its essential point, i.e. that, the time course of release is controlled by the voltage-dependent $T \rightleftharpoons S$ transformation. The differential equations corresponding to scheme (1) including the equations describing the influx and removal of Ca^{2+} are as follows.

Regulation of C is provided by

$$\frac{\mathrm{d}C}{\mathrm{d}t} = Y(t) - R(t) \qquad (2)$$

where $Y(t)$ describes the voltage-dependent influx of Ca^{2+}, given by

$$Y(t) = \frac{Ch_o(\varphi)\overline{Y}C_o}{K_y(\varphi) + C_o} \qquad (3)$$

$Ch_o(\varphi)$, the steady-state number of open Ca^{2+} channels at any membrane potential, φ, is obtained by solving the differential equations corresponding to,

$$Ch_c \underset{k_c^{(\varphi)}}{\overset{k_o^{(\varphi)}}{\rightleftharpoons}} Ch_o$$

$\overline{Y}$ denotes the maximal possible change in C obtained at high extracellular Ca^{2+} concentrations, C_o. $K_y(\varphi)$ is a Michaelis-Menten type constant, but here is voltage-dependent. The rate of removal $R(t)$ is given by

$$R(t) = \frac{\overline{R}C}{K_R + C} \qquad (4)$$

where $\overline{R}$ and K_R stand for the maximal rate and the half-saturation constant, respectively. For details and justification of Eqns. (2–4) see Lustig et al., (1990).

The differential equations associated with the main part of scheme (1) are

$$\frac{\mathrm{d}T}{\mathrm{d}t} = k_{-1}^{(\varphi)} S - k_1^{(\varphi)} T \qquad (5)$$

$$\frac{\mathrm{d}S}{\mathrm{d}t} = k_1^{(\varphi)} T + k_{-2} Q - k_{-1}^{(\varphi)} S - k_2 SC \qquad (6)$$

$$\frac{\mathrm{d}Q}{\mathrm{d}t} = k_2 SC - k_{-2} Q - nk_3 V_e Q^n + nk_{-3} V_e^* \qquad (7)$$

$$\frac{\mathrm{d}V_e^*}{\mathrm{d}t} = nk_3 V_e Q^n - nk_{-3} V_e^* - k_4 V_e^* \qquad (8)$$

$$\frac{\mathrm{d}L}{\mathrm{d}t} = k_4 V_e^* \qquad (9)$$

and the conservation law,

$$T + S + Q + nV_e^* + nL = T_o \qquad (10)$$

Limited in providing insight as it may be, it is the quantal content, L_{tot} rather than the time course of release ($\mathrm{d}L/\mathrm{d}t$), which is most often measured experimentally. To assess L_{tot}, Eqn. (11) must be solved.

$$L_{tot} = \int_0^{\tau} \frac{\mathrm{d}L}{\mathrm{d}t}\mathrm{d}t \qquad (11)$$

where τ is the duration of evoked relaese.

As our interest here is in obtaining a succinct analytical expression, we shall not elaborate further on the details of Eqns. (2–11). Such discussions may be found in Lustig et al. (1989, 1990) and in H. Parnas et al. (1986). It is important to note that, had we wished to describe the terminal with all its detailed processes, we would have needed to solve all the differential equations (2–11). The terminal, however, is only one component of the entire neuron, and other components would presumably involve no fewer details, and hence a similar number of differential equations. Motivated by this consideration, we now outline briefly a number of ways of condensing the differential Eqns. (2–11) into a single analytical expression.

An analytical expression to represent the nerve terminal

Lustig et al. (1989) derived an approximated analytical expression based on Eqns. (2–11). This expression provides the time course ($\mathrm{d}L/\mathrm{d}t$) as well as the total amount of transmitter being released, L_{tot}. The following is a brief account of the considerations that permitted such an approximation. Simulations performed by Lustig et al. (1989) show that the model faithfully captures the main experimetal results reported in the literature. In particular, as in the experiments (see above for references), the calculated time course of release is invariable to changes in the temporal distribution of C. Moreover, as in the experiments (Dudel, 1984a,b; I. Parnas et al. 1986; Arechiga et al., 1990), the only treatment that changed the time course of release had been an application of a brief hyperpolarizing pulse immediately or shortly after the depolarizing one. Such a post-pulse hyperpolarization reduced the amount of release, with a concomitant shortening of the duration of evoked release. The close agreement between the model described in scheme (1) and experimental findings not only provides assurance for the model's plausibility, but also suggests that the values for the various rate constants in the model are well chosen. Therefore, with the aid of simulations of Eqns. (2–11) Lustig et al. (1989) could highlight the important processes in Eqns. (2–11) and disregard all others in the derivation of the approximated analytical expression.

The main steps in arriving at such expressions are outlined below. The experimental results shown in Fig. 8 exemplify the striking independence of the time course of transmitter release on either C or membrane depolarization. It is only the amount of release that depends from either factor. The simulations of Eqns. (2–11) capture precisely this key feature, and they also show that Q (the complex formed when S binds C) has the same shape as $\mathrm{d}L/\mathrm{d}t$. We may, therefore, define the magnitude of Q at time t, $Q(t)$, by its peak amplitude Q_{m} times a shape-function $q(t)$ whose max is unity.

Our goal is now more clearly defined. We wish to find approximate expressions for both Q_{m} and $q(t)$, since it is their product that will yield the desired expression for L_{T} (total release) and $\mathrm{d}L/\mathrm{d}t$ (the time course of release). To do so, we make the following approximations. Reconsidering the independence of the time course of release of C (Fig. 8) allows us to consider C as a constant, $\overline{C}$ during the duration of the release process. By doing so, we need not continuously solve the differential equations associated with the regulation of C. Instead, we simply evaluate the constant $\overline{C}$ by calculating the Ca^{2+} influx once, under the appropriate experimental conditions.

We now turn to finding approximate expersions for describing the time course of S. To do so, we reconsider the experimental result according to which the only experimental manipulation that succeeded in altering the time course of release was the administration of a brief hyperpolarizing pulse immediately after the depolarizing one (see above for references). This implies (in the context of the Ca-Voltage hypothesis) that, during the depolarizing pulse, the main reaction concerning S is the transition $T \rightarrow S$, while upon repolarization it is mainly the reverse that occurs, i.e. $S \rightarrow T$. These observations allow us to derive a simple expression for S during the depolarizing pulse (at the end of which S arrives at its maximal level S_{m} as well as following termination of the pulse.

In our efforts to arrive at expressions for Q_{m} and $q(t)$ we still need to make approximations concerning the differential equations associated with Q. For this, we take advantage of simulations indicating that Q is small in comparison with the expression responsible for its formation, k_2SC (see Eqn. 7). This allows us to neglect altogether the term $nk_3V_{\mathrm{e}}Q^n$ from Eqn. (7) and, consequently to arrive at a solvable differential equation for Q.

The three approximations described above are the main steps taken to arrive at an approximated expression for Q_{m} and $q(t)$. The resulting final approximated expression for L_{tot} and $\mathrm{d}L/\mathrm{d}t$ is

$$L_{\mathrm{tot}} = k_3V_{\mathrm{e}}f\left(\frac{k_2\overline{C}S_{\mathrm{m}}}{k_{-1}^{(\varphi)} + k_2\overline{C}}\right)^n$$

$$\times\left(1 - e^{(k_{-1}^{(\varphi)}+k_2C)(t_{\mathrm{m}}-t_{\mathrm{p}})}\right)^n \qquad (12)$$

$$f = \int_0^{\tau} [q(t)]^n \mathrm{d}t \qquad (13)$$

and

$$q(t) = \frac{Q(t)}{Q_m} = \frac{e^{-k_2(t-t_p)} - e^{-b(t-t_p)}}{e^{-k_2(t_m-t_p)} - e^{-b(t_m-t_p)}} \quad (14)$$

where $b = k_{-1}^{(\varphi)} + k_2\overline{C}$

Since both L_{tot} in Eqn. (12), and the time course of release given by $q(t)$ in Eqn. (14) are determined by a combination of only a few rate constants, there is no need to solve the many differential equations (2–11). Examination of Eqn. (14) shows that the time course of release depends almost solely on only two rate constants, d_2 and $k_{-1}^{(\varphi)}$, all the others (in Eqns. 2–11) being of only secondary importance. The total amount of release, L_{tot} depends also on the (constant) level of C, $\overline{C}$, and on the characteristics of the depolarizing pulse (amplitude and width), which will in turn determine S_m. Derivation of the analytical expression obviously permits a considerable saving in computer time; more important, however, is the clarification of precisely which steps are of key importance for neurotransmitter release. For a more complete description of the method of approximation see Lustig et al. (1989).

The TERMINAL in SONN contains Eqns. (12) and (14) by means of which L_{tot} and dL/dt are obtained. The 'memory' in TERMINAL, which served to arrive at facilitation, is the remaining level of C, $C(t)$, between pulses. $C(t)$ can also be evaluated employing an analytical expression. If the physiological rate of removal is far from saturation, it will depend exponentially on C. We have opted to describe $C(t)$ as

$$C(t) = \overline{C} - \beta t \quad (15)$$

The choice of Eqn. (15) is based on evidence indicating that the physiological rate of removal is in the saturated range (H. Parnas and Segel, 1980; H. Parnas et al., 1982; I. Parnas et al., 1982).

Discussion and Summary

The example presented here, i.e. the auditory communication system of the cricket, is only one of many which demonstrate the importance of considering real physiological synapses in the framework of neuronal network analysis. Aspects such as synaptic delay, facilitation, depression, and duration were theoretically confirmed to play an important role in the overall behavior of neuronal networks. These being inherent features of all biological synapses, it is only natural to try to ensure that the model synapse faithfully captures them. On the other hand, when developing a model synapse, care must be taken to ensure that only key processes are considered, for otherwise, further study becomes impractical.

Discussion in the present study was limited to the presynaptic nerve terminal. Obviously, many of the features shown to be important could have also been demonstrated via the complementary postsynaptic counterpart. Moreover, for a complete synapse to be efficiently yet faithfully represented, an approximate expression for the postsynaptic component must eventually be derived. Indeed, SONN includes Eqns. (12) and (14) in its TERMINAL unit, and different analytical expression in its POSTSYNAPSE unit. The latter expression has been derived using methodology similar to that described here. Based on a large number of experimental results a fully detailed model has been formulated. Simulations of the detailed model were then used as guidelines to what approximations are permissible. By this method, H. Parnas et al. (1989a) and Buchman and Parnas (1992) arrived at a (single) manageable analytical expression to replace the set of differential equations associated with the postsynaptic processes. As a result, in SONN, the complete synapse is represented by two analytical expressions, one summarizing presynaptic processes and the other postsynaptic ones.

We suggest that these two rather simple expressions should be employed in simulations of neuronal networks. These expressions are as simple and efficient as the arbitrary α function but, in contrast to the α and like functions, they possess all the inherent characteristics of a real biological synapse.

References

Andreu, R. and Barret, E.F. (1980). Calcium dependence of evoked transmitter release at very low quantal contents at frog neuromuscular junction. *J. Physiol. (London)*, 308: 79–97.

Arechiga, H., Cannone, A., Parnas, H. and Parnas, I. (1990). Blockage of synaptic release by brief hyperpolarizing pulses in the neuromuscular junction of the crayfish. *J. Physiol. (London)*, 430: 119–133.

Av-Ron, E., Parnas, H. and Segel, L.A. (1991). A minimal biophysical model for an excitable and osicllatory neuron. *Biol. Cybern.*, 65: 487–500.

Buchman, E. and Parnas, H. (1992). Sequential approach to describe the time-course of synaptic channel opening under constant transmitter concentration. *J. Theor. Biol.*, 158: 517–534.

Datyner, N.B. and Gage, P.W. (1980) Phasic secretion of acetylcholine at the mammalian neuromuscular junction. *J. Physiol. (London)*, 303: 299–314.

Dudel (1984a) Control of quantal transmitter release at frog's nerve terminals. I. Dependence on amplitude and duration of depolarization. *Pflügers Archi.*, 402: 225–234.

Dudel (1984b) Control of quantal transmitter release at frog's nerve terminals. II. Modulation by de- or hyperpolarizing pulses. Pflügers Arch., 402: 235–243.

Hochner, B., Parnas, H. and Parnas, I. (1991) Effects of intra-axonal injection of Ca^{2+} buffers on evoked release and on facilitation in the crayfish neuromuscular junction. *Neurosci. Lett.*, 125: 215–218.

Huber, F. and Thorson, J. (1985) Cricket auditory communication. *Sci. Am.*, 253 (Dec): 46–54.

Listig, C., Parnas, H. and Segel, L.A., (1989) Neurotransmitter release: development of a theory for total release based on kinetics. *J. Theor. Biol.*, 136: 151–170.

Lustig, C., Parnas, H. and Segel, L.A. (1990) Release Kinetics as a tool to describe drug effects on neurotransmitter release. *J. Theor. Biol.*, 144: 225–248.

Parnas, H. and Parnas, I., (1989) Kinetics of release as a tool to distinguish between models for neurotransmitter release. In: A. Goldbeter (Ed.), *Cell to Cell Signalling: From Experiments to Theoretical Models*, Academic Press, London, pp. 47–59.

Parnas, H. and Rudolph, L. (1991) *Representation of Synapses in Neuronal Networks*. Technical report. The Leibniz Center for Research in Computer Science, Hebrew University, Jerusalem, Israel.

Parnas, H. and Segel, L.A. (1980) A theoretical explanation for some effects of calcium on the facilitation of neurotransmitter release. *J. Theor. Biol.*, 84: 3–29.

Parnas, H., Dudel, J. and Parnas, I. (1982). Neurotransmitter release and its facilitation in crayfish. I. Saturation kinetics of release, and of entry and removal of calcium. *Pflügers Arch.*, 393: 1–14.

Parnas, H., Dudel, J. and Parnas, I. (1986). Neurotransmitter release and its facilitation in crayfish. VII. Another voltage-dependent process beside Ca entry controls the time-course of phasic release. *Pflügers Arch.*, 406: 121–130.

Parnas, H., Flashner, M. and Spira M. (1989a) Sequential model to describe the nicotinic synaptic current. *Biophys. J.*, 55: 875–884.

Parnas, H., Hovav, G. and Parnas, I. (1989b) Effect of Ca^{2+} diffusion on the time-course of neurotransmitter release. Biophys. J., 55: 859–874.

Parnas, H., Parnas, I. and Segel, L.A. (1990) On the contribution of mathematical models to the understanding of neurotransmitter release. *Int. Rev. Neurobiol.*, 32: 1–50.

Parnas, I., Parnas, H. and Dudel, J. (1982) Neurotransmitter release and its facilitation in crayfish. II. Duration of facilitation and removal processes of calcium from the terminal *Pflügers Arch.*, 393: 232–236.

Parnas, I., Parnas, H. and Dudel J. (1986) Neurotansmitter release and its facilitation in crayfish. VIII. Modulation of release by hyperpolarizing pulses. *Pflügers Arch.*, 406: 131–137.

J. van Pelt, M.A. Corner H.B.M. Uylings and F.H. Lopes da Silva (Eds.)
Progress in Brain Research, Vol 102

CHAPTER 12

Intrinsic neuronal physiology and the functions, dysfunctions and development of neocortex

Barry W. Connors

Department of Neuroscience, Brown University, Providence, RI 02912, U.S.A.

Introduction

Neurons encode information as a time series of action potentials, and then distribute them spatially according to the structure of their axons. It follows that the pattern of input to a neuron strongly determines the pattern of its output. But neurons can also transform the temporal patterns of their input, yielding outputs that may be slowed down or speeded up, that may adapt over time or greatly outlast the input, or that may become newly rhythmic. These transformations are effected, in large part, by the many voltage- and calcium-dependent ion channels of the neuron's membrane. The ion channels determine the 'intrinsic physiology' of a neuron, as I narrowly define the term here.

Experimental studies and hypotheses about brain development and function have concentrated strongly on the synaptic and connectional features of brain circuits. The intrinsic physiology of component neurons is often assumed to be constant, and with mundane features at that, but neuronal properties are neither constant nor mundane. Because the complement of ion channels expressed by a particular class of neuron varies, intrinsic physiology varies, often dramatically. Physiology also changes with development. The aim of this chapter is to ask whether, and to what extent, variations in the intrinsic physiology of neurons contribute to normal brain development and function, and to neuropathological processes. The answer may seem obvious, even trivial; intrinsic properties are so complex and diverse, yet so specific from one cell type to the next, that of course they must contribute to the functions of the brain. Why else would the brain go to the trouble? The problem is that very few studies have shown directly a relationship between a particular pattern of intrinsic physiology and the function of a brain circuit. It is relatively easy to document the properties of neurons in the brain, but much harder to assign ultimate functions to their traits.

I will focus my discussion on cells of the cerebral neocortex because an extensive functional taxonomy of it has been compiled, because it is a good example of neurotaxonomy in need of unifying theories with supporting evidence, and because the cortex fascinates me.

Diversity and specificity of intrinsic physiology

Biophysical studies have hinted for a long time that the variety of membrane ion channels in the brain is large; molecular biological studies are showing that they number in the many dozens (Hille, 1992). It is no surprise then that the intrinsic physiology of central neurons is also very diverse, due to variations in the type, density and placement of channels (Llinás, 1988). In the neocortex, where morphologically based taxonomy has long implied a variety of neuron types (DeFelipe and Jones, 1988), physiological studies also show diversity (reviewed by Connors and Gutnick, 1990; Amitai and Connors, in press). This is illustrated by the repetitive firing of cortical neurons in response to a simple stimulus, a step of depolarizing current injected across the membrane (Fig. 1). Most cells fire initially at a high rate, but fall to a much lower rate within tens of milliseconds, i.e. they adapt. A minority of cells can sustain their high firing rates without adaptation for long periods. Other cells fire in spike clusters, or bursts, and may even generate repeating bursts at regular rates. There are also variations on these basic themes. Adaptation may proceed more or less rapidly or completely, and cells may shift from burst-firing to single spike-firing depending upon conditions. The basis for physiological diversity in cortical neurons lies, of course, in their ion channels. In single pyramidal cells of layer 5 alone there may be upwards of 20 types of voltage-, calcium- and sodium-dependent channels (Schwindt, 1992). There are very few studies that

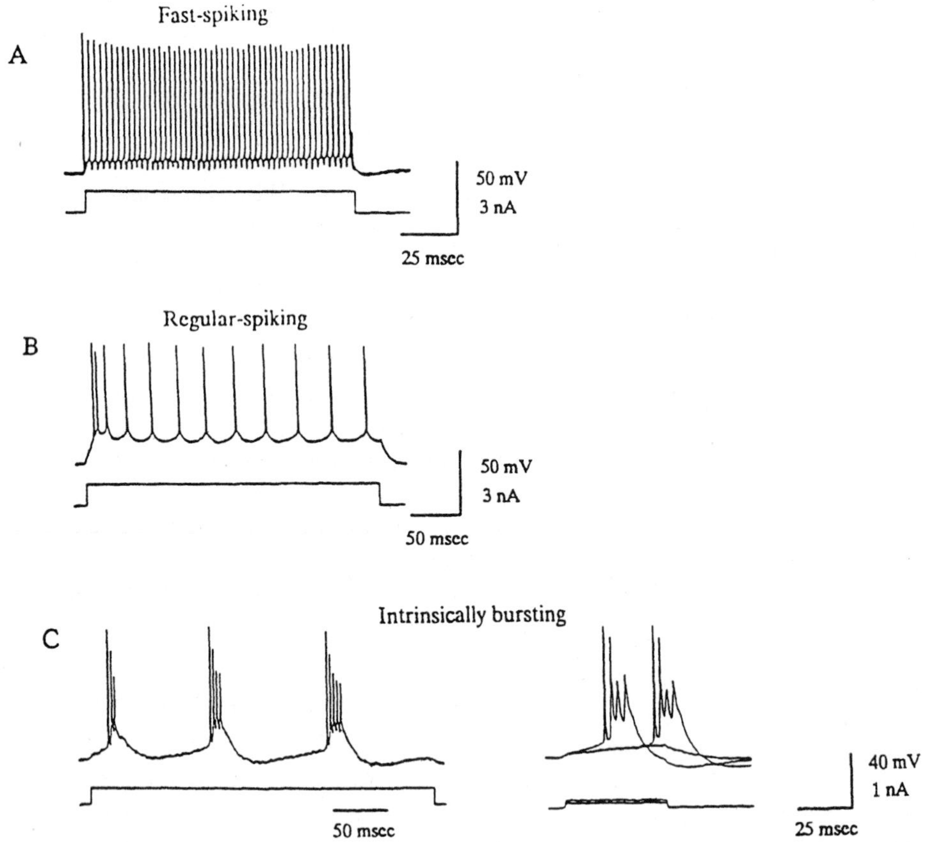

Fig. 1. Diversity of intrinsic firing patterns in neocortical neurons. (A) Fast-spiking neurons have brief action potentials, fire repetitively with little or no adaption, and are GABAergic interneurons. (B) Regular-spiking neurons have longer lasting action potentials, adapting patterns and are pyramidal or spiny stellate cells of layers 2 through 6. (C) Intrinsically bursting neurons generate clusters of spikes, sometimes in repeating patterns. They are pyramidal or spiny stellate cells primarily of layers 5, and occasionally 4. (Adapted from Connors and Gutnick, 1990.)

define the channel mechanisms underlying the diverse intrinsic firing properties in neocortex (Spain et al., 1991).

Variations in intrinsic physiology are not haphazardly distributed among cortical neurons, but show a high degree of specificity. In general, adapting and bursting patterns of firing occur in the spiny cells of the cortex, in particular the pyramidal cells (Fig. 2) (McCormick et al., 1985). Spiny cells are the excitatory neurons of the neocortex, and its exclusive source of output to the rest of the brain. Among them, intrinsic bursting occurs most frequently in a large subclass of pyramidal cells within layer 5; smaller layer 5 pyramids tend to be regular-spiking (Larkman and Mason, 1990; Chagnac-Amitai et al., 1990). The physiological dichotomy among layer 5 pyramidal cells extends even to their dendrites. Apical dendrites of large cells with elaborate dendritic morphology tend to have complex electrical properties with substantial calcium-dependent components, while smaller pyramids have simpler, purely sodium-dependent excitability (Kim and Connors, 1993). Intrinsically bursting cells of layer 5 project to major subcortical targets such as the pons and the superior colliculus (Wang and McCormick, 1993; Kasper et al., in press). On the other hand, layer 5 pyramidal cells that project to the contralateral cortex have non-bursting properties (Kasper et al., in press).

The other major class of neocortical neurons consists of the smooth (or sparsely spiny) inhibitory interneurons, which utilize γ-aminobutyric acid (GABA) as their transmitter. Some of these are the exclusive generators of the non-adapting, fast-spiking patterns (McCormick et al., 1985; Huettner and Baughman, 1988). A subclass of GABAergic neurons is not fast spiking, but gen-

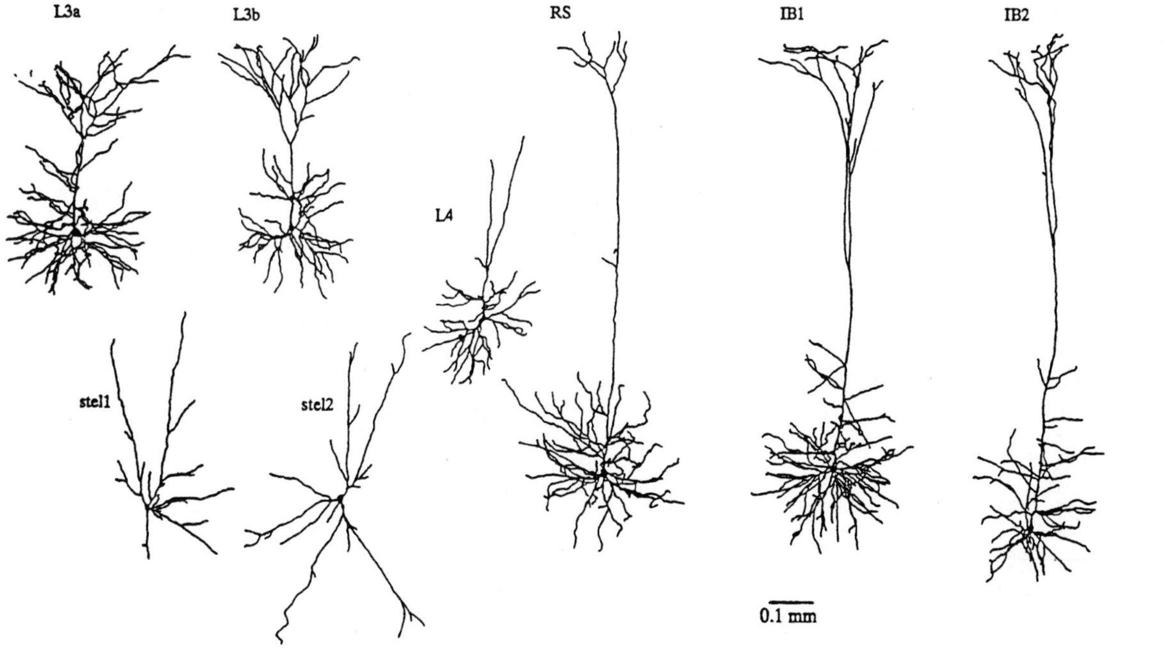

Fig. 2. Morphological correlates of intrinsic physiology. Examples of regular-spiking cells are the pyramidal neurons of layer 3 (L3a and L3b) and layer 5 (RS), and the spiny stellate neuron of layer 4 (L4). Intrinsically bursting cells include the largest pyramidal cells of layer 5 (IB1 and IB2). Fast-spiking cells are smooth or sparsely spiny interneurons (stel1 and stel2). (Adapted from Cauller and Connors, 1994)

erates voltage-dependent, low-threshold spiking patterns (Kawaguchi, 1993). There has been no laminar specificity for the intrinsic properties of smooth cells described so far. GABAergic neurons are distributed through all cortical layers, but there is laminar specificity to morphological subclasses (Somogyi, 1990) and we might expect physiological parallels to be found.

To summarize, the intrinsic physiology of cortical neurons is both diverse and specific. Cells can be classified on the basis of their structure, their function, and their biochemistry.

Development of membrane properties in neocortex

The intrinsic membrane properties of neocortical pyramidal cells change dramatically during the postnatal period in cats and rats (Purpura et al., 1965; Connors et al., 1983; Kriegstein et al., 1987; Hamill et al., 1991). In the rat, between the day of birth and over the next 2 weeks single action potentials increase their amplitude and their maximum rate of rise, and decrease their duration (Fig. 3) (McCormick and Prince, 1987). Over this period, pyramidal neurons also become more competent at repetitive firing, and can sustain increasingly higher rates for longer durations.

Voltage clamp studies of dissociated cells show that there are large postnatal increases in the density of voltage-dependent ion conductances, in both pyramidal and non-pyramidal neurons (Huguenard et al., 1988; Hamill et al., 1991).

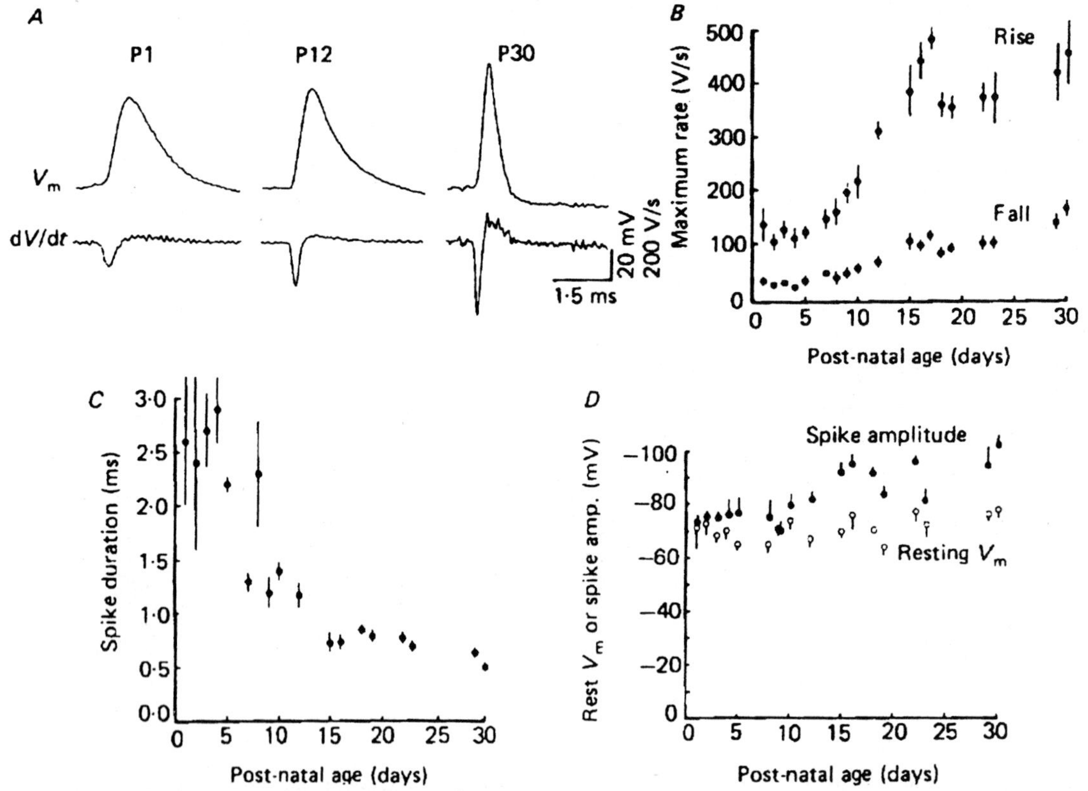

Fig. 3. Development of action potential properties in neocortical neurons. (A) Examples of single spikes (V_m) and their derivatives (dV/dt) at three postnatal ages. (B) Changes in the maximum rate of action potential rise and fall with age. (C) Age-dependence of the spike duration (measured at half-amplitude). (D) Changes in spike amplitude and membrane resting potential with age. (From McCormick and Prince, 1987.)

Interestingly, the kinetic properties of sodium currents do not change, even as their density increases 10-fold. The story for potassium currents is more complex, and as their density increases there are changes in the relative proportions of sustained and transient currents during the first 2 postnatal weeks. Calcium currents also increase postnatally; high threshold calcium currents increase approximately 3-fold in both pyramidal and non-pyramidal cells, while low-threshold calcium currents are undetectable at birth, but can be recorded several days later (Hamill et al., 1991).

This slow maturation of intrinsic physiology would be of only passing biological interest if the general functions of the cortex did not begin until the animal was 2 weeks old. This is certainly not the case, however. Afferent and efferent connections of the neocortex are established near or before the day of birth (Catalano et al., 1991), and excitatory synapses are functional (Agmon and O'Dowd, 1992; Kim et al., in press). The neurons of the cortex are also sensitive to manipulations of their input shortly after birth. For example, the sensory receptive fields of neurons in the rat somatosensory cortex are responsive to the removal of vibrissae during the first few postnatal days (Fox, 1992). The data imply that the relatively meager physiological properties of newborn rat cortical neurons are nevertheless enough to allow them to experience and respond to environmental stimuli. It has not been established whether the slow maturation of the physiology is relevant to development.

Pathology of intrinsic physiology

Ion channels are centrally important to every neuron in the brain, indeed to every cell in the body, so it seems likely that channel dysfunction (and thus altered intrinsic physiology) might be the basis for a variety of diseases. There are some clear examples of this outside the central nervous system. Cystic fibrosis is caused by faulty regulation of a chloride channel in the membranes of various secretory cells (Jentsch, 1993). A form of myotonia congenita is also associated with a disruption of chloride conductance, in the membrane of skeletal muscle (Adrian and Bryant, 1974). The latter results in a higher membrane resistivity and greater sensitivity to activity-dependent increases in extracellular potassium; the after-effects of a few action potentials can then lead to sustained depolarization, repetitive firing and tetanus. An analogous process is at least possible in the central nervous system. When chloride conductance is abolished in small central axons, a similar spontaneous, self-sustaining, potassium-dependent hyperexcitability can ensue (Connors and Ransom, 1984). The cyclic spontaneous firing diminishes as the axons mature and myelinate (Fig. 4).

There are few examples of brain disease that can be directly attributed to alterations in intrinsic membrane properties (I exclude from this statement disorders of ligand-gated, or neurotransmitter-associated, ion channels). One likely place to look would be disorders related to excitability, and epilepsy is the most obvious. Indeed, epileptologists have speculated for years that an increase in the intrinsic excitability of cerebral cortical neurons might underlie certain forms of

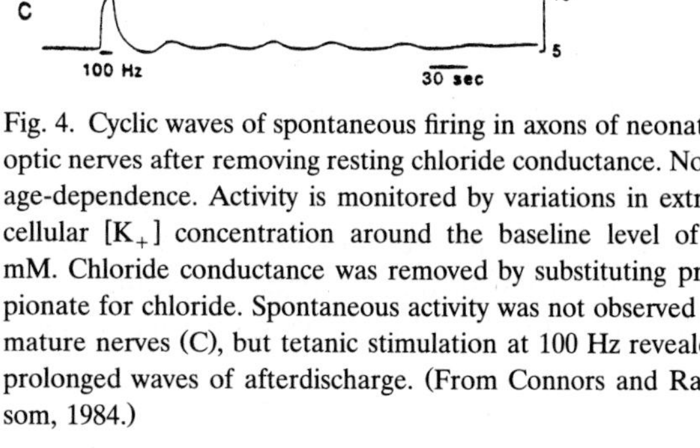

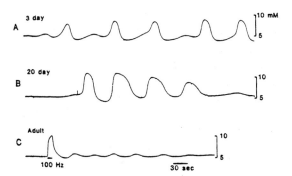

Fig. 4. Cyclic waves of spontaneous firing in axons of neonatal optic nerves after removing resting chloride conductance. Note age-dependence. Activity is monitored by variations in extracellular $[K_+]$ concentration around the baseline level of 5 mM. Chloride conductance was removed by substituting propionate for chloride. Spontaneous activity was not observed in mature nerves (C), but tetanic stimulation at 100 Hz revealed prolonged waves of afterdischarge. (From Connors and Ransom, 1984.)

seizures (Jackson, 1881; Prince, 1978; Schwartzkroin and Wyler, 1980; Connors and Amitai, 1993), but direct evidence has been lacking. In a recent study, Prince and Tseng (1993) showed changes in the intrinsic excitability of neurons in chronic epileptiform lesions caused by undercutting the white matter of the cortex. Layer 5 pyramidal cells of the epileptic tissue had a higher sensitivity to injected current, compared with non-epileptic control cells. The change in intrinsic excitability could be due to a variety of mechanisms, including altered channel types or densities or altered transmitter modulation of the same complement of channels. It is not at all clear whether the change in excitability is related to the seizures; the same tissue displayed changes in inhibitory synaptic functions, and perhaps the sprouting of local excitatory axons.

Spontaneous or engineered mutations offer a unique view of the role of intrinsic properties. As an example, the autosomal recessive mouse mutant *shiverer* has a severe deficit of central myelin due to deletions in its gene for myelin basic protein (Roach et al., 1985). By itself, eliminating myelin from an axon should greatly decrease or abolish excitability. However, *shiverer* mice have a *hyper*excitable phenotype, and exhibit tremor and seizures. The basis for this may be that the large-caliber central axons of *shiverer* develop a much higher than normal density of sodium channels (Noebels et al., 1991; Westenbroek et al., 1992), in effect overcompensating for their lack of myelin. An increased number of sodium channels has also been measured in the brains of E1 mice, a mutant strain of sensory-triggered epilepsy (Sashihara et al., 1992). Interestingly, there is a *decreased* density of sodium channels in hippocampal neurons of mice with trisomy 16, which is an animal model for Down's syndrome (human trisomy 21) (Galdzicki et al., 1993).

Shiverer is an example of how a severely impaired central nervous system (lacking myelin) can alter its constitution (increase sodium channel number) to avoid what should probably be a much more debilitating phenotype. Another mouse, called *reeler*, also compensates impressively. *Reeler* has a developmental defect that prevents cortical neurons from migrating into their proper layers. This results in disordered layers that are largely inverted compared with normal cortex (Caviness et al., 1988). *Reeler* has movement disorders related to disruption of its cerebellar circuitry, but its grossly malformed neocortex seems to compensate anatomically and physiologically in a remarkably effective way. Responses of visual cortical neurons are essentially normal (Simmons and Pearlman, 1983), and the intrinsic properties of cortical neurons are indistinguishable from control mice except that their laminar distribution is upside down (Fig. 5) (Silva et al., 1991). Intracellular injections into neurons of *reeler* cortex also demonstrate that the malformation of its neurons does not influence their ability to create an apparently normal intrinsic physiology (Silva et al., 1991).

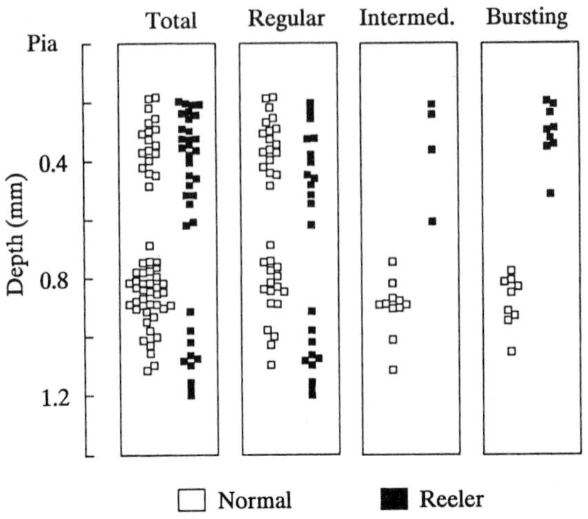

Fig. 5. Inversion of intrinsic physiological properties in mutant *reeler* cortex. The firing properties of neurons (classified as in Fig. 1) were assayed in various cortical layers. There were no differences in properties when neurons were grouped, and regular-spiking cells were found at all sampled depths in both genotypes. However intrinsically bursting cells, and a class of cells intermediate between bursting and regular-spiking, were seen only in layer 5 of normal cortex and in the upper (supragranular) layer of the *reeler*. (From Silva et al., 1991.)

While some of these studies suggest that ion

channel density, and thus intrinsic excitability, may be the basis for the expression of pathological behaviors, they are far from proof. Certainly developmental adjustments can mitigate the influence of a bad gene, while at the same time confusing an investigator. New methods allowing manipulation of specific channel genes, and their regulatory systems, in situ and on-demand will be helpful in designing more definitive studies.

Do variations of intrinsic physiology matter?

The likely answer is: yes, of course! The circumstantial evidence seems overwhelming. The intrinsic physiology of neurons varies immensely across the brain and within brain structures. It varies specifically and predictably. It changes systematically during development. It is altered in pathological states, often in ways that seem to explain the symptoms. In some cases the intrinsic properties of a class of neurons fit exquisitely well with the unique behaviors of their circuits; synchronous oscillations driven by intrinsically rhythmic neurons of the thalamus are a premier example from the vertebrate brain (Steriade and Llinás, 1988; McCormick, 1992). In the cerebral cortex, the intrinsic properties of different neurons may serve several distinct and important functions within the local circuitry (Amitai and Connors, in press).

An alternative answer to the question is that variations of intrinsic properties are an epiphenomenon, that they are incidental to the operations of the brain. It may suffice that neurons fire action potentials, and do so repetitively over a certain range, and that beyond these loose requirements the variations of their intrinsic rates of adaptation and rhythmicity do not matter. Indeed, descriptions of some computational models of neural circuits emphasize their 'robustness' in the face of changing neuronal properties; the large majority of models blithely assume a single simple form of intrinsic behavior and never explore the effects of varying it. Exploration of parameter space tends to focus on the structure and properties of synaptic connections, and the champion of current theories for developmental and pathological phenomena is synaptic plasticity. Indeed, neural circuitry can often accomplish what intrinsic membrane properties alone might do, as in the generation of rhythmic activities (Selverston, 1985; Getting, 1989; Grillner et al., 1991).

Unfortunately, the frank answer to the question in the heading is: we don't really know. Direct evidence for the importance of intrinsic properties is exceedingly hard to come by. It requires specifically altering the intrinsic properties of a particular population of neurons without also affecting synaptic mechanisms, and doing it under conditions in which the functions of the neural circuit can be assessed. Refined pharmacological and molecular biological methods offer hope. Formal modelling of neural circuits will be indispensable, but this requires a deep, quantitative knowledge of the modelled system that is so far difficult to achieve in the vertebrate nervous system.

Acknowledgements

The author's research is funded by the National Institutes of Health and the Office of Naval Research.

References

Adrian, R.H. and Bryant, S.H. (1974) On the repetitive discharge in myotonic muscle fibers. *J. Physiol. (London)*, 240: 505–515.

Agmon, A. and O'Dowd, D.K. (1992) NMDA receptor-mediated currents are prominent in the thalamocortical synaptic response before maturation of inhibition. *J. Neurophysiol.*, 68: 345–348.

Amitai, Y. and Connors, B.W. (in press) Intrinsic physiology and morphology of single neurons in neocortex. In: E.G. Jones and I.T. Diamond (Eds.), *Cerebral Cortex, Vol. 11, The Barrel Cortex of Rodents*, Plenum Press, New York.

Catalano, S.M., Robertson, R.T. and Killackey, H.P. (1991) Early ingrowth of thalamocortical afferents to the neocortex of the prenatal rat. *Proc. Natl. Acad. Sci. USA*, 88: 2999–3003.

Cauller, L.J. and Connors, B.W. (1994) Synaptic physiology of horizontal afferents to layer I of primary somatosensory cortex in rats. *J. Neurosci.*, 14: 751–752.

Caviness, V.S., Crandall, J.E. and Edwards, M.A. (1988) The

reeler malformation: implications for histogenesis. In: A. Peters and E.G. Jones (Eds.), *Cerebral Cortex, Vol. 7,* Plenum, New York, pp. 58–89.

Chagnac-Amitai, Y., Luhmann, H.J. and Prince, D.A. (1990) Burst generating and regular spiking layer 5 pyramidal neurons of rat neocortex have different morphological features. *J. Comp. Neurol.,* 296: 598–613.

Connors, B.W. and Amitai, Y. (1993) Generation of epileptiform discharge by local circuits of neocortex. In: P. Schwartzkroin (Ed.), *Epilepsy: Models, Mechanisms and Concepts,* Oxford University Press, New York, pp. 388–423.

Connors, B.W. and Gutnick, M.J. (1990) Intrinsic firing patterns of diverse neocortical neurons. *Trends Neurosci.,* 13: 99–104.

Connors, B.W. and Ransom, B.R. (1984) Chloride conductance and extracellular potassium concentration interact to modify the excitability of rat optic nerve fibres. *J. Physiol. (London),* 355: 619–633.

Connors, B.W., Benardo, L.S. and Prince, D.A. (1983) Coupling between neurons of the developing rat neocortex. *J. Neurosci.,* 3: 773–782.

Fox, K. (1992) A critical period for experience-dependent synapticplasticity in rat barrel cortex. *J. Neuroscience,* 12: 1824–1838.

DeFelipe, J. and Jones, E.G. (1988) *Cajal on the Cerebral Cortex,* Oxford University Press, New York, pp. 557–622.

Galdzicki, Z., Coan, E. and Rapaport, S.I. (1993) Cultured hippocampal neurons from trisomy 16 mouse, a model for Down's syndrome, have an abnormal action potential due to a reduced inward sodium current. *Brain Res.,* 604: 69–78.

Getting, P.A. (1989) Reconstruction of small neural networks. In: C. Koch and I. Segev (Eds.), *Methods in Neuronal Modelling. From Synapses to Networks,* MIT Press, Cambridge, MA, pp. 171–194.

Grillner, S., Wallén, P., Brodin, L. and Lansner, A. (1991) Neuronal network generating locomotor behavior in lamprey — circuitry, transmitters, membrane properties and simulation. *Annu. Rev. Neurosci.,* 14: 169–199.

Hamill, O.P., Huguenard, J.R. and Prince, D.A. (1991) Patch-clamp studies of voltage-gated currents in identified neurons of the rat cerebral cortex. *Cerebral Cortex,* 1: 48–61.

Hille, B. (1992) *Ionic Channels of Excitable Membranes,* 2nd ed., Sinauer Assoc., Sunderland, MA. 407 pp.

Huettner, J.E. and Baughman, R.W. (1988) The pharmacology of synapses formed by identified corticocollicular neurons in primary culture of rat visual cortex. *J. Neurosci.,* 8: 160–175.

Huguenard, J.R., Hamill, O. and Prince, D.A. (1988) Developmental changes in Na^+ conductances in rat neocortical neurons: appearance of a slowly inactivating component. *J. Neurophysiol.,* 59: 778–795.

Jackson, J.H. (1881), quoted in J. Taylor (Ed.), (1931) *Selected Writings of John Hughlings Jackson, Vol. 1, On Epilepsy and Epileptiform Convulsions,* Hodder and Stoughton, London.

Jentsch, T.J. (1993) Chloride channels. *Curr. Opin. Neurobiol.,* 3: 316–321.

Kasper, E., Larkman, A.U., Lübke, J. and Blakemore, C. (in press) Pyramidal neurons in layer 5 of the rat visual cortex. I. Correlation between cell morphology, intrinsic electrophysiological properties and axon targets. *J. Comp. Neurol.,*

Kawaguchi, Y. (1993) Groupings of non-pyramidal and pyramidal cells with specific physiological and morphological characteristics in rat frontal cortex. *J. Neurophysiol.,* 69: 416–431.

Kim, H.G. and Connors, B.W. (1993) Apical dendrites of the neocortex: correlation between sodium- and calcium-dependent spiking and pyramidal cell morphology. *J. Neurosci.,* 13: 5301–5311.

Kim, H.G., Fox., K. and Connors, B.W. (submitted) Properties of excitatory synaptic events in neurons of the primary somatosensory cortex of neonatal rats. *Cerebral Cortex,* in press.

Kriegstein, A.R., Suppes, T. and Prince, D.A. (1987) Cellular and synaptic physiology and epileptogenesis of developing rat neocortical neurons in vitro. *Dev. Brain Res.,* 431: 161–171.

Larkman, A.U., Mason, A. (1990) Correlations between morphology and electrophysiology of pyramidal neurons in slices of rat visual cortex. I. Establishment of cell classes. *J. Neurosci.,* 10: 1407–1414.

Llinás, R.R. (1988) The intrinsic electrophysiological properties of mammalian neurons: insights into central nervous system function. *Science,* 242: 1654–1664.

McCormick, D.A. (1992) Neurotransmitter actions in the thalamus and cerebral cortex and their role in neuromodulation of thalamocortical activity. *Prog. Neurobiol.,* 39: 337–388.

McCormick, D.A. and Prince, D.A. (1987) Post-natal development of electrophysiological properties of rat cerebral cortical pyramidal neurones. *J. Physiol. (London),* 393: 743–762.

McCormick, D.A., Connors, B.W., Lighthall, J.W. and Prince DA (1985) Comparative electrophysiology of pyramidal and sparsely spiny stellate neurons of the neocortex. *J. Neurophysiol.,* 54: 782–806.

Noebels, J.L., Marcom, P.K. and Jalilian-Tehrani, M.H. (1991) Sodium channel density in hypomyelinated brain increased by myelin basic protein gene deletion. *Nature,* 352: 431–434.

Prince, D.A. (1978) Neurophysiology of epilepsy. *Annu. Rev. Neurosci.,* 1: 395–415.

Prince, D.A., Tseng, G.-F. (1993) Epileptogenesis in chronically injured cortex: in vitro studies. *J. Neurophsyiol.,* 69: 1276–1291.

Purpura, D.P., Shofer, R.J. and Scarff, T. (1965) Properties of synaptic activities and spike potentials of neurons in immature neocortex. *J. Neurophysiol.,* 28: 925–942.

Roach, A., Takahashi, N., Pravtcheva, D., Ruddle, F. and Hood, L. (1985) Chromosomal mapping of mouse myelin basic protein gene and structure and transcription of the partially deleted gene in shiverer mutant mice. *Cell,* 42: 149–155.

Sashihara, S., Yanagihara, N., Kobayashi, H. and Izumi, F. (1992) Overproduction of voltage-dependent Na^+ channels in the developing brain of genetically seizure-susceptible E1 mice. *Neuroscience,* 48: 285–291.

Schwartzkroin, P.A. and Wyler, A.R. (1980) Mechanisms underlying epileptiform burst discharge. *Ann. Neurol.,* 7: 95–107.

Schwindt, P.C. (1992) Ionic currents governing input-output relations of Betz cells. In: T. McKenna, J. Davis and S.F. Zornetzer (Eds.), *Single Neuron Computation,* Academic Press, San Diego, pp. 235–258.

Selverston, A.I. (Ed.) (1985) *Model Neural Networks and Behavior,* Plenum, New York.

Silva, L.R., Gutnick, M.J. and Connors, B.W. (1991) Laminar distribution of neuronal membrane properties in neocortex of normal and reeler mouse. *J. Neurophysiol.,* 66: 2034–2040.

Simmons, P.A. and Pearlman, A.L. (1983) Receptive field properties of transcallosal visual cortical neurons in the normal and reeler mouse. *J. Neurophysiol.,* 50: 838–848.

Spain, W.J., Schwindt, P.C. and Crill, W.E. (1991) Post-inhibitory excitation and inhibition in layer V pyramidal neurones from cat sensorimotor cortex. *J. Physiol. (London),* 434: 609–626.

Somogyi, P. (1990) Synaptic organization of GABAergic neurons and GABAA receptors in the lateral geniculate nucleus and the visual cortex. In: D.M. Lam and C.D. Gilbert (Eds.), *Neural Mechanisms of Visual Perception,* Gulf Pub., Houston, pp. 35–62.

Steriade, M. and Llinás, L. (1988) The functional states of the thalamus and the associated neuronal interplay. *Physiol. Rev.,* 68: 649–742.

Wang, Z. and McCormick, D.A. (1993) Control of firing mode of corticotectal and corticopontine layer V burst-generating neurons by norepinephrine, acetylcholine, and 1S,3R–ACPD. *J. Neurosci.,* 13: 2199–2216.

Westenbroek, R.E., Noebels, J.L. and Catterall, W.A. (1992) Elevated expression of type II Na^+ channels in hypomyelinated axons of shiverer mouse brain. *J. Neurosci.,* 12: 2259–2267.

SECTION III

From Neuron to Network

A. Functional Activity and Other Formative Factors

J. van Pelt, M.A. Corner H.B.M. Uylings and F.H. Lopes da Silva (Eds.)
Progress in Brain Research, Vol 102

CHAPTER 13

Development of projection neurons of the mammalian cerebral cortex

Susan E. Koester and Dennis D.M. O'Leary

Molecular Neurobiology Laboratory, The Salk Institute, 10010 N. Torrey Pines Road, La Jolla, CA 92037, U.S.A.

Introduction

A major developmental issue unique to the nervous system is to define mechanisms underlying the establishment of specific axonal pathways and connections. Considerable progress has been made in this endeavor through studies of the simpler nervous systems of invertebrates. These studies have shown that molecular cues in the environment of the developing axon are crucial determinants of axon navigation (see Jacobson, 1991). A major question about which little is known is how neighboring neurons chose different pathways and innervate different targets. Again, work in invertebrates has led to the discovery of cell-fate determining genes (e.g. Chu-LaGraff and Doe, 1993), but how these genes relate to differentiation along anatomically or chemically defined phenotypic pathways remains unresolved. We are interested in eventually addressing similar questions with regard to the development of the efferent projections of the mammalian cerebral cortex. Requisite to addressing the molecular issues, it is important to define the normal development of the relevant populations of projection neurons. Here we summarize some of our recent studies on the development of two distinct classes of efferent projection neurons in the mammalian cortex: the callosal and subcortically projecting.

The cortex consists of six major layers which are distinguishable on the grounds of both cellular morphology and connectivity. The major efferent projections from the adult cortex are readily divisible based on anatomical criteria: the subcortical projection to targets in the brainstem and spinal cord and the callosal projection to the contralateral cortex. The subcortical projection arises from layer 6 (to thalamus) and layer 5 (to midbrain, hindbrain and spinal cord) while the callosal projection arises from all cortical layers except layer 1. Thus, both callosal and subcortically projecting neurons are found in layers 5 and 6. We have focused on the development of the populations within layer 5. Aside from residing in layer 5, these two types of projection neurons share other features: for example, both have a pyramidal morphology and they use the same neurotransmitters to convey information to their target neurons (Dinopoulos et al., 1989). It has even been reported that individual layer 5 cells extend axons both callosally and subcortically (Ramon y Cajal, 1894). However, electrophysiological and anatomical techniques have failed to uncover such a population. Several groups have used retrograde labeling with two fluorescent tracers in adult rats to show that individual cells cannot be retrogradely labeled from both the pyramidal decussation and the contralateral cortex (Catsman-Berrevoets et al., 1980; Wong and Kelly, 1981; Dreher et al., 1990; Koester and

O'Leary, 1993). In addition, individual layer 5 neurons cannot be antidromically stimulated from both the corticospinal tract and the contralateral cortex (Catsman-Berrevoets et al., 1980; Swadlow and Weyand, 1981). These data indicate that, in adult mammals, callosal and subcortically projecting cells are distinct populations, although intermingled in layer 5.

We have been interested in the questions of when and how this distinction comes about between callosal and subcortically projecting neurons. We have found that (1) the axon distinction between the two classes is present from very early stages of axogenesis, (2) a pronounced polarity of initial axon extension in cortex suggests that these two projection classes extend axons along their class-specific trajectories at different times of development, and (3) the two axon tracts are pioneered by distinct populations of early generated neurons. In contrast to these early distinctions in the development of the axonal projections between these two cell classes, their distinct dendritic arrangements are due in large part to a class-specific sculpting of an initial tall pyramidal morphology shared by both cell classes. Here we review these findings and consider possible mechanisms that could account for them.

Callosal and subcortically projecting neurons are developmentally distinct populations

Exuberant growth of axons, with subsequent elimination of functionally inappropriate axon segments and branches, is a widespread phenomenon in the developing nervous system and is a major mechanism for achieving the mature pattern of projections in the mammalian nervous system. Indeed, within both the callosal and subcortical projections, axon elimination has been shown to occur. As a population, layer 5 subcortically projecting cells in each neocortical area of rodents initially extend collateral projections to the complete set of layer 5 targets in the midbrain, hindbrain and spinal cord. Later, functionally inappropriate axon segments are eliminated, giving rise to the mature, area-specific patterns of projections. For example, layer 5 neurons in visual cortex loose their primary axon and collaterals distal to the basilar pons, while those in motor cortex loose axon collaterals to the superior colliculus (Stanfield et al., 1982; O'Leary and Stanfield, 1985; Stanfield and O'Leary, 1985; O'Leary and Terashima, 1988; O'Leary et al., 1990). Similarly, the mature restricted distribution of callosal projection neurons emerges from an early widespread distribution through the elimination of large numbers of callosal axon collaterals (Innocenti, 1981; Ivy and Killackey, 1981, 1982; O'Leary et al., 1981). In both instances, evidence suggests that the elimination process is influenced by functional activity (reviewed in O'Leary, 1992).

This phenomenon of early axonal exuberance followed by selective axon elimination raises the possibility that the adult distinction between these two projection populations (Fig. 1A) is achieved by the elimination of a callosal or subcortical axon by layer 5 neurons that initially extend both. To test this possibility, we have used a double retrograde labeling protocol, injecting Fast Blue (FB) into the pyramidal decussation, an easily accessible point in the subcortical projection pathway, and Diamidino Yellow (DY) into the contralateral cortex (Fig. 1B). A cell that is labeled with both tracers would have extended axon collaterals to both of the injection sites. This protocol used in adult rats results, predictably, in no double labeled cells. When the injections are made in newborn rats (a time prior to the phase of axon elimination described for each projection class) again, no cells are double labeled. Thus, the elimination of long projection axons cannot explain the projection distinction between the two classes.

An alternative is that cells whose final axons extend to one target initially extend axons toward both targets but not far enough to be detected by our retrograde tracer injections. To examine this possibility, we injected the lipophilic tracer 1,1-dioctadecyl-3,3,3′,3′-tetramethylindocarbocyanine perchlorate (DiI) into one hemisphere of cortex

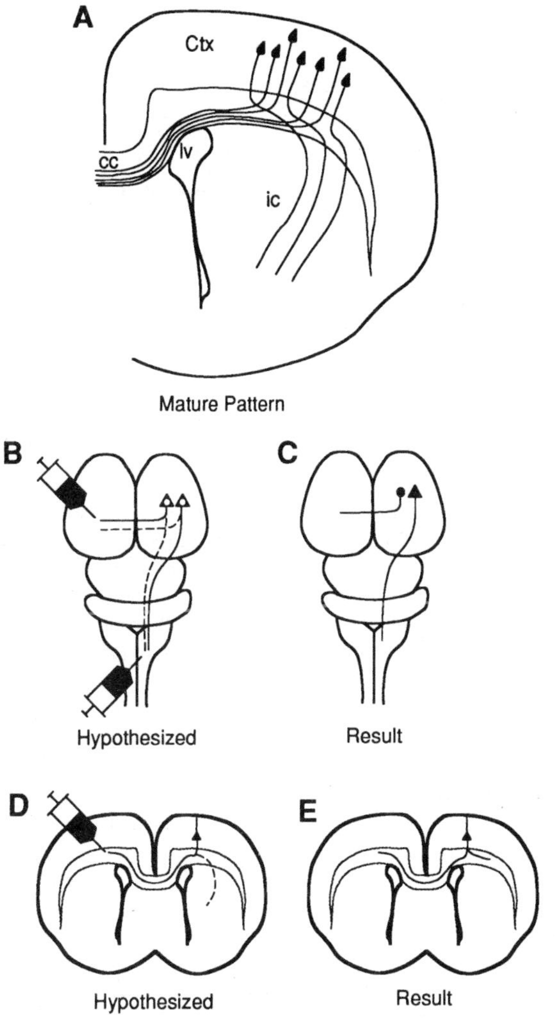

Fig. 1. Callosal and subcortically projecting neurons are distinct populations of projection neurons. (A) The mature pattern of projection neurons. Callosal neurons are found in both deep and superficial layers while subcortically projecting neurons are located exclusively in the deep layers. In layer 5, the two projection populations are intermingled. (B) To examine whether layer 5 projection neurons extend axon branches both callosally and subcortically, retrograde tracers were injected into contralateral cortex and the pyramidal decussation, an easily accessible point in the axon trajectory of subcortically projecting neurons. If a cell extended axons to both injection sites, it would be double labeled. (C) After double retrograde tracer injections in neonatal and adult animals, almost no cells were double labeled, indicating that in early postnatal animals, callosal and subcortically projecting neurons are distinct populations. (D) To determine whether callosal cells extended short axons into the internal capsule but not as far as our injection in the pyramidal decussation, we injected DiI into one hemisphere of the fixed brains of late embryonic rats. (E) Although axon branches were seen in the intermediate zone at these ages, virtually no axons were found in the internal capsule contralateral to the injection site.

of fixed embryonic rat brains to label retrogradely the callosal cells in the contralateral hemisphere (Fig. 1D). The earliest age at which callosal cells could be retrogradely traced from contralateral neocortex is embryonic day (E) 19. At this age, the axons of callosal cells extend radially to the intermediate zone (the future white matter) where they turn medially toward the corpus callosum and continue through it to the opposite cortical hemisphere. Frequently the axons of callosal cells branch in the intermediate zone, extending intracortical axon collaterals to other cortical areas. However, few or no labeled axons are found in the internal capsule contralateral to the injection site (Fig. 1E), indicating that callosal cells rarely if ever extend even short axons into the internal capsule, the pathway to subcortical targets.

Distinct populations of early generated neurons pioneer cortical efferent pathways

We next examined possible mechanisms underlying the class specific axon extension. Studies of axon guidance in simpler systems have provided examples of 'pioneer' neurons that are the first to extend axons along a particular trajectory and, in doing so, define an axonal pathway. These 'pioneer' cells extend axons at stages when the distances they must navigate are much shorter than later in development, and in some instances the terrain over which they grow is more readily traversed. Later growing axons could use the pioneer axons as a substrate for elongation and pathfinding. In some instances, pioneer axons have indeed been shown to be essential for the later development of the definitive axonal projection (e.g. Klose and Bentley, 1989). Possible roles for pioneer axons in the development of the efferent connections of the mammalian cortex have recently become a topic of great interest.

The first cells to become postmitotic in cerebral cortex come to overlie the neuroepithelium and at early stages form a transient layer termed the preplate. The preplate is subsequently split into a superficial marginal zone and a deep subplate layer by the cortical plate, which is formed by the migration and aggregation of later generated neurons within the preplate. The majority of the early generated preplate cells later reside in the subplate. McConnell et al. (1989) have shown in cats that subplate neurons extend the first axons from cortex along the subcortical trajectory through the internal capsule. DeCarlos and O'Leary (1992) confirmed that these cells also extend the earliest axons from cortex in rats, and demonstrated that the axons of subplate cells extend as far as the thalamus (the target of layer 6 axons) but do not continue along the subcortical trajectory taken by layer 5 axons into the hindbrain or spinal cord. Subplate neurons have been proposed to serve a crucial role in the establishment of the efferent connections of cortex. In support of such a role is the report that chemical ablation of subplate neurons disrupts the targeting of layer 6 axons in the thalamus (McConnell et al., 1994).

Given the potential importance of pioneering populations of axons, we have addressed the question of whether subplate cells also extend the first axons through the corpus callosum (Koester and O'Leary, 1994). The first step was to ascertain exactly when the first axons cross the midline of the corpus callosum in rats. Large DiI injections into dorsomedial cortex of the fixed brains of E17 rats anterogradely label growth cones at the midline (Fig. 2); this age is 1.5 days earlier than previously reported (Valentino and Jones, 1982). DiI injections into medial cortex of the contralateral hemisphere on E18 retrogradely label only a small number of neurons confined to a discrete cluster in the contralateral medial (cingulate) cortex. These retrogradely labeled cells that extend the first axons across the nascent corpus callosum have a complex morphology with

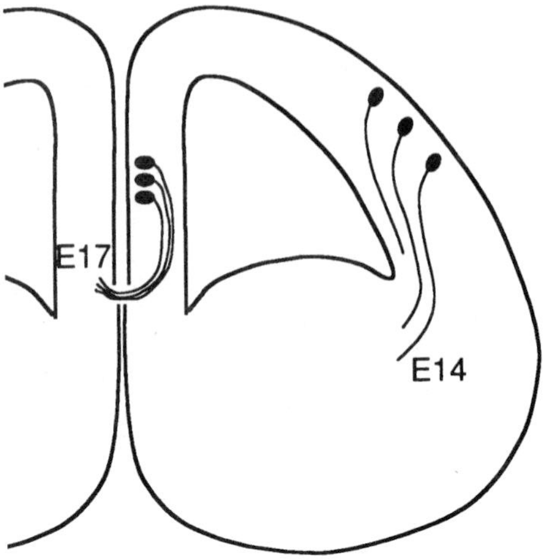

Fig. 2. The two efferent pathways from cortex are pioneered by distinct populations of early generated neurons at different times in development. The subcortical projection through the internal capsule is pioneered by neocortical subplate neurons. These axons first arrive in the internal capsule on E14. The callosal projection is pioneered by a small group of cortical plate neurons in cingulate cortex. The axons of these cells first cross the midline on E17.

extensively branched dendritic trees. Their axons run radially to leave the cortical plate, turn medially in the intermediate zone, and extend across midline.

We birth-dated the early callosal cells in cingulate cortex using $[^3H]$thymidine ($[^3H]$TdR) injected into timed pregnant rats at carefully chosen gestational stages. Fetuses were fixed at E18 for DiI retrograde labeling of the early callosal cells, followed by ^3H autoradiography. We found that the early callosal cells are generated predominantly on E14 with a very small fraction born on E13 and E15. This generation profile places these early callosal neurons amongst the earliest generated neurons in cingulate cortex (Bayer, 1990). Previous studies using histological methods have reported that cingulate cortex differs from neocortex in that it does not have a subplate (Bayer, 1990). However, in Nissl-stained sections of our material a subplate layer is clearly visible and is continuous from neocortical regions around to the most medial part of cingulate cortex, although in cingulate cortex the separation of the subplate from the cortical plate become progressively less distinct more medially. DiI injections into medial cortex label cells locally (ipsilaterally) with two distinct morphologies: radial/pyramidal typical of cortical plate neurons, and a population with more of the horizontal and multipolar morphologies that are suggestive of subplate neurons. Cells retrogradely labeled from contralateral cortex in cingulate regions of older rats are radially aligned with pyramidal morphologies, suggesting that most, if not all, of the early callosal cells are part of the cortical plate and not the subplate. This identification is consistent with our finding that injections of retrograde tracers such as FB, DY or rhodamine latex microspheres label at most only a few cells in the subplate, which are widely scattered across the contralateral cortical hemisphere. Thus, remarkably few subplate axons cross the corpus callosum, and these do so rather late in development. Based on these data, we conclude that a discrete population of cortical plate neurons within the cingulate cortex pioneer the callosal pathway (Fig. 2).

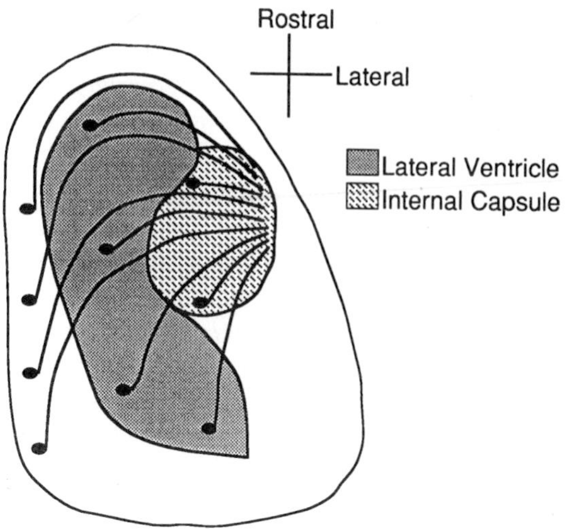

Fig. 3. At early stages of cortical axogenesis, essentially all axon extension is directed toward the internal capsule. Cells found in lateral regions of cortex extend axons laterally to reach the internal capsule. Those in more medial or caudal cortex extend axons both rostrally and laterally over the lateral ventricle to reach the internal capsule. This schematic shows a dorsal view of a flattened hemisphere of cortex.

Early axon extension in cortex is highly polarized

In rats, although the first axons extend into the internal capsule on E14 (De Carlos and O'Leary, 1992) the first callosal axons do not reach the midline of the corpus callosum until E17 (Fig. 2). What is the basis for this temporal disparity in the establishment of these two major cortical efferent pathways? Using the early neuronal marker TuJ1 (a monoclonal antibody to acetylated β-tubulin) (Moody et al., 1989), we have found that at early stages (E14–E16) the axonal pathway in the cortical intermediate zone becomes considerably thicker, moving from medial to lateral positions. This observation suggests that the majority of early axons are directed laterally, along the subcortical trajectory. To confirm this impression, we made focal DiI injections into various sites in the cortical wall of fixed embryonic rat brains. At E16 and earlier, virtually all retrogradely labeled cell bodies were found medial to the injection sites, indicating that they must have extended laterally directed axons. Cells in cingulate cortex, even those in the region where the early callosal neurons can later be labeled, also extend axons laterally toward the internal capsule. Based on these data, we conclude that initial axon extension in cortex is highly polarized, with virtually all axons directed laterally toward the internal capsule (Fig. 3).

By E17, the TuJ1 stained axonal pattern in the intermediate zone changes, suggesting that the marked polarity observed for early axon extension is breaking down. Indeed, DiI injections, made in cortex at E17 and later, resulted in retrogradely labeled cells both lateral and medial to the injection sites. Thus, by E17, cells in lateral cortex have begun to extend axons along the medial trajectory toward the corpus callosum and contralateral cortex. Since we know that the callosal and subcortical cells are co-generated (see below), from these data we conclude that callosal axon extension is delayed compared with subcortical axon extension, resulting in strong early polarity of axonal growth directed along the subcortical trajectory.

Three scenarios could explain the delay in callosal axon extension (Fig. 4). The first of these, and least likely in our opinion, is that initially all axon extension is directed laterally. Later, a subset of these axons is eliminated and collaterals are extended medially (Fig. 4A). However, we have seen no evidence supporting this possibility in our DiI labeled brains. A more likely possibility, therefore, is that axogenesis by callosal neurons is delayed compared with that by subcortical axons (Fig. 4B) or, alternatively, callosal neurons extend axons at the same time as the subcortically projecting neurons, but the callosal axons stall and wait before making their projection class-specific trajectory turn toward the midline (Fig. 4C).

Neuron class-specific dendritic elimination

In contrast to the early difference in axonal phenotype between callosal and subcortically projecting neurons, the characteristic distinct dendritic morphologies between these two projection classes do not become evident until much later in development. In adult mammals, subcortically projecting cells in layer 5 exhibit the classic tall pyramidal morphology with a tufted apical dendrite ramifying in layer 1, whereas callosal cells typically have apical dendrites that end in layer 4 (Games and Winer, 1988; Hübener and Bolz, 1988; Hallman et al., 1988). Thus, callosal cells have the short pyramidal morphology first described for a subset of layer 5 neurons by Lorente de Nó (1949). We have used DiI retrograde labeling in fixed rat brains to address whether these characteristic dendritic phenotypes are apparent early in dendritic growth or whether they emerge later through a process of selective dendritic elimination (Koester and O'Leary, 1992).

Surprisingly, we found that at postnatal day (P) 1 callosal neurons and subcortically projecting neurons (labeled from the superior colliculus, a major layer 5 subcortical target) have a very similar morphology: essentially all cells in both populations have an apical dendrite that extends to and arborizes in layer 1. The apical dendrites of both cell classes continue to develop until late in the first postnatal week. On P7, however, the dendrites of layer 5 callosal cells still extend beyond layer 4, but no longer reach layer 1. By P10, the apical dendrites of most layer 5 callosal neurons end in layer 4; and the dendritic distinctions between them and subcortically projecting neurons are similar to those seen in the adult. Measurements of the heights of the apical dendrites

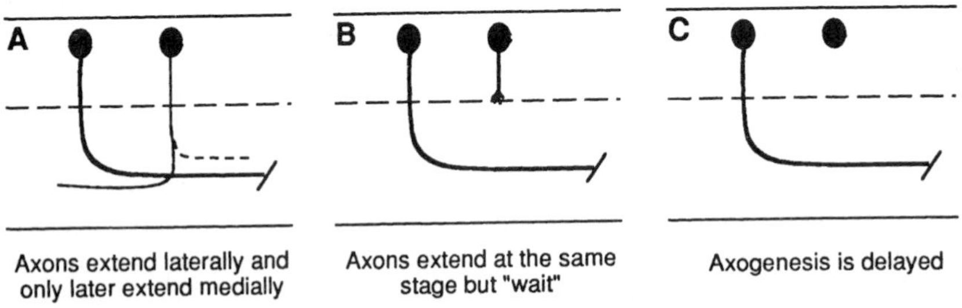
A
B
C
Axons extend laterally and only later extend medially
Axons extend at the same stage but "wait"
Axogenesis is delayed

Fig. 4. Three possible scenarios to explain the apparent delay in the extension of axons directed to the midline corpus callosum by callosal neurons. (A) Both callosal and subcortically projecting neurons initially extend axons laterally. Later callosal neurons eliminate their laterally directed axons (dotted line) and then axons medially. (B) Both projection neuron types extend axons at the same time but callosal axons 'wait' before they make their medially directed turn in the intermediate zone. (C) Callosal neurons do not extend axons at all until later in development, days after subcortically projecting neurons have extended their axons. In this schematic, medial (callosal) is to the left and lateral (subcortical) is to the right.

reveal a statistically significant decrease in the absolute height of layer 5 callosal cells between P4 and P7 and between P7 and P10, whereas the absolute heights of layer 5 subcortically projecting neurons increase over the same periods (Koester and O'Leary, 1992). These data show that the loss of the superficial portions of the apical dendrite by callosal neurons is due to an active elimination of the apical dendrite rather than a passive displacement due to subsequent growth and expansion of the superficial cortical layers. Interestingly, callosal cells in superficial layers permanently retain their branched apical dendrite in layer 1 indicating that the elimination process is a layer-specific phenomenon as well as a neuron class-specific one. Thus, in contrast to the axons extended by layer 5 callosal and subcortically projecting neurons, which grow along their class-specific trajectories from the earliest ages that we can identify them, the dendritic morphologies of these two projection classes are initially similar, and the class-specific laminar arrangements of their apical dendrites is not achieved until much later in development.

Callosal and subcortically projecting layer 5 neurons are co-generated

The data presented here define callosal and subcortically projecting neurons as distinct classes of projection neurons intermingled in layer 5. What mechanism determines this distinction? One possibility is that the two populations are generated at different times, allowing them to be exposed to different environmental cues that may control cell fate. This mechanism is similar to that proposed by McConnell and Kaznowski (1991) for the determination of laminar fate in the mammalian cortex. These investigators dissociated cells from the cortical neuroepithelium of ferrets, transplanted them into the cortical neuroepithelium of older hosts, and examined the final laminar location to which the transplanted cells migrated. They found that some cells migrated to deep layers, characteristic of the cells being generated in the donor cortex at the time the transplanted cells were removed, or to superficial layers, characteristic of the cells being generated in the host cortex at the time of transplantation (McConnell, 1985, 1988). The factor determining the layer to which the cells migrated was the environment in which the cells underwent their final phase of DNA synthesis: cells that underwent their final S phase in the host migrated to the superficial layers while those that became postmitotic in the donor migrated to the deep layers (McConnell and Kaznowski, 1991).

With these experiments in mind, we determined whether or not callosal and subcortically projecting neurons are generated at the same time. We exposed rat embryos to a single pulse of [^{3}H]TdR on E15, a time when large numbers of layer 5 cells are being generated (Koester and O'Leary, 1993). The animals were raised to adulthood and callosal and subcortically projecting neurons were retrogradely labeled in the same animals using two distinguishable tracers. We found that the same proportions of layer 5 callosal and subcortically projecting neurons were double labeled with [^{3}H]TdR, indicating that these two populations of neurons are co-generated. Therefore, a mechanism based primarily on temporal differences in cell generation cannot account for the distinctions seen later between the two projection populations.

Conclusions

Taking all these data together, we conclude that the two major layer 5 efferent projection classes of the neocortex, the callosal and the subcortical, are distinct populations of projection neurons generated at the same time in development. The evidence further suggests that the distinction between the two classes is determined early in development, either before or around the time that the neurons are generated. The projection class phenotype could be determined by cell-cell interactions such as those hypothesized to occur in the vertebrate retina (see Altschuler et al., 1991 for recent review). Alternatively, projection class could be determined by lineage dependent mechanisms. Lineage has been shown to be an important determinant of glial cell phenotype in cortex (e.g. Grove et al., 1993; Luskin et al., 1993) and is correlated with the differentiation of neurons into pyramidal and non-pyramidal morphologies (Parnavelas et al., 1991). Only direct analysis of the role of lineage in determining cortical projection phenotype will allow us to distinguish between these two mechanisms. These determination events are then played out during normal development as the two projection classes differentiate, elaborating their specific axonal and, later, dendritic phenotypes. Given the information that the callosal and subcortically projecting layer 5 neurons are developmentally distinct classes of cortical neurons, we are now in a position to address the molecular basis of determination of neuronal projection class in the rodent neocortex.

Acknowledgements

The work in the authors' laboratory summarized here was supported by NIH grant NS31558 and the Valley Foundation. Partial support for SEK was derived from an NSF predoctoral fellowship and from an Olin Fellowship administered through Washington University.

References

Altshuler, D., Turner, D. and Cepko, C. (1991) Specification of cell type in the vertebrate retina. In: D. M-K. Lam and C. Shatz (Eds.), *Development of the Visual System,* MIT Press, Cambridge, MA, pp. 37–58.

Bayer, S.A. (1990) Neurogenetic patterns in the medial limbic cortex of the rat related to anatomical connections with the thalamus and striatum. *Exp. Neurol.,* 107: 132–142.

Catsman-Berrevoets, C.E., Lemon, R.N., Verburgh, C.A., Bentivoglio, M. and Kuypers, H.G.J.M. (1980) Absence of callosal collaterals derived from rat corticospinal neurons. *Exp. Brain Res.,* 39: 433–440.

Chu-LaGraff, Q. and Doe, C.Q. (1993) Neuroblast specification and formation regulated by wingless in the *Drosophila* CNS. *Science,* 261: 1594–1597.

De Carlos, J.A. and O'Leary, D.D.M. (1992) Growth and targeting of subplate axons and establishment of major cortical pathways. *J. Neurosci.,* 12: 1194–1211.

Dinopoulos, A., Dori, I., Davies, S.W. and Parnavelas, J.G. (1989) Neurochemical heterogeneity among corticofugal and callosal projections. *Exp. Neurol.,* 105: 36–44.

Dreher, B., Dehay, C. and Bullier, J. (1990) Bihemispheric collateralization of the cortical and subcortical afferents to the rat's visual cortex. *Eur. J. Neurosci.,* 2: 317–331.

Games, K.D. and Winer, J.A. (1988) Layer V in rat auditory cortex: projections to the inferior colliculus and contralateral cortex. *Hearing Res.,* 34: 1–26.

Grove, E.A., Williams, B.P., Li, D.-Q., Hajihosseini, M., Friedrich, A. and Price, J. (1993) Multiple restricted lineages in the embryonic rat cerebral cortex. *Development,* 117: 553–561.

Hallman, L.E., Schofield, B.R. and Lin, C. (1988) Dendritic morphology and axon collaterals of corticotectal, corticopontine, and callosal neurons in layer V of the primary visual cortex of the hooded rat. *J. Comp. Neurol.,* 272: 149–160.

Hübener, M. and Bolz, J. (1988) Morphology of identified projection neurons in layer 5 of rat visual cortex. *Neurosci. Lett.,* 94: 76–81.

Innocenti, G.M. (1981) Growth and reshaping of axons in the establishment of visual callosal connections. *Science,* 212: 824–827.

Ivy, G.O. and Killackey, H.P. (1981) The ontogeny of the distribution of callosal projection neurons in the rat parietal cortex. *J. Comp. Neurol.,* 195: 367–389.

Ivy, G.O. and Killackey, H.P. (1982) Ontogenetic changes in the projections of neocortical neurons. *J. Neurosci.,* 2: 735–743.

Jacobson, M. (1991) *Developmental Neurobiology,* Plenum Press, New York.

Klose, M. and Bentley, D. (1989) Transient pioneer neurons are essential for formation of an embryonic peripheral nerve. *Science,* 245: 976–978.

Koester, S.E. and O'Leary, D.D.M. (1992) Functional classes of cortical projection neurons develop dendritic distinctions by class-specific sculpting of an early common pattern. *J. Neurosci.,* 12: 1382–1393.

Koester, S.E., O'Leary, D.D.M. (1993) Connectional distinction between callosal and subcortically projecting cortical neurons is determined prior to axon extension. *Dev. Biol.,* 160: 1–14.

Koester, S.E., O'Leary, D.D.M. (1994) Axons of early generated neurons in cingulate cortex pioneer the corpus callosum. *J. Neurosci.*, 14: in press.

Lorente de Nó, R. (1949) Cerebral cortex: architecture, intracortical connections, motor projections. In: J.F. Fulton (Ed.), *Physiology of the Nervous System,* Oxford University Press, New York, pp. 274–301.

Luskin, M.B., Parnavelas, J.G. and Barfield, J.A. (1993) Neurons, astrocytes, and oligodendrocytes of the rat cerebral cortex originate from separate progenitor cells: an ultrastructural analysis of clonally related cells. *J. Neurosci.,* 13: 1730–1750.

McConnell, S.K. (1985) Migration and differentiation of cerebral cortical neurons after transplantation into the brains of ferrets. *Science,* 229: 1268–1271.

McConnell, S.K. (1988) Fates of visual cortical neurons in the ferret after isochronic and heterochronic transplantation. *J. Neurosci.,* 8: 945–974.

McConnell, S.K. and Kaznowski, C.E. (1991) Cell cycle dependence of laminar determination in developing neocortex. *Science,* 254: 282–285.

McConnell, S.K., Ghosh, A. and Shatz, C.J. (1989) Subplate neurons pioneer the first axon pathway from the cerebral cortex. *Science,* 245: 978–982.

McConnell, S.K., Ghosh, A. and Shatz, C.J. (1994) Subplate pioneers and the formation of descending connections from cerebral cortex. *J. Neurosci.*, 14: 1892–1907.

Moody, S.A., Quigg, M.S. and Frankfurter, A. (1989) Development of the peripheral trigeminal system in the chick revealed by an isotype-specific anti-beta-tubulin monoclonal antibody. *J. Comp. Neurol.,* 279: 567–580.

O'Leary, D.D.M. (1992) Development of connectional diversity and specificity in the mammalian brain by the pruning of collateral projections. *Curr. Opin. Neurobiol.,* 2: 70–77.

O'Leary, D.D.M. and Stanfield, B.B. (1985) Occipital cortical neurons with transient pyramidal tract axons extend and maintain collaterals to subcortical but not intracortical targets. *Brain Res.,* 336: 326–333.

O'Leary, D.D.M. and Terashima, T. (1988) Cortical axons branch to multiple subcortical targets by interstitial axon budding: implications for target recognition and 'waiting periods'. *Neuron,* 1: 901–910.

O'Leary, D.D.M., Stanfield, B.B. and Cowan, W.M. (1981) Evidence that the early postnatal restriction of the cells of origin of the callosal projection is due to the elimination of axonal collaterals reather than to the death of neurons. *Dev. Brain Res.,* 1: 607–617.

O'Leary, D.D.M., Bicknese, A.R., De Carlos, J.A., Heffner, C.D., Koester, S.E., Kutka, L.J. and Terashima, T. (1990) Target selection by cortical axons: alternative mechanisms to establish axonal connections in the developing brain. Cold Spring Harbor Quant. Symp., 55: 453–468.

Parnavelas, J.G., Barfield, J.A., Franke, E. and Luskin, M.B. (1991) Separate progenitor cells give rise to pyramidal and nonpyramidal neurons in the rat telencephalon. *Cerebral Cortex,* 1: 463–468.

Ramon y Cajal, S. (1894) Les nouvelles idees sur la structure du systeme nerveux chez l'homme et chez vertebres. Reinwald, Paris.

Stanfield, B.B. and O'Leary, D.D.M. (1985) The transient corticospinal projection from the occipital cortex during the postnatal development of the rat. *J. Comp. Neurol.,* 238: 236–248.

Stanfield, B.B., O'Leary, D.D.M. and Fricks, C. (1982) Selective collateral elimination in early postnatal development restricts cortical distribution of rat pyramidal tract axons. *Nature,* 298: 371–373.

Swadlow, H.A. and Weyand, T.G. (1981) Efferent systems of the rabbit visual cortex: laminar distribution of the cells of origin, axonal conduction velocities and identification of axonal branches. *J. Comp. Neurol.,* 203: 799–822.

Valentino, K.L. and Jones, E.G. (1982) The early formation of the corpus callosum: a light and electron microscope study in foetal and neonatal rats. *J. Neurocytol.,* 11: 583–609.

Wong, D. and Kelly, J.P. (1981) Differentially projecting cells in individual layers of the rat auditory cortex: a double-labeling study. *Brain Res.,* 230: 362–366.

J. van Pelt, M.A. Corner H.B.M. Uylings and F.H. Lopes da Silva (Eds.)
Progress in Brain Research, Vol 102

CHAPTER 14

Naturally occurring and axotomy-induced motoneuron death and its prevention by neurotrophic agents: a comparison between chick and mouse

Lucien J. Houenou[1,2], Linxi Li[1,3], Albert C. Lo[1,2], Qiao Yan[4] and Ronald W. Oppenheim[1,2]

[1]*Department of Neurobiology and Anatomy,* [2]*The Neuroscience Program, and* [3]*Department of Opthalmology, Bowman Gray School of Medicine, Wake Forest University, Winston-Salem, NC 27157, and* [4]*Department of Neuroscience, Amgen, Inc., Thousand Oaks, CA 91320, U.S.A.*

Introduction

The normal development of the vertebrate neuromuscular system is characterized by the death of a substantial number (40–70%) of postmitotic motoneurons (MNs) during a period when these cells are forming synaptic connections with their target muscles (Oppenheim, 1981; Hamburger and Oppenheim, 1982; Clarke, 1985). Naturally occurring MN death has been described in many vertebrates, including chick (Hamburger, 1975; Chu-Wang and Oppenheim, 1978), mouse (Lance-Jones, 1982; Oppenheim et al., 1986), rat (Harris and McCaig, 1987), and human (Forger and Breedlove, 1987) embryos and fetuses. In mammals, the maintenance of MNs that survive the period of naturally occurring cell death is known to depend on continuous contact with the target for some time after birth, since transection of motor axons within the first postnatal week results in massive cell loss (Schmalbruch, 1984; Kashihara et al., 1987; Crews and Wigston, 1990; Pollin et al., 1991; Snider et al., 1992). Although naturally occurring MN death in avian embryos has been extensively studied (see e.g. Hamburger, 1975; Chu-Wang and Oppenheim, 1978; Pittman and Oppenheim, 1978, 1979), axotomy-induced MN death in this species has received little attention (Houthoff and Drukker, 1977).

There is increasing evidence suggesting that the survival of spinal MNs depends on trophic agents derived from peripheral targets (skeletal muscle) as well as from cells within the central nervous system (for reviews, see Oppenheim, 1989, 1991). It has recently been shown, for example, that members of the neurotrophin family (mRNA and protein) are present in muscle targets and can be retrogradely transported by MNs during the period of programmed cell death (Henderson et al., 1993; Homma et al., 1993; Koliatsos et al., 1993; McKay et al., 1993). Moreover, several neurotrophic agents, including some of the neurotrophins, have been recently shown to promote the in vivo survival of spinal MNs during the period of programmed cell death in avian embryos (Oppenheim et al., 1991, 1992, 1993; Neff et al., 1993), as well as following axotomy in chick embryos (Neff et al., 1993), in neonatal rats

(Sendtner et al., 1991, 1992a; Yan et al., 1992; Koliatsos et al., 1993), and in the mouse mutant *progressive motor neuropathy* (Sendtner et al., 1992b). However, whether neurotrophic agents can promote the in vivo survival of mammalian MNs during the period of programmed cell death, i.e. during fetal development, and how axotomy-induced MN death in avian species compares with that in mammals are still largely unanswered questions.

We review here our own findings on (1) the time-course of axotomy-induced death of lumbar spinal MNs in chick and mouse, and (2) the survival effects of a wide variety of known and putative trophic or growth factors on chick and mouse MNs, during naturally occurring cell death and following axotomy. We show that (1) chick MNs remain vulnerable to axotomy up to about embryonic day 12, whereas mouse MNs are still sensitive to axonal section on postnatal day 5, and (2) both avian and mouse MNs can be rescued by a variety of trophic agents, including the neurotrophins, during the period of programmed cell death as well as following axotomy.

Methodological considerations

The following purified neurotrophic agents and growth factors were examined for their effects on MN survival: mouse nerve growth factor (NGF, a gift from Eugene Johnson), brain-derived neurotrophic factor (BDNF, human recombinant protein Amgen, Inc.), neurotrophin-3 (NT-3, human recombinant protein from Amgen, Inc.), neurotrophin-5 (NT-4/5, human recombinant protein from A. Rosenthal, Genentech, Inc.), epidermal growth factor (EGF, a gift from Ralph Bradshaw), acidic fibroblast growth factor (aFGF, human recombinant protein from Ralph Petterson), basic fibroblast growth factor (bFGF, human recombinant protein from California Biotechnology, Inc.), ciliary neurotrophic factor (CNTF, human recombinant protein from Synergen, Inc.), insulin-like growth factor 1 and 2 (IGF-1 and IGF-2, human recombinant proteins from Cephalon, Inc.), transforming growth factor beta (TGF-ß, human recombinant protein from Bristol Myers Squibb), platelet-derived growth factor (PDGF-AB, human recombinant protein from Amgen, Inc.), S-100ß (recombinant protein provided by Linda van Eldik), heparin-binding growth-associated molecule (HB-GAM, recombinant protein from Heikki Rauvala—see Raulo et al., 1992), cholinergic differentiation factor/leukemia inhibitory factor (CDF/LIF, human recombinant protein from Amgen, Inc.), and interleukin 6 (IL-6, a gift from Amgen, Inc.).

Naturally occurring motoneuron cell death

White Leghorn chicken eggs were obtained as a gift from Hubbard Farms, Inc. (Statesville, NC) and were incubated in the laboratory at 37°C and 60% humidity. On embryonic (E) day 5 (E5), a small window was made in the shell, which was then sealed with a piece of tape, after which eggs were returned to the incubator. Embryos were treated once daily with trophic agents from E6 to E9, i.e. the period of programmed MN death in the chick (see e.g. Pittman and Oppenheim, 1978, 1979), as described previously (Oppenheim et al., 1991). Embryos were killed on E10. Their spinal cords were dissected, fixed in Carnoy's fluid overnight and processed for serial paraffin sections and histology. Motoneuron cell counts were performed as described (Oppenheim et al., 1990, 1993).

BALB/c ByJ mice (the Jackson Laboratory, Bar Harbor, Maine) were bred in our Medical School animal facility. On gestation day 14 (the morning a vaginal plug was observed is designated E0), pregnant females were anesthetized with ether, and partial laparatomy was performed under sterile conditions. One uterus (3–5 embryos) was exposed and each embryo was injected with 5 or 10 μg of a trophic agent in 5 or 10 μl of saline (0.9% NaCl, pH 7.2) using a modified 10 μl gauge Hamilton microsyringe, as described previously (Houenou et al., 1990). Injections were made into the amniotic fluid, embryos were replaced and the mother was allowed to recover after the abdomen was sutured closed. Fetuses from the contralateral uterus were used as controls. On

E18, mice were killed with an overdose of ether, and fetuses were collected by cesarean section. Spinal cords from control and trophic/growth factor-treated embryos were dissected out, fixed in Carnoy solution, and processed as described above for chicken tissues (see also Oppenheim et al., 1986).

Axotomy-induced motoneuron death

Chicken eggs were set and processed as described above. On E10, E12, E14, and E16, the right hindlimb of xylocaine anesthetized embryos was tightly sutured above the knee and the limb was amputated distal to the suture. Operated embryos were killed at 1- or 2-day intervals, their spinal cords were dissected, fixed in Bouin's fluid and processed for histology as described above. In another series of experiments, sciatic nerve axotomy (not limb amputation) was performed on posthatching day 1 (PH1) and operated animals were examined for lumbar MN survival at regular intervals up to postaxotomy (PAx) day 33. Since a significant number of MNs die 4 days after surgery performed on E12 (see below), and because operated embryos survived better at this age (versus surgery done at earlier ages), we used this operation paradigm (surgery on E12 and sacrifice on E16) to examine the effects of trophic/growth factors on axotomized chick MNs. Operated embryos were treated once daily from E12 to E15 with trophic agents as described previously (Houenou et al., 1991; Oppenheim et al., 1991). Embryos were killed on E16 and their special cords were dissected, fixed in Bouin's and processed for histology as described above. MN cell counts were performed as described previously (see e.g. Oppenheim et al., 1993).

Neonatal mice were anesthetized by hypothermia and the right sciatic nerve was sectioned in the mid-thigh region on postnatal (PN) days 2, 5, 10, 15 and 25. A 2- to 3-mm piece of the nerve was removed to prevent reinnervation. Following surgery, animals were sacrificed by an overdose of ether at 5-day intervals. Lumbar spinal cords with attached vertebra and dorsal root ganglia were dissected out and fixed by immersion in Bouin's fixative for 2–3 weeks (this procedure also results in decalcification). Tissues were embedded in paraffin, serially sectioned at 12 μm, and stained with hematoxylin and eosin. MN counts were made as described previously (Oppenheim et al., 1986). The unoperated contralateral side of the spinal cord was used as control.

In order to examine the effects of neurotrophic agents on axotomized mammalian motoneurons, mice were operated on PN5, and two mm^3 pieces of gelfoam (Upjohn Co., Kalamazoo, MI), soaked in 5-μl solutions of neurotrophic agents (1 μg/μl), were implanted at the axotomy site. Four days later, 5 μg of the same trophic agent (in 5 μl) were injected (using a micro syringe) into the musculature surrounding the site of axotomy. Animals were killed on PN12 (i.e. 7 days post-surgery) and their spinal cords were processed for histology as described above.

Empirical findings and interpretation

Trophic agents and programmed MN death

We have tested a wide variety of trophic/growth factors of which BDNF, NT-4/5, CNTF, IGF-1, IGF-2, S-100ß, PDGF-AB, TGF-ß, CDF/LIF, and IL-6 showed a significant survival-promoting activity on avian MNs during the period of naturally occurring cell death (Table I). We have also examined the in utero effects of NGF, BDNF, and CNTF on embryonic mouse MNs. Both BDNF and CNTF, but not NGF, were able to prevent MN death in the embryonic mouse (Fig. 1). In both chick and mouse, NGF was ineffective in rescuing MNs from programmed cell death (Table I and Fig. 1). This is in aggrement with the previously reported lack of activity of this neurotrophin on embryonic chick MNs (Oppenheim et al., 1982). In order to ascertain that the NGF used by us was biologically active, we have examined the survival of dorsal root ganglion (DRG) neurons (which also undergo programmed cell death between E6 and E10 (Hamburger and Oppenheim, 1982; Bhattacharyya et al., 1992)), in the same embryos

TABLE I

Motoneuron numbers (means ± S.E.M.) in the lumbar spinal cords of E10 chick embryos treated daily from E6 to E9 with various growth factors or neurotrophic agents[1]

Treatment (daily μg)	Motoneurons (N)
Control (Saline)	11,977 ± 108 (108)
NGF (5 μg)	12,108 ± 285 (15)
BDNF (5 μg)	15,221 ± 238 (20)[c]
NT-3 (5 μg)	12,674 ± 340 (10)
NT-4/5 (5 μg)	13,800 ± 481 (10)[a]
PDGF-AB (5 μg)	14,025 ± 301 (8)[b]
TGF-β (5 μg)	13,554 ± 343 (7)[a]
EGF (1-3 μg)	12,429 ± 305 (10)
IGF-1 (5 μg)	15,106 ± 234 (9)[c]
IGF-2 (5 μg)	14,166 ± 471 (6)[b]
IL-6 (5 μg)	14,760 ± 350 (6)[d]
CDF/LIF (5 μg)	13,645 ± 474 (8)[a]
CNTF (10 μg)	16,718 ± 331 (16)[d]
S-100β (15 μg)	15,919 ± 163 (12)[d]
HB-GAM (5 μg)	12,973 ± 398 (5)
bFGF (5 μg)	12,095 ± 464 (5)
aFGF (5 μg)	12,175 ± 355 (6)

[1]From Oppenheim et al. (1993).
N = sample sizes.
[a]$P < 0.05$.
[b]$P < 0.01$.
[c]$P < 0.003$.
[d]$P < 0.0001$ vs. control, t-test.

examined for MN survival. NGF increased neuron numbers by 40% in embryonic chick L3 DRG (control: 9021 ± 272, n = 76; NGF: 12 731 ± 275, n = 9; $P < 0.01$) and by 65% in fetal mouse S1 DRG (control: 3556 ± 329, n = 5; NGF: 5863 ± 192, n = 6; $P < 0.0001$).

The data presented here are the first to demonstrate that the neurotrophic agents, BDNF and CNTF (but not NGF), can rescue mammalian MNs from naturally occurring cell death in vivo (in utero). The recent demonstrations that both avian (Homma et al., 1993) and mammalian (DiStephano et al., 1992; Koliatsos et al., 1993; Yan et al., 1993) motoneurons retrogradely transport the neurotrophins employed in the present study, indicates that these agents could play an important role in normal MN survival and differentiation. However, in contrast to the other neurotrophins, NGF does not promote MN survival during the period of naturally occurring cell death, either in chick (Table I) (see also Oppenheim et al., 1982, 1992) or in mouse (Fig. 1).

Axotomy-induced death of MNs and rescue by trophic agents

Avian MNs were vulnerable to axotomy performed on or before E12 (Fig. 2). Axotomy on E14 or at older ages did not significantly affect cell survival (Fig. 2). Chick embryos survived better following axotomy on E12 versus, e.g., E9 or E10 (L.J.H., personal observation) and exhibited significantly decreased MN numbers (see Fig. 2), when examined 4 days later on E16. Since not all lumbar (L1–L8) motor axons contribute to the sciatic nerve (see e.g. Lance-Jones and Diaz, 1991), we also examined MN survival on E16 (operated on E12) in one lumbar segment (i.e. L4) in which all MNs project in the sciatic nerve. Four days after axotomy on E12, approximately 50% of L4 MNs die in the ipsilateral lateral motor column (LMC) when compared with the contralateral (control) side of the spinal cord (control: 3227 ± 82, n = 15; axotomy: 1592 ± 62,

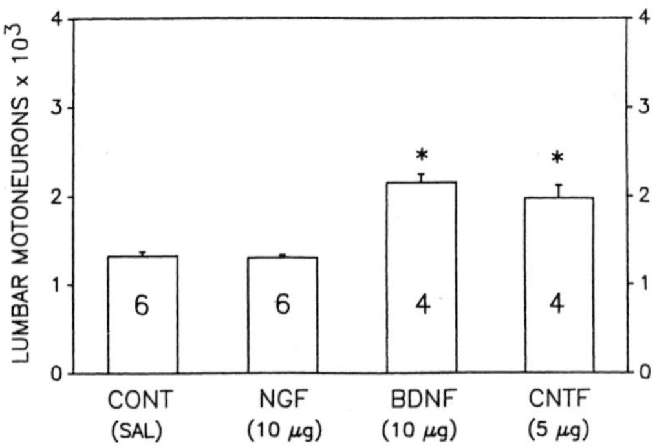

Fig. 1. Motoneuron numbers (means ± S.E.M.) in the lumbar spinal cord of mouse embryos injected on E14 with either saline (SAL), NGF, BDNF, or CNTF, and examined on E18. Numbers in bars are sample sizes. *$P < 0.001$ vs. control, t-test.

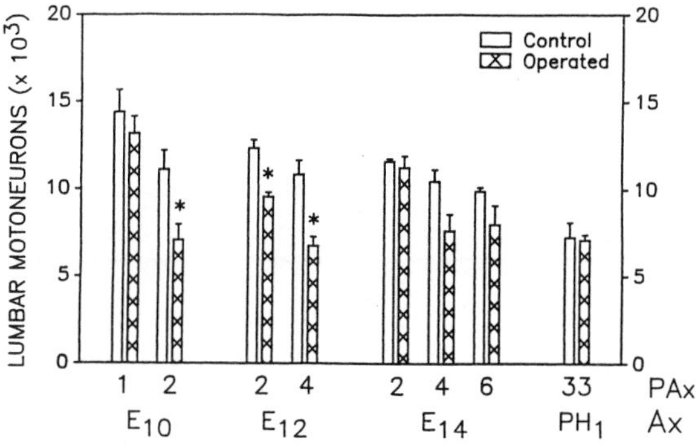

Fig. 2. Motoneuron numbers (means ± S.E.M.) in the lateral motor column (LMC) of chick embryos and posthatched chickens following axotomy (Ax) on embryonic days (E) 10, 12, 14, or 16 and examined 1, 2, 4, or 6 days later, or animals were operated or on posthatching (PH) day 1, and examined on postaxotomy (PAx) day 33. $^{*}P < 0.05$ vs. control (contralateral side), *t*-test. Each point involved three to five embryos.

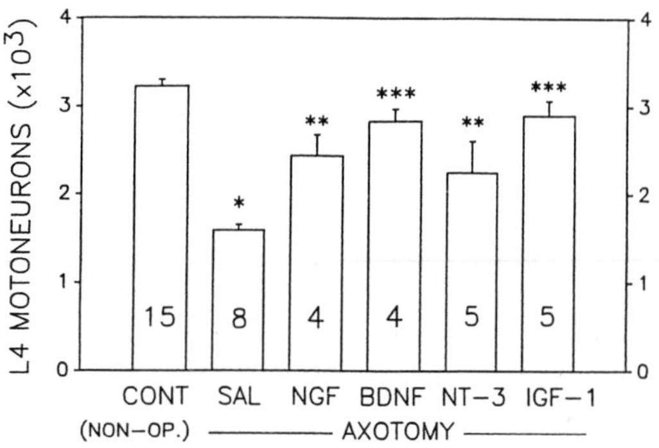

Fig. 3. L4 motoneuron numbers (means ± S.E.M.) in the spinal cord of E16 chick embryos following unilateral axotomy (hindlimb amputation) on E12 and treatment with either 0.2 ml of saline (SAL, 0.9% NaCl, pH7.2) or various trophic factors. Operated embryos were treated with 10 μg NGF, on E12 and E14, or they were injected daily from E12 to E15 with 5 μg BDNF, 5 μg NT-3, or 10 μg IGF-1 in 0.2 ml of saline. Cell counts of MNs contralateral to the side of axotomy were pooled (since no significant differences were observed between the different groups of embryos) and used as non-operated controls. Numbers in bars are sample sizes. $^{*}P < 0.0001$ vs. CONT; $^{**}P < 0.04$ vs. 'SAL' and $P < 0.001$ vs. CONT; $^{***}P < 0.0001$ vs. 'SAL' and $P > 0.05$ vs. CONT; *t*-test.

$n = 8$; $P < 0.0001$). We therefore, used this time course of axotomy (i.e. operation on E12 and sacrifice on E16) to test the effects of trophic agents on axotomized avian MNs. Treatment of operated embryos with either NGF, BDNF, NT-3, or IGF-1 significantly rescued L4 MNs from axotomy-induced death (Fig. 3). As shown in Fig. 4, CNTF was rescued L4 MNs from axotomy-induced death. However, the extent of this effect was dependent on repeated injection on CNTF from the time of axotomy to that of sacrifice (see legend to Fig. 4 for details), which was critical because of the reported short half-life of CNTF (Sendtner et al., 1992a).

When axotomy in the mouse was performed on or before PN5, there was a 33% decrease in total lumbar motoneuron numbers 5 days after axotomy and, by 10 days after surgery, MN numbers had declined to about 50% of controls. This reduction of motor neuron number appears to be stabilized by 10 days after surgery, since no further cell loss was detected between 10 and 25 days after axotomy (Fig. 5). In contrast, when axotomy was perfomed on PN10 mice, no significant cell loss was found 5 days post-axotomy, compared with controls. Further, axotomy per-

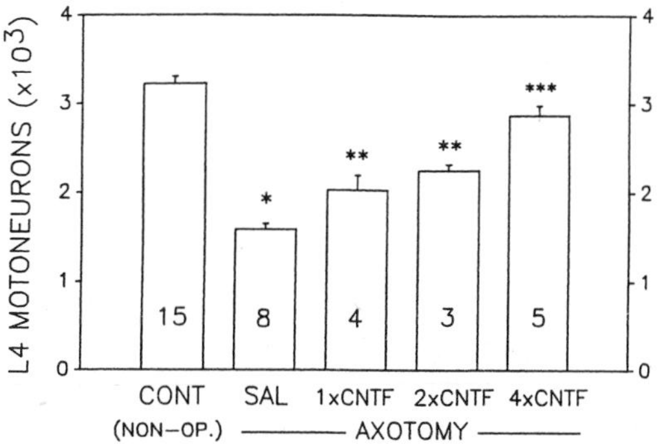

Fig. 4. L4 motoneuron numbers (means ± S.E.M.) in the spinal cord of chick embryos on E16 following unilatral sciatic nerve axotomy on E12. Operated embryos were treated either with saline (SAL), or with a total of 10 μg CNTF injected either *once 10μg*, i.e. on E12 (1 × CNTF), or *twice 5μg*, i.e. on E12 and E14 (2 × CNTF), or *four times 2.5μg*, i.e. on E12, E13, E14, and E15 (4 × CNTF). $^{*}P < 0.001$ vs. CONT.; $^{**}P < 0.001$ vs. 'Axotomy + SAL'; $^{***}P < 0.001$ vs. '1 × CNTF' or '2 × CNTF'. Note tht 4 × CNTF completely blocked cell death ($P > 0.1$ vs. CONT).

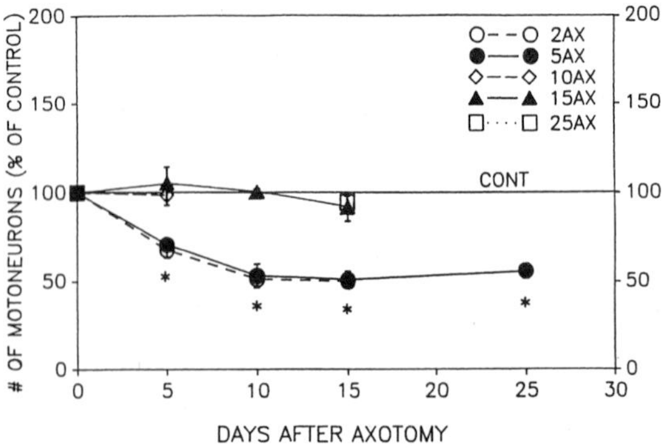

Fig. 5. Lumbar motoneuron numbers (% of controls) in the lateral motor column (LMC) of neonatal mice following unilateral sciatic nerve axotomy (Ax) on PN2, 5, 10, 15, and 25, and examined 5, 10, 15, or 25 days after surgery. $^*P < 0.05$ vs. control, *t*-test. Each point represents the mean ± S.E.M. of at least four animals.

formed on PN25 also failed to result in motoneuron death 15 days after surgery (Fig. 5). Since not all lumbar motoneurons send axons to the sciatic nerve, we examined the extent to which axotomy affects cell loss in each lumbar segment. As shown in Fig. 6, axotomy on PN2 affected 66% of L4 and 35% of L3 motoneurons, when animals were examined 10 days post-operation. Approximately 45% of L4 MNs died 7 days following axotomy performed on PN5 (Table II). Therefore, in subsequent studies, MNs were examined in the L4 segment of animals operated on PN5 and killed at least 7 days after surgery. As shown in Table II, NGF, BDNF, NT-3, NT-4/5, and IGF-1 rescued significant numbers of axotomized mouse MNs from cell death. Of these agents, BDNF, NT-3, and IGF-1 completely prevented axotomy-induced death of mouse MNs, whereas NGF, CNTF, and NT-4/5 were less effective (Table II).

These findings confirm and extend previous reports on the vulnerability of avian MNs to injury (Houthoff and Drukker, 1977). Previous studies have shown that avian isthmo-optic neurons also lose their target dependency at the end of the period of naturally occurring cell death (Catsicas and Clarke, 1987). The present data are also in agreement with previous demonstrations that mammalian MNs respond to axotomy performed within the first postnatal week by massive cell death (e.g. Schmalbruch, 1984; Pollin et al., 1991; Snider et al., 1992, review). Although rodents and chickens have very different life histories, with chickens being precocial and mice altricial at birth, the proximate cause of the difference in the time-course of MN sensitivity to axotomy between chick and mouse may be due primarily to differences in the time when nerve myelination occurs in the two species. In the chick spinal cord, myelination begins between E13 and E15, whereas this process occurs during postnatal development in the mouse. It is conceivable that myelination indicates a state of maturation when glial cells can secrete sufficient trophic agents to prevent neuronal cell death after axonal injury. Alternatively, this difference in MN sensitivity in the two species might be accounted for by a different developmental pattern of expression of trophic factor(s) in the local (CNS) environment of MNs (Koliatsos et al., 1993). Finally, it is also possible that spinal MNs are no longer dependent on trophic factors for their survival after E14 in the chickens and after PN10 in the mouse.

Although NGF did not prevent naturally occurring MN death (see Fig. 1 and Table I), it rescues MNs from deafferentation-induced death in the chick embryo (Oppenheim et al., 1992; Yin et al.,

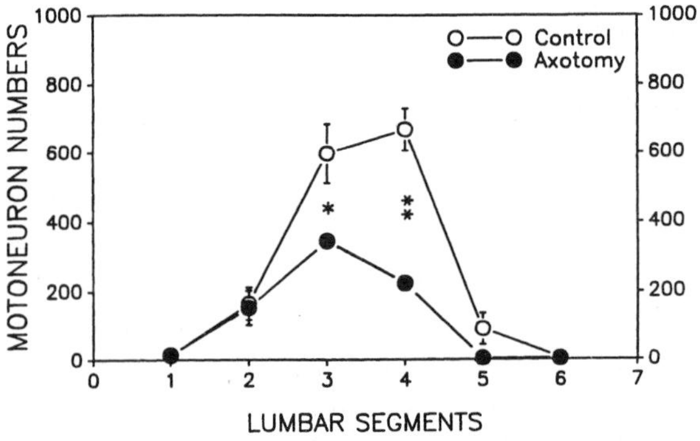

Fig. 6. Motoneuron numbers in the LMC of lumbar segments 1–6 of neonatal mice unilaterally axotomized (sciatic nerve) on day 2 and examined 10 days after surgery. $^*P < 0.05$ vs. control, *t*-test. $N = 5$ for each point.

TABLE II

Motoneuron numbers (means ± S.E.M.) in the lumbar spinal cord (L4) of 12 day-old mice axotomized on postnatal day 5 and treated with either saline or neurotrophic/growth factors

Treatment (*n*)	L4 motoneurons	Percentage
Contralateral (11)	921 ± 21	100
Ipsilateral		
Saline (4)	514 ± 45[a]	56
NGF (11)	764 ± 26[b]	83
BDNF (5)	935 ± 40[b]	101
NT-3 (5)	939 ± 52[b]	102
NT-4/5 (5)	742 ± 21[c]	81
CNTF (5)	651 ± 27[d]	71
IGF-1 (4)	868 ± 37[b]	94
PDGF-AA (4)	580 ± 19	63
PDGF-AB (5)	446 ± 19	48
IL-6 (4)	601 ± 64	65
bFGF (6)	630 ± 44	68

Operated mice received a total of 10 μg of the appropriate trophic agent at the site of axotomy between the day of surgery (PN5) and the day of sacrifice (PN12) as described under Methodological considerations. N = sample sizes.
[a]$P < 0.001$ vs. contralateral (control).
[b]$P < 0.001$ vs. 'saline'.
[c]$P < 0.01$ vs. 'saline'.
[d]$P < 0.05$ vs. 'saline'; *t*-test.

1994) and from axotomy-induced degeneration in both the chick (Fig. 3) and mouse (Table II). The ability of NGF to rescue MNs from axotomy-induced death in both species, as described in the present paper is consistent with a recent report (Yan et al., 1993), but is at odds with reports showing a lack of effect of this neurotrophin on axotomized rat MNs (Yan et al., 1988; Sendtner et al., 1992b; Koliatsos et al., 1993). These differences may be due to the difference in species, MN pools, or experimental methods. Taken at their face value, our results suggest that deafferented avian MNs (Oppenheim et al., 1992; Yin et al., 1994) and axotomized avian and mouse MNs (present data) express biologically active NGF receptors. It has been shown that developing MNs express $p75^{NGFR}$, the low affinity neurotrophin receptor (Yan and Johnson, 1988; Henderson et al., 1993) and very low levels of *trk*A, i.e. the tyrosine kinase receptor that mediates the biological activity of NGF (Henderson et al., 1993). Furthermore, developing MNs retrogradely transport NGF (Yan et al., 1988; Homma et al., 1993). However, it is not yet known whether axotomy upregulates the expression of *trk*A by MNs. It is also conceivable that the rather high local concentrations of NGF (due to our method of delivery) induce biological activity in a non-specific manner by binding to one of the other members of the *trk* family of receptors. Finally, we also cannot exclude the possibility that NGF in these situations rescues MNs indirectly, e.g. via a direct rescue effect on sensory neurons or interneurons.

Comparative aspects and prospects

Some of the agents (i.e. PDGF-AB and IL-6) utilized in the present study exhibited positive effects on avian but not on mouse MNs, either during the period of naturally occurring cell death or following injury. Another agent, NT-3, had a positive effect on *axotomized* avian and mouse MNs, but not on chick MNs during the period of naturally occurring cell death. Therefore, positive effects of a trophic/growth factor on MNs cannot necessarily be generalized from one species to another. However, several of the agents, e.g. BDNF, NT-4/5, IGF-1 and CNTF, were in fact found to promote the survival of mouse and chick MNs, not only during the period of naturally occurring cell death but also following axotomy. Of these agents, BDNF (Henderson et al., 1993; McKay et al., 1993), NT-4/5 (Henderson et al., 1993), and IGF-1 (Bondy et al., 1990; Ralphs et al., 1990; Streck et al., 1992) appear to be expressed in developing skeletal muscle. Furthermore, BDNF is up-regulated in muscle following denervation (Koliatsos et al., 1993). Taken together, these observations suggest that these trophic/growth factors could play a role in MN survival and differentiation during developoment.

In conclusion, the data presented here indicates that the time-course of axotomy-induced

MN death is different in chick and mouse spinal cord. In the chick, MNs become independent from their target for survival at the end of the period of programmed cell death. In the mouse, MNs remain vulnerable to the loss of their target for up to 2 weeks after the cessation of naturally occurring cell death. The data also show that a number of trophic/growth factors, including CNTF, IGF-1 and the neurotrophins, can promote the survival of avian and mouse MNs both during the period of naturally occurring cell death and following axotomy. Thus, these agents may be biologically important molecules during normal MN differentiation. Their effectiveness in preventing normal as well as injury-induced cell death also suggest that these agents may prove to be of value in the treatment of patients with motor neuron diseases. In fact, clinical trials are currently being conducted with amyotrophic lateral sclerosis (ALS) patients using CNTF and IGF and these will soon be extended to include BDNF.

Summary

Neuronal cell death is an important regressive event during the normal development of the peripheral and central nervous systems of many vertebrate and invertebrate species. Furthermore, when neurons are deprived of their target following axonal injury (axotomy) during embryonic, fetal, or early postnatal development, they undergo massive cell death. Both naturally occurring and axotomy-induced neuronal cell death can be prevented by treatment with growth factors or neurotrophic agents. Naturally occurring cell death of spinal MNs has been extensively studied in both avians and mammals. However, compared with mammals, there is little information on the effects of axotomy in avian species and it is not known whether trophic agents can modify axotomy-induced death in avian MNs. It is also not known whether trophic/growth factors can promote the in vivo survival of mammalian MNs during the period of naturally occurring cell death. We have examined (1) the time course of axotomy-induced death of lumbar spinal MNs in chick and mouse, and (2) the survival-promoting activity of a number of previously characterized growth and trophic factors on both programmed and axotomy-induced MN death in these two species. We show that axotomy performed on, or prior to, E12 in the chick results in a rapid decrease (i.e. 50%) in MN numbers within 3–4 days postsurgery, whereas these cells were able to survive for up to 1 week following axotomy on E14. By contrast, mouse MNs remained vulnerable to axotomy for at least 5 days after birth. Furthermore, we show that a number of trophic/growth factors rescue both avian and mouse MNs (albeit to different extents) from programmed and/or axotomy-induced cell death. These include NGF, BDNF, NT-3, NT-4/5, CNTF, IGF-1 and IGF-2, S-100ß, TGF-ß, PDGF, CDF/LIF, and IL-6.

Acknowledgements

The authors thank Ming Lei and David Prevette for technical assistance and Mary Ann Hayner for help in manuscript preparation. This work was supported by NIH grants HD29435 (L.J.H.) and NS20402 (R.W.O.) and grants from the Muscular Dystrophy Association, the International Research Institute for Paraplegia, and Synergen, Inc., (L.J.H.), and from the Amyotrophic Lateral Sclerosis Association, Amgen, Inc., and Cephalon, Inc. (R.W.O.).

References

Bhattacharyya, A., Oppenheim, R.W., Prevette, D., Moore, B.W., Brackenbury, R. and Ratner, N. (1992) S100 is present in developing chicken neurons and Schwann cells and promotes motor neurons survival in vivo. *J. Neurobiol.*, 23: 451–466.

Bondy, C.A., Werner, H., Roberts, C.T. and LeRoith, D., (1990) Cellular pattern of IGF-1 and type 1 IGF receptor gene expression in early organogenesis: comparison with IGF-2 gene expression. *Mol. Endocrinol.*, 4: 1386–1398.

Catsicas, S. and Clarke, P.G.H. (1987). Abrupt loss of dependence of retinoceptal neurons on their target cells, as shown by intraocular injection of kainate in chick embryos *J. Comp. Neurol.*, 262: 523–534.

Chu-Wang, I.W. and Oppenheim, R.W. (1978) Cell death of motoneurons in the chick embryo spinal cord. *J. Comp. Neurol.*, 177: 33–86.

Clarke, P.G.H. (1985) Neuronal death during development in isthmo-optic nucleus of the chick: sustaining role of afferents from the tectum. *J. Comp. Neurol.*, 234: 365–379.

Crews, L.L. and Wigston, D.J. (1990) The dependence of motor neurons on their target muscle during postnatal development of the mouse. *J. Neurosci.*, 10: 1643–1653.

DiStephano, P.S., Friedman, B., Wiegand, S.J. and Lindsay, R.M. (1992) The neurotrophins BDNF, NT-3 and NGF display distinct patterns of retrograde axonal transport of peripheral and central neurons. *Neuron*, 8: 893–993.

Forger, N.G. and Breedlove, S.M. (1987) Motoneuronal death during human fetal development. *J. Comp. Neurol.*, 264: 145–157.

Hamburger, V. (1975) Cell death in the development of the lateral motor column of the chick embryo. *J. Comp. Neurol.*, 160: 535–546.

Hamburger, V. and Oppenheim, R.W. (1982) Naturally occurring neuronal death in vertebrates. *Neurosci. Comment.*, 1: 39–55.

Harris, A.J. and McCraig, C.D. (1987) Motoneuron death and motor unit size during embryonic development of the rat. *J. Neurosci.*, 4: 13–24.

Henderson, C.E., Camu, W., Mettling, C., Gouin, A., Poulsen, K., Karihaloo, M., Rullamas, J., Evans, T., McMahon, S., Armanini, M.P., Berkemeier, L., Phillips, H.S. and Rosenthal, A. (1993) Neurotrophins promote motor neuron survival and are present in embryonic limb bud. *Nature*, 363: 266–270.

Homma, S., Prevette, D., Yan, Q. and Oppenheim, R.W. (1993) Binding sites and retrograde transport of neurotrophins in the chick embryo nervous system. *Soc. Neurosci. Abstr.*, 19: 1303.

Houenou, L.J., Pincon-Raymond, M., Garcia, L., Harris, A.J. and Rieger, F. (1990) Neuromuscular development following tetrodotoxin-induced inactivity in mouse embryos. *J. Neurobiol.*, 21: 1249–1261.

Houenou, L.J., McManaman, J.L. Prevette, D. and Oppenheim, R.W. (1991) Regulation of putative muscle-derived neurotrophic factors by muscle activity and innervation: *in vivo* and *in vitro* studies. *J. Neurosci.*, 11: 2829–2837.

Houthoff, H.J. and Drukker, J. (1977) Changing patterns of axonal reaction during neuronal development. A study in the developing chick nervous system. *Neuropathol. Appl. Neurobiol.*, 3: 441–451.

Kashihara, Y., Kuno, M. and Miyata, Y. (1987) Cell death of axotomized motoneurones in neonatal rats, and its prevention by peripheral reinnervation. *J. Physiol.*, 386: 135–148.

Koliatsos, V.E., Clartterbuck, R.E., Winslow, J.W., Cayouette, M.H. and Price, D.L. (1993) Evidence that brain-derived neurotrophic factor is a trophic factor for motor neurons in vivo. *Neuron*, 10: 359–367.

Lance-Jones, C. (1982) Motoneuron cell death in the developing spinal cord of the mouse. *Dev. Brain Res.*, 4: 473–479.

Lance-Jones, C. and Diaz, M. (1991) The influence of presumptive limb connective tissue on motoneuron axon guidance. *Dev. Biol.*, 143: 93–110.

McKay, S.E., Herzog, K.-H., Garner, A., Tucker, R.P., Oppenheim, R.W. and Large, T. (1993) Expression of BDNF and *trkB* during the development of the neuromuscular system in the chick embryo. *Soc. Neurosci. Abstr.*, 19: 252.

Neff, N.T., Prevette, D., Houenou, L.J., Lewis, M.E., Glicksman, M.A., Yin, Q.-W. and Oppenheim, R.W. (1993) Insulin-like growth factors: putative muscle-derived trophic agents that promote motoneuron survival. *J. Neurobiol.*, 24: 1578–1588.

Oppenheim, R.W. (1981) Cell death of motoneurons in the chick embryo spinal cord. V. Evidence on the role of cell death and neuromuscular function in the formation of specific peripheral connections. *J. Neurosci.*, 1: 141–151.

Oppenheim, R.W. (1989) The neurotrophic theory and naturally occurring motoneuron death. *Trends Neurosci.*, 12: 252–255.

Oppenheim, R.W. (1991) Cell death during development of the nervous system. *Annu. Rev. Neurosci.*, 14: 453–501.

Oppenheim, R.W., Maderdrut, J.L. and Wells, D.J. (1982) Cell death of motoneurons in the chick embryo spinal cord. VI. Reduction of naturally occurring cell death in the thoracolumbar column of terni by nerve growth factor. *J. Comp. Neurol.*, 210: 174–189.

Oppenheim, R.W., Houenou, L., Pincon-Raymond, M., Powell, J.A., Rieger, F. and Standish, L.J. (1986) The development of motoneurons in the embryonic spinal cord of the mouse mutant, muscular dysgenesis (*mdg/mdg*): survival, morphology and biochemical differentiation. *Dev. Biol.*, 114: 426–436.

Oppenheim, R.W., Prevette, D., Tytell, M. and Homma, S. (1990) Naturally occurring and induced neuronal death in the chick embryo *in vivo* requires protein and RNA synthesis: evidence for the role of cell death genes. *Dev. Biol.*, 138: 104–113.

Oppenheim, R.W., Prevette, D., Yin, Q.-W., Collins, F. and MacDonald, J. (1991) Control of embryonic motoneuron survival in vivo by ciliary neurotrophic factor. *Science*, 257: 1616–1618.

Oppenheim, R.W., Yin, Q.-W., Prevette, D. and Yan, Q. (1992) Brain-derived neurotrophic factor rescues developing avian motoneurons from cell death. *Nature*, 360: 755–757.

Oppenheim, R.W., Prevette, D., Haverkamp, L.J., Houenou, L., Yin, Q.-W. and McManaman, J. (1993) Biological studies of a putative avian muscle-derived neurotrophic factor that prevents naturally occurring motoneuron death in vivo. *J. Neurobiol.*, 24: 1065–1079.

Pittman, R. and Oppenheim, R.W. (1978) Neuromuscular

blockade increases motoneuron survival during cell death in the chick embryo. *Nature*, 271: 364–366.

Pittman, R. and Oppenheim, R.W. (1979) Cell death of motoneurons in the chick embryo spinal cord. IV. Evidence that functional neuromuscular interaction is involved in the regulation of naturally occurring cell death and the stabilization of synapses. *J. Comp. Neurol.*, 187: 425–446.

Pollin, M.M., McHanwell, S. and Slater, C.R. (1991) The effect of age on motoneuron death following axotomy in the mouse. *Development*, 112: 83–89.

Raulo, E., Julkunen, I., Merenmies, J., Pihlaskari, R. and Rauvala, H., (1992) Secretion and biological activities of heparin binding growth-associated molecule. *J. Biol. Chem.*, 267: 11408–11416.

Ralphs, J.R., Wylie, L. and Hill, D.J. (1990) Distribution of insulin-like growth factor peptides in the developing chick embryo. *Development*, 109: 51–58.

Schmalbruch, H. (1984) Motor neuron death after sciatic nerve section in newborn rats. *J. Comp. Neurol.*, 224: 252–258.

Sendtner, M., Arakawa, Y., Stockli, K.L., Kreutzberg, G.W. and Thoenen, H. (1991) Effect of ciliary neurotrophic factor (CNTF) on motoneuron survival. *J. Cell Sci. Suppl.*, 15: 103–109.

Sendtner, M., Schmalbruch, H., Stockli, K.A., Carroll, P., Kreutzberg, G.W. and Thoenen, H. (1992a) Ciliary neurotrophic factor prevents degeneration of motor neurons in mouse mutant progressive motor neuropathy. *Nature*, 358: 502–504.

Sendtner, M., Holtmann, B., Kolbeck, R., Thoenen, H. and Barde, Y.-A. (1992b) Brain-derived neurotrophic factor prevents the death of motoneurons in newborn rats after nerve section. *Nature*, 360: 757–759.

Snider, W.D., Elliott, J.L. and Yan, Q. (1992) Axotomy-induced neuronal death during development. *J. Neurobiol.*, 23: 1231–1246.

Streck, R.D., Wood, T.L., Hsu, M.S. and Pintar, J.E. (1992) Insulin-like growth factor I and II and insulin-like growth factor binding protein-2 RNAs are expressed in adjacent tissues within rat embryonic and fetal limbs. *Dev. Biol.*, 151: 586–596.

Yan, Q., and Johnson, Jr. E.M. (1988) An immunohistochemical study of the nerve growth factor receptor in developing rats. *J. Neurosci.*, 8: 3481–3498.

Yan, Q., Snider, W.D., Pinzone, J.J. and Johnson, Jr., E.M. (1988) Retrograde transport of nerve growth factor (NGF) in motoneurons of developing rats: assessment of potential neurotrophic effects. *Neuron*, 1: 335–343.

Yan, Q., Elliott, J.L. and Snider, W.D. (1992) Brain-derived neurotrophic factor rescues spinal motor neurons from axotomy-induced cell death. *Nature*, 360: 753–755.

Yan, Q., Elliott, J.L., Matheson, C., Sun, J., Zhang, L., Mu, X., Rex, K.L. and Snider, W.D. (1993) Influence of neurotrophins on mammalian motorneurons in vivo. *J. Neurobiol.*, 24: 1555–1577.

Yin, Q.-W., Johnson, J., Prevette, D. and Oppenheim R.W. (1994) Cell death of spinal motoneurons in the chick embryo following deafferentation: Rescue effects of tissue extracts, soluble proteins and neurotrophic agents. *J. Neurosci* (In press).

J. van Pelt, M.A. Corner H.B.M. Uylings and F.H. Lopes da Silva (Eds.)
Progress in Brain Research, Vol 102

CHAPTER 15

Synaptic development of the cerebral cortex: implications for learning, memory, and mental illness

Pasko Rakic[1], Jean-Pierre Bourgeois[2] and Patricia S. Goldman-Rakic[1]

[1]*Section of Neurobiology, Yale University School of Medicine, New Haven, CT, U.S.A. and* [2]*Molecular Neurobiology Laboratory, Pasteur Institute, Paris, France*

Introduction

The formation of appropriate synapses is the ultimate step in the construction of brain circuitry. Synaptic architecture defines the limits of individual mental capacity and provides the framework for comprehending the major psychiatric disorders. In the past two decades, developmental brain research has been dedicated to understanding the principles of neuronal specificity and the rules followed by growing axons as they discover their life-long partnerships with other cells. It is surprising, therefore, how little we know about synaptogenesis in the cerebral cortex and how poor our understanding is of the relationship between synaptic maturity and functional development. Are synapses added as we learn, as many have argued? Are there more synapses in some cortical areas than in others? Are there gender differences in synaptic density and do we lose synapses as we age? If we lose synapses with age, what is the timing and rate of this dissolution? These are a few of the many questions that can be asked. Although many opinions on these matters are expressed often, hard data in the form of rigorous quantitative analysis at the ultrastructural level is rare.

The findings to be described below, collected over the past 8 years in our laboratories, have begun to address these issues in the rhesus monkey, sometimes with counterintuitive results. We have performed quantitative electronmicroscopic analysis so far for five major cortical areas (Fig. 1) of the non-human primate brain and have counted and marked over one million synapses beginning with the earliest observed in embryonic life from the 50th embryonic day (E50) and thereafter through over 20 years of postnatal age. The detailed course and kinetics of synaptogenesis will be illustrated here primarily in the visual cortex, but reference will be made to other cortical areas as well. The study of major structural and functional subdivisions of the cortex over the primate lifespan offers a particularly comprehensive view of synapse formation. We believe that the effort and difficulty of obtaining such data is justified because the issues are both theoretically significant and socially important. Knowledge of the principles and limits of synaptic plasticity during childhood and adolescence could have enormous implications for educational theory and practice as well as for clinical treatment of childhood brain disorders.

Phase of rapid synaptic production

Most developmental neurobiologists would agree that synapse formation is a prolonged and rela-

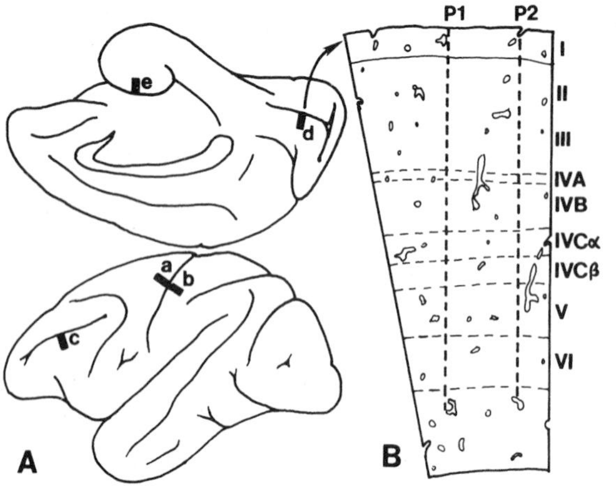

Fig. 1. (A) The lateral surface of the left cerebral hemisphere (bottom) and the medial surface, inverted (top), show the five cortical areas examined: (a) motor cortex (Brodmann's area 4) in precentral gyrus; (b) somatosensory cortex (area 1) in the postcentral gyrus; (c) prefrontal cortex (area 46) in the upper bank of the principal sulcus; (d) visual cortex (area 17) in the upper bank of the calcarine fissure; (e) molecular layer in the dentate gyrus (area 34). The blocks were post-fixed in osmium, embedded in Epon-Araldite, and 700-Å sections were cut across the entire width of the cortex. (B) An outline of an ultrathin section across the visual cortex (d), as an example. The two vertical lines (P1 and P2) indicate the localization of two probes each yielding about 100 electron micrographs that were printed at a final magnification of × 14,000. Similar probes were prepared for other cortical areas except the dentate gyrus, where probes were taken only across the width of the molecular layer of the suprapyramidal and infrapyramidal limbs (from Rakic et al., 1986).

tively late event marking the culmination of a series of steps beginning with the birth of neurons at the ventricular surface and ending, in the case of corticogenesis, with the completion of cell migration across the intermediate zone in an inside-out settlement of neurons to form the cortical layers (Rakic, 1972, 1974, 1988). In the visual cortex, the first apposed cytological profile containing round vesicles that could meet the criteria of a synapse was found in the marginal zone (prospective layer 1) in a rhesus monkey embryo at embryonic day 50 (E50), when the neurons of only the subplate zone and deep cortical layer V1 have been generated. Synapses continued to accrue slowly, following the inside-out pattern, as more superficial neurons arrive at the cortical plate. By E112, when the developing cortex has its full complement of neurons and lamination is clear, synapses were distributed throughout the depth of the cortex, albeit in low density. As is evident from Fig. 2, the most rapid accumulation of synapses in the macaque visual cortex starts at E112 (Bourgeois and Rakic, 1993). Similar rapid accumulation has been observed in somatosensory, motor, and prefrontal association cortex (Zecevic et al., 1989; Zecevic and Rakic, 1991; Bourgeois et al., 1994). For the visual cortex, the exponential growth of synapses occurs approximately 2 weeks after the end of the genesis of neurons destined for area 17 (Rakic, 1974) and not long after the completion of neuronal migration in the visual cortex (Rakic, 1974). Between 2 months before and 2 months after birth, the density of synapses in the striate cortex increases ten times faster (17-fold) (Bourgeois and Rakic, 1993) than the increase in the total volume of this area (1.7-fold) (Williams, R. and Rakic, P., unpublished data). The expansion of the population of synapses is not limited by the volume of cortical tissue since the synaptic boutons represent only 1.5% of the total volume of the striate cortex (Bourgeois and Rakic, 1993). Actually, the rate of accumulation of synapses during this exponential phase is strikingly high. We estimated that around 10,000 to 40,000 synapses are formed every second in each striate cortex of the macaque during this exponential phase. This high rate continues through birth and infancy and begins to subside only around the third postnatal month (Bourgeois and Rakic, 1993).

It is noteworthy that the course of synaptogenesis does not seem to change significantly at term as a consequence of birth or exposure to visual stimulation. We have conducted two experiments designed to determine if either the rate or final level of synaptogenesis could be altered by environmental events. In one experiment, (Bourgeois et al., 1989), we delivered monkey fetuses 3 weeks before term by Cesarean section in order to ad-

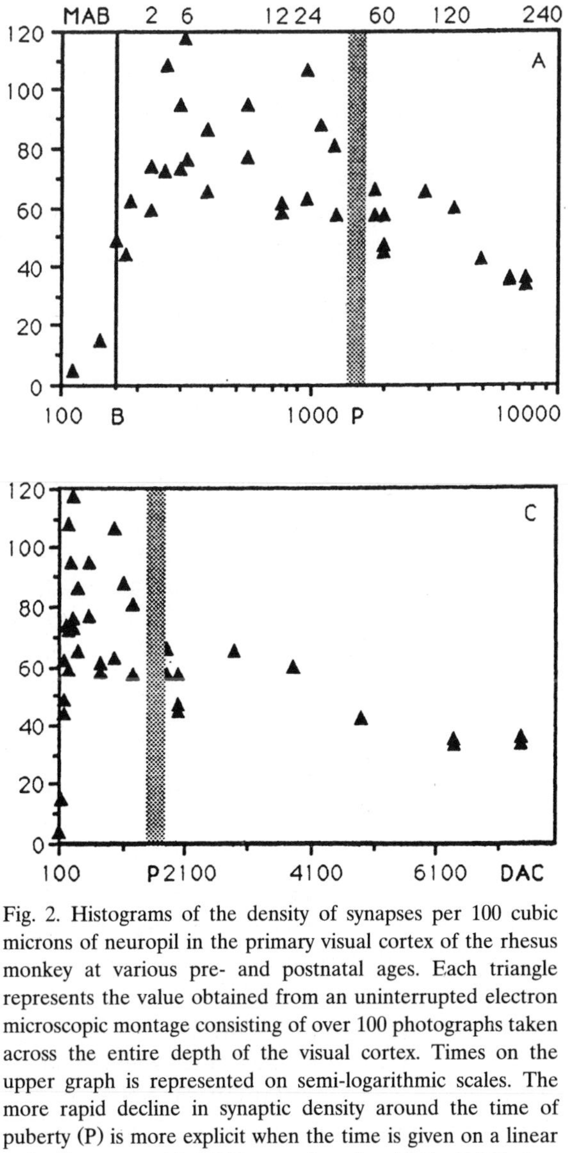

Fig. 2. Histograms of the density of synapses per 100 cubic microns of neuropil in the primary visual cortex of the rhesus monkey at various pre- and postnatal ages. Each triangle represents the value obtained from an uninterrupted electron microscopic montage consisting of over 100 photographs taken across the entire depth of the visual cortex. Times on the upper graph is represented on semi-logarithmic scales. The more rapid decline in synaptic density around the time of puberty (P) is more explicit when the time is given on a linear scale (lower graph). MAB: months after birth; DAC: days after conception; B: birth; P: puberty (from Bourgeois and Rakic, 1993).

vance and extend their exposure to visual stimulation compared with monkeys delivered at term. Subsequent analysis of synaptic density in pairs of monkeys matched for conceptual age, but differing in the prematurity of their birth, failed to reveal any difference in synaptic density in any layers of the visual cortex. In a second experiment, the visual cortex of monkeys that were enucleated bilaterally as fetuses was analyzed for changes in synaptic density relative to age-matched controls (Bourgeois and Rakic, 1987; Bourgeois, J.-P. and Rakic, P., in preparation). Again, there were no significant differences in synaptic density between the visual cortex of experimental or control subjects. These findings indicate that the timing and rate of synaptic production may be genetically programmed. This conclusion, however, does not contradict the finding of discrete differences in maturational rates of specific classes of synapses, which occur after birth in response to diverse functional states (e.g. Hubel et al., 1977a,b; Lund and Holbach, 1991). Our data only suggest that these influences probably act predominantly on the survival of existing, rather than on the production of new synapses (Bourgeois et al., 1989).

Rearrangement of synaptic pattern during synaptogenesis

During the phase of synaptogenesis, the junctions that are formed are not immutable or static. There is considerable evidence that reorganization of synapses could occur soon after they are formed (Plummer Lord and Behan, 1993; Missler et al., 1993). Previous anatomical (Rakic, 1976, 1977; LeVay and Stryker, 1979; Shatz, 1983) and physiological (Hubel and Wiesel, 1977a) studies have indicated that the segregation of the geniculo-cortical afferents into ocular dominance columns in the visual cortex during normal development results from rearrangement, and perhaps elimination, of synaptic terminals from the flanking domains of overlapping terminal fields. It is, therefore, of interest that the segregation of ocular dominance columns coincides with the time of increase, rather than decrease, in the number of synapses situated on dendritic spines and the spines themselves (Boothe et al., 1979; Lund and Holbach, 1991; Bourgeois and Rakic, 1993). In fact, the onset of the rapid phase of synaptogenesis at E112 begins about 1 month earlier than the onset of visible segregation of the ocular dominance domains (E144 according to Rakic, 1976, 1977). A similar timetable of maturation of dendritic spines in primate visual cortex has been observed in Golgi preparations (Lund and Holbach, 1991; Lund et al., 1991).

The period of rapid synaptogenesis after birth coincides with the so-called 'sensitive period' of both sensory and association areas. For example, in monocularly deprived monkeys this is the age when both anatomical and functional recovery is still possible if the initially deprived eye is reopened and the normal eye occluded (Hubel et al., 1977b; Blakemore et al., 1978, 1981; Lund et al., 1991). Our data on synaptogenesis in the cerebral cortex are consistent with the working hypothesis that the magnitude, rate, and timing of the phase of synaptic production is determined intrinsically, while the selective stabilization of only some of these synapses may be achieved through epigenetic mechanisms during postnatal development (Bourgeois et al., 1989). Comparable ultrastructural data for the other cortical areas are not available and this subject deserves the attention of researchers.

High synaptic density during adolescence

As mentioned, the rapid phase of synapse formation in the visual cortex subsides around 3 months of age. This ascending phase that culminates with supernumerary synaptic junctions has been observed in all studies which have examined various areas of macaque cerebrum in our and other laboratories (Bourgeois and Rakic, 1993; Bourgeois et al., 1994; Zecevic and Rakic, 1989, 1991; O'Kusky and Colonnier, 1982; Zielinski and Hendrickson, 1992). Thereafter, synaptic density remains at higher than adult levels for a period of several years (Fig. 3). At an earlier stage of our investigations, with fewer probes, and fewer cases, and a different measure (number of synapses per unit area of neuropil), it appeared that synaptic elimination to adult levels began immediately after 2–4 months of age (Rakic et al., 1986). However, in our latest analysis of visual cortex in a larger number of cases, and both with and without stereological corrections, the highest density of synapses in motor, somatosensory, and visual cortex is reached at 2–3 months of age and then declines, at first gradually (up to 3 years) and then more sharply after that point (Zecevic and Rakic, 1991; Bourgeois and Rakic, 1993; Bourgeois et al., 1994). A somewhat different time course is emerging for the prefrontal cortex, where the highest level of synaptic density appears to be sustained from the second postnatal month until the third year after birth before declining very gradually to the adult level (Bourgeois et al., 1994). Since we are dealing with trends on a life span scale of over 20 years, additional specimens would undoubtedly clarify whether the plateau observed in the prefrontal cortex is the general rule that applies for the entire cortical mantle or whether synaptic density begins a gradual decline after the peak is reached, as recorded in primary motor and sensory cortices.

The period of high synaptic density observed during the period of adolescence in all higher primates, including humans, is characterized by intense behavioral change and an immense capacity for learning. This is the period in human development that coincides with intellectual growth and the achievement of adult level competence in language and logical thought (e.g. Karmiloff-Smith, 1979; Yates, 1991), mathematics (e.g. Gelman and Gallistal, 1978). Many modifications in the morpho-functional maturation of the cerebral cortex have been proposed to explain dramatic behavioral change in this time frame, including reorganization and turnover of synaptic circuitry (Purves, 1988), synaptic vesicle proliferation (Miller, 1988), axonal myelination (Gibson, 1970), and biochemical differentiation (Goldman-Rakic and Brown, 1982). Most of these cellular events occur during the plateau phase of synaptogenesis and any and all of these parameters could contribute to the functional maturation of the cerebral cortex. Nevertheless, this phase of intense and highly efficient learning and rapid behavioral modification does not appear to be paralleled by either significant net accretion or net loss of synapses. The existing methods available do not allow determination of the extent to which synapses are stable and to what extent subject to constant turnover. In the human, the plateau phase for prefrontal cortex would correspond to the entire period of childhood through puberty, roughly 13 years (Fig. 4) (Huttenlocher, 1979). The findings obtained in non-human primates seem to indicate that factors which improve synaptic efficiency may be more critical than shifts in synaptic density as a neural basis for behavioral maturation and plasticity during this prolonged postnatal period.

Decrease in density of synapses

During the third year of age and possibly throughout adulthood and aging of the macaque monkey, there is a slow but steady decrease in the density of synapses in all cortical layers in each area examined (Fig. 3). In the prefrontal cortex, as indicated, the rate is less abrupt during sexual maturation than in the visual cortex and the decline seems to be more evenly distributed throughout the entire lifespan up to 20 years of age (Bourgeois et al., 1994); whereas, in the visual cortex there seems to be no further decline after 4 years (Bourgeois and Rakic, 1993). The evidence that these decreases in synaptic density do not represent a dilution of a constant number of

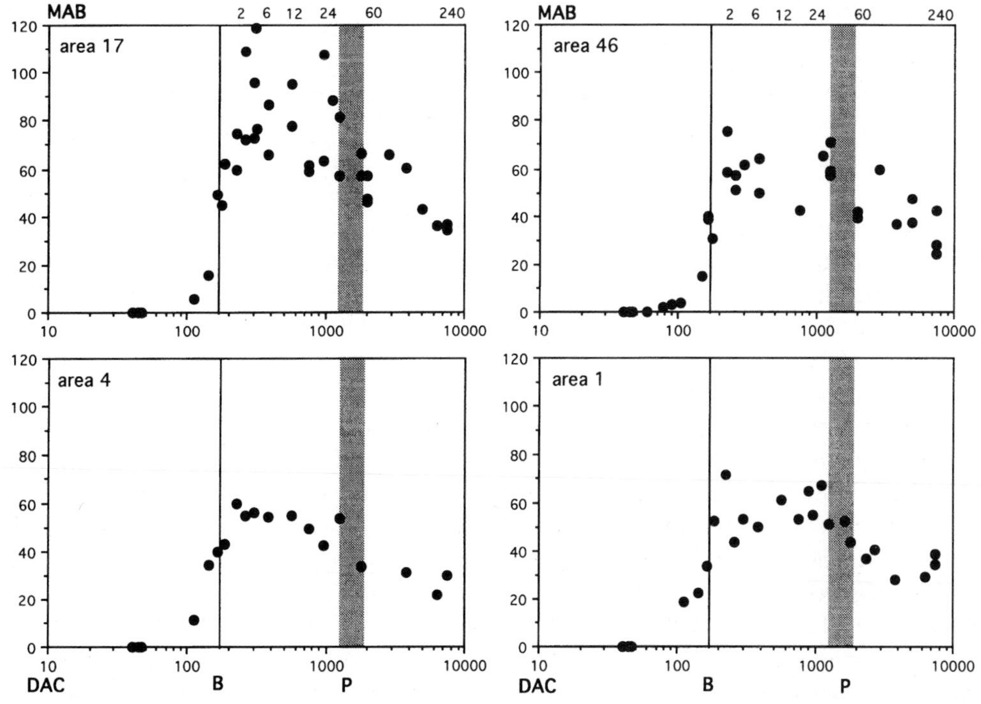

Fig. 3. Histograms of the density of synapses per 100 μm^3 of neuropil in the primary visual (area 17), prefrontal association (area 46), primary motor (area 4) and primary somatosensory (area 1) at various ages. Each black circle represents the value obtained from a single electronmicroscopic probe (see Fig. 1), and data are corrected and transferred into N per volume according to the Anker and Cragg formula. The time is represented on a semi-logarithmic scale. MAB: months after birth; DAC: days after conception; B: birth; P: puberty. (Data compiled from Zecevic et al., 1989; Zecevic and Rakic, 1991; Bourgeois and Rakic, 1993; Bourgeois et al., 1994.)

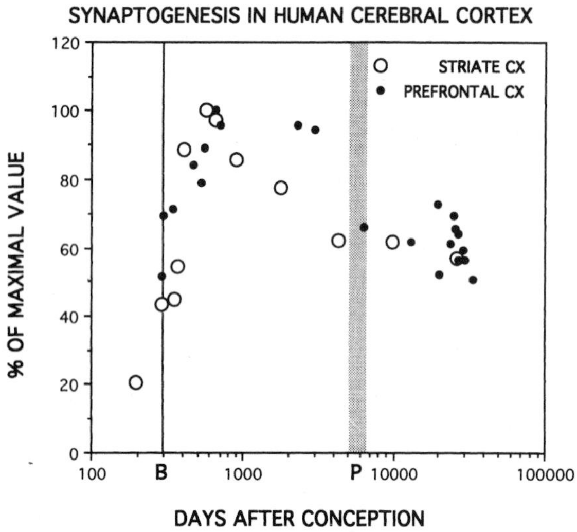

Fig. 4. Synaptic density in the human striate area (open, large circles) and prefrontal cortex (black, small circle) based on the studies of Huttenlocher and de Courten, 1987 and Huttenlocher, 1979, respectively. We replotted each point from their published data on a semilogarithmic scale as a function of conceptual age, and normalized to the maximum value of the curve. Under these conditions, both curves overlap as they do in monkeys. For further explanation see text.

synapses in an expanding volume of the cortex or neuropil, but a genuine attrition, has been addressed in detail in prior publications (Rakic et al., 1986; Zecevic et al., 1989; Zecevic and Rakic, 1991; Bourgeois and Rakic, 1993). An important part of the argument is that in all areas of the cerebral cortex examined to date, the decline of synaptic density is primarily due to a selective change in the number of asymmetric junctions situated on dendritic spines, while symmetric synapses situated on shafts remain relatively constant. This uneven loss of two basic types of synapses cannot be explained by a simple dilution. Furthermore, the change in synaptic density is not related to the differential increase in the percentage of neuropil, which actually decreases slightly during the course of postnatal life. Nor can it be attributed to neuronal loss, which occurs mainly before birth in both the human and monkey visual cortex (Leuba and Kraftsik, 1994; Williams and Rakic, unpublished data) and its magnitude is not sufficient to account for the observed changes in synaptic density (Bourgeois and Rakic, 1993). It should also be emphasized that we deliberately expressed synaptic density per unit of neuropil to assure that our measure would be impervious to increases in the size of neural and glial cells or other elements like blood vessels. Since the percentage of neuropil is relatively constant during adolescence and adulthood, most, if not all, change in synaptic density over that time appears due to actual synaptic loss.

The *total number of synapses* lost has been assessed so far only for the primary visual cortex (Bourgeois and Rakic, 1993). Our calculations show that as many as 1.9×10^{11} synapses may be lost between adolescence and adulthood in area 17 of each hemisphere. An average of 2450 synapses may be lost per second in each striate cortex, provided that they are being eliminated equally over a 24 h cycle between 2.5 and 5 years of age (Bourgeois and Rakic, 1993). Our estimates are of the same order of magnitude as those reported by O'Kusky and Colonnier (1982) for *Macaca fascicularis* and slightly smaller than that reported for marmoset monkey (Missler et al., 1993) (see also review in Peters, 1987). Although our data represent only mean values from a highly heterogeneous population of neurons, they reveal a 45–48% loss of synapses per neuron during puberty, close to the 46% loss observed in area 17 by O'Kusky and Colonnier (1982). Provided that the other areas lose a similar number of synapses per unit time, and visual cortex in macaque is only about 15% of the total cortical volume, as much as 30,000 synapses may be lost per second in the entire cerebral cortex of both hemispheres during the period of sexual maturation. These large changes in the density of synapses have very little effect either on the volume or the surface of the cortex because the total volume of synaptic boutons is less than 1.5% of the total cortical volume (Bourgeois and Rakic, 1993). Indeed, there is a small but statistically significant decrease in the volume of cortical gray matter between childhood and adulthood in humans (Jernigan et al., 1991).

In all cortical areas studied, the decrease in synaptic density is due mainly to the elimination of asymmetrical junctions situated on dendritic spines (see Bourgeois et al., 1994). Since this class of synapses is considered to be primarily excitatory, their massive loss is likely to be reflected in the overall excitability of the cerebral hemispheres. Indeed, between the ages of 10 and 14 years, when synaptic loss occurs in humans (Huttenlocher, 1979), marked decreases occur in the amplitude of the EEG and in the pattern and duration of the various stages of sleep (Feinberg et al., 1967). In addition, in both the human and monkey, there is a sharp decline in cortical metabolic activity during the same period (Chugani et al., 1987; Jacob et al., 1994).

Concurrent synaptogenesis and biochemical maturation in diverse cortical areas in monkey and human

Analysis of synaptogenesis in visual (Bourgeois and Rakic, 1993), motor (Zecevic et al., 1989), somatosensory (Zecevic and Rakic, 1991), and prefrontal (Bourgeois et al., 1994) cortex in macaque monkey indicates that a basic course of synaptogenesis, and, in particular, the ascending phase, occurs synchronously in all cortical areas examined (Fig. 3). Contrary to our initial expectations, we found that the entire population of synaptic junctions in the association cortex, like Brodmann's area 46, matures concurrently with, rather than after, the sensory and motor areas. We do not contest the observation that during this exponential phase there is a sequential accretion of synapses from diverse origins in each cortical area (see for example Lund and Holbach, 1991). We propose that most of these sequences of synaptogenesis are framed within a 'window of time', centered to this exponential phase of synaptogenesis in all areas of the neocortex within one hemisphere. The same may seem to be true in the human, although data at first appear to suggest a delay in the rapid phase of synaptogenesis in the prefrontal, as compared with that in the striate, area (Huttenlocher, 1979; Huttenlocher and De Courten, 1987). It now seems likely that technical and procedural factors could explain this apparent discrepancy. First, unlike in the human, all neocortical areas examined in macaque monkey were dissected from the same set of animals, which reduces sampling error. Second, the difference in timing and tempo of synaptogenesis observed between the striate and the prefrontal cortex in the human could result from the different sampling procedures used in the two studies of humans (Huttenlocher, 1979; Huttenlocher and De Courten, 1987). In the visual cortex, the data on synaptic density were expressed per unit volume of the *entire* cortex, and, therefore, were subject to differential regional changes in growth of other cellular elements, including myelination, which is more pronounced in the visual area, and this could affect the timetable of changes in the density of synapses per volume of tissue. In contrast, the study in the prefrontal cortex was confined to layer III. Third, studies in humans were done using phosphotungstic acid staining of synaptic junctions, a method which does not allow clear identification and measurements of pre- and post-synaptic cellular structures and precludes a correction for synaptic size. Finally, data in the human study were not available between 5 and 15 years of age, the period prior to and during puberty, when the plateau and elimination phases of synaptogenesis are observed in monkeys.

In spite of these procedural differences, the course of synaptogenesis in the human and monkey are remarkably similar across the entire cortex. Recently we re-examined the reported human data and expressed them in a manner that is more comparable with the work done in monkeys. In particular, when the data on synaptogenesis obtained from the human prefrontal (Huttenlocher, 1979), and human visual cortex (Huttenlocher and De Courten, 1987) were normalized and replotted to the maximum value of the curve on a semi-logarithmic scale, both curves were found to overlap, as they do in the monkey (Fig. 4). Moreover, by means of linear regression analysis, the two sets of values were fitted to a straight line (correlation coefficients: 0.90, $n = 5$, $P < 0.05$, for striate cortex, and 0.92, $n = 7$, $P < 0.05$ for prefrontal cortex). A statistical comparison between the two regression lines failed to reveal a significant difference in their slopes. It, therefore, appears that synaptogenesis both in the human and monkey proceeds synchronously in visual and prefrontal cortex.

The time course of synaptogenesis in different cortical areas corresponds well with the changes occurring simultaneously in cerebral metabolism in both the human and monkey cerebral cortex (Chugani et al., 1987; Jacob et al., 1994). In the human, the use of fluorodeoxyglucose, which indicates the level of metabolic activity in positron emission tomography, reveals that, after birth, metabolic activity also increases concurrently in prefrontal, motor, somatosensory, and visual cortex (Chugani et al., 1987). These studies lend support to the idea that maturation of diverse cortical areas in both the monkey and human occurs simultaneously rather than in pronounced sequential order. Further, the synchrony in synaptogenesis observed in the non-human primate is in harmony with biochemical and functional data on cortical maturation in the same species. Biochemical studies (Goldman-Rakic and Brown, 1982) suggest that the concentrations of dopamine, noradrenaline, and serotonin increase rapidly in the cortex of the macaque over the first 2 months, and approach adult levels by the fifth month after birth. Recent studies of the accumulation of major neurotransmitter receptor sites in different cortical areas show that their maximum density is also reached between 2 and 4 months after birth (Lidow et al., 1991; Lidow and Rakic, 1992). The curves of the increase in receptor density are very similar to those of synaptogenesis in all areas examined (Fig. 5). However, the phase of decline of receptor density appears to slightly precede the phase of synaptic decline. These observations from several divergent cortical areas suggest that the initial formation and maintenance of synapses, as well as their biochemical maturation, may be determined by intrinsic signals, which are common to the entire cortical mantle. The nature of these hypothetical genetic and epigenetic mechanisms remains to be elucidated.

The concurrent course of synaptogenesis and formation of neurotransmitter receptors in functionally different areas of the cerebrum is at variance with the tacit assumption, widespread in the literature, that development of cerebral cortex follows a hierarchical sequence of structural and functional development from the sensory to motor and, finally, to association cortex (originated by Flechsig, 1920; confirmed by Yakovlev and Lecours, 1967; reviewed in Greenfield, 1991). We argue that myelination is not the best criterion of functional maturation, and that hematoxylin staining of sections used to measure the amount of myelin is not a very precise quantitative method. Furthermore, there are clear exceptions to the hierarchical rule even in these studies. For example, the corticospinal motor system is among the last to myelinate and becomes stained by hemotyoxylin only during the second year of life (Yakovlev and Lecours, 1967). On the other hand, concurrent structural (synaptic) and biochemical (neurotransmitters, receptors) maturation of the cortical mantle appears reasonable in light of the intricate network of cortical connections that are essential for integration of sensory, motor, limbic, and associative components underlying even the simplest cortical functions (Goldman-Rakic, 1987a,b, 1988a,b). This concept of concurrency becomes even more compelling when one considers the coupling between the early maturation of the neuronal circuits of the cortex and the first expression of cognitive capacity that requires more than one cortical area. The synchronous synaptogenesis is also in harmony with contemporary ideas of early cognitive maturation in infants (Spelke, 1994).

Individual variability

The size of the cortex and its areas in macaque monkey, like that of the human, displays considerable individual differences (Purves and LaMantia, 1993). However, individual variability in the density of synaptic contacts during the rapid phase of production in the macaque monkey is

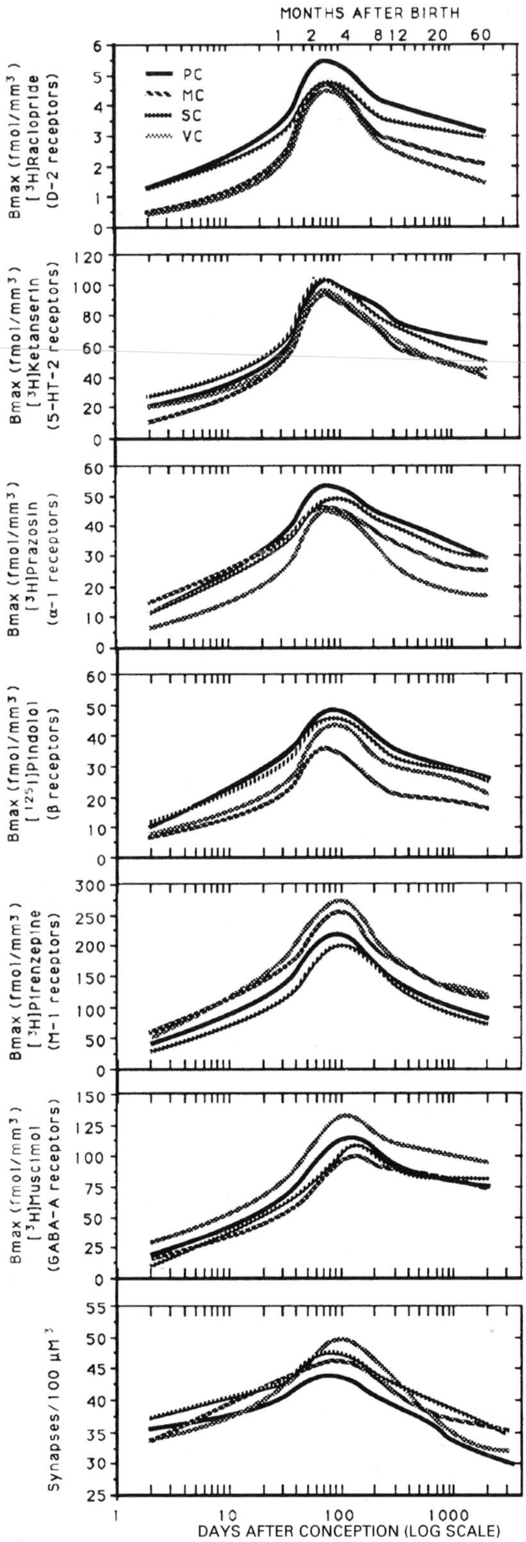

relatively small in all areas examined (Fig. 3). In contrast, in the phase of synaptic elimination, the densities of synaptic junctions begin to display considerable intra- and interindividual variability. It is possible that the rapid phase of synaptogenesis yields more homogeneous samples because it is controlled by a common intrinsic mechanism (Rakic et al., 1986). In contrast, the phases of synaptic elimination may display higher individual variability because of the diverse epigenetic events and variety of environmental factors that become more pronounced postnatally in the visual cortex (Bourgeois et al., 1989). The same variability was observed also in the other cortical areas at the same ages (Zecevic et al., 1989; Zecevic and Rakic, 1991; Bourgeois et al., 1994; Granger et al., 1995). It seems to be unrelated to the thickness of sections or the number of probes examined, and it is amplified by the stereological corrections. Use of the vertical probes in the present study does not appear to be the source of this variability, since a comparable level of individual variability was found in the anterior cingulate cortex when horizontal probes confined to each cortical layer were made out of the blocks obtained from

---

Fig. 5. Developmental changes in the overall (across all layers) density of the specific binding of radioligands labeling representative selection of neurotransmitter receptor subtypes in the prefrontal (PC), primary motor (MC), somatosensory (SC), and primary visual cortical (VC) regions. The figure also includes developmental changes in synaptic density in the same regions (adapted from Rakic et al., 1986). For receptor densities the curves were obtained by locally weighted least squares fit with 50% smoothing (KALEIDA GRAPH, Synergy Software, Reading, PA) based on mean B_{max} values obtained from the measurements of the entire cortical thickness in at least two animals at birth and at 1, 2, 4, 8, 12, 36, and 60 months of age. Age is presented in postnatal days on a logarithmic scale. The original data on synaptogenesis by Rakic et al. (1986) was presented as the number of synapses per unit area of neuropil. We recalculated the data per volume of tissue, which did not alter the timing of events and allowed us to compare the developmental course of cortical synapses with that of neurotransmitter receptors obtained in the present study. The curves representing synaptogenesis per area of neuropil (Rakic et al., 1986) were fitted in the same manner as those for neurotransmitter receptors. (From Lidow et al., 1991.)

the very same set of animals (Granger et al., 1995) (see also Zecevic and Rakic, 1991). The density of the different types of synapses also varies between the neighboring sample points along the lamina IV in the primary visual cortex of the cat (Winfield, 1983), as well as the lamina II of the rat (Wolff, 1976).

An amplitude of variability, similar to the one reported here, was also observed during the development of axo-dendritic spines (Boothe et al., 1979; Lund and Holbach, 1991) and axo-somatic contacts in layer IVC in area 17 of the same species (Mates and Lund, 1983; Lund and Harper, 1991). Finally, high variability has also been found in the number of neurons in the macaque neocortex (Vincent et al., 1989; Williams R. and Rakic, P., unpublished data), in size of cytoarchitectonic areas (Sarkisov et al., 1949; Rajkowska and Goldman-Rakic, 1995), as well as in the concentrations of several neurotransmitters (Wenk et al., 1989) and neurotransmitter receptors (Rakic et al., 1988). Thus, there is little doubt that there are considerable quantitative differences of synaptic density in the cerebral cortex of individual macaque monkeys at any age, as well as pronounced differences in the rate of their elimination. However, the significance of these differences encountered during the phase of elimination remains to be determined. A similar magnitude of variability can be expected in the developing human brain. Therefore, samples based on small number of specimens may give misleading results. Even gross morphological features, such as the shape of gyral pattern and their size in the human, are very different from individual to individual (e.g. Rademacher et al., 1993; Loftus and Gazzaniga, 1994). The functional significance of these individual variations is not clear, but competitive interactions seem to play an important role.

Regulation of the level and type of synapse retention and elimination

The balance between overproduction and elimination of neurons, axons, and synapses ultimately determines the size of a given pathway or the extent of territories devoted to a given terminal field in the mature cortex (Rakic, 1986; O'Leary and Koestler, 1993). Although a number of mechanisms have been proposed to regulate the specificity and therefore the selection of specific synaptic connections in the brain, this problem is still not resolved at the level of cellular and molecular processes. Indeed, a number of 'factors' studied by developmental neurobiologists could all be important. Studies in a variety of species, including the human, support the hypothesis that competitive interactions between two or more populations of neurons play a significant role in selective elimination of excess synapses before the adult level is reached. At the same time, the temporal register of the period of synaptic loss with the period of sexual maturation points logically towards the possible role of steroid hormones, which have been reported to influence synaptogenesis both in vivo and in vitro (e.g. Toran-Allerand, 1984; Arai and Matsumato, 1987). Another factor is experience, especially in the primate where synaptic loss occurs postnatally during young adulthood when the broadest of interactive experiences take place. In this regard, a recent longitudinal study carried out in the cat suggests that the period of plasticity in response to monocular deprivation lasts longer than previously thought and actually ends only during puberty, which occurs around 1 year of age in this species (Daw et al., 1992). Such studies emphasize the important role of sensory and motor experience, training, and, in the case of the human, education.

The first step should be to determine which connections and synapses are lost and whether this loss can be influenced by training. The methods presently available reveal only the emergence of developing synapses and the sites of their termination, but do not expose their origin. Therefore, we have only preliminary ideas as to the basis for the temporal correlation of various developmental events and the neural systems involved. It should be underscored, however, that the observed changes in synaptic densities occur well after all cortical neurons have been generated and settled in their appropriate layers (Rakic, 1974, 1982), and after thalamocortical (Darian-Smith et al., 1990) and callosal (Killackey and Chalupa, 1986; Schwartz and Goldman-Rakic, 1991) projections have established their region-specific topographic terminations. In fact, the largest wave of elimination of callosal axons in this species occurs during the first 2 postnatal months (LaMantia and Rakic, 1990), the period when synaptic density is still rapidly increasing. Paradoxically, however, changes in synapses are unlikely to be related to the 'exuberance' of callosal connections (Innocenti, 1981) because in this species they establish their adult pattern before birth (Killackey and Chalupa, 1986; Schwartz and Goldman-Rakic, 1991). Furthermore, since most synaptic classes, like callosal or thalamocortical terminals, represent only a very small fraction of the total innervation of the cortex, individually they cannot be responsible for the massive loss observed between infancy and adulthood. Therefore, it is likely that synapses belonging to several projection systems as well as to local circuits must be *simultaneously* involved in widespread competitive interactions that are triggered by a common regulatory mechanism (Rakic et al., 1986; Bourgeois et al., 1989).

The mode of normal development of geniculocortical projections provides a model for examining possible cellular mechanisms involved in formation of synapses and the territorial distribution of axonal terminals within the cerebral cortex. For example, the issue of whether competition between the two eyes plays a role in formation of segregated input to the cortex can be resolved by removing one eye prenatally and examining the cortex a few months after birth. In such cases ocular dominance stripes fail to develop and LGd axon terminals within layer IV form two continuous and uniform horizontal lines confined to layers IVA and IVC (Rakic, 1981). Since geniculocortical projections subserving two eyes initially overlap, these results can be explained, at least in part, by a failure of terminals subserving the remaining eye to retract from the territory that normally becomes occupied by terminals subserving the other (enucleated) eye. Additional evidence for competition is that binocular connections fail to withdraw to appropriate territories after blockade of electrical activity by tetrodotoxin, which prevents the influx of sodium ions across the membrane (Dubin et al., 1986; Stryker and Harris 1986; Shatz, 1990). There are some reports that NMDA receptors may also be essential (Shatz, 1990). However, recent examination of developmental events in the transgenic mouse, in which the R1 subunit of this receptor is 'knocked out' (Yuqing et al., 1993) indicate that a simple and straightforward explanation will not suffice to understand this complex phenomenon. The monocular enucleation model emphasizes the interplay between competing inputs but it is not clear that similar interplay is at work in all areas of the cortex nor is it clear how it applies to the maintenance of synapses.

Another model — binocular enucleation in the embryo — implicates genetic factors. In this model, the basic laminar pattern, including sublamination, develops normally in primate visual cortex after the removal of both eyes at early embryonic stages, prior to the formation of central connections, and prior to the genesis of the granular and supragranular layers in the visual cortex (Rakic, 1988; Rakic et al., 1991). In these cases, the density of synapses in the primary visual cortex is the same as in age-matched controls (Bourgeois and Rakic, 1987; Bourgeois and Rakic, in preparation). Furthermore, a variety of neurotransmitter receptors (Rakic and Lidow, 1994) as well as cytochrome oxidase patches and interpatch areas, that in normal animals presumably subserve color and non-color vision, develop well in the early enucleates (Kuljis and Rakic, 1990). These findings indicate that the pattern and quantity of certain classes of synapses can develop and be maintained in the total absence of visual stimulation. Yet, the area occupied by visual cortex in these animals is smaller, suggesting that genetic and epigenetic factors regulate different aspects of cortical development and specializations (for detailed discussion, see

Rakic, 1988). Thus, at present, competitive interaction should be viewed as an additional cellular mechanism that has emerged and been elaborated during the course of vertebrate evolution and has, perhaps, become most pronounced in primates (Easter et al., 1985). A provisional schema is that early developmental events, such as cell proliferation, neuronal migration, and the initial outgrowth of axons and even cell death may proceed in an orderly way in the absence of activity-dependent cues. In contrast, later phases of development, when the elimination of axons and synapses predominates, and the final circuits or topographical maps are shaped, may be regulated by the state of functional activity within the system. The theory of competitive elimination, reminiscent of Darwinian selection of the fittest, contributes to the specification of synaptic connectivity. This basic scheme has been used to model various aspects of development and learning (Barlow, 1975; Changeux and Danchin, 1976; Edelman and Finkel, 1984; Edelman, 1987; Changeux, 1993). However, since qualitative or quantitative change in synaptic architecture undoubtedly involves gene expression in response to activity (Changeux and Konishi, 1987), any distinction between genetic and epigenetic factors becomes an artificial or academic issue.

Steady state of synaptic density in adulthood

The final stages of synaptogenesis in the adult cerebral cortex are not clear. So far, the number of sexually mature monkeys that we have analyzed quantitatively is relatively small, and individual variability is still too large to make a definitive statement on the 'normal' number of synapses. The data collected to this point indicate that, from sexual maturity (3 years) to senescence (over 25 years), the density of synapses per unit volume of neuropil remains relatively steady or shows only small decline (Figs. 2, 3). Since the percentage of neuropil also remains steady during this period (Zecevic and Rakic, 1991; Bourgeois and Rakic, 1993), we can assume that the absolute number of synapses is relatively stable, and, if anything, their number does not increase. This stands in contrast to the studies reported in rodents, where learning and acquisition of new memory, or even exposure of rats to a single task, increases significantly the number of synapses, volume of neurons, number and size of spine, glial cells and endothelial cells (Greenough, 1992). Since, if anything, primates probably learn more during their much longer lifespan than do rodents, one would expect significant increases, not only of the number of synapses, but also of total brain weight. This evidently does not occur in primates.

We have argued elsewhere that a prolonged period of interaction with the environment, as pronounced as it is in all primates, especially human, requires a stable set of neurons and synapses to retain acquired experiences in the pattern of their synaptic connectivity (Rakic, 1985). It is in contrast to the continuing turnover and large addition of neurons in the non-mammalian vertebrates (e.g. Nottebohm and Alvarez-Buylla, 1993) and to the rodent brain, which apparently continues to receive a steady stream of new neurons in the hippocampus (Altman and Bayer, 1993), and possibly a small number even in the neocortex of the 3-month-old rat (Kaplan, 1985). Such addition may be due to a slow increase in the size of the rodent brain throughout a lifespan of about 2 years (Purves, 1988). We reason that humans can remember events that occurred 70 or 80 years earlier perhaps because their experiences and memories are 'imprinted' in stable neuronal assemblies that are equally old or older. Newly generated neurons and axons that replace older sets would conceivably render us eternal infants incapable of retaining learned behavior. Neurogenesis in the primate cerebral cortex does not occur after sexual maturity (Rakic, 1985; Rakic and Kornack, 1993), nor does neurogenesis contribute to the recovery of function following trauma sustained after the normal production of neurons ceases (reviewed in Rakic and Kornack, 1993). Of course, one can devise a hypothesis about how acquired information is transferred from dying neurons to newly generated ones, but such a complex scheme seems far less likely to account for prolonged memory retention in primates than the relatively simple maintenance of stable cell populations. Furthermore, we maintain that prolonged maintenance of the same set of neurons and connections throughout a lifetime has become an evolutionary advantage in primates as it enables the retention and updating of acquired information essential for survival (Rakic, 1985; Rakic and Kornack, 1993). Thus, our major goal is to find ways of preserving the ones already present in the adult brain.

Relevance for major psychosis

It is well established that many major psychoses, including schizophrenia, become manifest at the end of adolescence and in early adulthood. In addition, a growing number of studies of post-mortem tissue samples from schizophrenics and from magnetic resonance imaging of the brains of living patients have provided a solid factual basis for morphological signs, such as enlarged ventricles, which predate the illness (Weinberger, 1987). Finally, evidence for a genetic predisposition to schizophrenia is incontrovertible (Kety et al., 1968). Such findings have naturally led investigators to consider the developmental antecedents of this major psychosis and several hypotheses have invoked events related to synaptogenesis. Feinberg (1982) proposed that schizophrenia may be caused by a fault in programs of synaptic elimination during adolescence. In the absence of systematic data, Feinberg allowed that the defect could be either one of excess pruning or the opposite, failure of the normal pruning mechanisms such as described here. More recently, Hoffman and McGlashan have elaborated on the idea of excessive pruning and have used computational modeling to generate and test predictions relevant to particular symptoms in schizophrenia (Hoffman and McGlashan, 1993).

More direct evidence for the possibility of excessive pruning comes from the recent studies of non-invasive methods in living patients (Willamson et al., 1991; Pettegrew et al., 1993). These studies have employed ^{31}P magnetic resonance spectroscopy to examine the structural integrity of prefrontal cortex in drug naive schizophrenics. Their studies reveal a reduction in phosphomonesters indicative of phosolipid turnover in neuronal membranes. They interpreted their findings as evidence of excessive elimination of dendrites and axons in the patient population. Complementing the in vivo studies, Selemon et al., (1993) have used quantitative morphometry to examine schizophrenic prefrontal areas and found that there is a higher density of neurons per unit volume in these areas compared with controls — a finding also compatible with the loss of neuropil in schizophrenia. Glantz (1993) has reported preliminary evidence of reduced synaptophysin labeling in prefrontal cortex of schizophrenics. Shenton et al. (1992) have obtained MRI evidence of significant volume loss in temporal neocortex. These various studies indicate a pathological process at work in the cortical regions of patients — findings that are compatible with a neurodevelopmental perspective. Yet, we would caution that there are many gaps in the evidence that these changes occur at inflection points in the synaptogenesis curve and are causative rather than reflecting the results of the psychotic break or a more precipitous change occurring in response to more proximal causes. It is eminently clear, however, that knowledge of the normal course and mechanisms of synapse formation, the influence of various exogenous and endogenous events upon synapse stability and turnover, are essential prerequisites to determining the locus and timing of etiological factors in diseases that affect the cortex and alter cognitive function. This is the ultimate justification for studies of the type reported here in non-human primates, where conditions are experimentally controlled and monitored in a species that is unexcelled as an animal model of human brain development.

Acknowledgment

This research was supported by the U.S. Public Health Service Grants NS22807, NS14841 and

EY02593. Rhesus monkeys were obtained from breeding colonies at Yale University School of Medicine and the New England Regional Primate Research Center, Southborough, MA, which is supported by grant P51550017.

References

Altman, J. and Bayer, S.A. (1993) Are new neurons formed in the brain of adult mammals? In: A.C. Cuello (Ed.), *Restorative Neurology, Vol. 6, Neuronal Cell Death and Repair,* Elsevier, Amsterdam, pp. 203–225.

Arai, Y. and Matsumoto, A. (1987) Gonadal steroid control of synaptogenesis in the neuroendocrine brain. In: P.C.K. Leung, D.T. Armstrong, K.B. Ruf, and H.G. Fresen (Eds.), *Endocrinology and Physiology of Reproduction,* Plenum, New York, pp. 13–21.

Barlow, H.B. (1975) Visual experience and cortical development. *Nature,* 258: 199–204.

Blakemore, C., Garey, L.J.V. and Vital-Durand, F. (1978) The physiological effects of monocular deprivation and their reversal in the monkey's visual cortex. *J. Physiol.,* 283: 223–262.

Blakemore, C., Garey, L.J.V. and Vital-Durand, F. (1981) Recovery from monocular deprivation in the monkey. I. Reversal of physiological effects in the visual cortex. *Proc. R. Soc. London (B),* 213: 399–423.

Boothe, R.G., Greenough, W.T., Lund, J.S. and Wrege, K. (1979) A quantitative investigation of spine and dendrite development of neurons in visual cortex (area 17) of *Macaca nemestrina* monkeys. *J. Comp. Neurol.,* 186: 473–490.

Bourgeois, J.-P. and Rakic, P. (1987) Distribution, density and ultrastructure of synapses in the visual cortex in monkeys devoid of retinal input from early embryonic stages. *Abstr. Soc. Neurosci.,* 13: 1044.

Bourgeois, J.-P. and Rakic, P. (1993) Changing of synaptic density in the primary visual cortex of the rhesus monkey from fetal to adult stage. *J. Neurosci.,* 13: 2801–2820.

Bourgeois, J.-P., Jastreboff, P. and Rakic, P. (1989) Synaptogenesis in the visual cortex of normal and preterm monkeys: evidence for intrinsic regulation of synaptic overproduction. *Proc. Natl. Acad. Sci. USA,* 86: 4297–4301.

Bourgeois, J.-P., Goldman-Rakic, P.S. and Rakic, P. (1994) Synaptogenesis in the prefrontal cortex of rhesus monkey. *Cerebral Cortex,* 4: 78–96.

Changeux, J.-P. (1993) A critical view of neuronal models of learning and memory. In: P. Anderson (Ed.), *Memory Concepts,* Elsevier, Amsterdam, pp. 413–433.

Changeux, J.-P. and Danchin, A. (1976) Selective stabilization of developing synapses as a mechanism for the specification of neural network. *Nature,* 264: 705–712.

Changeux, J.-P. and Konishi, M. (Eds.) (1987) *The Neural and Molecular Bases of Learning,* Dahlem Konferenzen, Wiley, New York.

Chugani, H.T., Phelps, M.E. and Mazziotta, J.C. (1987) Positron emission tomography study of human brain functional development. *Ann. Neurol.,* 22: 487–497.

Darian-Smith, C., Darian-Smith, I. and Chema, S.S. (1990) Thalamic projections to sensorimotor cortex in the newborn monkey. *J. Comp. Neurol.,* 299: 47–63.

Daw N.W., Fox, K., Sato, H. and Czepita D. (1992) Critical period for monocular deprivation in the cat visual cortex. *J. Neurophysiol.,* 67: 197–202.

Dubin, M., Stark, L.A. and Archer, S.M. (1986) A role for action-potential activity in the development of neuronal connections in the kitten retinogeniculate pathway. *J. Neurosci.,* 6: 1021–1036.

Easter, Jr., S.S., Purves, D., Rakic, P. and Spitzer, N.C. (1985) The changing views of neuronal specificity. *Science,* 230: 507–511.

Edelman, G.M. (1987) *Neural Darwinism,* Basic Books, New York, 371 pp.

Edelman, G.M. and Finkel, L.J. (1984) Neuronal group selection in the cerebral cortex. In: G.M. Edelman, W.E. Gale and W.M. Cowan (Eds.), *Dynamic Aspects of Neocortical Function,* Wiley, New York, pp. 653–695.

Feinberg, I. (1982) Schizophrenia: Caused by fault in programmed synaptic elimination during adolescence? *J. Psychiatr. Res.,* 17 (Suppl. 4): 319–334.

Feinberg, I., Korssko, R.L. and Heller, N. (1967) EEG sleep patterns as a function of norma. and pathological aging in man. *J. Psychiatr. Res.,* 5: 107–144.

Flechsig, P. (1920) *Anatomie des Menschlichen Gehirns und Ruckenmarks auf Myelogenetischer Grundlage,* Thieme, Leipzig.

Gelman, R. and Gallistal, C. (1978) *The Child's Understanding of Numbers,*Harvard University Press, Cambridge, MA.

Gibson, K.R. (1970) Sequence of myelinization in the brain of *Macaca mulatta.*. Ph.D. dissertation, University of California, Berkeley.

Glantz, L.A. (1993) Synaptophysin immunoreactivity is selectively decreased in the prefrontal cortex of schizophrenic subjects. *Soc. Neurosci. Abstr.,* 19: 201.

Goldman-Rakic, P.S. (1987a) Circuitry of primate prefrontal cortex and regulation of behavior by representational memory. *Handb. Physiol.,* 5 (1): 373–417.

Goldman–Rakic, P.S. (1987b) Development of cortical circuitry and cognitive functions. *Child Dev.,* 58: 642–691.

Goldman-Rakic, P.S. (1988a) Topography of cognition: parallel distributed networks in primate association cortex. *Annu. Rev. Neurosci.,* 11: 137–156.

Goldman-Rakic, P.S. (1988b) Changing concepts of cortical connectivity: parallel distributed cortical networks. In: P. Rakic and W. Singer (Eds.), *Neurobiology of the Neocortex,* John Wiley, New York, pp. 177–202.

Goldman-Rakic, P.S. and Brown, R.M. (1982) Postnatal development of monoamine content and synthesis in the cerebral cortex of rhesus monkeys. *Dev. Brain Res.*, 4: 339–349.

Granger, G., Tekaia, F., LeSourd, A.M., Rakic, P. and Bourgeois, J.-P., LeSourd, A.M. and Rakic, P. (1995) Tempo of neurogenesis and synaptogenesis in the macaque cingulate mesocortex: comparison with the neocortex submitted.

Greenfield, P.M. (1991) Language, tools and brain: the ontogeny and phylogeny of hierarchically organized sequential behavior. *Behav. Brain Sci.*, 14: 531–595.

Greenough, W.T. (1992) Induction of brain structure by experience: substrates for cognitive development. *Minn. Symp. Child Psychol.*, 24: 155–200.

Hoffman, R.E. and McGlashan, T.H. (1993) Parallel distributed processing and the emergence of schizophrenic symptoms. *Schizophr. Bull.*, 19: 119–140.

Hubel D.H. and Wiesel, T.N. (1977a) Functional architecture of macaque monkey visual cortex. *Proc. Roy. Soc. London B*, 198: 1–59.

Hubel, D.H., Wiesel, T.N. and LeVay, S. (1977b) Plasticity of ocular dominance columns in monkey striate cortex. *Phil. Trans. Roy. Soc. London B*, 278: 377–409.

Huttenlocher, P.R. (1979) Synaptic density in human frontal cortex-developmental changes and effects of aging. *Brain Res.*, 163: 195–205.

Huttenlocher, P.R. and de Courten, C. (1987) The development of synapses in striate cortex of man. *Hum. Neurobiol.*, 6: 1–9.

Innocenti, G.M. (1981) Growth of reshaping of axons in the establishment of visual connections. *Science*, 212: 824–827.

Jacob, B., Chugani, H.T., Allada, V., Chen, S., Phelps, M.E., Pollack, D.B. and Raleigh, M.S. (1994) Metabolic brain development in rhesus Macaques and Vervet monkeys: a positron emission tomography study. *Cerebral Cortex*, in press.

Jernigan, T.L., Trauner, D.A., Hesselink, J.R. and Tallal, P.A. (1991) Maturation of human cerebrum observed in vivo during adolescence. *Brain*, 114: 2037–2049.

Kaplan, M.S. (1985) Formation and turnover of neurons in young and senescent animals: An electronmicroscopic and morphometric analysis. *Ann. NY Acad. Sci.*, 457: 173–192.

Karmiloff-Smith, A. (1979) Language development after five. In: P. Fletcher and M. Garman (Eds.), *Language Acquisition*, Cambridge University Press, Cambridge.

Kety, S.S., Rosenthal, D., Wender, P.H. and Schulsinger, F. (1968) The types and prevalence of mental illness in biological families of adopted schizophrenics. In: D. Rosenthal and S.S. Kety (Eds.) *Transmission of Schizophrenia*, Pergamon, Oxford, pp. 345–362.

Killackey, H. and Chalupa, L. (1986) Ontogenetic changes in the distribution of callosal projection neurons in the postcentral gyrus of the fetal rhesus monkey. *J. Comp. Neurol.*, 244: 331–384.

Kuljis, R.O. and Rakic, P. (1990) Hypercolumns in primate visual cortex develop in the absence of cues from photoreceptors. *Proc. Natl. Acad. Sci. USA*, 87: 5303–5306.

Leuba, G. and Kraftsik, R. (1984) changes in surface area, volume, 3-D shape and total numbers of neurons of the human primary visual cortex from midgestation until aging. *Anat. Embryol.*, in press.

LaMantia, A.S. and Rakic, P. (1990) Axon overproduction and elimination in the corpus callosum of the developing rhesus monkey. *J. Neurosci.*, 10: 2156–2175.

LeVay, S. and Stryker, M.P. (1979) The development of ocular dominance columns in the cat. In: J.A. Fernandelli (Ed.), *Aspects of Developmental Neurobiology*, Society for Neuroscience Symposia, Soc. Neurosci., Washington, DC., pp. 83–98.

Lidow, M.S. and Rakic, P. (1992) Scheduling of monoaminergic neurotransmitter receptor expression in the primate neocortex during postnatal development. *Cerebral Cortex*, 2: 401–416.

Lidow, M.S., Goldman-Rakic, P.S. and Rakic, P. (1991) Synchronized overproduction of neurotransmitter receptors in diverse regions of the primate cerebral cortex. *Proc. Natl. Acad. Sci. USA*, 88: 10218–10221.

Loftus, W.E. and Gazzaniga, M.S. (1994) Interindividual variations in left-right asymmetry of regional cortical surface area. *Cerebral Cortex*, in press.

Lund, J.S. and Harper, T.R. (1991) Postnatal development of thalamic recipient neurons in the monkey striate cortex. III. Somatic inhibitory synapse acquisition by spiny stellate neurons of layer 4C. *J. Comp. Neurol.*, 309: 144–149.

Lund, J.S. and Holbach, S.M. (1991) Postnatal development of thalamic recipient neurons in the monkey striate cortex: I. Comparison of spine acquisition and dendritic growth of layer 4C alpha and beta spiny stellate neurons. *J. Comp. Neurol.*, 309: 115–128.

Lund, J.S., Holbach, S.M. and Chung, W.W. (1991) Postnatal development of thalamic recipient neurons in the monkey striate cortex. II. Influence of afferents driving on spine acquisition and dendritic growth of layer 4C spiny stellate neurons. *J. Comp. Neurol.*, 309: 129–140.

Mates, S.L. and Lund, J.S. (1983) Developmental changes in the relationship between type 1 synapses and spiny neurons in the monkey visual cortex. *J. Comp. Neurol.*, 221: 91–97.

Miller, M.W. (1988) Development of projection and local circuit neurons in neocortex. In: A. Peters and E.G. Jones (Eds.) *Cerebral Cortex*, Vol. 7, Plenum Press, New York, pp. 133–175.

Missler, M., Wolff, A., Merker, H.-J. and Wolff, J.R. (1993) Pre- and postnatal development of the primary visual cortex of the common marmoset. II. Formation, remodeling, and elimination of synapses as overlapping process. *J. Comp. Neurol.*, 333: 53–67.

Nottebohm, F. and Alvarez-Buylla, A. (1993) Neurogenesis and neuronal replacement in adult birds. In: A.C. Cuello

(Ed.), *Restorative Neurology, Vol. 6, Neuronal Cell Death and Repair,* Elsevier, Amsterdam, pp. 227–236.

O'Kusky, J. and Colonnier, M. (1982) A laminar analysis of the number of neurons, glia and synapses in the visual cortex (area 17) of the adult macaque monkey. *J. Comp. Neurol.,* 210: 278–290.

O'Leary, D.M. and Koestler, S.E. (1993) Development of projection neuron types, axon pathways and patterned connections of the mammalian cortex. *Neuron,* 10: 991–1006.

Peters, A. (1987) Number of neurons and synapses in primary visual cortex. In: A. Peters and E.G. Jones (Eds.) *Cerebral Cortex,* Vol. 6, Plenum Press, New York, pp. 267–294.

Pettegrew, J.W., Keshavan, M.J. and Minshew, N.J. (1993) ^{31}P nuclear magnetic resonance spectroscopy: neurodevelopment and schizophrenia. *Schizophr. Bull.,* 19: 35–53.

Plummer, K. and Behan, M. (1993) Development of corticotectal synaptic terminals in the cat: a quantitative electronmicroscopic analysis. *J. Comp. Neurol.,* 338: 458–474.

Purves, D. (1988) *A Trophic Theory of Neuronal Organization,* Harvard University Press, Cambridge, MA., 231 pp.

Purves, D. and LaMantia, A.S. (1993) Development of blobs in the visual cortex of macaques. *J. Comp. Neurol.,* 334: 169–175.

Rademacher, J., Caviness, V.S., Steinmetz, H. and Galaburda, A.M. (1993) Topographical variation of the human primary cortices: implication neuroimaging, brain mapping and neurobiology. *Cerebral Cortex,* 3: 313–329.

Rajkowska, G. and Goldman-Rakic, P.S. (1995) Cytoarchitectonic definition of prefrontal areas in the normal human cortex: variability in locations of areas 9 and 46 and relationship to the Talairach coordinate system. *Cerebral Cortex,* in press.

Rakic, P. (1972) Mode of cell migration to the superficial layers of fetal monkey neocortex. *J. Comp. Neurol.,* 145: 61–84.

Rakic, P. (1974) Neurons in the monkey visual cortex: systematic relation between time of origin and eventual disposition. *Science,* 183: 425–427.

Rakic, P. (1976) Prenatal genesis of connections subserving ocular dominance in the rhesus monkey. *Nature,* 261: 467–471.

Rakic, P. (1977) Prenatal development of the visual system in the rhesus monkey. *Phil. Trans. R. Soc. London B,* 278: 245–260.

Rakic, P. (1981) Development of visual centers in primate brain depends on binocular competition before birth. *Science,* 214: 928–931.

Rakic, P. (1982) Early developmental events: cell lineages, acquisitions of neuronal positions, and areal and laminar development. *Neurosci. Res. Prog. Bull.,* 20: 439–451.

Rakic, P. (1985) Limits of neurogenesis in primates. *Science,* 227: 154–156.

Rakic, P. (1986) Mechanisms of ocular dominance segregation in the lateral geniculate nucleus. Competitive elimination hypothesis. *Trends Neurosci.,* 9: 11–15.

Rakic, P. (1988) Specification of cerebral cortical areas. *Science,* 241: 170–176.

Rakic, P. and Kornack D.R. (1993) Constraints on neurogenesis in adult primate brain: An evolutionary advantage? In: A.C. Cuello (Ed.), *Neuronal Death and Regeneration, Restorative Neurology 6,* Elsevier, Amsterdam, pp. 257–266.

Rakic, P. and Lidow, M.S. (1994) Distribution and density of neurotransmitter receptors in the absence of retinal input from early embryonic stages. *J. Neurosci.,* in press.

Rakic, P., Bourgeois, J.-P., Eckenhoff, M.F., Zecevic, N. and Goldman-Rakic, P.S. (1986) Concurrent overproduction of synapses in diverse regions of the primate cerebral cortex. *Science,* 232: 232–235.

Rakic, P., Gallager, D. and Goldman-Rakic, P.S. (1988) Areal and laminar distribution of major neurotransmitter receptors in the monkey visual cortex. *J. Neurosci.,* 8: 3670–3690.

Rakic, P., Suner, I. and Williams, R.W. (1991) A novel cytoarchitectonic area induced experimentally within the primate visual cortex. *Proc. Natl. Acad. Sci. USA,* 88: 2083–2087.

Sarkisov, S.A., Filimonov, I.N. and Preobrazenskaya, N.S. (1949) *Cytoarchitecture of the Cerebral Cortex.* Moscow, Medgiz.

Schwartz, M.L. and Goldman-Rakic, P.S. (1991) Prenatal specification of callosal connections in rhesus monkey. *J. Comp. Neurol.,* 307: 144–162.

Selemon, L.D., Rajkowska, G. and Goldman-Rakic, P.S. (1993) Cytologic abnormalities in area 9 of the schizophrenic cortex. *Soc. Neurosci. Abstr.,* 19: 367.

Shatz, C.J. (1983) Prenatal development of cats retinogeniculate pathway. *J. Neurosci.,* 3: 482– 499.

Shatz, C. (1990) Impulse activity and the patterning of connections during CNS development. *Neuron,* 5: 1–10.

Shenton, M.E., Kikinis, R., Jolesz, F.A., Pollak, S.D., LeMay, M., Wible, C.G., Hokama, H., Martin, J., Metcalf, D., Coleman, M. et al. (1992) Abnormalities of the left temporal lobe and thought disorder in schizophrenia. A quantitative magnetic resonance imaging study. *N. Engl. J. Med.,* 327: 604–612.

Spelke, E. (1994) Object perception, object-directed action, and physical knowledge in infancy. In: M. Gazzaniga, (Ed.), *Cognitive Neuroscience,* MIT Press, in press.

Stryker, M.P. and Harris, W.A. (1986) Binocular impulse blockade prevents the formation of ocular dominance columns in cat visual cortex. J. Neurosci. 6: 2117–2133.

Toran-Allerand, D. (1984) On the genesis of sexual differentiation of the central nervous system: morphogenetic consequences of steroidal exposure and possible role of α-Feto protein. In: G. de Vries, J. de Bruin, N. Uylings, and M. Corner (Eds.), *Progress in Brain Research, Vol. 61, Sex Differences in the Brain,* Elsevier, Amsterdam, pp. 63–78.

Vincent, S.L., Peters, A. and Tigges, J. (1989) Effects of aging on the neurons within area 17 of rhesus monkey cerebral cortex. *Anat. Rec.,* 223: 329–341.

Weinberger, D.R. (1987) Implications of normal brain development for the pathogenesis of schizophrenia. *Arch. Gen. Psychiatry,* 44: 464–669.

Wenk G.L., Pierce, D.J., Struble, R.G., Price, D.L. and Cork, L.C. (1989) Age-related changes in multiple neurotransmitter systems in the monkey brain. *Neurobiol. Aging,* 10: 11–19.

Williamson, P., Drost, D., Stainly, J., Carr, T., Morrison, S. and Merskey, H. (1991) Localized phosphorus 31 magnetic resonance spectroscopy in chronic schizophrenic patients and normal controls. *Arch. Gen. Psychiatry,* 48: 578.

Winfield, D.A. (1983) The postnatal development of synapses in the different laminae of the visual cortex in the normal kitten and in kittens with eyelid suture. *Dev. Brain Res.,* 9: 155–169.

Wolff, J.R. (1976) Stereological analysis of the heterogeneous composition of the central nervous tissue: synapses of the cerebral cortex. In Proc. 4th Internat. Congr. Stereol. *Nat. Bur. Stand. (U.S.), Spec. Publ.,* 431: 331–334.

Yakovlev, P.I. and Lecours, A.R. (1967) The myeloarchitectonic cycles of regional maturation of the brain. In: A. Minkowsky (Ed.), *Regional Development of the Brain in Early Life,* Blackwell Scientific, Oxford, pp. 3–70.

Yates, T. (1991) Theories of cognitive development. In: M. Lewis (Ed.), *Child and Adolescent Psychiatry,* Chapter 10, Williams and Wilkins, Baltimore, pp. 109–129.

Yuqing, L., Messersmith, E., Shatz, C. and Tohegawa, S. (1993) Development of CNS structure and connections in the NMDA receptor deficient mouse. *Soc. Neurosci. Abstr.,* 19: 1272.

Zecevic, N. and Rakic, P. (1991) Synaptogenesis in monkey somatosensory cortex. *Cerebral Cortex,* 1: 510–523.

Zecevic N., Bourgeois, J.-P. and Rakic, P. (1989) Changes in synaptic density in motor cortex of rhesus monkey during fetal and postnatal life. *Dev. Brain Res.,* 50: 11–32.

Zielinski, B.S. and Hendrickson, A. (1992) Development of synapses in macaque monkey striate cortex. *Visual Neurosci.,* 8: 491–504.

J. van Pelt, M.A. Corner H.B.M. Uylings and F.H. Lopes da Silva (Eds.)
Progress in Brain Research, Vol 102

CHAPTER 16

Activity-dependent neurite outgrowth and neural network development

A. van Ooyen and J. van Pelt

Graduate School of Neuroscience and Netherlands Institute for Brain Research, Amsterdam, The Netherlands

Introduction

In the course of development neurons become assembled into functional neural networks. Among the many factors shaping the ultimate structure of these networks, electrical activity plays a pivotal role (for review see Fields and Nelson, 1992). Many processes that determine synaptic connectivity and neuronal form are modulated by electrical activity, e.g. neurite outgrowth and growth cone behaviour (e.g. Cohan and Kater, 1986; Fields et al., 1990a; Schilling et al., 1991), naturally occurring cell death (e.g. Ferrer et al., 1992), production of trophic factors (e.g. Thoenen, 1991), synaptogenesis (e.g. Constantine-Paton, 1990), elimination of synapses (e.g. Shatz, 1990; Van Huizen et al., 1985, 1987a), changes in synaptic strength (e.g. Madison et al., 1991; Tsumoto, 1992), and functional maturation and differentiation of neurons (e.g. Spitzer, 1991; Corner and Ramakers, 1992). As a result of these activity-dependent processes, a reciprocal influence exists between the formation of neuronal form and synaptic connectivity on the one hand, and neuronal and network activity on the other hand. Thus, a given network may generate activity patterns which modify the organization of the network, leading to altered activity patterns which further modify structural or functional characteristics, and so on. Such a feedback loop must be expected to have major implications not only for the mature network and neurons, but also for their ontogenetic stages. In this article we will address the possible implications of one of these activity-dependent processes, namely activity-dependent neurite outgrowth.

A number of studies have demonstrated that electrical activity can directly affect neurite outgrowth (for review see Mattson, 1988). Electrical activity of the neuron reversibly arrests neurite outgrowth or even produces retraction (Cohan and Kater, 1986; Fields et al., 1990a; Schilling et al., 1991; Grumbacher-Reinert and Nicholls, 1992). Similarly, depolarizing media and neurotransmitters affect neurite outgrowth of many cell types (e.g. Sussdorf and Campenot, 1986; McCobb et al., 1988; Lipton and Kater, 1989; Mattson and Kater, 1989; Todd, 1992; Neely, 1993), with excitatory neurotransmitters inhibiting outgrowth and inhibitory ones stimulating outgowth (many of the effects of neurotransmitters on neuron outgrowth are related to their effects on electrical activity (Mattson, 1988)).

The morphological responses to neurotransmitters and electrical activity are probably mediated by changes in intracellular calcium levels (Cohan et al., 1987; Mattson, 1988; Kater et al., 1988; Kater and Guthrie, 1990; Fields et al., 1990b; Kater et al., 1990., Kater and Mills, 1991). High levels of activity, resulting in high intracellular calcium concentrations, would cause neurites to retract, whereas low levels of activity and consequently low calcium concentrations would allow further outgrowth (Kater et al., 1990).

From these studies on activity-dependent neurite outgrowth, the realization is growing that electrical activity and neurotransmitters are not only involved in information coding, but also play an important role in shaping neuronal form and in defining the structure of the networks in which they operate (Mattson, 1988; Lipton and Kater, 1989). Using simulation models, we will explore and elucidate the possible implications of activity-dependent outgrowth and locally interacting excitatory and inhibitory cells for neuronal morphology and network development (Van Ooyen and Van Pelt, 1992, 1993, 1994; Van Ooyen et al., in press)

The model

We use a distributed network, with neuron dynamics governed by the following set of (dimensionless) equation (shunting model):

$$\frac{dX_i}{dT} = -X_i + (1 - X_i)\sum_k^N W_{ik} F(X_k) - (H + X_i)\sum_l^M W_{il} F(Y_l)$$
$$\frac{dY_j}{dT} = -Y_j + (1 - Y_j)\sum_k^N W_{jk} F(X_k) - (H + Y_j)\sum_l^M W_{jl} F(Y_l) \quad (1)$$

where X_i is the (mean) membrane potential of excitatory cell i, Y_j is the membrane potential of inhibitory cell j, N and M are the total number of excitatory and inhibitory cells, respectively, H is the ratio of the inhibitory saturation potential to the excitatory one and W_{ik}, W_{il}, W_{jk}, W_{jl} represent the connection strengths (all $W \geq 0$; k and l are the indices of the excitatory and inhibitory driver cells, respectively; i and j are the indices of the excitatory and inhibitory target cells, respectively). The firing rate function F is taken to be sigmoidal:

$$F(u) = \frac{1}{1 + e^{(\theta - u)/\alpha}} \quad (2)$$

where α determines the steepness of the function and θ represents the firing threshold. The (low) firing rate when the membrane potential is subthreshold may be considered as representing spontaneous activity.

Growing cells are modelled as expanding circular areas, which can be conceived of as neuritic fields. When two such fields overlap, the corresponding neurons become connected with a strength proportional to the area of overlap:

$$W_{ij} = A_{ij} S \quad (3)$$

where $A_{ij} = A_{ji}$ is the amount of overlap ($A_{ii} = 0$) and S is a constant of proportionality. A_{ij} may be regarded as representing the total number of synapses formed reciprocally between neurons i and j, while S would represent the average synaptic strength. Strength may depend on the type of connection. We distinguish S^{ee}, S^{ei}, S^{ie}, and S^{ii}, which are constants representing the excitatory-to-excitatory, inhibitory-to-excitatory, excitatory-to-inhibitory, and inhibitory-to-inhibitory synaptic strengths, respectively (in S^{ei}, for example, e represents the target and i the driver cell). In the model, the outgrowth of each individual neuron, whether excitatory or inhibitory, depends in an identical way upon its own level of electrical activity.

$$\frac{dR_i}{dT} = \rho G(F(X_i)) \quad (4)$$

where R_i is the radius of the (circular) neuritic field of neuron i, and ρ determines the rate of outgrowth. The outgrowth function G is defined as

$$G(F(X_i)) = 1 - \frac{2}{1 + e^{(\epsilon - F(X_i))/\beta}} \quad (5)$$

where ϵ is the value of $F(X_i)$ for which $G = 0$, and β determines the steepness of the function (see Fig. 1). Depending on the firing rate, a neuritic field will grow out ($G > 0$ when $F(X_i) < \epsilon$), retract ($G < 0$ when $F(X_i) > \epsilon$) or remain con-

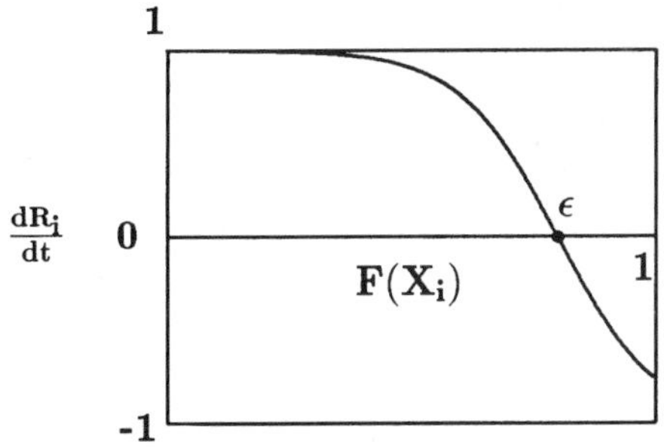

Fig. 1. Outgrowth function G (Eqn. (5)).

stant ($G = 0$ when $F(X_i) = \epsilon$). Equation (5) is thus simply a phenomenological description of the theory of Kater et al. (Kater et al., 1988, 1990; Kater and Guthrie, 1990) to the effect that, via calcium influx, the neuron's electrical activity affects its outgrowth.

Since in most CNS neural tissues there are more excitatory than inhibitory cells, we took $M/(N = M)$ mostly in the range of 0.1–0.2. For the rest, all the parameter values were the same for excitatory and inhibitory cells. In most of the simulations, we took $H = 0.1$, $\theta = 0.5$, $\alpha = 0.10$, $\beta = 0.10$ and $\epsilon = 0.60$. The results, however, do not depend on the exact choices of the parameters.

Overshoot

A general feature of nervous system development, in vivo as well as in vitro, is that virtually all structural elements show an initial overproduction, followed by an elimination during further development. These so-called overshoot phenomena occur, for example, with respect to synapse number (e.g. Purves and Lichtman, 1980; O'Kusky, 1985; Lnenicka and Murphy, 1989; in vitro: Van Huizen et al., 1985, 1987a), number of dendrites (Miller, 1988), number of axons (e.g. Heathcote and Sargent, 1985; Schreyer and Jones, 1988), and total dendritic length (Uylings et al., 1990).

In our model of a developing neural network, we found that such a transient overproduction with respect to connections or synapses can arise as the result of activity-dependent outgrowth in combination with a neuronal response function possessing some form of firing threshold. To explain this, first consider a purely excitatory network ($M = 0$) (Van Ooyen and Van Pelt, 1992, 1994). For a given connectivity **W** the network has convergent activation dynamics (Hirsch, 1989); the equilibrium points are solutions of

$$0 = -X_i + (1 - X_i) \sum_k^N W_{ik} F(X_k) \qquad \forall i \quad (6)$$

If the variations in X_i are small (relative to $\overline{X}$, the average membrane potential of the network), we find:

$$0 \simeq -\overline{X} + (1 - \overline{X})\overline{W}F(\overline{X}) \qquad (7)$$

Based on this approximation, the average connection strength $\overline{W}$ can be written as a function of $\overline{X}$:

$$\overline{W} = \frac{\overline{X}}{(1 - \overline{X})\, F(\overline{X})} \qquad 0 \leq \overline{X} < 1 \qquad (8)$$

which gives the equilibrium manifold of $\overline{X}$ ($d\overline{X}/dT = 0$) as depending on $\overline{W}$ (Fig. 2). The equilibria are stable on the branches *ABC* and *DEF*, and unstable on *CD*. This hysteresis loop underlies the emergence of overshoot in the model. The presence of hysteresis hinges upon

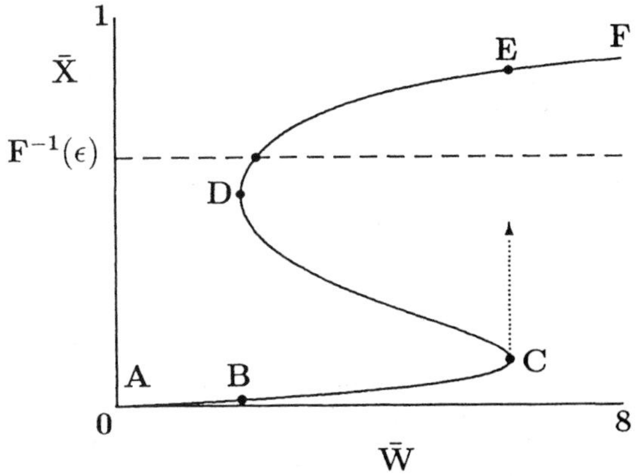

Fig. 2. Hysteresis between average connectivity ($\overline{W}$) and average membrane potential ($\overline{X}$) in a purely excitatory network. Shown is the steady state dependence of $\overline{X}$ ($d\overline{X}/dT = 0$) on $\overline{W}$ ($\overline{W} = 1/N\ \Sigma_{i,k}^N\ W_{ik}$) according to Eqn. (8).

the firing-rate function having some form of threshold along with low, but non-zero, values for sub-threshold membrane potentials (i.e. spontaneous activity; also see Pakdaman et al., 1994).

The size of a neuritic field remains constant if $X_i = F^{-1}(\epsilon)$, where F^{-1} is the inverse of F. Thus, at the intersection point of the line $\overline{X} = F^{-1}(\epsilon)$ with the equilibrium manifold, $\overline{W}$ remains constant. Outgrowth of neurons is on a time scale of days or weeks, so that connectivity can be regarded as quasi-stationary on the time scale of membrane potential dynamics. Thus, for changing $\overline{W}$, $\overline{X}$ follows the curve of Fig. 2. Initially, $\overline{W} = 0$, $dR_i/dT > 0$ and $\overline{W}$ increases, whereby $\overline{X}$ follows the branch ABC until it reaches w_2, where it jumps to the upper branch, thus exhibiting a transition from quiescent to activated state. In the activated state, however, $F(X_i) > \epsilon$, as a result of which the neuritic fields begin to retract ($dR_i/dT < 0$) and $\overline{W}$ to decrease, whereby $\overline{X}$ now moves along the upper branch from E to the intersection point. Thus, in order to arrive at an equilibrium point on DE, a developing network has to go through a phase in which $\overline{W}$ is higher than in the final situation, thus exhibiting a transient overshoot in $\overline{W}$. In the presence of inhibition, overshoot still occurs, and can even be enhanced (Fig. 3). To counterbalance the effect of

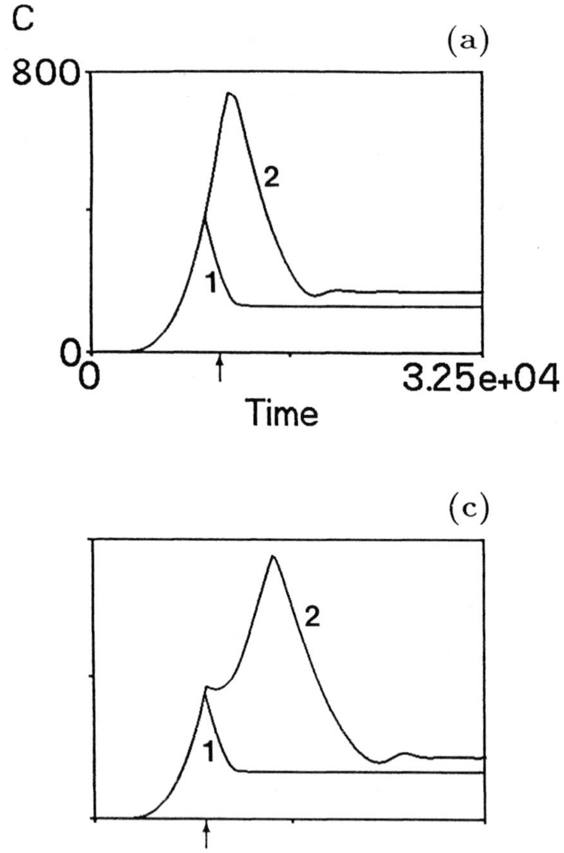

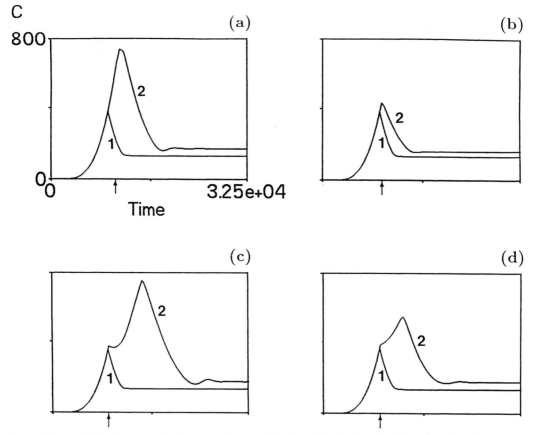

Fig. 3. Effect of inhibition on the development of connectivity. In all figures $N = 32$, $M = 4$, and $S^{ee} = S^{ie} = 0.6$. Total connectivity ($C = \Sigma_{p,q}^{N+M} A_{pq}$) in network (1) without inhibitory transmission ($S^{ei} = 0$, $S^{ii} = 0$) and (2) with inhibition ($S^{ei} \neq 0$, $S^{ii} \neq 0$). Arrows indicate the onset of network activity in the networks with inhibition. (a) $S^{ei} = 1.4$, $S^{ii} = 0.6$. (b) $S^{ei} = 1.0$, $S^{ii} = 0.6$. (c), (d) $S^{ei} = 1.4$, $S^{ii} = 0.6$ but with different and less regular spatial distribution of inhibitory cells.

inhibition, a higher excitatory connectivity is necessary for the network to reach the point at which the average connectivity starts declining. Also the excitatory connectivity level in the stable network must be higher. If, however, inhibition is too strong (many inhibitory cells or a high value of S^{ei}) the electrical activity in the network will remain so low that the cells continue to grow out.

With respect to the development of electrical activity and connectivity, the model shows similarities with what has been observed in developing in vitro cultures of dissociated nerve cells: an initial phase of neurite outgrowth and synapse formation while electrical activity is low, a rather abrupt onset of network electrical activity when the synapse density reaches a critical value, followed by a phase of neurite retraction (Purkinje cells in cerebellar cultures: Schilling et al., 1991) and/or elimination of synapses (cerebral cortex cells: Van Huizen et al., 1985, 1987a; Van Huizen, 1986). As in the model, blockade of electrical activity results in continued neurite outgrowth in both types of culture (Schilling et al., 1991; Van Huizen and Romijn, 1987) and prevents synapse elimination in cortex cultures (Van Huizen et al., 1985).

Onset of pruning

In purely excitatory model networks with a more or less homogeneous cell density, the decrease in connectivity begins shortly after the onset of network activity (Van Ooyen and Van Pelt, 1994). In mixed networks, however, the decline in overall connectivity can be considerably delayed relative to the onset of network activity (Fig. 3). In parts of the network with many inhibitory cells, excitatory cells can still be growing out, while in parts with fewer inhibitory cells they are already retracting, so that average connectivity can still increase markedly after the onset of network activity. This is in fact what was observed in neocortical cell cultures (Van Huizen et al., 1985): generalized electrical activity was readily detectable after about 12 days in vitro, whereas the overall decline in synapse numbers started only after about 18 days in vitro.

Delayed inhibition

The development of inhibition may lag that of excitation (Jackson et al., 1982; Barker and Harrison 1988; Corner and Ramakers, 1992; Rörig and Grantyn, 1993). If the development of inhibition in the model is delayed by giving the inhibitory cells a lower outgrowth rate, the growth curve of the number of *inhibitory* connections no longer exhibits overshoot (Fig. 4). The inhibitory cells develop within a network that has already attained a more or less stable activity level, and will therefore simple grow out until their overlap is such that $F(X_i) = \epsilon$. The observation that in tissue cultures of dissociated cerebral cortex cells the putative inhibitory synapses (synapses on shafts: Shepherd, 1990) show no pronounced overshoot during development, while the synapses on spines (which are mostly excitatory: Shepherd, 1990) show a clear overshoot (Van Huizen et al., 1985) would thus be consistent with a progressive increase in the ratio of effective inhibitory to excitatory synaptic activity during development.

Critical period

An interesting observation is that networks grown under conditions in which the generation of electrical activity is blocked, thus inducing an abnormally high connectivity, do not necessarily prune their connections after the block has been removed (Fig. 5). Although a restoration of activity occurs (possibly in the form of oscillations) this causes no reduction in connectivity. On the contrary, connectivity increases still further, as the average firing rate is below ϵ for most of the time due to inhibition. The ability for the network to prune its connections after it has been silenced appears to depend on the level of connectivity attained, and therefore on the time it spent under conditions of electrical silence. If this is longer than a certain 'critical period', elimination of connections can no longer take place, even though the cells themselves have not changed in their ability to do so (for details see Van Ooyen et al., in press). On the other hand, blocking the activity

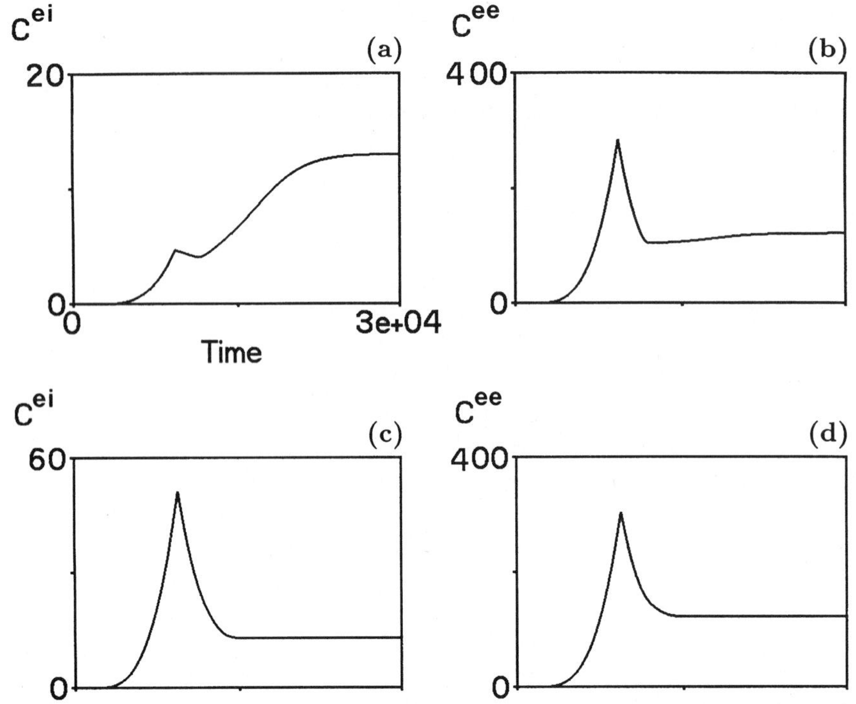

Fig. 4. Effect of delayed development of inhibition. Same network as in Fig. 3a, but with $S^{ei} = 0.6$. In (a), (b) the inhibitory cells have a lower outgrowth rate than the excitatory cells, $\rho = 0.00003$ and $\rho = 0.0001$, respectively. In (c), (d) both type of cells have the same growth rate, $\rho = 0.0001$. In (a), (c) the inhibitory-excitatory connectivity is shown ($C^{ei} = \Sigma_i^N \ \Sigma_l^M \ A_{il}$), in (b), (d) the excitatory-excitatory connectivity ($C^{ee} = \Sigma_{i,k}^N \ A_{ik}$)

in a normally developed network in which the process of overshoot and elimination of connections have already occurred, should always result in an increase in connectivity. These results, too, show similarities with findings in cultures of disscociated cerebral cortex cells (Van Huizen et al., 1987b). Thus, if cultures in which electrical activity had been chronically blocked during development—resulting, as in the model, in an enhanced neurite outgrowth (Van Huizen and Romijn, 1987) and a prevention of synapse elimination (Van Huizen et al., 1985)—were then placed in control medium, no elimination of synapses occurred even over a period of several weeks, although electrical activity was restored. On the other hand, blocking the activity in normally developed and highly mature cultures in which the process of synaptic elimination had already occurred resulted in a substantial increase in synapse density (as in the simulation model), pointing to a process of either terminal or collateral sprouting.

Compensatory sprouting

Various brain regions may lose neurons with aging (e.g. Curcio et al., 1982). When cells in a mature network are progressively deleted, the average neuritic field size is found to increase with the number of deleted neurons (Fig. 6). After excitatory cell loss, the average firing rate drops below ϵ, and cells (especially those in the neighbourhood of the deleted cells) will begin to grow out until they all have the same activity level as before. To compensate for the lost cells, a larger neuritic (dendritic) field is now necessary in order to receive sufficient input from the surviv-

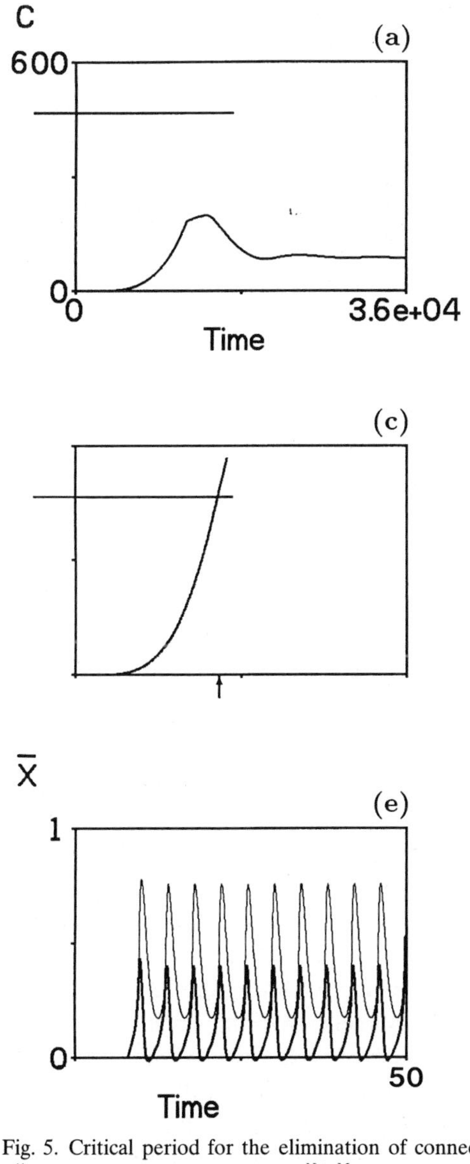

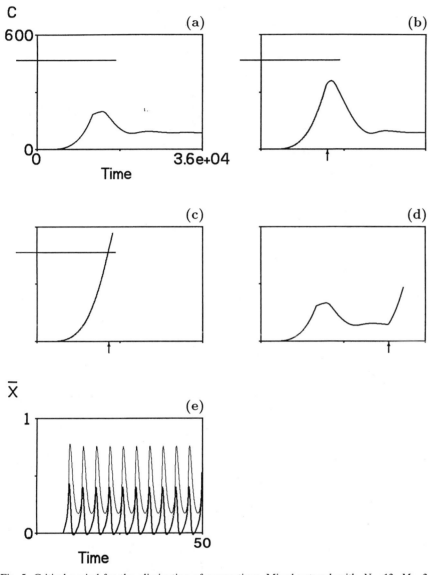

Fig. 5. Critical period for the elimination of connections. Mixed network with $N = 13$, $M = 3$, $S^{ee} = 0.6$, $S^{ei} = 1.1$, $S^{ie} = 0.6$, and $S^{ii} = 0$. C = total connectivity = $\Sigma_{p,q}^{N+M} A_{pq}$. (a) Normal development. (b), (c) The generation of electrical activity is blocked till the time indicated by the arrow. The horizontal line indicates the connectivity level above which connectivity can no longer decrease when activity returns. Below this line activity returns without oscillations or in the form of oscillations that gradually change, as connectivity changes, into a high 'steady' activity, followed by a normal decrease in activity and connectivity. Above this line, oscillations change into low 'steady' activity (network becomes inhibited and cells keep growing out) as connectivity further increases. Starting at still higher positions above this line, the network comes directly in the inhibited state, without a transient oscillatory phase. (d) Activity is blocked in a normally developed network at the time indicated by the arrow. (e) The average membrane potential ($\overline{X}$) of the excitatory (thick line) and inhibitory population (thin line) shown just after removing the blockade in (c).

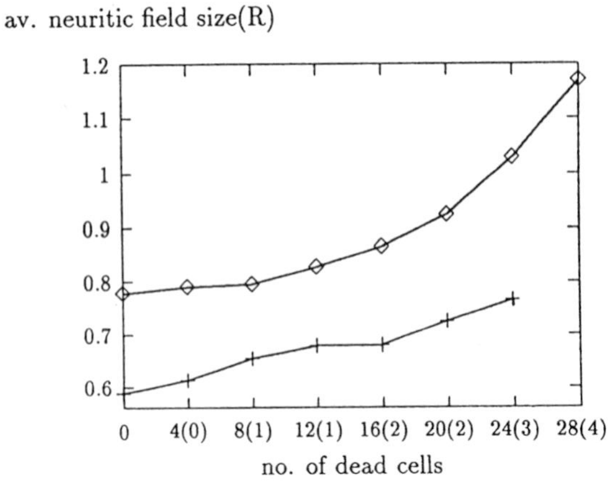

Fig. 6. Compensatory sprouting in response to cell death. Same network as in Fig. 3a ($N = 32$, $M = 4$) but with $S^{ee} = S^{ei} = S^{ie} = S^{ii} = 0.6$. Average fieldsize of excitatory (◇) and inhibitory cells (+) against the total number of deleted cells (in parentheses the number of deleted inhibitory cells). Cells were deleted at random. Note that the effect of cell death becomes relatively larger as the number of remaining cells decreases.

ing cells. Indeed, it was found in human cortex that the dendritic extent per neuron increased steadily through old age (Beull and Coleman, 1979; Coleman and Flood, 1986). These results were interpreted to indicate that dendritic proliferation is a compensatory response to the death of neighbouring neurons (Curcio et al., 1982; Coleman and Flood, 1986).

Network and neuritic field size

In purely excitatory networks we showed that, with a relatively low synaptic strength, cells develop into a single interconnected network, whereas a high synaptic strength yields a number of more or less loosely connected sub-networks (Van Ooyen and Van Pelt, 1994). The presence of inhibitory cells increases the degree of connectivity within the network: excitatory cells will need to grow larger neuritic fields in order to receive sufficient excitatory input to stop growing; as a result, more cells make mutual contacts, and sub-networks that otherwise would have been relatively disconnected, now become tightly linked. Thus, by inducing outgrowth, inhibitory cells can help to selectively connect different parts of a neural network structure.

In a purely excitatory network, all cells are identical except for their position. Local variations in cell density, however, suffice to generate a great variability among individual cells, with respect both to their neuritic field size at equilibrium and to the developmental course of their field size and firing behaviour. Cells surrounded by a high number of neighbouring cells tend to become small, since a small neuritic field will already give sufficient overlap with other cells such that $F(X_i) = \epsilon$. In contrast, relatively isolated cells must grow large neuritic fields in order to contact a sufficient number of cells (see Fig. 9). One might say that the neuritic fields adapt to the avaiblable space so as to cover it optimally.

In the presence of inhibition, such variability arises even without any local differences in cell density. In mixed networks, even though there are no intrinsic differences in growth properties, the neuritic field of an inhibitory cell will tend to become smaller than that of an excitatory cell. These differences emerge solely as the result of simple outgrowth rules and cell interactions. To illustrate how this can come about, consider a one-dimensional string of cells containing one inhibitory cell (Fig. 7). During the initial period of development, all cells have the same size, but the moment the network becomes activated, cells will differentiate, in such a way that the inhibitory cell ends up having the smallest neuritic field, adjacent to two excitatory cells with large fields. The influence of an inhibitory cell is not restricted to its direct neighbours but, rather, percolates through the network so that a characteristic distribution of cell sizes is induced. The presence of one inhibitory cell in a string of excitatory cells gives rise to a pattern of alternating small and large cells which gradually damps out (Fig. 8). A similar, though more complex, situation is obtained in the two-dimensional case: a characteris-

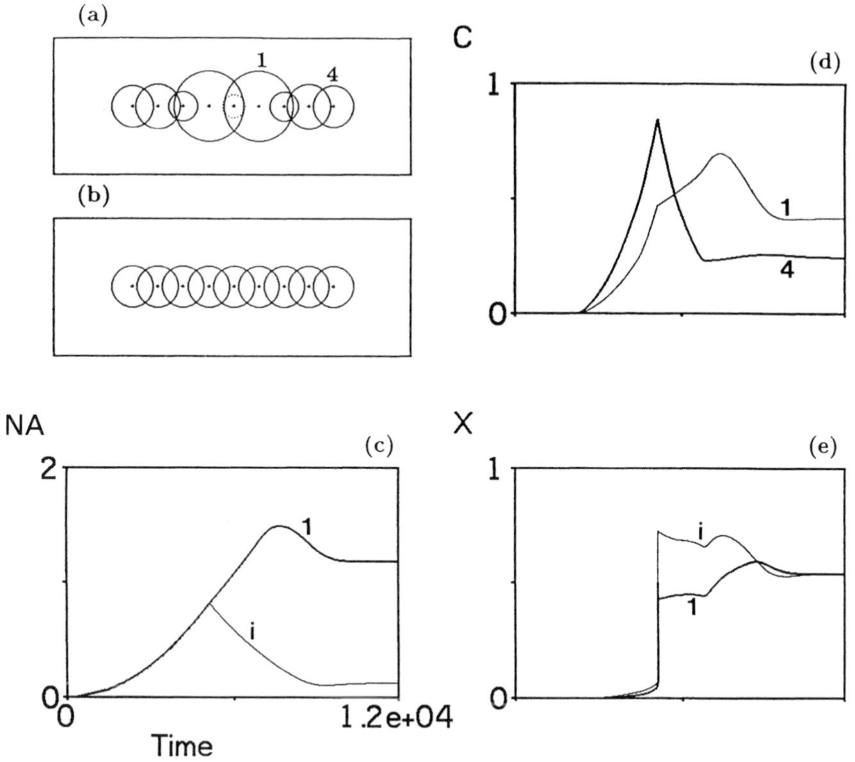

Fig. 7. Development of cell size differences. String of cells with $N = 8$, $M = 1$, $S^{ee} = S^{ei} = S^{ie} = 8.0$, and torus boundary conditions. (a) Network at equilibrium. Central cell (dashed line) is inhibitory. (b) Same network without inhibition ($N = 9$, $M = 0$). (c)Neuritic field area (*NA*) of the inhibitory cell (i) and its directly neighbouring excitatory cell (cell 1). (d) The overlap of cell 1 with other excitatory cells ($\Sigma_k^N A_{1k}$), and of cell 4 with other excitatory cells ($\Sigma_k^N A_{4k}$). Note that cell 1, the one getting inhibition, has the largest overlap at equilibrium. (e) Membrane potential of the inhibitory cell (i) and cell 1.

tic distribution of cell sizes—a kind of damping wave—is generated in the region surrounding each inhibitory cell (Fig. 9). With more than one inhibitory cell, interference patterns are generated, resembling those observed in other wave phenomena.

The mechanism causing cell sizes to differ is as follows. An excitatory cell that receives inhibition needs more excitation (i.e. a greater overlap with other excitatory cells) than does a cell that is not inhibited (see Fig. 7d), and must therefore produce a larger neuritic field (assuming a more or less homogeneous distribution of cells). As a consequence, an inhibitory cell will become surrounded by large excitatory cells, whereas—since the same growth rules apply to inhibitory cells as well—the inhibitory cell itself can remain small because even a small neuritic field will yield a large overlap with its surrounding cells. In other words, an inhibitory cell becomes small by causing the size of its direct neighbours to increase.

In the standard model we used neuritic fields without making a distinction between axons ('sending') and dendrites ('receiving'). If this distinction is introduced, it is the *dendritic* field of inhibitory cells that become large, since a cell regulates its own activity only by adapting the size of its receiving, i.e. dendritic field.

In the cerebral cortex the dendritic (but also axonal) fields of inhibitory neurons are indeed smaller, on the whole, than those of excitatory neurons. Two main types of neurons can be dis-

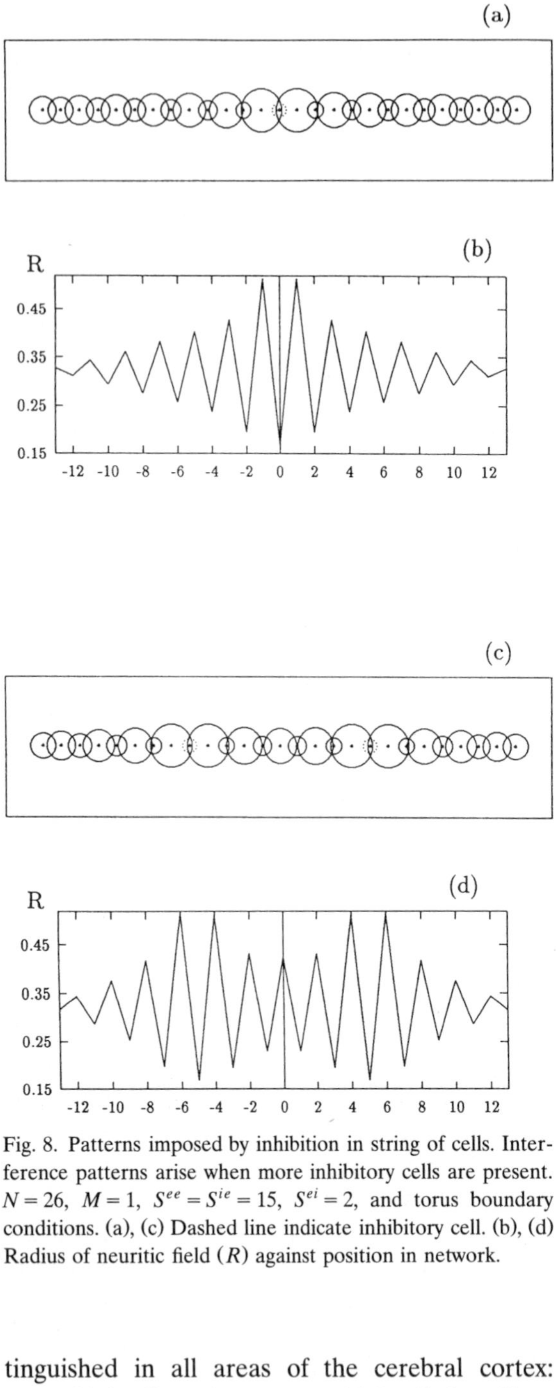

Fig. 8. Patterns imposed by inhibition in string of cells. Interference patterns arise when more inhibitory cells are present. $N = 26$, $M = 1$, $S^{ee} = S^{ie} = 15$, $S^{ei} = 2$, and torus boundary conditions. (a), (c) Dashed line indicate inhibitory cell. (b), (d) Radius of neuritic field (R) against position in network.

tinguished in all areas of the cerebral cortex: pyramidal cells and non-pyramidal cells (e.g. Kandel et al., 1991; Abeles, 1991). Pyramidal cells are excitatory and have large apical dendrites that often cross several layers; their axons are long, terminating in other areas of the cortex, and each axon has many collaterals which make synapses on neighbouring cells. The non-pyramidal cells, most of which are inhibitory, usually have smaller cell bodies with dendritic and axonal branches that extend only locally.

Self-inhibition

If the spatial distribution of inhibitory cells is such that they can contact each other, the effect of self-inhibition will become important (Fig. 10). As a consequence of the outgrowth rules, the ultimate level of inhibition will be higher with than without self-inhibition. When an inhibitory cell is electrically inhibited, its outgrowth is stimulated, so that more excitatory cells will eventually be contacted, and with a larger overlap. In this way, not only the number of inhibitory cells but also their distribution is crucial in determining the level of synaptic inhibition. Note that long-range inhibition will be obtained when inhibitory cells occur in a clustered fashion, so that they are in a position to stimulate each other's outgrowth.

Discussion

Many processes that play a role in the development of neurons into functional networks depend upon electrical activity. We have made a start at unravelling the possible implications for network ontogeny and neuron morphology of activity-dependent neurite outgrowth. Several interesting properties arise as the result of interactions among outgrowth, excitation and inhibition:

- a transient overproduction ('overshoot') during development with respect to synapse number or connectivity, both in purely excitatory and in mixed networks;
- the neuritic fields of inhibitory and excitatory cells become differentiated, whereby those of inhibitory cells tend to become smaller;
- the distribution of inhibitory cells becomes important in determining the ultimate level of inhibition;

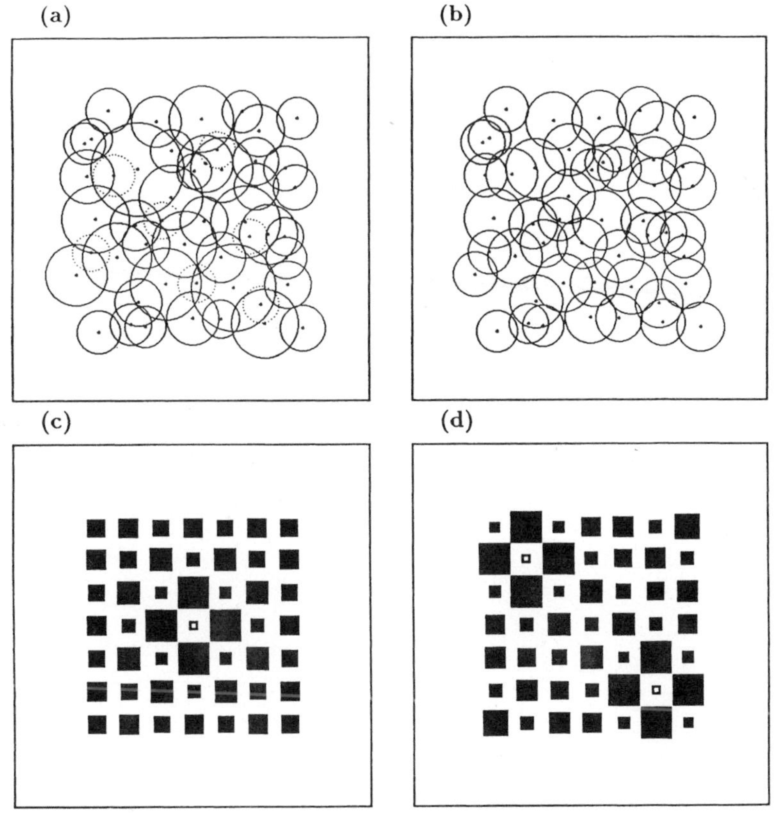

Fig. 9. Patterns imposed by inhibition. $S^{ee} = S^{ie} = 3$, $S^{ei} = 5$, $S^{ii} = 0$. Torus boundary conditions. (a) $N = 42$, $M = 7$. Cells on 'noisy' grid positions. Dashed line indicate inhibitory cell. (b) Same placing of cells as in (a), but all former inhibitory cells are now excitatory ($M = 0$). (c) $N = 48$, $M = 1$. Cells on grid positions. Diameter of square is proportional to area of neuritic field. Scaled to maximum area. Cell with white dot is inhibitory. (d) Same as in (c) but with $M = 2$. Notice interference patterns.

- inhibitory cells, by inducing outgrowth, can help to selectively connect sub-networks;
- pruning of connections can no longer take place if the network has grown without proper electrical activity for longer than a certain time ('critical period');
- excitatory cell death will be accompanied by an increased neuritic field of surviving neurons ('compensatory sprouting').

The results do not depend critically on precise choices of the parameter values.

The model may account for various (seemingly unrelated) phenomena observed in developing cultures of dissociated cells: (i) a sudden transition from a quiescent state to one of network activity; (ii) a transient overproduction of synapses; (iii) effects observed after totally blocking electrical activity, and after selective blocking of inhibitory transmission; (iv) different growth curves for synapses on shafts and on spines; (v) a delayed onset of the pruning phase relative to the onset of network activity; (vi) a critical period for synapse elimination but not for synapse formation; (vii) size differences between inhibitory and excitatory neurons. Experimental studies are now needed for testing explicitly whether activity-dependent outgrowth indeed plays an important role in the mechanisms underlying these phenomena.

Neuronal form results from both intrinsic

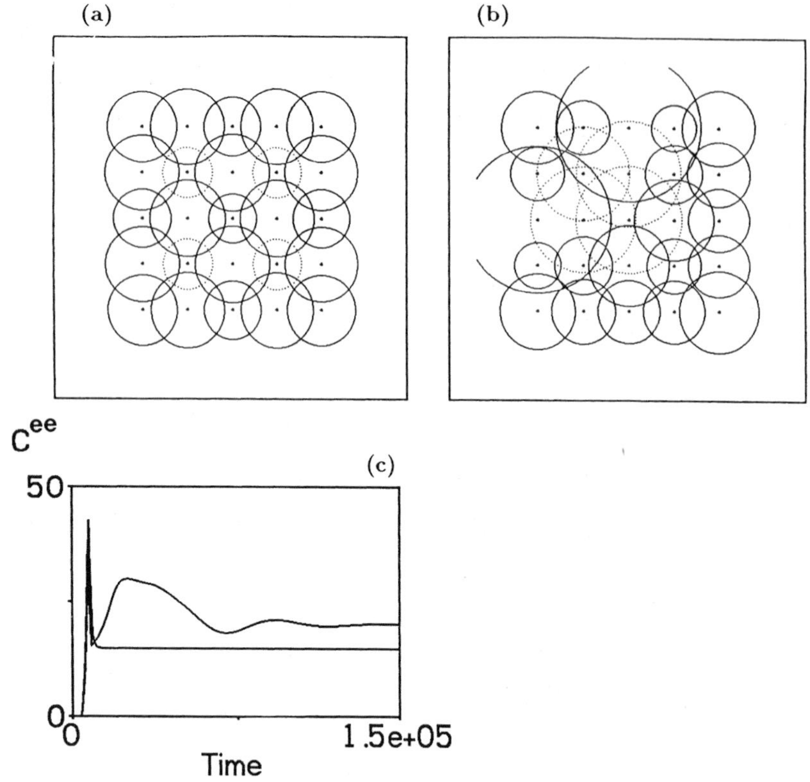

Fig. 10. Effect of distribution of inhibitory cells. Network on grid with $N=21$, $M=4$, $S^{ee}=S^{ie}=3$, $S^{ei}=1$, $S^{ii}=2$. Dashed line indicate inhibitory cell. (a) Inhibitory cells regularly distributed. (b) Inhibitory cells clustered. (c) The connectivity from excitatory-to-excitatory ($C^{ee}=\Sigma_{i,k}^{N} A_{ik}$) cells is higher at equilibrium when cells are clustered and self-inhibition plays a role.

growth properties and interactions with other cells. This study demonstrates how, as the result of cell interactions, a differentiation with respect to size could arise between excitatory and inhibitory cells that all have the same intrinsic growth properties. The observation in the model that pruning of connections is no longer possible if the network has grown without electrical activity for longer than a certain time serves to illustrate that a 'critical period' need not be the result of a predetermined cellular time schedule, but can arise as a result of non-linear neuron properties and cell interactions.

The hysteresis loop in the activity vs. connectivity curve underlies the emergence of overshoot in the model. Hysteresis may thus provide a mechanism by which overshoot phenomena in general can be understood. To illustrate this, consider the following variant of the model. Suppose that cell death is more likely at high levels of electrical activity, and that neurons (whose neuritic fields are now taken to be of constant size) are produced at a given source, transported, and progressively incorporated into a network during development. As before, neurons become connected when their neuritic fields overlap, and at a given cell density, the network will become activated and start losing cells if the resulting activity is too high. Because of the hysteresis effect, a higher cell density will be necessary to activate the network than to sustain it once activity has been initiated, thus giving rise to a reduction in cell numbers.

Two patterns of dendritic development have

been described in literature, a pattern of initial exuberant dendritic growth followed by retraction ('overshoot'), but also a pattern where dendritic arbors simply increase until their adult length is attained (e.g. Petit et al., 1988; Ulfhake et al., 1988). In the model, both patterns can be observed within one and the same network. A cell connecting to a structure that is not yet electrically active will exhibit overshoot in its growth curve, whereas a cell growing into structrue that has a stable level of activity will fail to show overshoot. Such a difference could in itself explain the existence of these two modes of dendritic development.

To explain how neuronal loss might induce dendritic proliferation in the surviving neurons, Coleman and Flood (1986) have suggested that death of a neuron brings about the release of a trophic factor which induces the proliferation of neighbouring dendrites. Alternatively, cell death could result in a reduced competition among the surviving neurons for afferent supply which, in return, would allow dendrites to proliferate (Coleman and Flood, 1986). The mechanism emerging from the present model provides a simpler alternative. Following (excitatory) neuronal death, the level of electrical activity drops, thus permitting outgrowth until all the cells have the same activity level as before. Considering that neurotransmitter release and electrical activity are closely coupled, this mechanism is essentially the same as the one put forward by Mattson (1988). He proposed that loss of inputs as a consequence of cell death will result in reduced availibility of neurotransmitter (which normally stabilizes dendritic morphology), leading to resumed outgrowth until the dendrites encounter another terminal that releases transmitter; dendritic outgrowth would again be stabilized, and existing circuitry consolidated at that point.

In the model inhibitory cells impose a structure on neighbouring excitatory cells. Interestingly, Lund et al. (1993) have found indications that inhibitory neurons may help to shape the patchy and stripe-like connectivity patterns which are found in several areas of macaque monkey cerebral cortex. Along the same lines, DeFelipe et al. (1990) have presented data suggesting that the 'double bouquet' cell (which is probably GABAergic) is an important factor in the mechanism by which a micro-columnar organization is imposed upon the cerebral cortex.

For some values of ϵ, the model can generate sustained oscillations in overall activity (and connectivity). The period of these oscillations is determined by the rate of outgrowth of neuritic fields. A network in which ϵ is distributed over a range of values can display complex behaviour, with oscillations of different frequencies.

As in tissue cultures, the model cells have no external input. External excitatory input to a single cell diminishes its neuritic field size and reduces overshoot, whereas inhibitory input does the opposite.

In summary, experimental results have indicated that neurotransmitters and associated electrical activity, by means of their effects on neurite outgrowth, have considerable potential for controlling the development of neuronal form and circuitry. In this study we have begun to explore this potential, and have been able to show that activity-dependent neurite outgrowth in a network of interacting excitatory and inhibitory cells can indeed have profound and complex effects on both neuronal morphology and network development.

References

Abeles, M. (1991) *Corticonics. Neural Circuits of the Cerebral Cortex*. Cambridge University Press, Cambridge. pp. 6–18.

Barker, J.L. and Harrison, N.L. (1988) Outward rectification of inhibitory postsynaptic currents in cultured rat hippocampal neurons. *J. Physiol.*, 403: 41–55.

Beull, S.J. and Coleman, P.D. (1979) Dendritic growth in the aged human brain and failure of growth in senile dementia. *Science*, 206: 854–856.

Codhan, C.S. and Kater, S.B. (1986) Suppression of neurite elongation and growth cone motility by electrical activity. *Science*, 232: 1638–1640.

Cohan, C.S., Connor, J.A. and Kater, S.B. (1987) Electrically and chemically mediated increases in intracellular calcium in neuronal growth cones. *J. Neurosci.*, 7: 3588–3599.

Coleman, P.D. and Flood, D.G. (1986) Dendritic proliferation in the aging brain as a compensatory repair mechanism. *Prog. Brain Res.*, 70: 227–237.

Constantine-Paton, M. (1990) NMDA receptor as a mediator of activity-dependent synaptogenesis in the developing brain. *Cold Spring Harbor Symposia on Quantitative Biology, Volume LV*, Cold Spring Harbor Laboratory Press, Cold Spring Harbor, NY, pp. 431–443.

Corner, M.A. and Ramakers, G.J.A. (1992) Spontaneous firing as an epiginetic factor in brain development—physiological consequences of chronic tetrodotoxin an picrotoxin exposure on cultured rat neocortex neurons. *Dev. Brain Res.*, 65: 57–64.

Curcio, C.A., Buell, S.J. and Coleman, P.D. (1982) Morphology of the aging nervous system. In: J.A. Mortimer, F.J. Pirozzolo and G.J. Maletta (Eds.), *The Aging Motor System, Advances in Neurogerontology, Vol.* 3, Praeger Publ., New York, pp. 7–35.

DeFelipe, J., Hendry, S.H.C., Hashikawa, T., Molinari, M. and Jones, E.G. (1990) A microcolumnar structure of monkey cerebral cortex revealed by immunocytochemical studies of double bouquet cell axons. *Neuroscience*, 37: 655–673.

Ferrer, I., Soriano, E., Del Rio, J.A., Alcantara, S. and Auladell, C. (1992) Cell death and removal in the cerebral cortex during development. *Prog. Neurobiol.*, 39: 1–43.

Fields, R.D. and Nelson, P.G. (1992) Activity-dependent development of the vertebrate nervous system. Int. Rev. Neurobiol., 34: 133–214.

Fields, R.D., Neale, E.A. and Nelson, P.G. (1990a) Effects of patterned electrical activity on neurite outgrowth from mouse neurons. *J. Neurosci.*, 10: 2950–2964.

Fields, R.D., Guthrie, P.B., Nelson, P.G. and Kater, S.B. (1990b) Calcium homeostatic capacity is regulated by patterned electrical activity in growth cones of mouse DRG neurons. *Soc. Neurosci. Abstr.*, 16: 457.

Grumbacher-Reinert, S. and Nicholls, J. (1992) Influence of substrate on retraction of neurites following electrical activity of leech Retzius cells in culture. *J. Exp. Biol*, 167: 1–14.

Heathcote, R.D. and Sargent, P.B. (1985) Loss of supernumary axons during neuronal morphogenesis. *J. Neurosci.*, 5: 1940–1946.

Hirsch, M.W. (1989) Convergent activation dynamics in continuous time networks. *Neural Networks*, 2: 331–349.

Jackson, M.B., Lecar, H., Brenneman, D.E., Fitzgerald, S. and Nelson, P.G. (1982) Electrical development in spinal cord cell culture. *J. Neurosci.*, 2: 1052–1061.

Kandel, E.R., Schwartz, J.H., Jessel, T.M. (1991) *Principles of Neural Science*, Prentice-Hall, Englewood Cliffs, NJ. pp. 292–779.

Kater, S.B. and Mills, L.R. (1991) Regulation of growth cone behaviour by calcium. *J. Neurosci.*, 11: 891–899.

Kater, S.B. and Guthrie, P.B. (1990) Neuronal growth cone as an integrator of complex environmental information. *Cold Spring Harbor Symposia on Quantitative Biology, Volume LV*, Cold Spring Harbor Laboratory Press, Cold Spring Harbor, NY, pp. 359–370.

Kater, S.B., Mattson, M.P., Cohan, C. and Connor, J. (1988) Calcium regulation of the neuronal growth cone. *Trends Neurosci.*, 11: 315–321.

Kater, S.B., Guthrie, P.B. and Mills, L.R. (1990) Integration by the neuronal growth cone: a continuum from neuroplasticity to neuropathology. In: P.D. Coleman, G.A. Higgins and C.H. Phelps (Eds.), *Progress in Brain Research, Vol.* 86, *Molecular and Cellular Mechanisms of Neuronal Plasticity in Normal Aging and Alzheimer's Disease*, Elsevier Science Publishers, Amsterdam, pp. 117–128.

Lipton, S.A. and Kater, S.B. (1989) Neurotransmitter regulation of neuronal outgrowth, plasticity and survival. *Trends Neurosci.*, 12: 265–270.

Lnenicka, G.A. and Murphy, R.K. (1989) The refinement of invertebrate synapses during development. *J. Neurobiol.*, 20: 339–355.

Lund, J.S., Yoshioka, T. and Levitt, J.B. (1993) Comparison of intrinsic connectivity in different areas of macaque monkey cerebral cortex. *Cerebral Cortex*, 3: 148–162.

Madison, D.V., Malenka, R.C. and Nicoll, R.A. (1991) Mechanisms underlying long-term potentiation of synaptic transmission. *Annu. Rev. Neurosci.*, 14: 379–397.

Mattson, M.P. (1988) Neurotransmitters in the regulation of neuronal cytoarchitecture. *Brain Res. Rev.*, 13: 179–212.

Mattson, M.P. and Kater, S.B. (1989) Excitatory and inhibitory neurotransmitters in the generation and degeneration of hippocampal neuroarchitecture. *Brain Res.*, 478: 337–348.

McCobb, D.P., Haydon, P.G. and Kater, S.B. (1988) Dopamine and serotonin inhibition of neurite elongation of different identified neurons. *J. Neurosci. Res.*, 19: 19–26.

Miller, M.W. (1988) Development of projection and local circuit neurons in neocortex. In: A. Peters and E.G. Jones (Eds.), *Cerebral Cortex. Development and Maturation of the Cerebral Cortex, Vol.* 7, Plenum Press, New York, pp. 133–175.

Neely, M.D. (1993) Role of substrate and calcium in neurite retraction of leech neurons following depolarization. *J. Neurosci.*, 13: 1292–1301.

O'Kusky, J.R. (1985) Synapse elimination in the developing visual cortex: a morphometric analysis in normal and dark-reared cats. *Dev. Brain Res.*, 22: 81–91.

Pakdaman, K., Van Ooyen, A., Houweling, A.R. and Vibert, J.-E. (1994) Hysteresis in a two neuron-network: basic characteristics and physiological implications. In: M. Marinaro and P.G. Morasso, (Eds.) *Artificial Neural Networks*, 1, *Proc. ICANN'94*, Springer Verlag, New York, pp. 162–165.

Petit, T.L., LeBoutillier, J.C., Gregorio, A. and Libstug, H. (1988) The pattern of dendritic development in the cerebral cortex of the rat. *Dev. Brain Res.*, 41: 209–219.

Purves, D. and Lichtman, J.W. (1980) Elimination of synapses in the developing nervous system. *Science*, 210: 153–157.

Rörig, B. and Granty, R. (1993) Glutamergic and GABAergic synaptic currents in ganglion cells from isolated retinae of pigmented rats during postnatal development. *Dev. Brain Res.*, 74: 98–110.

Schilling, K., Dickinson, M.H., Connor, J.A. and Morgan, J.I. (1991) Electrical activity in cerebellar cultrues determines Purkinje cell dendritic growth patterns. *Neuron* 7: 891–902.

Schreyer, D.J. and Jones, E.G. (1988) Axon elimination in the developing corticospinal tract of the rat. *Dev. Brain Res.*, 38: 103–119.

Shatz, C.J. (1990) Impulse activity and the patterning of connections during CNS development. *Neuron*, 5: 745–756.

Shepherd, G.M. (1990) *The Synaptic Organization of the Brain*. Oxford University Press, New York, Oxford. pp. 392–420.

Spitzer, N.C. (1991) A developmental handshake: neuronal control of ionic currents and their control of neuronal differentiation. *J. Neurobiol.*, 22: 659–763.

Sussdorf, W.S. and Campenot, R.B. (1986) Influence of the extracellular potassium environment on neurite growth in sensory neurons, spinal cord neurons and sympathetic neurons. *Dev. Brain Res.*, 25: 43–52.

Thoenen, H. (1991) The changing scene of neurotrophic factors. *Trends Neurosci.*, 14: 165–170.

Todd, R.D. (1992) Neural development is regulated by classical neurotransmitters: dopamine D2 receptor stimulation enhances neurite outgrowth. *Biol. Psychiatry*, 31: 794–807.

Tsumoto, T. (1992) Long-term potentiation and long-term depression in the neocortex. *Prog. Neurobiol.*, 39: 209–228.

Ulfhake, B., Cullheim, S. and Franson, P. (1988) Postnatal development of cat hind motoneurons. I. Changes in length, branching structure and spatial distribution of dendrites of cat triceps surae motoneurons. *J. Comp. Neurol.*, 278: 69–87.

Ulyings, H.B.M., Van Eden, C.G., Parnavelas, J.G. and Kalsbeek, A. (1990) The prenatal and postnatal development of rat cerebral cortex. In: B. Kolb and R.C. Trees (Eds.), *The Cerebral Cortex of the Rat*, MIT Press, Cambridge, MA. pp. 35–76.

Van Huizen, F. (1986) Significance of bioelectric activity for synaptic network formation. PhD Thesis, University of Amsterdam.

Van Huizen, F. and Romijn, H.J. (1987) Tetrodotoxin enhances initial neurite outgrowth from fetal rat cerebral cortex cells in vitro. *Brain Res.*, 408: 271–274.

Van Huizen, F., Romijn, H.J. and Habets, A.M.M.C. (1985) Synaptogenesis in rat cerebral cortex is affected during chronic blockade of spontaneous bioelectric activity by tetrodotoxin. *Dev. Brain Res.*, 19: 67–80.

Van Huizen, F., Romijn, H.J., Habets, A.M.M.C. and Van den Hooff, P. (1987a) Accelerated neural network formation in rat cerebral cortex cultures chronically disinhibited with picrotoxin. *Exp. Neurol.*, 97: 280–288.

Van Huizen, F., Romijn, H.J. and Corner, M.A. (1987b) Indications for a critical period for synapse elimination in developing rat cerebral cortex cultures. *Dev. Brain Res.*, 13: 1–6.

Van Ooyen, A. and Van Pelt, J. (1992) Phase transitions hysteresis and overshoot in developing neural networks. In: I. Aleksander and J. Taylor (Eds.), *Artificial Neural Networks*, 2, *Proc. ICANN'92*, Elsevier, Amsterdam, pp. 907–910.

Van Ooyen, A. and Van Pelt, J. (1993) Implications of activity-dependent neurite outgrowth for developing neural networks. In: S. Gielen and B. Kappen (Eds.), *Artificial Neural Networks*, *Proc. ICANN'93*, Springer Verlag, New York, pp. 177–182.

Van Ooyen, A. and Van Pelt, J. (1994) Activity-dependent outgrowth neurons and overshoot phenomena in developing neural networks. *J. Theor. Biol.*, 167: 27–43.

Van Ooyen, A., Van Pelt, J. and Corner, M.A. Implications of activity-dependent neurite outgrowth for neuronal morphology and network development. *J. Theor. Biol.*, in press.

J. van Pelt, M.A. Corner H.B.M. Uylings and F.H. Lopes da Silva (Eds.)
Progress in Brain Research, Vol 102

CHAPTER 17

γ-Aminobutyric acid (GABA): a fast excitatory transmitter which may regulate the development of hippocampal neurones in early postnatal life

Y. Ben-Ari, V. Tseeb, D. Raggozzino, R. Khazipov and J.L. Gaiarsa

INSERM U29, 123 Bd de Port-Royal, 75014 Paris, France

Introduction

In adult central nervous systems, γ-aminobutyric acid (GABA) is the main inhibitory neurotransmitter, acting on two main classes of receptors: $GABA_A$ and $GABA_B$ receptors (Sivilotti and Nistri, 1991). The bicuculline sensitive $GABA_A$ receptors activate a channel permeable to chloride, underlying the fast IPSP in various brain structures (Ben-Ari et al., 1981; Sivilotti and Nistri, 1991), and is potentiated by benzodiazepines and barbiturates (Study and Barker, 1981). $GABA_B$ receptors can be subclassified according to their signal-transduction mechanisms; activation of postsynaptic $GABA_B$ receptors leads to an increase of the potassium conductance that underlies the slow IPSP (Newberry and Nicoll, 1984; Dutar and Nicoll, 1988a) or to a decrease of voltage dependent conductances (Dolphin and Scott, 1986, 1987). Both effects are mediated through activation of a pertussis toxin sensitive G-protein (Dolphin and Scott, 1987; Dutar and Nicoll, 1988b). In addition to these postsynaptic effects, $GABA_B$ receptor agonists also depress glutamatergic and GABAergic synaptic transmission, acting via presynaptically localised receptors (Bowery et al., 1980; Howe et al., 1987) which may have different pharmacological properties than the postsynaptic ones (Thompson and Gähwiler, 1992; Dutar and Nicoll, 1988b).

The excitatory drive is mediated primarily by glutamatergic receptors; AMPA/kainate subtypes of receptors are responsible for the fast component of the excitatory responses while NMDA receptors mediate the slow component (Mayer and Westbrook, 1987), which is thought to play an important role in neuronal plasticity (Collingridge and Bliss, 1987; Ben-Ari et al., 1992).

In the adult hippocampus, as in other brain structures, GABAergic interneurones inhibit neuronal firing and prevent synchronised paroxysmal activity. Indeed, application of $GABA_A$ antagonists generate spontaneous and evoked paroxysmal activity (Johnston and Brown, 1984) and benzodiazepines, which potentiate $GABA_A$ mediated transmission, are clinically antiepileptic and anxyolytic agents.

In the hippocampus, as in neocortex, GABAergic interneurones become postmitotic before the pyramidal and granule cells do (Amaral and Kurz, 1985; Lübbers et al., 1985), and biochemical studies have shown that the GABAergic system is well developed at birth (Coyle and Enna, 1976; Rozenberg et al., 1989). In contrast, excitatory inputs progressively develop during the first postnatal week of life (Amaral and Dent, 1981; Pokorny and Yamamoto, 1981; Richter and Wolf, 1990). This and other lines of evidence (Ben-Ari et al., 1990; Cherubini et al., 1991) suggest that GABAergic interneurones play a unique role dur-

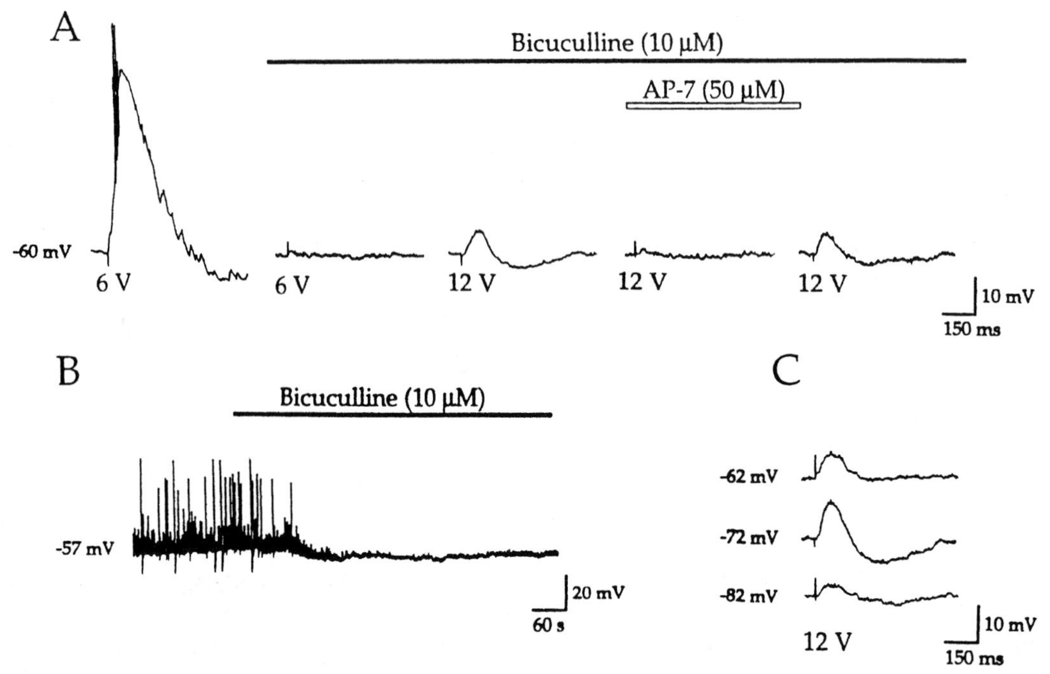

Fig. 1. $GABA_A$ receptors provide the excitatory drive in neonatal neurones. Current-clamp recording with KCl filled electrode of a CA3 pyramidal neurone at P2. (A) Bath application of bicuculline (10 μM) blocked evoked synaptic activity. Increasing the strength of stimulation revealed an AP-7 sensitive EPSP in the presence of bicuculline. (B) Continuous trace of the same cell showing that bicuculline (10 μM) blocked the spontaneous synaptic activity, reduced the synaptic noise and induced a membrane hyperpolarisation. (C) Voltage depency of the AP-7 sensitive EPSP in the presence of bicuculline (10 μM).

ing the early postnatal period. In the present report, we review recent observations on the properties of neonatal GABAergic interneurones and their functional role in hippocampal development.

Experimental overview

Neonatal $GABA_A$ receptors mediate excitatory drive

Intracellular recordings of CA3 pyramidal neurones on hippocampal slices revealed the presence of spontaneous and evoked giant depolarising potentials (GDPs) in over 85% of the neurones recorded between P0 and P4 (Fig. 1A) (Ben-Ari et al., 1989). These GDPs (300–500 msec duration, 10–50 mV in amplitude) were network driven events mediated by $GABA_A$ receptors. Spontaneous and evoked GDPs reversed polarity at the same potential as the responses evoked by exogenous application of GABA or isoguvacine ($GABA_A$ agonist) (i.e. -50 mV with potassium-methylsulfate filled electrodes and -20 mV with KCl filled electrodes) (Ben-Ari et al., 1989). Furthermore the evoked and spontaneous GDPs were blocked by bicuculline (10 μM, Figs. 1A, 1B), which induced a small membrane hyperpolarisation (Fig. 1B) and failed to induce any paroxysmal activity. Starting from the end of the first postnatal week, GDPs disappeared progressively and were replaced by spontaneous large hyperpolarising potentials (LHPs). At this period of development, bicuculline (as in adult tissue) generates paroxysmal activity mediated by glutamate, while exogenously applied GABA or $GABA_A$ agonist (isoguvacine) hyperpolarised CA3 pyramidal neurones. These GABA responses reversed polarity at around -65 mV, again as in adults (Ben-Ari et al., 1989).

These data therefore indicate that, during the first postnatal week of life, GABA (acting on $GABA_A$ receptors) depolarises the immature CA3 pyramidal neurones and, in fact, provides most of the excitatory drive at early stages of development.

Developmental profile of glutamatergic synaptic potential

As stressed above, bicuculline completely blocked spontaneous and evoked hippocampal synaptic activity between P0 and P4. Interestingly, in the presence of bicuculline, an EPSP was recorded in 60% of the cells ($n = 14$) when the strength or frequency of stimulation was increased. In four out of eight neurones the EPSPs were fully blocked by NMDA antagonists (APV or AP-7, 50 μM, Fig. 1A) and exhibited a voltage dependency similar to that of NMDA-mediated synaptic events in adult tissue (Fig. 1C). In two other neurones, the bicuculline-insensitive EPSPs were not affected by NMDA antagonists but were blocked by the non-NMDA antagonist CNQX (10 μM). In the two remaining cells, the bicuculline-insensitive EPSPs were blocked by concomitant application of CNQX (10 μM) and APV (50 μM).

The properties of the evoked synaptic responses were also studied using patch-clamp recording in the whole-cell configuration. In control ACSF, the amplitude of the evoked synaptic current increased with hyperpolarisation, and the I-V relations were linear over a wide range of membrane potentials (not shown). Addition of CNQX (20 μM) did not markedly change either the amplitude of the evoked synaptic currents or its I-V relation (Figs. 2A and 2B, top). However, further addition of bicuculline (20 μM) strongly reduced the amplitude of the synaptic currents evoked at holding potential (-50 mV) from 48 ± 9 pA to 19 ± 5 pA ($n = 5$). The bicuculline-resistant components represent NMDA synaptic currents, since (i) they were blocked by APV (50 μM) and (ii) I-V relations in the range of -30 to -90 mV showed a negative slope, presumably reflecting the voltage-dependent inhibition of these channels by Mg^{2+} (Nowak et al., 1984).

The properties of postsynaptic glutamatergic receptors were further studied in the presence of TTX (1 μM). Bath application of glutamatergic agonists (AMPA, NMDA, quisqualate) induced inward currents which, in immature CA3 pyramidal neurones, had the same characteristics as those recorded in adult cells in terms of amplitude, pharmacology, and voltage dependency (Gaiarsa et al., 1990, 1991). Furthermore, recordings in the cell-attached configuration revealed that NMDA receptor-gated channels are present on neonatal hippocampal neurones and have the expected properties even at this early stage of development.

These data therefore show that during the first postnatal week of life, when GABA depolarises CA3 pyramidal neurones, the excitatory glutamatergic inputs are weakly developed or even quiescent. Furthermore, in immature CA3 pyramidal neurones, EPSPs are primarily mediated by NMDA receptors which already have the expected Mg^{2+} dependent block.

Developmental study of benzodiazepine modulation of IPSPs

Benzodiazepines enhance neuronal $GABA_A$ responses via two main types of receptors (BZ1 and BZ2). In immature cells only BZ2 are present, and BZ1 develops only after P21 (Lippa et al., 1981). We investigated the effects of benzodiazepines on $GABA_A$ responses of CA3 pyramidal cells.

Monosynaptic $GABA_A$ IPSPs were intracellularly recorded from CA3 cells in the presence of APV (50 μM) and CNQX (10 μM) to block excitatory input. Midazolam (300 nM), a BZ1 and BZ2 agonist, increased the amplitude and duration of the IPSPs in young and adult rats (Fig. 3A). The effects of midazolam on the IPSPs rarely recovered in full after wash-out (10–30 min) but the addition of the antagonist Ro 15-1788 (10 μM) reversed the effect of midazolam (Fig. 3A). As expected from a late developmental onset of

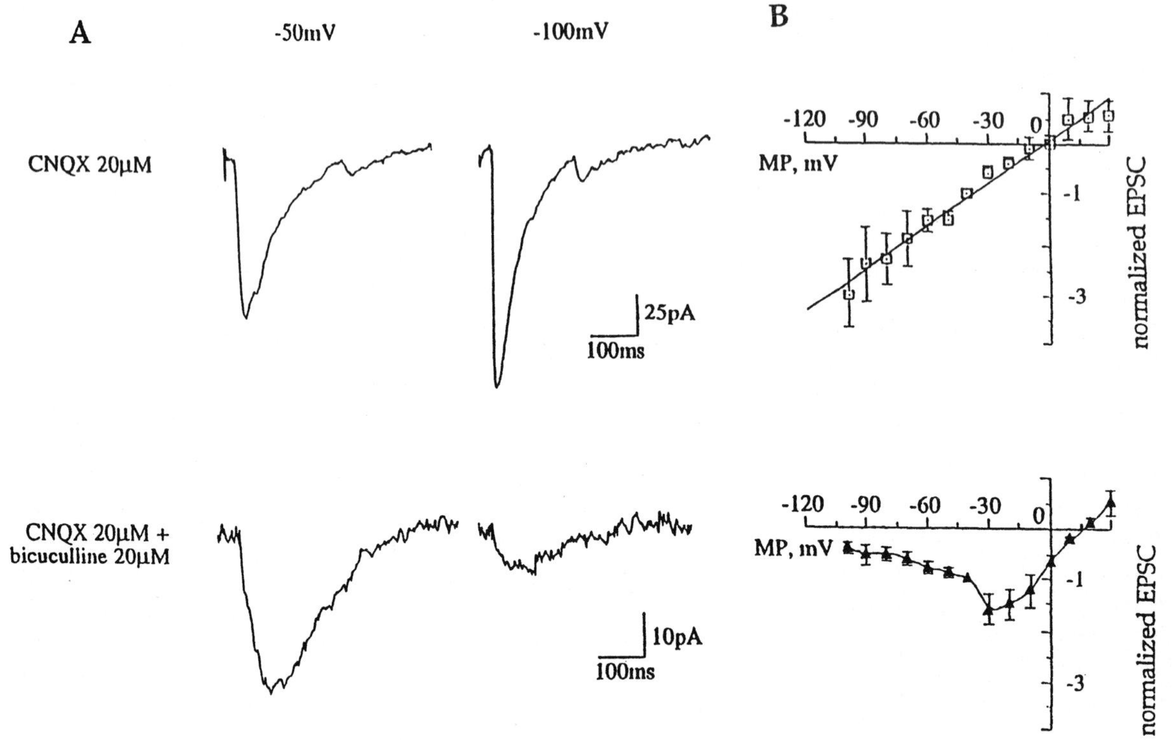

Fig. 2. NMDA-mediated synaptic currents in immature CA3 pyramidal cells. (A) Whole-cell currents were recorded from CA3 pyramidal neurone in hippocampal slice from immature (P6) rat. Excitatory synaptic currents (EPSCs) induced by stimulation of the stratum radiatum in the presence of CNQX (20 μM, top). Note elevated amplitude of the EPSC at −100 mV. Addition of bicuculline (20 μM) to the medium reduced the amplitude of the evoked EPSCs (bottom). Note the smaller amplitude of the EPSCs when the cell was hyperpolarised. (B) I-V relations of the EPSCs in the presence of CNQX (top) and in the presence of CNQX + bicuculline (bottom). The data represent an average made from six pyramidal neurones. Note that bicuculline revealed a region of negative slope at hyperpolarised potential.

BZ1, the specific BZ1 agonist zolpidem (100 nM) did not increase the IPSPs before P21 (Rovira and Ben-Ari, 1991, 1993). In contrast to the IPSPs, the current responses evoked by exogenous GABA were not increased by midazolam before P21 (Fig. 3B).

These observation suggest that (i) BZ2 receptors are involved in the potentiation of the IPSPs, and (ii) these BZ2 receptors are localised on GABAergic interneurones, their activation leading to enhanced GABA release probably via calcium channels. To test this hypothesis, the effects of BZs on barium potentials were studied in young (P7–P12) and in adult rats in the presence of bicuculline (10 μM), CNQX (10 μM) and AP5 (50 μM). In adult and neonatal CA3 pyramidal neurones, the duration of barium spikes was increased by midazolam (300 nM) and decreased by the inverse agonist (DMCM, 10 μM) (Fig. 4). In contrast, zolpidem shortened barium spikes in adults (100 nM; 66%) but not in young cells.

These observations suggest that BZ2 ligands modulate calcium current in immature cells, an effect compatible with a subsequent increase in GABA release. In contrast, BZ1 ligands are effective only in adult cells, where they cause a decrease in calcium entry.

Developmental differences between presynaptic and postsynaptic $GABA_B$ inhibition

Starting from P6, as in adulthood, a biphasic IPSP is elicited by electrical stimulation in the

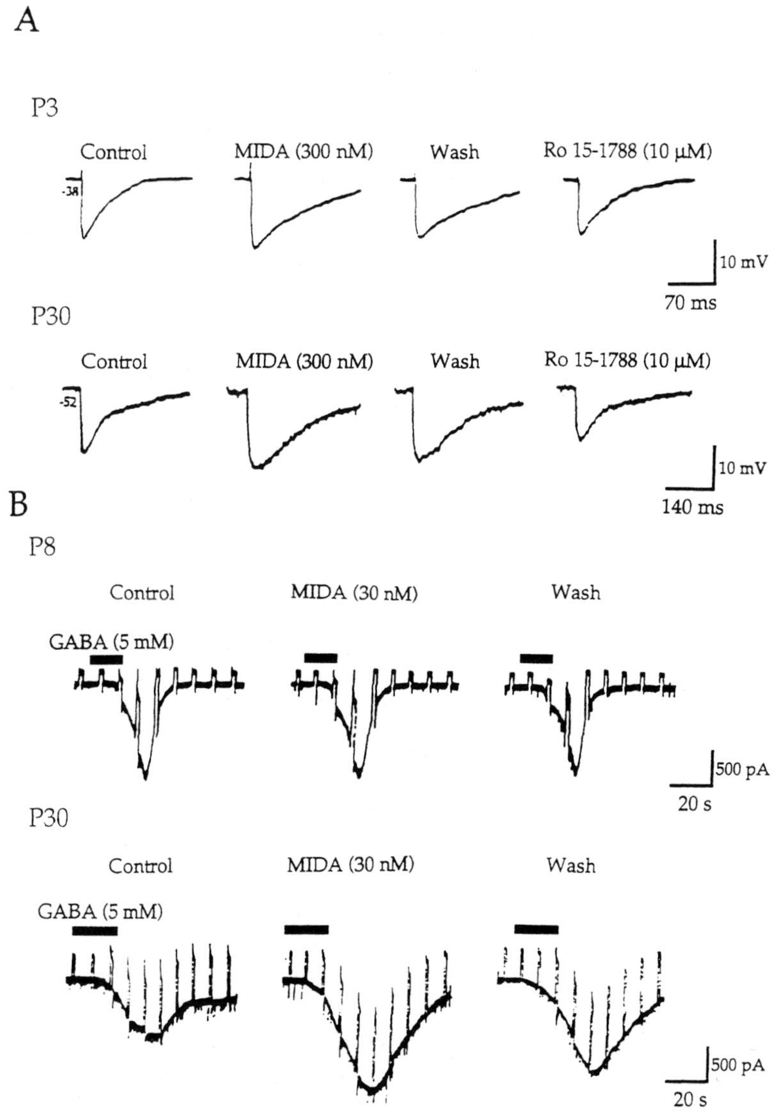

Fig. 3. Benzodiazepine increases the $GABA_A$-IPSP of immature cells without enhancing the GABA current. (A) Current-clamp recordings with potassium-metylsulfate filled electrode of CA3 pyramidal neurones. Midazolam increases the monosynaptic IPSPs in pup (P5, top traces) and in adult (bottom traces). The IPSPs are evoked after hilar stimulation in the presence of CNQX (10 μM) and APV (50 μM). The benzodiazepine antagonist Ro 15-1788 (10 μM) reverses the effects of midazolam. (B) Voltage-clamp recordings with KCl filled electrode of CA3 pyramidal neurones. In contrast to adult rat (bottom traces), midazolam does not increase the GABA current in cells of young rats (P12, top traces). Tetrodotoxin (1 μM) was superfused throughout the experiment, GABA and midazolam were bath applied. The upward deflection on the current traces are responses to the 20 mV voltage step used to monitor the input conductance of the cell.

presence of excitatory amino acid receptor blockers (CNQX, 10 μM; APV, 50 μM); the fast component is mediated by $GABA_A$ receptors, while the slow component is mediated by $GABA_B$ receptors. In contrast, in P0–P3 slices, only a depolarising $GABA_A$ potential was recorded under the same conditions in most neurones (seven out of nine CA3 pyramidal neurones), suggesting

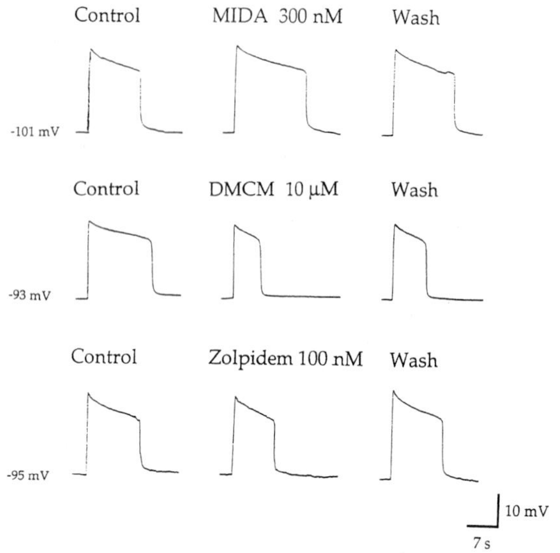

Fig. 4. $GABA_A$-receptor unrelated effects of BZs on the duration of the barium spikes in young and adult cells. TTX (1 μM), TEA (25 mM), 4AP (5 mM), CNQX (10 μM), AP5 (50 μM) and bicuculline (10 μM) were added to the basic ACSF wich contains 2mM barium instead of calcium. Cells were recorded with CsCl electrodes. Barium potentials were evoked by depolarising pulses (2 nA, 10–15 msec, 0.04 Hz) at the membrane potential indicated in front of the traces in each row. Recordings from different adult cells with CsCl electrodes are shown in each row.

that $GABA_B$ mediated IPSPs appear at a later stage of development. In keeping with this, application of baclofen (a $GABA_B$ agonist) induced an outward currents in P6 CA3 pyramidal neurones but not earlier (Fig. 5A).

In contrast to the postsynaptic $GABA_B$ mediated inhibition, hippocampal presynaptic $GABA_B$ inhibition is well developed at birth. Indeed, when using a cesium-filled electrode to block postsynaptic $GABA_B$ receptors, bath application of baclofen reversibly depressed glutamatergic and GABAergic transmission (Fig. 5B). These effects are antagonised by CGP 35 348 (500 μM); they must be due to activation of presynaptic receptors since baclofen has no effect either on input resistances and membrane potentials or on postsynaptic isoguvacine- and glutamate-mediated currents.

These data, therefore, show that postsynaptic $GABA_B$ inhibition is not present at birth, but appears during the first postnatal week of life. In contrast, presynaptic $GABA_B$ inhibition is already operative at birth on both glutamatergic and GABAergic terminals.

Morphological development of the GABAergic network

Our electrophysiological data indicate that GABA provides most of the excitatory drive in the hippocampus during the first postnatal week of life, while excitatory inputs are still poorly developed or quiescent. To examine the relationship between electrophysiological and morphological changes in the GABAergic network, we have intracellularly stained interneurones with biocytine using the whole cell patch-clamp configuration. The cells had resting potentials from −50 to −70 mV and showed the fast firing activity typical for interneurones. In 14 interneurones injected in the strata oriens and radiatum of CA3-CA2 regions at postnatal days 2 and 6, axons emerging from the cell body or proximal part of dendrites were smooth, and had a relatively uniform diameter in comparison with the dendrites. The axonal network of these interneurones was already as complex at P2 as at P6, with a wide innervation extending to most of the hippocampal formation (Fig. 6A). In contrast, the axonal network of the CA3 pyramidal neurones is clearly much less developed: while Schaffer collaterals are often observed, immature CA3 pyramidal neurones exhibit only a few recurrent collaterals within the CA3 field (Fig. 6B) (Gaiarsa et al., 1992).

$GABA_A$ receptors modulate the outgrowth of hippocampal neurones

Several experiments indicate that neurotransmitters can affect the morphological development of neurones grown in culture (Mattson, 1988). For instance, glutamate has been shown to regulate dendritic outgrowth of hippocampal pyramidal neurones (Mattson, 1988), and recent observations suggest that this effect is mediated by membrane depolarisation leading to an in-

A

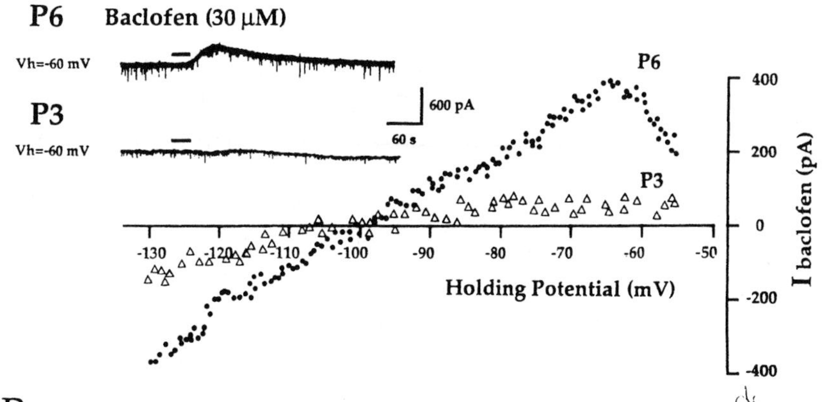

B

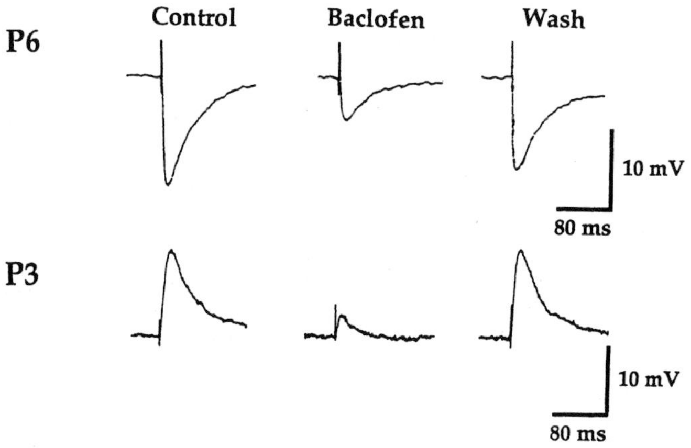

Fig. 5. Postnatal developement of baclofen mediated currents. (A) Intracellular recordings (CsCl filled electrode) of postnatal days 6 and 3 (P6, P3) CA3 pyramidal neurones in single electrode voltage clamp mode in the presence of TTX (1 μM). Baclofen (30 μM) was bath applied at a holding potential of -60 mV. The I-V plot was computed by digital subtraction of the steady state currents resulting from the voltage ramp (500 msec duration, from -120 to -40 mV) in the absence and in the presence of baclofen. The current generated by baclofen at P3 is clearly smaller compared with the current generated at P6 (KCl filled electrode). (B) Current-clamp recordings of CA3 pyramidal neurones. In the presence of CNQX (10 μM) and APV (50 μM), to block excitatory synaptic input, and using a cesium-methylsulfate filled electrode to block postsynaptic $GABA_B$ mediated responses, electrical stimulation induced an hyperpolarising $GABA_A$ IPSP at P6 (V_{mb} = y $-$ 57 mV) and a depolarising $GABA_A$ IPSP at P3 ($V_{mb} = -60$ mV). Baclofen (10 μM) reversibly depressed the evoked synaptic response at P6 and P3.

crease in intracellular calcium concentration (Kater et al., 1988). In order to determine the functional role of depolarising GABA responses in early postnatal development, we have tested the effect of bicuculline on the outgrowth of hippocampal neurones in culture. Cells were grown for 3 days at low density in serum-free medium. The presence of functional $GABA_A$ receptors on cultured neurones was verified using patch-clamp recording. In neurones grown for 3 days in the presence of bicuculline (200 μM), the neuritic outgrowth of the hippocampal neurones

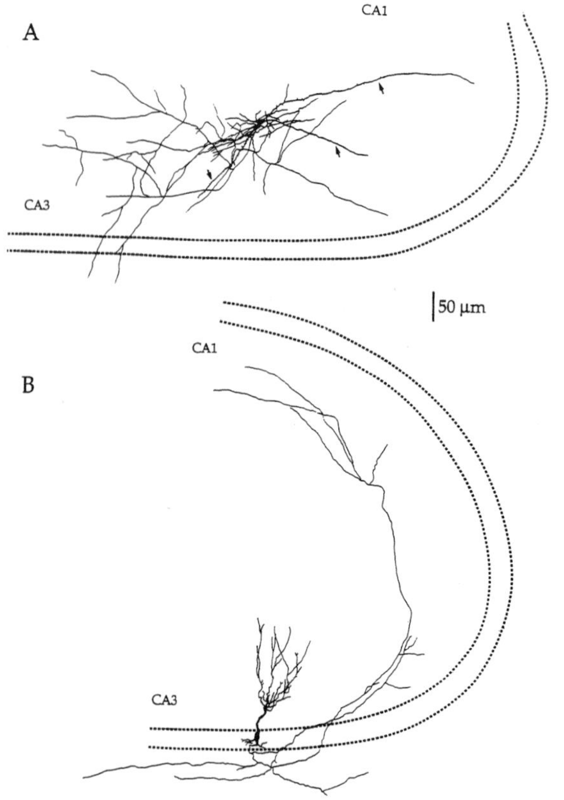

Fig. 6. Morphological development of interneurones and CA3 pyramidal neurones. Camera lucida reconstructions of interneurone (A) and CA3 pyramidal neurone (B) injected with biocytine at P5 using the whole cell patch technique. Dendrites of the interneurone are noted with arrows. Note the extent and complexity of the axonal network of the interneurone compared with the CA3 pyramidal neurone.

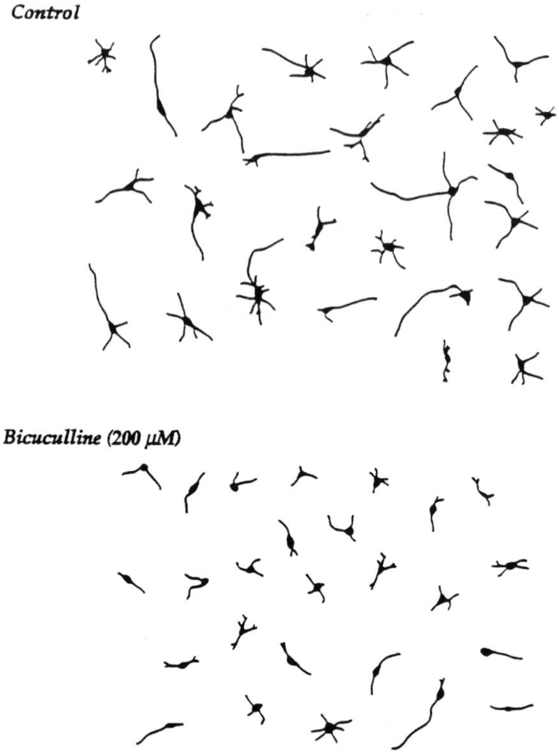

Fig. 7. Morphological alterations of the hippocampal neurones in culture induced by bicuculline. Cultures were grown for 3 days in the absence and in the presence of bicuculline (200 μM) and fixed for silver staining. For each experimental condition 25 neurones taken at random were drawn.

was profoundly altered (Fig. 7) (Barbin et al., 1993): there was a noticeable reduction in the total dendritic length (50%) as well as in the number of primary neurites (25%) and branching points (60%). This study therefore suggests that activation of $GABA_A$ receptors exerts a trophic action on the morphological development of hippocampal neurones (see also Prassad and Barker, 1990).

Theoretical considerations

Our studies show that: (1) activation of $GABA_A$ receptors, either by exogenously applied GABA or by direct electrical stimulation of the GABAergic interneurones, leads to membrane depolarisation in immature hippocampal pyramidal cells; (2) benzodiazepines potentiate the immature GABAergic transmission by acting on type-2 receptors, presumably localised on GABAergic interneurones; (3) postsynaptic $GABA_B$ inhibition is absent or poorly developed at early stages of postnatal development, whereas presynaptic inhibition is already present at birth; (4) $GABA_A$ receptors appear to play a trophic role, as well as providing the excitatory drive required for optimal pyramidal cell growth.

Depolarising action of $GABA_A$-receptors

Depolarising effects of $GABA_A$ receptors were confirmed in the CA3 region by other groups

(Fiszman et al., 1990; Xie and Smart, 1991), and recently extended to the developing CA1 region (Michelson and Lothman, 1989; Zhang et al., 1991) and immature visual cortex (Luhmann and Prince, 1991; Yuste and Katz, 1991), suggesting that similar processes occur in different brain structures during the maturation of GABAergic transmission.

Two hypotheses can be suggested to explain the depolarising responses of GABA. First, neonatal $GABA_A$-activated channels have a different ionic permeability than do the adult ones, most notably to bicarbonate ions (HCO_3^-) (Bormann et al., 1987; Kaila et al., 1989), which have an equilibrium potential more depolarised that the normal resting potential (Chesler, 1990). However, we found no change in the reversal potential of the exogenous GABA responses when the slices were superfused with HCO_3^--free ACSF, and several observations suggest that neonatal $GABA_A$ mediated responses depend strongly upon the chloride gradient (Ben-Ari et al., 1989). An alternative hypothesis is that the intracellular Cl^- concentration is higher in immature than in adult CA3 pyramidal neurones. This could be due either to (i) a delayed maturation of potassium/chloride co-transport mechanisms involved in chloride extrusion (Alvarez-Leefmans, 1990), as occurs in the CA1 region (Zhang et al., 1991), or (ii) to the presence at early stages of development of co-transport mechanisms involved in Cl^- accumulation, as in dorsal root ganglion cells (Alvarez-Leefmans et al., 1988) and rat sympathetic neurones (Ballany and Grafe, 1985), where GABAergic responses are also depolarising.

Modulation of GABA release by benzodiazepines

In young rats, benzodiazepine agonists such as midazolam and zolpidem do not enhance the responses evoked by exogenous application of GABA, suggesting that immature $GABA_A$ receptors differ in their subunit composition, and might not be associated to the γ subunit which confers benzodiazepine sensitivity to the $GABA_A$ receptor (Pritchett et al., 1989). The hypothesis of a different subunit composition for immature $GABA_A$ receptors is also suggested by the developmental loss of the zinc blocking action (Smart et al., 1991).

Our study also shows that midazolam, a BZ1 and BZ2 agonist, but not zolpidem (a selective BZ1 agonist), enhances the duration of the IPSP in young rats, thus indicating that only the BZ2 receptors are involved. Furthermore, midazolam does not potentiate the responses to exogenously applied GABA, suggesting that these BZ2 receptors are localised on GABAergic interneurones. Our working hypothesis was that midazolam regulates the release of GABA presynaptically through an action on calcium channels and we have effectively shown that BZ-R ligands affect calcium entry. Indeed, midazolam increases the duration of barium spikes in young and adult rats, while zolpidem reduces the barium spike duration in adult but has no effect in young rats. This observation cannot be reasonably attributed to general changes in cell excitability, since: (1) BZ-R ligands did not change the cell input resistance, (2) the barium spike duration, but not the spike height, was affected (which contrasts with the voltage-dependent inactivation), and (3) preliminary experiments using the voltage-clamp technique confirm these observations. Furthermore, since they were not supressed by bicuculline, the effects of BZ-R ligands do not appear to be mediated through a $GABA_A$-R.

In young rats, midazolam (a BZ1 and BZ2 agonist) increases the duration of the barium spikes and IPSPs while zolpidem (a selective BZ1 agonist) has no effect, indicating that BZ2 receptors are involved. These data thus support the interpretation that midazolam increases the duration of the IPSPs in young rats by modulating calcium entry on GABAergic terminals. Furthermore, these effects on calcium entry in immature neurones could be of special relevance in regard to the trophic role of calcium dependent processes (see below). Zolpidem and midazolam have distinct modes of action in adult rats. Midazolam increases calcium entry while zolpidem decreases it. However, the block of barium spikes by zolpidem could be due to a release of calcium from an intracellular pool, as is known to occur with kainate (Nistri and Cherubini, 1991). Such a bicuculline resistant reduction in the calcium spike duration by midazolam or diazepam was also observed in the CA1 adult pyramidal cells (Carlen et al., 1983) and in myenteric neurons (Cherubini and North, 1985) and it involves a high affinity BZ-R (nM range). Thus, there would exist in the hippocampus a high affinity BZR site not associated to a $GABA_A$-R but rather coupled to calcium channels.

Early development of pre (but not post) synaptic $GABA_B$ mediated responses

In the hippocampal CA3 region, postsynaptic $GABA_B$ inhibition is very poorly developed at early postnatal stages of development whereas presynaptic $GABA_B$ inhibition is already present at birth. A delayed maturation of the postsynaptic $GABA_B$ inhibition was also observed in different stuctures (Harrisson, 1990; Luhmann and Prince, 1991; Blanton and Kreigstein, 1992). In adult brains, activation of $GABA_B$ receptors leads to an increase in postassium conductance throught activation of a pertussis toxin sensitive G-protein. Therefore, the absence of postsynaptic $GABA_B$ inhibition could be due to an absence of $GABA_B$ receptors or potassium channels, but also to a delayed maturation of the second messenger system coupling the receptor to its channel. Interestingly, we have observed a similar developmental profile for the postsynaptic responses mediated by serotonin which, through $5HT_{1a}$ receptors, activates the same potassium channel (Andrade et al., 1986). Furthermore, a delayed development of postsynaptic $GABA_B$ receptors seems unlikely since the presynaptic receptors are present on both glutamatergic and GABAergic terminals, and a recent autoradiographic study has shown that $GABA_B$ binding-site levels peak on P3 in the CA3 region (Turgeon and Albin, 1992).

Morphological development of the GABAergic axonal network

At an early stage of development, GABAergic interneurones provide most of the excitatory synaptic drive onto immature CA3 pyramidal cells. Indeed, bicuculline fully blocked all spontaneous and evoked synaptic activity, while conventional excitatory synaptic events could be observed only when the strength or the frequency of stimulation was greatly increased. These observations suggest that the GABAergic network is well developed at an early stage of development, whereas the glutamatergic one is quiescent. In keeping with these observations, our morphological study shows that the axonal GABAergic network at P2 is already as complex as at P6–P8. Indeed, at early stages of development, axons from these interneurones extend widely in the stratum radiatum of the CA3 region, and often send collaterals into the CA3 stratum oriens and into the CA1 region. In contrast, the recurrent axonal network of the pyramidal neurones is weakly developed: even when collaterals were observed, they were directed to the CA1 fields (the Schaffer collaterals) or to the hilus, with very few recurrent collaterals within the CA3 fields.

Physiological function of the GABAergic interneurones at early stage of development

Our data show that GABA, acting on $GABA_A$ receptors, can affect the morphological development of hippocampal neurones. Ealier studies have also shown that GABA can promote neurite outgrowth and differentiation on differents types of neurones in culture (Spoerri and Wolff, 1981; Spoerri, 1988; Prasad and Barker, 1990) as well as increasing the density of organelles involved in protein synthesis and facilitating synaptogenesis in vivo (Spoerri, 1988). Furthermore, several studies have shown that growth cone motility and neurite outgrowth strongly depend on cytosolic calcium free concentration (Kater et al., 1988). An elevation of intracellular calcium concentration may originate from an influx through voltage sensitive channels or through receptor linked channel complex, as well as by release from intracellular stores (Miller, 1988). We therefore propose that, by depolarising immature CA3 pyramidal neurones, GABA will induce a rise in intracellular calcium concentration through activation of voltage sensitive calcium channels. In agreement, recent studies have shown that exogenous GABA increases intracellular calcium concentrations in immature cortical neurones (Yuste and Katz, 1991) and oligodendrocytes (Kirchhof and Kettenmann, 1992) while depolarising them. The GABA-induced rise in intracellular calcium concentration was blocked by voltage sensitive calcium channel blockers. Alternatively, however, GABA might also elevate intracellular calcium concentration via release from internal calcium stores, as it was recently shown to do on type I astrocytes in primary cell culture (Nilsson et al., 1993).

Summary and conclusion

The properties of neonatal GABAergic synapses were investigated in neurones of the hippocampal CA3 region. GABA, acting on $GABA_A$ receptors, provides most of the excitatory drive on immature CA3 pyramidal neurones at an early stage of development, whereas glutamatergic synapses (in particular, those mediated by AMPA receptors) are mostly quiescent. Thus, during the first postnatal week of life, bicuculline fully blocked spontaneous and evoked depolarising potentials, and $GABA_A$ receptor agonists depolarised CA3 pyramidal neurones. $GABA_A$ mediated currents also had a reduced sensitivity to benzodiazepines. In the presence of bicuculline, between P0 and P4, increasing the stimulus strength reveals an excitatory postsynaptic potential which is mostly mediated by NMDA receptors. During the same developmental period, pre- (but not post) synaptic $GABA_B$ inhibition is present. Intracellular injections of biocytin showed that the axonal network of the GABAergic interneurones is well developed at birth, whereas the pyramidal recurrent collaterals are only beginning to develop. Finally, chronic bicuculline treatment of hippocampal neurones in culture reduced the extent of neuritic arborisation, suggesting that GABA acts as a trophic factor in that period.

In conclusion, it is suggested that during the first postnatal week of life, when excitatory inputs are still poorly developed, $GABA_A$ receptors provide the excitatory drive necessary for pyramidal cell outgrowth. Starting from the end of the first postnatal week of life, when excitatory inputs are well developed, GABA (acting on both $GABA_A$ and $GABA_B$ receptors) will hyperpolarise the CA3 pyramidal neurones and, as in the adult, will prevent excessive neuronal discharges.

Our electrophysiological and morphological studies have shown that hippocampal GABAergic interneurones are in a unique position to modulate the development of CA3 pyramidal neurones. Developing neurones require a certain degree of membrane depolarisation, and a consequent rise in intracellular calcium, for stimulating neurite outgrowth; the GABAergic network, which develops prior to the glutamatergic one, appears to provide this depolarisation. Starting from the end of the first postnatal week of life, at a time when excitatory pathways are developing, GABA (acting on both $GABA_A$ and $GABA_B$ receptors) would reverse its action, and start to play its well-known role as an inhibitory neurotransmitter.

References

Alvarez-Leefmans, F.J. (1990) Intracellular Cl^- regulation and synaptic inhibition in vertebrate and invertebrate neurones. In: J.F. Alvarez-Leefmans and J.M. Russel (Eds.), *Chloride Channels and Carriers in Nerve, Muscle and Glial Cells,* Plenum Press, New York, pp. 109–150.

Alvarez-Leefmans, F.J., Gamino, S.M., Giraldez, F. and Nogureon, I. (1988) Intracellular chloride regulation in amphibian dorsal root ganglion neurones studied with ion-selective microelectodes. *J. Physiol.,* 406: 225–246.

Amaral, D.G. and Dent, J.A. (1981) Development of the mossy fibers of the dentate gyrus: I.A. light and electron microscopic study of the mossy fibers and their expansions. *J. Comp. Neurol.,* 195: 51–86.

Amaral, D.G. and Kurz, J. (1985) The time of origin of cells demonstrating glutamic acid decarboxylase-like immunoreactivity in the hippocampal formation of the rat. *Neurosci. Lett.,* 59: 33–39.

Andrade, R., Malenka, R.C and Nicoll, R. A. (1986) A G protein couples serotonine and $GABA_B$ receptors to the same channels in hippocampus. *Science,* 234: 1261–1265.

Ballany, K. and Grafe, P. (1985) An intracellular analysis of γ-aminobutyric-acid associated ion movements in rat sympathetic neurones. *J. Physiol.,* 365: 41–58.

Barbin, G., Pollard, H., Gaiarsa, J.L. and Ben-Ari, Y. (1993) Involvement of $GABA_A$ receptors in the outgrowth of cultured hippocampal neurons. *Neurosci. Lett.,* 152: 150–154.

Ben-Ari, Y., Krnjevic, K., Reiffenstein, R.J. and Reinhardt, W. (1981) Inhibitory conductance changes and action of γ-aminobutyrate in rat hippocampus. *Neuroscience,* 6: 2445–2463.

Ben-Ari, Y., Cherubini, E., Corradetti, R. and Gaiarsa, J.L. (1989) Giant synaptic potentials in immature rat CA3 hippocampal neurones. *J. Physiol.,* 416: 303–325.

Ben-Ari, Y., Rovira, C., Gaiarsa, J.L., Coradetti, R., Robain, O. and Cherubini, E. (1990) GABAergic mechanims in the CA3 hippocampal region during early postnatal life. In: J. Storm-Mathisen, J. Zimmer and O.P. otterson (Eds.), *Progress in Brain Research, Vol. 83, Understanding the Brain through the Hippocampus,* Elsevier, Amsterdam, pp. 313–321.

Ben-Ari, Y., Anikstejn, L. and Bregetovski, P. (1992) Protein kinase C modulation of NMDA currents: an important link for LTP induction. *Trends Neurosci.,* 15: 333–338.

Blanton, M.G. and Kreigstein, A.R. (1992) Properties of amino acids neurotransmitter receptors in embryonic cortical neurons when activated by exogenous and endogenous agonists. *J. Neurophysiol.,* 67: 1185–1200.

Bormann, J., Hamill, O.P. and Sackmann, B. (1987) Mechanism of anion permeation through channels gated by glycine and γ-aminobutyric acid in mouse cultured spinal neurones. *J. Physiol.,* 385: 243–286.

Bowery, N. G., Hill, D. R., Hudson, A. L., Doble, A., Middlemiss, D. N., Shaw, J. and Turnbull, M. (1980) (−)Baclofen decreases neurotransmitter release in the mammalian CNS by an action at a novel GABA receptor. *Nature,* 283: 92–94.

Carlen, P. L. Gurevich N. and Polc, P. (1983) Low-dose benzodiazepine neuronal inhibition: enhanced Ca^{2+}-mediated K^+-conductance. *Brain Res.,* 271: 358–364.

Chesler, M. (1990) The regulation and modulation of pH in the nervous system. *Prog. Neurobiol.,* 34: 410–427. Cherubini, E. and North, R. A. (1985) Benzodiazepines both enhance γ-aminobutyrate responses and decrease calcium action potentials in guinea-pig myenteric neurones. *Neuroscience,* 14: 309–315.

Cherubini, E., Gaiarsa, J.L. and Ben-Ari, Y. (1991) GABA: an excitatory transmitter in early postnatal life. *Trends Neurosci.,* 14: 515–519.

Collingridge, G.L. and Bliss, T.V.P. (1987) NMDA receptors — their role in long-term potentiation. *Trends Neurosci.,* 10: 289–293.

Coyle, J.T. and Enna, S.J. (1976) Neurochemical aspects of the ontogenesis of GABAergic neurons in the rat. *Brain Res.,* 111: 119–133.

Dolphin, A. C. and Scott, R. H. (1986) Inhibition of calcium currents in cultured rat dorsal root ganglion neurones by (−)-baclofen. *Br. J. Pharmacol.,* 88: 213–220.

Dolphin, A. C. and Scott, R. H. (1987) Calcium channel currents and their inhibition by (−)-baclofen in rat sensory neurones: modulation by guanine nucleotides. *J. Physiol.,* 386: 1–17.

Dutar, P. and Nicoll, R. A. (1988a) A physiological role for $GABA_B$ receptors in the central nervous system. *Nature,* 332: 156–158.

Dutar, P. and Nicoll, R. A. (1988b) Pre- and postsynaptic $GABA_B$ receptors in the hippocampus have a different pharmacological properties. *Neuron,* 1: 585–591.

Fiszmann, M., Novotny, E.A., Lange, G.D. and Barker, J.L. (1990) Embryonic and early postnatal hippocampal cells respond to nanomolar concentration of muscimol. *Dev. Brain Res.,* 53: 186–193.

Gaiarsa, J.L., Coradetti, R., Cherubini, E. and Ben-Ari, Y. (1990) the allosteric glycine site on the *N*-methyl-D-aspartate receptor modulates GABAergic-mediated synaptic events in neonatal rat CA3 hippocampal neurons. *Proc. Natl. Acad. Sci.,* 87: 343–346.

Gaiarsa, J.L., Coradetti, R., Cherubini, E. and Ben-Ari, Y. (1991) Modulation of GABA-mediated synaptic potentials by glutamatergic agonists in neonatal CA3 rat hippocampal neurons. *Eur. J. Neurosci.,* 3: 301–309.

Gaiarsa, J.L., Beaudoin, M. and Ben-Ari, Y. (1992) Effect of neonatal degranulation on the morphological development of rat CA3 pyramidal neurons: inductive role of mossy fibers on the formation of thorny excrescences. *J. Comp. Neurol.,* 321: 612–625.

Harrisson, N.L. (1990) On the presynaptic action of baclofen at inhibitory synapses between cultured rat hippocampal neurons. *J. Physiol.,* 422: 433–446.

Howe, J. R., Sutor, B. and Zieglgänsberger, W. (1987) Baclofen reduces post-synaptic potentials of rat cortical neurons by an action other than its hyperpolarising action. *J. Physiol.,* 384: 539–569.

Johnston, D. and Brown, T.M. (1984) Mechanisms of neuronal burst generation. In: P.A. Schwartzkroin and H.V. Wheal (Eds.), *Electrophysiology of Epilepsy,* Academic Press, New York, pp. 227–301.

Kaila, K., Pasternack, M., Saarikoski, J. and Voipio, J. (1989) Influence of GABA-gated bicarbonate conductance on potential, current and intracellular chloride in crayfish muscle fibers. *J. Physiol.,* 330: 163–165.

Kater, S.B., Mattson, M.P., Cohan, C. and Connor, J. (1988) Calcium regulation of the neuronal growth cone. *Trends Neurosci.,* 11: 315–321.

Kirchhoff, F. and Kettenmann, H. (1992) GABA triggers a $[Ca^{2+}]_i$ increase in murine precursor cells of the oligodendrocyte lineage. *Eur. J. Neurosci.,* 4: 1049–1058.

Lippa, A.S., Beer, B., Sano, M.C., Vogel, R.A. and Meyerson, L.R. (1981) Differential ontogeny of type 1 and type 2 benzodiazepine receptors. *Life Sci.,* 28: 2343–2347.

Lübbers, K., Wolf, J.R. and Frotscher, M. (1985) Neurogenesis of GABAergic neurons in the rat dentate gyrus: a combined autoradiographic and immunocytochemical study. *Neurosci. Lett.,* 62: 317–322.

Luhmann, H.J. and Prince, D.A. (1991) Postnatal maturation of GABAergic system in rat neocortex. *J. Neurophysiol.,* 65: 247–263.

Mattson, M.P. (1988) Neurotransmitters in the regulation of neuronal cytoarchitecture. *Brain Res. Rev.,* 13: 179–212.

Mayer, M.L. and Westbrook, G.L. (1987) The physiology of excitatory amino acids in the vertebrate central nervous system. *Prog. Neurobiol.,* 28: 197–276.

Michelson, H.B. and Lothman, E.W. (1989) An in vivo electrophysiological study of the ontogeny of excitatory and inhibitory processes in the rat hippocampus. *Dev. Brain Res.,* 47: 113–122.

Miller, R.J. (1988) Calcium signaling in neurons. *Trends Neurosci.,* 11: 415–419.

Newberry, N. R. and Nicoll, R. A. (1984) Direct hyperpolarising action of baclofen on hippocampal pyramidal cells. *Nature,* 308: 450–452.

Nilsson, M., Erikson, P.S., Rönnbäck, L. and Hnsson, E. (1993) GABA induces Ca^{2+} transients in astrocytes. *Neuroscience,* 53: 605–614.

Nistri, A. and Cherubini, E. (1991). Depression of a sustained calcium current by kainate in rat hippocampal neurones in vitro. *J. Physiol.,* 435: 465–481.

Nowak, K.L., Bregestovski, P., Ascher, P., Herbert, A. and Prochiantz, A. (1984) Magnesium gates glutamate activated channels in mouse central neurons. *Nature (London)* 307: 462–464.

Pokorny, J. and Yamamoto, T. (1981) Postnatal ontogenesis of the hippocampal CA1 area in rats. II. Development of ultrastructure in stratum lacunosum and molecular. *Brain Res. Bull.,* 7: 121–130.

Prassad, A. and Barker, J.L. (1990) Functional $GABA_A$ receptors are critical for survival and process outgrowth of embryonic chick spinal cord cells. *Soc. Neurosci.,* 16: 647.

Pritchett, D.B., Sontheimer, H., Shivers, B.D., Ymer, S., Kettenmann, H., Schofield, P.R. and Seeburg, P.H. (1989). Importance of a novel $GABA_A$ receptor subunit for benzodiazepine pharmacology. *Nature,* 338: 582–585.

Rovira, C. and Ben-Ari, Y. (1991) Benzodiazepines do not potentiate GABA responses in neonatal hippocampal neurons. *Neurosci. Lett.,* 123: 265–268.

Rovira, C. and Ben-Ari, Y. (1993) Developmental study of benzodiazepine effects on monosynaptic $GABA_A$-mediated IPSPs of rat hippocampal neurons. *J. Neurophysiol.,* 70: 3: 1076–1085.

Richter, K. and Wolf, G. (1990) High-affinity glutamate uptake of rat hippocampus during postnatal development: a quantitative autoradiographic study. *Neuroscience,* 34: 49–55.

Rozenberg, F., Robain, O., Jardin, L. and Ben-Ari, Y. (1989) Distribution of GABAergic neurons in late and early postnatal rat hippocampus. *Dev. Brain Res.,* 50: 177–187.

Sivilotti, L. and Nistri, A. (1991) GABA receptor mechanism in the central nervous system. *Prog. Neurobiol.,* 36: 35–92.

Smart, T.G., Moss, S.J., Xie, X. and Hganir, R.L. (1991) $GABA_A$-receptors are differentially sensitive to zinc: dependance on subunit composition. *Br. J. Pharmacol.,* 103: 1837–1839.

Spoerri, P.E. (1988) Neurotrophic effects of GABA in culture of embryonic chick brain and retina. *Synapse,* 2: 11–22.

Spoerri, P.E. and Wolff, J.R. (1981) Effect of GABA administration on murine neuroblastoma cells in culture. *Cell Tissue Res.,* 218: 567–579.

Study, R.E. and Barker, J.L. (1981) Diazepam and (−)pentobarbital: fluctuation analysis reveals different mechanisms for potentiation of γ-aminobutyric acid responses in cultured central neurons. *Proc. Natl. Acad. Sci. USA,* 78: 7180–7184.

Thompson, S.M. and Gähwiler, B.H. (1992) Comparison of the actions of baclofen at pre- and postsynaptic receptors in the rat hippocampus in vitro. *J. Physiol.,* 451: 329–345.

Turgeon, S.M. and Albin, R.L. (1992) Ontogeny of $GABA_B$ binding in rat brain. *Soc. Neurosci.,* 18: 279.15.

Xie, X. and Smart, T.G. (1991) A physiological role for endogenous zinc in rat hippocampal synaptic neurotransmission. *Nature,* 349: 521–524.

Yuste, R. and Katz, L.C. (1991) Control of postsynaptic Ca^{2+} influx in developing neocortex by excitatory and inhibitory neurotransmitters. *Neuron,* 6: 333–344.

Zhang, L., Spigelman, I. and Carlen, P.L. (1991) Development of GABA-mediated, chloride-dependent inhibition in CA1 pyramidal neurones of immature rat hippocampal slices. *J. Physiol.,* 444: 25–49.

SECTION III

From Neuron to Network

B. Sensory Stimulation and Cortical Maturation

J. van Pelt, M.A. Corner H.B.M. Uylings and F.H. Lopes da Silva (Eds.)
Progress in Brain Research, Vol 102

CHAPTER 18

Regulation of *N*-methyl-D-aspartate (NMDA) receptor function during the rearrangement of developing neuronal connections

Magdalena Hofer and Martha Constantine-Paton

Department of Biology, Yale University, New Haven, CT 06511, U.S.A.

Introduction

The development of mature synaptic connectivity in the central nervous system is the result of a multistage process. Cell-surface cues and initial synaptic contacts set the stage for a dynamic phase of synaptogenesis during which contacts within a developing neuropil are rearranged in a pattern partially imposed by neuronal activity. The final sharpening and fine-tuning of synaptic contacts involves elimination of incorrect and redundant terminals and stabilization of appropriate ones.

Donald Hebb first proposed, as part of his theory for associative learning, that synapses can undergo strengthening whenever activity in presynaptic cells occurs simultaneously with that in the postsynaptic cell (Hebb, 1949). The *N*-methyl-D-aspartate (NMDA) type of glutamate receptor is believed to play an important role in the process of activity-dependent synaptic restructuring (Constantine-Paton et al., 1990). As it gets activated only when the postsynaptic cell membrane is already depolarized (Mayer et al., 1984; Mayer and Westbrook, 1985; MacDermott et al., 1986) it is capable of detecting correlated afferent activity. Ca^{2+} influx through the integral ion channel may initiate the selective stabilization of simultaneously active synapses by a Hebbian-like mechanism (Constantine-Paton et al., 1990; Shatz, 1991; Cline and Tsien, 1991; Yuste and Katz, 1991).

The potential of synapses to reorganize is, in most neuronal systems, restricted to a short period of time in early life. The question therefore arises: what are the developmental mechanisms that permit synaptic plasticity and that terminate it? What is the molecular basis for the critical difference between the pronounced structural plasticity of the developing brain and the less plastic properties of many regions in the mature brain? The answer to these questions is largely unknown. Accumulating evidence, however, suggests that the molecular properties of NMDA receptors become altered during development, and these changes may define periods of developmental plasticity. This paper reviews some of the data illustrating developmental changes in the properties of the NMDA receptor complex that may reflect alterations in the ability of synapses to rearrange. It will focus on the visual system of amphibia and rodents for a discussion of cellular and molecular changes that may be involved in the mutability of synaptic connections.

Developmental changes in NMDA receptors and synaptic plasticity

If NMDA receptors are involved in developmental stabilization and sorting of synapses, then some aspects of their function should change in parallel with anatomical and physiological alterations in developmental plasticity. Although transient increases in glutamate receptor function and density appear to be common during development, the correlation with periods of enhanced synaptic plasticity is less clear. One of the systems in which a relationship between NMDA receptor efficacy and some form of synaptic plasticity was demonstrated is the cat visual cortex (reviewed by Fox and Daw, 1993). A decrease of the NMDA receptor component of the visual response occurs with increasing age in layers IV, V and VI of cat visual cortex (Tsumoto et al., 1987; Fox et al., 1989). This down-regulation of NMDA receptor efficacy correlates with the end of the period of segregation of geniculate axon terminals into ocular dominance columns in layer IV (LeVay et al., 1978) but not with the reduced plasticity of layer IV as measured in the monocular deprivation paradigm (Olson and Freeman, 1980). The relationship between NMDA receptor protein levels, as measured in receptor binding assays, and critical periods for monocular deprivation in cat visual cortex also remains perplexing. Increases in the levels of NMDA receptor binding appear to parallel the critical period for monocular deprivation (Bode-Greuel and Singer, 1989; Reynolds and Bear, 1991; Gordon et al., 1991), but they do not correspond to the period of sensitivity of the visual response of neurons in the deep cortical layers to NMDA receptor antagonists (Fox et al., 1989).

Rearing kittens in the dark, a procedure which prolongs physiological plasticity (Cynader and Mitchell, 1980) but not the anatomically analyzed formation of ocular dominance columns (Mower et al., 1985), does not delay the decrease in NMDA binding sites (Bode-Greuel and Singer, 1989; Gordon et al., 1991). This treatment does, however, prevent the decline in the NMDA receptor potency (Fox et al., 1991). These discrepancies may, in part, be caused by technical difficulties associated with receptor binding studies. For example, [^{3}H]MK-801 binding of cortical homogenates addresses changes in the number of receptors as well as in agonist affinity. However, this technique does not resolve individual cortical layers (Reynolds and Bear, 1991; Gordon et al., 1991). Layer by layer changes in NMDA receptor binding have been resolved by receptor autoradiography using [^{3}H]glutamate displaced by APV (Bode-Greuel and Singer, 1989). It seems likely that some of these inconsistencies will be reconciled as the subunit composition of NMDA receptors in different layers and at different ages become defined.

Examples from other systems indicate that developmental changes in the efficacy of the receptor, their density and their function are widespread. In hippocampus, the effectiveness of the NMDA receptor is enhanced in young animals compared with adult (Hamon and Heinemann, 1988; McDonald et al., 1988). NMDA receptor number as assayed by ligand binding was shown to increase transiently during postnatal days 6–10 and then decrease to adult levels by day 13 in rat hippocampus (Tremblay et al., 1988). However, NMDA receptor-mediated hippocampal long-term potentiation (LTP: a long lasting change in synaptic efficacy) persists throughout adulthood. In cerebellum, a transiently enhanced sensitivity of Purkinje and granule cell responses to NMDA corresponds to the period of synapse elimination during development (Dupont et al., 1987; Garthwaite et al., 1987). Other functional properties of the NMDA receptor are changed during development. The Mg^{2+} sensitivity and voltage dependence of the NMDA receptor have been shown to be lower in hippocampal neurons from immature rats than those from adult ones (Ben-Ari and Cherubini, 1988; Bowe and Nadler, 1990; Morrisett et al., 1990; Kleckner and Dingledine, 1991), whereas the sensitivity for glycine is higher in young rats than in adult ones (Kleckner and Dingledine, 1991). These studies suggest that NMDA receptor from immature animals are more easily excitable and may pass more current in comparison with adult animals.

Recently, a developmental change in the duration of NMDA receptor mediated currents was observed in collicular neurons: in 10- to 15-day-old animals the NMDA currents are several times longer than in 23- to 33-day-old animals (213 $\pm$ 57 msec vs. 85 $\pm$ 37 msec) (Hestrin, 1992). A similar change in NMDA channel open time was observed in the rat visual cortex. Moreover, this change is apparently activity-dependent, as it can be delayed by either dark rearing or TTX treatment (Carmignoto and Vicini, 1992). This change in NMDA receptor channel kinetics is similar to the observation at the neuromuscular junction where the time course of the embryonic end-plate current is slow compared with that of the adult (Sakmann and Brenner, 1978). In the muscle nicotinic acetylcholine receptor, a subunit replacement results in the transition from the fetal to the adult form, leading to a decrease in the mean open time of the ion channel (Mishina et al., 1986). An analogous alteration in the subunit composition of the NMDA receptor may define its channel open time in a similar manner.

In an attempt to obtain information about the postulated changes in the NMDA receptor subunit composition, Williams et al. (1993) measured the affinity of rat brain NMDA receptors to the non-competitive antagonist ifenprodil, that may modulate polyamine sites. They found a uniformly high affinity in neonatal animals whereas, in 21-day-old or adult animals, a second population of receptors with a 100-fold lower affinity appears. Complementary voltage-clamp recording of *Xenopus* oocytes expressing various stoichiometric configurations of NMDA receptor subunits revealed potent inhibition by ifenprodil of responses of homomeric NR1 and heteromeric NR1/NR2B receptors, but not of NR1/NR2A receptors, suggesting that differences in subunit composition are the basis for the observed change in antagonist affinity (Williams et al., 1993). In addition, in situ hybridization studies have revealed differential developmental regulation of the spatial and temporal expression pattern of mRNA coding for various NMDA receptor subunits, supporting the notion that changes in the subunit composition of the NMDA receptor channel complex take place during development (Watanabe et al., 1992). Taken together, these findings from many different brain regions make clear that a number of aspects of NMDA receptor function are developmentally regulated. Whether these alterations correspond to structural changes and the ability of synapses to reorganize still remains to be demonstrated.

Regulation of NMDA receptor efficacy in the frog retinotectal system

The visual system of cold-blooded vertebrates has served as an excellent model for studying the chemospecific and activity-dependent processes that lead to a precisely organized projection. The retinotectal projection in amphibians and fish retains structural plasticity well into maturity as retinal terminals are constantly shifting over the tectal surface in order to maintain retinotopy throughout larval development, despite asymmetric patterns of retinal and tectal cell proliferation. A preparation that anatomically assays the operation of the mechanisms that mediate synaptic competition has been developed using *Rana pipiens* (leopard frog) embryos. By implanting a third eye primordium into a tadpole embryo and forcing two afferent projections to share one tectal lobe, a series of alternating eye-specific afferent termination zones or stripes is created in the optic tectum (Constantine-Paton and Law, 1978; Law and Constantine-Paton, 1981). NMDA receptors have been shown to be required for the formation and maintenance of these eye-specific stripes, as chronic treatment of the tecta of three-eyed tadpoles with the specific NMDA receptor blocker APV for 4–6 weeks results in the desegregation of the normal striped pattern (Cline et al., 1987). In contrast, chronic treatment of three-eyed tadpoles with NMDA (the agonist) sharpens stripe boundaries (Cline and Constantine- Paton, 1990). Even though the initial formation of the retinotectal projection is not activity-dependent in cold-blooded animals (Harris, 1980; Harris, 1984; Stuermer et al., 1990), treatment of normal two-eyed tadpoles with APV dramatically disrupts the topography of the retinotectal map (Cline and Constantine-Paton, 1989).

Complementing these anatomical studies, Debski et al. (1991) used a physiological approach in order to investigate the state of NMDA receptors in animals chronically treated with NMDA, and found that chronic treatment with NMDA decreases the electrophysiologically measured sensitivity of the optic tectum to applied NMDA (Debski et al., 1991). Receptor binding autoradiography on similarly treated tissue failed to reveal any change in binding site density. Thus, there appears to be a change in the effectiveness of individual NMDA receptors rather than a decrease in receptor number. The lowered sensitivity to NMDA is a long-term change, and remains stable for hours even after the agonist has been removed, thus differing from the rapid desensitization reported by others (Trussell et al., 1988; Mayer et al., 1989). This experimentally induced decrease in NMDA receptor effectiveness may be analogous to the naturally occurring reduction in receptor function that is correlated with the end of some periods of visual plasticity. It may be reflected anatomically by the sharpening of stripes, that represent a further restriction of the intermingling of axon branches from the two eyes. A similar limiting of synapse stabilization to areas where afferent activity is most highly correlated must occur as a young brain matures in order to ensure that only highly correlated, efficient contacts persist into adulthood. If NMDA receptor activation is indeed the critical trigger for such synapse stabilization, then a decrease in its effectiveness with age would be sufficient to restrict the stabilization of new contacts in older brain. Thus, new contacts would be less likely to be established in mature brains, and synaptic plasticity would be reduced.

Recent whole-cell patch clamp analysis of tadpole tectal neurons have demonstrated that these neurons have both non-NMDA and NMDA receptor-mediated currents, and that NMDA receptors on these tectal neurons are not responsible for the bulk of normal excitatory transmission (Hickmott and Constantine-Paton, 1993) . The same study also examined GABAergic inhibitory currents in tectal neurons and suggested that inhibition might modulate NMDA receptor-mediated excitatory responses. Similarly, in rat cortex a transient manifestation of strong NMDA receptor-mediated potentials is a consequence of the relative immaturity of GABA-mediated inhibition. Further development is accompanied by a decrease in NMDA receptor mediated excitation, thus providing a temporal window of enhanced sensitivity for LTP induction (Luhmann, 1990; Luhmann and Prince, 1990). These observations have implications for the control of synaptic plasticity during development: the relative contribution of NMDA and non-NMDA receptors to postsynaptic responses appears to change depending upon alterations in inhibitory input and these changes may affect the ability of synapses to rearrange during development.

Regulation of NMDA receptors at the mRNA level in the rat superior colliculus

To get insight into the mechanisms of topographic map formation in warm-blooded animals, we have studied the development of retinocollicular projections in rats. The period of rearrangement of retinal terminals in the superior colliculus is restricted to the first 2 weeks of postnatal development, and several clearly defined stages of this process have been described (Simon and O'Leary, 1992) (see Fig. 1). The developing retinocollicular projection therefore provides a more suitable system for studying factors that may control the onset and the termination of plasticity than does the amphibian retinotectal projection. Developing rat retinal axons do not grow directly to their topographically appropriate location in the SC, but initially mistarget widely (Simon and O'Leary, 1992). This is in contrast to the development of retinal projections in cold-blooded vertebrates where a precisely organized retinotopic map is present from early stages, and coherence of this projection is maintained throughout a subsequent period of synapse sorting and retinal and tectal growth (Gaze et al., 1979; Holt and Harris, 1983; Constantine-Paton and Reh, 1985; Stuermer, 1988). Disruption of activity patterns interferes with the maintenance and regeneration of these maps in amphibia and fish, but it does not affect the initial organization of retinal projections in these species (Harris, 1980, 1984; Stuermer et al., 1990).

In rats, the establishment of retinocollicular order requires a major remodeling of the early, diffuse retinal projection prior to eye opening and pattern vision. A large-scale elimination of aberrantly positioned branches and arbors, along with an increase in branching and arborization at topographically correct locations in the SC, occurs during the first 2 postnatal weeks (Simon and O'Leary, 1992). This refinement of the retinotopic map appears to depend upon normal NMDA receptor function: chronic blockade of NMDA receptors in the SC from birth disrupts the remodeling of the retinocollicular projection so that mistargeted axons remain and arborize at topographically incorrect sites (Simon et al., 1992). Since NMDA receptors are involved in retinotopic map formation in the SC, we asked whether developmental changes in their properties would parallel this process. As a first step in addressing this issue, we investigated the regulation of mRNA coding for NMDA and non-NMDA receptor subunits in the developing rat retinocollicular system in order to see whether NMDA receptor expression and the period of topographic map refinement in the SC are temporally correlated.

Figure 2 shows the temporal expression pattern of mRNA coding for various glutamate receptor subunits in the developing rat superior colliculus. It is obvious that the mRNAs coding for the NMDA receptor subunits NR1 and NR2B, the AMPA receptor subunit GluR2 and the metabotropic receptor mGR1 are regulated differentially, suggesting different roles in the development of the retinocollicular projection. The levels of mRNA coding for the NMDA receptor subunit NR1 are low during the first postnatal week (Fig. 1a) with a marked rise of NR1 mRNA occurring between postnatal day (P) P6 and P12. This increase in NR1 gene expression coincides with the completion of the refinement of the retino- and corticocollicular projections. It also seems to parallel an increase in synaptic density within the superficial layers of the SC (Warton and McCart, 1989) and the development of electrophysiological activity in the SC (Molotchnikoff and Itaya, 1993). If NMDA receptors are involved in the process of retinocollicular map development, we may expect their density to be elevated during this period. However, we find that NR1 mRNA expression reaches high levels only during the final stages of synaptic rearrangements of afferent projections to the SC. Assuming that mRNA levels of the NR1 subunit reflect the density of the NMDA receptor complex, our results may indicate that NMDA receptors are less important for the early stages of map development, which involve widespread collateralization and the onset of synaptogenesis, than for the later stages when mistargeted projections are completely withdrawn.

The expression pattern of the NMDA receptor subunit NR2B (Monyer et al., 1992) differs from the NR1 mRNA (Fig. 1b). Fairly high levels are present at early postnatal ages followed by a decrease to adult levels. It is conceivable that the high NR2B mRNA levels during the period of retinocollicular map refinement provide a condition permissive for synapses to rearrange, and that the drop in NR2B mRNA levels restricts this ability. In contrast to NR1 and NR2B mRNA, the mRNA expression pattern of the non-NMDA receptor subunit GluR2 in the developing SC was

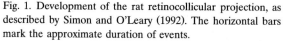

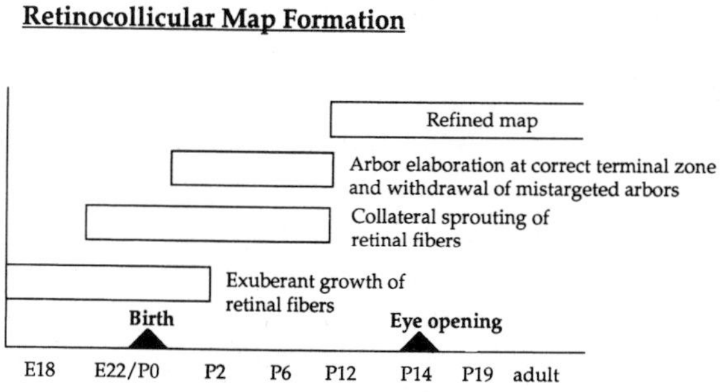

Fig. 1. Development of the rat retinocollicular projection, as described by Simon and O'Leary (1992). The horizontal bars mark the approximate duration of events.

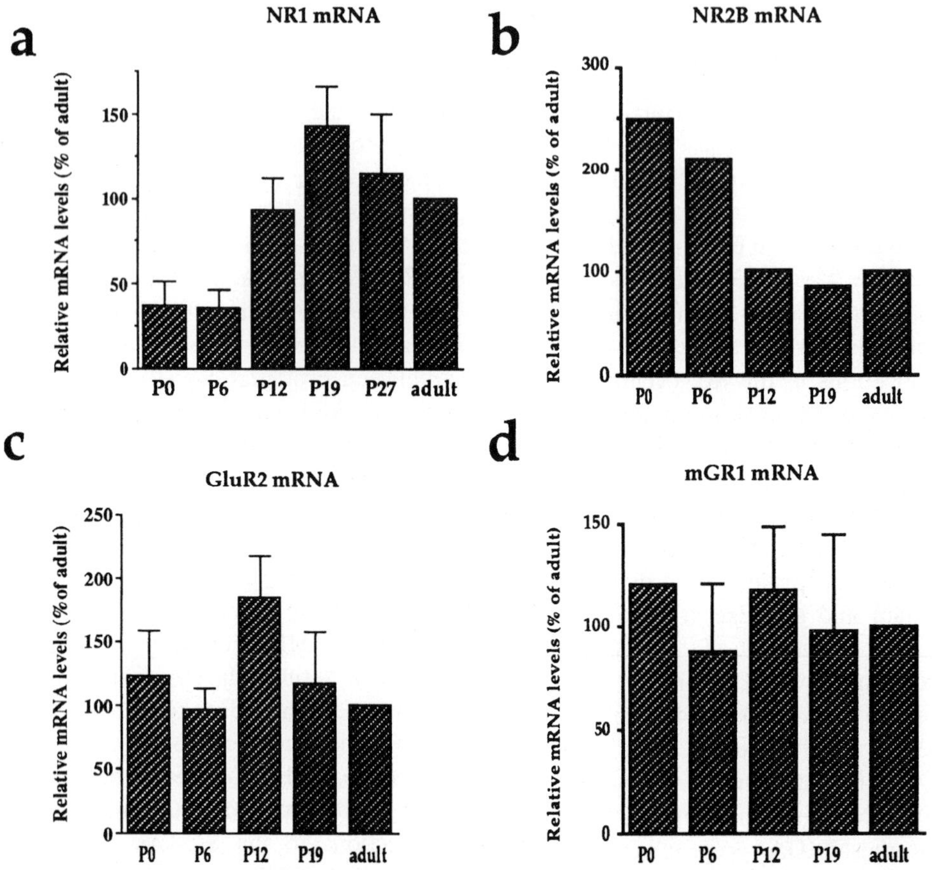

Fig. 2. Expression patterns of mRNA coding for the NMDA receptor subunits NR1 (a) and NR2B (b), the AMPA receptor subunit GluR2 (c) and the metabotropic receptor mGR1 (d). RNA from the superficial superior colliculus was analyzed by Northern blots and quantitative measurements of mRNA levels were obtained by densitometric scanning of autoradiographic signals.

not closely correlated with synaptic changes (Fig. 1c). We chose to examine the expression of GluR2 mRNA because GluR2, when coexpressed with the AMPA receptor subunits GluR1 or GluR3 in *Xenopus* oocytes, decreases the Ca^{2+} permeability of the channel formed by these subunits (Hollman et al., 1991). Thus, low levels of GluR2 subunits could endow non-NMDA ionotropic receptors with a high Ca^{2+}-permeability similar to that of NMDA receptors, but without the voltage-gating property that is presumably essential for detection of correlated activity (Constantine-Paton et al., 1990; Bourne and Nicoll, 1993). The observation that the time course of GluR2 mRNA expression appears to be unrelated to the anatomically defined periods of synaptic rearrangement and topographic map refinement in the SC suggests that this subunit is involved more generally in synaptic transmission than is the more tightly regulated NMDA receptor subunit NR1. A similar result was found with the regulation of the mRNA coding for a metabotropic glutamate receptor, mGR1 (Masu et al., 1991). There was no clear correlation between the mRNA expression pattern of mGR1 in developing SC and synaptogenic events (Fig. 1d).

After it was shown that chronic APV-blockade of NMDA receptors in the SC disrupts the development of the retinotopic map in SC (Simon et al., 1992) we examined whether NR1 mRNA levels are affected by this treatment as well. Our results showed that NR1 mRNA levels are dramatically reduced at P12 and P19 after chronic APV application. In fact, NR1 mRNA levels in treated animals were similar to the levels measured in normal P0 or P6 animals. Thus, chronic NMDA receptor blockade appears to prevent the developmental increase of NR1 mRNA levels. If the NR1 subunit is a common element of the native NMDA receptor, as suggested by its widesread distribution (Kutsuwada et al., 1992; Monyer et al., 1992), and if its transcript levels reflect the density of the whole NMDA receptor complex, our findings indicate that chronic exposure to APV causes a reduction in the number of NMDA receptors. During early postnatal development of the SC, neuronal activity, mediated through NMDA receptors, may therefore be necessary to trigger the developmental rise of NR1 mRNA, while disruption of activity may prevent the normal developmental increase of NR1 mRNA expression. Our result indicates that the suppression of the normal increase of the subunit NR1 may, in part, account for the disruption of the refinement of the retinocollicular projection. Thus, the current data support the hypothesis that the NMDA receptor is specifically involved in structural modifications of the developing nervous system, and that activity-dependent regulation of NMDA receptor subunits at the mRNA level may be an important factor in this process.

Summary

There is evidence from a number of studies that the molecular and biophysical properties of NMDA receptors are altered during normal development. A temporal correlation with changes in NMDA receptor efficacy and periods of synaptic plasticity has been demonstrated in several systems, suggesting that NMDA receptors have a critical function in determining periods of synaptic plasticity. Data from our laboratory demonstrate reduced NMDA sensitivity of the tectal evoked potential following chronic application of NMDA to the tadpole tectum, a treatment that may mimic a naturally occurring mechanism for limiting neuronal plasticity to certain stages of development. Our analysis of the expression pattern of mRNA coding for various glutamate receptor subunits in the rat retinocollicular system establishes that differential regulation of NMDA receptor subunits at the mRNA level could be a molecular basis for changes in biophysical and pharmacological properties of the NMDA receptor complex. However, even though the NMDA receptor is the best studied candidate to function as a 'plasticity switch', there are large gaps in our understanding of the complete set of factors that control the ability of synapses to rearrange during development.

Acknowledgments

The authors would like to thank Dr. J. Boulter for the GluR2 clone, Dr. S. Nakanishi for the NR1 and mGR1 cDNA and Dr. P. Seeburg for the NR2B cDNA. This work was supported by EMBO longterm fellowship to M.H., and by NEI grant EY06039 to M.C.-P.

References

Ben-Ari, Y. and Cherubini, E. (1988) Changes in voltage dependence of NMDA currents during development. *Neurosci. Lett.*, 94: 88–92.

Bode-Greuel, K. and Singer, W. (1989) The development of *N*-methyl-D-aspartate receptors in cat visual cortex. *Dev. Brain Res.*, 46: 197–204.

Bourne, H.R. and Nicoll, R. (1993) Molecular machines integrate coincident synaptic signals. *Cell / Neuron*, 72/10 (Suppl.): 65–75.

Bowe, M.A. and Nadler, J.V. (1990) Developmental increase in the sensitivity to magnesium of NMDA receptors on CA1 hippocampal pyramidal cells. *Dev. Brain Res.*, 56: 55–61.

Carmignoto, G. and Vicini, S. (1992) Activity-dependent decrease in NMDA receptor responses during development of the visual cortex. *Science*, 258: 1007–1011.

Cline, H.T. and Constantine-Paton, M. (1989) NMDA receptor antagonists disrupt the retinotectal topographic map. *Neuron*, 3: 413–426.

Cline, H.T. and Constantine-Paton, M. (1990) NMDA receptor drug treatment alters RGC terminal morphology in vivo. *J. Neurosci.*, 10: 1197–1216.

Cline, H.T. and Tsien, R.W. (1991) Glutamate-induced increases in intracellular Ca^{2+} in cultured frog tectal cells mediated by direct activation of NMDA receptor channels. *Neuron*, 6: 259–267.

Cline, H.T., Debski, E. and Constantine-Paton, M. (1987) NMDA receptor antagonist desegregates eye specific strips. *Proc. Natl. Acad. Sci. USA*, 84: 4342–4345.

Constantine-Paton, M. and Law, M.I. (1978) Eye-specific termination bands in tecta of three-eyed frogs. *Science*, 202: 639–641.

Constantine-Paton, M. and Reh, T. (1985) Dynamic synaptic interactions during the formation of a retinotopic map. In: *Neurobiology: Molecular Biological Approaches to Understanding Neuronal Function and Development*. E.P. O'League. (Ed.), New York: Alan R. Liss, Inc., pp. 151–168.

Constantine-Paton, M., Cline, H.T. and Debski, E.A. (1990) Patterned activity, synaptic convergence and the NMDA receptor in developing visual pathways. *Annu. Rev. Neurosci.*, 13: 129–154.

Cynader, M. and Mitchell, D.E. (1980) Prolonged sensitivity to monocular deprivation in dark-reared cats. *J. Neurophysiol.*, 43: 1026–1040.

Debski, E.A., Cline, H.T., McDonald, J.W. and Constantine-Paton, M. (1991) Chronic application of NMDA decreases the NMDA sensitivity of the evoked potential in the frog. *J. Neurosci.*, 11: 2947–2957.

Dupont, J.L., Gardelle, R. and Crepel, F. (1987) Postnatal development of the chemo-sensitivity of rat cerebellar Purkinje cells to excitatory amino acids: an in vitro study. *Dev. Brain Res.*, 34: 59–68.

Fox, K. and Daw, N.W. (1993) Do NMDA receptors have a critical function in visual cortical plasticity? Trends Neurol. Sci., 16: 116–122.

Fox, K., Sato, H. and Daw, N. (1989) The location and function of NMDA receptors in cat and kitten visual cortex. *J. Neurosci.*, 9: 2443–2454.

Fox, K., Daw, N., Sato, H. and Czepita, D. (1991) Dark-rearing delays the loss of NMDA receptor function in the kitten visual cortex. *Nature*, 350: 342–344.

Garthwaite, G., Yamini, B.J. and Garthwaite, J. (1987) Selective loss of Purkinje and granule cell responsiveness to *N*-methyl-*D*-apartate in rat cerebellum during development. Dev. *Brain Res.*, 36: 288–292.

Gaze, R.M., Keating, M.J. Ostberg, A. and Chung, S.H. (1979) The relationship between retinal and tectal growth in larval *Xenopus*. Implication for the development of the retinotectal projection. *J. Embryol, Exp. Morphol.* 53: 103–143.

Gordon, B., Daw, N.W. and Parkinson, D. (1991) The effect of age on binding of MK-801 in the cat visual cortex. *Dev. Brain Res.*, 62: 61–68.

Hamon, B. and Heinemann, V. (1988) Developmental changes in neuronal sensitivity to excitatory amino acids in area CA1 of the rat hippocampus. *Dev. Brain Res.*, 38: 286–290.

Harris, W.A. (1980) The effect of eliminating impulse activity on the development of the retinotectal projection in salamanders. *J. Comp. Neurol.*, 194: 303–317.

Harris, W.A. (1984) Axonal pathfinding in the absence of normal pathways and impulse activity. *J. Neurosci.*, 4: 1153–1162.

Hebb, D.O. (1949) *Organization of Behavior*, John Wiley, New York.

Hestrin, S. (1992) Developmental regulation of NMDA receptor-mediated synaptic currents at a central synapse. *Nature*, 357: 686–689.

Hickmott, P.W. and Constantine-Paton, M. (1993) The contribution of NMDA, non-NMDA and GABA receptors to postsynaptic responses in neurons of the optic tectum. *J. Neurosci.* 13: 4339–4353.

Hollman, M., Hartley, M. and Heinemann, S. (1991) Ca^{2+} permeability of KA-AMPA-gated glutamate receptor channels depends on subunit composition. *Science*, 252: 851–853.

Holt, C.E. and Harris, W.A. (1983) Order in the initial tectal map in *Xenopus*: a new technique for labeling growing nerve fibres. *Nature*, 301: 150–152.

Kleckner, N.W. and Dingledine, R. (1991) Regulation of hippocampal NMDA receptors by magnesium and glycine during development. *Mol. Brain Res.*, 11: 151–159.

Kutsuwada, T., Kashiwabuchi, N., Mori, H., Sakimura, K., Kushiya, E., Araki, K., Meguro, H., Masaki, H., Kumanishi, T., Arakawa, M. and Mishina, M. (1992) Molecular diversity of the NMDA receptor channel. *Nature*, 358: 36–41.

Law, M.I. and Constantine-Paton, M. (1981) Anatomy and physiology of experimentally produced striped tecta. *J. Neurosci.*, 1: 741–759.

LeVay, S., Stryker, M.P. and Shatz, C.J. (1978) Ocular dominance columns and their development in layer IV of the cats visual cortex: a quantitative study. *J. Comp. Neurol.*, 179: 233–244.

Luhmann, H.J. (1990) Control of NMDA receptor-mediated activity by GABAergic mechanisms in mature and developing rat neocortex. *Dev. Brain Res.*, 54: 287–290.

Luhmann, H.J. and Prince, D.A. (1990) Transient expression of polysynaptic NMDA receptor-mediated activity during neocortical development. *Neurosci. Lett.*, 111: 109–115.

MacDermott, A.B., Mayer, M., Westbrook, G.L., Smith, S.J. and Barker, J.L. (1986) NMDA-receptor activation increases cytoplasmic calcium concentration in cultured spinal cord neurons. *Nature*, 321: 519–522.

Masu, M., Tanabe, Y., Tsuchida, K., Shigemoto, R. and Nakanishi, S. (1991) Sequence and expression of a metabotropic glutamate receptor. *Nature*, 349: 760–765.

Mayer, M.L. and Westbrook, G.L. (1985) The action of *N*-methyl-D-aspartate on mouse spinal neurones in culture. *J. Physiol. (London)*, 361: 65–90.

Mayer, M.L., Westbrook, G.L. and Guthrie, P.B. (1984) Voltage-dependent block by Mg^{2+} of NMDA responses in spinal cord neurons. *Nature,* 309: 261–263.

Mayer, M.L., Vyklicky, L.J. and Clements, J. (1989) Regulation of NMDA receptor desensitization in mouse hippocampal neurons by glycine. *Nature,* 338: 425–427.

McDonald, J.W., Silverstein, F.S. and Johnston, M.V. (1988) Neurotoxicity of *N*-methyl-D-aspartate is markedly enhanced in developing rat central nervous systems. *Brain Res.,* 459: 200–203.

Mishina, M., Kurosaki, T., Tobimatsu, T., Morimoto, Y., Noda, M., Yamamoto, T., Terao, M., Lindstrom, J., Takahashi, T., Kuno, M. and Numa, S. (1986) Expression of functional acetylcholine receptor from cloned cDNAs. *Nature,* 307: 604–608.

Molotchnikoff, S. and Itaya, S.K. (1993) Functional development of the neonatal rat retinotectal pathway. *Dev. Brain Res.,* 72: 300–304.

Monyer, H., Sprengel, R., Schoepfer, R., Herb, A., Higuchi, M., Lomeli, H., Burnashev, N., Sakmann, B. and Seeburg, P.H. (1992) Heteromeric NMDA receptors: molecular and functional distinction of subtypes. *Science,* 256: 1217–1221.

Morrisett, R.A., Mott, D.D., Lewis, D.V., Wilson, W.A. and Swartzwelder, H.S. (1990) Reduced sensitivity of the *N*-methyl-D-aspartate component of synaptic transmission to magnesium in hippocampal slices from immature brain. *Dev. Brain Res.,* 56: 151–262.

Mower, G.D., Caplan, C.J., Christen, W.G. and Duffy, F.H. (1985) Dark rearing prolongs physiological but not anatomical plasticity of the cat visual cortex. *J. Comp. Neurol.,* 235: 448–466.

Olson, C.R. and Freeman, R.D. (1980) Profile of the sensitive period for monocular deprivation in kittens. *Exp. Brain Res.,* 39: 17–21.

Reynolds, I.J. and Bear, M.F. (1991) Effects of age and visual experience on [^{3}H]MK-801 binding to NMDA receptors in kitten visual cortex. *Exp. Brain Res.,* 85: 611–615.

Sakmann, B. and Brenner, H.R. (1978) Change in synaptic channel gating during neuromuscular development. *Nature,* 276: 401–402.

Shatz, C. (1991) Impulse activity and the patterning of connections during CNS development. *Neuron,* 5: 745–756.

Simon, D.K. and O'Leary, D.D.M. (1992) Development of topographic order in the mammalian retinocollicular projection. *J. Neurosci.,* 12: 1212–1232.

Simon, D.K., Prusky, G.T., O'Leary, D.D.M. and Constantine-Paton, M. (1992) NMDA receptor antagonists disrupt the formation of a mammalian neural map. *Proc. Natl. Acad. Sci. USA,* 89: 10593–10597.

Stuermer, C.A.O. (1988) Retinotopic organization of the developing retinotectal projection in the zebrafish embryo. *J. Neurosci,* 8: 4513–4530.

Stuermer, C.A.O., Rohrer, B. and Münz, H. (1990) Development of the retinotectal projection in zebrafish embryos under TTX-induced neural-impulse blockade. *J. Neurosci.,* 10: 3615–3626.

Tremblay, E., Roisin, M.P., Represa, A., Charriaut-Marlangue, C. and Ben Ari, Y. (1988) Transient increased density of NMDA binding sites in developing rat hippocampus. *Brain Res.,* 461: 393–396.

Trussell, L.O., Thio, L.L., Zorumski, C.F. and Fischbach, G.D. (1988) Rapid desensitization of glutamate receptors in vertebrate central neurons. *Proc. Natl. Acad. Sci. USA,* 85: 2834–2838.

Tsumoto, T., Hagihara, K., Sato, H. and Hata, Y. (1987) NMDA receptors in the visual cortex of young kittens are more effective than those of adult cats. *Nature,* 327: 513–514.

Warton, S.S. and McCart, R. (1989) Synaptogenesis in the stratum griseum superficiale of the rat superior colliculus. *Synapse,* 3: 136–148.

Watanabe, M., Inoue, Y., Sakimura, K. and Mishnia, M. (1992) Developmental changes in distribution of NMDA receptor channel subunit mRNAs. *Neuroreport* 3: 1138–1140.

Williams, K., Russell, S.L., Shen, Y.M. and Molinoff, P.B. (1993) Developmental switch in the expression of NMDA receptors occurs in vivo and in vitro. *Neuron,* 10: 267–278.

Yuste, R. and Katz, L.C. (1991) Control of post-synaptic Ca^{2+} influx in developing neocortex by excitatory and inhibitory neurotransmitters. Neuron, 6: 333–344.

J. van Pelt, M.A. Corner H.B.M. Uylings and F.H. Lopes da Silva (Eds.)
Progress in Brain Research, Vol 102

CHAPTER 19

Role of the visual environment in the formation of receptive fields according to the BCM theory

C. Charles Law, Mark F. Bear and Leon N Cooper

Departments of Physics and Neuroscience, Institute for Brain and Neural Systems, Brown University, Providence, RI 02912, U.S.A.

Introduction

Modification of synaptic effectiveness between neurons in cortex is widely believed to be the physiological basis of learning and memory; further, there is now evidence that similar synaptic plasticity occurs in many areas of mammalian cortex (Kirkwood et al., 1992). In 1982, Bienenstock, Cooper and Munro (BCM) proposed a concrete synaptic modification hypothesis in which two regions of modification (Hebbian and anti-Hebbian) were stabilized by the addition of a sliding modification threshold.

There are two ways to test a theory like that of Bienenstock, Cooper and Munro. One is to compare its consequences with experiment; the other is to directly verify its underlying assumptions. Recently two such avenues of research have supported this model of plasticity. Physiological experiments have verified some of its basic assumptions, while analysis and simulations have shown that the theory can explain existing experimental observations of selectivity and ocular dominance plasticity in kitten visual cortex in a wide variety of visual environments and make testable predictions.

The BCM theory was originally created to explain the development of orientation selectivity and binocular response of neurons in various visual environments in kitten striate cortex, one of the most thoroughly studied areas in neuroscience. The research philosophy of our laboratory is to keep our model of the cortex as simple as possible, and add details after behavior and consequences have been thoroughly understood. In this paper we will present a more realistic representation of the previous simplified visual environment. Effects on our previous findings, and the additional ways the extension allows further comparisons with visual cortex will be examined.

Research in this area began with Nass and Cooper (1975) who explored a model in which the modification of visual cortical synapses was Hebbian; i.e. a change to a synapse was based on the multiplication of the pre- and postsynaptic activities, and stabilization of the synaptic weights was produced by stopping modification when the cortical response reached a specified maximum — thus tying local modifications to the total cortical response. The idea that the sign of the modification should be based on whether the postsynaptic response is above or below a threshold was incorporated by Cooper et al. (1979) (see Fig. 1) to explain variations in selectivity with different visual environments. To stabilize the synapses without having to impose external constraints on them, the threshold was allowed, by Bienenstock et al. (1982), to slide as a non-linear function of the recent history of the cell's postsynaptic response.

The original 'single-cell' theory, which modeled only afferents from the lateral geniculate nucleus (LGN) that synapsed on a single cortical neuron, was extended by Scofield and Cooper (1985) to a network of interconnected neurons, such as that in kitten striate cortex. The fully connected network was later simplified by Cooper and Scofield (1988) with the introduction of a mean-field theory, which in effect replaces all of the individual but redundant connections to a 'cortical' neuron from every other cell in the intracortical network by a single set of 'effective synapses' that conveys to each cell the average activity of all of the others. Cooper and Scofield showed that the evolution of LGN-cortical synapses is similar for either a single neuron receiving only LGN input or a neuron embedded in a mean-field network.

In the BCM theory, a change in the weight of a synapse is equal to the product of the presynaptic activity and a certain function (ϕ) of the postsynaptic activity. The qualitative consequences of the theory follow from a few properties of the ϕ function:

- when postsynaptic activity is above the modification threshold Θ, ϕ is positive, (i.e. an active synapse will be potentiated);
- when postsynaptic activity is below the threshold Θ, ϕ is negative, (i.e. an active synapse will be depressed);
- there is no modification ($\phi = 0$) when postsynaptic activity is equal either to the spontaneous firing rate, or to the threshold Θ;
- the value of Θ is not fixed, but rather 'slides' as a function of the postsynaptic activity.

One possible modification function which satisfies these constraints is displayed in Fig. 1.

A recent investigation (Dudek and Bear, 1992) using rat hippocampal slices provides evidence for the plausibility of the modification function hypothesized in the BCM theory. In this experiment, the Schaffer collateral projection to area CA1 is stimulated with 900 electrical pulses at frequencies ranging from 0.5 to 50 Hz, and the resulting modification is measured as a change in

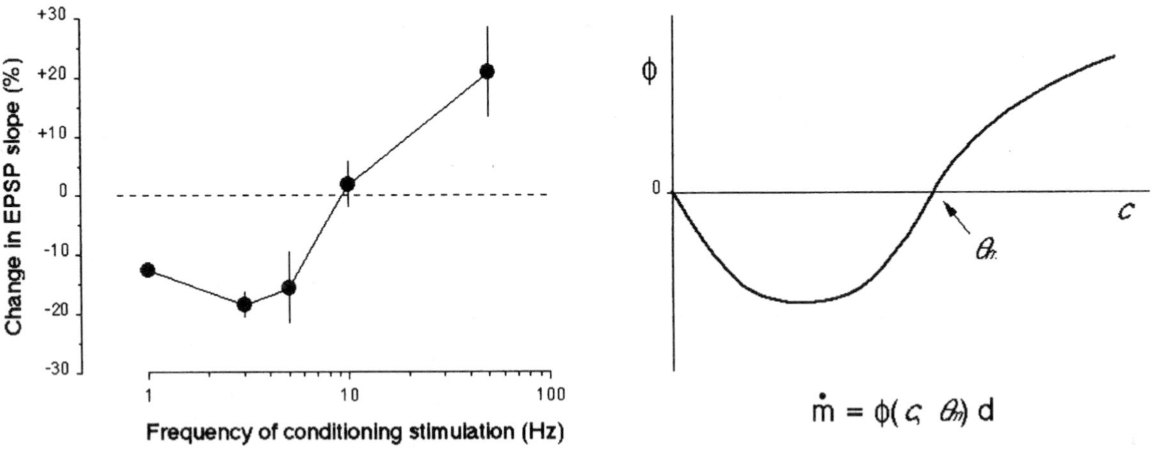

Fig. 1. Comparison of the theoretical ϕ function with a 'modification function' measured with physiological experiments (Dudek and Bear, 1992). The left graph shows the relationship between the frequency of the stimulation of the Schaffer collaterals, and the resulting change in the initial slope of the population EPSP in area CA1; 900 stimulus pulses were used at each frequency. The right graph shows the modification function hypothesized by the BCM theory. The variable c represents the postsynaptic activity, and the value of ϕ is the direction and magnitude of synaptic modification.

the initial slope of the population EPSP. The magnitude of the change is plotted as a function of the stimulation frequency, and is compared with the BCM theoretical modification function in Fig. 1. The observed plasticity has many properties required by the BCM theory. Thus, stimulation frequencies below 0.5 Hz produce no significant change in the population EPSP, whereas, when the Schaffer collaterals were stimulated at frequencies ranging from 1 to 5 Hz, a small but significant long-term depression of the response is observed. At stimulation frequencies above 10 Hz potentiation of the population EPSP is observed. In this investigation, the modification threshold, Θ, would be the cross-over between synaptic depression and potentiation at approximately 10 Hz. Whether this threshold moves, and is dependent on the history of the postsynaptic activity, is the subject of ongoing investigations.

The similarity between the BCM and the physiological modification functions is rather striking, assuming that (i) altering stimulation frequency is comparable with changing postsynaptic activity, and (ii) synaptic plasticity in rat hippocampus is similar to plasticity in kitten visual cortex. Recent experiments (Kirkwood et al., 1992) have shown that the later assumption is valid for superficial layers of kitten striate cortex. They report that stimulus protocols that can induce long-term potentiation (LTP) or depression (LTD), respectively in area CA1 of the hippocampus are also effective for inducing LTP and LTD in the connections from layer IV to superficial layers of visual cortex.

Visual cortical plasticity

Clothiaux, Bear and Cooper (CBC) (1991) show that simulations based on the BCM synaptic modification theory, using a fixed set of parameters, reproduce both the kinetics and equilibrium states of experience-dependent modifications observed experimentally in kitten striate cortex. During a sensitive period of post-natal development, the response properties of neurons in striate cortex of the kitten can be modified by manipulating the visual experience of the animal. The rearing conditions which have been simulated include normal rearing (NR), monocular deprivation (MD), reverse suture (RS), strabismus (ST), and binocular deprivation (BD).

The development of orientation selectivity: NR, dark rearing and correlated activity

Two important characteristics of the visual response of most neurons in cat striate cortex are that they are binocular and show a strong preference for contours of a particular orientation (Hubel and Wiesel, 1962). Although some orientation selectivity exists in striate cortex prior to visual experience, maturation to adult levels of specificity and responsiveness requires normal contour vision during the first 2 months of life (Barlow, 1975; Bonds, 1975; Blakemore and Van Sluyters, 1975; Fregnac and Imbert, 1978; Albus and Wolf, 1984; Fregnac and Imbert, 1984; Movshon and Van Sluyters, 1984; Braadstad and Heggelund, 1985).

When a kitten is reared in the dark, selectivity in striate cortex develops normally for the first 3 post-natal weeks, but after this period the number of selective cells decrease and the number of non-specific cells increase (Buissert and Imbert, 1976). It has been suggested that the development of selectivity prior to visual experience is due to correlated activity generated spontaneously in the retina (Miller et al., 1989). Two types of correlated retinal activity which could influence this development have been observed in the absence of visual stimuli. Meister et al. (1991) found that cells in the retina of a fetal cat and newborn ferrets spontaneously fire in bursts of activity which propagate slowly across the retina over a duration of several seconds. Mastronarde (1989) observed that activity of ganglion cells with overlapping receptive field centers have correlated activity at various ambient light levels.

The ocular dominance of a neuron is a measure of how effectively the neuron can be driven through the left and right eyes, respectively. Visual signals that originate in the retina are relayed through the LGN which project to layer IV of striate cortex. Until the level of the LGN, visual information originating from the two eyes is segregated in separate pathways, so LGN neurons are visually responsive to stimulation through only one eye. Striate cortex is thus the first site where individual cells receive afferent projections from both left and right eyes, and in normal kitten striate cortex most of the visually responsive cells are binocular. In cortical layer IV, which receives direct synaptic connections from the LGN, physiological recordings from individual cells indicate that in varying degrees approximately 60% of the visually responsive neurons can be activated through either eye (Shatz and Stryker, 1978).

One of the most dramatic examples of cortical plasticity is the alteration of ocular dominance in kitten striate cortex with a rearing procedure called monocular deprivation, during which one eye is deprived of patterned stimuli (by either suturing the eyelid closed or using an eye patch). In such an imbalanced visual environment, cells in the kitten's striate cortex change from mostly binocular to almost exclusively monocular: in less than 24 h most cells lose their responsiveness to stimulation through the deprived eye and can only be driven through the eye which remains open (Wiesel and Hubel, 1963; Hubel and Wiesel, 1970; Blakemore and Van Sluyters, 1974; Olson and Freeman, 1975; Movshon and Dursteler, 1977). The change induced by monocular deprivation is reversible. The rearing condition called reverse suture follows an initial period of monocular deprivation: after the cortical neurons have become monocular, the deprived eye is opened and the other eye closed. In this situation the cortical neurons soon lose responsiveness to the newly closed eye, and become responsive to the newly opened eye (Blakemore and Van Sluyters, 1974). Acute studies indicate that, as the ocular dominance of cortical responsiveness shifts from one eye to the other, there is rarely a period when cells can be strongly activated by both eyes (Movshon, 1976; Mioche and Singer, 1989).

Modeling a realistic visual environment

One of the significant assumptions used by Clothiaux et al. (1991) in the simulations of the deprivation experiments described above concerns the effects of visual experience on the activity of neurons in the LGN. Contoured stimuli are modeled with an abstract set of patterns, representing the activity of LGN afferents resulting from stimuli with different orientations. LGN activity in a normal visual environment is modeled as patterned input distorted by noise. In the absence of visual contours, LGN-cortical input activity is assumed to be uncorrelated noise distributed around a constant level of spontaneous activity. The justification for this assumption is the fact that neurons in primary visual cortex have small visual fields, so that visual contours when viewed through a small aperture are either noise or edges distorted by noise. This allows visual input to be represented by a pattern set, with a single variable representing the orientation of the stimulus.

To examine the validity of this representation of the visual environment used by CBC, we have run simulations with more realistic training patterns. Biologically realistic inputs are generated by processing natural images with a model of the retina and LGN. This model has many advantages over the previous model, and makes fewer assumptions about the activity of the LGN neurons. In the previous model, the same input patterns as were used for training were also used to test the selectivity of the BCM neuron. With the realistic input, none of the patterns used for testing the model were explicitly used in the training procedure. The BCM neuron is tested with separate sets of bar stimuli, spot stimuli and sinusoidal gratings. Bars of light test for orientation specificity, spots of light generate two-dimensional maps of the receptive field, and sinusoidal gratings test for spatial frequency selectivity. The greater diversity of test stimuli allows a more direct comparison to be made between selectivity developed by the model and that seen in the visual cortex.

Circular regions from the left and the right

retinas, covering the same visual space, are used to generate input to a single BCM neuron. The LGN is assumed to simply relay the signal generated by the retina to the visual cortex. Each retina includes an array of ganglion cells spaced one unit apart, and an array of receptors which are also spaced one unit apart. Only ganglion cells, the receptive-field midpoints of which fall within a circular visual area with radius $R_r^v = 5.0$ units, are included in the model. Each ganglion cell has an antagonistic center-surround receptive field which approximates a difference of two 'Gaussians'. The standard deviation of the center Gaussian is 1 unit, while that of the surround Gaussian is 3 units. This creates a receptive field center with a diameter of 4.4 units. The receptive field of each ganglion cell is balanced so that uniform illumination of any intensity results in spontaneous activity. A non-linearity in the form of a threshold restricts the absolute ganglion cell activity to be positive.

The activity c, of the cortical neuron is defined as $c = \sigma(\mathbf{m} \cdot \mathbf{d})$, where the vector **m** represents the strength of the synapses which connects the input, **d**, to the BCM neuron, and the function σ is a threshold which keeps the absolute cell activity from becoming negative. For each iteration of the simulation, the modification of the synaptic weights is determined by the equation $\dot{m}_k = \eta\phi(c, \Theta)d_i$, where η is the learning speed constant, and the function σ is defined as a simple parabolic function satisfying the general requirements as stated by BCM, $\phi(c, \Theta) = (1/\Theta)c(c - \Theta)$. (The extra factor $1/\Theta$ in effect makes the learning rate a function of Θ. This does not affect results qualitatively but makes the simulations somewhat more stable.) Following Intrator and Cooper (1992), the sliding threshold Θ is defined as the square of the average cell activity. This average is computed using an exponential time window,

$$\Theta = \tau^{-1} \sum_{t'=-\infty}^{t} c^2 e^{(t'-t)\tau^{-1}},$$

where τ is the time-constant of the sliding threshold's memory. The set of parameters used for these simulations is summarized in Table I.

The visual environment of the model consists of 24 gray scale images with dimensions 256 × 256 pixels (Fig. 2). For each cycle of the simulation, the activity of the receptors in the retina is determined by randomly picking one of the 24 images, and randomly shifting the image on the model retina. The shift is restricted so that none of the ganglion cell receptive field centers fall within ten units of the image border. The activity of each receptor in the model is determined by the intensity of a pixel in the image. This method

TABLE I

Parameters of the model

Parameter	Significance	Value
R_r^v	Radius of the retinal patches	5.0
STD_c	Standard deviation of a ganglions excitatory center Gaussian	1.0
STD_s	Standard deviation of a ganglions inhibitory surround Gaussian	3.0
d_s	Level of ganglion spontaneous activity	2.0
c_s	Level of BCM spontaneous activity	1.0
$\overline{n^2}$	Average square of the ganglion noise from a sutured eye	0.03
mj (0)	Starting value of the synaptic weights for normal rearing	(0.0, 0.01)
η	Modification step size	0.0001
τ	Time constant in the definition of Θ	1000

Fig. 2. Some of the 24 natural images used for training the model.

generates a very large training set because of the many shifts which are possible. Ganglion cell activity generated by a sutured eye is taken to be uniformly distributed noise.

At selected times during the simulations, the properties of the BCM neurons are tested separately through the left and right eyes. Two-dimensional maps of the receptive field are generated by 'shining' small spots of light on the retina at many locations, and recording the cortical neurons activity generated for each spot. This is similar to the process used by Jones and Palmer (1987) to generate two-dimensional receptive field profiles of simple cells in cat striate cortex. Orientation tuning of the neuron is determined by presenting bars of light. For each orientation, several bar stimuli are generated at different spatial locations. The best response to the set of stimuli determines the amplitude of the tuning curve for that orientation. The maxima of the left eye and the right eye tuning curves are then used to determine the binocularity of the BCM neuron.

In simulations of NR, both eyes receive correlated pattern input from the same natural images. At the beginning of the simulation of normal rearing, the weights are initialized to small random values. Even though at this stage the cortical neuron is relatively unresponsive to visual stimuli, the initialization of the weights adds a small bias which can influence the final selectivity of the neuron. Figure 3 shows the results of several simulations which are initialized with different random weights. At the end of normal rearing, the BCM neuron is binocular and has properties similar to simple cells found in striate cortex. Its receptive field has adjacent excitatory and inhibitory bands which produce selectivity to the orientation of bar stimuli. In our simulations, the BCM neuron most often becomes selective to horizontal or vertical orientations. The preference for horizontal stimuli is a property of the natural images, not a spurious bias introduced by the retinal or cortical models. When the environment is rotated by rotating the natural images, the orientation preference of the model also shifts. Figure 4 shows orientation preference histograms generated with two models: the first has the normal visual environment, the second a visual environment rotated by 45 degrees.

The results of these simulations are very robust, and do not depend on the retinal properties. Changing the parameters of the retina does not significantly influence the stable fixed points of the simulations. In fact, even completely bypassing the retina has little effect on the selectivity developed by the BCM neuron. Figure 5 shows the receptive field that developed when the BCM neuron was trained directly from the natural images. Further, the simulations are not dependent on the size of the retinal patches which converged on the cortical neuron. When simulations are run with much larger retinas, the BCM neuron simply disconnects itself from most of the ganglion cells. Figure 6 shows the receptive field when the retinal diameter is four times larger than the diameter of the retina in the control simulations. Enlarging the size of the retinas, which project to the BCM neurons, does not significantly change the BCM neuron's receptive-field size. The major difference in this simulation is the increased orientation selectivity due to the slight elongation of the receptive field.

Training with correlated noise

It has been observed that the activities of neighboring ganglion cells are correlated in the absence of contour vision (Mastronarde, 1989). To examine the significance of these observations to our results, the deprivation experiments were simulated with deprived eye correlations generated by the receptive fields of the ganglion cells. Uncorrelated noise is presented to the model and the antagonistic center-surround receptive fields in the retina produce positive correlation at small distances (less than 2.2 units), and negative correlations at medium distances (over 2.2 units). This addition altered the results of the deprivation simulations in minor ways. The most notable change occurred in the simulation of reverse suture. Correlations in the deprived eye delay the

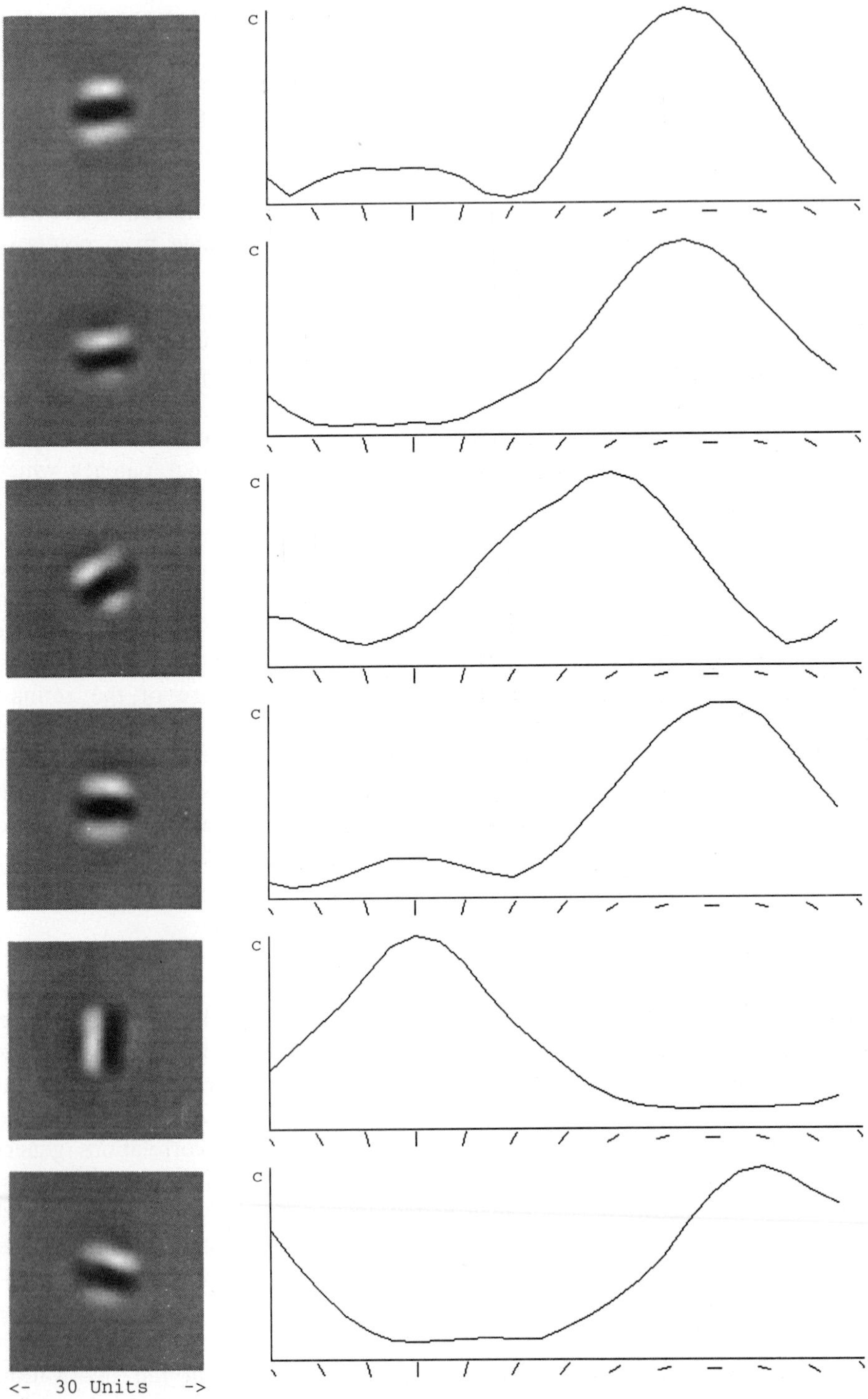

Fig. 3. These receptive fields were developed by BCM neurons trained with natural images. The eight simulations had $R_r^v = 5.0$ units. The only difference between the simulations was the initialization of the random number generator. (1) The light areas of the receptive field maps represent excitatory regions, and the dark areas represent inhibitory regions. (2) Orientation tuning curves were generated with bars of light.

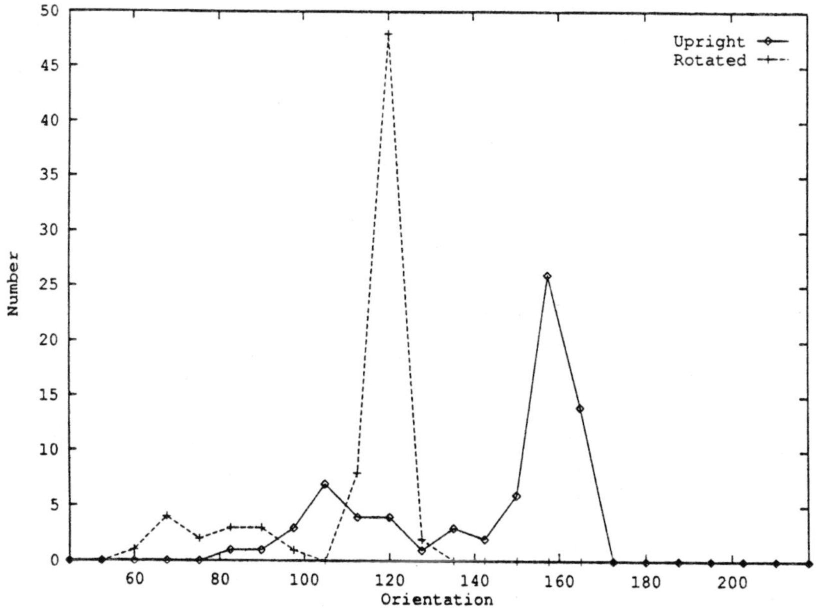

Fig. 4. These histograms show the orientation preferences developed by 100 individual simulations of normal rearing trained with natural images, and 100 simulations trained with natural images rotated by 45 degrees.

recovery of the newly opened eye. This dependency suggests that the results of reverse suture may be dependent on the amount and types of correlation in a deprived eye. It is likely that various forms of visual deprivation, i.e. lid suture, eye patch, TTX or diffuse lens, will produce different forms of correlations, because the level of ambient light affects the retinal spontaneous correlations. It is possible that with more simulations and analysis, our model would uncover significant dependencies between quantitative results of deprivation experiments and the method by which the deprivation was effected.

Conclusion

Generalization of the CBC visual environment to include retinal preprocessing and a naturalistic rearing environment did not significantly alter the CBC results during simulations of binocular de-

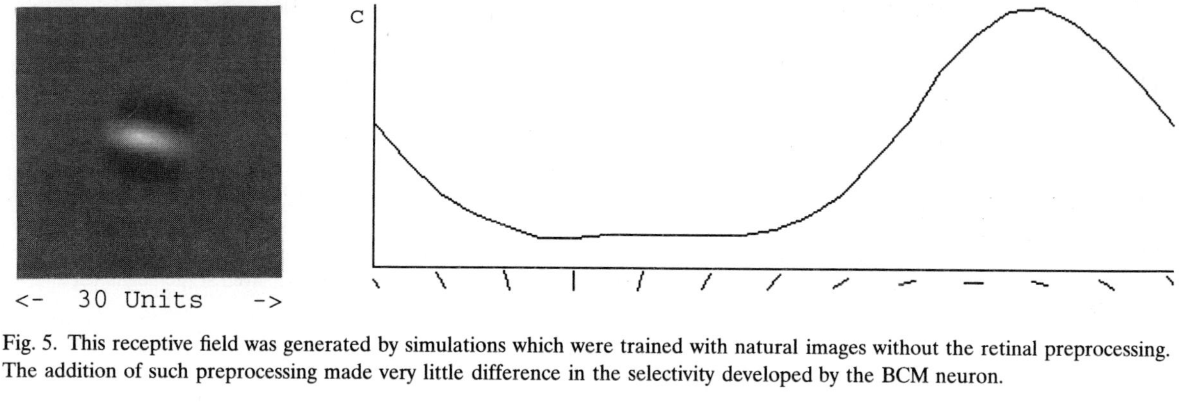

Fig. 5. This receptive field was generated by simulations which were trained with natural images without the retinal preprocessing. The addition of such preprocessing made very little difference in the selectivity developed by the BCM neuron.

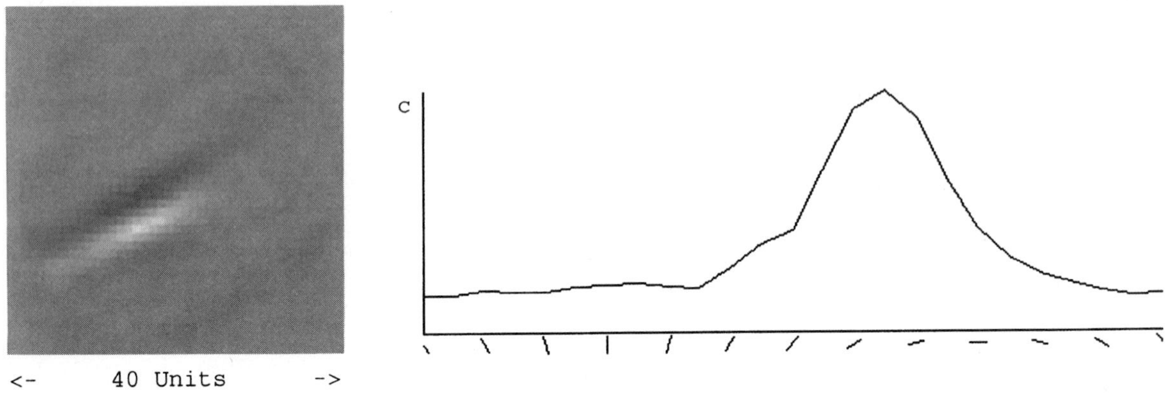

Fig. 6. This receptive field is the result of a simulation, where the size of the retinal patches that connected to the BCM neuron was four times larger than previous simulations. Except for an increase in selectivity, the properties of the BCM neuron are the same as simulations with input with a smaller spatial extent.

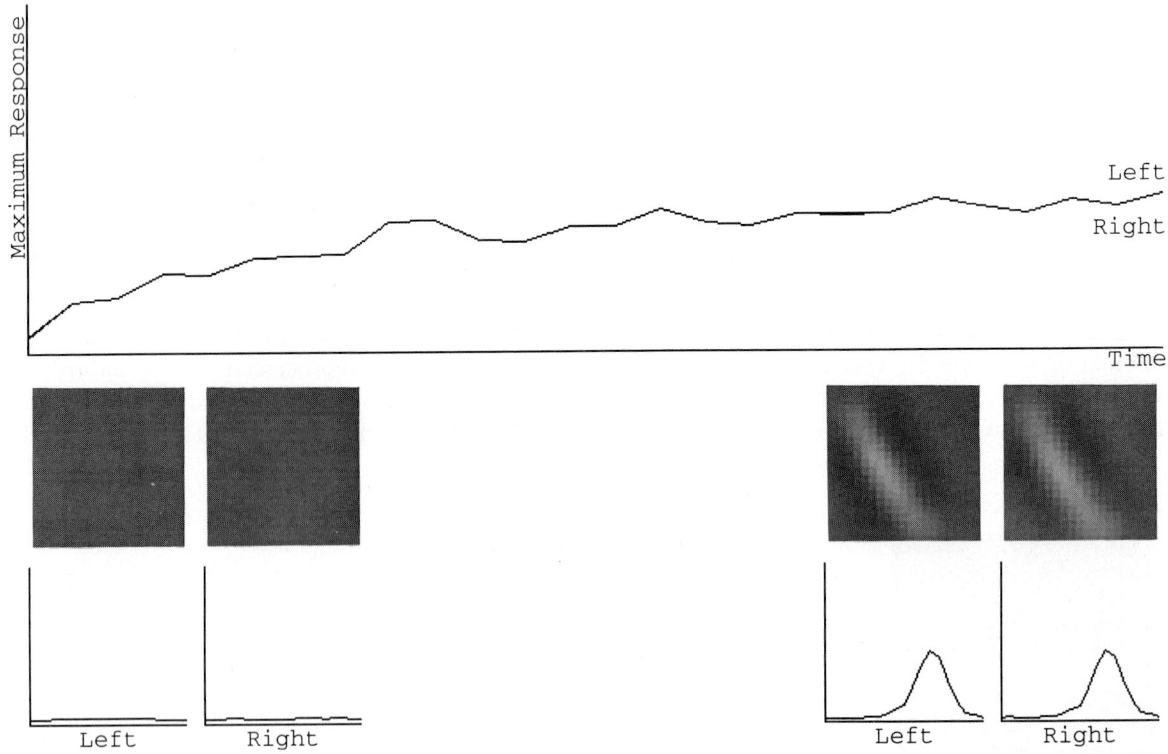

Fig. 7. This figure displays the results of the simulation of normal rearing. The graph shows the responsiveness of the BCM neuron to stimulation through the left and right eyes. The left and right eye receptive field properties are displayed at the beginning and the end of the simulation.

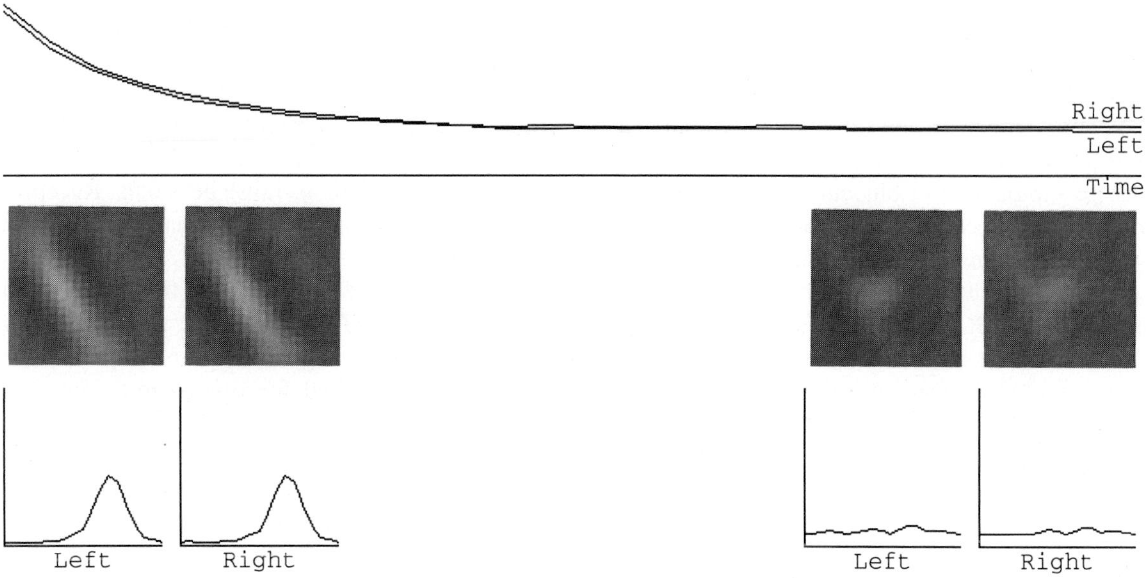

Fig. 8. This figure displays the results of the simulation of binocular deprivation.

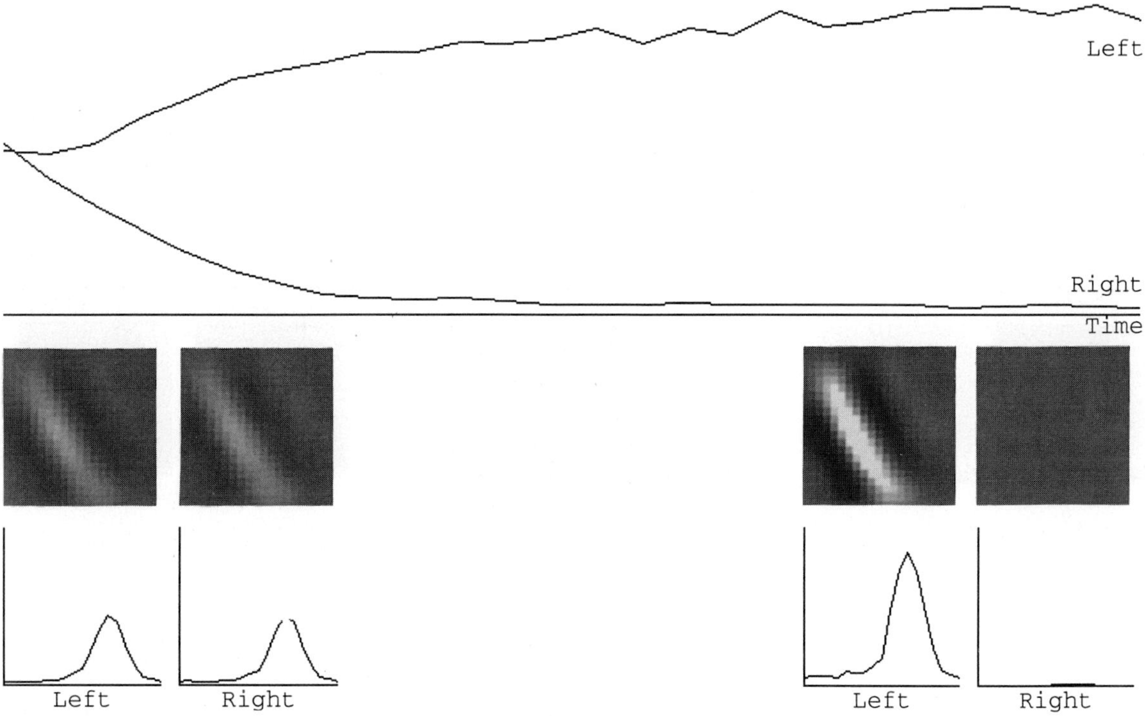

Fig. 9. This figure displays the results of the simulation of monocular deprivation.

privation, monocular deprivation, reverse suture and strabismus. Figures 7–10 show the results of deprivation simulations using natural images. As in the CBC simulations, normal rearing produces a selective binocular neuron which is equally responsive to stimulation through the left and right eyes. In simulations of binocular deprivation, the BCM neuron becomes less responsive and selective, but remains binocular. In simulations of monocular deprivation, the sutured eye disconnects from the BCM neuron, while in simulations of reverse suture the newly closed eye disconnects from the BCM neuron before the newly open eye reconnects. These results suggest that the abstract patterns of CBC were adequate representations of visual experience for these simulations of visual deprivation experiments.

Our simulations have also shown that the BCM theory can be consistent with the hypothesis that correlated retinal spontaneous activity is responsible for the development of selectivity prior to visual experience. Crude orientation selectivity develops in the absence of contoured visual experience, when the model is trained exclusively with correlated noise. However, to obtain this effect the learning speed, η, must be small. Receptive fields somewhat similar to normally reared neurons develop.

Figure 11 shows the results of three such simulations. The BCM neurons have adjacent excitatory and inhibitory regions in their receptive fields, but are poorly tuned for orientation. These results are consistent with our previous assumption that, although some selectivity may develop in the absence of contour vision, normal visual experience is necessary for full selectivity to develop.

Although simulations of the deprivation experiments using the realistic inputs produce qualita-

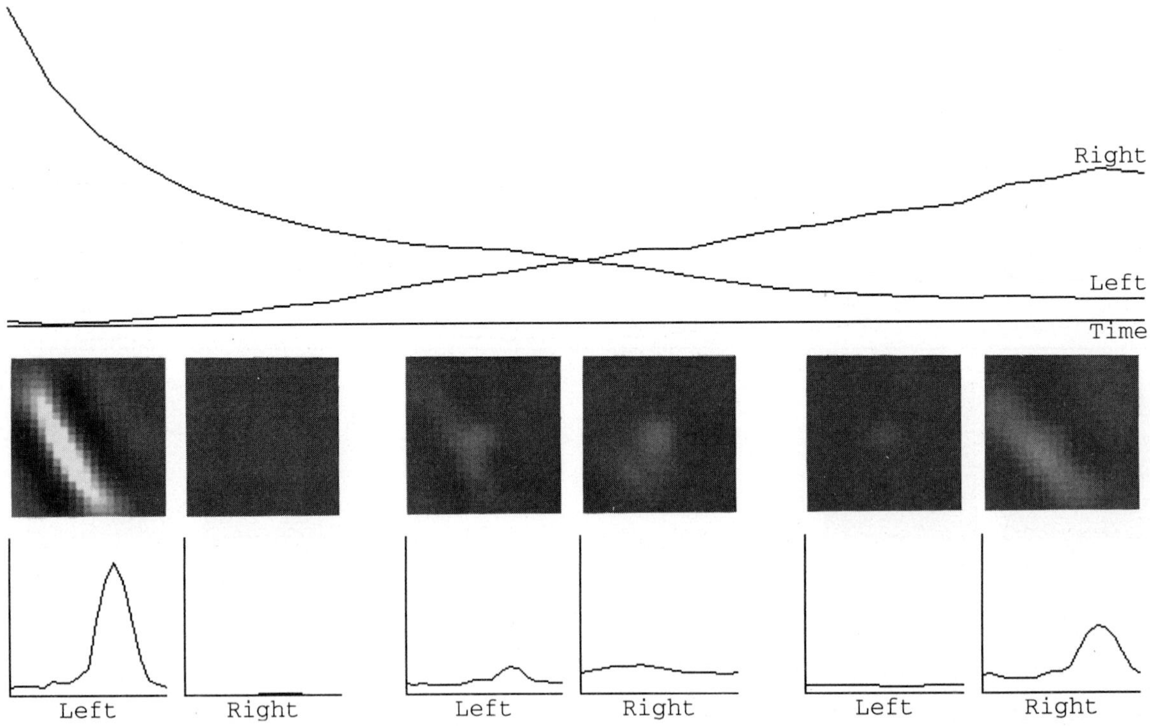

Fig. 10. This figure displays the results of the simulation of reverse suture. The graph displays the responsiveness of the BCM neuron to stimulation through the left and right eyes. The left and right eye receptive field properties are displayed at the beginning, middle and end of the simulation.

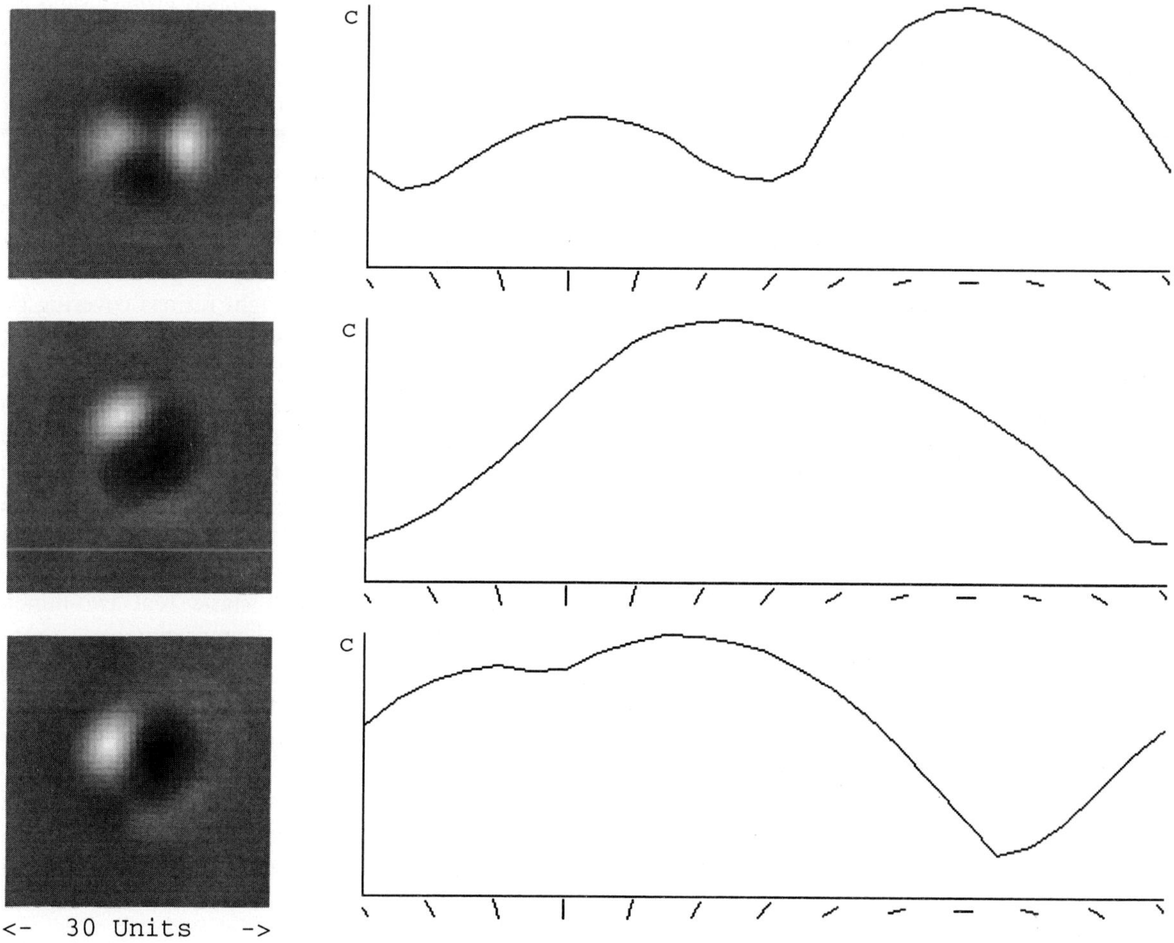

Fig. 11. This figure show the three results of simulations of single cell trained with white noise images. The only difference between the three simulations was the initialization of the random number generators.

tively similar results to CBC, in some cases the detailed behavior agreed more closely with experimental observations (Mioche and Singer, 1989). In simulations of RS, realistic inputs eliminate a possible discrepancy between experimental findings and the CBC simulations. It was noted by Mioche and Singer (1989) that the recovery of the newly opened eye is slow and incomplete following reverse suture, while the CBC model turns out to have a fast and robust recovery. As suggested in CBC (1991) this has several plausible explanations. However, without any additional assumptions, the present simulation of reverse suture displayed in Fig. 10 agrees more closely with the experimental findings: the BCM neuron does recover responsiveness to the newly opened eye, but recovery is slow.

In the simulation of monocular deprivation there is possibly a small discrepancy in the single cell simulation of MD, the response generated by the open eye quickly grows to offset the loss of input from the closed eye, and the weights connecting the open eye to the BCM increase. When Mioche and Singer (1989) chronically recorded from cells during monocular deprivation no increase in the cells response to stimulation through the open eye occurred until the closed eye was completely disconnected. We do not consider this discrepancy serious for several reasons: in network simulations of MD the increase is almost eliminated, and the increased responsiveness seen in the single cell model is parameter dependent; specifically, it is greatly influenced by any non-linearity in the definition of neuron activity.

Discussion and Conclusions

Our results support the contention that BCM synaptic modification is being implemented in cortex. Physiological experiments are consistent with some of the fundamental assumptions of the BCM theory, and simulations in realistic visual environment have reproduced results of normal rearing and the various visual deprivation experiments. Thus, we have successfully tested consequences of the theory as well as some of its underlying assumptions. The theory makes various predictions (discussed in detail elsewhere) that should be tested; in addition a question of increasing importance is the molecular basis for BCM modification.

Although some of this work is in an early state, it provides an excellent illustration of the benefit of the interaction of theory with experiment. Theory enables us to follow a long chain of arguments and connect, in a fairly precise way, various hypotheses with their consequences. It forces us to refine our language so that questions can be formulated with clarity and precision. Experiment focuses our attention on what is real; it separates what might be from what is; it tells us what must be explained and what is possible among explanations.

As is illustrated in this article, the theoretician who develops his arguments with close attention to the experimental results may create a concrete structure of sufficient clarity so that new questions, of great interest and amenable to experimental verification, become apparent.

The BCM (1982) theory of synaptic plasticity models the development of orientation selectivity and binocular interaction in primary visual cortex, and has successfully reproduced kitten visual deprivation experiments (Clothiaux et al., 1991). Additional support for the underlying assumption of this theory comes from recent physiological experiments (Dudek and Bear, 1992).

In order to better compare the consequences of the BCM theory with experiment, the original abstraction of the visual environment is replaced in this work by real visual images. Circular regions from the left and right retinas covering the same visual space are used to generate input to a single BCM neuron. The LGN is assumed to simply relay signals generated in the retina to the visual cortex. Each ganglion cell has an antagonistic 'center-surround receptive field' generated by a difference of two Gaussians, and ganglion cell activity is restricted to be positive.

This extension allows the BCM neuron to be trained and tested with static real two-dimensional images. The visual environment is represented by 24 gray scale natural images, which can be shifted across the artificial retinas. In this environment, the BCM neuron develops receptive similar to simple cells found in primary visual cortex. It displays adjacent excitatory and inhibitory bands, when tested with spot stimuli, and orientation selectivity when tested with bar stimuli.

Acknowledgments

This research was supported by the Office of Naval Reasearch, and the Army Research Office.

References

Albus, K. and Wolf, W. (1984) Early post-natal development of neuronal function in the kitten's visual cortex: a laminar analysis. *J. Physiol. (Lond.)* 110: 463–480.

Barlow, H.B. (1975) Visual experience and cortical development. *Nature (London)*, 258: 199–205.

Bienenstock, E.L., Cooper, L.N. and Munro, P.W. (1982) Theory for the development of neuron selectivity: orientation specificity and binocular interaction in visual cortex. *J. Neurosci.*, 2: 32–48.

Blakemore, C. and van Sluyters, R.C. (1974) Reversal of the physiological effects of monocular deprivation in kittens:

further evidence for a sensitive period. *J. Physiol. (London),* 237: 195–216.

Blakemore, C. and van Sluyters, R.C. (1975) Innate and environmental factors in the development of the kitten's visual cortex. *J. Physiol. (Lond.)* 248: 663–716.

Bonds, A.B. (1975) Development of orientation tuning in the visual cortex of kittens. In: R.D. Freeman (Ed.), *Developmental Neurobiology of Vision,* Plenum, New York, pp. 663–716.

Braadstad, B.O. and Heggelund, P. (1985) Development of spatial receptive field organization and orientation selectivity in kitten striate cortex. *J. Neurophysiol.,* 53: 1158–1178.

Buissert, B.O. and Imbert, M. (1976) Visual cortical cells: their developmental properties in normal and dark reared kittens. *J. Physiol. (London),* 255: 511–525.

Clothiaux, E.E., Bear, M.F. and Cooper, L.N. (1991) Synaptic plasticity in visual cortex: comparison of theory with experiment. *J. Neurophysiol.,* 66: 1785–1804.

Cooper, L.N. and Scofield, C.L. (1988) Mean-field theory of a neural network. *Proc. Natl. Acad. Sci.,* 85: 1973–1977.

Cooper, L.N., Liberman, F. and Oja, E. (1979) A theory for the acquisition and loss of neuron specificity in visual cortex. *Biol. Cybernet.,* 33: 9–28.

Dudek, S.M. and Bear, M.F. (1992) Homosynaptic long-term depression in area CA1 of hippocampus and the effects of NMDA receptor blockade. *Proc. Natl. Acad. Sci. USA,* 89: 4363–4367.

Fregnac, Y. and Imbert, M. (1978) Early development of visual cells in normal and dark-reared kittens: relationship between orientation selectivity and ocular dominance. *J. Physiol.,* 278: 27–44.

Fregnac, Y. and Imbert, M. (1984) Development of neuronal selectivity in primary visual cortex of cat. *Physiol. Rev.,* 64: 325–434.

Hubel, D.H. and Wiesel, T.N. (1962) Receptive fields, binocular interaction and functional architecture in the cat's visual cortex. *J. Physiol.,* 160: 106–154.

Hubel, D.H. and Wiesel, T.N. (1970) The period of susceptibility to the physiological effects of unilateral eye closure in kittens. *J. Physiol.,* 206: 419–436.

Intrator, N. and Cooper L.N (1992) Objective function formulation of the BCM theory of visual cortical plasticity: statistical connections, stability conditions. *Neural Networks,* 5: 3–17.

Jones, J.P. and Palmer, L.A. (1987) The two-dimensional spatial structure of simple receptive fields in cat striate cortex. *J. Neurophysiol.,* 58: 1187–1211.

Kirkwood, A., Aizenman, C.D. and Bear, M.F. (1992) Common forms of plasticity in hippocampus and visual cortex in vitro. *Neurosci. Abstr.,* 18: 1499.

Mastronarde, D.N. (1989) Correlated firing of cat retinal ganglion cells. *Trends Neurosci.,* 12: 75–80.

Meister, M., Wong, R.O., Baylor, D.A. and Shatz, C.J. (1991) Synchronous bursts of action potentials in ganglion cells of the developing mammalian retina. *Science Wash. DC,* 252: 939–943.

Miller, K.D., Keller, J.B. and Stryker, M.P. (1989) Ocular dominance column development: analysis and simulation. *Science (Wash. DC),* 252: 939–943.

Mioche, L. and Singer, W. (1989) Chronic recordings from single sites of kitten striate cortex during experience-dependent modifications of receptive-field properties. *J. Neurophysiol.,* 63: 185–197.

Movshon, J.A. (1976) Reversal of the physiological effects of monocular deprivation in the kitten's visual cortex. *J. Physiol. (London),* 261: 125–174.

Movshon, J.A. and Dursteler, M.R. (1977) Effects of brief periods of unilateral eye closure on the kitten's visual system. *J. Neurophysiol.,* 40: 1255–1265.

Movshon, J.A. and van Sluyters, R.C. (1984) Visual neural development. *Annu. Rev. Psychol.,* 40: 1255–1265.

Nass, M.M. and Cooper, L.N (1975) A theory for the development of feature detecting cell in visual cortex. *Biol. Cybernet.,* 19: 1–18.

Olson, C.R. and Freeman, R.D. (1975) Progressive changes in kitten striate cortex during monocular vision. *J. Neurophysiol.,* 38: 26–32.

Scofield, C.L. and Cooper, L.N., (1985) Development and properties of neural networks. *Contemp. Phys.,* 26: 125–145.

Shatz, C.J. and Stryker, M.P. (1978) Ocular dominance in layer IV of the cat's visual cortex and the effects of monocular deprivation. *J. Physiol.,* 281: 267–283.

Wiesel, T.N. and Hubel, D.H. (1963) Single cell responses in striate cortex of kittens deprived of vision in one eye. *J. Neurophysiol.,* 26: 1003–1017.

J. van Pelt, M.A. Corner H.B.M. Uylings and F.H. Lopes da Silva (Eds.)
Progress in Brain Research, Vol 102

CHAPTER 20

Models of activity-dependent neural development*

Kenneth D. Miller

Departments of Physiology and Otolaryngology,, W.M. Keck Foundation, Center for Integrative Neuroscience, and Neuroscience Graduate Program, University of California, San Francisco, CA 94143-0444, U.S.A.

Introduction

In the development of many vertebrate neural systems, an initially rough connectivity pattern refines to a precise, mature pattern through activity-dependent synaptic modification or rearrangement (reviewed in Miller, 1990a; Constantine-Paton et al., 1990). What is known about these processes is generally consistent with a hypothetical rule first proposed by Hebb (1949): synapses are strengthened if there is temporal correlation between their pre- and postsynaptic patterns of activity. The development often appears to be competitive: for a given pattern of activation, a correlated group of inputs may lose strength when competing with a more strongly activated correlated input group, yet retain or gain strength when competitors are absent (Wiesel and Hubel, 1965; Guillery, 1972; Miller and MacKay, 1994).

The classic example of such correlation-based, competitive development is the formation of ocular dominance columns in the mammalian visual cortex (reviewed in Miller and Stryker, 1990; Shatz, 1990). Visual inputs from the lateral geniculate nucleus (LGN) to the visual cortex terminate in separate stripes or patches consisting largely or entirely of terminals serving a single eye (Fig. 1). There is a regular, periodic alternation across the cortex of patches dominated by each eye. This segregated projection develops in an activity-dependent manner from a diffuse, overlapping initial projection in which inputs serving the two eyes project roughly equally throughout cortical layer 4.

Orientation columns are another striking feature of visual cortical organization (Hubel and Wiesel, 1962; reviewed in LeVay and Nelson, 1991). Most cortical cells are orientation selective, responding selectively to light/dark edges over a narrow range of orientations. The preferred orientation of cortical cells varies regularly and periodically across the cortex. It has long been a popular notion that orientation columns, like ocular dominance columns, may develop by a process of activity-dependent synaptic competition. However, evidence has been lacking, because orientation selectivity generally develops before animals are born or have functioning vision (Wiesel and Hubel, 1974; Sherk and Stryker, 1976; Chapman and Stryker, 1993). This does not rule out activity-dependent development, because experiments have shown that there is spontaneous neural activity, locally correlated within each eye, in the absence of vision and in the fetus (Mastronarde, 1989; Maffei and Galli-Resta, 1990; Meister et al., 1991; Wong et al., 1993), and that this activity is sufficient to guide activity-dependent ocular dominance segregation (Shatz and Stryker, 1988; Shatz, 1990; Miller and Stryker, 1990).

*This article is an updated version of an article that first appeared in *Seminars in the Neurosciences*, Vol. 4, No. 1 (1992), pp. 61–73. With permission of Academic Press.

The rules of synaptic modification that drive cortical development are presumably local to the environment of each synapse and its pre- and postsynaptic cells. We would like to understand these rules, yet often what can be observed is only a large-scale developmental outcome, such as ocular dominance segregation. At the same time, we would like to understand the mechanisms underlying these large-scale outcomes: which synaptic plasticity rules could produce them, what other factors are required and how can alternatives be tested? To address these questions and others, various models have been devised for the development of columnar systems through local activity-dependent synaptic modification.

The Von der Malsburg model of column development

Von der Malsburg first formulated such a model for the development of visual cortical columns through 'self-organization' (Von der Malsburg, 1973; Von der Malsburg and Willshaw, 1976; Willshaw and Von der Malsburg, 1976) (see also related models developed at about the same time: Wilson and Cowan, 1973; Nass and Cooper, 1975; Perez et al., 1975; Grossberg, 1976). This model established many of the elements of current models, and so is worth examining in detail. Von der Malsburg assumed that synapses of LGN inputs onto cortical neurons are modified by a Hebbian rule, and that the process is competitive, so that some synapses are strengthened only at the expense of others. He enforced the competition by holding constant the total strength of synapses converging on each cortical cell (conservation rule). He assumed further that inputs tend to be activated in clusters or patterns, so that there are correlations in the firing of the inputs; and that cortical cells also tend to be activated in clusters due to the intrinsic connectivity of the cortex, e.g. short-range horizontal excitatory connections and longer-range horizontal inhibitory connections.

The results expected from this model were

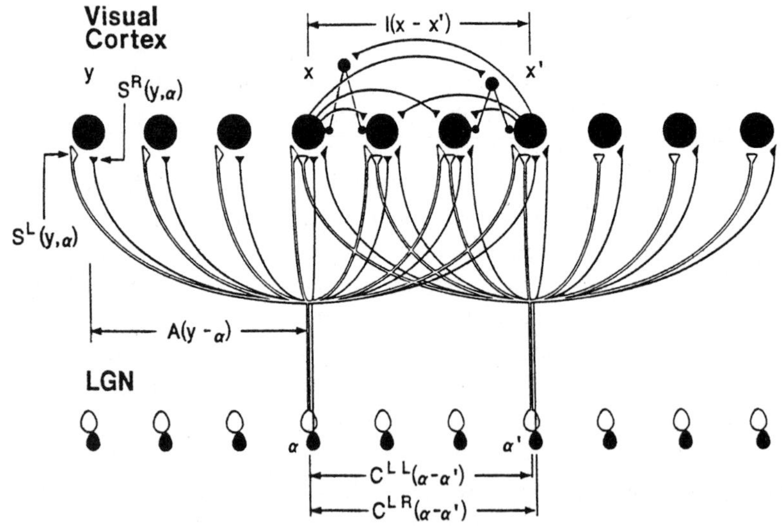

Fig. 1. Elements of the mature visual projection. Retinal ganglion cells from the two eyes project to separate laminae of the lateral geniculate nucleus (LGN). LGN neurons serving the two eyes in turn project to separate patches or stripes, known as ocular dominance columns, within layer 4 of the primary visual cortex. Binocular regions receiving inputs from both LGN layers are shown at the borders between ocular dominance columns; such binocular regions exist in some species, such as cats, but not in others, such as monkeys. The cortex is shown in cross-section, so that layers 1 through 3 are above and layers 5 and 6 below the layer 4 projection region. Reprinted by permission from Miller et al. (1989). © 1989 by the AAAS.

described as follows. The synaptic conservation rule should

'... (lead) to positive or negative interference between fibres connected to the same target cell. If such fibres are correlated in their activity ... their training effects are mutually reinforcing. On the other hand, synapses made on the same target cell by fibres which are anticorrelated never get reinforced.'

'After some time the modifiable synapses will have been rearranged so that each possible pattern of activity in the presynaptic sheet will evoke one of the possible patterns of activity in the postsynaptic sheet, thereby establishing associations between pairs of patterns. For discovering which patterns become linked together we use the crucial fact that each cell in a sheet belongs to several overlapping patterns. Under this condition the final projection has the property that overlapping stimuli will evoke overlapping responses and the responses to non-overlapping stimuli will not overlap. The geometrical interpretation of the resulting projections depends on the structure of the patterns of activity allowed.' (Von der Malsburg and Willshaw, 1976).

Von der Malsburg applied this model to the development of orientation columns (Von der Malsburg, 1973), the development of ocular dominance columns (Von der Malsburg and Willshaw, 1976) and, with Willshaw, to the development of topography (refinement of the retinotopic map) (Willshaw and Von der Malsburg, 1976). Computer simulations demonstrated that the models could work in each case. For orientation columns, inputs were activated in oriented patterns of all possible orientations. Individual cortical cells developed selective response to one such oriented pattern, with nearby cortical cells preferring nearby orientations. In the case of ocular dominance columns, inputs were activated in monocular patterns: each activation pattern was a localized cluster of cells from a single eye, and the two eyes were never simultaneously active. Individual cortical cells came to be driven exclusively by a single eye, and clusters of cortical cells came to be driven by the same eye. The tendency of overlapping cortical clusters to prefer the same eye resulted in a final arrangement of stripes of cortical cells preferring a single eye. 'The spacing between the stripes is determined by the range of inhibition' (Von der Malsburg and Willshaw, 1976), i.e. by the diameter of the intrinsic clusters of cortical activity.

Thus, a synaptic conservation rule was used that forced a Hebbian rule to become competitive. This led individual cells to become selective for a single correlated pattern of inputs. Combined with the idea that the cortex was activated in intrinsic clusters, this suggested an origin for cortical columns: coactivated cells in a cortical cluster would tend to become selective for similar, coactivated patterns of inputs.

A more general framework for analyzing cortical development

This model provided many of the elements of a model of cortical development, and it included a fundamental experimental prediction: the width of an ocular dominance column should be determined by measurable intracortical inhibition. But otherwise, the study of the model was largely a set of demonstrations that the ideas could work. The range of conditions under which the model *would* work, and the dependence of the model results on quantities a neurobiologist could measure, were largely unexamined. Such knowledge provides the means of testing a model or distinguishing it from alternative mechanisms.

Consider the model for ocular dominance columns: how do the results depend on the correlations in activities in the input layers? If the two eyes are sometimes coactivated, either because they are activated independently or because they are partially correlated by vision, can ocular dominance segregation occur? If inputs are correlated over a larger or smaller distance within an eye, how will this alter the outcome of development? How will the fact that inputs from the LGN have localized arborizations ('arbors') in cortex affect development? Can the size of these arbors, or the distance over which inputs are correlated, influence the width of the final patches? Can columns develop under alternative patterns of intracortical connectivity, for example, if the intracortical inhibition is weak or non-existent during development? How would these same questions be answered under alternative models of column development?

To address questions like these, we developed a similar model of ocular dominance column development that allowed analytical characterization of the outcome of development (Miller et al., 1989; Miller, 1990a,b; Miller and Stryker, 1990). This analysis gives a general understanding of the framework for column formation proposed by Von der Malsburg, and extends this framework to include alternative plasticity mechanisms, alternative structures of connectivity and of correlation, and alternative implementations of the postsynaptic conservation rule that ensures a competition. A detailed discussion of the mathematical and biological assumptions necessary for this analysis is given in Miller, 1990b.

We showed that a large class of developmental models could be expressed in terms of three measurable elements (Fig. 2): (1) a set of correlation functions, describing the correlation between the activities of input pairs as a function of their eyes of origin and their retinotopic separation; (2) an arbor function that describes the spread of input arborizations in cortex allowed by retinotopy; (3) a cortical interaction function describing interactions within the cortex by which activity at one location influences the effectiveness of correlated synapses on different cortical cells at nearby locations. In the absence of such intracortical influences, the competition occurring at each cortical cell would be independent, and hence development of large-scale clustering of cortical properties, such as ocular dominance, would not be expected.

The cortical interaction function is particularly dependent on the biological mechanism proposed to underly plasticity. In a Hebbian mechanism, as used in Von der Malsburg's model, the cortical

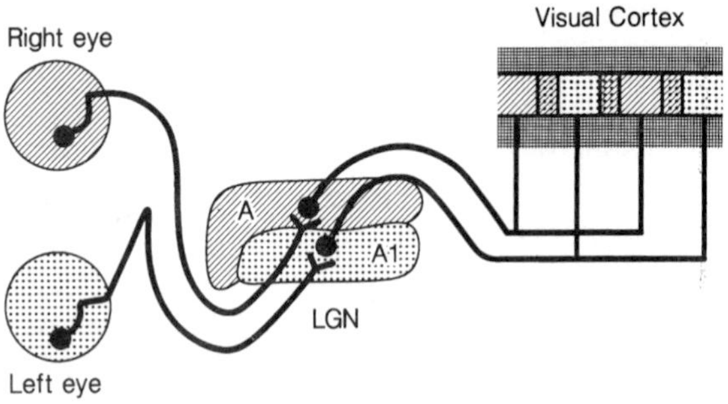

Fig. 2. Elements of a model of cortical development. Left (white) and right (black) input cells innervate cells in layer 4 of the visual cortex. α and α' label positions in the input layer (LGN) and x and x' label the retinotopically corresponding positions in the output layer (cortical layer 4); y labels an additional cortical position. The layers are taken to be two-dimensional, that is x, α, etc. are two-dimensional variables. The afferent correlation functions C^{LL} and C^{LR} measure, respectively, the correlation in activity between two left-eye afferents, and the correlation in activity between a left-eye and a right-eye afferent, as a function of the retinotopic separation of the two afferents. The arbor function A measures anatomical connectivity from a geniculate location to a cortical location, as a function of the retinotopic separation between the two locations. The cortical interaction function I measures the effect of neural activity at one cortical location (x') on the development of synapses at a lateral cortical site (x); in a Hebbian model, this function summarizes the effects of intracortical synaptic connections, by which activation at x' influences postsynaptic activation at x. The left-eye and right-eye synaptic strengths S^{L} and S^{R} represent the total physiological strength with which a given afferent activates a given cortical cell. These synaptic strengths are the dynamical (time-varying) variables in the model. The same framework can be applied to a competition between ON-center and OFF-center inputs from a single eye, rather than between left-eye and right-eye inputs: for example the correlation functions would then be $C^{\mathrm{ON\text{-}ON}}$, $C^{\mathrm{ON\text{-}OFF}}$, etc. Reprinted by permission from Miller et al. (1989). © 1989 by the AAAS.

interaction function is determined by intracortical synaptic connections. It is positive over distances at which cortical cells tend to excite one another and negative over distances at which cortical cells tend to inhibit one another. For alternative mechanisms that involve the activity-dependent release and uptake of a trophic or modification factor, the cortical interaction function also incorporates the spread of influence across the cortex due to diffusion.

Predicting the outcome of development

The outcome of development under mechanisms that can be expressed within this framework can be predicted as follows (Fig. 3) (Miller et al., 1989; Miller, 1990a; Miller and Stryker, 1990). We define the *arbor radius* as the radius of the arbor function, i.e. the radius of the retinotopically allowed arborization of an input to cortex. First, cortical cells tend to develop receptive fields consisting of a subset of inputs that are as correlated as possible (Fig. 3A). The most correlated subset is determined by the correlation functions but also by the arbor function, which restricts the set of inputs that can potentially innervate a cortical cell. This subset will be from a single eye, and hence cortical cells will tend to develop monocular receptive fields, provided that (1) inputs within each eye are locally correlated, and (2) at all separations within an arbor radius, inputs are better (or no worse) correlated with inputs from their own eye than with inputs from the opposite eye (Fig. 3A, top). Second, cortical cells tend to develop receptive fields that are as correlated as possible with other cortical receptive fields at excitatory distances across the cortex, but as anticorrelated as possible with other cortical receptive fields at inhibitory distances (Fig. 3B, top left). For ocular dominance, correlated receptive fields are those representing the same eye. Thus, as suggested by Von der Malsburg, cortical cells should be arranged in patches such that the patch width of a single eye corresponds approximately to the diameter of an excitatory region in cortex. More precisely, the width of a left-eye patch plus a right-eye patch corresponds to the spatial period that maximizes the Fourier transform of the cortical interaction function.

If intracortical inhibition is absent, then large clusters of cortical cells dominated by a single eye will form (Fig. 3B, top right). It is, however, possible for periodically alternating ocular dominance columns to develop in the absence of intracortical inhibition if an additional rule is invoked: the total synaptic strength made by each *input* cell must be approximately conserved (Fig. 3B, bottom). Note that this presynaptic conservation rule is distinct from the conservation rule that ensures competition, which is applied to each postsynaptic cell. The arbor function limits the synapses of each input to an arbor radius from its 'best' cortical location. Thus, presynaptic conservation limits the width of a patch of inputs from one eye to be no wider than the arbor radius, so that inputs from both eyes can form their synapses within the diameter of any arbor. In this sense, the presynaptic conservation rule has an effect similar to intracortical inhibition on a scale of about half an arbor radius. If inhibition is present on such a scale or finer, the rule does not alter the unconstrained course of development, but in the absence of such inhibition, the rule leads to development of ocular dominance columns with a width set by the arbors.

When ocular dominance segregation does occur, our analysis also allows prediction of the relative degree of segregation. In cats, where ocular dominance segregation in cortical layer 4 is not complete, there is some binocular overlap at the borders between the patches of the two eyes. In monkeys, where there is complete segregation in layer 4, there is no overlap. To understand such alternative outcomes, it is helpful to think of the development as a competition between monocular and binocular patterns of input. Because ocular dominance segregation does tend to occur, the fastest-growing pattern of inputs to a single cortical cell is a monocular pattern (Fig. 3A, top); but how much faster does it grow than a binocular pattern consisting of inputs from one eye in half of the receptive field, and inputs from the other eye in the other half (Fig. 3A, bottom right)? If the correlations within each eye are weak and occur only over input separations that are very small compared with an arbor radius, then the binocular pattern grows nearly as well as the monocular one. In this case the binocular pattern has slightly less total correlation than the monocular one because of the boundary between left-eye and right-eye inputs, but this loss is small because the correlations are weak and do not extend very far. Alternatively, if the correlations

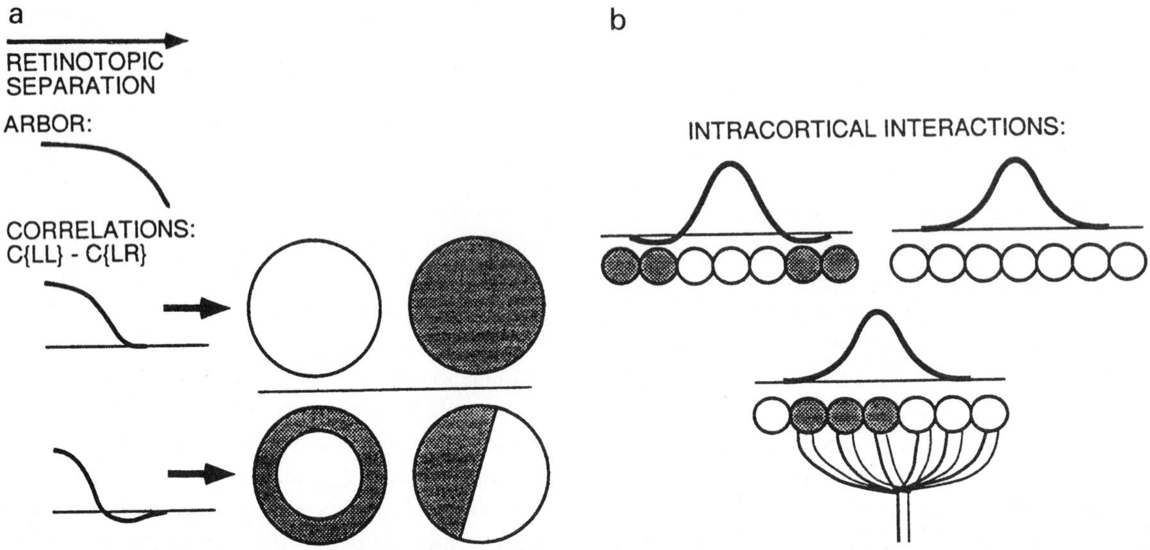

Fig. 3. Determinants of the outcome of cortical development as described by the model. (A) The correlations among input activities determine receptive field (RF) structure. Each large circle indicates the receptive field of a single cortical cell. White indicates left-eye inputs, shading indicates right-eye inputs. RF structure is determined by the degree to which two same-eye inputs with a given retinotopic separation are better correlated with one another than with an opposite-eye input at the same separation. This is described by the difference correlation function, $C^{D} \equiv C^{LL} - C^{LR}$, where we have assumed equality of the two eyes, $C^{LL} = C^{RR}$. TOP: If C^{D} is always positive within an arbor radius, ocular dominance segregation occurs: a cortical cell receptive field comes to receive only left-eye or only right-eye input. BOTTOM: If C^{D} were to change sign, so that at small separations same-eye inputs were best correlated, but at larger separations within the arbor radius opposite-eye inputs were best correlated, then receptive fields would develop segregated left-eye and right-eye subregions. Two alternative possible arrangements of such subregions are indicated. The correlation structure alone does not strongly distinguish between these; but the intracortical interactions lead to oriented arrangements, such as that at the lower right. The examples shown may be generalized: receptive fields tend to form with an alternation of left- and right-eye subregions whose spatial frequency maximizes the Fourier transform of C^{D}. (B) Intracortical interactions determine the arrangement of cortical cells. Only the case of ocular dominance segregation (A, top) is illustrated. TOP: Cells become arranged to best match the interactions. If the interactions are purely excitatory, arbitrarily large regions will become dominated by a single eye (top right). Note that interactions are assumed to be a function of distance, identical for every cell, although the figure shows them only for one cell. BOTTOM: If total synaptic strength is conserved over each presynaptic arbor, then ocular dominance columns can develop in the absence of intracortical inhibition. Initial arbors are shown stretching over six cortical cells. For each eye's inputs to retain approximately equal strength in the final arrangement, as a presynaptic conservation rule requires, eye patches can be no larger than half the arbor diameter (three cells wide); so patches of this size will develop in the absence of inhibition. This description of the determinants of development applies equally to a competition between ON- and OFF-center inputs from a single eye: in that case, $C^{D} \equiv C^{ON\text{-}ON} - C^{ON\text{-}OFF}$, white indicates ON-center inputs, shading indicates OFF-center inputs.

are strong and extend throughout the arbor, then inputs in one half of the receptive field cooperate with correlated inputs in the other half and the monocular pattern greatly outgrows the binocular one. In both cases ocular dominance segregation will occur, but in the first it will occur weakly: binocular patterns will develop on some cells, particularly those at the boundaries between the patches from the two eyes. In the second case, there will be complete segregation.

The method used to ensure competition can alter the outcome of development, by altering the minimal correlation structure needed for development of ocular dominance (Miller et al., 1989; Miller and MacKay, 1994). Recall that competition is modeled by conservation of total synaptic strength over each cortical cell. If this conservation is subtractive, i.e. implemented by reducing each synaptic strength by the same amount, then ocular dominance can develop when the inputs from one eye are even slightly more correlated with their neighbors than with those from the other eye, as described above (Fig. 3A). If, however, this conservation is multiplicative (as was Von der Malsburg's), implemented by multiplying all synaptic strength by a renormalizing constant, then the tendency to develop ocular dominance can be spoiled if the two eyes sometimes fire together, even randomly. Thus, if conservation is multiplicative, the inputs must be strictly monocular in order for ocular dominance to develop (the exact requirement depends on details of the Hebbian rule but this is a reasonable intuitive characterization). Ocular dominance begins to develop in the fetus in some animals, and continues its development after birth in the presence of vision, despite the fact that the two eyes may fire together randomly in the fetus and in darkness (Mastronarde, 1989; Maffei and Galli-Resta, 1990; Meister et al., 1991) and are partially correlated by vision after birth. This suggests that subtractive rather than multiplicative mechanisms may be more appropriate for modeling the competitive nature of a Hebbian rule in the visual cortex.

Simulations have confirmed that this framework is sufficient to account for many aspects of visual cortical development, including the development of monocular cells and their organization into periodic ocular dominance patches, the degree of monocular segregation, the restriction of afferent arbors to periodic patches of cortical innervation, and the effects of monocular deprivation including a critical period (Miller et al., 1989; Miller, 1990a; Miller and Stryker, 1990).

Application to the development of orientation selectivity

In cats, the cortical cells in the layers receiving the LGN inputs are primarily 'simple cells' (Hubel and Wiesel, 1962; Bullier and Henry, 1979): orientation-selective cells whose receptive fields consist of oriented, spatially segregated subregions receiving exclusively ON-center or OFF-center excitatory input. The understanding we have achieved of the determinants of development raises the possibility that oriented cortical simple cells could result from a competition between ON-center and OFF-center inputs, very much as ocular dominance segregation results from a competition between left-eye and right-eye inputs (Miller, 1992, 1994).

The parameter regime in which ocular dominance segregation does not develop (Fig. 3A, bottom) results in receptive fields reminiscent of simple cells, with segregated subregions each receiving a different class of input. In particular, oriented receptive fields develop (as in Fig. 3A, lower right) if development occurs in the presence of intracortical interactions. For a competition between left- and right-eye inputs, the correlation structure that yields this outcome is not biologically reasonable, but this correlation structure, in which sign changes as a function of distance, is plausible for a competition between ON- and OFF-center inputs (Fig. 4).

Thus, orientation selectivity could develop through competition between ON-center and OFF-center inputs. The outcome of such competition is the formation of receptive fields like those of Fig. 3A, lower right, but with segregated subregions of ON-center and of OFF-center inputs rather than of left-eye and of right-eye inputs. Such receptive fields strongly resemble cortical simple cells. The parameter regime leading to this result (Fig. 3A, bottom) is that in which ON-center cells are best correlated with other ON-center cells at small retinotopic separations, but are best correlated with OFF-center cells at larger retinotopic separations within an arbor radius (Fig. 4). Experimental measurement will be necessary to determine whether such a correlation structure is actually present during development.

Simulations have demonstrated that competition between ON- and OFF-center inputs under these conditions leads both to the development of oriented simple cells and to their continuous ar-

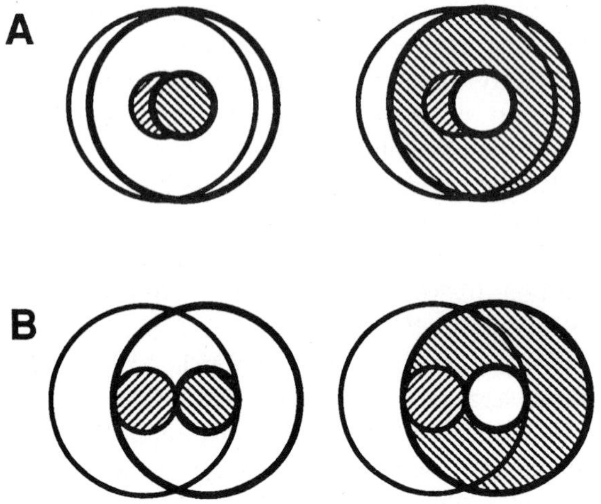

Fig. 4. Plausibility of a correlation structure between ON- and OFF-center inputs that would lead to development of orientation selectivity. Striped regions represent ON regions, and white regions represent OFF regions. Orientation selectivity develops in the dark (Wiesel and Hubel, 1974; Sherk and Stryker, 1976; Chapman and Stryker, 1993), so we consider correlations in spontaneous activity, in the absence of vision, due to common input from photoreceptors (Mastronarde, 1989). (A) Correlations at small retinotopic separations, where receptive field (RF) centers overlap. Left: Two ON-center RFs have overlapping ON-centers and overlapping OFF-surrounds, and hence would be expected to frequently receive common input and thus to be well correlated. Right: For an ON-center and an OFF-center RF at similar separations, ON-center overlaps OFF-center, and ON-surround overlaps OFF-surround, so anticorrelation is expected. Such correlations and anticorrelations at separations where centers overlap have been observed in dark activity of retinal ganglion cells (Mastronarde, 1989). (B) Correlations at larger retinotopic separations, where RF center overlaps RF surround. Left: Two ON-center RFs. The ON-center of each RF overlaps the OFF-surround of the other, and so the two would be expected to rarely receive common input, and thus to be poorly correlated or anticorrelated. Right: An ON-center and an OFF-center cell. The ON-center of one RF overlaps the ON-surround of the other, and similarly OFF-center overlaps OFF-surround, so better correlation might be expected than at left. While such correlations have not been observed, measurements have only been made in the retina; the LGN has stronger RF surrounds compared to the retina, so this effect might originate in or become enhanced in the LGN.

rangement in periodic orientation columns resembling those seen experimentally (Miller, 1992, 1994) (Fig. 5). The determinants of development follow the same general framework discovered for ocular dominance. The receptive fields consist of a subset of inputs that are as correlated as possible. Because of the change in correlations with distance, this subset consists of segregated, adjacent subfields of ON- and OFF-center inputs. For example, ON-inputs in an ON- subfield are, given this correlation structure, better correlated with OFF-inputs in an adjacent subfield than they would be with ON-inputs at that adjacent location. The intracortical interactions determine that the receptive fields of cortical cells are as correlated as possible at excitatory distances and as anticorrelated as possible at inhibitory distances. This both determines that individual receptive fields will be oriented (Fig. 3A bottom right rather than bottom left), and determines the arrangement of the orientation map. The determination of the width of the orientation columns by this mechanism is complicated, involving intracortical interactions, correlations and arbors. The fact that the width of orientation columns is determined somewhat differently than the width of ocular dominance columns is consistent with experiment: in both cats and monkeys, the periods of these two systems differ by 20–30% (Hubel et al., 1978; Loewel et al., 1988). A novel finding of the model is that the spatial phase of cortical simple cells — i.e. the spatial location within the receptive field of the ON-regions and of the OFF-regions — is a critical variable in determining the orientation maps. Hence, the maps of orientation and of spatial phase must be measured simultaneously to understand the origin of the orientation map alone.

Note that these ideas have diverged from those of Von der Malsburg. He conceived of orientation selectivity arising through a process by which competing oriented patterns of inputs become associated with different clusters of coactivated cortical cells. In the present model, there are no oriented patterns of inputs. Rather, competing ON- and OFF-center inputs converge onto cortical cells but become segregated within receptive fields. Given intracortical interactions, this leads to the emergence of orientation selectivity and its organization across cortex despite the absence of oriented patterns in the inputs.

These ideas demonstrate the potential power of modeling. We began with the belief, based on work in ocular dominance columns, that a Hebbian or similar mechanism is operating in the development of visual cortex. A general analysis of such competition demonstrated an unexpected outcome: convergence of two competing input populations onto postsynaptic cells but segregation within individual receptive fields. Previously, Hebbian competition between competing input populations was assumed to produce segregation between postsynaptic cells, as in ocular dominance segregation (but see Linsker, 1986a,b,c, discussed below). Identification of this novel parameter regime with cortical simple cells suggested an unexpected explanation for the origin of orientation selectivity in visual cortex. This explanation is testable: experiment can determine whether the postulated correlation structure is present during development, and experimental manipulations that would disrupt the ON/OFF correlation structure, for example forcing all inputs to fire in synchrony, should prevent the development of orientation selectivity if the hypothesis is correct.

Strengths and limitations of this approach

This framework allows understanding of the results of development based on the spatial arrangements of correlations, arbors, and intracortical interactions. This is the same level of detail as that studied in many experiments on column formation (Miller and Stryker, 1990): measurement or perturbation may be made of the correlations between competing sets of inputs, the arborizations of those inputs in the cortex, or the types and interconnections of cells in cortex, as well as of the biochemical mechanisms underlying plasticity. The model thus systematically connects experimentally measurable and perturbable quantities with expected developmental outcome at similar levels of detail. It also demonstrates that formulation of the problem at this level is sufficient to account for a wide range of observed phenomena, and it makes a number of novel predictions, for example, that the development of orientation selectivity may depend on the ON/OFF correlation structure.

Our analysis focuses on predictions that are independent of the details of biological non-linearities, because the nature of these is largely unknown. We accomplish this by concentrating on the initial development of a pattern of the difference between two similar input projections; this difference is initially small, allowing linearization of the equations (Miller, 1990b). We restrict our analysis to elements that develop in the early, linear regime of the model and thus are very robust to implementation details. These include the column width, and the very facts that cortical cells become monocular or orientation-selective. We do not address questions like the detailed layout of ocular dominance columns (i.e. long straight stripes vs. irregular patches), that depend on non-linearities and hence may vary with details of model implementation. The results must also be robust to noise, meaning that they develop from essentially any randomly chosen initial condition. Similarly, we respect biological constraints such as the exclusively excitatory nature of the LGN input to cortex.

This framework does not address the details of neuronal structure or the temporal structure of activation and plasticity. The conjunctions of activation that lead to Hebbian plasticity depend in a complex way on the temporal and spatial distributions of input activities, on interactions through the dendritic structure of the postsynaptic cell, and on the biophysical and learning mechanisms at the synapse. Yet in our model this is expressed simply through a correlation function that summarizes the ability of inputs to cooperate in achieving Hebbian plasticity on any postsynaptic cell, and the postsynaptic cell is collapsed to a point without dendritic structure. In a sense we are begining with the answer: we know that the competition ultimately leads an entire cell to become dominated by a single pattern of inputs. Thus, competition is ultimately integrated over the cell as a whole, so we ignore finer levels of competition.

Our framework also ignores cell- and cell-type specific connectivity. Given the complexity and

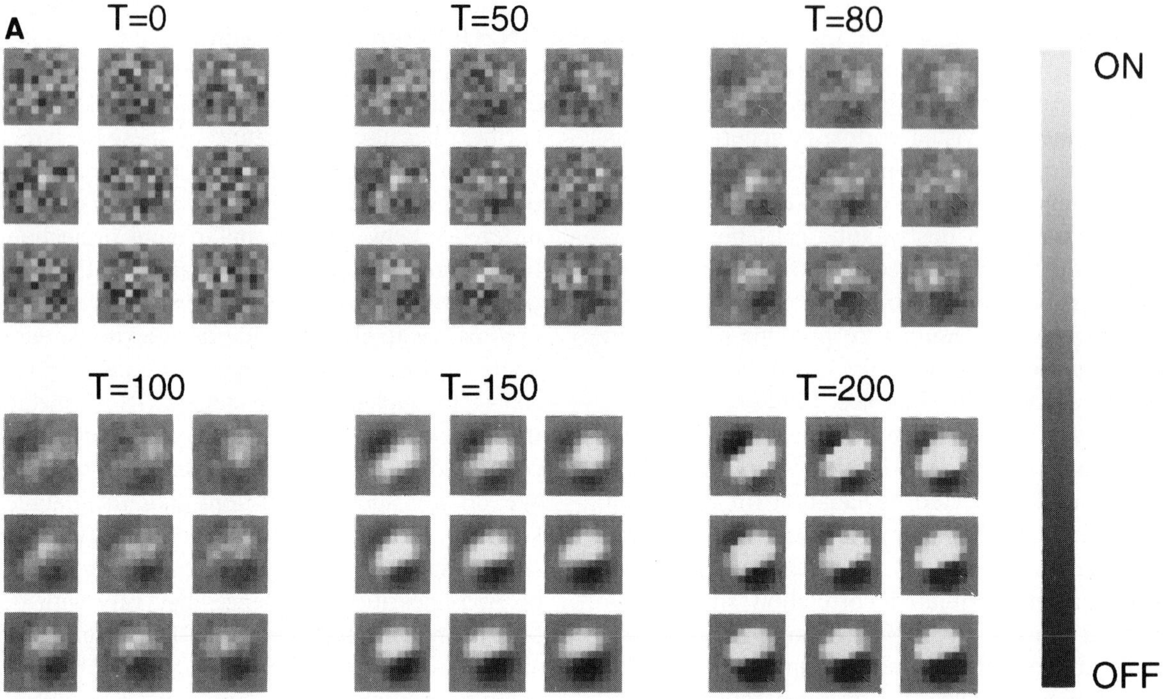

Fig. 5. Development of orientation selectivity through competition between ON- and OFF-center inputs. Results from simulation of development in a 31 by 31 array of cortical cells. (A) Development of orientation selectivity in receptive fields of a 3 by 3 group of cortical cells. Difference between ON-center and OFF-center synaptic strength at each point in each receptive field is shown, at six time steps ranging from a randomly assigned initial condition ($T = 0$) to the final state. (B) Time development of map of preferred orientation from the same simulation. Each pixel shows the 'best' orientation of one cortical cell. Orientation selectivity

specificity of cortical connections, why should the intracortical interactions be described simply as a function of the horizontal separation of two cortical sites? There is at present no compelling answer for this question based on realistic models of cortical circuitry, although simplified circuits with this property can be constructed (Von der Malsburg, 1973; Wilson and Cowan, 1973; Pearson et al., 1987), and there are bits of evidence for the real cortex (Miller and Stryker, 1990). Again, we begin with the solution: without an interaction function that depends on separation, it is difficult to account for such regular spatial properties as periodic columns. Should the cortical interaction function and correlation functions be independent of one another? It may be that the average distance over which intracortical interactions are excitatory depends on the average size of a coactivated set of inputs, and not simply on intrinsic circuitry in cortex. By beginning with the separate role of each of these functions, we reach an understanding that will serve us even if it turns out that one function partly determines the other.

The role of realistic models

Realistic modeling may help bridge the gap between biophysically realistic neuronal properties and the level at which our modeling begins. If the plasticity rule is coupled to local postsynaptic voltage, why should the competition on a single cortical cell result in the left eye or right eye winning the entire cell, rather than in an independent competition for each dendrite? The answer will ultimately be determined by detailed single-cell modeling (Lytton and Wathey, 1992), which is

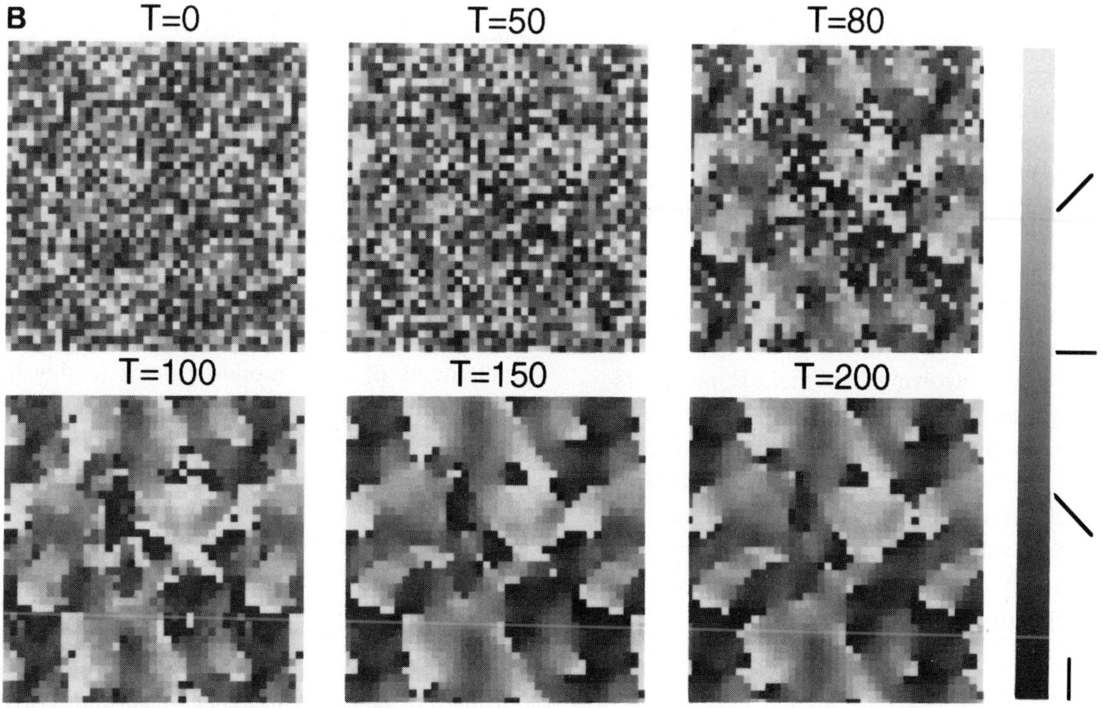

corn(preference for best orientation) is initially extremely slight, as seen in (A). Note that, because orientation is a circular variable, black and white extremes both represent vertical. If one grid interval is regarded to represent about 100 μm, the final map shows reasonable resemblence to maps measured in cats. This scale is biologically plausible, corresponding for this simulation to arbor diameters of about 1 mm, purely excitatory intracortical interactions over about 200 μm, and a change in sign in ON/OFF correlations (Fig. 3A bottom) at a retinotopic separation corresponding to about half an arbor radius.

just beginning to be used to explore synaptic plasticity (Holmes and Levy, 1990; Zador et al., 1990; Brown et al., 1991). Using realistic but linear membrane biophysics, realistic synaptic conductances, and a Hebbian plasticity rule that depends on local postsynaptic voltage, it has been found that there are conditions in which one or a small number of patterns of synaptic inputs 'takes over' an anatomically detailed hippocampal pyramidal cell (Brown et al., 1991). The parameters that give this result are still under investigation, but are difficult to characterize due to the complexity of the model. Should intracortical interactions be described simply as a function of intracortical separation? This could be investigated by constructing biophysically and anatomically realistic models of activation in cortical networks, using methods like those described in Traub and Miles (1992), in order to determine the cortical interactions that emerge on a time scale appropriate to affect plasticity in a correlated group of inputs.

Such piece-by-piece use of biophysical modeling to evaluate or alter key assumptions in a simplified model should be contrasted with an attempt to directly attack a developmental problem using a biophysically detailed simulation. The latter approach faces two major hurdles. First, it must necessarily use a very simplified version of the biophysics, for practical reasons. It is at present computationally impossible to simulate detailed model networks like those, say, of Traub and Miles (Traub and Miles, 1992) for the hours or days of real time that are the minimum needed to study development. Second, despite such simplification a biophysically detailed model of plasticity necessarily has a large number of parameters that are not constrained by experimental knowledge. Some means of managing this complexity must be found.

Inclusion of biophysical detail does not necessarily render a model more realistic (see related discussion in Sigvardt and Williams, 1992). For example, one recent model of development in somatosensory cortex (Pearson et al., 1987) used a plasticity rule involving hypothetical receptors with complicated kinetics and 14 arbitrary parameters, and rules for cortical cell activation requiring specification of 10–15 additional parameters. But, under the conditions studied, the plasticity rule reduced to a Hebbian rule in which plasticity occurs only when the postsynaptic cell is depolarized, and the pattern of cortical activation was simply short-distance excitation and longer-distance inhibition (Miller and Stryker, 1990). Thus, our analysis could be applied so that the results could be understood from the correlations in the input patterns used, the arbors, and the intracortical interactions as specified by the range of excitation and inhibition (Miller and Stryker, 1990). Interesting developmental results were obtained, but understanding was obscured rather than enhanced by using biologically unconstrained complexity to implement a very simple model. In particular, because the authors did not understand what determined the width of the cortical patches that developed in their model, they did not realize that certain aspects of their results could be ruled out by existing physiological observations (Miller and Stryker, 1990).

In sum, models must be informed and constrained by biological knowledge at their particular level of realism or abstraction. An increase in detail does not imply an increase in realism; unrealistic or unconstrained details may obscure rather than enhance understanding. Genuine study of the problem of plasticity at deeper levels of biological realism will extend and transform our knowledge, but this will likely need to be guided by, and will in turn guide, the understanding we gain at a more simplified level of description.

Abstract models of activity-dependent development

Models of development may also be formulated at an abstract level. These models explore dynamical or optimization principles that may lead to patterns like those observed in development. The elements of these models cannot readily be identified with biologically measurable quantities; but the simplified setting may expose principles that can carry over to a biological setting.

One set of models characterizes the Fourier transforms of biological patterns, either directly (Rojer and Schwartz, 1990) or by positing developmental rules that are simple in Fourier space (Swindale, 1980, 1982). Swindale proposed that ocular dominance at one cortical site influences the growth of ocular dominance at another cortical site according to a simple function of distance, and that these influences add linearly; the biological nature of the influence was not specified (Swindale, 1980). Such a rule selects periodic patterns of ocular dominance whose spatial period corresponds to the peak of the Fourier transform of the influence function. A similar rule, proposing that orientations influence one another's growth according to a function of distance (Swindale, 1982), selects a periodic pattern of preferred orientations in the same way.

The strength of these models is that they accurately describe observed maps of orientation and ocular dominance in a very simple way. Furthermore, the simplicity of the models allows interesting reasoning about issues such as monocular deprivation (Swindale, 1980) or the spacing of orientation singularities (Swindale, 1982). The problems with these models are that they cannot obviously be distinguished from more complex models that would yield a similar final distribution of power in Fourier space, and they do not propose a biological mechanism by which the posited influences might be exerted.

In the case of the ocular dominance model, there is a simple bridge to a biologically identifiable model. In the limit in which each eye fires as a unit, so that correlations are completely specified by eye of origin without need to specify retinotopic position, the model we proposed for ocular dominance reduces to one in terms of an influence function. The influence function is then specified in terms of biologically identifiable factors: correlations, arbors, cortical interactions, and presynaptic conservation rules if any. However, in the case of orientation selectivity our model does not provide such a connection: in our model, the interaction between two cells of the same orientation can vary from maximally positive to maximally negative depending on the spatial phases of the two receptive fields, so an influence function cannot be derived that depends on orientation difference alone.

Another set of models posit simplified versions of Hebbian rules. Linsker used linear Hebbian rules with synapses that could be either positive or negative, and a constraint that fixed the total percentage of positive and of negative synaptic strength in a receptive field (Linsker, 1986a,b,c; analyzed in Miller, 1990b; MacKay and Miller, 1990a,b). Variation of those percentages and of the input correlations revealed circumstances in which interesting receptive field structures arise. Correlations that oscillate led to spatially structured receptive fields with segregated regions of positive and negative synaptic input. Symmetry could spontaneously break, so that circularly symmetric inputs and arbors led to oriented receptive fields. However, the orientation-selective cells developed only when the constraint was finely tuned, and then only in the non-linear regime of development (MacKay and Miller, 1990a). Most of the results similarly depended on the constraint and its tuning, on the ability of synapses to become either positive or negative (or more generally, on the assumption of complete indistinguishability of positive and negative synapses (Miller, 1990b)), and on the particular choice of saturating synaptic non-linearities (Miller, 1990b; MacKay and Miller, 1990a,b). Thus, this model demonstrated previously unknown dynamical outcomes that could arise from Hebbian rules, but it did not demonstrate biologically plausible or robust ways to achieve those outcomes.

Other simplified Hebbian models use the self-organizing feature maps of Kohonen (Kohonen, 1989) or similar algorithms (Durbin and Mitchison, 1990) to model the development of ocular dominance and orientation columns (Durbin and Mitchison 1990; Obermayer et al., 1990, 1991, 1992). The Kohonen mappings are an abstraction of a Hebbian rule. The major abstractions are that, for any given input pattern, the output of cortical cells is determined without intracortical interactions, and reinforcement of active inputs then occurs only on the most activated cortical cell and its near neighbors, a representation of localized clusters in cortical activation. Multiplicative constraints conserve the sum of squares of the synaptic weights over each cortical cell. These mappings are amenable to analysis and have a number of interesting features. They lead to continuous maps in which all inputs gain equal representation in cortex if activated equally often, nearby inputs develop nearby cortical representations, and a constant distance across the cortex corresponds to a roughly constant distance in 'input space'. If the input space has more than the cortex's two dimensions — for example, the five-dimensional space of ocular dominance, preferred orientation, orientation selectivity, and 2D retinotopic position — this means that, when one feature is changing rapidly across cortex, the others will be changing slowly. The algorithms produce realistic maps of orientation (Durbin and Mitchison, 1990; Obermayer et al., 1990) and of ocular dominance (Kohonen, 1989; Obermayer et al., 1991, 1992), and reproduce many of the experimentally observed relationships between orientation and ocular dominance columns (Obermayer et al., 1991, 1992).

These mapping algorithms have several problems as biological models: they do not predict the column widths in any straightforward way from biologically identifiable parameters; as implemented thus far, they cannot break symmetry to develop oriented responses from non-oriented input patterns, and more generally, the biological interpretations of the various abstractions used are not clear. But these models demonstrate simple rules that can account for complicated aspects of cortical maps, and their formal closeness to Hebbian rules suggests that they may serve as a guide for more biologically based models.

The abstract models demonstrate that differing dynamical mechanisms may converge on similar results resembling those seen biologically. Conversely, very different biological mechanisms may hide similar underlying dynamics. Furthermore some aspects of the results of a model may vary with details of implementation and thus not be firmly tied to the mechanism being studied. Thus, resemblance of model and biological results should not be viewed as a validation of a proposed model; rather it is a demonstration that a proposed mechanism can be involved, under some circumstances, in achieving results like those found biologically.

Summary and conclusions

What makes a useful model of neural development? One important contribution of modeling is to demonstrate that proposed biological mechanisms can be sufficient to account for experimental results. The Von der Malsburg model is a classic example. But such demonstrations alone do not provide tools to experimentally distinguish one mechanism from another. To draw such distinctions, the connection between measurable biological quantities and developmental outcomes must be established.

Perhaps the most important task for the future of developmental modeling is to deepen the connection between theory and experiment. Experimentally, this requires detailed and difficult measurements or experimental perturbations of the correlations among inputs and the intracortical connectivity existing during development. Simultaneous measurement of the maps of spatial phase and orientation of mature simple cells will provide important information for the understanding of orientation column development.

Theoretically, the number of open problems is enormous. How will inclusion of additional plasticity mechanisms, such as sprouting and retraction of synapses or plasticity of intracortical connections, alter the analytical understanding thus far achieved? What precisely determines the width of orientation columns in the model presented here? Can the relationship between ocular dominance and orientation columns be understood from developmental rules in a testable way? The existing framework may be extended to a three-dimensional cortex and to more complex models of intracortical connectivity. It may also be applied to other developmental phenomena including the development of lamination in the LGN (Shatz and Stryker, 1988; Hahm et al., 1991), the formation of visual maps in experimentally altered auditory cortex (Roe et al., 1990, 1992), and the mapping of visual and auditory maps in the optic tectum (Knudsen and Brainard, 1991; Brainard and Knudsen, 1993). For each system the goal is to develop testable predictions as to the patterns of activity and connectivity that could or could not lead to the results observed given a proposed mechanism of plasticity. Incorporation of deeper levels of biophysical realism will extend, deepen, and perhaps fundamentally alter the framework presented here. An important goal for the future will be to understand the computational and functional significance of developmental rules.

Activity-dependent, competitive mechanisms of synaptic plasticity appear to play an important role in many processes of late neural development, where an initially rough connectivity pattern refines to a precise, mature pattern. A prominent example is the formation of ocular dominance columns in the visual cortex of many mammals. These processes may be modeled at several levels. Simple models use abstract neurons and assume synaptic modification according to a Hebbian or similar correlation-based rule. These models incorporate biological constraints and attempt to predict large-scale developmental patterns from the combination of synaptic-level plasticity rules and measurable biological patterns of activation and connectivity. More detailed models attempt to incorporate various levels of biophysical realism, including membrane and channel properties and dendritic geometry. Abstract models examine the connectivity patterns that may result if biological development follows certain dynamical or other abstract rules, without concern for how such rules might be implemented at the synapse. The strengths and weaknesses of these approaches have been examined in the present review through study of models for the development of ocular dominance and of orientation selectivity in the visual cortex.

References

Brainard, M.S. and Knudsen, E.I. (1993) Experience-dependent plasticity in the inferior colliculus: a site for visual calibration of the neural representation of auditory space in the barn owl. *J. Neurosci.,* 13: 4589–4608.

Brown, T.H., Zador, A.M., Mainen, Z.F. and Claiborne, B.J. (1991) Hebbian modifications in hippocampal neurons. In: J. Davis and M. Baudry (Eds.), *Long Term Potentiation: A Debate Cf Current Issues,* MIT Press, Cambridge, MA, pp. 357–389.

Bullier, J. and Henry, G.H. (1979) Laminar distribution of first-order neurons and afferent terminals in cat striate cortex. *J. Neurophysiol.,* 42: 1271–1281.

Chapman, B. and Stryker, M.P. (1993) Development of orientation selectivity in ferret visual cortex and effects of deprivation. *J. Neurosci.,* 13: 5251–5262.

Constantine-Paton, M., Cline, H.T. and Debski, E. (1990) Patterned activity, synaptic convergence and the NMDA receptor in developing visual pathways. *Annu. Rev. Neurosci.,* 13: 129–154.

Durbin, R. and Mitchison, G. (1990) A dimension reduction framework for understanding cortical maps. *Nature,* 343: 644–647.

Grossberg, S. (1976) On the development of feature detectors in the visual cortex with applications to learning and reaction-diffusion systems. *Biol. Cybern.,* 21: 145–159.

Guillery, R.W. (1972) Binocular competition in the control of geniculate cell growth. *J. Comp. Neurol.,* 144: 117–130.

Hahm, J.O., Langdon, R.B. and Sur, M. (1991) Disruption of retinogeniculate afferent segregation by antagonists to NMDA receptors. *Nature,* 351: 568–570.

Hebb, D.O. (1949) *The Organization of Behavior,* John Wiley, New York.

Holmes, W.R. and Levy, W.B. (1990) Insights into associative long-term potentiation from computational models of NMDA receptor-mediated calcium influx and intracellular calcium concentration changes. *J. Neurophysiol.,* 63: 1148–1168.

Hubel, D.H. and Wiesel, T.N. (1962) Receptive fields, binocular interaction and functional architecture in the cat's visual cortex. *J. Physiol.,* 160: 106–154.

Hubel, D.H., Wiesel, T. and Stryker, M.P. (1978) Anatomical demonstration of orientation columns in macaque monkey. *J. Comp. Neurol.,* 177: 361–380.

Knudsen, E.I. and Brainard, M.S. (1991) Visual instruction of the neural map of auditory space in the developing optic tectum. *Science,* 253: 85–87.

Kohonen, T. (1989) *Self-Organization and Associative Memory,* Springer-Verlag, Berlin, 3rd edition.

LeVay, S. and Nelson, S.B. (1991) The columnar organization of visual cortex. In: A.G. Leventhal (Ed.), *The Neural Basis of Visual Function,* CRC Press, Boca Raton, pp. 266–315.

Linsker, R. (1986b) From basic network principles to neural architecture: emergence of orientation columns. *Proc. Natl. Acad. Sci. USA,* 83: 8779–8783.

Linsker, R. (1986a) From basic network principles to neural architecture: emergence of orientation-selective cells. *Proc. Natl. Acad. Sci. USA,* 83: 8390–8394.

Linsker, R. (1986c) From basic network principles to neural architecture: emergence of spatial-opponent cells. *Proc. Natl. Acad. Sci. USA,* 83: 7508–7512.

Loewel, S., Bischof, H.-J., Leutenecker, B. and Singer, W. (1988) Topographic relations between ocular dominance and orientation columns in the cat striate cortex. *Exp. Brain Res.,* 71: 33–46.

Lytton, W.W. and Wathey, J.C. (1992) Realistic single-neuron modeling. *Semin. Neurosci.,* 4: 15–25.

MacKay, D.J.C. and Miller, K.D. (1990a) Analysis of Linsker's applications of Hebbian rules to linear networks. *Network,* 1: 257–298.

MacKay, D.J.C. and Miller, K.D. (1990b) Analysis of Linsker's simulations of Hebbian rules. *Neural Comput.,* 2: 173–187.

Maffei, L. and Galli-Resta, L. (1990) Correlation in the discharges of neighboring rat retinal ganglion cells during prenatal life. *Proc. Natl. Acad. Sci. USA,* 87: 2861–2864.

Mastronarde, D.N. (1989) Correlated firing of retinal ganglion cells. *Trends Neurosci.,* 12: 75–80.

Meister, M., Wong, R.O.L., Baylor, D.A. and Shatz, C.J. (1991) Synchronous bursts of action-potentials in ganglion cells of the developing mammalian retina. *Science,* 252: 939–943.

Miller, K.D. (1990a) Correlation-based models of neural development. In: M.A. Gluck and D.E. Rumelhart (Eds.), *Neuroscience and Connectionist Theory,* Lawrence Erlbaum Ass., Hillsdale, NJ, pp. 267–353.

Miller, K.D. (1990b) Derivation of linear Hebbian equations from a non-linear Hebbian model of synaptic plasticity. *Neural Comput.,* 2: 321–333.

Miller, K.D. (1992) Development of orientation columns via competition between ON- and OFF-center inputs. *NeuroReport,* 3: 73–76.
Miller, K.D. (1994) A model for the development of simple cell receptive fields and the ordered arrangement of orientation columns through activity-dependent competition between ON- and OFF-center inputs. *J. Neurosci.,* 14: 409–441.
Miller, K.D. and MacKay, D.J.C. (1994) The role of constraints in Hebbian learning. *Neural Comput.,* 6: 98–124.
Miller, K.D. and Stryker, M.P. (1990) The development of ocular dominance columns: mechanisms and models. In: S.J. Hanson and C.R. Olson (Eds.), *Connectionist Modeling and Brain Function: The Developing Interface,* MIT Press/Bradford, Cambridge, MA, pp. 255–350.
Miller, K.D., Keller, J.B. and Stryker, M.P. (1989) Ocular dominance column development: analysis and simulation. *Science,* 245: 605–615.
Nass, M.M. and Cooper, L.N. (1975) A theory for the development of feature detecting cells in visual cortex. *Biol. Cybern.,* 19: 1–18.
Obermayer, K., Ritter, H. and Schulten, K. (1990) A principle for the formation of the spatial structure of cortical feature maps. *Proc. Natl. Acad. Sci. USA,* 87: 8345–8349.
Obermayer, K., Schulten, K. and Blasdel, G.G. (1991) A neural network model for the formation and for the spatial structure of retinotopic maps, orientation- and ocular dominance columns. Theoretical Biophysics Technical Report TB-91-09, The Beckman Institute, University of Illinois Urbana-Champaign.
Obermayer, K., Blasdel, G.G. and Schulten, K. (1992) A statistical mechanical analysis of self-organization and pattern formation during the development of visual maps. *Phys. Rev. A,* 45: 7568–7589.
Pearson, J.C., Finkel, L.H. and Edelman, G.M. (1987) Plasticity in the organization of adult cerebral cortical maps: a computer simulation based on neuronal group selection. *J. Neurosci.,* 7: 4209–4223.
Perez, R., Glass, L. and Shlaer, R. (1975) Development of specificity in the cat visual cortex. *J. Math. Biol.,* 1: 275–288.
Roe, A.W., Pallas, S.L., Hahm, J.-O. and Sur, M. (1990) A map of visual space induced in primary auditory cortex. *Science,* 250: 818–820.
Roe, A.W., Pallas, S.L., Kwon, Y.H. and Sur, M. (1992) Visual projections routed to the auditory pathway in ferrets: receptive fields of visual neurons in primary auditory cortex. *J. Neurosci.,* 12: 3651–3664.
Rojer, A.S. and Schwartz, E.L. (1990) Cat and monkey cortical columnar patterns modeled by bandpass-filtered 2D white noise. *Biol. Cybern.,* 62: 381–391.
Shatz, C.J. (1990) Impulse activity and the patterning of connections during CNS development. *Neuron,* 5: 745–756.
Shatz, C.J. and Stryker, M.P. (1988) Tetrodotoxin infusion prevents the formation of eye-specific layers during prenatal development of the cat's retinogeniculate projection. *Science,* 242: 87–89.
Sherk, H. and Stryker, M.P. (1976) Quantitative study of cortical orientation selectivity in visually inexperienced kitten. *J. Neurophysiol.,* 39: 63–70.
Sigvardt, K.A. and Williams, T.L. (1992) Models of central pattern generators as oscillators: the lamprey locomotor CPG. *Semin. Neurosci.,* 4: 37–46.
Swindale, N.V. (1980) A model for the formation of ocular dominance stripes. *Proc. R. Soc. London B.,* 208: 243–264.
Swindale, N.V. (1982) A model for the formation of orientation columns. *Proc. R. Soc. London B.,* 215: 211–230.
Traub, R.D. and Miles, R. (1992) Modeling hippocampal circuitry using data from whole cell patch clamp and dual intracellular recordings in vitro. *Semin. Neurosci.,* 4: 27–36.
Von der Malsburg, C. (1973) Self-organization of orientation selective cells in the striate cortex. *Kybernetik,* 14: 85–100.
Von der Malsburg, C. and Willshaw, D.J. (1976) A mechanism for producing continuous neural mappings: ocularity dominance stripes and ordered retino-tectal projections. *Exp. Brain Res.,* 1(Suppl.): 463–469.
Wiesel, T.N. and Hubel, D.H. (1965) Comparison of the effects of unilateral and bilateral eye closure on cortical unit responses in kittens. *J. Neurophysiol.,* 28: 1029–1040.
Wiesel, T.N. and Hubel, D.H. (1974) Ordered arrangement of orientation columns in monkeys lacking visual experience. *J. Comp. Neurol.,* 158: 307–318.
Willshaw, D.J. and Von der Malsburg, C. (1976) How patterned neural connections can be set up by self- organization. *Proc. R. Soc. London B.,* 194: 431–445.
Wilson, H.R. and Cowan, J.D. (1973) A mathematical theory of the functional dynamics of cortical and thalamic nervous tissue. *Kybernetik,* 13: 55–80.
Wong, R.O., Meister, M. and Shatz, C.J. (1993) Transient period of correlated bursting activity during development of the mammalian retina. *Neuron,* 11: 923–938.
Zador, A., Koch, C. and Brown, T.H. (1990) Biophysical model of a Hebbian synapse. *Proc. Natl. Acad. Sci. USA,* 87: 6718–6722.

J. van Pelt, M.A. Corner H.B.M. Uylings and F.H. Lopes da Silva (Eds.)
Progress in Brain Research, Vol 102

CHAPTER 21

Role of chaotic dynamics in neural plasticity

Walter J. Freeman

Department of Molecular and Cell Biology, LSA 129, University of California at Berkeley, CA 94720-3200, U.S.A.

Introduction

Neurobiologists are encountering intractable difficulties in pursuing their interests owing to the complexity of the systems under study. By tradition (Sherrington, 1906) the nervous system is viewed hierarchically in terms of nerve cells and their parts, organelles, and macromolecules in the reductive direction, and in terms of local assemblies and networks, areas of cortex and nuclei, sensory and motor systems, and functioning brains in the integrative direction. In past decades the concept of 'simpler systems' arose (Fentress, 1976) by which problems of neural function were to be analyzed using isolated neurons, ganglia, and intact segments of the lower invertebrates and vertebrates. This is the route by which, for example, long-term potentiation (LTP) came to light and to be pursued from the snail to the hippocampus as a model for associative learning (Alkon, 1992).

What has become apparent in recent years is that every level of the neural hierarchy reveals unbounded complexity, and that what is learned about a synapse or a neuron cannot easily be applied upwardly to an area of cortex or a brain system. Reductionism is no panacea. It is also obvious that biologists increasingly cannot understand their new data or fit them together into meaningful patterns without the use of models (Freeman, 1993). In retrospect it should have become clear when the Hodgkin-Huxley model was introduced in 1952 that mathematical formalisms are essential for significant progress to be made in the biology of complex systems. However, the construction of the requisite abstractions is an exceedingly difficult task that requires facility in both experiment and theory (Freeman, 1975).

Artificial neural networks (ANN) form a class of models already at hand, which has grown explosively in the past decade. Despite their variety and breadth of applications, they have seen little use by neurobiologists. There are some investigators who have taken great interest in them as the basis for a fresh approach to studies of the function of nervous systems in the contexts of computational neuroscience, neurocomputation, and neural information processing. But the majority of biologists cite the lack of relevance of the new models to biological neural networks (BNN) as they know them, and appear to be offended upon hearing a neuron analogized to a transistor or an area of cortex to a multilayered perceptron. Certainly the components, connectivities, architectures and dynamics differ substantially between ANN and BNN, but the important differences lie much more deeply. Their origins are to be found in the misconceptions that biologists sustain in their own views about how brains function. Despite their disdain for cognitive connectionists, or perhaps because of it, biologists seem unaware that ANN are not the independent creations of engineers. They are accurate reflections of how most biologists conceive of the operations of nervous systems. Most of these conceptions derive from studies in anesthetized or paralyzed animals or those inactivated by surgical incisions to disconnect or destroy the cerebrum. In such states the neurons and brain systems by design become incapable of goal directed behavior, and they perform reflex actions only upon delivery of input, just as a neural network or a PC sits quietly and awaits instructions before springing into action. Neurobiologists usually require that their BNN maintain a stable baseline state to which they return after each test or trial, as the basis for getting reproducible results (Freeman, 1987a). Thereby they seek to describe an analytic operation by which a sensory input is transformed to a bounded output within a brief time frame. This is exactly the mode in which ANN are operated. A deviation from asymptotic approach to a steady state in a BNN is tolerated only if the neuron or network enters into periodic oscillation at a time-invariant frequency. The BNN can then be modeled with an oscillator, which is also a stable and predictable device. If a neuron or a network begins to generate aperiodic activity, which is unpredictable because it is not periodic, the BNN is commonly thought to be 'deteriorating', in much the same way that an ANN which performs unpredictably is thought to be defective, perhaps having noisy components that should be replaced.

BNN in the normal waking state are very different from most ANN as currently conceived (Table 1). They are ceaselessly active in generating so-called 'spontaneous' activity, which has the form of random pulse trains generated by axons and broad spectrum EEG activity generated by the dendrites of cortical neurons in large populations. Brains manifest continual changes in state, sometimes at the prompting of the investigator, but more often outside experimental control, implying that they are highly unstable, in strong contrast to ANN, which maintain silence when not addressed and become active on command, but within a state or a succession of states carefully controlled by the engineer or computer scientist. ANN wait like anesthetized animals for the arrival of input and then perform the equivalent of a conditioned reflex, whereas BNN actively seek input as in sniffing, feeling, looking and listening. ANN are designed to perform selected tasks, whereas BNN are oriented toward the achievement of self-organized, internally formulated goals.

TABLE I

Comparison of artificial and biological neural networks

ANN	BNN
Stationary	Unstable
Input-driven	Self-organizing
Task-oriented	Goal-oriented
Computational	Dynamic
bits	flows
symbols	patterns
information	meaning
Memory	Remembering
An object	A process
Representations	Trajectories
Gradient descent	State transition
Retrieval	Construction
Test by matching	Test by motor act

Of course, in one sense a brain is stable throughout a lifetime unless it becomes epileptic, and in another sense a computer has a new state whenever it is loaded with a new program, but these interpretations of the words miss a crucial difference between computational and dynamical systems. Most importantly, the operations of BNN are by dynamic flows and state changes, not by computation. To compute is to calculate, reckon, or estimate by use of numbers. Symbols in languages and mathematics are constructed in human brains by neural dynamics. BNN lack numbers or other formal codes and the agencies for attaching meaning to symbols that are manipulated according to logical rules. A perceptual act requires the construction of a neural activity pattern, not the retrieval of a fixed pattern stored in a memory bank. That which is constructed is not an image of the stimulus that initiated the construction (Bartlett, 1932). It is an expression of the meaning or significance of the stimulus for the subject, in the light of previous experience with the stimulus by the subject. Whereas ANN can be built to manipulate symbols in codes that convey information in the sense defined by Shannon and Weaver (Lucky, 1989), who divorced information from meaning (Dreyfus, 1979), BNN offer a dynamics of meaning, according to which the patterns constructed in sensory cortex serve to destabilize and drive motor systems that initiate and maintain search behavior for new stimuli.

Since biological information cannot be defined or measured, the phrase 'information processing' serves only as a metaphor to describe the process of self-induced modification in a BNN using its environment as a source of input, and it cannot be regarded as describing a set of laws governing throughput and transformation of measured quantities as in a telephone line. Moreover, a central activity pattern of a BNN can be described metaphorically as a 'representation' of the outside world, but the term cannot be used in the senses that a digit represents a number or a painting represents a scene, because within the animal brain there is no means for an observer to verify what is being represented by which part to which other part of the brain (Skarda and Freeman, 1987). With respect to meaning as opposed to stimulus energy, BNN are closed systems. The internal activity patterns must ultimately bear some relation to the outside world, because they organize the firing patterns of neurons by the hundreds of millions, leading to coordinated activity of a subject into the world for the incorporation of new stimulation, but the meanings are a private, self-organized set of relations within the brain of each subject. We cannot know but can only infer what others, even our most intimate friends, are perceiving and thinking.

Those kinds of ANN that derive from passive networks are capable of little more than homeostatic reflexes that constitute error correction by gradient descent, but other kinds of models have been developed as goal-seeking automata, which have much more in common with BNN, such, for example, as the 'turtles' of Walter (1956), the 'homeostats' of Ashby (1956), and the 'vehicles' of Braitenberg (1984). These models demonstrate the feasibility of constructing dynamical systems having the self-organizing properties of animals without requiring the computational machineries of ANN that rely upon codes, symbols and other forms of representation. The review here is focused upon the sensory side of animals regarded as free-roving automata, in particular those stages which constitute the interface between the infinite complexity of the outside world and the finite inner machinery for predictive control of behavior, and which have been described by the visionary poet William Blake as 'The Doors of Perception' (Freeman, 1991).

The questions to be considered are: What do patterns of perceptual activity in sensory cortices look like? How are they constructed through instabilities of chaotic dynamics? And what properties of the cortical neurons are necessary for the instabilities of the cortices to arise?

What do patterns of perceptual activity look like?

Neural activity exists in many forms of electrical and chemical energy, which can be observed by means of a variety of techniques. In the approach described here, reliance is placed on recording the electrical signs of activity of neurons in behaving animals, with emphasis on the cerebral cortex. These signs are in two classes (Fig. 1). Axons generate pulse trains that are accessed by microelectrodes one or a few neurons at a time. Dendrites generate electric current that determines the pulse frequencies of axons. The current of each neuron flows in a closed loop in accordance with Kirchoff's Current Law intracellularly from the synaptic 'battery' to the axon initial segment and back extracellularly. The intracellular limb is private to the neuron. The extracellular limb is across the cortical tissue where it sums with the other neurons using the same current path, giving rise to a field potential (electroencephalogram,

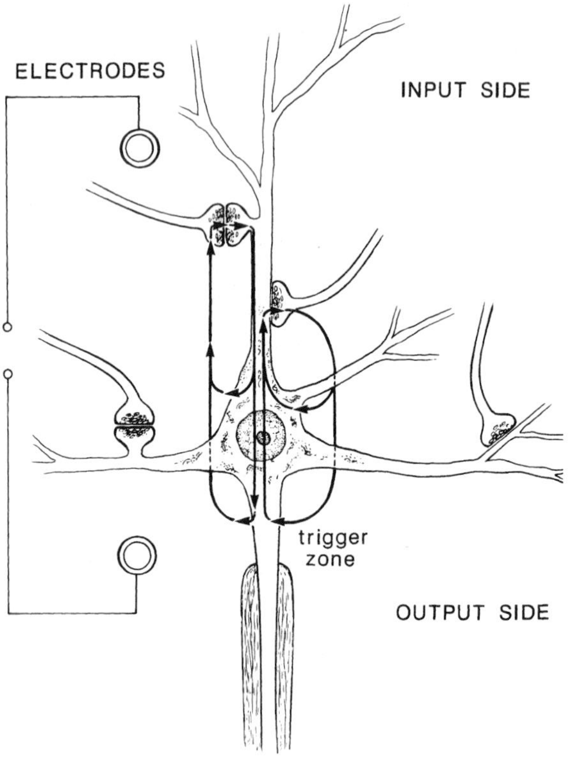

Fig. 1. A schematic drawing of a neuron shows the dendrites radiating upwardly and laterally from the cell body and the axon below. Each synapse acts as a small battery, which drives electric current in a closed loop. The excitatory current is inward at the synapse and outward at the initial segment of the axon where the output pulse is triggered. The inhibitory current flows in the opposite direction. The potential differences of the currents for the single cell sum at the trigger zone. The potential differences for the same current across the extracellular tissue resistance sum for all active neurons in the neighborhood. This EEG is recorded with transcortical electrode pairs.

EEG) that is accessed with macroelectrodes. When the positions of the macroelectrodes are properly specified with respect to the geometries and cytoarchitectures of neuron populations, the EEG provides a measure of the local mean field intensities of macroscopic activity of neural populations, as distinct from the microscopic activity of single neurons (Freeman, 1992).

Because the axons and dendrites of neurons and populations have two principal modes of function, pulses and waves, a mode of a neuron must have at least two state variables, one to represent dendritic current as a continuously varying amplitude and the other to represent axonal pulse frequency. A model of a population must likewise have two macroscopic state variables one to represent dendritic current density and the other to represent pulse density. Transforms at the microscopic level are needed to describe the conversion at synapses between the modes of pulse frequency and current amplitude, and the inverse at trigger zones between the modes of current amplitude and pulse frequency. Comparable but not identical transforms are needed to describe the relations between pulse density input, current density, and pulse density output at the macroscopic level (Freeman, 1975). Furthermore, two transforms are needed to describe the transfers of events between the two levels. Because transmission in the brain is by axons and is observed in pulse trains, whereas integration is by dendrites and is observed as wave amplitudes, it is logical to assert that tranformation of microscopic sensory input is at synapses, whereas the inverse transformation of macroscopic cortical activity observed in dendritic waves to output pulse patterns is at trigger zones.

Sensory events at the microscopic level and perceptual events at the macroscopic level co-exist as space-time patterns of neural activity in cortex in this hierarchy. Neural analysis of sensory input, which leads to feature extraction, is observed by recording the action potentials of single neurons and analyzing them as point processes. Neural synthesis of input with past experience and expectancy of future action is observed by recording the activity of populations in local mean fields, and describing the dendritic integration by constructing and solving differential equations. Both levels of activity are found to coexist in olfactory, auditory, somesthetic and visual cortices, each preceding and then following the other. The transformation of sensory input in the pulse mode to perceptual output in the dendritic mode is enacted by a state transition of a cortical population, in whch the microscopic activity in both pulse and wave modes just prior to the transition determines the selection of a macroscopic pattern, and thereafter that pattern in both wave and pulse modes 'enslaves' the microscopic activity of single neurons within the population (Haken and Stadler, 1990). The state transition observed in the EEG is a property of a population, whereas the state transition between firing and not firing is a property of an axon. These state transitions should not be confused with tranformations between pulse and wave modes or between different hierarchical levels.

An example of EEGs from the olfactory system is shown in Fig. 2. Each recording shows the electrical potentials established by the flow of dendritic current from neurons within a radius on the order of 1 mm, which are simultaneously active and therefore contribute anonymously to the sum of excitatory current in one direction and inhibitory current in the other direction. These waves show the typical irregularity and aperiodicity of oscillations in macroscopic activity found at sites throughout the brain. They also reflect a degree of influence by both sensory and motor activity, by which inhalation instigated by the limbic system through the respiratory centers activates olfactory receptors in the nose, which excite the bulb and cause it to change state. The state change is not readily apparent in the single EEG traces, which are continually changing, but it can be detected by simultaneously recording the EEGs from 64 electrodes in an 8 × 8 array. This method shows that the aperiodic wave form is shared across the array of traces, that it differs in amplitude from point to point forming a pattern, and that the spatial patterns can be made to

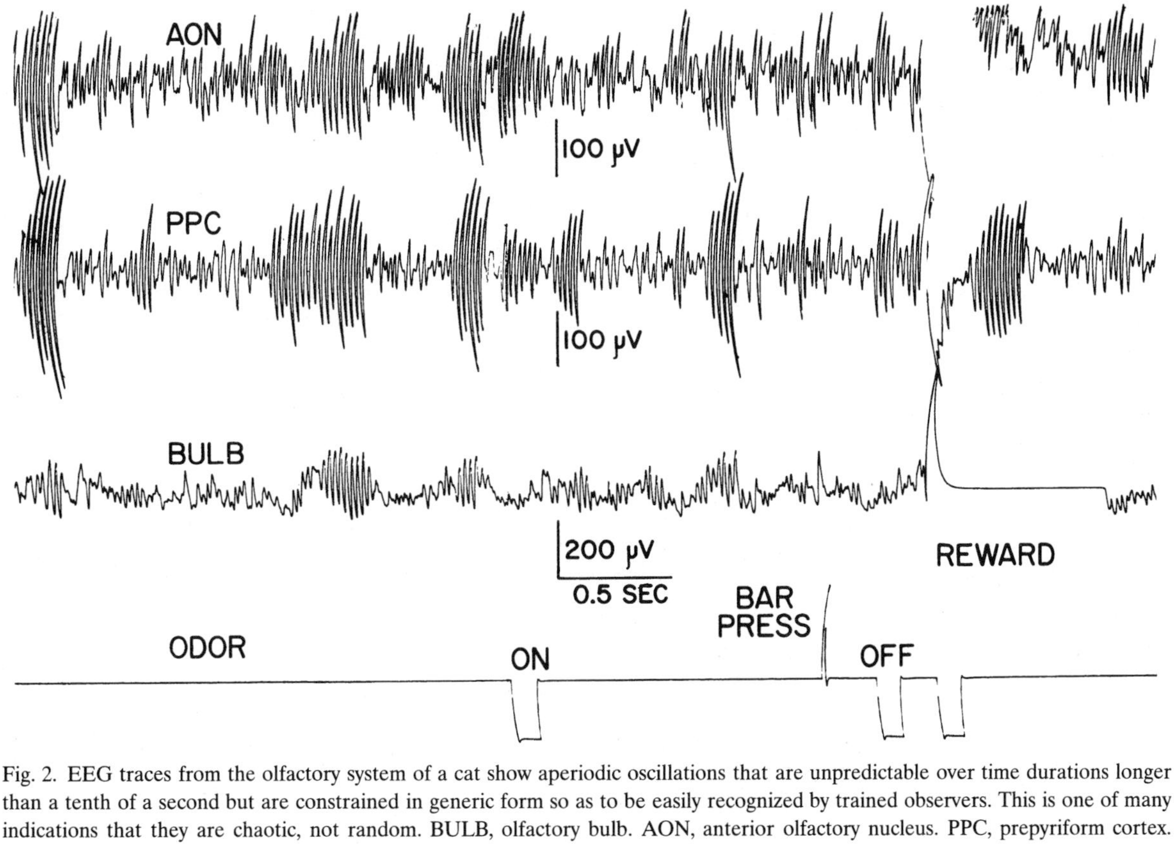

Fig. 2. EEG traces from the olfactory system of a cat show aperiodic oscillations that are unpredictable over time durations longer than a tenth of a second but are constrained in generic form so as to be easily recognized by trained observers. This is one of many indications that they are chaotic, not random. BULB, olfactory bulb. AON, anterior olfactory nucleus. PPC, prepyriform cortex.

differ by training subjects to identify and respond to olfactory stimuli.

These spatial patterns can be displayed by means of contour plots (Fig. 3) that show the amplitude of the 64 waves during brief time segments lasting on the order of one tenth of a second and corresponding to single inhalations. In this example records were taken just after the rabbit was placed in the recording situation (A) and then again after several weeks of familiarization (B) to achieve a steady baseline. The change with learning to adapt to the recording conditions turned out to be typical of the rabbits. When a novel odorant stimulus was given, the steady pattern was temporarily replaced by wildly irregular activity with no definable pattern (Fig. 4), and the subject oriented to the stimulus. If the novel odorant was not paired with a rewarding or punishing unconditional stimulus, the steady pattern returned, and the orienting response abated as the subject habituated to the odorant. If the novel stimulus was paired with a reinforcement, the steady spatial pattern changed (C, 'sawdust'), and it did so with each new odorant (D, E) in series (Freeman and Davis, 1990). When the sawdust odorant was re-introduced, the pattern again changed to a new form (F) and not to the preceding shape (C).

This lack of invariance with respect to the stimulus, which is shown here in generalization during recursive serial conditioning, was found to hold in a variety of forms of learning, including over-training and contingency reversal in discriminative learning. The latter is illustrated in

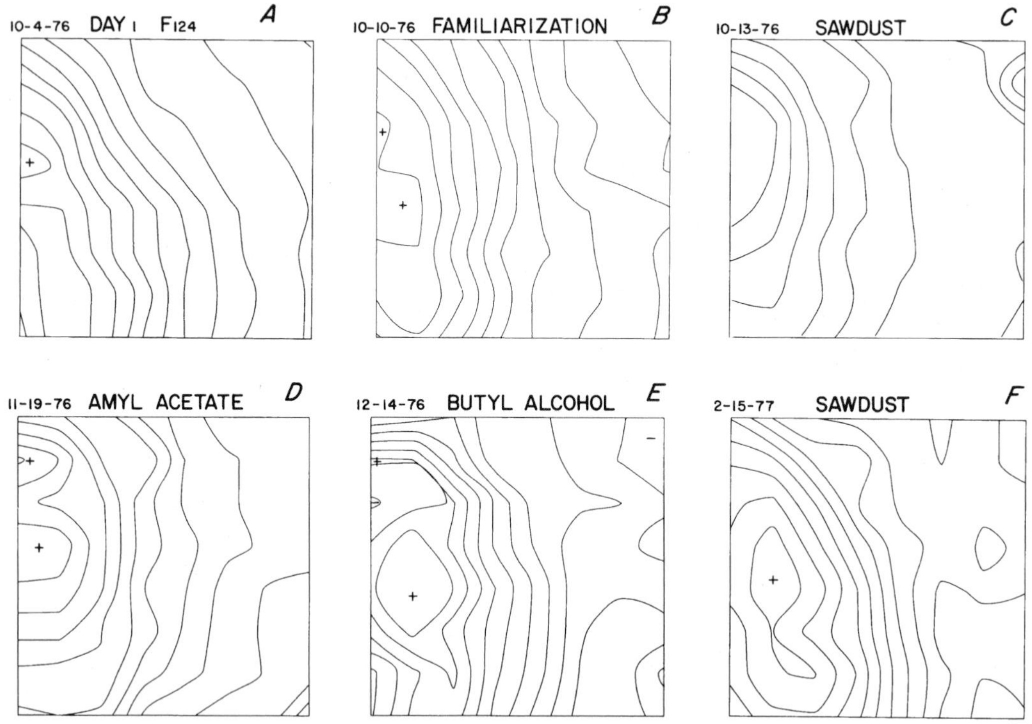

Fig. 3. Contour plots show the root mean square (r.m.s) amplitude of the EEG as recorded simultaneously at 64 sites with an 8×8 array of electrodes placed onto the surface of the olfactory bulb of a rabbit. The spatial patterns of amplitude modulation of a common wave form change with serial conditioning under reinforcement.

graphical form in Fig. 5 in which three stimuli, the background control C, an odorant paired with water (CS +), and another not so paired (CS −), are shown to elicit each of three spatial patterns at time *t*. They are followed by each of three responses (do nothing, lick, or sniff). When the reinforcement is switched, the stimuli and the responses are the same, but the central patterns for the odorants are different at time *t* + 1. The pattern with the control also changes (Freeman and Grajski, 1987). The conclusion is that the central patterns are not the computational transforms of the stimuli, but that they are constructs based on past experience with the stimuli and the contexts of the experience. Thereby they reflect the meaning of the stimuli for the subjects. It appears, further, that the meaning of a stimulus is changed by a change in the context, as should be expected in a truly associative memory.

What is the role of chaos in construction of patterns?

The finding that the central EEG patterns are not computable from the spatial patterns of the stimuli forces consideration of the mechanisms whereby masses of neurons might construct them. An answer proposed here in brief and developed elsewhere at greater length (Freeman, 1975, 1987a,b, 1992) is that each sensory cortex maintains a global chaotic attractor. It is manifested by the aperiodic activity seen in the basal state of the EEG between stimuli, which destabilize the bulb and lead to high amplitude oscillations known as 'bursts' (Fig. 6). The performance of the system is shown to advantage by plotting each variable against another in two or three dimensions in such fashion that, in conditions of au-

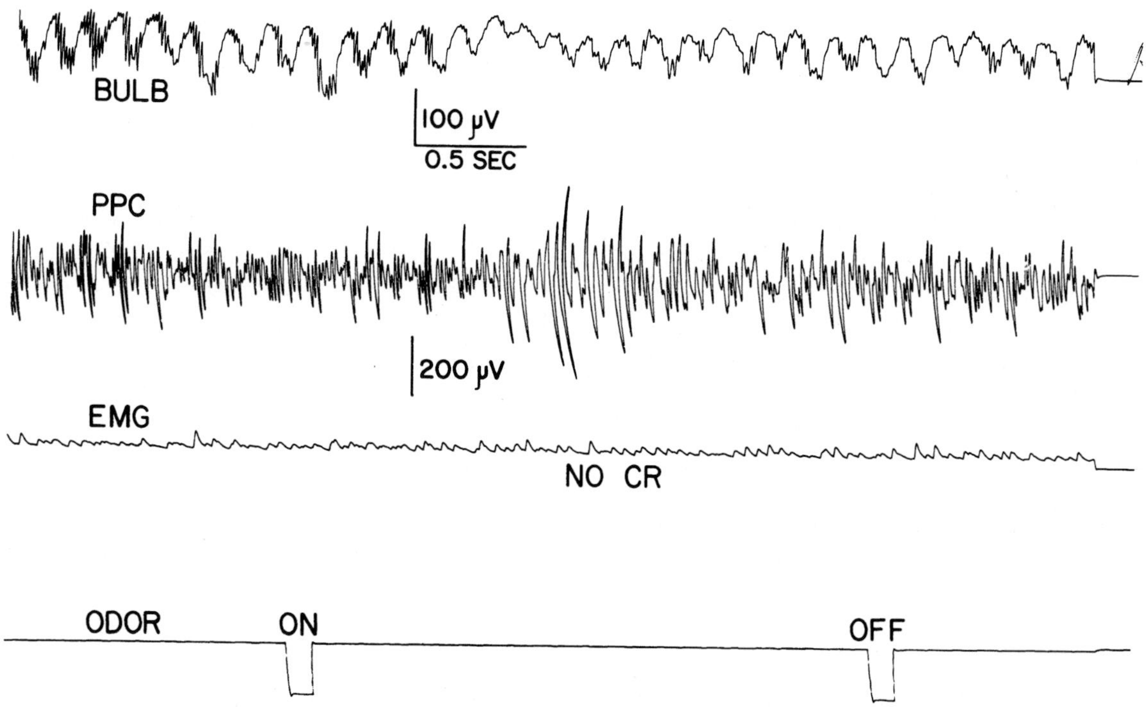

Fig. 4. The EEG reaction is shown for a novel stimulus, which induces an orienting response. The activity is unpatterned, which suggests that it is the basis for the construction of a new wing of a global attractor in Hebbian learning.

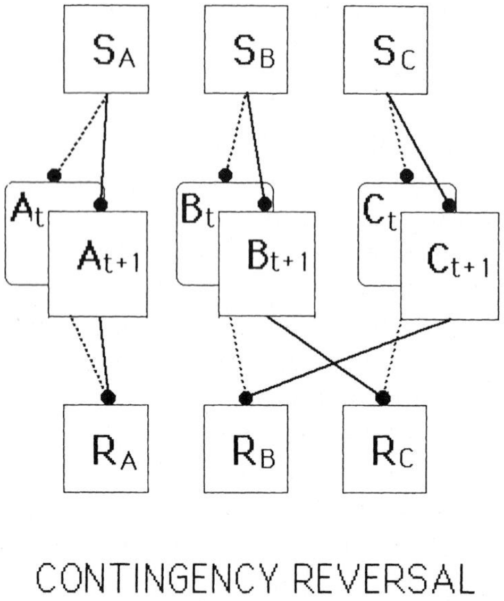

Fig. 5. The finding is indicated in schematic form that merely changing the reinforcement between two stimuli, one reinforced and the other not, is accompanied by changes in spatial patterns for both stimuli and for the control state as well, thus illustrating the lack of invariance of the central 'memory store' of BNN.

tonomy and stationarity (Freeman, 1992) the existence of a preferred pattern of activity, known as an attractor is revealed. The attractor for the basal state reveals little or no identifiable structure (Fig. 7), but the bursts provide a basis for visualizing the structure in the larger context of receiving stimuli, in which the cortex is neither autonomous nor stationary. Its global attractor is seen to have multiple side lobes or 'wings' in analogy with the two wings of the Lorenz attractor with the dihedral shape of a butterfly. Each wing is accessed by a learned odorant stimulus, and it determines a spatial pattern of cortical activity (Freeman, 1992). The access takes place when the presentation of the stimulus destabilizes the cortex, so that it leaves the basal state and is confined to the appropriate wing for the duration of the central effect of the stimulus. The confinement is expressed in the spatial pattern. When a new odorant is learned, a new wing is formed with a new spatial pattern, and all of the other patterns change slightly by about 7% of the mean variance for each wing (Freeman and Grajski, 1987), which expresses the modification of the context by the new experience. In this manner the cortical system during a perceptual act goes directly to an expression of the meaning of a stimulus without doing a computational search and match, followed by the attachment of a meaning to the selected pattern.

The dynamics of neurons and of populations can be modeled in both software and hardware embodiments with networks of coupled ordinary differential equations. Examples of basic models include the RC passive membrane, the cable equation, and the Hodgkin-Huxley equations (Freeman, 1992), which are components of cortical models. The simpler models employ point and limit cycle attractors to specify stable solution sets. The more realistic models are designed to function also in deterministic chaotic modes (Thompson and Stewart, 1988). An example of the connectivity diagram for the KIII model that suffices for the olfactory system is shown in Fig. 8. Each of the three main parts (bulb, nucleus and prepyriform) can be modeled as an oscillator, owing to the reciprocal interactions between excitatory and inhibitory neurons. Yet the characteristic frequencies of the three parts are incommensurate, so that entrainment at one frequency is not possible. The failure to agree leads to chaos. If the parts are surgically isolated, each then goes to a point attractor with non-oscillatory output (Freeman, 1975), showing that the chaotic activity is a global property of the system, not of its parts or their component neurons. Thus it is that when the olfactory system changes its pattern upon a state transition, it does so over its entire extent virtually simultaneously (Freeman and Baird, 1987). This rapid spread of the state transition is required to enable the bulb to express its output in the form of spatial patterns that extend over the entire extent of the surface of each structure, yet that can be spatially integrated without loss owing to phase and frequency dispersion.

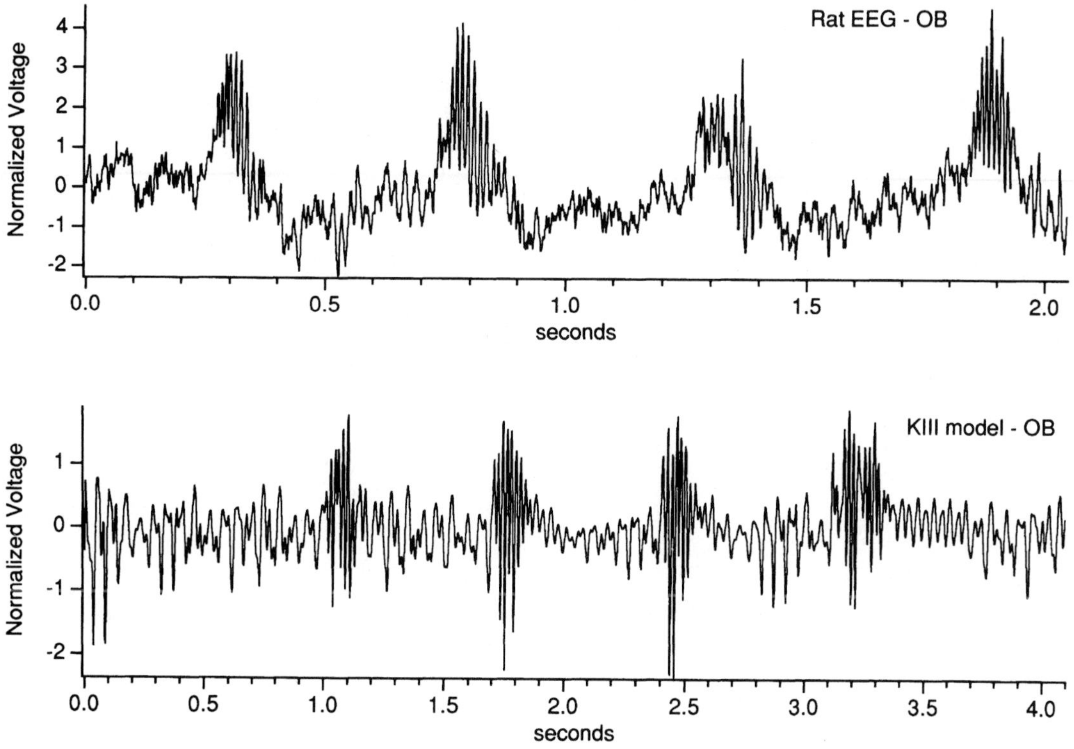

Fig. 6. The EEG of a rat recorded from the olfactory bulb during normal respiration is simulated with the KIII model of the olfactory system. From Kay et al. (1993).

During associative learning on the basis of examples under reinforcement, the classes of sensory input are formed and stored by modified synaptic connections. The modification is based on the correlation of pre- and postsynaptic activity at individual synapses, which is a form of Hebbian learning. Here enters a crucial role for chaotic activity as seen in Fig. 4 for the response to a novel stimulus, whether or not it is reinforced. Hebbian learning requires that there be activity in both neurons forming a synapse. However, in order that a new attractor be formed, the activity cannot conform to any pre-existing attractor, or there would be merely strengthening of that output pattern with no emergence of a new pattern. The pattern of Hebbian activity must be unlike any that has occurred previously. An effective solution is the chaotic generator, for which the pattern is not only unpredictable but is not reproducible for initial conditions which are varying. A learned stimulus constitutes a known initial condition with a basin of attraction, whereas a novel stimulus is by definition unknown. Hence it is the sensitivity to initial conditions as well as the capacity to act as an information source that enables a chaotic system to create novel patterns of output. Perception and recognition are by a state transition from one basin of attraction to another of the global chaotic attractor, that is by Hebbian learning and habituation. Neural models having these dynamic properties show robust capacities for amplification and for correct classification of noisy and incomplete patterns that correspond to sensory inputs to biological nervous systems in attentive and motivated animals (Yao and Freeman, 1990).

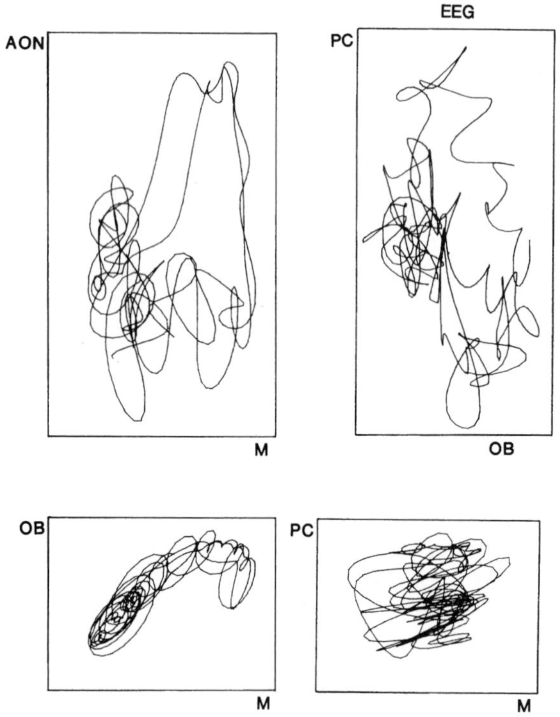

Fig. 7. The upper right frame shows an example of a phase portrait of the olfactory system made by plotting the amplitude of the OB EEG on the ordinate against that of the OB EEG on the abscissa from a 5 sec record. Simulations of the portraits by the KIII model are seen from different 'views' of the basal chaotic attractor. From Freeman (1987b).

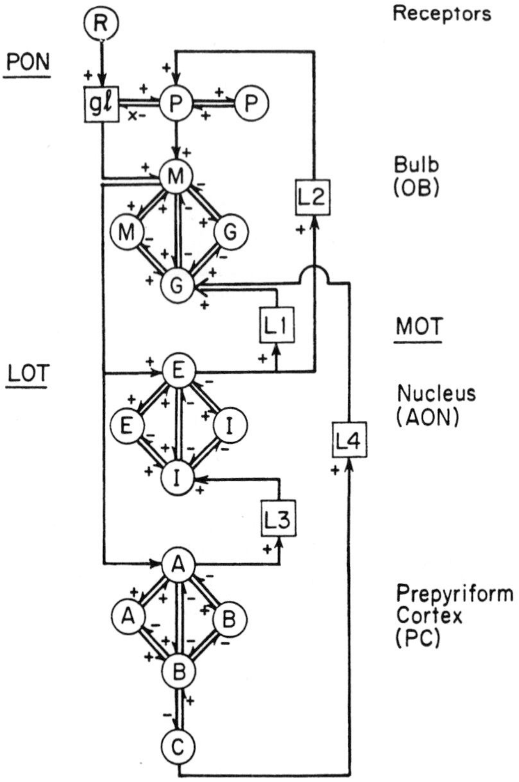

Fig. 8. A flow diagram of the olfactory system includes the receptors (R) transmitting to the glomeruli (gl) by the primary olfactory nerve (PON) and ending on excitatory (+) periglomerular (P) and mitral (M) cells, the latter interacting with the inhibitory (−) granule cells (G). Mitral cells project by way of the lateral olfactory tract (LOT) to the AON and PC, and these project back by the medial olfactory tract (MOT). There are distributed delays in each feedback path (L1 to L4). From Freeman (1987b).

How is instability controlled by learning?

Emphasis has been placed here on the instability of brain systems. Two kinds of state transition have been characterized with the aid of the models. One kind is based on a change of an internal parameter such as a synaptic weight that is expressed as a gain coefficient of a connection in a model. This state transition is a bifurcation, and the weight then becomes a bifurcation parameter. The other kind depends not on an internal parameter change but on the delivery of input. This type of change has no bifurcation parameter, because the input should not be regarded as a parameter, being outside the model. Time varying input causes repeated state transitions in a process that has been labeled 'chaotic itinerancy' (Kaneko, 1990; Tsuda, 1991), both in view of the tendency always to be moving on in a life-long evolution, and in respect to the return close to previously experienced patterns in the manner of a migratory worker coming back seasonally to the same or similar settings.

The instability of BNN in contrast to the stability of conventional ANN can be ascribed to two properties of the basic components of BNN that differ from those of ANN. First, the kernel for the spatiotemporal integration by the dendritic

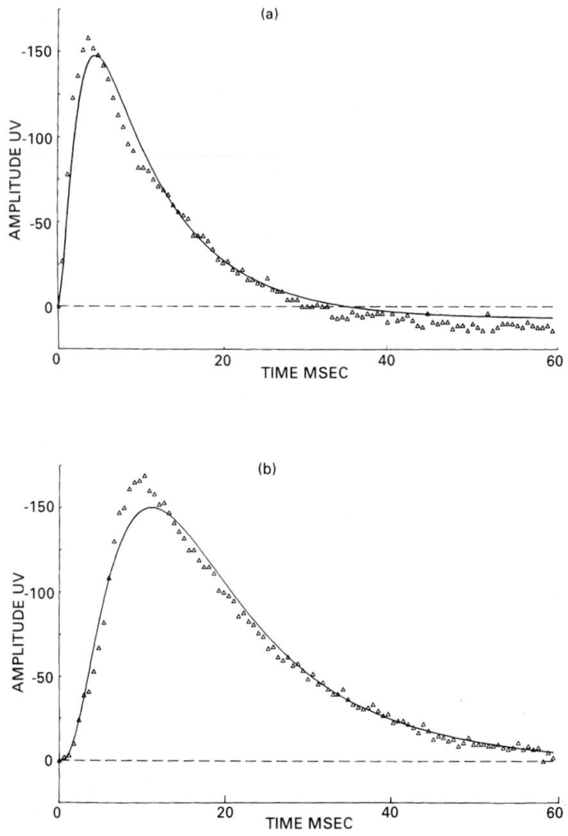

Fig. 9. Open loop impulse responses are shown for the granule cells (G) on electrical stimulation of the LOT (above) and of the PON (below). The impulse input for the latter is first passed through the mitral cells (M), which accounts for the slow rise time of the granule cell response. From Freeman (1972).

trees of neurons is defined by the open loop impulse response (Freeman, 1975) under deep anesthesia. This event conforms to the extracellular compound postsynaptic potential of the cortical neurons receiving the input from a single electrick shock. The rise time is finite (Fig. 9) owing to synaptic and cable delays in series with the passive membrane decay, so that the minimal representation is a second-order ordinary differential equation (ODE). Most ANN rely on a first-order ODE. The significance becomes apparent when two populations are interconnected in a negative feedback loop. With increasing feedback gain the first-order kernel shows an increase in frequency but no decrease in the negative real part of the closed loop rate constants. It is stable for all values of feedback gain. The loop with the second-order kernel reveals increasing frequency and decreasing decay rate of the impulse response with increasing feedback gain, and at some modest level of gain the real part becomes positive, and the system becomes unstable (Freeman, 1975, 1992). In brief, the synaptic and cable delays of neurons contribute to the destabilization of cortex in oscillatory modes, owing to the existence of both limit cycle and chaotic basins of attraction.

Second, the relationship in the single neuron between membrane potential and the probability of axonal firing is not linear. When a neuron is depolarized by dendritic current, the probability of firing increases exponentially as threshold is approached. Yet if the neuron does fire, its ability to do so again is limited by the after-effects of the firing, so that it approaches a maximum asymptotically (Freeman, 1979a). Most neurons spend most of their lives just below their thresholds and are in the firing state less than 1% of the time. Their firing is uncorrelated with the firings of other neurons in their neighborhoods. Yet they depend on those others by virtue of the reciprocal re-excitation they receive through short and long pathways. The feedback path can be modeled using a one-dimensional diffusion equation (Freeman, 1964), so that the distribution of their firing times in an interval histogram is Poisson. The outcome of the random firing is that the relation between dendritic depolarization and the density of firing in a cortical population is a sigmoid curve, but the curve is asymmetric as seen in Fig. 10. The slope of the curve increases with excitation, owing to the exponential increase in firing probability of the component neurons. Because the slope expresses the forward gain of the population at the trigger zones, the feedback gain between two population is increased by input. When this input-independent gain is combined with the second-order kernel of integration, it follows that input to cortex can cause a state

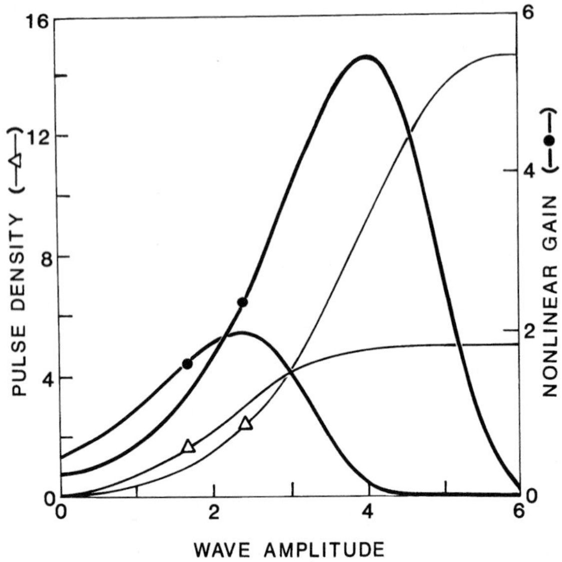

Fig. 10. The sigmoid curve of cortical neural populations facilitates the destabilization of cortex in two ways. First, it is asymmetric, so that excitatory input not only gives increased output, but it also increases the gain. In a mutually excitatory population this effect on gain can lead to an explosive growth in activity that is seen as a 'burst'. Second, the steepness of the curve increases with arousal and motivation, such as by food or water deprivation. Both effects on input-dependent gain are important for perception. Adapted from Freeman (1979a) and Eeckmann and Freeman (1990, 1992).

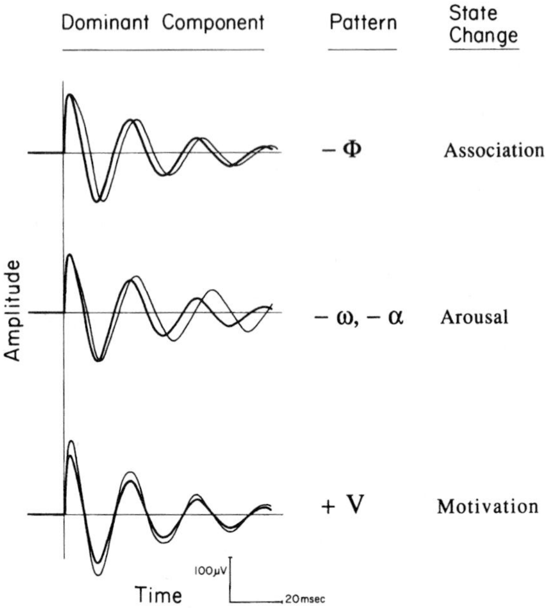

Fig. 11. Sensory cortex is selectively destabilized by learning to identify an electrical stimulus, when the stimulus used to evoke an impulse response is paired with reinforcement in a conditioning paradigm. The instability is revealed by a reduction in the decay rate of the damped sine wave impulse response. It can be simulated only by an increase in synaptic strength between excitatory neurons that are co-activated by the stimulus (Hebbian learning). From Emery and Freeman (1969).

transition from a point attractor under perturbation by noise or a low energy chaotic attractor to an oscillatory limit cycle or high energy chaotic attractor (a 'burst').

Impulse responses of the olfactory system are shown schematically in Fig. 11 and by the reproduction of averaged evoked potentials from a cat in Fig. 12. These illustrate the role of learning in facilitating or defacilitating selective instability of cortex. When associative learning occurs by training a subject to respond to the electrical shock that is used for the conditioned stimulus, the wave form of the impulse response changes in a characteristic way (Emery and Freeman, 1969). The initial amplitude does not increase, showing that the site of synaptic change is not at the synapses between the incoming axonal volley and the dendrites of the cortical neurons, as has been predicted in studies of LTP and modeled in perceptron-like ANN. Instead, the duration of the first half cycle increases, showing that the synaptic change is an increase in the strength of mutual excitation among the cortical excitatory neurons having both short and long range connections. As confirmed by modeling the network (Freeman, 1975, 1979b) the decay rate of the oscillatory impulse response decreases with associative learning, showing that the cortex is made less stable selectively to the learned input. Conversely, if the subject is given a stimulus without reinforcement, or if the reward is withheld so that the conditioned response is extinguished, then the impulse response to the electric shock serving as the conditioned stimulus undergoes an increase in decay rate, showing that habituation selectively

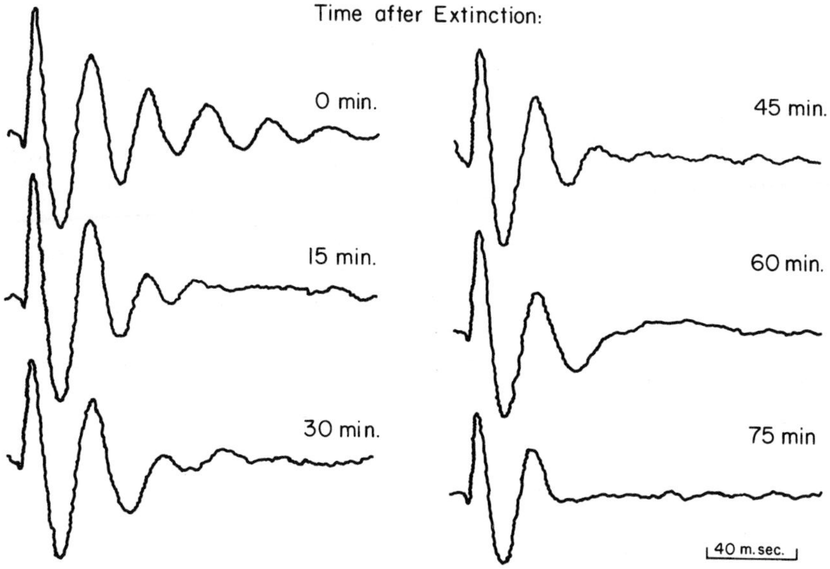

Fig. 12. Habituation to the electrical stimulus used to evoke the cortical impulse response is undertaken by presenting the same stimulus repeatedly and without reinforcement, so that eventually the subject fails to respond behaviorally. During this process the decay rate of the damped sine wave impulse response is increased, showing that the cortex is made more stable selectively in respect to the unwanted stimulus. This change is simulated in the KIII mdoel by decreasing the forward connections to other neurons of the excitatory neurons that are receiving the unwanted input, whether those other neurons are excitatory and inhibitory (Freeman, 1979b). From Freeman (1962).

stabilizes the cortex to the unwanted input. This is modeled by decreasing the forward gain of the cortical excitatory neurons that are receiving the input, and not by decreasing the strength of the synapses carrying the input to the cortical neurons (Freeman, 1979b).

Summary and conclusions

Mathematical models are essential for the study of complex neural systems at all levels of the hierarchy from macromolecules through neurons to brain systems. ANN are readily available, but most of them are inappropriate for modeling brain function in normal behavior, because they stem from studies of neural systems in anesthetized or paralyzed animals which are capable only of reflex output. That class of models lacks the goal-directed, self-organizing properties of neural systems in behaving animals. In contrast to the stability of ANN and their reliance on asymptotic convergence to steady states (point attractors) and periodic oscillations (limit cycle attractors), BNN are intrinsically unstable. They continuously generate 'spontaneous' aperiodic activity that manifests the operations of chaotic dynamics undergoing repeated state transitions. Observations on the activity patterns of sensory cortex reveal that the perceptual outputs of BNN are by construction of spatial patterns and dynamic trajectories and not by computation using symbolic representations. Chaotic dynamics plays essential roles both in the construction of perceptions and in the continuing update of cortical populations, which requires selective synaptic modification during associative learning and habituation.

Simultaneous multichannel recording from the olfactory bulb and cortex has given the following experimental results. (1) The cortical activity that relates to the perception of a sensory stimulus is carried macroscopically by populations of neurons, not microscopically by a small number of single neurons e.g. 'units', 'feature detectors'. (2) The macroscopic activity reflects the meaning and significance of the stimulus for the experimental subject and not the stimulus as it is known to the observer. (3) The activity carries the meaning in spatial patterns, not in time series (the difference between a phonograph or radio and movie or TV). (4) The spatial patterns of activity that accompany previously learned stimuli or responses are changed by the introduction of new stimuli and also by modifications in reinforcement contingencies. There is no invariance in the memory store within the populations. (5) The patterns of activity are created by dynamic neural interactions in sensory cortex, not by registration or filtering of stimuli. There is no evidence for storage, retrieval, cross-correlation or logical tree search. (6) The dynamics is chaotic, not merely noisy, so that each act of perception involves a new construction by the cortex and not mere information processing. From these findings we infer that chaotic dynamics plays a crucial role in the construction of the associational contexts comprising the memory systems of experimental subjects.

Acknowledgements

This work was supported by grants from the National Institute of Mental Health and the Office of Naval Research.

References

Alkon, D.L. (1992) *Memory's Voice. Deciphering the Mind-Brain Code*, Harper Collins, New York.

Ashby, W.R. (1956) *Design for a Brain: The origin of Adaptive Behavior*. Revised, 2nd ed., Wiley, New York.

Bartlett, F. (1932) *Remembering*, Cambridge University Press, Cambridge.

Braitenberg, V. (1984) *Vehicles. Experiments in Synthetic Psychology*, The MIT Press, Cambridge, MA.

Dreyfus, H.L. (1979) *What Computers Can't Do. The Limits of Artificial Intelligence*, revised edition, Harper Colophon, New York.

Eeckman, F.H. and Freeman, W.J. (1990) Correlations between unit firing and EEG in the rat olfactory system. *Brain Res.*, 528: 238–244.

Eeckman, F.H. and Freeman, W.S. (1991) Asymmetric sigmoid nonlinearity in the rat olfactory system. *Brain Res.*, 528: 238–244.

Emery, J.D. and Freeman, W.J. (1969) Pattern analysis of cortical evoked potential parameters during attention changes. *Physiol. Behav.*, 4: 67–77.

Fentress, J.C. (Ed.) (1976) *Simple Networks and Behavior*, Sinauer and Assoc., Sunderland, MA.

Freeman, W.J. (1962) Phasic and long-term excitability changes in prepyriform cortex cats. *Exp. Neurol.*, 5: 500–518.

Freeman, W.J. (1964) A linear distributed feedback model for prepyriform cortex. *Exp. Neurol.*, 10: 525–547.

Freeman, W.J. (1972) Measurement of open-loop responses to electrical stimulation in olfactory bulb of cat. *J. Neurophysiol.*, 35: 745–761.

Freeman, W.J. (1975). *Mass Action in the Nervous System*, Academic Press, New York.

Freeman, W.J. (1979a) Nonlinear gain mediating cortical stimulus-response relations. *Biol. Cybern.*, 33: 237–247.

Freeman, W.J. (1979b) Nonlinear dynamics of paleocortex manifested in the olfactory EEG. *Biol. Cybern.*, 35: 21–37.

Freeman, W.J. (1987a) Techniques used in the search for the physiological basis of the EEG. In: A. Gevins and A. Remond (Eds.), *Handbook of Electroencephalography and Clinical Neurophysiology*, Vol 3A, Part 2, Ch. 18, Elsevier, Amsterdam, pp. 583–664.

Freeman, W.J. (1987b) Simulation of chaotic EEG patterns with a dynamic model of the olfactory system. *Biol. Cybern.*, 56: 139–150.

Freeman, W.J. (1991) The physiology of perception. *Sci. Am.*, 264: 78–85.

Freeman, W.J. (1992) Tutorial on neurobiology: from single neurons to brain chaos. *Int. J. Bifurcation Chaos*, 2: 451–482.

Freeman, W.J. (1993) Chaos in psychiatry. *Biol. Psychiatry*, 31: 1079–1081.

Freeman, W.J. and Baird, B. (1987) Relation of olfactory EEG to behavior: spatial analysis: *Beh. Neurosci.*, 101: 393–408.

Freeman, W.J. and Davis, G.W. (1990) Olfactory EEG changes under serial discrimination of odorants by rabbits. In: D. Schild (Ed.), *Chemosensory Information Processing*, Springer-Verlag, Berlin, NATO ASI Series, Vol. H39, pp. 375–391.

Freeman, W.J. and Grajski, K.A. (1987) Relation of olfactory EEG to behavior: factor analysis. *Behav. Neurosci.*, 101: 766–777.

Haken, H. and Stadler, M. (Eds.) (1990) *Synergetics of Cognition*, Springer-Verlag, Berlin.

Kaneko, K. (1990) Clustering, coding, switching, hierarchical ordering and control in network of chaotic elements. *Physica D*, 41: 137–142.

Kay, L., Schimoide, K. and Freeman, W.J. (1993) Comparison of EEG time series from rat olfactory system with model composed of nonlinear coupled oscillators. In preparation.

Lucky, R.W. (1989) *Silicon Dreams: Information, Man, and Machines*, St. Martin's Press, New York.

Sherrington, C.S. (1906) *The Integrative Action of the Nervous System*, Yale University Press, New Haven.

Skarda, C.A. and Freeman, W.J. (1987) How brains make chaos in order to make sense of the world. *Behav. Brain Sci.*, 10: 161–195.

Thompson, J.M.T. and Steward, H.B. (1988) *Nonlinear Dynamics and Chaos*, Wiley, New York.

Tsuda, I. (1991) Chaotic itinerancy as a dynamical basis of hermeneutics in brain and mind. *World Futures*, 32: 167–184.

Walter, W.G. (1963) *The Living Brain*, Norton, New York.

Yao, Y. and Freeman, W.J. (1990) Model of biological pattern recognition with spatially chaotic dynamics. *Neural Networks*, 3: 153–170.

J. van Pelt, M.A. Corner H.B.M. Uylings and F.H. Lopes da Silva (Eds.)
Progress in Brain Research, Vol 102

CHAPTER 22

Neural networks in the brain involved in memory and recall

Edmund T. Rolls and Alessandro Treves*

Department of Experimental Psychology, University of Oxford, South Parks Road, Oxford, OX1 3UD, U.K.

Introduction

Damage to the hippocampus and related structures leads to anterograde amnesia, i.e. an inability to form many types of memory. Old memories are relatively spared. Recent memories, formed within the last few weeks or months, may be impaired (Squire, 1992). The learning tasks that are impaired include spatial and some non-spatial tasks in which information about particular episodes, such as where a particular object was seen, must be remembered. Further, some hippocampal neurons in the monkey respond to combinations of visual stimuli and places where they are seen (Rolls, 1990a,b, 1991; Rolls and O'Mara, 1993). It is also known that inputs converge into the hippocampus, via the adjacent parahippocampal gyrus and entorhinal cortex, from virtually all association areas in the neocortex, including areas in the parietal cortex concerned with spatial function, temporal areas concerned with vision and hearing, and the frontal lobes (Fig. 1). An extensively divergent system of output projections enables the hippocampus to feed back into most of the cortical areas from which it receives inputs. On the basis of these and related findings the hypothesis is suggested that the importance of the hippocampus in spatial and other memories is that it can rapidly form 'episodic' representations of information originating from many areas of the cerebral cortex, and act as an intermediate term buffer store. In this paper, analyses of how the architecture of the hippocampus could be used as such a buffer store are considered, and then a hypothesis on how recent memories could be recalled from this store back into the cerebral cortical association areas, to be used as needed in the formation of long-term memories, is presented (see Marr, 1971; Rolls, 1989a,b, 1991; Treves and Rolls, 1994).

The hippocampus

Hippocampal CA3 circuitry (see Fig. 1)

Projections from the entorhinal cortex reach the granule cells (of which there are 10^6 in the rat) in the dentate gyrus (DG) via the perforant path (pp). The granule cells project to CA3 cells via the mossy fibres (MF), which provide a *sparse* but possibly powerful connection to the 3×10^5 CA3 pyramidal cells in the rat. Each CA3 cell receives approximately 50 mossy fibre inputs, so that the sparseness of this connectivity is thus 0.005%. By contrast, there are many more—pos-

* Present address: S.I.S.S.A. - Biophysics, via Beirut 2–4, 34013 Trieste, Italy.

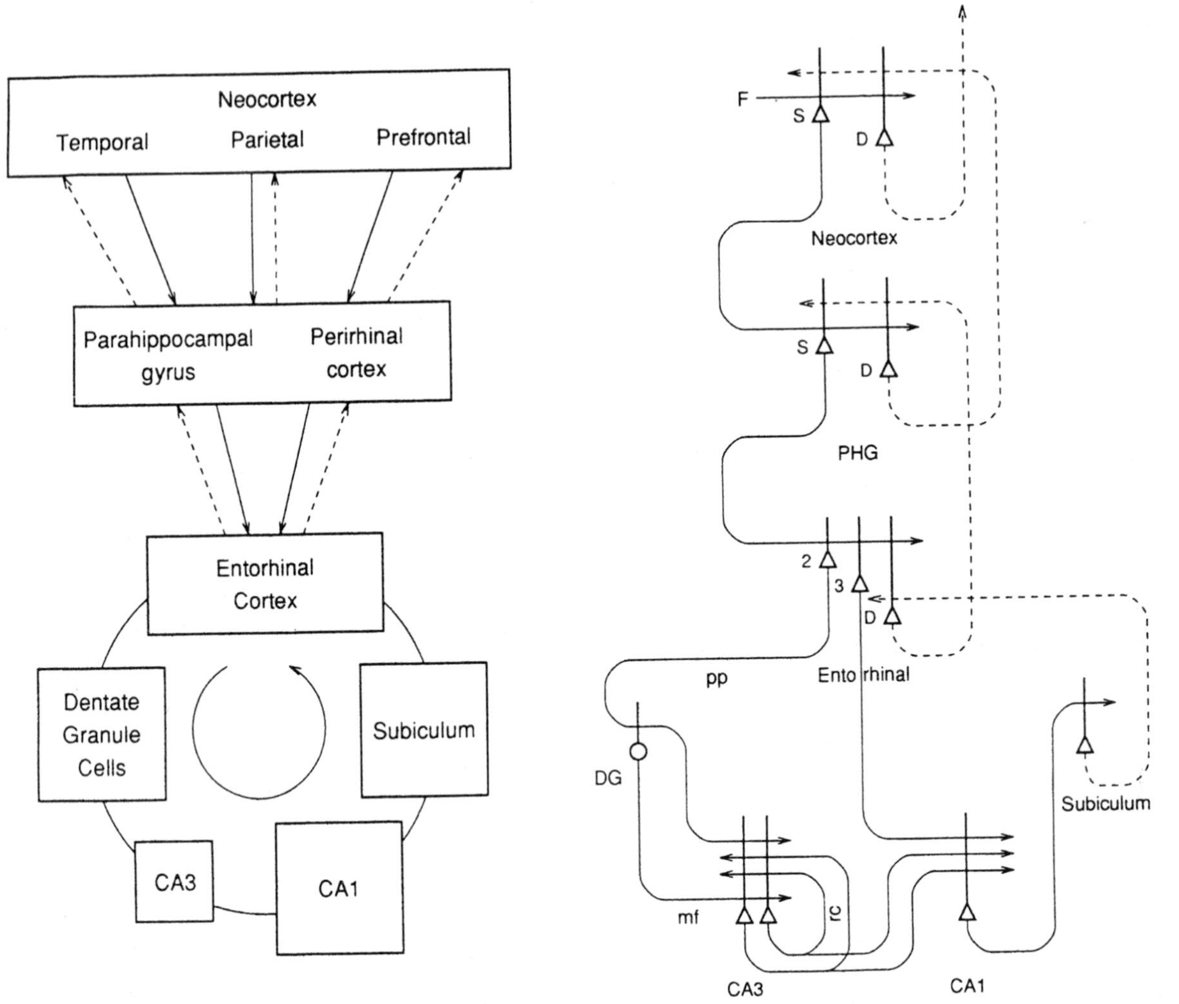

Fig. 1. Forward connections (solid lines) from areas of cerebral association neocortex via the parahippocampal gyrus and perirhinal cortex, and entorhinal cortex, to the hippocampus, and back-projections (dashed lines) via the hippocampal CA1 pyramidal cells, subiculum, and parahippocampal gyrus to the neocortex. There is great convergence in the forward connections down to the single network implemented in the CA3 pyramidal cells, and great divergence again in the back-projections. Left: block diagram. Right: more detailed representation of some of the principal excitatory neurons in the pathways. Abbreviations—D: deep pyramidal cells; DG: dentate granule cells; F: forward inputs to areas of the association cortex from preceding cortical areas in the hierarchy; mf: mossy fibres; PHG: parahippocampal gyrus and perirhinal cortex; pp: perforant path; rc: recurrent collateral of the CA3 hippocampal pyramidal cells; S: superficial pyramidal cells; 2: pyramidal cells in layer 2 of the entorhinal cortex; 3: pyramidal cells in layer 3 of the entorhinal cortex. The thick lines above the cell bodies represent the dendrites.

sibly weaker—direct perforant path inputs onto each CA3 cell, in the rat of the order of 40×10^3. The largest number of synapses (about 1.2×10^4 in the rat) on the dendrites of CA3 pyramidal cells is, however, provided by the (recurrent) axon collaterals of CA3 cells themselves (rc). It is remarkable that the recurrent collaterals are distributed to other CA3 cells throughout the hippocampus (Amaral and Witter, 1989; Amaral et al., 1990; Ishizuka et al., 1990), so that effectively the CA3 system provides a single network, with a connectivity of approximately 4% between the different CA3 neurons.

CA3 as an autoassociation memory

Many of the synapses in the hippocampus show associative modification as shown by long-term potentiation, and this synaptic modification appears to be involved in learning (Morris, 1989). On the basis of the evidence summarized above, Rolls (1987, 1989a,b,c, 1990a,b, 1991) has suggested that the CA3 stage acts as an autoassociation memory which enables episodic memories to be formed and stored for an intermediate term in the CA3 network, and that subsequently the extensive recurrent collateral connectivity allows for the retrieval of a whole representation to be initiated by the activation of some small part of the same representation (the cue). We have therefore performed quantitative analyses of the storage and retrieval processes in the CA3 network (Treves and Rolls, 1991, 1992). We have extended previous formal models of autoassociative memory (see Amit, 1989) by analysing a network with graded response units, so as to represent more realistically the continuously variable rates at which neurons fire, and with incomplete connectivity (Treves, 1990; Treves and Rolls, 1991). We have found that in general the maximum number p_{max} of firing patterns that can be (individually) retrieved is proportional to the number C^{RC} of (associatively) modifiable RC synapses per cell, by a factor that increases roughly with the inverse of the sparseness a of the neuronal representation. The sparseness is defined as

$$a = \frac{<\eta>^2}{<\eta^2>} \quad (1)$$

where $<\cdot>$ denotes an average over the statistical distribution characterizing the firing rate η of each cell in the stored patterns. Approximately,

$$P_{max} \simeq \frac{C^{RC}}{a\ln(1/a)}k \quad (2)$$

where k is a factor that depends weakly on the detailed structure of the rate distribution, on the connectivity pattern, etc., but is roughly in the order of 0.2–0.3 (Treves and Rolls, 1991).

The main factors that determine the maximum number of memories that can be stored in an autassociative network are thus the number of connections on each neuron devoted to the recurrent collaterals, and the sparseness of the representation. For example, for $C^{RC} = 12{,}000$ and $a = 0.02$ (realistic estimates for the rat), p_{max} is calculated to be approximately 36,000.

We have also indicated how to estimate I, the total amount of information (in bits per synapse) that can be retrieved from the network. I is defined with respect to the information i_p (in bits per cell) contained in each stored firing pattern, by subtracting the amount i_l lost in retrieval and multiplying by p/C^{RC}:

$$I \equiv \frac{p}{C^{RC}}(i_p - i_l). \quad (3)$$

The maximal value I_{max} of this quantity was found (Treves and Rolls, 1991) to be in several interesting cases around 0.2–0.3 bits per synapse, with only a mild dependency on parameters such as the sparseness of coding a.

The requirement of the input systems to CA3 for the efficient storage of new information

By calculating the amount of information that would end up being carried by a CA3 firing pattern produced solely by the perforant path input and by the effect of the recurrent connections, we have been able to show (Treves and Rolls, 1992) that an input of the perforant path type, alone, is unable to direct efficient information storage. Such an input is too weak, it turns out, to drive the firing of the cells, as the ‘dynamics’ of the network is dominated by the randomizing effect of the recurrent collaterals. This is the manifestation, in the CA3 network, of a general problem affecting storage (i.e. learning) in *all* autoassociative memories. The problem arises when the system is considered to be activated by a set of input axons making synaptic connections that have to compete with the recurrent connections, rather than having the firing rates of the neurons artificially clamped into a prescribed pattern.

In an argument developed elsewhere, we hypothesize that the mossy fibre inputs force efficient information storage by virtue of their strong and sparse influence on the CA3 cell firing rates (Rolls, 1989a,b; Treves and Rolls, 1992).

A different input system is needed to trigger retrieval

An autoassociative memory network needs afferent inputs also in the other mode of operation, i.e. when it retrieves a previously stored pattern of activity. We have shown (Treves and Rolls, 1992) that if the cue available to initiate retrieval is rather small, one needs a large number of associatively modifiable synapses. The number needed is of the same order as the number of concurrently stored patterns p. For such reasons we suggest that the direct perforant path system to CA3 (see Fig. 1) is the one involved in relaying the cues that initiate retrieval.

The dentate granule cells

The theory is developed elsewhere that the dentate granule cell stage of hippocampal processing which precedes the CA3 stage acts to produce during learning the sparse yet efficient (i.e. non-redundant) representation in CA3 neurons which is required for the autoassociation to perform well (Rolls, 1989a,b,c, 1991; see also Treves and Rolls, 1992). One way in which it may do this is by acting as a competitive network to remove redundancy from the inputs producing a more orthogonal, sparse, and categorized set of outputs (Rolls, 1987, 1989a,b,c, 1990a,b). A second way arises because of the very low contact probability in the mossy fibre–CA3 connections, which helps to produce a sparse representation in CA3 (Treves and Rolls, 1992). A third way is that the powerful dentate granule cell–mossy fibre input to the CA3 cells may force a new pattern of firing onto the CA3 cells during learning.

The CA1 organization

The CA3 cells are connected to the CA1 cells. It is suggested that the CA1 cells, given the separate parts of each episodic memory which must be separately represented in CA3 ensembles, can allocate neurons, by competitive learning, to represent at least larger parts of each episodic memory (Rolls, 1987, 1989a,b,c, 1990a,b). This implies a more efficient representation, in the sense that when eventually after many further stages, neocortical neuronal activity is recalled (as discussed below), each neocortical cell need not be accessed by all the axons carrying each component of the episodic memory as represented in CA3, but instead by fewer axons carrying larger fragments.

Dynamics and the temporal dimension

The analysis described above of the capacity of a recurrent network such as the CA3 considered steady state conditions of the firing rates of the neurons. The question arises of how quickly the recurrent network would settle into its final state. With reference to the CA3 network, how long does it take before a pattern of activity, originally evoked in CA3 by afferent inputs, becomes influenced by the activation of recurrent collaterals? In a more general context, recurrent collaterals between the pyramidal cells are an important feature of the connectivity of the cerebral neocortex. How long would it take these collaterals to contribute fully to the activity of cortical cells. If these settling processes took in the order of hundreds of milliseconds, they would be much too slow to contribute usefully to cortical activity, whether in the hippocampus or the neocortex (Rolls, 1992). As this question is crucial to understanding cortical function, we have analysed the time taken for neocortical pyramidal cells in the inferior temporal cortex of macaques to settle into a stable pattern of firing in response to a visual stimulus. We have found that within 20–40 msec of first firing, that is within 1–3 spikes from the fastest firing cells, single neurons have settled into stable firing rates which provide approximately 50% of the maximal information which can be obtained from the spike train of the neuron (Tovee et al., 1993). Could recurrent collaterals contribute to settling as rapid as this?

A partial answer to this question can be inferred from a recent theoretical development based on the analysis of the collective dynamical properties of realistically modelled neuronal units (Treves, 1993; Treves et al., 1994). The method incorporates the biophysical properties of real cell membranes, and considers the dynamics of a network of integrate-and-fire neurons, laterally connected through realistically modelled synapses. The analysis indicates that the model network will attain a stable distribution of firing rates over time scales determined essentially by synaptic and intrinsic conductance inactivation times. Some of these (e.g. the conductance time constants associated with excitatory synapses between pyramidal cells) are very short, less than 10 msec, implying that the activation of recurrent collaterals between pyramidal cells will contribute to determine the overall firing pattern within a period of a very few tens of msec (for further details see Treves, 1993; Treves et al., 1994). With respect to the CA3 network, the indication is thus that retrieval would be rapid, indeed fast enough for it to be biologically plausible.

Backprojections to the neocortex

The hippocampus as a buffer store

It is suggested that the hippocampus is able to recall the whole of a previously stored episode for a period of days, weeks or months after the episode, when even a fragment of the episode is available to start the recall. This recall from a fragment of the original episode would take place particularly as a result of completion produced by the autoassociation implemented in the CA3 network. It would then be the role of the hippocampus to reinstate in the cerebral neocortex the whole of the episodic memory. The cerebral cortex would then, with the whole of the information in the episode now producing firing in the correct sets of neocortical neurons, be in a position to incorporate the information in the episode into its long-term store in the neocortex.

We suggest that during recall, the connections from CA3 via CA1 and the subiculum would allow activation of at least the pyramidal cells in the deep layers of the entorhinal cortex (see Fig. 1). These neurons would then, by virtue of their backprojections to the parts of the cerebral cortex that originally provided the inputs to the hippocampus, terminate in the superficial layers of those neocortical areas, where synapses would be made onto the distal parts of the dendrites of the cortical pyramidal cells (see Rolls, 1989a,b,c).

Our understanding of the architecture with which this would be achieved is shown in Fig. 1. The feed-forward connections from association areas of the cerebral neocortex (solid lines in Fig. 1), show major convergence as information is passed to CA3, with the CA3 autoassociation network having the smallest number of neurons at any stage of the processing. The back-projections allow for divergence back to neocortical areas. The way in which we suggest that the back-projection synapses are set up to have the appropriate strengths for recall is as follows (see also Rolls, 1989a,b). During the setting up of a new episodic memory, there would be strong feed-forward activity progressing towards the hippocampus. During the episode, the CA3 synapses would be modified, and via the CA1 neurons and the subiculum, a pattern of activity would be produced on the back-projecting synapses to the entorhinal cortex. Here the back-projecting synapses from active back-projection axons onto pyramidal cells being activated by the forward inputs to the entorhinal cortex would be associatively modified. A similar process would be implemented at preceding stages of the neocortex, that is in the parahippocampal gyrus/perirhinal cortex stage, and in association cortical areas.

Quantitative constraints on the connectivity of back-projections

How many back-projecting fibres does one need to synapse on any given neocortical pyramidal cell, in order to implement the mechanism outlined above? Clearly, if the theory were to produce a definite constraint of the sort, quantitative anatomical data could be used for verification or falsification.

Consider a polysynaptic sequence of back-projecting stages, from hippocampus to neocortex, as a string of simple (hetero-)associative memories in which, at each stage, the input lines are those coming from the previous stage (closer to the hippocampus). Implicit in this framework is the assumption that the synapses at each stage are modifiable and have been indeed modified at the time of first experiencing each episode, according to some Hebbian associative plasticity rule. A plausible requirement for a successful hippocampo-directed recall operation, is that the signal generated from the hippocampally retrieved pattern of activity, and carried backwards towards the neocortex, remain undegraded when compared with the noise due, at each stage, to the interference effects caused by the concurrent storage of other patterns of activity on the same back-projecting synaptic systems. That requirement is equivalent to that used in deriving the storage capacity of such a series of heteroassociative memories, and it was shown in Treves and Rolls (1991) that the maximum number of independently generated activity patterns that can be retrieved is given, essentially, by the same formula as (2) above

$$p \simeq \frac{C}{a\ln(1/a)}k' \qquad (2')$$

where, however, a is now the sparseness of the representation at any given stage, and C is the average number of (back-)projections each cell of that stage receives from cells of the previous one. (k' is a similar slowly varying factor to that introduced above.) If p is equal to the number of memories held in the hippocampal buffer, it is limited by the retrieval capacity of the CA3 network, p_{max}. Putting together the formula for the latter with that shown here, one concludes that, roughly, the requirement implies that the number of afferents of (indirect) hippocampal origin to a given neocortical stage (C^{HBP}), must be $C^{HBP} = C^{RC}a_{nc}/a_{CA3}$, where C^{RC} is the number of recurrent collaterals to any given cell in CA3, the average sparseness of a neocortical representation is a_{nc}, and a_{CA3} is the sparseness of memory representations in CA3 (Treves and Rolls, 1993).

One is led to a definite conclusion: a mechanism of the type envisaged here could not possibly rely on a set of monosynaptic CA3-to-neocortex back-projections. This would imply that, to make a sufficient number of synapses on each of the vast number of neocortical cells, each cell in CA3 has to generate a disproportionate number of synapses (i.e. C^{HBP} times the ratio between the number of neocortical cells and the number of CA3 cells). The required divergence can be kept within reasonable limits only by assuming that the back-projecting system is polysynaptic, provided that the number of cells involved grows gradually at each stage, from CA3 back to neocortical association areas (see Fig. 1).

The theory of the recall of recent memories in the neocortex from the hippocampus provides a clear view about why backprojections should be as numerous as forward projections in the cerebral cortex. The reason suggested for this is that as many representations may need to be accessed by back-projections for recall and related functions (see Rolls, 1989a,b,c) in the population of cortical pyramidal cells as can be accessed by the forward projections, and this limit is given by the number of inputs onto a pyramidal cell (and the sparseness of the representation), irrespective of whether the input is from a forward or a back-projection system.

Summary

We have considered how the neuronal network architecture of the hippocampus may enable it to act as an intermediate term buffer store for recent memories, and how information may be recalled from it to the cerebral cortex using modified synapses in back-projection pathways from the hippocampus to the cerebral cortex. The recalled information in the cerebral neocortex could then be used by the neocortex in the formation of long-term memories, which is severely impaired by damage to the hippocampus.

Acknowledgements

Different parts of the research described here were supported by the Medical Research Council,

PG8513790, by an EEC BRAIN grant, by the MRC Oxford Research Centre in Brain and Behaviour, by the Oxford McDonnell-Pew Centre in Cognitive Neuroscience, and by a Human Frontier Science program grant.

References

Amaral, D.G., Ishizuka, N. and Claiborne, B. (1990) Neurons, numbers and the hippocampal network. *Prog. Brain Res.*, 83: 1–11.

Amaral, D.G. and Witter, M.P. (1989) The three-dimensional organization of the hippocampal formation: a review of anatomical data. *Neuroscience*, 31: 571–591.

Amit, D.J. (1989) *Modelling Brain Function*. Cambridge University Press, New York.

Ishizuka, N., Weber, J. and Amaral, D.G. (1990) Organization of intrahippocampal projections originating from CA3 pyramidal cells in the rat. *J. Comp. Neurol.*, 295: 580–623.

Marr, D. (1971) Simple memory: a theory for archicortex. *Phil. Trans. R. Soc. Lond. B*, 262: 24–81.

Rolls, E.T. (1987) Information representation, processing and storage in the brain: analysis at the single neuron level. In: J.-P. Changeux and M. Konishi (Eds.), *The Neural and Molecular Bases of Learning,* Wiley, Chichester, pp. 503–540.

Rolls, E.T. (1989a) Functions of neuronal networks in the hippocampus and neocortex in memory. In: J.H. Byrne and W.O. Berry (Eds.), *Neural Models of Plasticity: Experimental and Theoretical Approaches*, Ch. 13, Academic Press, San Diego, pp. 240–265.

Rolls, E.T. (1989b) The representation and storage of information in neuronal networks in the primate cerebral cortex and hippocampus. In: R. Durbin, C. Miall and G. Mitchison (Eds.), *The Computing Neuron*, Ch. 8, Addison-Wesley, Wokingham, England, pp. 125–159.

Rolls, E.T. (1989c) Functions of neuronal networks in the hippocampus and cerebral cortex in memory. In: R.M.J. Cotterill (Ed.), *Models of Brain Function*, Cambridge University Press, Cambridge, pp. 15–33.

Rolls, E.T. (1990a) Theoretical and neurophysiological analysis of the functions of the primate hippocampus in memory. *Cold Spring Harbor Symp. Quant. Biol.,* 55: 995–1006.

Rolls, E.T. (1990b) Functions of the primate hippocampus in spatial processing and memory. In: D.S. Olton and R.P. Kesner (Eds.), *Neurobiology of Comparative Cognition*, Ch. 12, Lawrence Erlbaum, Hillsdale, NJ, pp. 339–362.

Rolls, E.T. (1991) Functions of the primate hippocampus in spatial and non-spatial memory. *Hippocampus,* 1: 258–261.

Rolls, E.T. (1992) Neurophysiological mechanisms underlying face processing within and beyond the temporal cortical visual areas. *Phil. Trans. R. Soc.,* 335: 11–21.

Rolls, E.T. and O'Mara, S. (1993) Neurophysiological and theoretical analysis of how the hippocampus functions in memory. In: T. Ono, L.R. Squire, M. Raichle, D. Perrett and M. Fukuda (Eds.), *Brain Mechanisms of Perception: From Neuron to Behavior,* Oxford University Press, New York pp. 276–300.

Squire, L.R. (1992) Memory and the hippocampus: a synthesis from findings with rats, monkeys and humans. *Psychol. Rev.,* 99: 195–231.

Tovee, M.J., Rolls, E.T., Treves, A. and Bellis, R.P. (1993) Information encoding and the responses of single neurons in the primate temporal visual cortex. *J. Neurophysiol.,* 70: 640–654.

Treves, A. (1990) Graded-response neurons and information encodings in autoassociative memories. *Phys. Rev. A,* 42: 2418–2430.

Treves, A. (1993) Mean-field analysis of neuronal spike dynamics. *Network*, 4: 259–284.

Treves, A. and Rolls, E.T. (1991) What determines the capacity of autoassociative memories in the brain? *Network*, 2: 371–397.

Treves, A. and Rolls, E.T. (1992) Computational constraints suggest the need for two distinct input systems to the hippocampal CA3 network. *Hippocampus*, 2: 189–199.

Treves, A. and Rolls, E.T. (1994) A computational analysis of the role of the hippocampus in memory. *Hippocampus,* in press.

Treves, A., Rolls, E.T. and Tovee, M.J. (1994) On the time required for recurrent processing in the brain. *Proc. Natl. Acad. Sci. USA*, in press.

J. van Pelt, M.A. Corner H.B.M. Uylings and F.H. Lopes da Silva (Eds.)
Progress in Brain Research, Vol 102

CHAPTER 23

Categories of cortical structure

Dale Purves, David R. Riddle, Leonard E. White, Gabriel Gutierrez-Ospina and Anthony-Samuel LaMantia

Department of Neurobiology, Duke University Medical Center, Durham, NC 27710, U.S.A.

Introduction

The organization of the mammalian cortex is usually discussed in terms of maps and modules. Maps are topographic cortical (or subcortical) representations in which the geometry of the body is laid out in a more or less continuous fashion that reflects the arrangement of receptor surfaces or musculature. Modules are iterated structures within such maps whose significance and genesis has remained mysterious (see, for example, Purves et al., 1992). These traditional distinctions have sometimes made it difficult to compare diverse cortical systems. For example, the primary somatic sensory map is not easily equated with the cortical map of visual space. Further, iterated modules of similar size and appearance in various cortical regions are not necessarily members of the same class. More useful categories for comparison are cortical structures generated by induction, recognition and subsequent trophic interactions, and iterated cortical units generated by the effects of neural activity on neuronal growth. This analysis of cortical development considers barrels in the somatic sensory cortex and entities such as the primary visual cortex to be members of the same category, and distinguishes barrels from activity-dependent structures such as ocular dominance columns.

Organization of the sensory cortices

A more detailed consideration of the somatic sensory system illustrates some of the difficulties encountered when one tries to compare different cortical regions. The primary somatic sensory cortex is the first cortical station for processing information derived from the body's mechanosensory receptors (i.e. sensory receptors that respond to touch, pressure, stretch and vibration). In a number of mammals, this region is characterized by striking cytoarchitectonic and metabolic units called barrels and barrel-like-structures (Fig. 1) (Woolsey and Van der Loos, 1970; Wallace, 1987; Dawson and Killackey, 1987; Riddle et al., 1992). Each of these units appears to be the cortical representaiton of a peripheral sensory specialization such as a whisker on the face, or a digit or digital pad on the paw (Woolsey and Van der Loos, 1970; Dawson and Killackey, 1987; Killackey, et al., 1990; see also Catania et al., 1993). Because these cytoarchitectonic elements collectively delineate the entire body, they are often considered to be a somatic sensory map. Barrels and the space between them do not, however, constitute a continuous representation of receptors as in the primary visual and auditory cortices. Thus, the area between barrels in layer IV is not devoted to processing information arising from the body surface between adjacent mystacial vibrissae; on the contrary, the skin and fur between whiskers is represented separately (Welker, 1971, 1976; Verley and Pidoux, 1981; Chapin and Lin, 1984; Nussbaumer and Van der Loos, 1985; Killackey et al., 1990). Nor do the regions between

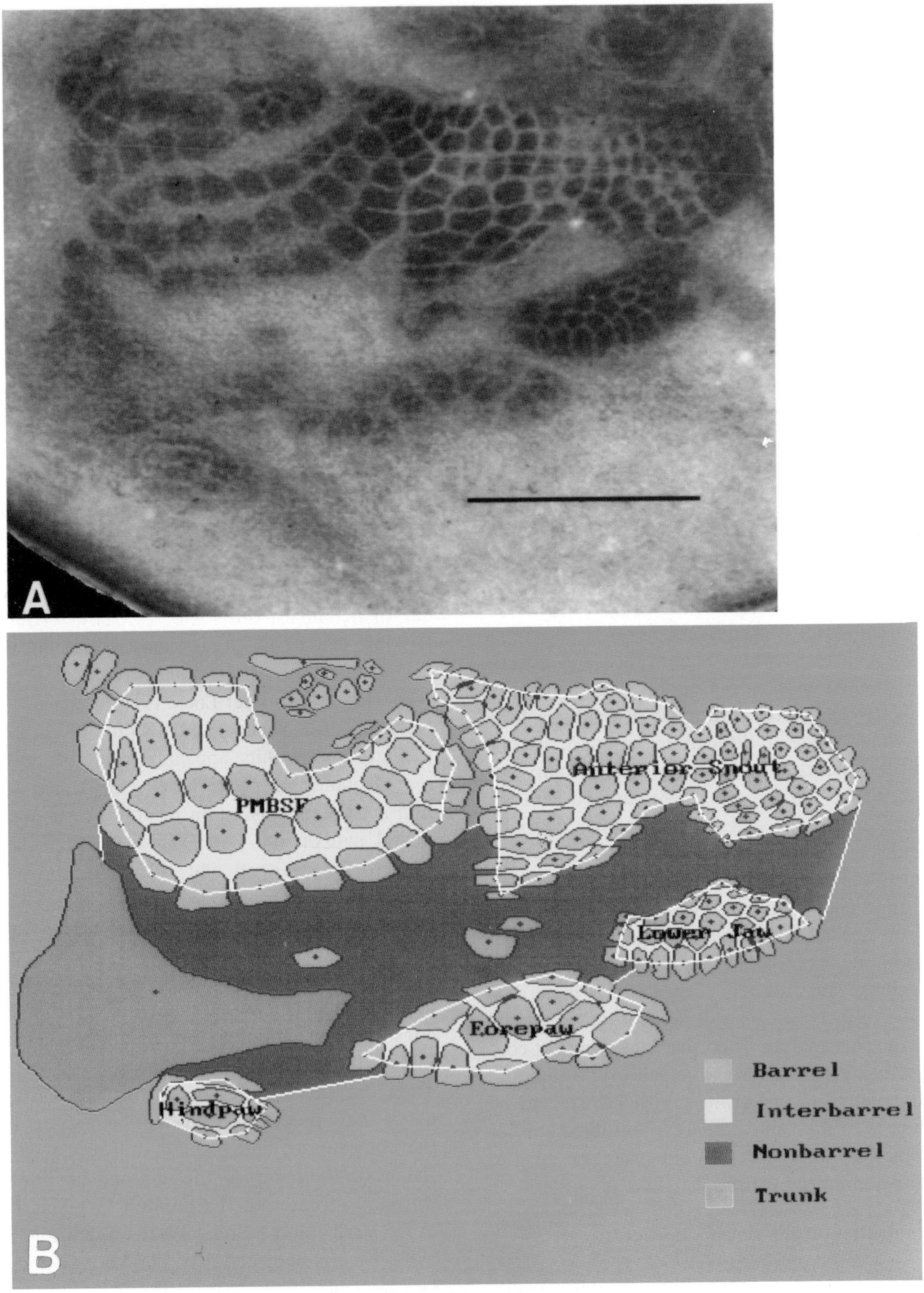
A
PMBSF
Anterior Snout
Lower Jaw
Forepaw
Hindpaw
Barrel
Interbarrel
Nonbarrel
Trunk
B

cortical barrels receive their major input from the ventral basdal complex of the thalamus (as do barrels); rather, they are driven from a separate thalamic nucleus (the posterior nucleus; see Welker et al., 1984; Koralek et al., 1988; Fabri and Burton, 1991) and from the contralateral cortex via the corpus callosum (Ivy et al., 1978; Koralek and Killackey 1990; Killackey et al., 1990). Such discontinuities in the primary somatic representation are not limited to rodents. The skin of the primate hand is also discontinuously mapped, the representation of the palmar surface with its specialized ridges being separated from the representation of the adjacent skin on the dorsum of the hand and fingers (Merzenich et al., 1978). In contrast, maps in the primary visual and auditory cortices are continuous representations of visual space or the audible frequency spectrum. These differences among sensory systems presumably reflect the nature and arrangement of various peripheral receptors. In those instances in which peripheral receptors are contiguous, as in the retina or the cochlea, the cortical representation of the receptors is a continuous map, but where the peripheral receptors occur as discrete populations, as in whisker follicles or digital pads, the corresponding cortical representation is likewise discontinuous (see also Killackey et al., 1990).

Adding to the difficulty of rationalizing the organization of different sensory cortices is the fact that many regions of the mammalian brain comprise repeating elements, the size and appearance of which are similar to barrels in the somatic sensory system (Fig. 2) These elements are most often referred to as modules or columns. Such structures are generally thought to represent a basic unit of cortical structure and function (Szentagothai, 1977; Mountcastle, 1978; Eccles, 1981; although see Towe, 1975; Crick and Asanuma, 1986; Swindale, 1990; Purves et al., 1992). Because these entities may look very much like barrels, one might assume that they belong to the same category and that their development proceeds according to the same principles. In what follows, we suggest that this assumption conflates cortical structures that are quite different in their cell biology, molecular underpinnings and, in all probability, their purpose. Accordingly, it may be more appropriate to consider each of the diverse cortical and subcortical entities shown in Fig. 2 as members of one of two basic ontogenetic classes: (1) structures that arise primarily from induction, intercellular recognition, and later trophic interactions, best exemplified by the barrels of the rodent somatosensory cortex, and (2) structures that arise primarily from the effects of activity on neuropil development, best exemplified by ocular dominance columns in the primate visual cortex.

Contributions of induction, intercellular recognition, and trophic interations

Induction

Most studies of cortical development are limited to interactions that occur after the majority of cortical neurons have been generated. This restriction is partly a consequence of technical limitations and partly the result of a tendency to consider the development of neural circuits separately from the developmental history of their precursor cells. In fact, neuronal precursor cells necessarily acquire information that influences mature patterns of neural circuitry (e.g. Horvitz, 1981; Goodman and Bate, 1981; Purves, 1982; Thomas et al., 1984; McConnell and Kaznowski, 1991; Yamada et al., 1993).

Much of this information derives from neighboring cells by means of induction. More specifically, induction occurs when one developing tissue initiates the differentiation of another via intercellular signaling (reviewed by Gurdon, 1992;

Fig 1. The primary somatic sensory cortex and its constituent parts. (A) Digitized image of a single SDH-stained tangential section of the rat primary somatic sensory cortex. Scale bar = 2 mm. (B) Complete S1 map in the same animal reconstructed from multiple sections. Barrels are gray, interbarrel areas are yellow, and non-barrel S1 areas are red (from Riddle et al., 1993).

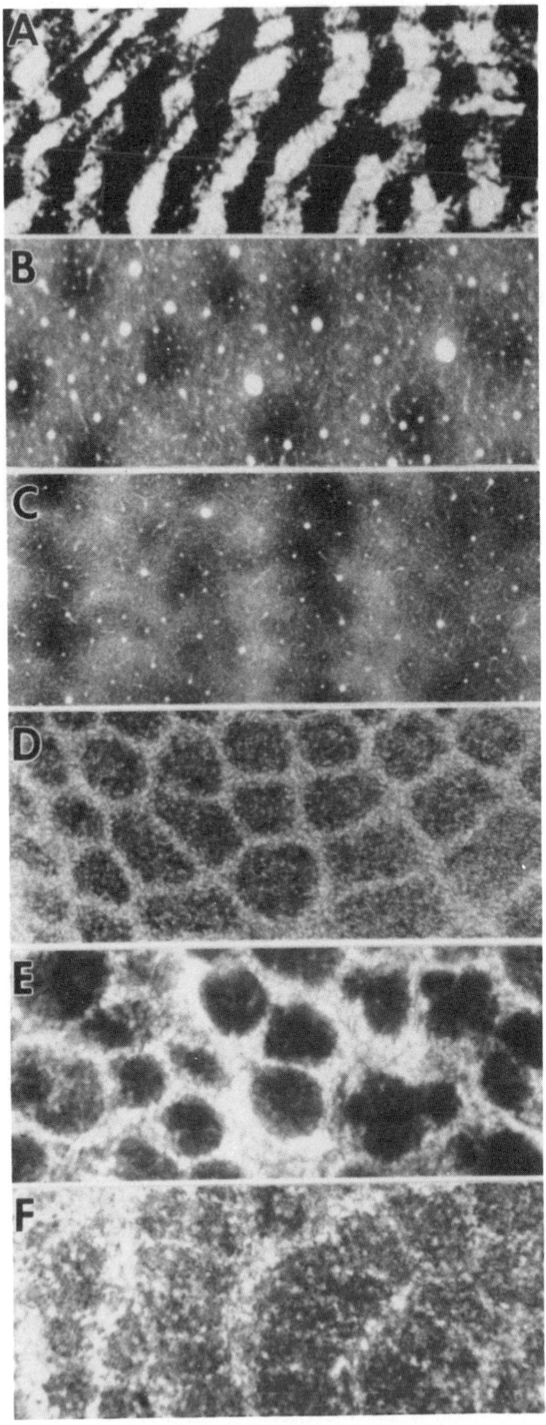

Fig. 2. Examples of modular circuitry in the mammalian brain. (A) Ocular dominance columns in layer IV of the primary visual cortex (V1) of a rhesus monkey (autoradiography after injection of radioactive proline into one eye). (B) Blobs in layers II–III of the primary visual cortex of a squirrel monkey (cytochrome oxidase histochemistry). (C) Stripes in layers II–III in the secondary visual cortex (V2) of a squirrel monkey (cytochrome oxidase histochemistry). (D) Barrels in layer IV in the primary somatic sensory cortex of a rat (succinic dehydrogenase histochemistry). (E) Glomeruli in the olfactory bulb of a mouse (Sudan Black staining). (F) Barreloids in the ventroposterior medial nucleus (VPM) in the thalamus of a rat (succinic dehydrogenase histochemistry) (from Purves et al., 1992.

Jessell and Melton, 1992; Slack, 1993). Such interactions are mediated by signaling molecules that modify gene expression, most often by regulating the activity of transcription factors and other DNA binding proteins. Peptide hormones like fibroblast growth factor and activin (Kimelman and Kirschner, 1987; Hemmati-Brinvalou and Melton, 1992; Yamada et al., 1993), and steroid-like molecules such as retinoic acid (Evans, 1988; Linney and LaMantia, 1994) are the best understood examples of inductive signals. Although such molecules act through specific receptors to modify gene expression in distinct regions and cell types (Hemmati-Brinvalou and Melton, 1992; Mangelsdorf and Evans, 1992), the results of induction may be remote in both time and space from the initial interaction. In mammals, induction establishes territories that eventually give rise to sensory and motor neurons in the spinal cord (Yamada et al., 1993), cranial nerve nuclei in the brainstem (Lumsden and Keynes, 1989; Kessel, 1993), and olfactory pathways in the forebrain (Stout and Graziadei, 1980; LaMantia et al., 1993). Of particular relevance to the present argument is the fact that regional patterns arising from inductive interactions prefigure patterns of connectivity in the mature brain (Bovolenta and Dodd, 1991; LaMantia et al., 1993).

In the mammalian somatic sensory system, potential sites for induction exist at each level of the developing trigeminal pathway. Thus inductive interactions underlie the early formation of the fronto-nasal region of the head (Jacobson, 1963), as well as the formation of individual whisker follicles (Wrenn and Wessells, 1984; Horne et al., 1986). There is also evidence that induction establishes positional information and regional identity in the parts of the neural tube that ultimately participate in barrel formation. In particular, distinct sequences and patterns of regional differentiation in the hindbrain are correlated with the emergence of specific domains of putative transcription factor expression (Wilkinson et al., 1989; Kessel and Gruss, 1991). In short, region-specific transcriptional regulation as a consequence of inductive signaling occurs at all of the levels of the central nervous system where barrels and their subcortical relays (called barrelettes in the brainstem, and barreloids in the thalamus) eventually appear.

Despite continuing uncertainty about the precise mechanisms involved, induction is clearly an early step in the sequence of conditional instructions for the developing somatic sensory (or any other) pathway. Although inductive events do not establish cortical barrels, they provide early positional instructions upon which the ultimate pattern of these cortical units is built. The adult pattern will be realized only if subsequent intercellular recognition and trophic signaling proceeds normally.

Intercellular recognition

Patterns of organization initiated by the inductive interactions that bias cells toward assuming a particular position and fate must be supported by subsequent intercellular interactions that influence cell movement and adhesion. This generalization is especially germane to the nervous system, where cell position and local interaction generate representations of sensory, motor and ultimately cognitive information. It is generally assumed that these aspects of neuronal development (i.e. cell migration, axon outgrowth and synapse formation) are controlled by cell–cell recognition and are mediated by specific molecules that guide cell movement and direct growing axons to their targets.

This idea was formally proposed by J.N. Langley in the late 19th century (Langley, 1895), and again by R.W. Sperry in the middle of this century (Sperry, 1963); both sought to explain in this way the orderly functional mapping that occurs during development and nerve regeneration. The phenomenon of target recognition by axons has since been documented in a variety of neural regions including peripheral ganglia, spinal cord, cerebellum, thalamus, and the cerebral cortex (Purves and Lichtman, 1978; Lance-Jones and Landmesser, 1981a,b; Mason and Gregory, 1984; Shatz and Kirkwood, 1984; Rakic, 1977, 1988; Dodd and Jessell, 1988). A number of cell adhesion molecules have been identified that appear at appropriate places and times (and have appropriate effects on cell behavior) to mediate directed cell migration, axon outgrowth and synaptogenesis (Edelman, 1986; Jessell, 1988; Hynes and Lander, 1992).

In particular, several aspects of the developing somatosensory system indicate the influence of intercellular recognition. Ingrowing trigeminal axons are precisely directed to their targets (Moody and Heaton, 1983), as are relay axons to the somatosensory thalamus, and the thalamic axons that subsequently project to the rudimentary somatosensory cortex (Senft and Woolsey, 1991; Catalano et al., 1991). All these events are accompanied by the expression of subsets of cell surface and extracellular matrix adhesion molecules (Steindler et al., 1989; Scarisbrick and Jones, 1993). Intercellular recognition in the developing barrel pathway therefore acts as a further conditional step to constrain populations of growing axons to their appropriate targets in somatosensory relays, thus influencing the initial formation of the barrel pattern.

Trophic interactions

Trophic interactions are long-term interdependent relationships of neurons and the cells they innervate, mediated by one of a growing number of neurotrophic factors (Purves and Lichtman, 1985; Barde, 1989; Thoenen, 1991; Loughlin and Fallon, 1993). Unlike inductive interactions, trophic feedback is not generally influential until differentiated neurons contact their targets (see, for example, Lumsden and Davies, 1983). Trophic interactions have been studied most intensively in the peripheral nervous system, where the action of a particular neurotrophin—nerve growth factor—has served as a paradigm for understanding the manifold consequences of these relationships (reviewed in Purves and Lichtman, 1985; Purves, 1988; Purves et al., 1988). More recently, trophic interactions have been recognized and investigated in the central nervous system, primarily in the context of neuron survival and the possibility of a neurotrophic basis for several neurodegenerative diseases (Barde, 1989; Thoenen, 1991). Here we propose that trophic interactions are also involved in the formation and maintenance of cortical structures that are initiated by early inductive events. By the same token, the abrogation of trophic support may explain the results of a number of experimental manipulations in developing sensory systems.

In the peripheral nervous system, several phenomena have been taken as evidence of trophic interplay between innervation and target. Interruption of the link between neurons and target causes a series of changes in the innervating neurons that are conveyed trans-synaptically to other stations in the relevant pathway. Such changes are reversible if neuron–target relationships are re-established promptly, whereas long-term interruption leads to the death of the trophically dependent neurons. Most of these phenomena are apparent in the developing somatic sensory systems. For example, the trigeminal ganglia are the rostral counterparts of the dorsal root ganglia, in which trophic interactions were first demonstrated (see Hamburger and Levi-Montalcini, 1949). Nerve transecton or ablation of the target tissue in early life causes dramatic changes in the trigeminal ganglion cells, leading to the eventual death of many of these neurons (Savy et al., 1981; Waite and Cragg, 1982; Waite, 1984; Klein et al., 1988; Woolsey, 1990). Numerous experiments have also shown that the effects of target ablation are transmitted centrally, as evidenced by changes in the brainstem, thalamus, and somatic sensory cortex after follicle cautery or nerve section (reviewed in Woolesy, 1990).

These central effects of early peripheral intervention in the rodent somatic sensory system have usually been interpreted as demonstrating the ability of whisker follicles to *instruct* the establishment of a central pattern (e.g. Van der Loos and Dörfl, 1978; Andres and Van der Loos, 1985; Killackey et al., 1990). In support of this view, the central patterns in the barrel pathway appear sequentially in brainstem, thalamus, and cortex (reviewed in Killackey et al., 1990). Furthermore, critical periods end progressively later as one moves centrally in the system (Woolsey, 1990; Killackey et al., 1990). Another relevant finding is that strains of mice that have an anomalous number of whiskers invariably have a corresponding number of cortical barrels (Yamakado and Yohro, 1979; Van der Loos et al., 1984). It should be recognized, however, that these observations provide only circumstantial evidence for an instructive role of the vibrissal follicles in the formation of their central representations in the brainstem, thalamus, and cortex. Another explanation of the effects of peripheral deprivation is that such manipulations interrupt the trophic chain of support that sustains the development of the central patterns. In this way, vibrissal follicles do not instruct the formation of barrelettes, barreloids, and barrels; the follicles (or other peripheral receptors) are instructive only in the sense that their innervation serves to validate and sustain the neural architecture initiated by earlier inductive and recognitory events. The vibrissal follicles are but one in a set of elements whose inductive histories and trophic interdependencies eventually lead to the formation of modular cytoarchitectonic and metabolic structures in the various stations of the trigeminal pathway.

Contribution of neural activity

The development of some modular patterns in the visual system of carnivores and primates clearly depends upon neural activity (Shatz, 1990; Constantine-Paton et al., 1990; Katz, 1993; Hockfield and Kalb, 1993). Perhaps the most compelling observation in this regard is that blocking activity in the retina with tetrodotoxin prevents the formation of ocular dominance columns (Stryker and Harris, 1986). Most workers agree that this iterated pattern in the visual cortex reflects the influence of asynchronous neural activity on competitive interactions among neurons (Miller et al., 1989; Shatz, 1990).

In contrast, neural activity is not decisive for the normal formation of the barrel pattern in S1 and its subcortical relays (Chiaia et al., 1992; Henderson et al., 1992; see also Schlagger and O'Leary, 1993). Thus, silencing the neonatal cortex (or the relevant peripheral nerves) for a few days after birth with tetrodotoxin or by other means does not prevent the appearance of apparently normal cortical barrels (or the corresponding subcortical structures in the thalamus and brainstem). Nor does functional deprivation (clipping or plucking whiskers without damaging the peripheral nerve or its terminal processes) have any obvious effect on this process (Woolsey and Wann, 1976; Simons and Land, 1987). Activity does, however, appear to influence the size of barrels during normal development (Riddle et al., 1992, 1993—see below). Activity can also influence the extent of the electrophysiologically determined representation of whiskers following certain manipulations (Kossut and Hand, 1984; Simons and Land, 1987; Kossut et al., 1988; Junti et al., 1991). For instance, plucking all but one of the mystacial vibrissae on the day of birth results in an enlargement of the electrophysiological representation of the remaining whisker, and in some cases a modest expansion in the apparent size of the corresponding barrel determined histochemically (Fox, 1992). Thus, whereas the initial formation of the barrel pattern is largely immune to altered activity, subsequent barrel growth (whether normal or in response to manipulation) can be affected in this way.

The most plausible interpretation of these observations is that neural activity affects *both* competitive interactions at the level of individual target cells *and* the growth of cortical neuropil more generally. Evidence that activity affects competitive interactions is well established (see, for example, Purves and Lichtman, 1985). Evidence that activity also influences neuropil growth, though less fully understood, comes from several different sorts of observations. The brains of all mammals, including man, increase in size progressively throughout the period of postnatal development (e.g. Dekaban and Sadowsky, 1978; Purves, 1994). In parallel, both synaptic numbers and neuropil complexity increase steadily (e.g. Armstrong-James and Johnson, 1970; Pomeroy et al., 1990; Bourgeois and Rakic, 1993). In consequence, the area of maps and modules also increases; for example, the cross-sectional area of barrels approximately double during postnatal maturation (Riddle et al., 1992). During this process, the growth of metabolically more active regions of cortex is greater than that of the adjoining cortex, which is less active during development (Riddle et al., 1993; see also Purves and LaMantia, 1993). Thus, barrels in the somatic sensory cortex, which are distinguished by their high levels of neural activity, grow about twice as much postnatally as the intervening non-barrel cortex. Additional evidence for the participation of electrical activity in the modulation of cortical growth comes from studies of binocularly enucleated rats. In this case, increased cortical metabolism (and presumably neural activity) is correlated with greater than normal growth of S1 barrels (D. Zheng and D. Purves, unpublished observations; see also Rauschecker et al., 1990; Bronchti et al., 1992). A more direct link between electrical activity and neuronal growth has been reported in the rodent hypothalamus, where natural cycles of activity associated with lactation elicit corresponding changes in the magnocellular neurons of the paraventricular and supraoptic nuclei (Hatton, 1985; Theodosis and Poulain, 1992). Nursing increases hypothalamic activity and foments growth of the relevant neurons, whereas weaning has the opposite effect. Taken together, these observations indicate that neural activity and neuropil growth are tightly linked in mammalian brain development.

Comparison of S1 and V1

Failure to distinguish inductive/trophic influences and neural activity as different avenues to the formation of cortical structure may explain why it has been so difficult to understand the effects of deprivation in the somatic sensory system. Numerous workers, ourselves included, have attempted to rationalize the plasticity of S1 and its critical period in early life by analogy with the visual system, despite the fact that altered activity has quite different effects on the formation of barrels and ocular dominance columns.

The likely explanation of these otherwise puzzling observations is simply that the wrong things are being compared. Barrels in the somatosensory system are discrete cytoarchitectonic entities isomorphically related to particular collections of peripheral sensors; ocular dominance columns are not. On the other hand, somatosensory barrels and V1 as a whole *are* quite similar: both constitute inductive/ trophic imprints of distinct collections of functionally related peripheral sensors—a vibrissal-follicular complex in the case of a barrel, and the retina in the case of V1. This line of reasoning suggests that V1 as a whole is the structure in visual cortex most appropriately compared with a barrel. This counterintuitive conclusion is supported by the fact that binocular enucleation at an early stage in primates blurs the area 17/18 border (Rakic, 1988; Rakic et al., 1991), in much the same way that ablation of peripheral receptors blurs barrel borders (which tend to merge into rows instead of maintaining discrete boundaries after this manipulation—e.g. Woolsey, 1990). In contrast, neither the boundaries of barrels nor of V1 are much influenced by altered activity. The power of developmental observations in the visual system, together with superficial structural similarities, has long militated for placing barrels and ocular dominance columns in the same developmental category; while conceptually attractive, this assignment is unwarranted.

Although these ontogenetic distinctions between structures initiated by induction and those initiated by activity are straightforward at a macroscopic level, it should be emphasized that the same neurons necessarily participate in the developmental strategies of both these cortical categories. For instance, activity-generated structures such as ocular dominance columns are present within V1, a structure that is, we argue, comparable with barrels in the somatic sensory cortex. At the cellular level, all neurons must respond to inductive/trophic signals *and* to activity-mediated modulation. Indeed, some aspects of trophic interactions may themselves be modulated by activity (Purves, 1988). That neurons can participate simultaneously in both these major categories of cortical structures is therefore not surprising; nor does this fact diminish the importance of making the distinctions we have outlined or of determining the cellular mechanisms underlying the formation and maintenance of each category.

Predictions

The characteristics of cortical structures initiated by induction and those initiated by activity are summarized in Table I. This interpretation of cortical organization makes a number of predictions:

1. The borders of structures initiated by induction (e.g. barrel/interbarrel boundaries or the 17/18 border) will distinguish populations of neurons that have different inductive histories and trophic dependencies.
2. The borders of structures initiated by activity (e.g. ocular dominance columns) will distinguish populations of neurons that support different temporal patterns of neural activity.
3. The borders of structures initiated by induction will be especially sensitive to interference with normal trophic interactions during development.
4. The borders of structures initiated by activity will be especially sensitive to interference with normal patterns of electrical activity during development.
5. The borders of structures initiated by induction will be coextensive with cytoarchitectonic boundaries.

TABLE I

Characteristics of the two major categories of cortical structure discussed

Cortical structures initiated by induction	Cortical structures initiated by activity
1. Isomorphic with peripheral sensory specialization; pattern therefore relatively constant from animal to animal	Not isomorphic with any peripheral specialization; pattern therefore variable from animal to animal
2. Initiated by signals that influence subsequent gene expression	Initiated by effects of activity on neuropil growth
3. Dependent upon trophic interactions	Dependent upon temporal patterns of activity
4. Boundaries delineate different inductive/trophic domains	Boundaries delineate domains with different temporal patterns of activity
5. Coextensive with cytoarchitectonic boundaries	Not coextensive with cytoarchitectonic boundaries
6. Critical period reflects decline in sensitivity to inductive and/or trophic influences on neuronal growth	Critical period reflects decline in sensitivity to effects of neural activity on neuronal growth
7. Relatively insensitive to activity deprivation during critical period	Relatively insensitive to trophic deprivation during critical period
8. Form relatively early in development	Form relatively late in development

6. The borders of structures initiated by activity will not be defined by cytoarchitectonic boundaries.
7. Because of their peripheral isomorphism, structures initiated by induction will remain constant in number during development.
8. Because of their lack of peripheral isomorphism, structures initiated by activity may change in number as development proceeds.
9. Critical periods for both induction-initiated and activity-initiated structures will be dictated by the effects of diminished trophic interactions or sensitivity to patterns of activity, respectively, on neuronal growth.
10. Since neural activity and trophic interactions both influence growth, a variety of growth-abating influences will ultimately be found to underlie critical periods in various regions of the brain.

How a variety of other cortical structures (e.g. orientation columns in V1, the primary auditory cortex (A1), binaural columns in A1, Mountcastle's somatic sensory columns, callosal stripes) fit the categories we have come to call 'trophons' (structures arising from inductive and trophic interactions that are relatively independent of activity) and 'actons' (structures arising from the influences of activity that do not depend, except in a permissive way, on induction, recognition and trophic interaction) remains to be seen. Indeed, trophons and actons may be the extremes of a continuum.

Conclusion

Two major classes of cortical structures can be distinguished on the basis of their developmental history: those that represent the imprint of early inductive events and later trophic support (exemplified by somatic sensory barrels in rodents), and those that represent the effects of activity on neuropil growth (exemplified by ocular dominance columns in the carnivore and primate visual cortex). Recognition of these categories may facilitate comparison of different cortical regions and the identification of appropriate homologies among their diverse structural features.

Acknowledgments

We are grateful to Larry Katz and David Fitzpatrick for helpful criticisms.

References

Andres, F.L. and van der Loos, H. (1985) From sensory periphery to cortex: the architecture of the barrel field as modified by various early manipulations of the mouse whiskerpad. *Anat. Embryol.*, 172: 11–20.

Armstrong-James, M. and Johnson, F.R. (1970) Quantitative studies of postnatal changes in synapses in rat superficial motor cerebral cortex. An electron microscopial study *Z. Zellforsch. Mikros. Anat.*, 110: 559–568.

Barde, Y.-A. (1989) Trophic factors and neuronal survival. *Nature*, 2: 1525–1534.

Bourgeois, J.-P. and Rakic, P. (1993) Changes of synaptic density in the primary visual cortex of the macaque monkey from fetal to adult stage. *J. Neurosci.*, 13: 2801–2820.

Bovolenta, P. and Dodd, J. (1991) Perturbation of neuronal differentiation and axon guidance in the spinal cord of mouse embryos lacking a floor plate: analysis of Danforth's short-tail mutation. *Development*, 113: 625–639.

Bronchti, G., Schöenberger, N., Welker, E., Van der Loos, H. (1992) Barrelfield expansion after neonatal eye removal in mice. *NeuroReport*, 3: 489–492.

Catalano, S.M., Robertson, R.T. and Killackey, H.P. (1991) Early ingrowth of thalamocortical afferents to the neocortex of the prenatal rat. *Proc. Natl. Acad. Sci. USA*, 88: 2999–3003.

Catania, K.C., Northcutt, R.G., Kaas, J.H. and Beck, P.D. (1993) Nose stars and brain stripes. *Nature*, 364; 493.

Chapin, J.K. and Lin, C.-S (1984) Mapping the body representation in the SI cortex of anesthetized and awake rats. *J. Comp. Neurol.*, 229: 199–213.

Chiaia, N.L., Fish, S.E., Bauer, W.R., Bennett-Clarke, C.A. and Rhoades, R.W. (1992) Postnatal blockade of cortical activity by tetrodotoxin does not disrupt the formation of vibrissa-related patterns in the rat's somatosensory cortex. *Dev. Brain. Res.*, 66: 244–250.

Constantine-Paton, M., Cline, H.T. and Debski, E. (1990) Patterned activity, synaptic convergence, and the NMDA receptor in developing visual pathways. *Annu. Rev. Neurosci.*, 13: 129–154.

Crick, F.H.C. and Asanuma, C. (1986) Certain aspects of the anatomy and physiology of the cerebral cortex. In: J.L. McCelland and D.E. Rumelhart (Eds.), *Parallel Distributed Processing: Explorations in the Microstructure of Cognition*, Chapter 20, Wiley, New York, pp. 333–371.

Dawson, D.R. and Killackey, H.P. (1987) The organization and mutability of the forepaw and hindpaw representations in the somatosensory cortex of neonatal rat. *J. Comp. Neurol.*, 25: 246–256.

Dekaban, A.S. and Sadowsky, D. (1978) Changes in brain weights during the span of human life: relation of brain weight to body heights and body weights. *Ann. Neurol.*, 4: 345–356.

Dodd, J. and Jessell, T.M. (1988) Axon guidance and the patterning of neuronal projections in vertebrates. *Science*, 242: 692–699.

Eccles, J.C. (1981) The modular operation of the cerebral neocortex considered as the material basis of mental events. *Neuroscience*, 6: 1839–1856.

Edelman, G.M. (1986) Cell adhesion molecules in the regulation of animal form and tissue pattern. *Annu. Rev. Cell Biol.*, 2: 81–116.

Evans, R.M. (1988) The steroid and thyroid hormone receptor superfamily. *Science*, 240: 889–895.

Fabri, M. and Burton, M. (1991) Topography of connections between primary somatosensory cortex and posterior complex in rat: a multiple fluorescent tracer study. *Brains Res.*, 538: 351–357.

Fox, K. (1992) A critical period for experience-dependent synaptic plasticity in rat barrel cortex. *J. Neurosci.*, 12: 1826–1838.

Goodman, C. and Bate, M. (1981) Neuronal development in the grasshopper. *Trends Neurosci.*, 4: 163–169.

Gurdon, J.B. (1992) The generation of diversity and pattern in animal development. *Cell*, 68: 185–199.

Hamburger, V. and Levi-Montalcini, R. (1949) Proliferation, differentiation and degeneration in the spinal ganglia of the chick embryo. *J. Exp. Zool.*, 111: 457–501.

Hatton, G.I. (1985) Reversible synapse formation and modulation of cellular relationships in the adult hypothalamus under physiological conditions. In: Cotman, C.W. (Ed.), *Synaptic Plasticity*, Guilford Press, New York, pp. 373–404.

Hemmati-Brinvalou, A. and Melton, D.A. (1992) A truncated activin receptor inhibits mesoderm induction and formation of axial structures in *Xenopus* embryos. *Nature*, 359: 690–614.

Henderson, T.A., Woolsey, T.A. and Jacquin, M.F. (1992) Infraorbital nerve blockade from birth does not disrupt central trigeminal pattern formation in the rat. *Dev. Brain Res.*, 66: 146–152.

Hockfield, S. and Kalb, R.G. (1993) Activity-dependent structural changes during neuronal development. *Curr. Opin. Neurobiol.*, 3: 87–92.

Horne, K.A., Jahoda, C.A.B. and Oliver, R.F. (1986) Whisker growth induced by implantation of cultured vibrissa-dermal papilla cells in the adult rat. *J. Embryol. Exp. Morphol.*, 97: 111–124.

Horvitz, H.R. (1981) Neuronal cell lineages in the nematode *Caenorhabditis elegans*. In: D. Garrod and J. Feldman (Eds.), *Development of the Nervous System*, Cambridge University Press, Cambridge, pp. 331–34.

Hynes, R.O. and Lander, A.D. (1992) Contact and adhesive specificities in the associations, migrations and targeting of cells and axons. *Cell*, 68: 303–322.

Ivy, G.O., Akers, R.M. and Killackey, H.P. (1978) Differential distribution of callosal projection neurons in the neonatal and adult rat. *Brains Res.*, 173: 532–537.

Jacobson, A.G. (1963) The determination and positioning of

the nose, lens and ear. I. Interactions within the ectoderm, and between the ectoderm and underlying tissues. *J. Exp. Zool.*, 154: 273–284.

Jessell, T.M. (1988) Adhesion molecules and the hierarchy of neural development. *Neuron*, 1: 3–13.

Jessell, T.M. and Melton, D.A. (1992) Diffusible factors in vertebrate embryonic induction. *Cell*, 68: 257–270.

Junti, Y., Merzenich, M.M., Woodruff, T.J., Jenkins W.M. (1991) Functional reorganization of rat somatosensory cortex induced by synchronous afferent stimulation. *Acta Acad. Med. Sin.*, 13: 278–282.

Katz, L.C. (1993) Cortical space race. *Nature* 364: 578–579.

Kessel, M. (1993) Reversal of axonal pathways from rhombomere 3 correlate with extra Hox expression domains. *Neuron*, 10:379–393.

Kessel, M. and Gruss, P. (1991) Homeotic transformations of murine vertebrae and concomitant alteration of Hox codes induced by retinoic acid. *Cell*, 67: 89–104.

Killackey, H.P., Jacquin, M.F. and Rhoades, R.W. (1990) Development of somatosensory structures. In: J.R. Coleman (Ed.), *Development of Sensory Systems in Mammals*, Wiley, New York, pp. 403–429.

Kimelman, D. and Kirschner, M. (1987) Synergistic induction of mesoderm by FGF and TGF-β and the identification of an mRNA coding for FGF in the early *Xenopus* embryo. *Cell*, 51: 869–877.

Klein, G.B., Renehan, W.E., Jacquin, M.F. and Rhoades, R.W. (1988) Anatomical consequences of neonatal infraorbital nerve transection upon the trigeminal ganglion and vibrissa follicle nerves in the adult rat. *J. Comp. Neurol.*, 268: 469–488.

Koralek, K.A. and Killackey, H.P. (1990) Callosal projections in rat somatosensory cortex are altered by early removal of afferent input. *Proc. Natl. Acad. Sci. USA*, 87: 1396–1400.

Koralek, K.A., Jensen, K.F, and Killackey, H.P. (1988) Evidence for two complementary patterns of thalamic input to the rat somatosensory cortex. *Brain Res.*, 463: 346–351.

Kossut, M. and Hand, P. (1984) The development of vibrissal cortical columns: a 2-deoxyglucose study. *Neurosci. Lett.*, 46: 1–6.

Kossut, M., Hand, P.J., Breenberg, J. and Hand, C.L. (1988) Single vibrissal cortical column in SI cortex of rat and its alterations in neonatal and adult vibrissa-deafferented animals: a quantitative 2DG study. *J. Neurophysiol.*, 60: 829–852.

LaMantia, A.-S., Colbert, M.C. and Linney, E. (1993) Retinoic acid induction and regional differentiation prefigure olfactory pathway formation in the mammalian forebrain. *Neuron*, 10: 1035–1048.

Lance-Jones, C. and Landmesser, L. (1981a) Pathway selection by chick lumbosacral motoneurons during normal development. *Proc. R. Soc. Lond.* (*Biol*), 214: 1–18.

Lance-Jones, C. and Landmesser, L. (1981b) Pathway selection by embryonic chick motoneurons in an experimentally altered environment. *Proc. R. Soc. Lond.* (*Biol*), 214: 19–52.

Langley, J.N. (1985) Note on regeneration of preganglionic fibres of the sympathetic. *J. Physiol.* (*Lond.*), 18: 280–284.

Linney, E. and LaMantia, A.-S (1994) Retinoid signalling in mouse embryos. *Adv. Dev. Biol.*, in press.

Loughlin, S.E. and Fallon, J.H. (Eds.) (1993) *Neurotrophic Factors*, Academic Press, Inc., Harcourt Brace Jovanovich, Publishers, San Diego, CA.

Lumsden, A.G.S. and Davies, A.M. (1983) Earliest sensory nerve fibers are guided to peripheral targets by atractans other than nerve growth factor. *Nature*, 306: 787–788.

Lumsden, A. and Keynes R. (1989) Segmental patterns of neuronal development in the chick hindbrain. *Nature*, 337: 424–428.

Mangelsdorf, D.J. and Evans, R.M. (1992) Retinoid receptors as transcription factors. In: S.L. McKnight and K.R. Yamamoto (Eds.), *Transcription Regulation*, Cold Spring Harbor Laboratory Press, Cold Spring Harbor, NY, pp. 1137–1167.

Mason, C.A. and Gregory, E. (1984) Postnatal maturation of cerebellar mossy and climbing fibers: transient expression of dual features of single axons. *J. Neurosci.*, 4: 1715–1735.

McConnell, S.K. and Kaznowski, C.E. (1991) Cell cycle dependence of laminar determination in developing neocortex. *Science*, 254: 282–285.

Merzenich, M.M., Kaas, J.H., Sur, M. and Lin, C.S. (1978) Double representation of the body surface within cytoarchitectonic areas 3b and 1 in "SI" in the owl monkey (aotus trivirgatus). *J. Comp. Neurol.*, 181: 41–73.

Miller, K.D., Keller, J.B. and Stryker, M.P. (1989) Ocular dominance column development: Analysis and simulation. *Science*, 245: 605–615.

Moody, S.A. and Heaton, M.B. (1983) Developmental relationships between trigeminal ganglia and trigeminal motoneurons in chick embryos. III. Ganglion perikarya direct motor axon growth in the periphery. *J. Comp. Neurol.*, 213: 350–364.

Mountcastle, V.B. (1978) An organizing principle for cerebral function: The unit module and the distributed system. In: G.M. Edelman and V.B. Mountcastle (Eds.), *The mindful Brain: Cortical Organization and the Group-Selective Theory of Higher Brain Function*, MIT Press, Cambridge, MA, pp. 7–50.

Nussbaumer, J.C. and Van der Loos, H. (1985) An electrophysiological and anatomical study of projections to the mouse cortical barrel field and its surroundings. *J. Neurophysiol.*, 53: 686–698.

Pomeroy, S.L., LaMantia, A.-S. and Purves, D. (1990) Postnatal construction of neural circuitry in the mouse olfactory bulb. *J. Neurosci.* 10: 1952–1966.

Purves, D. (1982) Guidance of axons during development and after nerve injury. In: J.G. Nicholls (Ed.), *Repair and Regeneration in the Nervous System*. Dahlem Konferenzen, Berlin, pp. 107–125.

Purves, D. (1988) *Body and Brain: A Trophic Theory of Neural Connections*, Harvard University Press, Cambridge, MA.

Purves, D. (1994) *Neural Activity and the Growth of the Brain*. Cambridge University Press, Cambridge, UK.

Purves, D. and LaMantia, A.-S. (1993) The development of blobs in the visual cortex of macaques. *J. Comp. Neurol.*, 334: 169–175.

Purves, D. and Lichtman, J.W. (1978) The formation and maintenance of synaptic connections in autonomic ganglia. *Physiol. Re.*, 58: 821–863.

Purves, D. and Lichtman, J.W. (1985) *Principles of Neural Development*, Sinauer Associates, Inc., Sunderland, MA.

Purves, D., Snider, W.D. and Voyvodic, J.T. (1988) Trophic regulation of nerve cell morphology and innervation in the autonomic nervous system. *Nature*, 336: 123–128.

Purves, D., Riddle, D. and LaMantia, A.-S. (1992) Iterated patterns of brain circuitry (or how the cortex gets its spots). *Trends Neurosci.*, 15: 362–368.

Rakic, P. (1977) Prenatal development of the visual system in rhesus monkey. *Phil Trans. R. Soc. Lond. B*, 278: 245–260.

Rakic, P. (1988) Specification of cerebral cortical areas. *Science*, 240: 170–176.

Rakic, P., Suñer, I. and Williams, R.W. (1991) A novel cytoarchitectonic area induced experimentally within the primate visual cortex. *Proc. Natl. Acad. Sci. USA*, 88: 2083–2087.

Rauschecker, J.P., Tian, B., Korte, B., Egert, U. (1990) Crossmodal changes in the somatosensory vibrissa/barrel system of visually deprived animals. *Proc. Natl. Acad. Sci. USA*, 89: 5063–5067.

Riddle, D., Richards, A., Zsuppan, F. and Purves, D. (1992) Growth of the rat somatic sensory cortex and its constituent parts during postnatal development. *J. Neurosci.*, 12: 3509–3524.

Riddle, D.R., Gutierrez, G., Zheng, D., White, L., Richards, A. and Purves, D. (1993) Differential metabolic and electrical activity in the somatic sensory cortex of the developing rat. *J. Neurosci.*, 13: 4193–4213.

Savy, C., Margules, S., Farkas-Bargeton, E. and Verley, R. (1981) A morphometric study of the mouse trigeminal ganglion after unilateral destruction of vibrassae follicles at birth. *Brain Res.*, 217: 265–277.

Scarisbrick, I.A. and Jones, E.G. (1993) NCAM immunoreactivity during major developmental events in the rat maxillary nerve-whisker system. *Dev. Brain Res.*, 71: 121–135.

Schlaggar, B.L. and O'Leary, D.D.M. (1993) Postsynaptic control of plasticity in developing somatosensory cortex. *Nature*, 364: 623–626.

Senft, S.L. and Woolsey, T.A. (1991) Growth of thalamic afferents into mouse barrel cortex. *Cerebral Cortex*, 1: 308–335.

Shatz, C.J. (1990) Impulse activity and the patterning of connections during CNS development. *Neuron*, 5: 745–756.

Shatz, C.J. and Kirkwood, P.A. (1984) Prenatal development of functional connections in the cat's retinogeniculate pathway. *J. Neurosci.*, 4: 1378–1397.

Simons, D.J. and Land, P.W. (1987) Early experience of tactile stimulation influences organization of somatic sensory cortex. *Nature*, 326: 695–697.

Slack, J.M.W. (1993) Embryonic induction. *Mech. Dev.*, 41: 91–107.

Sperry, R.W. (1963) Chemoaffinity in the orderly growth of nerve fiber patterns and connections. *Proc. Natl. Acad. Sci. USA*, 50: 703–710.

Steindler, D.A., Cooper, N.G.F., Faissner, A. and Schachner, M. (1989) Boundaries defined by adhesion molecules during development of the cerebral cortex: The J1/tenascin glycoprotein in the mouse somatosensory cortical barrel field. *Dev. Biol.*, 131: 243–260.

Stout, R.P. and Graziadei, P.C.P. (1980) Influence of the olfactory placode on the development of the brain in *Xenopus laevis* (*Daudin*). I. Axonal growth and connections of the transplanted olfactory placode. *Neuroscience*, 5: 2175–2186.

Stryker, M.P. and Harris, W.A. (1986) Binocular impulse blockade prevents the formation of ocular dominance columns in cat visual cortex. *J. Neurosci.*, 6: 2117–2133.

Swindale, N.V. (1990) Is the cerebral cortex modular? *Trends Neurosci.*, 13: 487–492.

Szentagothai, J. (1977) The neuron network of the cerebral cortex: a functional interpretation. *Proc. R. Soc. Lond. B*, 201: 219–248.

Theodosis, D.T. and Poulain, D.A. (1992) Neuronal-glia and synaptic plasticity of the adult oxytocinergic system. *Ann. NY Acad. Sci.*, 652: 303–325.

Thoenen, H. (1991) The changing scene of neurotrophic factors. *Trends Neurosci.*, 14: 165–170.

Thomas, J.B., Bastiani, M.J., Bate, M. and Goodman, C.S. (1984) From grasshopper to Drosophila: a common plan for neuronal development. *Nature*, 310: 203–207.

Towe, A.L. (1975) Notes on the hypothesis of columnar organization in somatosensory cerebral cortex. *Brain Behav. Evol.*, 11: 16–47.

Van der Loos, H. and Dörfl, J. (1978) Does the skin tell the somatosensory cortex how to construct a map of the periphery? *Neurosci. Lett.*, 76: 23–30.

Van der Loos, H., Dörfl, J. and Welker, E. (1984) Variation in the pattern of mystacial vibrissae: a quantitative study of mice of ICR stock of several untried strains. *J. Hered.*, 75: 326–336.

Verley, R. and Pidoux, B. (1981) Electrophysiological study of the topography of the vibrissae projections to the tactile thalamus and cerebral cortex in mutant mice with hair defects. *Exp. Neurol.*, 72: 475–485.

Waite, P.M.E. (1984) Rearrangement of neuronal responses in the trigeminal system of the rat following peripheral nerve section. *J. Physiol.*, 352: 425–445.

Waite, P.M.E. and Cragg, B.G. (1982) The peripheral and central changes resulting from cutting or crushing the afferent nerve supply to the whiskers. *Proc. R. Soc. Lond. B*, 214: 191–211.

Wallace, M.N. (1987) Histochemical demonstration of sensory maps in the rat and mouse cerebral cortex. *Brain Res.*, 418: 178–182.

Welker, C. (1971) Microelectrode delineation of fine grain somatographic organization of SmI cerebral neocortex in albino rat. *Brain Res.*, 26: 259–275.

Welker, C. (1976) Receptive field of barrels in the somatosensory neocortex of the rat. *J. Comp. Neurol.*, 166: 173–190.

Welker, W., Sanderson, K.J. Shambes, G.M. (1984) Patterns of afferent projections to transitional zones in the somatic sensorimotor cerebral cortex of albino rats. *Brain Res.*, 292: 261–267.

Wilkinson, D.G., Bhatt, S., Cook, M., Boncinelli, E. and Krumlauf, R. (1989) Segmental expression of Hox-2 homeobox-containing genes in the developing mouse hindbrain. *Nature*, 341: 405–409.

Woolsey, T. (1990) Peripheral alteration and somatosensory development. In: J.R. Coleman (Ed.), *Development of Sensory Systems in Mammals*, Wiley, New York.

Woolsey, T.A. and Van der Loos, H. (1970) The structural organization of layer 4 in the somatic sensory region (SI) of mouse cerebral cortex. *Brain Res.*, 17: 205–242.

Woolsey, T.A. and Wann, J. (1976) Areal changes in mouse cortical barrels following vibrissal damage at different postnatal ages. *J. Comp. Neurol.*, 170: 53–66.

Wrenn, J.T. and Wessells, N.K. (1984) The early development of mystacial vibrissae in the mouse. *J. Embroyl. Exp. Morphol.*, 83: 137–156.

Yamada, T., Pfaff, S.L., Edlund, T. and Jessell, T.M. (1993) Control of cell pattern in the neural tube: motor neuron induction by diffusible factors from notochord and floor plate. *Cell*, 73: 673–686.

Yamakado, M. and Yohro, T. (1979) Subdivision of mouse vibrissae on an embryological basis, with descriptions of variations in the number and arrangement of sinus hairs and cortical barrels in BALB/c (nu/+; nude, nu/nu and hairless (hr/hr) strains. *Am. J. Anat.*, 155: 153–174.

SECTION IV

Neural Network Dynamics

J. van Pelt, M.A. Corner H.B.M. Uylings and F.H. Lopes da Silva (Eds.)
Progress in Brain Research, Vol 102

CHAPTER 24

Dynamics of local neuronal networks: control parameters and state bifurcations in epileptogenesis

Fernando H. Lopes da Silva[1], Jan-Pieter Pijn[2] and Wytse J. Wadman[1]

[1]*Institute of Neurobiology, Graduate School of Neurosciences, University of Amsterdam, Faculty of Biology, Kruislaan 320, 1098 SM Amsterdam, The Netherlands and* [2] *Instituut voor Epilepsiebestrijding, 'Meer en Bosch', De Cruquiushoeve, P.O. Box 21, 2100 AA Heemstede, The Netherlands*

Introduction

In this overview, our objective is to present and discuss some aspects of the general behavior of local neuronal networks (LNNs) considered as dynamic systems. We focus here on the CA1 area of the rat hippocampus because this area has been studied in detail. After a brief description of some of the main structural and physiological properties of the neuronal elements and the synaptic circuits of the CA1 area, we discuss the dynamics of the LNNs considered as systems, i.e. their modes of activity under different conditions. The basic theoretical concepts of non-linear dynamics are treated briefly in order to introduce notions such as attractors, dimensions, bifurcations and chaos. Next, we examine the analogies between dynamic non-linear systems and LNNs, considering the changes in correlation dimension under different conditions. Finally, we consider how disturbances of the control parameters, in particular the balance between excitatory and inhibitory processes (E/I balance) can give rise to bifurcations and pathological behavior such as in epilepsy.

Local neuronal networks: general principles of organization and a case study — the CA1 area of the hippocampus

In general terms, neuronal networks of the central nervous system (CNS) consist of interconnected excitatory (E) and inhibitory (I) neurones. In theoretical studies, it is common to consider the behavior of neurones in a stereotyped way, as wave-to-pulse and pulse-to-wave transformers, with static non-linear properties, namely threshold and saturation (Hopfield and Tank, 1986). However, neurons have a variety of intrinsic membrane properties that allow them to exhibit complex firing behavior with highly non-linear dynamic characteristics, such as regular spiking, burst firing at different frequencies (Steriade and Llinás, 1988).

In terms of interconnectivity, the current view (Freeman, 1987) is that within a LNN both excitatory and inhibitory synaptic connections exist in all possible combinations: excitatory synapses on inhibitory neurons and vice versa, and also excitatory synapses on excitatory neurons as well as inhibitory synapses on inhibitory neurons. Although in principle all the different types of connections indicated above are indeed possible, there is a large variability in their relative importance depending on brain area and on the stage of brain development. To make this description more explicit, we will summarize in more detail the basic elements of the brain structure, the hippocampus CA1 area, where we obtained most of our experimental results.

The CA1 area of the hippocampus

As a basic and relatively simple cortical structure, we chose the hippocampus, since we know a good deal in terms of neuronal elements and connections, of its basic physiological properties, as summarized in Lopes da Silva et al. (1990) and also of its pathophysiological behavior in epilepsy. In the Ammon's horn (CA1 field), we can distinguish the pyramidal cells that form the main excitatory neurons having glutamate as neurotransmitter, and different types of interneurons (INs) that in most cases have GABA as neurotransmitter. The intrinsic properties of the main cell groups in this cortex have been described by several authors (Wong and Prince 1978, 1981; Gustafsson et al., 1982; Nakajima et al., 1986; Numan et al., 1987), and especially several types of interneurons have been described in the CA1 area (for details see Lacaille et al., 1987; Lacaille and Schwartzkroin, 1988; Lacaille and Williams, 1990).

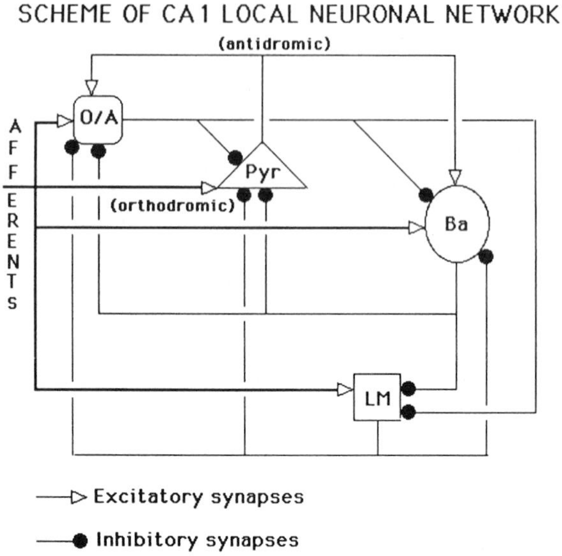

Fig. 1. Summary of the main local circuits of the CA1 area of the hippocampus of the rat. In this scheme we can distinguish: (i) afferent (Aff) or extrinsic excitatory synapses on pyramidal and on interneurons; (ii) pyramidal neurons (Pyr) form excitatory synapses on basket (Ba) and on the interneurons of the stratum oriens/alveus (O/A cells) but not on the interneurons of the stratum lacunosum-moleculare (LM cells); (iii) all types of interneurons (Ba, O/A, LM cells) make inhibitory synapses on pyramidal neurons; (iv) Ba cells make synapses also on O/A cells; O/A cells make synapses also on Ba cells and possibly also on LM cells; it is likely that the latter also make inhibitory synapses on Ba cells. (Adapted from Lacaille and Schwartzkroin, 1988.)

If one wishes to analyze the dynamics of such a network, it is important to consider which circuits are present (these are summarized in Fig. 1). The excitatory synapses are glutamatergic. Different types of receptors present in the CA1 area are usually known by the name of the corresponding agonists, AMPA and kainate, NMDA, t-ACPD, and can be characterized by their molecular composition, as reviewed in Gasic and Hollmann (1992). Although glutamate is the endogeneous ligand for these receptors, a wealth of subtle differences in kinetics may result from the variety of receptor subunits, the functional importance of which has still to be determined. Some of the interneurons receive inputs from outside the CA1 area, and assure feedforward inhibition, whereas others constitute essential elements of the local feedback inhibitory loops, and receive inputs from axon collaterals of the pyramidal cells, i.e. recurrent collaterals.

Two types of inhibitory synaptic activities mediated by GABA must be distinguished: one will open chloride (Cl^-) channels, through $GABA_A$ receptors, causing a relatively fast IPSP, while the other one will open potassium (K^+) channels, through $GABA_B$ receptors, causing a much slower, but deeper, hyperpolarization (Knowles and Schwartzkroin, 1981; Lacaille et al., 1987; Lacaille and Schwartzkroin, 1988). Recently, Pearce (1993) presented physiological evidence for two distinct GABAA synaptic actions in CA1: a fast type that decays within 3–8 msec and a slow type that decays within 30–70 msec. These two types of currents could subserve distinct functional roles: the fast type would shunt depolarizing currents from the spike-initiating zone and would counteract the effect of an EPSP; the slow type causing dendritic inhibition would correspond to the conventional early IPSP that underlies the intensity dependent phenomenon of paired-pulse depression.

Local neuronal network as dynamic non-linear systems — theoretical considerations

Point and chaotic attractors and correlation dimension

LNNs are composed of neuronal elements with non-linear properties such as threshold, saturation and hysteresis, and these elements are interconnected by different types of synaptic processes and delay lines. Therefore, they form complex non-linear systems that can display a wide variety of behaviors. Let us assume that N independent variables are needed for the description of the behavior of the system. In mathematical terms, this implies that the state of the system is represented by one point in a N-dimensional space, and that the time evolution of the system corresponds to a trajectory in this space.

Systems of this kind, when perturbed tend to relax to a defined state, called the *attractor* of the system. The dynamic characteristics of the system can best be analysed by studying the topology of the dynamic trajectories of the momentary states of the system in phase space, i.e. a space that is spanned by the variables that characterize the system. In practice, we do not know the system's variables and thus we have to plot in phase space a variable taken from the observed time series. In general terms, a perturbation will evoke a transient response after which the system relaxes to a stable attracting solution that occupies a subset of the phase space. Trajectories starting from closely placed starting points will tend towards the same attractor. Three of these attractors can be distinguished. (1) The simplest is the case where all trajectories spiral towards one point in phase space: the *point attractor*. (2) When the trajectories converge to a periodic solution, i.e. an oscillation of fixed period, the system has a *periodic or limit cycle attractor*. (3) Finally, a more recently discovered type is the *strange or chaotic attractor*, where the ensemble of trajectories converges to a non-periodic solution that has an infinitely fine structure but occupies only a small proportion of the phase space.

In the case of real systems, the value of N is unknown, but it is always interesting to estimate it. According to a theorem of Takens (1981), it is possible to reconstruct the attractor given a single signal generated by the system. The problem is that we do not know, in general, the dimension that the space should have in which the attractor must be constructed. The solution is, first, to reconstruct the attractor in a lower dimensional subspace of dimension m. This is called the embedding of the attractor in m-dimensional space, where $m < N$, and it means that an m-dimensional projection of the attractor has to be reconstructed. The 'density' averaged over the whole attractor corresponds to the correlation integral $C(r,m)$. There is a power law relationship between $C(r,m)$ and r. The exponent of this power relationship is called the correlation dimension D_2. D_2 is therefore the local slope of the plot of log $C(r,m)$ against log (r). Since $m < N$, such a projection will be a compressed version of the attractor, and therefore it will have a higher 'density' than the original attractor.

The 'density' represents the probability of two points of the attractor lying within a unit volume. The key to the solution is to estimate this 'density' for various values of m. This procedure is repeated for different values of m until the 'density' does not change anymore for increasing values of m. To estimate the 'density', the concept of 'correlation dimension D_2' is used. The procedure generally employed can be described intuitively as follows (for a quantitative description, see Pijn et al., 1991): around each point of the m-dimensional embedded attractor an m-dimensional sphere with radius r is constructed. The 'density' is calculated as function of r and this procedure is repeated for increasing values of m. The estimate D_2 is obtained when these plots saturate for a given m. In Figs. 2 and 3, D_2 is plotted against log r; if the dynamics of the system correspond to an attractor a plateau should be evident in such plots. This does not occur in the case of filtered noise. Figure 2 shows a number of signals that are the solutions of the Duffing's equation with forcing, to illustrate the occurrence of different types of attractor and the procedures used in the corresponding analysis. The Duffing equation was chosen because it is a well known example of a non-linear oscillator that has been extensively studied with respect to chaotic dynamics (Thompson and Stewart, 1986).

Control parameters, operating point and bifurcations

Depending on the system's parameters, a dynamic non-linear system may have different attractors. The ensemble of all starting conditions leading to a stable attractor solution forms the *basin of attraction*. It is interesting to analyse how such a system behaves as function of a number of parameters. This means that the location in phase space of the attractors and basins has to be found for a family of parameters. We may consider

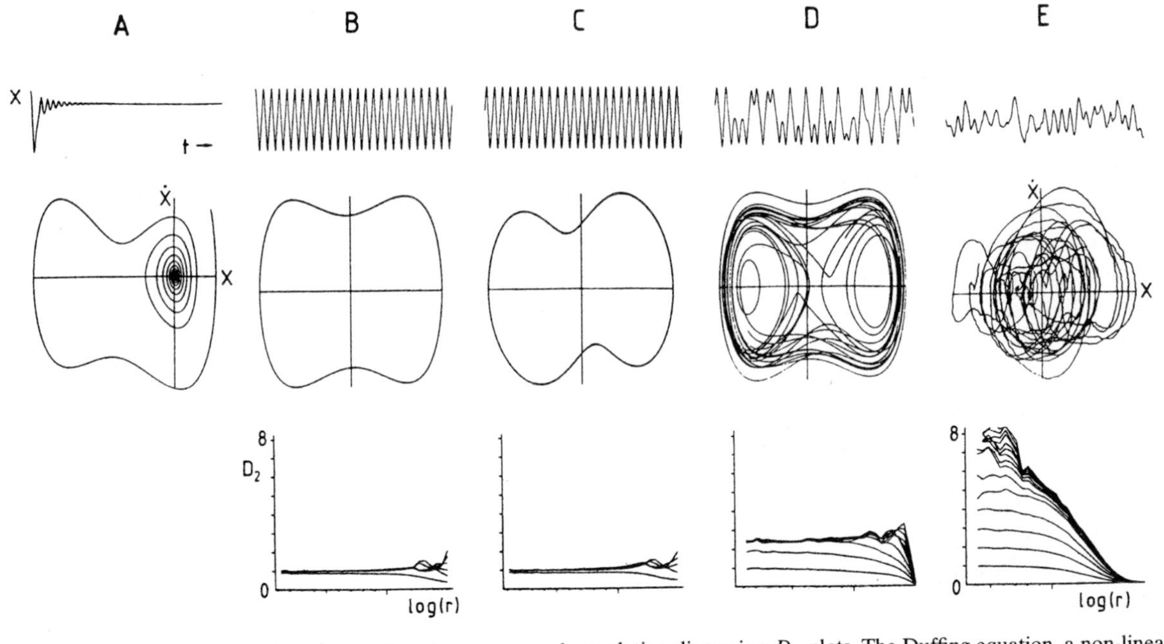

Fig. 2. Examples of signals with corresponding phase-space and correlation dimension D_2 plots. The Duffing equation, a non-linear differential equation was chosen as an example: $x + \delta x \; - \; x + x^3 = \gamma$ cos ωt, with $\delta = 0.15$ and $\gamma = 0.3$ for signals B and D, and $\gamma = 0$ for signal A. The starting values of x and t differed for signals B and D. This non-linear differential equation was solved in a digital computer for a number of conditions to demonstrate that this system can have multiple attractors and to illustrate the corresponding phase-space and D_2 plots. Three conditions are shown: point attractor (A), limit cycle (B) and chaotic attractor (D). To test the hypothesis whether the signal D is a linear correlated noise, a surrogate signal was created by the following procedure. The Fourier Transform (amplitude and phase spectrum) of the original signal was computed, then the phase angle of each frequency component was replaced by a random number and, finally, an inverse transformation was carried out to create the surrogate signal. The latter is a random signal (E). The same procedure was carried out for signal B, resulting in signal C. In this case, the limit cycle and the correlation dimension are similar for B and C. B is a periodic signal that has a line spectrum; after the transformation, the resulting signal C has the same characteristics, and both show a value of $D_2 = 1$, as expected. On the contrary, D and E differ since D corresponds to low-dimensional chaos ($2 < D_2 < 3$) and E to a random signal that has a relatively high dimension (> 10). Therefore, the hypothesis that the signal D would be correlated noise can be rejected. The calculations were performed on signals with 3.2 times the length of the examples shown, totalling 16,384 samples.

these as *control parameters* of the system. The control parameters define the state at which the system operates, i.e. the *operating point* of the system. As the settings of the control parameters change, the operating point also changes and the system may converge to different attractors, for example it may *bifurcate*, i.e. a rapid state change may take place, for example from a periodic to a chaotic attractor, or from a high- to a low-dimensional attractor. Indeed, in non-linear systems, multiple attractors are common. A given set of values of the control parameters will lead to a qualitative change, or *bifurcation*, between two stable states of the system's dynamics (for a tutorial overview of the theory, the reader is referred to Thompson and Stewart, 1986).

Behavior of a LNN (CA1 area) as a dynamic system

It is sometimes assumed that knowledge of the elementary properties of individual neurons is sufficient to understand the behavior of the whole neuronal population, namely the occurrence of oscillations or of spatio-temporal patterns of activity. However, the fact that different groups of neurons with non-linear properties are interconnected, forming complex LNNs, can give rise to emergent properties and to cooperative actions which are not simply deductible from elementary properties. In view of the difficulty of deducing the dynamics of the whole population from the properties of neuronal elements, we have to approach this problem using a macroscopic analysis. A general approach is to estimate the correlation dimension D_2 of a realization of the LNN, i.e. of an EEG signal generated by the system under different conditions.

Here we studied the dynamics of the CA1 area under three conditions: restful wakefulness, exploratory behavior and during a kindled epileptic seizure. As shown in Fig. 3, the two first signals have high D_2 values that are not distinguishable from those of band-limited noise. However, the epileptiform seizure had a low value in the range $2 < D_2 < 4$ and differed conspicuously from that of noise. This means that under such condition the signal may be considered to be generated by a LNN with a chaotic attractor. We should add that under the resting conditions (i.e. before the application of the series of tetani that induces kindling), a single tetanus causes only a short transient without an afterdischarge, such that the system returns to the initial state in a very short time. If this state would be characterized by an EEG signal of constant amplitude, the corresponding correlation dimension, D_2, would be equal to zero. However, the EEG of the rest condition does not have a constant amplitude but is characterized by on-going oscillations. The corresponding correlation dimension is very high, which indicates that the LNN behaves apparently in a random way. We may state that the system fills a small subspace of a high-dimensional space. This does not necessarily imply that the underlying process is stochastic in nature, however, merely that the process is of such complexity that it is undistinguishable from noise.

We may conclude that during the kindling procedure, the operating point of the LNN shifts progressively and comes close to a bifurcation point, i.e. it may undergo a rapid state change. The system comes close to a condition where it can switch between two states: a high dimensional random state and a low dimensional chaotic attractor (Fig. 4). The shift in operating point renders the LNN more sensitive to the same input, i.e. a tetanus. This is a typical property of non-linear systems that may have multiple attractors.

Changes in control parameters of CA1 LNN during kindling

Our working hypothesis is that the basic control parameter of the dynamics of the CA1 LNN is the balance between excitatory (E) and inhibitory (I) drives. We examine here the experimental evidence for taking this possibility seriously.

Impulse responses: single and paired-pulse

The E/I balance should be apparent in the *impulse response* of the LNN. A problem is how to

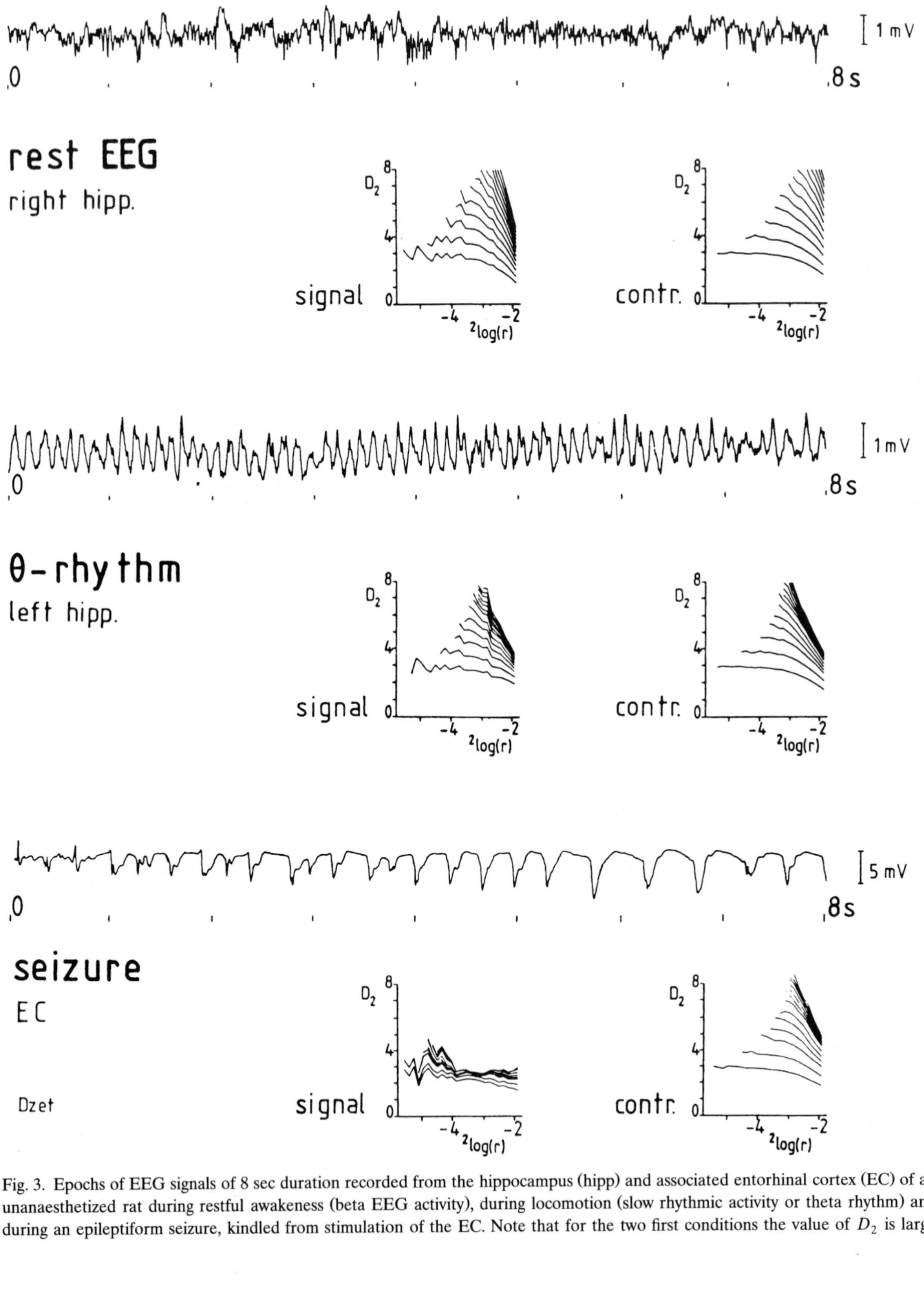

Fig. 3. Epochs of EEG signals of 8 sec duration recorded from the hippocampus (hipp) and associated entorhinal cortex (EC) of an unanaesthetized rat during restful awakeness (beta EEG activity), during locomotion (slow rhythmic activity or theta rhythm) and during an epileptiform seizure, kindled from stimulation of the EC. Note that for the two first conditions the value of D_2 is large

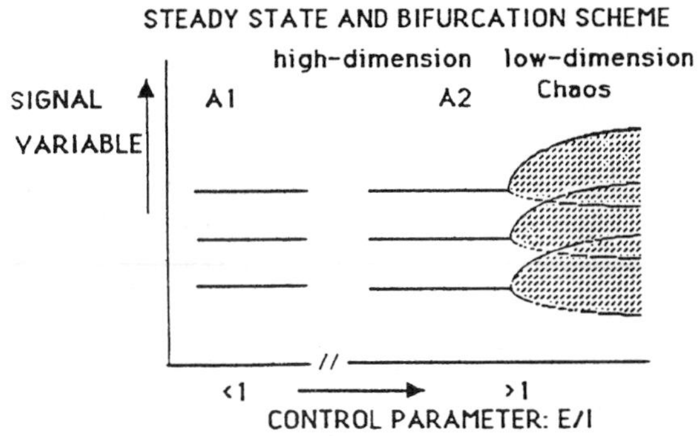

Fig. 4. Qualitative scheme of a model of the dynamics of the local neuronal network (LNN). The diagram represents a map of the behavior of the various steady states that trace out paths in space and can show bifurcations, as function of a control parameter. The latter is assumed to be the ratio between excitatory (E) and inhibitory (I) processes and is plotted along the horizontal axis. The vertical axis represents an output variable of the LNN, for example the variance of the signal amplitude. Region A1 represents the case that E/I has a small value (i.e. < 1) where different steady-state paths can be found, depending on levels of awareness/drowsiness; these states correspond to high-dimensional processes like those illustrated in the two first traces of Fig. 2. Region A2 represents the case where the E/I is > 1 and a bifurcation takes place to steady states characterized by low-dimensional chaotic attractors, such as during seizures as shown in the lower trace of Fig. 2. Note that under normal conditions the system behaves as in A1. Due to a change of E/I (either acute during a strong tetanus for example, or chronically due to the kindling process) the system can behave as in A2, and it can easily be attracted into the basin of a low-dimensional chaotic attractor.

choose a representative response of the LNN. A typical choice is to record the field evoked potential (EP) generated by the jointly and coordinated activity of the neuronal elements. In general, all types of membrane currents, intrinsic as well as synaptic, will contribute to the potential recorded from the surrounding field. However, the main contribution to the local EP is given by the post-synaptic currents flowing into the extracellular medium.

The most characteristic impulse response of the CA1 LNN is the EP elicited by a single pulse applied to the Schaffer collateral/commissural fibers (Leung, 1979). More insight into the nature of the dynamics of the LNN can be obtained by using *paired-pulse stimulation* (Fig. 5). In general, the response to the second pulse of a pair is conditional on the response to the first; accordingly, the first stimulus is usually called the conditioning stimulus (CS), and the second the test stimulus (TS). Two parameters are of especial importance in this protocol: the interstimulus interval (ISI) and the stimulus intensity.

In general, at low stimulus intensities and at ISI shorter than 500 msec, the response to the second stimulus of the pair (test response or TR) is facilitated with respect to the first response (CR). This phenomenon is called 'paired-pulse facilitation' (PPF). The PPF reflects the enhancement of the release of glutamate to the TS, probably due to the fact that $[Ca^{2+}]$ remains elevated in the pre-synaptic terminal for tens of milliseconds after the CS (Zucker, 1989). In addition, it is also likely that a reduction of feed-forward inhibition may contribute to the PPF. This may occur due to the fact that the activity of the GABAergic interneurons responsible for this feedforward inhibition, can be depressed by the GABA released during the CR, via pre-synaptic GABAB (auto)receptors (Nathan et al., 1990). With a progressive increase in stimulus intensity, the PPF gradually declines and a 'paired-pulse depression' (PPD) becomes apparent. This effect is rapidly encountered 'in vivo' (Kamphuis et al., 1988) but it is much more difficult to elicit in slices in vitro (Nathan et al., 1990). The PPD is caused mainly by IPSPs on the pyramidal neu-

---

and similar for the EEG signals and for the surrogate (or control, contr.) signals, indicating that the underlying process cannot be distinguished from random noise. On the contrary, the third trace (i.e. the epileptiform EEG) has a low value (between 2 and 4) that differs strongly from that of the surrogate signal, indicating the existence of a chaotic attractor of low dimension during an epileptiform seizure. (Adapted from Pijn et al., 1991.)

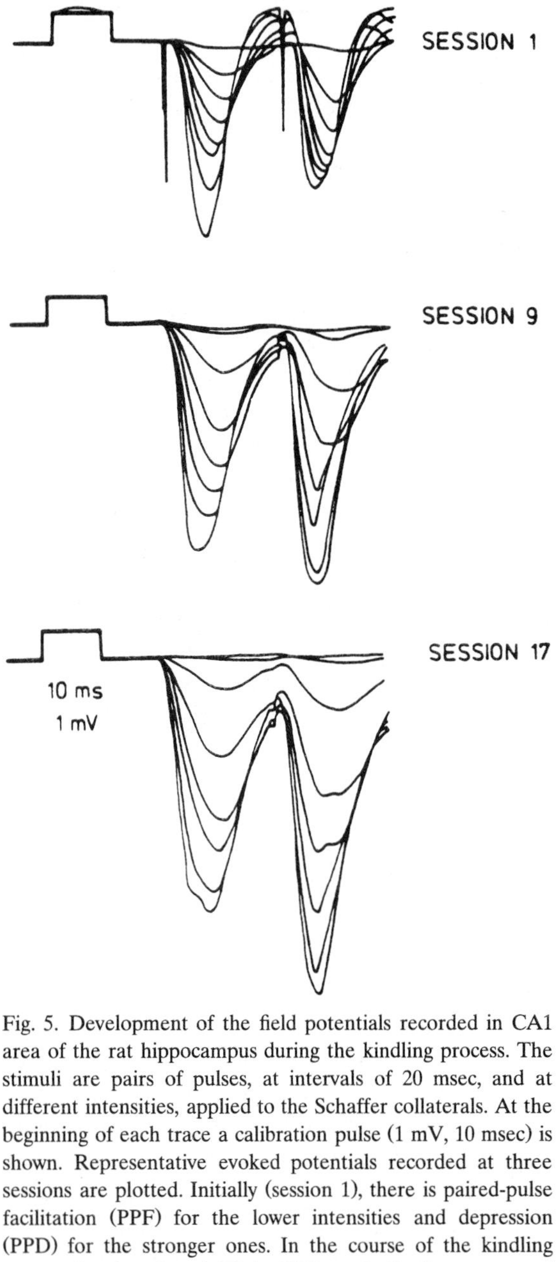

Fig. 5. Development of the field potentials recorded in CA1 area of the rat hippocampus during the kindling process. The stimuli are pairs of pulses, at intervals of 20 msec, and at different intensities, applied to the Schaffer collaterals. At the beginning of each trace a calibration pulse (1 mV, 10 msec) is shown. Representative evoked potentials recorded at three sessions are plotted. Initially (session 1), there is paired-pulse facilitation (PPF) for the lower intensities and depression (PPD) for the stronger ones. In the course of the kindling process (sessions 9 and 17) the PPD gradually disappears, and only PPF can be found.

rons. These IPSPs are most likely mediated by $GABA_{A,slow}$ synaptic actions, taking into consideration that this component decays within 30–70 msec (Pearce, 1993), which corresponds to the ISI at which PPD is usually encountered.

From the above, we may conclude that there are several excitatory (E) and inhibitory (I) processes in the CA1 area of the hippocampus. In the following, we will examine how this balance may change under different conditions. In addition to the parameters described above, we must also consider that the intrinsic membrane characteristics of CA1 pyramidal neurons will also contribute to a change in the dynamics of the LNN. CA1 pyramidal neurons, isolated from a kindled focus possess an increased calcium conductance that will considerably increase the intracellular calcium concentration, and thus may contribute to the enhanced excitability found after kindling (Vreugdenhil and Wadman, 1992).

Bifurcation in activity mode: the kindling phenomenon

In order to examine whether, and if so how, the balance between excitation and inhibition is affected during kindling, we (Kamphuis et al., 1988, 1992; Huisman and Wadman, 1990) and others (Lothman et al., 1992) have examined this question both in vivo and in vitro.

In vivo, we found a characteristic long-term change of the local field potentials evoked by stimulation of the Schaffer collaterals using the paired-pulse paradigm: (i) a decrease in the slope of the decaying phase of the EP recorded from the stratum radiatum; (ii) a decrease in paired-pulse depression; (iii) the emergence of a population spike in the field potentials recorded from stratum oriens and/or pyramidale. Furthermore, the phenomenon of long-term potentiation, LTP, defined as an increase in the amplitude of the response to the conditioning stimulus, was also found but usually only in the early phases of kindling. The first change is probably caused by an attenuation of the early IPSP, either the recurrent and/or the feedforward component. The second is a most conspicuous and robust change that is invariably seen in CA1 area during kindling of the Schaffer collaterals and probably determined by the decrease of the inhibitory (feedback or recurrent) component; this can be put in evidence by plotting the paired-pulse index,

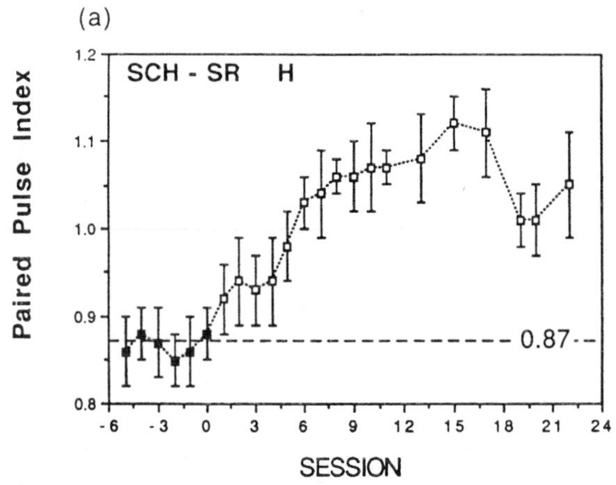

Fig. 6. Plot of the paired-pulse index (PPI) as function of kindling session. The PPI is calculated as the ratio between the amplitude of the response to the second stimulus of a pair, and that of the first, measured in evoked potentials as those of Fig. 5. The first six closed symbols represent the measurements in the pre-kindling period and the open symbols indicate the recordings during the kindling period. The horizontal broken line indicates the mean pre-kindling value of PPI. Values represent means ± S.E.M. ($n = 8$). (Adapted from Kamphuis et al., 1993.)

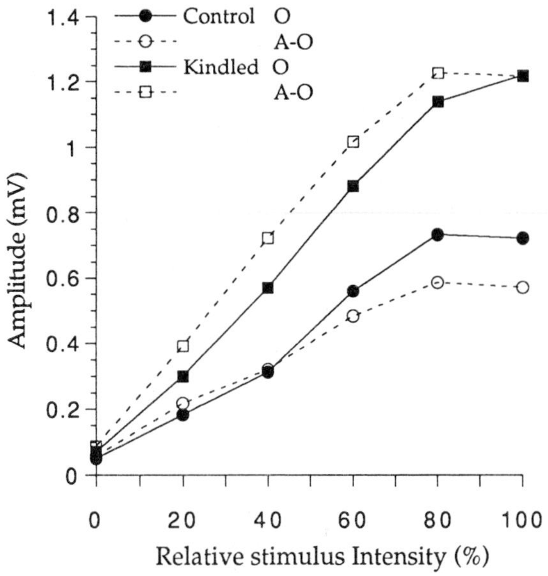

Fig. 7. Plot of the input-output relations of the responses recorded in CA1 pyramidal area to stimulation of the afferent fibers (Schaffer collaterals). Two conditions were studied for controls and for kindled rats: in one condition the stimulus was applied only to the afferents (see Fig. 1, orthodromic stimulation, O). In the other an antidromic stimulus was applied to the outgoing fibers 20 ms before the O stimulus. The antidromic stimulus activates not only pyramidal neurons but also, through the axon collaterals, inhibitory interneurons after one synaptic relay (AO). In the control situation it can be seen that the amplitude of the evoked population spike, as a function of stimulus intensity, is depressed relative to the O stimulation. This is not the case after kindling, indicating that the recurrent inhibition is less effective.

i.e. the ratio between the amplitude of the response to the second stimulus (TR) and that of the first (CR), as function of kindling session (Fig. 6). The third is most likely due to an impairment of the feedforward IPSP (possibly the $IPSP_{A,slow}$) since the population spike appears, and progressively increases in amplitude, in the field potential elicited by the conditioning stimulus when the latter is relatively strong.

The nature of the underlying changes was analyzed in more detail in hippocampal slices obtained from kindled animals and maintained in vitro. Using the combination of antidromic and orthodromic stimulations described above, we observed (Fig. 7) that the inhibitory recurrent component, elicited by the antidromic stimulation of the alvear tract, was decreased after kindling (Huisman and Wadman, 1990).

Taking both the in vivo and the in vitro experimental results into consideration, we may conclude that both the recurrent and the feedforward inhibitory components are impaired by the kindling procedure. This has been supported by other experimental facts: (i) a reduction in CA1, by almost 50%, of a subpopulation of GABA-immunoreactive somata that do not co-localize with the calcium-binding protein parvalbumin (Kamphuis et al., 1989); (ii) a reduction of the GABAergic inhibitory action on the firing rate of CA1 pyramidal neurons, studied iontophoretically (Kamphuis et al., 1991); (iii) a decrease of the GABAergic binding sites in CA1 studied with the [^{3}H]muscimol agonist (Titulaer et al., 1994). In addition, there are also indications that this change in balance may be further amplified by strengthening of excitatory processes, since in the latter study the responsiveness of CA1 neurons to iontophoretic application of glutamate was long-lastingly increased after the last kindled convulsion. We have limited ourselves in this description to the CA1 area of the hippocampus. However, it may be noted that in the kindling model of the dentate gyrus also, the balance between excitation and inhibition changes in favour of the former, but this appears to be due, in this case, to an increase in the function of the NMDA receptor/channels (Mody et al., 1992).

Conclusions

The physiological analyses described above allow us to draw the conclusion that the balance between excitatory (E) and inhibitory (I) processes, during kindling epileptogenesis in the CA1 area of the hippocampus, shifts in the direction of a prevalence of the former. The operating point of the LNN changes accordingly. The consequences of this change for the dynamics of the network can be perceived when we take into consideration that a LNN behaves as a non-linear dynamic system that may have multiple attractors, including a low-dimensional chaotic attractor. Indeed, we may conclude from the experimental measurements described above that the hippocampal networks may present bifurcations between different states characterized by different correlation dimensions. When the LNN is within the basin of attraction of the chaotic attractor, it can generate complex oscillations that can last for a long time, say many seconds or even minutes, and can possess that combination of 'randomness and structure' that is nowadays called *chaos*. The network can fall more easily into the basin of this attractor when its operating point is close to a bifurcation, as it occurs during epileptogenesis. In this case, even a weak stimulus can be sufficient to switch the dynamics of the LNN to chaos.

This is in essence the main difference between a normal and an epileptogenic network: the former can also produce epileptiform discharges but only after very strong stimulations (e.g. electroshock, injection of convulsants such as pentylenetetrazol) since its operating point is quite far from a bifurcation leading to chaos; in the epileptogenic network, on the contrary, the distance between operating and bifurcation points is much reduced.

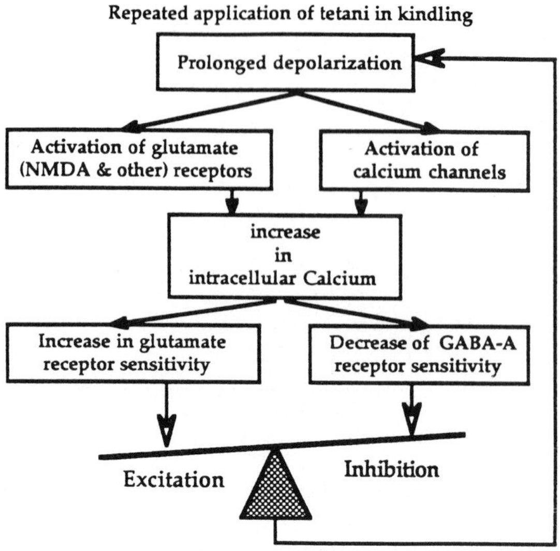

Fig. 8. Scheme representing the main changes at the cellular/molecular level that occur during the process of kindling epileptogenesis of the CA1 area of the hippocampus. The repeated depolarization, elicited by the tetani, causes the increase of Ca^{2+} conductances and the stimulation of different types of glutamate receptors. At long-term, changes occur in the number/sensitivity of these receptors, and of the Ca^{2+} voltage-dependent channels, possibly due to modification of the corresponding molecular composition. This results in tipping the balance between excitation and inhibition in favor of the former.

We should note that, according to this perspective, the essence of the epileptiform behavior is any change in operating point of the LNN that brings it close to a bifurcation leading to chaos, independently of the exact mechanism at the microscopic level that may underlie this change in the E/I balance. In other words, different cellular/molecular mechanisms can lead to the same change in dynamics of the LNN, on condition that they induce the same type of change in E/I. This implies, of course, that a variety of factors may cause epileptogenesis. In the case of kindling of the CA1 area of the rat hippocampus the main cellular and molecular changes that lead to the change in the control parameter E/I are summarized in Fig. 8. This figure shows in a schematic way how the successive depolarizations caused by the application of the tetani lead to changes in calcium conductances (Vreugdenhil and Wadman, 1992), and to changes both of the $GABA_A$ and the glutamate receptors (Kamphuis et al., 1991, 1994; Titulaer et al., 1994) that tip the E/I balance in favour of the excitatory processes.

Summary

The aim of this overview is to present evidence that local neuronal networks (LNNs) are functionally organized in such a way that they behave as *dynamic non-linear systems* that can exhibit multiple types of *attractor* and can present *bifurcations* between different attractors, depending on *control parameters*. To begin with, some of the theoretical concepts of non-linear dynamics and chaos are briefly presented. As a case study, we described the CA1 area of the hippocampus and the changes that the corresponding LNNs undergo during *kindling epileptogenesis*. During epileptic seizures, evidence exists for the presence of low-dimensional chaos, since the *correlation dimension* estimated from the corresponding EEG signals decreases dramatically from a large value, characteristic of the resting state, to a low value typical of *deterministic chaos*. We propose that, among other things, an important control parameter of the dynamics of this brain area is the *balance between excitatory (E) and inhibitory (I) processes*. We assume that this balance can be experimentally estimated by using a paired-pulse paradigm. Accordingly, we demonstrate that the paired-pulse response changes during kindling epileptogenesis in the sense that the E/I ratio increases in the course of the establishment of a kindled epileptogenic focus. This change in E/I leads to a shift in the *operating point* of the LNN moving it close to a bifurcation where a rapid state change takes place. In this way, the LNN dynamics can change more readily to the basin of attraction of a chaotic attractor than under normal conditions. This is in essence what makes the behavior of the LNN more sensitive to tetanus, and predicts the facilitated occurrence of epileptic seizures during kindling.

Acknowledgements

The expert secretarial help of Trúc Ngô-Hà and Cristine Knaap-Cabi is gratefully acknowledged. This work was supported by a grant from the National Epilepsy Fund CLEO-Alog.

References

Freeman, W.J. (1987) Techniques used in the search for the physiological basis of the EEG. In: A. Gevins and A. Rémond (Eds.), *Handbook of Electroencephalography and Clinical Neurophysiology (revised edition)*, Vol. 3A, Part 2, C. 18, Elsevier, Amsterdam, pp. 583–664.

Gasic, G.P. and Hollmann, M. (1992) Molecular neurobiology of glutamate receptors. *Annu. Rev. Physiol.*, 54: 507–536.

Gustafsson, B.H, Galvan, P., Grace, P. and H. Wigström (1982) A transient outward current in a mammalian central neuron blocked by 4-aminopyridine. *Nature London*, 299: 252–254.

Hopfield, J.J. and Tank, D.W. (1986) Computing with neural circuits: a model. *Science*, 233: 625–633.

Huisman, E. and Wadman, W.J. (1990). Excitation and inhibition determined by double pulse protocols in rat hippocampal slices. *Pfluegers Arch.*, 418: R147.

Kamphuis, W., Lopes da Silva, F.H. and Wadman, W.J. (1988) Changes in local evoked potentials in the rat hippocampus (CA1) during kindling epileptogenesis. *Brain Res.*, 440: 205–215.

Kamphuis, W., Huisman, E., Wadman, W.J., Heizmann, C.W. and Lopes da Silva, F.H. (1989) Kindling induced changes in parvalbumin immunoreactivity in rat hippocampus and its relation to long-term decrease in GABA-immunoreactivity. *Brain Res.*, 479: 23–34.

Kamphuis, W., Gorter, J.A. and Lopes da Silva, F.H. (1991) A long-lasting decrease in the inhibitory effect of GABA on glutamate responses of hippocampal pyramidal neurons induced by kindling epileptogenesis. *Neuroscience*, 41: 425–431.

Kamphuis, W., Gorter, J.A. and Lopes da Silva, F.H. (1992) Inhibitory and excitatory synaptic processes in hippocampal kindling of the rat. In: E.-J. Speckmann and M.J. Gutnick (Eds.), *Epilepsy and Inhibition*, Urban and Schwarzenberg, Baltimore, pp. 363–374.

Knowles, W.D. and P.A. Schwartzkroin (1981) Local circuit synaptic interactions in hippocampal brain slice. *J. Neurosci.*, 1: 318–322.

Lacaille, J.C. and Schwartzkroin, P.A. (1988) Stratum lacunosum-moleculare interneurons of hippocampal CA1 region. II. Intrasomatic and intradendritic recordings of local circuit synaptic interactions. *J. Neurosci.,* 8: 1411–1424.

Lacaille, J.C. and Williams, S. (1990) Membrane properties of interneurons in stratum oriens-alveus of the CA1 region of rat hippocampus in vitro. *Neuroscience,* 36: 349–359.

Lacaille, J.C., Mueller, A.L., Kunkel, D.D. and Schwartzkroin, P.A. (1987) Local circuit interactions between oriens alveus interneurons and CA1 pyramidal cells in hippocampal slices: electrophysiology and morphology. *J. Neurosci.,* 7: 1979–1993.

Leung, L.-W.S. (1979) Orthodromic activation of hippocampal CA1 region of the rat. *Brain Res.,* 176: 49–63.

Lopes da Silva, F.H., Witter, M.P., Boeijinga, P.H. and Lohman, A.H.M. (1990) Anatomical organisation and physiology of the limbic cortex. *Physiol. Rev.,* 70: 453–511.

Lothman, E.W., Bertran, E.H. and Stringer, J.L (1992) Functional anatomy of hippocampal seizures. *Prog. Neurobiol.,* 37: 1–82.

Mody, I., Otis, Th. S., Stanley, K.J. and Köhr, G. (1992) The balance between excitation and inhibition in dentate granule cells and its role in epilepsy. *Epilepsy Res.,* Suppl. 9: 331–339.

Nakajima, Y., Nakajima, S., Leonar, R.J. and Yamaguchi, K. (1986) Acetylcholine raises the excitability by inhibiting the fast transient potassium current in cultured hippocampal neurons. *Proc. Natl. Acad. Sci. USA,* 83: 3022–3026.

Nathan, T., Jensen, M.S. and Lambert, J.D.C. (1990) $GABA_B$ receptors play a major role in paired-pulse facilitation in area CA1 of the rat hippocampus. *Brain Res.,* 531: 55–65.

Numan, R.E., Wadman, W.J. and Wong, R.K.S. (1987) Outward currents of single hippocampal cells obtained from the adult guinea-pig. *J. Physiol. London,* 393: 331–353.

Pearce, R.A. (1993) Physiological evidence for two distinct $GABA_A$ responses in rat hippocampus. *Neuron,* 10: 189–200.

Pijn, J.P., van Neerven, J., Noest, A. and Lopes da Silva, F.H. (1991) Chaos or noise in EEG signals; dependence on state and brain site. *Electroenceph. Clin. Neurophysiol.,* 79: 371–381.

Steriade, M. and Llinás, R.R. (1988) The functional states of the thalamus and the associated neuronal interplay. *Physiol. Rev.,* 68: 649–742.

Takens, F. (1981) Detecting strange attractors in turbulence. Dynamical system and turbulence. In: D.A. Rand and L.J. Young (Eds.), *Lecture Notes in Mathematics, Vol. 898,* Springer, New York, pp. 365–381.

Thompson, J.M.T. and Stewart, N.B. (1986) Nonlinear dynamics and chaos. John Wiley, Chichester, 376 pp.

Titulaer, M.N.G., Kamphuis, W., Pool, C.W., van Heerikhuize, J.J. and Lopes da Silva, F.H. (1994) Kindling induces time-dependent and regional specific changes in the [^{3}H]muscimol binding in the rat hippocampus: a quantitative autoradiographic study. *Neuroscience,* 59: 817–826.

Vreugdenhil, M. and Wadman, W.J. (1992) Enhancement of calcium currents in rat hippocampal CA1 neurons induced by kindling epileptogenesis. *Neuroscience,* 49: 373–381.

Wong, R.K.S. and D.A. Prince (1978) Participation of calcium spikes during intrinsic burst firing in hippocampal neurons. *Brain Res.,* 159: 385–390.

Wong, R.K.S. and D.A. Prince (1981) Afterpotential generation in hippocampal pyramidal cells. *J. Neurophysiol.,* 45: 86–97.

Zucker, R.S. (1989) Short-term synaptic plasticity. *Annu. Rev. Neurosci.,* 12: 13–31.

J. van Pelt, M.A. Corner H.B.M. Uylings and F.H. Lopes da Silva (Eds.)
Progress in Brain Research, Vol 102

CHAPTER 25

Non-linear dynamics in neural networks

J.G. Taylor

Centre for Neural Networks, Department of Mathematics, King's College London, Strand, London WC2R 2LS, UK

Introduction

Non-linear analysis is being used ever more extensively to explore the properties of neural networks composed of increasingly sophisticated neurons. I wish here to describe recent results in applying non-linear methods, involving both dynamical systems and other branches of mathematics, to problems of information processing. The topics I will cover are more specifically: (a) non-linear principle component analysis and decorrelation of outputs, by means of extensions of the Hebb rule with stabilising terms, and its possible use in early visual processing; (b) temporal sequence storage by temporal neurons and recurrent nets, using an extension of Hebbian learning, and its relation to the problems of hippocampal, temporal and frontal lobe memory; (c) stochastic responses associated with quantal synaptic transmission; (d) the manner in which feed-forward synchronisataion is affected by stochasticity.

These results, and the more general problem areas in which they lie, also have great relevance to the fast-developing field of artificial neural networks. Since the topics under (a), (b), (c) and (d) cover an enormous area I will only indicate very briefly some of the details, and in some cases refer to the referenced papers, for more extensive discussion.

Principle components and decorrelation

It has been suggested (Barlow, 1961) that feature analysis in the visual system can arise through attempted decorrelation of inputs. Another approach to early visual processing is that of maximisation of throughput of information (Linsker, 1988), which is known to be closely related to principle component analysis (Plumbley, 1993a,b). In this section the ideas of principle component analysis will be explored, first for linear neurons, and then at the non-linear level.

Let us first review the Oja rule (Oja, 1982) for a linear neuron before discussing the extension to a quadratic neuron. The Oja neuron has n input lines each with a weight $w_j = 1,2,3...N$. We denote the weighted sum of inputs as

$$y = \sum_{j=1}^{N} w_j \sigma_j = \underline{w}^T \underline{\sigma}$$

where $\underline{\sigma}$ is a column vector of inputs drawn from some distribution on $\mathbf{R}^N$ and $\underline{w}^T$ is the transpose of the n weight column vector. The Oja learning rule consists of the usual Hebbian term plus a decay factor which ensures the eventual normalisation of $\underline{w}$:

$$\dot{w}_i(t) = r(t)y(t)(\sigma_i - y(t)w_i(t)) \quad (1)$$

where the dots denotes differentiation with respect to time and $r(t)$ is the learning rate. Averaging over the input distribution and denoting the components of the correlation matrix ($\mathbf{C}_2$) as $C_{ij} = < \sigma_i \sigma_j$ (where the brackets correspond to averaging over the set of input patterns) we obtain

$$\frac{1}{r(t)}\langle \dot{w}_i(t)\rangle = \sum_{j=1}^{N} C_{ij}\langle w_j(t)\rangle - \sum_{j,k=1}^{N} \langle w_j(t)\rangle \times C_{jk}\langle w_k(t)\rangle\langle w_i(t)\rangle \quad (2)$$

We have made the assumption that the weights are uncorrelated with the patterns and with themselves. After dropping the angle brackets, Eqn. (2) may be written in vector-matrix notation as

$$\frac{1}{r(t)}\dot{w} = \underline{\underline{C}}\,\underline{w} - \left(\underline{w}^{\mathrm{T}}\underline{\underline{C}}\,\underline{w}\right)\underline{w} \quad (3)$$

At a fixed point the right hand side of (3) becomes zero. This implies that $\underline{w}$ is an eigenvector of $\mathbf{C}_2$ with eigenvalue

$$\lambda = \underline{wt}\underline{\underline{C}}\,\underline{w} \quad (4)$$

The right hand side of (4) is equal to $\lambda\|\underline{w}\|^2$ and so the Euclidean norm of $\underline{w}$ is one. A local stability analysis (Krogh and Hertz, 1990) shows that the only stable solution of (3) is the eigenvector of $\mathbf{C}_2$ with the largest eigenvalue. Therefore the Oja rule obtains the normalised eigenvector of the correlation matrix of the input space associated with the largest eigenvalue. This is typically called the first principal component. The learning rule (3) will only converge to the first principal component under certain conditions on the learning rate (Ljung, 1977). These conditions are

(i) $\sum^{\infty} r(t) = \infty$

(ii) $\sum^{\infty} r^p(t) > \infty$ for some p

(iii) $r(t)$ is a decreasing sequence of positive real numbers;

(i) and (ii) set a limit on the rate of reduction of the learning step size. If $r(t)$ decreases faster than allowed, then there is some non-zero probability that the fixed point is not reached; (iii) simply implies that the learning rate falls off with time. We have investigated the dynamics of the Oja learning rule with the aim of finding suitable choices for the form of $r(t)$ in (Taylor and Coombes (1993), to which further discussion is referred.

For many purposes it is adequate to describe the Oja learning rule as an algorithm which maximises the variance of the output of a linear neuron, keeping the weights approximately normalised at all times (Sanger, 1989). In practice one finds this to be true especially if the initial weights are normalised. Although the Oja learning rule does not perform gradient descent on an energy function, it is a simple matter to show that the variance $F(t)$ is an increasing function of time t, since $F(t+1) - F(t) = 2r[(\underline{\underline{C}}\underline{w})^2 - (\underline{w}^t\underline{\underline{C}}\,\underline{w})^2] + \Delta\underline{w}^{\mathrm{T}}\underline{\underline{C}}\Delta\underline{w} \geq 0$ where $\Delta\underline{w} = \underline{w}(t+1) - \underline{w}(t)$, if the weights are normalised at all times.

We shall now discuss the extension to quadratic neurons (Taylor and Coombes, 1993). Let the output of our neuron be $y(t) = \sum_{ij}(t)\sigma_i\sigma_j$ and the learning rule be

$$\dot{w}_{ij}(t) = r(t)y(t)\left(\sigma_i\sigma_j - y(t)w_{ij}(t)\right) \quad (5)$$

On averaging over the presented patterns we obtain

$$\frac{1}{r(t)}\langle \dot{w}_{ij}(t)\rangle = \sum_{l,m} C_{ijlm}\langle w_{lm}(t)\rangle - \sum_{l,m,r,s}\langle w_{rs}(t)\rangle C_{lmrs}\langle w_{rs}(t)\rangle \times \langle w_{ij}(t)\rangle \quad (6)$$

C_{ijkl} denote the components of the fourth rank correlation tensor ($\mathbf{C}_4$). Relabelling as $(i, j) = a,(l,m) = b,(r,s) = c$ Eqn. (6) becomes

$$\frac{1}{r(t)}\langle \dot{w}_a(t)\rangle = \sum_b C_{ab}\langle w_b(t)\rangle - \sum_{b,c}\langle w_b(t)\rangle C_{bc}\langle w_c(t)\rangle\langle w_a(t)\rangle \quad (7)$$

This is of the same form as the original Oja learning rule (2) and, as such, its behaviour is known. The output of a quadratic neuron is proportional to the energy function of a Hopfield net, and the link between the two is explained in the next section.

The Eqn. (7) has identical form to Eqn. (3), but now involves up to fourth-order statistics of the input patterns. It is evident that this process can be continued by adding higher order terms still into the right-hand side of the quadratic neurons output $y(t)$. If terms or order n are included, we then define the object $\Omega^T = W_i, W_{ij}..., W_{il}...i_n)$, as the extended weight vector whilst the correlation matrix C is enlarged to

$$\underline{\underline{C}} = \begin{pmatrix} C_2 & C_3 & \cdots & C_{n+1} \\ C_{n+1} & \cdots & \cdots & C_{2n} \end{pmatrix} \quad (8)$$

where C_n denotes the correlation of n patterns. Equation (7) remains unchanged, taking the form

$$\dot{\Omega} = \underline{\underline{C}}\Omega - \left(\Omega^T \underline{\underline{C}}\Omega\right)\Omega \quad (9)$$

Extensions may be given to learning eigen vectors corresponding to lower eigenvalues of C. These will be of importance in giving a better reconstructed image. However, we will not discuss these in any detail here.

In all of these cases the learning proceeds till Ω becomes the normalised eigenvector of $\underline{\underline{C}}$ with maximal eigenvalue. As in the linear case, principal component analysis may be shown to minimise the mean-squared reconstruction error, so leading again to important data compression. The reduction is not completely trivial, so we will give a brief demonstration here.

The optimal reconstruction is expected to take the form

$$\hat{X}_i^\alpha = y^\alpha G_i + y^\alpha y^\beta G_i^\beta + \ldots \quad (10)$$

where α is the pattern label. The value of $\hat{x}^\alpha$ can be expressed in vector form using the notation

$$\vec{Y}^\alpha(y)^T = (y^\alpha, y^\alpha y^\beta, y^\alpha y^\beta y^\gamma, \ldots) \quad (11a)$$

$$\underline{G}^T = \left(\underline{G}, \underline{G}^\beta, \vec{\underline{G}}^{\beta\gamma} \ldots\right)$$

$$= \left(\underline{a}, \underline{a}_j x_j^\beta, a_{j\underline{k}} x_j^{\beta_j} x_j^\gamma, \ldots\right) \quad (11b)$$

In terms of (11), $\underline{\hat{x}}^\alpha$ may be written as

$$\underline{\hat{x}}^\alpha = \vec{\underline{G}}.\vec{Y}^\alpha(y) \quad (12)$$

where the dot product in (12) is over the vector indices β, γ, etc. in (11). The mean square reconstruction error is defined as

$$\epsilon = E\left(\|\underline{x} - \underline{\hat{x}}\|^2\right) = E\left(\left\|\underline{x} - \vec{\underline{G}}.\vec{\underline{Y}}(y)\right\|^2\right) \quad (13)$$

where y is given by the non-linear expression, extending the quadratic neuron, as,

$$y = W_i x_i + W_{ij} x_i x_j + \ldots \quad (14)$$

If we denote combinations of indices $j_1 \ldots j_n$ by $\underline{j}$, $\gamma, \ldots, \gamma_m$ by $\underline{\gamma}$ and $x_{j_1}^{\beta_1} \ldots x_{j_n}^{\beta_n}$ by $x_{\underline{j}}^{\beta} x_{\underline{j}}^{\beta}$

$$\epsilon = \sum_{\alpha,i} \left[x_i^\alpha - y^\alpha y^\beta - a_{ij} x_{\underline{j}}^\beta\right]^2 \quad (15)$$

Variations of ϵ in (15) with respect to $a_{i\underline{j}}$ to minimise the reconstruction error ϵ of equations (15) leads to the matrix equation

$$a_{i\underline{j}} \Lambda_{j\underline{k}} = \Lambda_{i\underline{k}} \quad (16)$$

where

$$\Lambda_{i\underline{k}} = (C\Omega)_i (C\Omega)_{\underline{k}}$$

and

$$\Lambda_{\underline{kl}} = (\Omega^T C\Omega)(C\Omega)_{\underline{k}}(C\Omega)_{\underline{l}}$$

where

$$(C\Omega)\underline{k} = \prod_{r=1}^{n} (C\Omega)k_r$$

Then a solution of (16) is

$$a_{j\underline{k}} = (C\Omega)_i (\underline{C\Omega})_{\underline{k}} \left[\sum_{\underline{k}} \left(\underline{\Omega C^2 \Omega}\right)^{|\underline{k}|}\right]^{-1} \left(\Omega^T C\Omega\right)^{-1} \quad (17)$$

which is a non- linear extension of the linear version (and reduces to it in the linear case). The minimum error for this solution maybe obtained from (13) and (17) as

$$\frac{1}{2}\left[\sum_{\alpha} \left(\underline{x}^\alpha\right)^2 - \left(\Omega^T C^2 \Omega / (\Omega^T C\Omega)\right)\right] \quad (18)$$

This is minimised on the original weight vector when Ω is chosen as the principle eigenvector of C. In fact there are more general solutions $a_{i\underline{k}} = (C\Omega)_i f_{\underline{k}}(\Omega C^2 \Omega)$ to equation (16), but they all lead to the same minimum error (18), and hence also to the principal component choice. The errors (15) may be extended to approximating higher order statistics of the image, and leads to a minimal reconstruction error of such statistics by use of the principal components of $\underline{\underline{C}}$ of equation (8) (Plumbley and Taylor, 1993).

The quadratic extension appears to be the most relevant neurobiologically, since complex cells in V1 and V2 are regarded as performing a quadratic map on their activity (Pollen et al., 1989). Thus the above approach shows how a Hebbian learning rule on the quadratic (and linear) weights would lead to minimal reconstruction error on the signal up to and including the fourth-order statistics.

Temporal sequence storage

Storage of sequences of patterns must occur in the brain, since we can certainly produce such stored sequences when, for example, we recite a poem or sing a tune. There are numerous methods which have been attempted to achieve storage of such sequences. Here we will consider how temporal neurons aid such storage. The basic objects used are leaky integrators (LINs), since activity on their surface decays away over a certain time.

LINs act by storing activity coming on to their surface, this activity $A(t)$ at time t dying away with a certain time constant. Thus in discrete time

$$A(t+1) = (1-d)A(t) + I(t) \tag{19}$$

where $(1-d)$ is the reduction factor experienced at each time step, and $I(t)$ is the input at that time. Further temporality could be obtained by including non-trivial geometry for the neurons, but we will not pursue that here. Finally input channel dynamics can be included by replacing $(I(t)$ by a channel variable $C_2(t)$ which is coupled to another channel variable C_1 by an equation like (19), whilst C_1 instead of A, is driven by $I(t)$ by (19). The result is equivalent to convoluting the input with the alpha function $\alpha^2 te^{-\alpha t}$ of (Jack et al., 1975) before adding it on the right of (19). This gives an effect of delaying the maximum of the input till a time $1/\alpha$ of its original arrival at the synapse. It is relevant to note that some papers on biological memory consider LTP as a reduction in the latency $1/\alpha$, so should be added to the biologically realistic adaptive parameters of the net (unlike the time constant).

A simple way of looking at what the LIN type of node can achieve (Bressloff and Taylor, 1990) is to consider the task of learning a given binary classification of a set of sequences $\underline{S}^m(r)$ (r denoting time, m the pattern label). For LINs on each of the input line of a single layer perceptron, their activity at time t will be

$$\sum_{r=0}^{t} \underline{S}^m(r)(1-d)^{t-r} = \underline{\tilde{S}}^m(t) \tag{20}$$

The classification problem can be solved by a single layer threshold neuron, provided the set of patterns $\underline{\tilde{S}}^m(r)$ is linearly separable. This is a different problem from that of the linear separability of the set of patterns $\underline{S}^m(r)$. Consider, for example, the vectors

$$A = (1,0)^{\mathrm{T}}, \quad B = (0,1)^{\mathrm{T}}$$

and consider the task of learning the mappings $AB \to 1{,}0$ $BA \to 0{,}0$. This is an ordering problem, since the pattern A produces output 1 or 0 according as to whether it is before or after the pattern B; it could not be solved by a standard perceptron. But it can be learnt for the pattern $\tilde{S}$ formed by (20), since (with k = 1-d)

$$\tilde{A}(1) = (1,0)^{\mathrm{T}}, \quad \tilde{A}(2) = (k,1)^{\mathrm{T}}, \quad \tilde{B}(2) = (1,K)^{\mathrm{T}}$$

and the sets $\{\tilde{A}(1)\}$ and $\{\tilde{A}(2),\tilde{B}(1),\tilde{B}(2)\}$ are linearly separable in $\underline{R}^2$. We should add that choice of different values for d on the different input LINs, with $A = (1,1)^{\mathrm{T}}$, is not linearly separable if $d_1 = d_2$ but is so if $d_1 \neq d_2$. Learning the correct set of ds to make a task linearly separable is an interesting problem.

From temporal sequence classification let us turn to TSS, using what is predictive self-learning. The basic idea behind this approach (Reiss and Taylor, 1991) is to use a single layer perceptron net (in general with sigmoidal output) to produce the next input $\underline{P}(t+1)$ of a pattern sequence, the

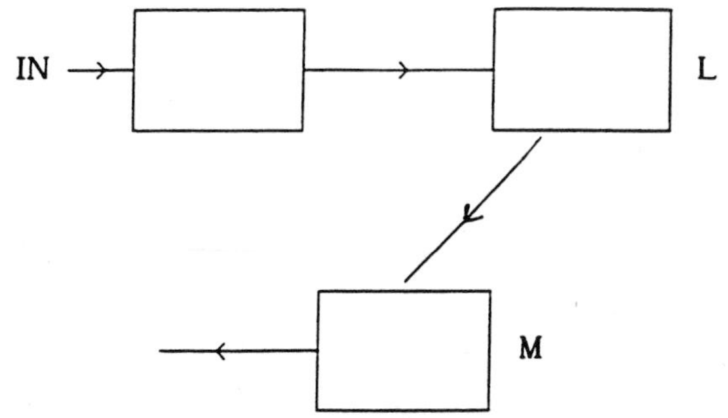

Fig. 1. The architecture of the learning system for predictive self-learning of TSS. Input is sent in a 1:1 manner to the bank L of LINs (with varying time constants) whose thresholded outputs appose on the net M of output neurons. The output of these neurons is compared with the inputs in the IN net at one step later in time (the nets M and N have the same number of neurons), and the appropriate weights onto the neuron of the M net are changed so that these two (M output and successive IN input) become the same. In the sequence generating phase the output from the M net is used for the succeeding input to the IN net after starting off with the initial sequence input to IN.

input being a stored history of the earlier patterns by means of a net of LINs. This may be achieved by the architecture of Fig. 1, where the input IN is fed in a 1:1 fashion into the net L of LINs with a range of time constants and channel variables. The output of these LINs, suitably thresholded, is fed onto the SLP net M. The weights of these latter inputs are modified in the teaching phase, so that the output of M is identical to the input on IN at one later time step (by the delta learning rule). On feeding the output of M back into IN, the stored sequence (provided it has been correctly learnt) will be generated from the net M from that time. Improvement of pattern storage has been achieved (Reiss and Taylor, 1991) by adding adaptive lateral connections to M so as to increase the size of basins of attraction for each of the patterns. The method of TSS seems very effective, and was explored in Reiss and Taylor (1991) for its dependence on the various channel and neuron parameters. In particular, the storage capacity was shown by simulation to be $0(N)$, where N is the number of neurons in L. This was also shown by replica symmetry using statistical mechanical techniques (Taylor, 1991).

A slightly different version of TSS (Bressloff and Taylor, 1990), still using the architecture of Fig. 1, guarantees convergence of the learning algorithm for suitable temporal sequences. This approach uses linear outputs of the LIN net L, and the nodes of M are thresholded which becomes a thresholded SLP. In terms of the transformed pattern space $\tilde{S}^m(t)$ one can use the perceptron learning algorithm for the change of the weight from the ith LIN to the jth neuron in M, with guarantee of convergence provided the set of patterns $\tilde{S}^m(t)$ is linearly separable. Moreover, an extension can be given to the optimal learning algorithm of (Gardner et al., 1988) with a non-zero stability parameter K. This has still to be explored by simulation. The learning of a function of time can also be obtained by these approaches.

Memory in humans appears to occur in temporal sequences. The hippocampus and its nearby (neo)cortex is the brain region which appears critical for such storage. The gross feedback architecture of Fig. 1 seems to relate closely to that of hippocampus, in which L is recognised as the dentate gyrus and M as $CA3$ and/or $CA1$ (Reiss and Taylor, 1991, 1992). The lateral connections in M, mentioned earlier, which make the storage clearer, are known to exist in $CA3$. A biologically realistic implementation of the learning rule is not obvious, but one solution has been worked out in Reiss and Taylor (1991) using a suitable course for the formation of long-term memory traces.

Since memory does not appear to be permanently laid down in the hippocampus, at least in primates, the temporal sequences stored initially in hippocampus must ultimately be stored elsewhere, most likely in the parahippocamal gyrus and nearby brain areas. Such regions do not have the obvious feedback-loop structure of Fig. 1, so another mode of TSS must be employed. That could arise most naturally from the temporal topographic map of Chappell and Taylor (1993), which it is reasonable to expect could occur somewhere in the cortex. The detailed implementation of this will be considered in detail elsewhere, but one point can be made immediately. It is that the topographic map process naturally adds activity in a region closest to where memories of a similar character were previously stored. In the case of TSS, this would be expected to correspond to allow for the concatenation of temporal sequences. In other words an earlier temporal sequence will capture a later one if that later one is a reasonable continuation of the earlier sequence. This appears to be a very efficient procedure.

One can add, to the above, approaches which use more of the three-dimensional structure of the neuron. In Bressloff and Taylor (1993a,b) a neural network model is constructed that includes details concerning the passive membrane properties of a neuron's dendrites, and the output response of the model to synaptic input patterns is analysed. The main results of the analysis are the following:

(i) A compartmental model of passive dendritic processing can be reformulated as a neural network model in which the activation state of the neuron is of the general form $V(t) = \Sigma_j \int X_j(t,s) W_j a_j(s) \mathrm{d}s$, where W_j is the weight of the jth input line, $a_j(s)$ is the input at time s, and X_j is a response function that incorporates details concerning the dendrites.

(ii) In the case of an infinite one-dimensional dendritic chain, one can derive a simple analytical expression for the response function of the neuron. In particular, one finds that for an input pattern of specific spatial frequency k across the chain, the model neuron is equivalent to a leaky-integrator neuron with effective decay rate $\epsilon(k) = a = b\mathrm{f}(1 - \cos k)$, where a, b are positive constants.

(iii) The activation state of the neuron forms an internal representation of an input sequence, thus allowing the neuron to extract temporal features such as the ordering patterns within the sequence. One of the advantages of taking into account dendritic structure is that it allows for a more flexible representation of input history than given by the decaying activity traces of time summating neurons, for example. A version of the perceptron learning-rule and convergence theorem can be constructed and, at least in simple cases, geometrical methods used to study the ability of the neuron to resolve temporal features of an input sequence.

(iv) The activation state of the model neuron is a non-linear function of the inputs when shunting effects are taken into account. At the network level, shunting inhibition can lead to states of self-sustained firing in which the individual neurons have low firing rates. Shunting inhibition could also provide a mechanism for modulating the internal representation of temporal input history discussed in (iii).

In conclusion, the model neuron constructed in the paper of Bressloff and Taylor (1993) should be of interest both to artificial neural modellers and to those involved with real neurons. For the first group, the model neuron, is in a form suitable for inclusion as a processing element in a neural network; the addition of compartmental structure can lead to enhanced temporal discrimination features. In the case of the latter group, our analytical expressions for compartmental neurons contains sufficient detail concerning the properties of real neurons to allow comparison with experimental results and to provide insight into principle features of neuronal information processing; some of these features have been specified in (i)–(iv) above.

Neuronal stochasticity

An important component of information processing in the central neurons system is the emission of amounts of chemical transmitter substance(s) on arrival of an action potential (Katz, 1969). The arrival of a nerve impulse at a synaptic 'bouton' causes a depolarisation of the presynaptic cell membrane which opens channels for calcium ion entry leading to the release of a number, n, of packets of chemical transmitters into the synaptic cleft. The distribution function $p^{(st)}(n)$ giving the probability of the stimulated emission of n packets has been taken variously as a Poisson distribution of mean ω, as a Binomial distribution of size M and probability p, which describes a 'one vesicle' model when $M = 1$, and as a compound Binomial-type distributions. The Bionomial distributions of size M are the most realistic, with $M \approx 1$–10^3–10^4 for lower animals and peripheral systems and $M \approx 1$–10 for central nervous systems (Korn and Faber, 1987). In the latter case $p \approx 0.25$–1 is often observed. There is also spontaneous release of transmitter at a synapse when no impulse arrives. The associated distribution function of vesicular release, denoted by $p^{(s)}$, will be of similar form to $p^{(st)}$ but usually with considerably smaller mean $\bar{n}$.

When a transmitter molecule flows across the synaptic cleft there is a certain probability, called the postsynaptic efficiency ϵ, that it binds to a receptor site of the postsynaptic membrane. If such a binding occurs, than a channel is opened allowing ions (Na^+, K^+, CP^-) to move in and out of the cell under the influence of concentration and potential gradients (Kuffler, et al., 1984). Such ionic currents lead to changes in membrane potential across the postsynaptic membrane, i.e. they produce PSPs. These PSPs may be positive or negative in sign corresponding, respectively, to excitatory or inhibitory transmitters. Each PSP flows decrementally (in space) down to the axon hillock which is the region of the neuron's cell body connected directly to the axon. Such flow may be modelled by a multi-compartmental model of the neuron as a connected set of leaky integrators joined by resistors, as was discussed in the previous section. We shall assume here, for simplicity, that the neuron is an idealised point processor (i.e. a single compartment) and will, therefore, only be concerned with the temporal aspects of PSP summation. We shall also neglect active transport of PSPs such as dendritic spiking. If the resulting membrane potential at the action hillock exceeds a certain threshold, then the neuron fires an action potential. Threshold noise may be included by taking the threshold κ to be a random variable with distribution function $p^{th}(\kappa)$. Note, however, that there is no clear neurophysiological evidence that such noise is actually present to any significant degree; the only solid evidence is that noise arises at the synapses due to quantal chemical transmission. In the absence of threshold noise, $p^{th}(\kappa) = \delta(\kappa - \kappa_0)$ for some fixed threshold κ_0. Once a neuron has initiated an action potential it requires a recovery time (refractory period) before it is capable of firing again. There is also a synaptic delay time between the arrival of an action pulse at the presynaptic membrane and the occurrence of a PSP at the axon hillock. A significant contribution to this delay arises from the time between depolarisation of the presynaptic membrane and transmitter release.

A simple mathematical model of the quantal release process at a synapse was developed some time ago (Taylor, 1972). This has since evolved, both into an extension of the random iterative function approach to chaos, in terms of the random iterative network (Bressloff and Taylor, 1990), and to a hardware-realisable neuron, the probabilistic RAM or pRAM (Gorse and Taylor, 1988, 1989), for which considerable further development has occurred (Gorse and Taylor, 1993).

The simplest approach to these developments, and to modelling stochastic quantal release, is to extend the Caianiello equations for the McCulloch-Pitts neurons

$$u_i(t+1) = \theta\left(\sum_{j=1}^{N} W_{ij}u_j(t) - s_i\right) \qquad (21)$$

where $u_i(t)$ is the activity of the ith neuron at time t, with $u_i(t) = 1$ if the neuron fires then, otherwise it is zero, s_i is the threshold for this response, W_{ij} is the strength of the connection from the jth to ith neuron and θ ( ) is the Heaviside or step function $\theta(x) = 1(x > 0), = 0(x < 0)$. The extension is achieved by expressing the weight W_{ij} as the product of presynaptic release quantity q_{ij} and a factor ϵ_{ij} for the efficiency of the update of the released chemical transmitter

$$W_{ij} = q_{ij}\epsilon_{ij} \tag{22}$$

The chemical transmitter amount q_{ij}, is now regarded as a random variable, with distribution function given by whatever is most appropriate, as noted above. Thus one may choose amongst:

$$\rho(x) = \delta(x - n_0 q_0) \quad (\textit{deterministic}) \tag{23a}$$

$$= \Sigma(\mu^n/n!)e^{-\mu}\delta(x - nq_0) \quad (\textit{Poisson}) \tag{23b}$$

$$= p\delta(x - q_0) + (1 - p)\delta(x) \quad (\textit{Bernoulli}) \tag{23c}$$

or other possibilities, such as the compound binomial. In (23), q_0 is the amount of transmitter contained in a single packet so that, in (23a) n_0 packets are released by the arrival of a nerve impulse, in (23b) μ is the mean number of packets released by the impulse, and in (23c) p is the probability of release of a single packet. The threshold s_i may also be made probabilistic, thus encompassing the spin glass approach to artificial networks with the distribution function for s_i as

$$\rho(s_i) = \frac{d}{ds_i}[1 + e^{-\beta s_i}]^{-1} \tag{23d}$$

In (23d), β^{-1} is the usual noise temperature. However, as was noted above, there is not much biological reality about the choice (23d). It is now possible to set up a Markov process for the time development of the network of such stochastic neurons. We may use Eqns. (21), (22) and (23) to deduce that:

$$\textit{Prob}\ (u_i(t + 1) = 1|u(t))$$

$$= \int\Pi\, d\underline{q}\rho_1(\underline{q})d\underline{s}\rho_2(\underline{s})\theta\left(\sum_j W_{ij}u_j(t) - s_i\right) \tag{24}$$

In the right hand side of (24), the distribution $\rho_1(\underline{q})$, $\rho_2(\underline{s})$ may be taken to be that of (23) or even more general expressions. It is possible to extend the random iterative net approach even further by using threshold functions of non-linear expressions of earlier inputs

$$u_i(t + 1) = \theta\left(\sum_j a_{ij}u_j(t) + \sum_{j,k} a_{ijk}u_j(t)u_k(t) + \ldots\right) \tag{25}$$

where all weight $a_{ij}, a_{ijk}, \ldots$ are random variables. In the case where powers of inputs up to $O(N)$ are considered (where N is the number of neurons in the net) this gives 2^N independent variables which are expected to contribute to the activity of the net by the expressions

$$\text{Prob}(u_i(t + 1) = 1|\underline{u}(t)) = \alpha^{(i)}_{\underline{u}(t)} \tag{26}$$

where, for each neuron i, the 2^N probabilities $\alpha^{(i)}_{\underline{u}(t)}$ are expressible using the integrals of a similar form (24) but now with the variables extended by non-linearity, as in (25). Conversely, given the set of 2^N probabilities $\alpha^{(i)}_{\underline{u}(t)}$, it is possible to determine (though not necessarily uniquely) distribution functions on the non-linear weights so that the expressions generalising (24) are valid.

The quantities (26) define (for each i) an N-input pRAM loaded with the probability $\alpha_{\underline{u}}$ at each binary address $\underline{u}$; $\alpha_{\underline{u}}$ is the probability of the pRAM outputting a spike (a one) when location $\underline{u}$ is accessed. Since there is maximal functionality with respect to the N binary inputs, the pRAM possesses maximal non-linearity, and can be shown to be the most general stochastic automaton operating in the binary domain $\{0,1\}^N$ (Gorse and Taylor, 1991). If the N binary inputs are replaced by N spike trains in which a 1 is present, with probability $x_i, i = 1 \ldots N$, then the pRAM is able to perform mappings $[0,1]^N \rightarrow [0,1]$, with the average output.

$$\sum_{\underline{u}} \alpha_{\underline{u}} X_{\underline{u}}(\underline{x}) \tag{27}$$

where $X_{\underline{u}}(\underline{x}) = \prod_{i=1}^{N}[x_i u_i + (1 - x_i)(1 - u_i)]$ gives the probability address $\underline{u}$ will be accessed at any given time by the N input spike trains. The expression (27) does not have maximal functionality in the real-valued domain $[0,1]^N$, for example, not containing items like x^2 but its biological realism is not reduced thereby.

The three main features of pRAMs, as expressed by Eqns. (26) and (27) are: (1) they have continuously variable 'weights' $\alpha_{\underline{u}}$ to which the classical neural network training algorithms are applicable (so that they are not really 'weightless' systems at all—although the same could be said for all RAM-based neurons, since even the simplest systems allow for parameters with two values {0,1}, not one); (2) they have ultimate non-linearity for binary inputs; (3) they are stochastic devices which emit spike trains (which for finite length have non-zero variance).

Item (1) above allows all of the continuous neural network training algorithms to be applied to pRAMs. Even error back-propagation can be developed and used for pRAMs. However such supervised learning rules are not easily implemented in hardware (although pRAM radial basis function nets have interesting features since they lead to non-linear functions on the Gaussian centres).

We have concentrated mainly on reinforcement learning, which we recognised early on as having the important advantage of being hardware realisable in pRAMs (Gorse and Taylor, 1990) and also have some biological realism. Thus, we will consider the learning algorithm

$$\Delta \alpha_{\underline{u}} = \rho\left[(a - \alpha_{\underline{u}})r + \lambda(1 - a - \alpha_{\underline{u}})p\right] \quad (28)$$

where ρ is the learning rate, r and p are the reward and penalty signals delivered by the environment for action a $\in$ {1,0} in context $\underline{u}$, and λ a parameter governing the amount of 'exploration' in the system (its tendency to try an alternative action when a given one is penalised). The reinforcement learning rule (28) ahs been analysed very extensively in Gorse and Taylor (1991). Unlike more heuristic training rules for RAM-based systems, a rule such as (28), which manipulates continuous quantitites, can be investigated with regard to convergence properties using conventional mathematical techniques (in this case from the theory of stochastic learning automata). The standard conditions on the time-dependence of the learning rate ($\rho(t) \to 0$ as $t \to \infty, \sum \rho_t = \infty, \sum \rho_t^2 < \infty$) may be shown to apply to (28), provided the penalty parameter $\lambda \neq 0$. If $\lambda = 0$, then there may be absorption by one of the boundary points, $\alpha_{\underline{u}} = 0$ or 1 (for some $\underline{u}$). Although the rule (28) is not guaranteed to converge unless ρ is decremented in the above way, in practice it is often sufficient to use a small (less than 0.1, say) but constant ρ, although there will in this case be some residual 'jitter' in the weight values.

Any training scheme may be modified by the addition of noise to the training data. This adds robustness to the learning system, and in the case of pRAMs is also a way of introducing generalisation (even for binary pattern learning). In the case of pRAMs, noise may be added in a very natural way by manipulating input spike trains (so that in the binary-input case, for example, and input-0 spike train now has some probability ϵ of containing a 1, and input-1 spike train a probability ϵ of containing a 0). When added to the above hardware-realisble reinforcement training scheme, this spike-train manipulation results in a very powerful and robust learning mechanism. Results on training in noise for patterns recognition have already been obtained, with an indication of its value in the biological context.

A model for latencies in the visual system

There are conflicting data on latencies in the visual system. Measurements based on electrical stimulations consistently indicate latencies of less than 2 msec par neuronal relay (Ferster and Lindstrom, 1983), but when visual simulation is used, and average latency of 10 msec per neuronal relay is a general accepted value (Thorpe and Imbert, 1989). It is observed, for instance, between layer 4 and layers 2–3 in the striate cortex (Maunsell and Gibson, 1992). What is the cause for these longer latencies with visual input?

It is assumed that electrical stimulations produce highly synchronized input spikes (Douglas and Martin, 1991) which cause unnaturally short latencies. With visual stimulation, input spikes arrive synchronously and it has been proposed that they must be integrated for appropriate intervals (Maunsell and Gibson, 1992). However, temporal integration is incompatible with the observed irregularity of the spike trains (Softky and Koch, 1993). On the contrary, coincidence detection based on very short integration times, or possibly a synaptic saturation mechanism (Bugmann, 1992) seems required. Coincidence detection corresponds to a multiplicative AND-type, function of the neuron which has been postulated for neurons in MST (Verri et al., 1992). If we assume neurons operating as coincidence detectors, and thereby discard the temporal integration hypothesis, how can the long latencies be explained?

We describe here (for more detail see Bugmann and Taylor, 1993) a simple model of information processing in the visual system, based on coincidence detection, which can reconcile the above-mentioned observations, producing long delays with visual stimulation and short ones with electrical stimulations. We make the following assumptions: (1) neurons in the visual cortex are part of a pyramidal multilayer network where each neuron receives spikes from m distinct neurons in the preceding layer; (2) neurons act as coincidence detectors, firing only if all m inputs provide a spike within a given time window; (3) neurons receive excitatory feedback from local-circuit neurons and stay in a state of persistent activity ('ON' state) after producing their first spike, as suggested by physiological observations (Douglas and Martin, 1991); (4) neurons in the self-sustained ON-state are silenced by inhibitory feedback from their target neurons in the next layer (this feature ensures that neurons fire only during the minimum necessary time, and prevents the loss of information from one layer to the next, although we will not discuss this aspect in detail here).

Both simulations and theory have been used to analyse the network response (Bugmann and Taylor, 1993). The simulations were performed with a pyramid of pRAM neurons. These neurons can act as coincidence detectors, like leaky integrate-and-fire neurons, and are easier to analyse mathematically. The pRAM operates in discrete time-steps Δt with spike trains defined as sequences of 1s (a spike) and 0s. Each pRAM was chosen to have $m = 4$ inputs, and was set for the coincidence detection mode, firing only when all four input spikes are present during time-step Δt. As soon as the first spike is produced, the pRAM is set to fire at each subsequent time step, with probability P_1. This produces a random spike train of frequency $f_1 = P_1/\Delta t$, characteristic of the ON-state, and simulates the effect of local exitatory feedback. Such a random firing is consistent with findings that, at all frequencies, biological neurons fire with near Poisson distributions of interspike intervals.

The pyramidal network has 64 neurons in its input layer (layer 0), divided into 16 groups of four neurons connected to 16 neurons in layer 1. These neurons are fivided into four groups of four neurons connected to four neurons in layer 2; these latter four neurons are connected to a single neuron in layer 3.

The inputs are initially set to fire randomly, with a frequency $f_0 = P_0/\Delta t$. However, as soon as their first spike is produced, their frequency is set to f_1. So, the only role of f_0 is to produce a jitter in the starting time of the neurons in the first layer.

For a neuron in layer $n + 1$ to fire, there are two conditions: (1) all m neurons in layer n must be in the 'ON' state, and (2) a coincidence has to occur (coincidence probability $P_c = P_1^m$). In biological neurons, coincidences are defined as spikes arriving in a given time window τ, and occur with a probability $P_c' = f_s^M \tau^m$ (Bugmann, 1991) where f_s is the biological sustained rate. If we assume the coincidence probabilities to be equal in both systems, this gives us the relation between the time-steps and frequencies used in simulations and those in biological systems to be $f_1 \Delta t = f_s \tau$.

We define the total latencies L_n as the time elapsed from the start of the simulation until a neuron in layer n has reached an average firing probability $P_1/2$. We define the relative latencies as $\Delta L_n = L_n - L_{n-1}$. The relative latencies all converge to the same minimum value $\Delta L_{\min}$ for large values of P_1. Theoretical analysis of the system predicts a $1/P_1^m$ dependence for ΔL_n. This is confirmed by simulations. The minimum latencies ΔL_{min} (determined at $P_1 = 1$) for various values of P_0, has the theoretical prediction $\Delta L_{\text{min}} = \log(1/m)/\log(1 - P_0)$. This again was in close agreement with the simulation.

In the model we have not included transmission delays, synaptic delays or background potential effects, so the model provides a lower bound to latencies, and is purely related to the performed computation. It shows two components to the computational latencies: (1) the initial jitter or time necessary for all m input neurons to be in the ON state: this depends on P_0 in the model; (2) the time for a coincidence to occur, which depends on P_1.

The initial jitter, corresponding physiologically to the fluctuations in onset times in retinal ganglion cells (Levick, 1973), determines the absolute minimum in per-layer computation time ΔL_{min}. The gain in per-layer computation time due to the level of sustained activity P_1 becomes smaller and smaller as P_1 is increased. With a value $P_1 = 0.6-0.8$, most of the gain is already made. Assuming a realistic biological time-window for coincidences of the order of 2 msec, sustained frequencies of 300–400 Hz (as observed in biological neurons) would be sufficient for the computation time to approach its smallest possible value. As ΔL_{min} increases with the number of inputs, the spread of latencies of neurons in the same layer (Maunsell and Gibson, 1992) could effect differences in fan-in. Preliminary investigations show that this model may also explain the psychometric curves obtained in visual masking experiments.

If the hypothesis of this model is valid, it indicates that the visual system uses maximum firing rates probably near to the optimum with regards to computation speed. The main limiting factor would therefore be the noise at the level of the sensory receptors, which determines the latencies in all subsequent processing layers. 'Noise' refers here to jitter in onset times, not to fluctuations in firing levels. A prediction of this model is that latencies in the visual system can be manipulated by modifying the onset-time jitter in retinal ganglion cells.

Summary

A general framework for the analysis of neurons as stochastic, three-dimensionally complex and non-linear units with a range of temporal properties is outlined, and a class of problems delineated. Some general mathematical properties of the resulting network are deduced, together with information-theoretic questions to be pursued. In particular examples of the relevance of the non-linear, temporal and stochastic properties of neurons in effective information processing are briefly outlined.

References

Barlow, H.B. (1961) Three points about lateral inhibitions. In: W. Rosenblith (Ed.), *Sensory Communication* MIT Press, Boston, MA, pp. 782–786.

Bressloff, P.C. and Taylor, J.G. (1990) Random iterative networks. *Phys Rev.,* A41: 1126–1137.

Bressloff, P.C. and Taylor, J.G. (1993a) Compartmental model response function for dendritic trees. *Biol. Cybern.,* 69: 199.

Bressloff, P.C. and Taylor, J.G. (1993b) Spatio-temporal pattern processing in a compartemental model neuron. *Phys Rev. E,* 47: 2899.

Bugmann, G. (1991), Summation and multiplication: two distinct operation domains of leaky integrate-and-fire neurons. *Network,* 2: 489–509.

Bugmann, G. (1992) Multiplying with neurons: compensation of irregular input spike trains by using time-dependent synaptic efficiencies. *Biol. Cybern.,* 68: 87–92.

Bugmann, G. and Taylor, J.G. (1993) A model for latencies in the visual system. In: S. Gielen and B. Kappen (Eds.), *Proc ICANN '93,* Springer-Verlag, London. pp. 165–168.

Chappell, G. and Taylor, J.G. (1993) The temporal kohonen map. *Neural Networks,* 6: 441–445.

Douglas, R.J. and Martin K.A.C. (1991) A functional microcircuit for cat visual cortex. *J. Physiol.,* 440: 735–769.

Ferster, D. and Lindstrom, S. (1983) An intracellular analysis of geniculo-cortical connectivity in area 17 of the cat. *J. Physiol.,* 342: 181–215.

Gardner, E., Gulfreurd, H. and Yekutieli, I. (1989) The phase space of interactions in neural networks with definite symmetry: *J. Phys.* A, 22, 1995–2008.

Gorse, D. and Taylor, J.G. (1988) On the equivalence and

properties of noisy neural and probabilistic RAM nets. *Phys. Lett.*, A131: 326–332.

Gorse, D. and Taylor, J.G. (1989) An analysis of noisy RAM and neural nets. *Physica*, D34: 90–114.

Gorse, D. and Taylor, J.G. (1990) A general model of stochastic neural processing. *Biol. Cybern.*, 61: 299–306.

Gorse, D. and Taylor, J.G. (1991) Universal associative stochastic learning automata. *Neural Network World*, 1: 193–202.

Gorse, D. and Taylor, J.G. (1993) A review of the theory of pRAMs. In: N.M. Allinson (Ed.) Proceedings of the *Weightless Neural Networks* Conference, York, UK, Publ. Dept of Electronics, Univ. of York.

Jack, J., Noble, D. and Tsien, R.W. (1975) *Electrical Current Flow in Excitable Cells*, 2nd Edn. Clarendon Press, Oxford.

Katz, B. (1969) *The Release of Neural Transmitter Substance*, Liverpool University Press, Liverpool.

Korn, H. and Faber, D.S. (1987) Regulations and significance of probabilistic release mechanism at central synapses. In: W.M. Edelman, W. Gall and W.M. Cowan (Eds.), *Synaptic Functions*, Wiley, New York, pp. 57–108.

Krogh, A. and Hertz, J.A. (1990) Hebbian Learning of Principal Components. In: *Parallel Processing in Neural Systems and Computers*, Elsevier Science Publishers, Amsterdam, pp. 183–186.

Kuffler, S.W., Niccols, J.G. and Martin, A.R. (1984), *From Neuron to Brain*, Sinauer, Sunderland MA.

Levick (1973) Variation in the response latency of cat retinal ganglion cells. *Vision Res.*, 13, 853–873.

Linsker, R. (1988) Self-organisation in a perceptual network. *IEEE Comput.*, 21(3): 105–117.

Ljung, L. (1977) Analysis of recursive stochastic algorithms. *IEEE Trans. Automatic Control*, AC-22: 551–575.

Maunsell, J.H.R. and Gibson, J. (1992) Visual response latencies in striate cortex of the macaque monkey. *J. Neurophysiol.*, 68: 1332–1344.

Oja, E. (1982) A simplified neuron model as a principal component analyser. *Math. Biol.*, 15: 267–273.

Plumbley, M.D. and Taylor, J.G. (1993) The information approach to neural networks. In: J.G. Taylor (Ed.), *Mathematical Approaches to Neural Networks*, Elsevier, Amsterdam. pp. 307–340.

Plumbley, M.D. (1993b) Efficient information transfer and anti- Hebbian neural networks, in press.

Pollen, D.A., Gaska, J.P. and Jacobson, L.D. (1989) Physiological constraints on models of vision. In: R.M.J. Cotterhill (Eds.) *Models of Brain Function*, Cambridge University Press, New York, pp. 115–136.

Reiss, M. and Taylor, J.G. (1991) Storing temporal sequences. *Neural Networks*, 4: 773–787.

Reiss, M. and Taylor, J.G. (1992) Does the Hippocampus store temporal sequences? *Neural Network World*, 2: 365–393.

Sanger, T.D. (1989) Optimal unsupervised learning in a single-layer linear feedforward neural network. *Neural Networks*, 2: 459–473.

Softky, W.R. and Koch, C. (1993) The highly irregular firing of cortical-cells is inconsistent with temporal integration of random EPSPs. *J. Neurosci.*, 13: 334–350.

Taylor, J.G. (1972) Spontaneous behaviour in neural networks. *J. Theor. Biol.*, 36: 513–528.

Taylor, J.G. (1991) On the capacity for temporal sequence storage. *Int. J. Neural Syst.*, 2: 47.

Taylor, J.G. and Coombes, S. (1993) Learning higher order correlations. *Neural Networks* 6 (3): 423–428.

Thorpe, S.J. and Imbert, M. (1989) Biological constraints on connectionist modelling. In: R. Pfeifer et al. (Eds.), *Connectionism in Perspective*, Elsevier, Amsterdam, pp. 63–92.

Verri, A., Straforini, M. and Torre, V. (1992) Computational aspects of motion perception in natural and artificial neural networks. In: G.A. Orban and H.H. Nagel (Eds.), *Artificial and Biological Vision Systems*, Springer, New York, pp. 71–92.

J. van Pelt, M.A. Corner H.B.M. Uylings and F.H. Lopes da Silva (Eds.)
Progress in Brain Research, Vol 102

CHAPTER 26

Are there unifying principles underlying the generation of epileptic afterdischarges in vitro?

Roger D. Traub[1] and John G.R. Jefferys[2]

[1]*IBM, T.J. Watson Research Center, Yorktown Heights, NY 10598, U.S.A. and Department of Neurology, Columbia University, New York, NY 10032, U.S.A. and* [2]*Department of Physiology, St. Mary's Hospital Medical School, Imperial College, London W2 1PG, U.K.*

Introduction

It has been exceedingly difficult to elucidate the basic mechanisms of human epilepsy. There are simply too many clinical and electrical seizure types, with widely differing (but virtually all poorly understood) underlying etiologies. Recording from individual neurons in the human brain *in situ* is difficult and impractical in the majority of patients; intracellular recording *in situ* is likely to prove impossible. Experimental models of epilepsy, particularly *in vitro* models, make it easier to record from the neurons and to control the chemical environment, but unfortunately it is not easy to extrapolate insights from *in vitro* experiments to the human disease state.

We discuss here one approach: the comparison of several distinct forms of experimental epilepsy *in vitro*. This approach is feasible in view of recent advances in understanding cellular mechanisms of afterdischarges in the hippocampal slice. Mechanisms that are shared by distinct experimental seizure types ought to apply to certain human seizure types as well. We are less likely, in this way, to be led astray by the peculiarities of one favorite experimental approach. The three epilepsy types that we shall consider are these: (1) disinhibition (specifically, blockade of $GABA_A$ receptors, as with the drugs bicuculline, picrotoxin or penicillin) (Traub et al., 1993a) or after injection of tetanus toxin *in vivo* (Empson and Jefferys, 1993); (2) bathing tissue in medium lacking Mg^{2+} ions ('low Mg epilepsy'), thereby increasing synaptic transmitterr release and membrane excitability and also increasing the conductance of *N*-methyl-D-aspartate (NMDA) receptor/channel complexes (Schneiderman and MacDonald, 1987; Mody et al., 1987; Neuman et al., 1989; Tancredi et al., 1990; Avoli et al., 1991); (3) 4-aminopyridine (4AP), a potassium channel blocker that likewise enhances intrinsic neuronal excitability, but that also increases and prolongs synaptic potentials (Galvan et al., 1982; Voskuyl and Albus, 1985; Avoli and Perreault, 1987; Rutecki et al., 1987; Ives and Jefferys, 1990; Perreault and Avoli, 1991). We shall analyze these epilepsies using data from the CA3 region of the rat or guinea pig hippocampal slice, supplemented by detailed network simulations. The CA3 region generates epileptic events quite readily, so that considerable data, both anatomical and physiological, are now available concerning its cellular and synaptic organization.

Salient features of the CA3 region

The organization of this region has been reviewed in detail elsewhere (Traub and Miles, 1991). Here we shall briefly review certain features of special relevance to epileptogenesis. We discuss also how these features influence the representation of CA3 in network simulations. Many technical details concerning the computer simulations can be found in recent publications (Traub et al., 1992a 1993a).

Pyramidal neurons

Pyramidal neurons are the principal cell type. These cells can fire single action potentials, doublets, or intrinsic bursts (recognizable *in vivo* as so-called complex spikes) consisting of closely spaced runs of three or more action potentials followed by a long-lasting (many seconds) afterhyperpolarization (AHP). During depolarization of the neuron with steady injected currents, three rhythmical modes of firing occur: (1) intrinsic bursts at a frequency of up to about 3 Hz are seen during somatic current injection; (2) larger somatic current injection then causes the cell to switch, tonically, to repetitive single action potentials at frequencies up to 40 Hz or more (Wong and Prince, 1981); (3) injection of depolarizing current into the apical dendrite elicits rhythmical slow action potentials, presumably Ca spikes, in the frequency range 5–8 Hz (Traub et al., 1993a). These different firing modes result from the interaction of a number of ionic conductances, including most importantly: (1) g_{Na}, (2) a high-threshold g_{Ca}, (3) one or more voltage-dependent K^+ conductances that repolarize Na^+ spikes, (4) a Ca^{2+}- and voltage-dependent K^+ conductance (the C-conductance) that contributes to repolarization of the Ca spike, and (5) the slow Ca-dependent conductance ($g_{K(AHP)}$) that yields the AHP. These conductances are apparently distributed non-uniformly across the soma-dendritic membrane. A model of the CA3 pyramidal neuron, incorporating voltage-clamp data (mostly from acutely isolated cells) suggests that this non-uniform distribution of conductances is, in fact, critical for the operation of the cell (Traub et al., 1991). This appears to be true also for cerebellar Purkinje cells. Dendritic g_{Ca}, in particular, is essential in forming intrinsic bursts and afterdischarges, and we predict (based on simulations) that dendritic g_{Ca} also provides an amplification that increases the potency of recurrent excitatory synapses in CA3. Refining this cell model is a continuing process.

Recurrent excitatory synapses

Recurrent excitatory synapses, an essential substrate of epilepsy (except during so-called field bursts (Haas and Jefferys, 1984; Taylor and Dudek, 1984)), are abundant in CA3. Dual intracellular recordings in slices indicate a probability of monosynaptic connection that is a few percent (Miles and Wong, 1986, 1987; Traub and Miles, 1991). The connection probability *in vivo* is expected to be much higher. Recurrent axonal branches are seen anatomically in both basilar and mid-apical dendritic regions (Ishizuka et al., 1990), suggesting that recurrent connections form in both locations. A current-source density analysis obtained during disinhibition-induced epilepsy is consistent with this notion (Swann et al., 1986), while a similar analysis obtained during low-Mg epilepsy suggests that the majority of NMDA receptors, in recurrent connections, lie in the mid-apical dendrites (J.G.R. Jefferys and M.A. Whittington, unpublished data).

Let us consider the admittedly indirect evidence concerning the excitatory amino-acid receptors involved in recurrent excitation. It appears that both α-amino-3-hydroxy-5-methyl-4-isoxazole propionic acid (AMPA) and NMDA receptors are involved: during disinhibition, 6-cyano-7-nitroquinoxaline-2,3-dione (CNQX), an AMPA blocker, almost always abolishes synchronous discharges (Lee and Hablitz, 1989). On the other hand, in low Mg solutions that enhance NMDA conductances (Ascher and Nowak, 1988; Jahr and Stevens 1990), synchronous activity occurs that is resistant to CNQX (Jefferys and Traub, 1992). In both disinhibition and low Mg epilepsy, a combination of CNQX and the NMDA blocker 5-aminophosphonovaleric acid (APV, AP-5) block all synchronous activity (Lee and Hablitz, 1989; Traub et al., 1993a).

The *functional* properties of CA3 recurrent excitatory synapses are extremely important. Most of the data on these synapses derive from a bathing medium with normal or increased Mg concentration, and without 4AP. Under such circumstances, a single presynaptic action potential induces an EPSP of roughly 1 mV and 5–12 msec time-to-peak, but very rarely a postsynaptic spike (Miles and Wong, 1986; Traub and Miles, 1991). In contrast, a presynaptic *burst* of three or more action potentials causes, with a probability somewhat under a half, a postsynaptic *burst*, typically with a latency of 10 msec or more, and lasting tens of milliseconds (Fig. 4 of Miles and Wong, 1987; Traub and Miles, 1991; Fig. 1). It is this ability of bursting to spread from cell to cell that allows the stimulation of a single neuron, under special conditions such as $GABA_A$ blockade, to lead eventually (with delays up to 200 msec) to synchronous firing of the entire neuronal population (Traub and Wong, 1982; Miles and Wong, 1983, 1987).

What patterns exist in the CA3 recurrent connectivity? Suppose that the recurrent excitatory synaptic connections form the basis of an associative memory (Rolls, 1989), or are used to store distances between the place fields of corresponding pyramidal neurons (Muller et al., 1991; Traub et al., 1992b). In such a case, then either the synaptic strengths are not all equal, or there is non-randomness in the synaptic connectivity, or both. To see this, suppose that recurrent synapses encode distance between place fields. Consider three connected neurons, A, B and C whose respective place fields are A′, B′ and C′. Let $\sigma(A,B)$ be the strength of the A → B connection and $d(A', B')$ be the distance between their place fields. We are assuming that there exists some function f such that σ (A, B) $= f[d(A', B')]$. Then, because of the geometry of the world, $d(A', B')$, $d(A',C')$ and $d(B', C')$ must be related. It follows that $\sigma(A,B)$, $\sigma(A,C)$ and $\sigma(B,C)$ must also then be related. But in fully random network the quantities $\sigma(A,B)$, $\sigma(A,C)$ and $\sigma(B,C)$ would be completely uncorrelated. We conclude that synaptic strengths cannot be distributed randomly in a CA3 network encoding distance between place fields.

At the moment, one type of structure to the CA3 connectivity is recognizable in longitudinal slices: there is finite average connection distance (roughly 1 mm), less than half the maximum extent of the connections, and much smaller than the 1 cm length of the longitudinal slice. In other words, the synaptic connectivity is not globally random. Beyond that, it has not yet proven possible experimentally to discover the information-bearing correlations in synaptic connectivity in vitro. We usually perform simulations of 1000 neuron networks, the minimum size of experimental mini-slices that can generate afterdischarges (Miles et al., 1984). This is a small enough system (500–700 μm along stratum pyramidale) to ignore the spatial fall-off in anatomical connection probability. For studies of epileptogenesis on this spatial scale, a uniform, random connectivity has proven acceptable. Our simulations use an excitatory connectivity of 2%, at the lower limit of the likely range of values. Connectivity in vivo is doubtless denser. In simulations of several thousand cells, spatial localization of synaptic connections allows apileptic discharges to propagate, just as they do in hippocampal slices (Miles et al., 1988; Traub et al., 1993b).

Inhibitory cells

Inhibitory cells in CA3 are found in all tissue layers, but have been best characterized physiologically in the stratum pyramidale (Miles and Wong, 1984; Miles, 1990a). These cells differ in their current-induced and spontaneous firing patterns (R. Miles, personal communication), in the magnitude of unitary IPSPs generated in nearby pyramidal cells (Miles, 1990b), and, very likely, in whether they primarily activate $GABA_A$ or $GABA_B$ receptors (Segal, 1987; Müller and Misgeld, 1991). Many (but by no means all) of the cells fire non-adapting trains of narrow action potentials, with deep and fast post-spike AHPs, and this is the firing pattern developed by our simulated inhibitory cells. Other inhibitory cells can fire in intrinsic bursts (Traub et al., 1987;

Kawaguchi and Hama, 1988) or can exhibit subthreshold oscillations involving low-threshold Ca channels (Fraser and MacVicar, 1991).

Network simulations have so far embodied connectivity data, as well as firing pattern data, from stratum pyramidale interneurons (Traub and Miles, 1991). There are several important differences in this connectivity compared with pyramidal cell connectivity. For example, CA3 pyramidal cell axons spread several mm (Tamamaki et al., 1984) and contact only a small fraction of cells in their territory of arborization, whereas inhibitory cell axons typically spread less than 1 mm but contact more than half of the cells in their territory (R. Miles, personal communication). Functional connectivity to and from interneurons is also distinctive. Thus, the pyramidal cell → interneuron EPSP is particularly rapid in onset and effective, often leading to an action potential after a single presynaptic spike (Miles, 1990a). Indeed, disynaptic inhibition (pyramidal cell → interneuron → pyramidal cell) leads to an IPSP peak several milliseconds earlier than the peak of a monosynaptic pyramidal cell EPSP (Miles and Wong, 1984, 1986).

The interneuron EPSP is largely attenuated by CNQX (Miles, 1990a) and so contains an AMPA component, but NMDA receptors also exist on interneurons (Hestrin et al., 1990; Jones and Bühl, 1993). Based on indirect evidence (synchronous bursts in low Mg and CNQX are prolonged by bicuculline; Traub et al., in press), we suspect that some of these interneuronal NMDA receptors are in local recurrent synapses, rather than all being located at afferent excitatory synapses. Interneurons form GABAergic synapses with each other, probably of both A and B types, and in complex bathing media (excitatory amino acid blockers plus 4AP), some of the $GABA_A$ synapses can actually be excitatory (Michelson and Wong, 1991). As the role of such effects in epilepsy is still uncertain, however, and in order to set a bound upon the complexity of our network models, we have omitted interneuron → interneuron synapses from most simulations; we do, however, include recurrent excitation of interneurons, with both AMPA- and NMDA-mediated postsynaptic conductances.

Modeling synaptic interactions

Epileptogenesis induced by pharmacological agents, or by selective changes in ionic composition of the medium, occurs so quickly (within minutes) that anatomical rewiring in the tissue is highly unlikely: the altered population behavior so induced must occur through functional changes either in the cells or in the synapses. Indeed, in all experimental epilepsies studied so far, synaptic alterations seem—in our opinion—to be of predominant importance. Thus, a reasonably accurate—but also computationally manageable—description of synaptic interactions is essential. We use a simple formalism that applies to the four types of synapse/receptor that are modeled: AMPA, NMDA, $GABA_A$, $GABA_B$ (Traub et al., 1992a). This formalism works as follows. (1) In each compartment, the synaptic conductance of a given type (say AMPA) is proportional to a time-independent scaling parameter C_{AMPA}, and to a time-varying variable S_{AMPA}. (2) S_{AMPA} is influenced by presynaptic activity, and it can be taken to represent the number of receptors in the respective compartment that are open at any given time. (3) In the absence of presynaptic activity, S_{AMPA} decays with first-order kinetics and time constant $\tau_{AMPA} = 2$ msec; τ for the other types of receptor is longer. We typically use $\tau_{NMDA} = 150$ msec, $\tau_{GABA(A)} = 7$ msec $\tau_{GABA(B)} = 100$ msec. (4) For a single presynaptic action potential, S_{AMPA} obeys a differential equation

$$\frac{dS_{AMPA}}{dt} = k_{AMPA}(t) - \frac{S_{AMPA}}{\tau_{AMPA}} \qquad (1)$$

In Eqn. (1), the forcing function $k_{AMPA}(t)$ is a square-wave, equal to 0 always, except for the interval T_{AMPA} when $k_{AMPA} = 1$; T_{AMPA} is usually equal to 2 msec. The parameter T_{AMPA} defines the time scale over which a presynaptic action potential will 'activate' postsynaptic receptors, thereby determining the time-to-peak of the postsynaptic conductance. The corresponding parameter for the other synaptic receptors is different: $T_{NMDA} = 5$ msec; $T_{GABA(A)} = 1$ msec; $T_{GABA(B)} = 150$ msec. (5) Further, multiple presynaptic action potentials, either as temporal sequences from single neurons, or from multiple different neurons, lead to addition of the (possibly time-shifted) respective forcing functions k_{AMPA} in Eqn. (1). (6) For the 'slow' receptors (i.e. NMDA and $GABA_B$), the respective S value is constrained to be less than or equal to a saturation value S_{max}. (Otherwise, these conductances become unreasonably large during epileptic states when large neuronal populations are firing simultaneously). The use of saturation appears reasonable, given that there must be a finite number of synaptic receptors.

For simulating the NMDA receptor/channel complex, an additional consideration is important. Specifically, we must take into account the effects of transmembrane potential and external $[Mg^{2+}]$ on the conductance. We ignore the kinetics of voltage and Mg interactions with the channel, and assume that, in each membrane compartment bearing NMDA receptors, the NMDA conductance is a product of the scaling parameter C_{NMDA}, the time-dependent term $S_{NMDA}(t)$ (constrained to be less than a saturation value), *and* a function $f(V,[Mg])$, where $0 \leq f \leq 1$. The function $f(V,[Mg])$ that we use derives directly from the data of Jahr and Stevens (1990) as formulated in their equation 4a and Table 1. $f(V,[Mg])$ is monotonic increasing in V for fixed [Mg], and monotonic decreasing in [Mg] for fixed V.

The simulated actions of convulsant drugs can be described in terms of this synaptic formalism. Thus, to simulate the effects of $GABA_A$ blockade (e.g. from bicuculline), we reduce the scaling parameter $C_{GABA(A)}$, usually to 0. To simulate APV, we set $C_{NMDA} = 0$. To simulate low Mg, we set the parameter [Mg] to 0 in the function $f(V,[Mg])$. Finally, to simulate the effects of 4AP, we increase T_{AMPA} from its usual value of 2 msec to 10 msec or more (ignoring the likely prolongation of $T_{GABA(A)}$, etc.). $GABA_A$ blockers are not known to affect intrinsic membrane properties, and we have not attempted to emulate the effects of low Mg and 4AP on membrane kinetics or voltage-dependent K conductances (e.g. the transient A-type of K conductance). Our reasons for believing that low Mg and 4AP actions on intrinsic cell properties may be of secondary importance for epileptogenesis are these: (1) synchronous activity occurs in low Mg media even when total divalent cation concentration is compensated by increasing Ca^{2+} concentration (Hamon et al., 1987); (2) synchronous activity resembling experimental recordings in 4AP can be simulated without inclusion of the effects on A currents (R. Traub, J.G.R. Jefferys, B.W. Strowbridge, II. Michelson and R.K.S. Wong, unpublished data).

A more difficult problem is to know what the 'normal' values of the synaptic scaling parameters should be. Our approach is an empiric one: to search for results that are both consistent with experiment and are valid over a reasonable range of the scaling parameters.

Synaptic noise

Synaptic noise, both EPSPs and IPSPs, occurs in CA3 pyramidal neurons (Brown et al., 1979; Miles, 1991). Spontaneous IPSPs reflect in part the spontaneous firing of interneurons as they often occur simultaneously, under resting conditions, in pairs of neurons (Miles and Wong, 1984). Spontaneous IPSPs are also recorded in pyramidal cells in the presence of blockers of excitatory amino acid neurotransmission (Miles, 1991). Spontaneous EPSPs likely reflect background firing of pyramidal cells, as well as spontaneous, presumably quantal transmitter release at mossy fiber and recurrent synapses; not all of the spontaneous EPSPs in CA3 pyramidal cells derive from mossy terminals, as some spontaneous EPSPs have an NMDA component (McBain and Dingledine, 1992). The frequency and duration of spontaneous EPSPs are incresed in low Mg (J.G.R. Jeffreys and M.A. Whittington, unpublished data), and background noise is greatly enhanced in 4AP (Ives and Jefferys, 1990). Background EPSPs appear to be of particular significance in the initiation of spontaneous synchronous bursts, at least in high potassium epilepsy (Chamberlin et al., 1990; Traub and Dingledine, 1990). It therefore is necessary to include background synaptic activity in a model of spontaneous epileptic events; yet, at the same time, one does not wish to introduce too many free parameters. For this reason, in simulations to be presented here, we have made a compromise: to keep the *frequency* of the noise constant in the different epilepsy models, but to vary the *amplitude and/or time course* of the spontaneous synaptic potentials according to the known or presumed actions of the respective epileptogenic agents. This assumption facilitates comparison of the different epilepsy models, although it also leads to false estimates of the frequency of spontaneous synchronized bursts. The average frequencies used here are, respectively, 10 Hz for mossy (AMPA) EPSPs and 10 Hz for recurrent (AMPA + NMDA) EPSPs on pyramidal cells, and 2 Hz for AMPA EPSPs on inhibitory cells, the latter in turn causing inhibitory neuronal firing and background IPSPs in pyramidal cells. The synaptic noise is Poisson-distributed and independent for the different cells.

Unitary excitatory synaptic potentials

Based on the above assumptions about how drugs and altered bathing media exert their actions, we can now consider (Fig. 1), at least qualitatively, how unitary EPSPs might appear. (So far as we are aware, the only experimental recordings of monosynaptic recurrent EPSPs have been obtained in media with normal or increased Mg concentrations, and without using 4AP or NMDA blockers; Miles and Wong, 1986, 1987; Miles, 1990a.) What we shall present should be regarded as a set of experimental predictions. Figure 1 illustrates simulations of the expected effect of a presynaptic action potential (left) or a presynaptic

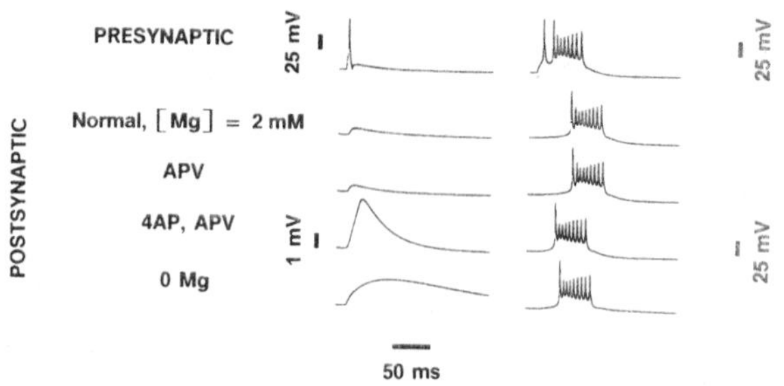

Fig. 1. Simulated unitary excitatory synaptic interactions, for single presynaptic spike (left: postsynaptic holding current = −0.2nA) and a presynaptic burst of action potentials (right: postsynaptic holding current = −0.05nA). Shown are an EPSP in normal medium; an AMPA-mediated EPSP ('APV', indicating blockade of the NMDA-mediated component); an AMPA-mediated EPSP as it might be amplified by the simulated action of 4AP (prolongation of the AMPA receptor 'on' duration, T_{AMPA}, after a presynaptic action potential from 2.0 msec to 17.5 msec); and an AMPA + NMDA EPSP wherein the NMDA component has been amplified by 0 Mg (see text), and the peak NMDA channel conductance is same as for the 'normal' EPSP. While the unitary EPSPs differ in shape and amplitude, stereotypic burst transmission is predicted to occur in each case, although with different latency (R.D. Traub and J.G.R. Jefferys, unpublished data).

burst (right) on a single postsynaptic pyramidal cell. The EPSP interacts with voltage-dependent conductances in the postsynaptic dendritic membrane. Experimentally, it has been shown that unitary EPSPs (Fig. 8 of Miles and Wong, 1986) and afferent stimuli (Wong et al., 1979; Masukawa and Prince, 1984) elicit voltage-dependent responses. While the voltage-dependent behavior of NMDA receptor/channels might contribute to such reponses, NMDA currents are not likely to be the sole generator: similar voltage-dependent responses can be elicited by current injection, via a micro-electrode, into the postsynaptic cell body or dendrite (Wong et al., 1979; Masukawa and Prince, 1984; Miles and Wong, 1986). Injection of a hyperpolarizing current reduces the voltage-dependent depolarizing currents associated with the EPSP, both in simulations and experimentally (Miles and Wong, 1986). Note that the normal EPSP and the pure AMPA-mediated EPSP ('APV') are prolonged and smoothed also by the electrotonic properties of the membrane. The simulated EPSP produced by simultaneous activation of AMPA and NMDA receptors has a depolarizing tail under baseline conditions, the amplitude of which increases with membrane depolarization (not shown). The membrane depolarization acts both on the NMDA conductance and on intrinsic Na and Ca channels. Prolongation of T_{AMPA} to 17.5 msec, simulating the effect of 4AP, increases the EPSP amplitude, delays its peak and prolongs the EPSP. Enhancement of the simulated NMDA component in low Mg not only increases the EPSP amplitude, but also produces a prolonged depolarizing tail—the predictable result of the slow decay of the NMDA conductance (τ_{NMDA} = 150 msec in this case).

Despite the striking differences in unitary EPSPs occurring under the illustrated conditions, our model predicts (Fig. 1) that burst transmission should occur, in a robust and stereotypic way, except for differences in latency, provided that the postsynaptic cell is not excessively hyperpolarized or shunted. Such burst transmission, with its relative sensitivity to membrane potential (at least when synaptic conductances are near threshold) is a consequence of the all-or-none quality of the fully regenerative dendritic Ca spike. The ability of burst transmission to occur via activation of AMPA receptors alone would be consistent with the experimental demonstration that synchronized bursts occur in disinhibited slices during NMDA blockade (Herron et al., 1985; Ashwood and Wheal, 1986; Dingledine et al., 1986; Lee and Hablitz, 1989).

The effect of one pyramidal cell on an indivudal other pyramidal cell is of critical importance in epileptogenesis, and during disinhibition is probably the single most important factor leading to synchronization of bursting (Traub and Wong, 1982). Nevertheless, this effect is not sufficient to explain how synchronization can sometimes occur even when inhibition is present, as it is known to be in the case of both low Mg (Tancredi et al., 1990; Jefferys et al., 1994) and in 4AP (Buckle and Haas, 1982; Rutecki et al., 1987; Perreault and Avoli, 1991). Nor are unitary interactions sufficient to explain the characteristic shaping of a sustained epileptic afterdischarge into a series of synchronous oscillations. The initiation process involves some large part of the population, possibly including interneurons, at a time when the overall population-averaged firing rate is low. In contrast, shaping of the afterdischarge depends on the response of pyramidal cells to multiple, high-frequency (perhaps receptor-saturating) synaptic inputs—excitatory and, in some epilepsy models, inhibitory as well. Network simulations have thus proven useful, as a complement to experiments, for understanding both initiation and shaping of epileptic events (Traub and Miles, 1991; Traub et al., 1993a).

Synaptic events just prior to synchronized firing

Figure 2 uses superimposed intracellular records (from experiments and network simulations) to contrast the initiation process in disinhibition, 4AP treatment and low Mg, respectively. During disinhibition, there is little perceptible build-up in synaptic activity just prior to the synchronized event. This is probably caused by the relative rapidity of spread of firing during disinhibition (Traub and Wong, 1982). On the other hand, in 4AP (Ives and Jefferys, 1990) and in low Mg, synchronous firing is preceded by a significant build-up in synaptic activity and spike firing, an effect similar to what occurs in high K-induced synchronization (Traub and Dingledine, 1990; Chamberlin et al., 1990). An interesting detail is the larger size of the synaptic potentials in 4AP than in low Mg, in both experiment and in simulations (even though NMDA conductances were blocked in the 4AP simulation). Why should synchrony be relatively delayed in low Mg and 4AP, even though unitary EPSPs are expected to be enhanced? We attribute this to the 'braking' effect of $GABA_A$ inhibition. Simulations are consistent with this notion, as synchrony develops with extreme rapidity when $GABA_A$ blockade is added to the presumed effects of 4AP or of low

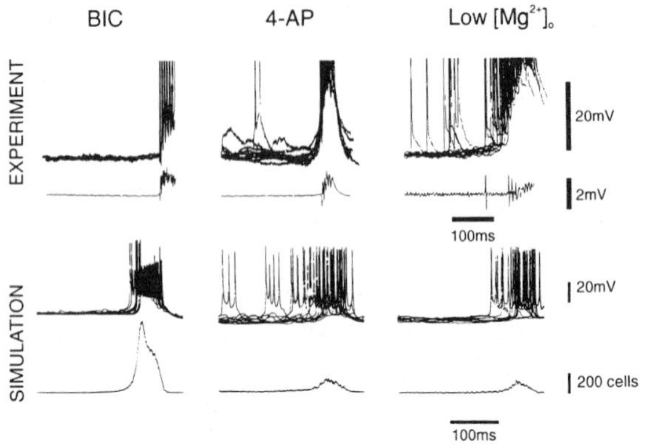

Fig. 2. Cellular activity just prior to synchronous firing in three types of epilepsy: $GABA_A$ blockade ('Bicuculline'), 0 Mg, and 4AP. Experiments above, simulations below (1000 pyramidal cells, 100 inhibitory cells). Note the relative absence of synaptic potentials and cell firing prior to synchrony in Bicuculline. In contrast, in the two models with significant baseline noise and with synaptic inhibition (0 Mg and 4AP), there is considerable activity prior to the synchronous firing. The experimental data were obtained by simultaneous intracellular and extracellular recording of spontaneous activity for many seconds, holding a single neuron with the intracellular electrode; records were then superimposed by aligning them with respect to the largest negativity in the field potential. The simulation figures were obtained by superimposing potentials from 11 different cells 'recorded' simultaneously (J.G.R. Jefferys, M.A. Whittington, A.E. Watts and R.D. Traub, unpublished data; see also Ives and Jefferys, 1990.) Experiments were performed in 400 μm thick transvers hippocampal slices of the rat at 33° C. Bicuculline methiodide concentration was 70 μM and 4AP concentration was 100 μM. NMDA conductances were blocked in the 4AP simulation.

Mg, and a similar effect is seen experimentally in low Mg (M.A. Whittington and J.G.R. Jefferys, unpublished data).

Predicted events during three types of afterdischarge

Figure 3 illustrates the shaping of simulated afterdischarges by large synaptic conductances. The common feature is a series of dendritic Ca spikes. During $GABA_A$ blockade, we predict (see also Traub et al., 1993a) the occurrence of a large plateau of NMDA conductance, while AMPA conductance fluctuates in phase with the population bursts. Experimental recordings of cells held near 0 mV (following intracellular cesium injection), during picrotoxin-induced afterdischarges, indicate that both slow and faster oscillating excitatory conductances occur (Fig. 2 of Miles et al., 1984), but it has not yet proven possible to separate these components by receptor sub-type. A suggestive piece of evidence is that the secondary bursts (corresponding to the predicted NMDA plateau during disinhibition) are abolished by NMDA blockers, but the initial burst is not (Lee and Hablitz, 1990; Traub et al., 1993a). In low Mg afterdischarges, we predict also the occurrence of a large NMDA plateau that drives dendritic Ca spikes (Fig. 3). In simulations, this NMDA plateau may be so intense that it is necessary to postulate an NMDA desensitization process (Benveniste et al., 1990) in order for the afterdischarge to terminate (Traub et al., in press).

Does this imply that the NMDA receptor underlies all afterdischarges? On experimental grounds, clearly not: synchronized bursts and afterdischarges can occur in 4AP even during NMDA blockade (Psarropoulou and Avoli, 1992; H. Michelson and R.K.S. Wong, personal communication). The 4AP simulation of Fig. 3, run with NMDA 'receptors' blocked, suggests that prolongation of T_{AMPA}—corresponding physically to prolonged glutamate release after each presynaptic action potential—might explain the synchrony and afterdischarges: very large and long-lasting waves of AMPA input can apparently evoke dendritic electrogenesis without the aid of NMDA receptors and in the presence of a large $GABA_A$ input. Under such conditions, AMPA receptors would be forced to behave, functionally, like NMDA receptors—i.e. to generate prolonged synaptic currents. In 4AP, the AMPA receptors would presumably lack the voltage-dependence of NMDA receptors, but that is not likely to be crucial: during low Mg epileptogenesis, the NMDA receptor/channels will have lost most of their voltage-dependence. Of course, for the postulated action of 4AP to be reasonable, AMPA channels must not completely desensitize (Tang

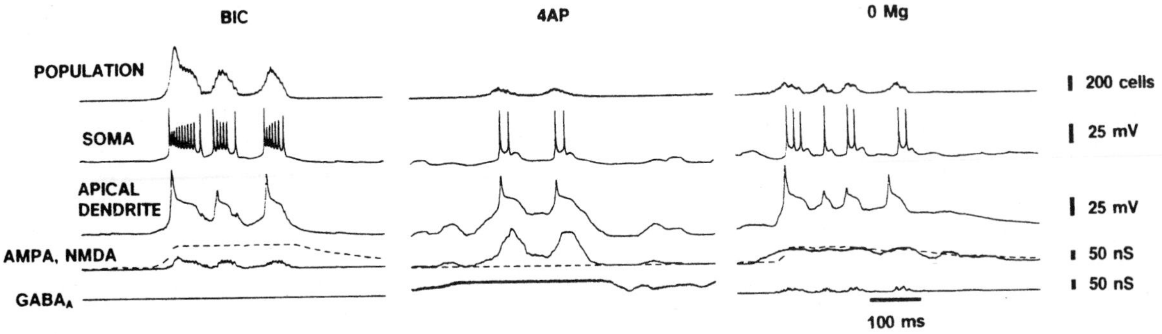

Fig. 3. Cellular events and synaptic inputs in three simulated afterdischarge types (1000 pyramidal cells, 100 inhibitory cells). Plotted from above downward are: population activity (the number of pyramidal cells with soma depolarized > 20 mV), the soma potential of a selected neuron, the apical dendritic potential (0.6 λ from the soma) of the same neuron, AMPA and NMDA conductances developing on the same neuron (NMDA shown with dashed line, AMPA with solid line), and the respective GABA$_A$ conductance. The NMDA input plotted is the ligand-gated component only, and does not reflect the multiplicative voltage- and Mg-dependent term. In the bicuculline case, a large NMDA input develops, eliciting dendritic Ca spikes, each corresponding to somatic (and population) bursts and, hence, corresponding to waves of AMPA input. A similar series of events occurs in 0 Mg, although GABA$_A$ inhibition is there present. The 4AP simulation was performed with NMDA receptors blocked (which does not prevent synchrony in physiological experiments: H. Michelson and R.K.S. Wong, unpublished data; Psarropoulou and Avoli, 1992). In the 4AP case, in the model, dendritic Ca spikes are driven by large waves of AMPA input. Transduction into somatic bursts is limited by the large GABA$_A$ input. (In the 0 Mg simulation, the NMDA conductance desensitizes, with time constant 350 msec, to 20% of its baseline value.)

et al., 1989). If the AMPA channels do indeed desensitize dramatically on a time scale of tens of milliseconds, then other mechanisms may need to be invoked in order to explain 4AP epileptogenesis.

Summary and Conclusions

To find general principles in the cellular mechanisms of epileptogenesis, one must analyze experimental epilepsy models and determine what exists in common between them. We consider here afterdischarges in hippocampal slices induced using either (1) GABA$_A$ blockade (e.g. with bicuculline), (2) a bathing solution lacking Mg^{2+} ions (low Mg-induced epilepsy), or (3) 4-aminopyridine (4AP). By 'afterdischarge' we mean an event that lasts hundreds of milliseconds or more, involving the synchronous firing of all the neurons in a population, shaped into a long initial burst and a series of one or more secondary bursts, and terminating in a prolonged afterhyperpolarization (AHP). We propose that the following features exist in common between these three experimental epilepsies: (1) recurrent excitatory synaptic connections; (2) sustained dendritic synaptic excitation, mediated by either AMPA or NMDA receptors, or both; (3) an intrinsic cellular response to sustained excitation, consisting of rhythmical dendritic bursts, primarily mediated by Ca spikes.

In conclusion, if the picture outlined here proves correct, then the stereotypic appearance of epileptic afterdischarges—consisting of synchronized population bursts in series, whatever the network alteration leading to seizures—does indeed reflect a common set of mechanisms. The mechanisms cannot, apparently, be formulated in simple terms of this receptor or that receptor. Rather, we suggest, the recurrent excitatory synapses are able, under diverse circumstances, collectively to produce sustained dendritic conductances in neuronal populations. Pyramidal neurons, by virtue of their normal intrinsic membrane properties, respond to such sustained conductances with rhythmical bursts. The recurrent synapses, in a dual role, serve to maintain the synchrony of these bursts, and so shape the activity into a synchronized oscillation.

Acknowledgements

We thank Drs. Richard Miles, Robert K.S. Wong, Hillary Michelson, Miles Whittington and Ben Strowbridge for helpful discussions and for sharing unpublished data. This work was supported by the IBM Corporation and the Wellcome Trust. JGRJ is a Wellcome Trust Senior Lecturer.

References

Ascher, P. and Nowak, L. (1988) The role of divalent cations in the *N*-methyl-D-aspartate responses of mouse central neurones in culture. *J. Physiol.*, 399: 247–266.

Ashwood, T.J. and Wheal, H.V. (1986) Extracellular studies on the role of *N*-methyl-D-aspartate receptors in epileptiform activity recorded from the kainic acid-lesioned hippocampus. *Neurosci. Lett.*, 67: 147–152.

Avoli, M. and Perreault, P. (1987) A GABAergic depolarizing potential in the hippocampus disclosed by the convulsant 4-aminopyridine. *Brain Res.*, 400: 191–195

Avoli, M., Drapeau, C., Louvel, J., Pumain, R., Olivier, A. and Villemure, J.-G. (1991) Epileptiform activity induced by low extracellular magnesium in the human cortex maintained in vitro. *Ann Neurol.*, 30: 589–596.

Benveniste, M., Clements, J., Vyklicky, Jr., L., and Mayer, M.L. (1990) A kinetic analysis of the modulation of *N*-methyl-D-aspartic acid receptors by glycine in mouse cultured hippocampal neurones. *J. Physiol.*, 428: 333–357.

Brown, T.H., Wong, R.K.S. and Prince, D.A. (1979) Spontaneous miniature synaptic potentials in hippocampal neurons. *Brain Res.*, 177: 194–199.

Buckle, P.J. and Haas, H.L. (1982) Enhancement of synaptic transmission by 4-aminopyridine in hippocampal slices of the rat. *J. Physiol.*, 326: 109–122.

Chamberlin, N.L., Traub, R.D. and Dingledine, R. (1990) Role of EPSPs in initiation of spontaneous synchronized burst firing in rat hippocampal neurons bathed in high potassium. *J. Neurophysiol.*, 64: 1000–1008.

Dingledine, R., Hynes, M.A. and King, G.L. (1986) Involvement of *N*-methyl-D-aspartate receptors in epileptiform bursting in the rat hippocampal slice. *J. Physiol.*, 380: 175–189.

Empson, R.M. and Jefferys, J.G.R. (1993) Synaptic inhibition in primary and secondary chronic epileptic foci induced by intrahippocampal tetanus toxin in the rat. *J. Physiol.*, 465: 595–614.

Fraser, D.D. and MacVicar, B.A. (1991) Low-threshold transient calcium current in rat hippocampal lacuonsum-moleculare interneurons: kinetics and modulation by neurotransmitters. *J. Neurosci.*, 11: 2812–2820.

Galvan, M., Grafe, P. and ten Bruggencate, G. (1982) Convulsive actions of 4-aminopyridine on the guinea-pig olfactory cortex slice. *Brain Res.*, 241: 75–86.

Haas, H.L. and Jefferys, J.G.R. (1984) Low-calcium field burst discharges of CA1 pyramidal neurones in rat hippocampal slices. *J. Physiol.*, 354: 185–201.

Hamon, B., Stanton, P.K. and Heinemenn, U. (1987) An *N*-methyl-D-aspartate receptor-independent excitatory action of partial reduction of extracellular [Mg^{2+}] in CA_1-region of rat hippocampal slices. *Neurosci. Lett.*, 75: 240–245.

Herron, C.E., Williamson, R. and Collingridge, G.L. (1985) A selective *N*-methyl-D-aspartate antagonist depresses epileptiform activity in rat hippocampal slices. *Neurosci. Lett.*, 61: 255–260.

Hestrin, S., Nicoll, R.A., Perkel, D.J. and Sah, P. (1990) Analysis of excitatory synaptic action in pyramidal cells using whole-cell recording from rat hippocampal slices. *J. Physiol.*, 422: 203–225.

Ishizuka, N., Weber, J. and Amaral, D.G. (1990) Organization if intrahippocampal projections originating from CA3 pyramidal cells in the rat. *J. Comp. Neurol.*, 295: 580–623.

Ives, A.E. and Jefferys, J.G.R. (1990) Synchronization of epileptiform bursts induced by 4-aminopyridine in the in-vitro hippocampal slice preparation. *Neurosci. Lett.*, 112: 239–245.

Jahr, C.E. and Stevens, C.F. (1990) Voltage dependence of NMDA-activated macroscopic conductances predicted by single-channel kinetics. *J. Neurosci.*, 10: 3178–3182.

Jefferys, J.G.R. and Traub, R.D. (1992) Synchronization of CA3 pyramidal neurons by NMDA-mediated excitatory synaptic potentials in hippocampal slices incubated in low-Mg^{2+} solutions. *J. Physiol.*, 452: 32P.

Jefferys, J.G.R., Whittington, M.A. and Traub, R.D. (1994) Depression of fast inhibitory postsynaptic currents in CA3 pyrimidal cells in rat hippocampal slices incubated in vitro in magnesium-free solutions; possible roles in epileptic activity. *J. Physiol.*, 476: 71P.

Jones, R.S.G. and Bühl, E.H. (1993) Basket-like interneurones in layer II of the entorhinal cortex exhibit a powerful NMDA-mediated synaptic excitation. *Neurosci. Lett.*, 149:35–39.

Kawaguchi, Y. and Hama, K. (1988) Physiological heterogeneity of nonpyramidal cells in rat hippocampal CA1 region. *Exp. Brain Res.*, 72: 494–502.

Lee, W.-L. and Hablitz, J.J. (1989) Involvement of non-NMDA receptors in picrotoxin-induced epileptiform activity in the hippocampus. *Neurosci. Lett.*, 107: 129–134.

Lee, W.-L. and Hablitz, J.J. (1990) Effect of APV and ke-

J. van Pelt, M.A. Corner H.B.M. Uylings and F.H. Lopes da Silva (Eds.)
Progress in Brain Research, Vol 102

CHAPTER 27

Synchronization in neuronal transmission and its importance for information processing

M. Abeles, Y. Prut, H. Bergman and E. Vaadia

Department of Physiology and the Center for Neural Computation, The Hebrew University of Jerusalem, P.O. Box 1172, Jerusalem 91-010, Israel

Introduction

The cognitive abilities of man and his extensive memorizing capabilities are assumed to rely on cortical processing. Yet, very little is known about the detailed mechanisms of these processes. Ever since Sherington's (1940) pictorial description of the cortical activity in the awake brain, it has been accepted that the most relevant parameters for describing and understanding cortical function are the spatio-temporal patterns of activity. In the present contribution we show that such patterns can be recorded in the cortex of behaving monkeys. We show that they span hundreds of milliseconds and maintain accuracy of a few milliseconds. The properties of these patterns suggest the nature of the neural circuitry that generates them, and therefore the mechanism that underlies cortical processing. By way of simulations we show how such circuits produce periodic, and non-periodic oscillations, and how activities in such circuits at different cortical regions can bind together to form compound entities.

In the following sections we summarize briefly the theoretical considerations and the experimental evidence that lead us to postulate the nature of the basic computing circuit in the cortex. Then we proceed to illustrate some of its properties as seen in simulations.

Theoretical consideration

The cerebral cortex is densely packed with synapses, axons, dendrites and neurons. In mammals all regions of neocortex contain approximately 800,000,000 synapses, 4 km of axons and 0.5 km of dendrites per cubic mm. Neuronal densities vary between animals, cortical regions and cortical layers. The range of neuronal densities is from 200,000 cells per cubic mm in layers II and III of the mouse to 10,000 cells per cubic mm in deep layers of the human motor cortex (Abeles, 1991 ch. 1; Braitenberg and Schuz, 1991). Approximately 90% of the cortical synapses are excitatory. It is estimated that half are derived from axons that reach the cortex through the white matter, while the other half are derived from axons that stay within the cortex and belong to neurons in the same region.

Each cortical neuron receives upon its dendrite 4000 to 80,000 excitatory synapses. On the average each of them must be very weak. It is usually postulated that internal cortical circuits are laid down by selective strengthening of some synapses. Particular attention is often paid to the synapses upon dendritic spines. Most of the synapses on spines are excitatory and most of the spines are on dendrites of excitatory neurons. Thus, computing circuits are built of excitatory neurons that become effectively coupled to each other by synaptic strengthening. In this sea of excitation, the inhibitory neurons maintain the important roles of preventing the recurrent excitation from building up into an epileptic feat, of isolating circuits from each other, and of providing a background of mutual inhibition upon which competition between several coexisting circuits can take place.

The high density of axons, dendrites, and synapses may give the impression that all neurons within a cortical region are connected to each other. However, both anatomical considerations and electrophysiological recordings show that the probability of two neighboring (within 300 μm) excitatory cells being synaptically coupled is approximately 10% (Braitenberg, 1978; Fetz et al., 1991). The strengths of the excitatory synapses are indeed very low. Even the strongest synapses ever found cannot support one to one transmission between two cortical neurons (Abeles, 1991 ch.3, Fetz et al., 1991).

With this type of anatomy and physiology, one must assume that cortical activation is maintained by neurons that excite each other in groups which are interconnected by multiple diverging/converging connections. The basic cellular properties of refractoriness and synaptic delays leads to the conclusion that the short range structure of the internal excitation in the cortex assumes the form of a feed-forward network composed of groups of neurons which are coupled through multiple diverging/converging connections. This conclusion is strengthened by the experimental findings that the strongest responses of most cortical neurons are well below 100/sec. Thus the activity profiles within 10 or so msec are essentially those of a feed-forward networks. Only with longer delays can one expect to see reactivation of the same neuron and feedback loops may become effective. A stylized diagram of such connections is given in Fig. 1.

Transmission of activity between two groups of cells that are coupled by multiple connections as in Fig. 1, can take two forms. Cells in one group may elevate their firing rates for some period, causing depolarization of the cells in the other group by spatial and temporal summation of synaptic potentials, which in turn produce an elevation of firing rates in the excited group. This form of transmission of excitation is termed *asynchronous* transmission. On the other hand if all the neurons in one group fire once in synchrony (within 1–3 msec), the synchronous volley of synaptic potentials which would arrive at the second group would evoke a synchronous spiking within the second group. This form of transmission is termed *synchronous* transmission.

In the regime of low firing rates, a neuron is much more sensitive to synchronous activation.

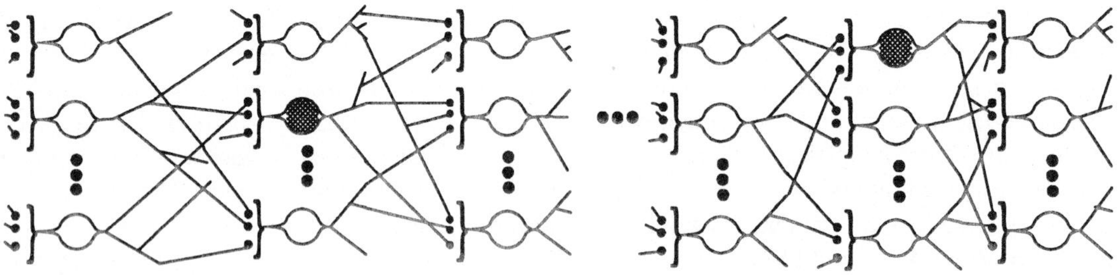

Fig. 1. A chain of diverging/converging connections. Locally these connections look like a multi-layered feed-forward network. However when a long chain is considered the same neuron may take part in more than one link. The neuron marked by gray shading takes part in two links.

This property has been analyzed quantitatively by Abeles (1982), and more recently be several other groups (e.g. Nelken, 1988; Koch et al., 1990; Douglas et al., 1991). Thus under the prevailing cortical conditions transmission through a chain of neuronal groups connected as in Fig. 1 is most likely to be in the synchronous form (see Abeles 1991, ch. 7 and Abeles et al., 1993b for further elaborations on this issue).

In summary, the basic anatomy and physiology of the cerebral cortex suggest that intracortical processing be carried by multiple diverging and converging connections. Analysis of the performance of links made by such diverging/converging connections shows that they preferably transmit synchronized volleys of action potentials. A network comprising groups of cells with diverging/converging connections between the groups is called a *syn-fire* chain.

Experimental evidence

Theoretical considerations lead us to postulate that cortical activity is processed and transmitted by syn-fire chains. The cortical connectivity and the strength of the intracortical connections suggest that each link in a syn-fire chain be composed of a few tens of neurons (see Abeles, 1991, chs. 6 and 7). The chances of obtaining simultaneous recording from several neurons in two successive links are meager. Therefore, it is not experimentally possible to observe directly the organization of activity in such chains. However, if one records from several neurons along a chain, then whenever the activity sweeps down the syn-fire chain, one expects to observe tight time locking (1–3 msec) among the recorded cells. Such time locking should be maintained even after long delays (hundreds of milliseconds).

Abeles and Gerstein (1988) described an algorithm that detects all excessively repeating patterns, irrespective of duration and neuronal composition. Recently, a search for such patterns was carried in data recorded from the frontal cortex of behaving monkeys. A large excess of such repeating patterns was found often (Abeles et al., 1993a).

More recently, a less global algorithm for detecting patterns was applied by one of us (Y.P.) to data from a different set of experiments. In this approach, which exploits the Joint Post Stimulus Time Histogram (Aertsen et al., 1989) method, only patterns composed of three spikes are looked for. This is done by using the spikes of one neuron as a trigger (instead of a stimulus), and counting the number of joint coincidences of spikes from two other neurons at delays (t_1, t_2) after this trigger. This method is equivalent to the construction of a three-fold correlation of spike trains (see Abeles et al., 1993b,c for more details).

While this method is limited to patterns of three spikes, it has the advantage of getting a better statistical estimate of what can be expected by chance and does not suffer from the masking effects that neurons with high firing rates had in the more global searching algorithm. By applying this new algorithm, excessively repeating patterns were found in most recording sessions.

Figure 2 illustrates one such pattern. It was composed of a spike from unit 12 followed by a spike from unit 2 after 151 ± 1 msec, and then, after 289 ± 1 msec by a spike from unit 13. This pattern (designated as (12,2,13); (0,151 ± 1,289 ± 1)) repeated 20 times an event whose odds to happen by chance were less than one in ten million. Figure 2 shows all the spikes of units 12, 2 and 13 from 200 msec before, up to 500 msec after, the first spike in the pattern. The spikes associated with successive occurrences of the patterns are plotted above each other. Note that the spikes which were associated with the pattern are not embedded in high frequency firing bursts, but stand in isolation. Patterns with slightly different delays (e.g. (12,2,13); (0,148 ± 1,289 ± 1), or (12,2,13); (0,151 ± 1,292 ± 1)) occurred at chance levels. We encountered hundreds of such events in our recordings.

Most of these events were associated with stimuli delivered to the monkey or with its motor responses. Very often a large fraction of the

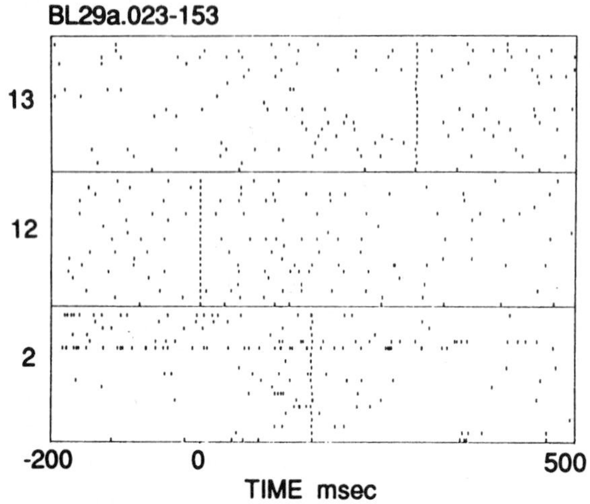

Fig. 2. Spatio-temporal firing patterns. The pattern (12,2,13); (0,151 $\pm$ 1,289 $\pm$ 1) repeated 20 times, an event which is highly unlikely to occur by chance. The activity around all 20 repetitions is plotted in rasters. The lower raster represents all the spikes around the first occurrence, above it are all the spikes around the second occurrence, and so on. All the rasters are aligned on the first spike in the pattern.

unit's responses was composed of spikes organized in well-timed patterns (Abeles et al., 1993a). All these properties support the hypothesis that cortical activity is generated by syn-fire chains.

The syn-fire structure as shown in Fig. 1 suggests a simple, multi-layered, feed-forward structure. If indeed cortical computations were performed by such a simple structure, then it is expected that there would be an almost one-to-one correspondence between the observed patterns and the behavior. Instead, we observed that any given behavior, or external event, was associated with several different patterns composed of spikes coming from the same neurons. Furthermore, often two or more of the spikes in a pattern came from the same single neuron.

In addition, some patterns could be preferentially associated with one condition, while other, having the same spike composition, could be associated with a different condition. For example, in one of the recording sessions the pattern (6,6,6); (0,56 $\pm$ 1,338 $\pm$ 1) appeared 17 times after the monkey was stimulated by a light blink from the left, and the monkey 'knew' that later, he would be required to respond to this stimulus. The same pattern appeared only four times when the same stimulus was delivered and the monkey 'knew' that he is asked to ignore the stimulus. On the other hand the pattern (6,6,6); (0,32 $\pm$ 1,152 $\pm$ 1) appeared 17 times when the monkey had to ignore this stimulus, and only three times when the monkey 'knew' that later he would be required to respond to it (Abeles et al., 1993b).

These observed properties did not support the notion that activity is being generated by simple feed-forward syn-fire chains. The mere fact that the same single unit can participate several times in the pattern, means that it may take part in more than one link of the syn-fire chain. For instance, assume that the neurons marked by gray shading in Fig. 1 are one and the same, then whenever activity sweeps through the syn-fire chain this neuron would emit two spikes with a fixed interval in between.

If all the neurons took part in all the links, the network would become a fully connected feedback network (as in Hopfield, 1982). In such a fully connected network there would be no preferred firing sequences. Thus, the existence of repeated patterns with the same neuron participating at several points, means that we are dealing with a neuronal structure which is between the fully recurrent feedback net and the multilayered feed-forward net. Activity in such a network may re-excite itself like in a feedback attractor, yet maintain well-timed sequences. This type of self-maintained activity is what physiologists mean when they speak about *reverberations*. The coexistence of several repeating spatio-temporal pattens suggests that activity may reverberate in several different preferred pathways. The accurate timing of the observed patterns suggests that they were generated by non-periodic reverberations in syn-fire chains. Examination of how syn-fire chains with repeated participation of some of its neurons may reverberate, was examined by way of simulations as described in the next section.

In summary, if cortical activity is produced by syn-fire chains, then whenever one succeeds in measuring the activity of several neurons from the same chain, one should observe the repeated appearance of well-timed firing patterns among these neurons. By recording through several microelectrodes in the cortex of behaving monkeys we found large numbers of such patterns. They may last hundreds of milliseconds yet maintain a timing accuracy of less than 3 msec. The properties of the patterns suggested that a given neuron may participate in more than one link along the chain. A syn-fire chain with this structure can maintain several modes of reverberations as have been observed experimentally.

Simulations

Computer simulations of networks of 'integrate and fire' neurons were used to gain insight into properties of syn-fire chains. The networks were composed of 1000 or more excitatory neurons each having the following properties: The 'cell body' exhibited passive capacity and leakage conductivities as well as active Na and K conductivities. The membrane potential of the cell body behaved according to:

$$C\frac{dV}{dt} = g_k(V - V_k) + g_{cl}(V - V_{cl}) + g_{Na} \times (V - V_{Na}) + g_{leak}V + I_{den}$$

where the resting potential is considered zero and I_{den} is the dendritic current. The dendritic current represented all the synaptic currents induced by the network and a background noisy current representing inputs produced by 'other brain regions' which are not associated with the processes studied in the simulation. Few simulated cells could also be excited by external currents, which represented a 'sensory input' to the network.

The excitatory synaptic currents had a rise time of 1 msec and a decay time of 5 msec. Two types of inhibitory currents were simulated. Both currents had a rise time of 1 msec. Half the inhibitory synapses had a decay time of 5 msec and the other half had a decay time of 20 msec. The noise current had a Gaussian amplitude distribution and was randomly fluctuating with a time constant of 5 msec. The noise of the various cells was uncorrelated.

The 'resting' membrane potential (i.e. the potential when no synaptic inputs were provided) was considered as 0. The threshold was set at 10 mV above that potential. Slow adaptation was introduced by allowing the threshold to fluctuate with the membrane potential. At every simulation step 1% of the membrane-potential deviation from the resting level was added to the threshold. These fluctuations accumulated with a time constant of 20 msec. When the membrane potential hit the threshold and the cell was not refractory it 'fired' an action potential. This was achieved by increasing the Na conductivity for 1 msec and then setting it back to its low resting level. The cell could not fire again for 3 msec (absolute refractoriness). K conductivity was turned on 1 msec after the cell fired, and decayed back with a time constant of 20 msec. One millisecond after the cell fired, the currents at its post-synaptic targets were activated.

The network contained two types of cells: excitatory and inhibitory. Each of the excitatory cells was connected at random to 20% of the other excitatory cells with very weak synapses. Some excitatory cells also formed strong conections with few other excitatory cells. Every excitatory cell excited all the inhibitory neurons. Every inhibitory neuron inhibited all the excitatory cells as well as all the inhibitory cells (including itself).

The strong synaptic connections were selected to form a variety of syn-fire chains. In the illustrations below each chain was composed of 75 links and each link was 12 neurons wide. The first three links (comprising 36 neurons) served as the input layer, which could be excited by a weak, continuous depolarizing current.

Parameters were selected to assure a low firing rate (less than 1/sec) when no external current was applied, and to prevent the explosion of activity when the network was stimulated externally.

The simulations showed clearly that a chain of neuronal groups connected by diverging/converging links (as in Fig. 1) will become preferentially active in the synchronous mode. This preference is so strong that the diffuse activation of the first links by continuous weak current will be converted by the network into synfire activity (as illustrated in Fig. 3B). When the neurons of the syn-fire chain took part in more than one link, reverberations could be achieved easily. It is enough to have half the neurons participate in two links at random to enable such reverberations.

The simulations also showed that the same neurons may form two different syn-fire chains. In these conditions the activity within the same cell group can reverberate in more than one mode (Abeles et al., 1993c). This property can account for the experimental observation (as described in the previous section) that the same neuron can generate one type of firing pattern in one behavioral state and a different pattern in another state. The experimental findings and the results of the simulations suggest that information about the nature of a stimulus (or about internal brain processes) be coded not just by elevated firing rates of some neurons, but also by the fine time structure of the firing cells.

Thus, different reverberating modes may qualify the nature of the information being processed beyont that which can be deduced by considering the question "which neurons elevated their firing rates?". This finding suggests that such modes can be used to bind syn-fire activities in various brain regions. The feasibility fo this idea was studied by a simulation of two cell assemblies; each could reverberate in two modes (1 and 2). A Hebb-like learning process which took place while both assemblies were reverberating in mode 1, or when both were reverberating in mode 2 was used to create cross-assembly linkage. After a few sessions the two assemblies became bound so that activating one of them elicited the complementary activity in the other. Furthermore, when the two assemblies were activated in incompatible modes (say, one assembly in mode 1, while the other was in mode 2), one assembly changed its reverberating mode to conform with the other.

After further strengthening of the inter-assembly synapses, when both assemblies were reverberating in compatible modes, coherent oscillations of the two assemblies were observed. These could take the form of either periodic, or non-periodic oscillations depending on the network parameters.

An example of the behavior of such coupled assemblies is given in Fig. 3, which illustrate binding of activity between two cell assemblies. Each assembly was composed of 720 excitatory cells, with the same properties as described above. Two syn-fire chains were embedded within each assembly. The structure of the chains is shematically illustrated in Fig. 3A, where the top panel shows the two syn-fire chains of cell assembly 1 and the bottom of cell assembly 2. Each neuron is represented by a dot, and each link by a column of 12 dots. Each syn-fire chain comprised 506 neurons, and was constructed from 75 pools each

Fig. 3. Binding of activity between two syn-fire chains. Excerpts from a simulation in which activities in two cell assemblies were bound by learned cross-connections are shown. (A) The syn-fire structure of the network. The top two panels show the two syn-fire chains embedded in cell assembly 1, and the lower two panels show the syn-fire chains of cell assembly 2. Connectivity is between adjacent columns of cell from left to right. Neurons that participate only once in each assembly are marked by thin dots, those that participate in two or more links are marked by heavy dots. Beyond the multiple enrollment of the neurons in each syn-fire chain, some of the neurons participated in both chains. (B) Syn-fire activity was initiated in both assemblies by applying a weak depolarizing current to the input stages of both assemblies at time 0. The evolvement of activity during the first six simulation steps (one step simulated 1 msec) is shown. The four rectangles in each panel represent the four syn-fire chains shown in (A), but only the neurons that fired are plotted. Activity started to appear in syn-fire 2 of assembly 2 immediately after stimulation, but reached synchronicity only at steps 3 and 4. Activity in syn-fire 2 of assembly 1 started only at time step 3. It became fully synchronous only at time step 6. From then on, syn-fire activity propagated coherently in both assemblies. Note that each neuron took part in several links and could participate in both syn-fire chains 1 and 2. Therefore, a synchronous firing constellation shows both as a vertical line (e.g. the lower right rectangle at step 4) and as scattered dots in both syn-fire chains. (C) Firing constellations at four instances in

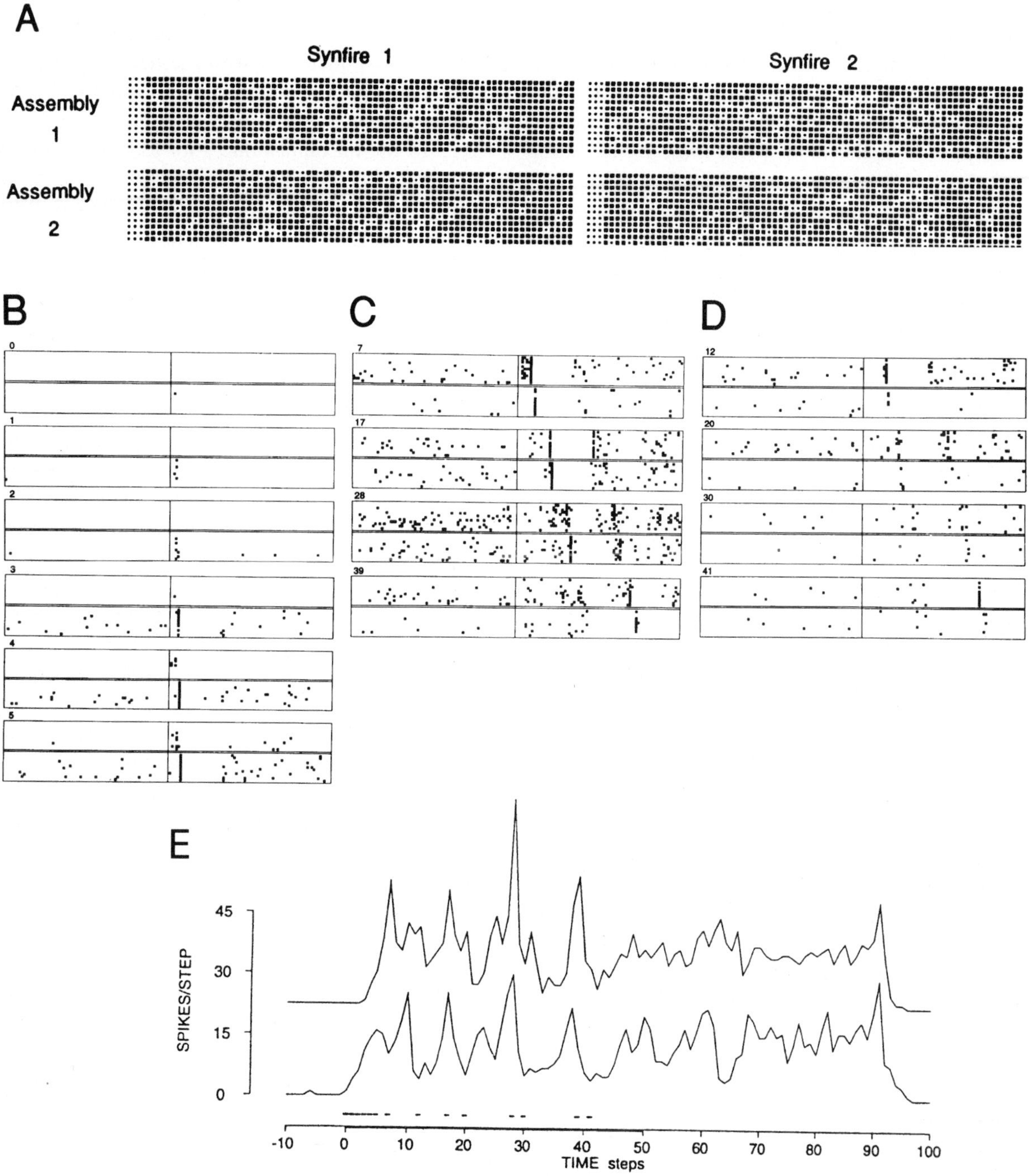

which the activity in assembly 1 reached local maxima. (D) Firing constellations at four instances in which the firing constellations of assembly 2 reached local minima. (E) The total level of activity expressed in number of firing cells per step for both assemblies as a function of time. The trace for assembly 1 is displaced up to prevent confusion between two traces. Stimulation time (0–50) is marked by a heavy bar under the abscissa. The short bars above the abscissa mark the times at which activity snapshots are presented in B, C and D.

of which contained twelve neurons. Connectivity is from left to right. The first three links to the left were 'input' neurons which could be stimulated by weak depolarizing current. Neurons that take part in only one link of each syn-fire chain are plotted as thin dots, while those participating in two (or three) links are plotted as thick dots. Beyond the multiple enrollment of the neurons in each syn-fire chain, some of the neurons participated in both chains. Nine of the 36 input neurons of one chain took part also in the input stage of the other chain, and 295 of the 470 neurons in the rest (non-input) of one chain took part also in the other chain.

Beside the internal connections, as described above, the two assemblies were cross-connected by weak excitatory connections. Every cell was initially connected at random to half the excitatory cells in the other assembly. Each syn-fire chain could be started by applying a week excitatory current to its first three pools for 50 msec. Figure 3B shows the initiation of syn-fire activity when the input stages of syn-fire 2 in both assemblies were stimulated. Note that initially the depolarizing current generated diffuse excitation among the input neurons, but within a few steps this activity is converted into syn-fire activity (i.e. sequential synchronous activation of the links).

With the initial structure, each cell assembly responded to its excitatory inputs without much regard to what was happening in the other assembly. The random cross-linking between assemblies was then modified by the following procedure: both assemblies were excited consistently, that is either syn-fire 1 or syn-fire 2 were started in both assemblies. Whenever the activity reached the end of both syn-fire chains successfully, the appropriate cross-assembly connections were modified according to the following (Hebb like) rule. All the cross-connections of the cells that fired were reduced by a fixed value (5% of the initial strength) and multiplied by 0.98. If by this decrement the value became negative it was clamped to zero and no further modification of this connection was allowed. Non-zero connections, for which the presynaptic cell was active shortly before the post-synaptic cell, were strengthened by a fixed value. The level of this fixed increment decayed exponentially with the delay between pre and post-synaptic activity time. Strengthening was maximal for a delay of 1 msec, and decayed to $1/e$ of the maximal value after a delay of 5 msec.

After ten training sessions the activity of the two assemblies became strongly associated. When both assemblies were stimulated consistently (as in Fig. 3), their activity evolved in a cooperative way. Typically the delay between the propagating activity waves in both assemblies was maintained at a stable (non-zero) level. When only one assembly was started, its activity evoked after some time its companion activity in the other assembly. When the assemblies were excited by conflicting inputs, the activity started in the two 'non-compatible' syn-fire chains. After approximately 10–20 steps activity was aroused also in the non-stimulated syn-fire chains. The two configurations competed for a while until one configuration took over, and from then on, the activity coherently propagated in both assemblies.

After a few more training sessions the binding between the two assemblies became very marked. At this stage coherent oscillations of the activity levels in both assemblies were observed. Figure 3E shows the total activity level of the two asssemblies as a function of time after the stimulation onset. During the first 40 steps the activity levels oscillate in a quasi-periodic way and they were coherent in both assemblies. The firing constellations near the peaks and troughs of these oscillations are plotted in Figs. 3C and 3D.

Although in general a syn-fire progression along the chain is observed, more complex propagation patterns can also be seen. For instance at time steps 17 and 28 there were two regions of synchronized activities. Between time steps 28 and 39 activity 'jumped' over 20 links in the middle of the chain. The total duration of activity was longer then 75 synaptic delays, suggesting that the activity reverberated for a while within the chain.

During the first 40 steps the total activity levels tend to oscillate periodically. However, careful examination of Fig. 3 shows that the repeated undulations of firing levels in the cell assemblies were not associated with repeated activation of the same neurons. Nor were these oscillations associated with periodic reverberations, as can be seen from examining the firing constellations at the peaks (C) or the troughs (D) of the activity waves. The activity wave undulations of both assemblies were coherent for the first 40 steps and much less later, although activity in both syn-fire chains continued to propagate with a fixed delay between the two assemblies throughout.

Thus, the activity level in a region may show periodic oscillations while none of the excitatory neurons in the region fire periodically! Inhibitory neurons would show periodicities because the undulations of activity levels are caused by the interplay between the positive feedback provided by mutual excitation and the negative inhibitory feedback. The coherency is caused by the cross-assembly connections.

With further strengthening of the connections between the two assemblies, non-periodic undulations of activity were observed. These could appear either in a coherent mode in both assemblies, or in a non-coherent mode. Such phenomena mimic well the coherent activity as was described for many compound brain processes. These simulations illustrate a case where coherent oscillation mark the fact that the activities in the two cell assemblies are bound together, but in themselves they are not the mechanisms of binding, nor carriers of information about what is being 'computed' by the cell assemblies. Both the binding process and the information are carried by the fine spatio-temporal organization of activities in the assemblies.

Conclusion

Anatomy, physiology, experimental results and simulations lead to the conclusion that accurate spatio-temporal firing patterns are a significant aspect of information processing in the cortex, and that these patterns are generated by reverberations in syn-fire chains. Syn-fire chains are also sensitive to the time structure of incoming information. Thus, syn-fire reverberations in different brain regions can enhance or interfere with each other according tot he compatibility of the sequences of activation in the chains. These properties may serve as the neuronal substrate for the 'compositionality' of language (Bienenstock, 1991). The oscillations observed in many brain regions while processing information (e.g. Eckhorn et al., 1988; Sheer, 1989; Gray and Singer, 1989) may be a by product of the excitatory-inhibitory imbalance in an activated region. Coherent oscillations among regions are a natural outcome of binding of syn-fire activities between the regions. Both phenomena may serve as an indication for the experimenter on activation and binding, but the internal signals of these processes for the brain itself are the accurate spatio-temporal firing patterns of the involved neurons.

Acknowledgements

The authors are indebted to E. Ahissar and I. Nelken who helped in developing the experimental setup, to I. Haalman, E. Margalit, and H. Slovin who helped in carrying out the experiments, and to V. Sharkansky for help in the artwork. This research was supported in part by grants from the United-States Israeli Binational Science Foundation (BSF) and the Basic Research Fund administered by the Israel Academy of Sciences and Humanities.

References

Abeles, M. (1982) Role of cortical neuron: integrator or coincidence detector? *Isr. J. Med. Sci.*, 18: 83–92.

Abeles, M. (1991) *Corticonics: Neural Circuits of the Cerebral Cortex*, Cambridge University Press, New York. 280 pp.

Abeles, M. and Gerstein, G.L. (1988) Detecting spatiotemporal firing patterns among simultaneously recorded single neurons. *J. Neurophysiol.*, 60: 909–924.

Abeles, M., Bergman, H., Margalit, E. and Vaadia, E. (1993a) Spatio-temporal firing patterns in the frontal cortex of behaving monkeys. *J. Neurophysiol.*, 70: 1629–1638.

Abeles, M., Prut, Y., Bergman, H., Vaadia, E. and Aertsen A. (1993b) Integration, synchronicity and periodicity. In: A.

Aertsen (Ed.), *Brain Theory: Spatio-temporal Aspects of Brain Function*, Elsevier, Amsterdam, pp. 149–181.

Abeles M., Vaadia E., Bergman H., Prut Y., Haalman I. and Slovin H. (1993c) Dynamics of neuronal interactions in the frontal cortex of behaving monkeys, Submitted.

Aertsen, A.M.H.J., Gerstein, G.L., Habib, M.K. and Palm, G. (1989) Dynamics of neuronal firing correlation: modulation of "effective connectivity". *J. Neurophysiol.* 61: 990–917.

Bienenstock, E. (1991) Notes on the growth of a "composition machine". In: D. Andler, E. Bienenstock and B. Laks (Eds.), *Contributions to Interdisciplinary Workshop on Compositionality in Cognition and Neural Networks* pp. 25–43.

Braitenberg, V. (1978) Cell assemblies in the cerebral cortex. In: R. Heim and G. Palm (Eds.), *Lecture Notes in Biomathematics, Vol.* 21, *Theoretical Approaches to Complex Systems*, Springer-Verlag, Berlin, pp. 171–188.

Braitenberg, V. and Schuz, A. (1991) *Anatomy of the Cortex. Statistics and Geometry*, Springer-Verlag, Berlin.

Douglas, R.J., Martin, K.A.C. and Whitteridge D. (1991) An intracellular analysis of the visual responses of neurones in cat visual cortex. *J. Physiol.* (*London*), 440: 659–696.

Eckhorn, R., Bauer, R., Jorden, W., Brosch, M., Kruse, W., Munk, M. and Reitboeck, H.J. (1988) Coherent oscillations: a mechanism for feature linking in the visual cortex? *Biol. Cybern.*, 60: 121–130.

Fetz, E., Toyama, K. and Smith W. (1991) Synaptic interactions between cortical neurons. In: A. Peters (Ed.), *Cerebral Cortex*, Vol. 9, Plenum, New York. pp. 1–47.

Gray, C.M. and Singer, W. (1989) Stimulus-specific neuronal oscillations in orientation columns of cat visual cortex. *Proc. Natl. Acad. Sci. U.S.A.* 86: 1698–1702.

Hopfield, J.J. (1982) Neural networks and physical systems with emergent collective computational abilities. *Proc. Natl. Acad. Sci. U.S.A.*, 79: 2554–2558.

Koch, C., Douglas, R.J. and Wehmeier, U. (1990) Visibility of synaptically induced conductance changes: theory and simulations of anatomically characterized cortical pyramidal cells. *J. Neurosci.*, 10: 1728–1744.

Nelken, I. (1988) Analysis of the activity of single neurons in stochastic settings. *Biol. Cybern.*, 59: 201–215.

Sheer, D.E. (1989) Focused arousal and the cognitive 40-Hz event-related potentials: Differential diagnosis of Alzheimer disease.*Prog.Clin. Biol. Res.*, 317: 79–94.

Sherrington, C.S. (194) *Man on his Nature*, Cambridge Univ. Press, London, pp. 166–183.

J. van Pelt, M.A. Corner H.B.M. Uylings and F.H. Lopes da Silva (Eds.)
Progress in Brain Research, Vol 102

CHAPTER 28

Oscillatory and non-oscillatory synchronizations in the visual cortex and their possible roles in associations of visual features

Reinhard Eckhorn

Department of Biophysics, Philipps-University, Renthof 7, 35032-Marburg, Germany

Introduction

Stimuli composed of coherent features are integrated by our sensory systems into perceptual entities, even if the features are dispersed among different sensory modalities. We can perceive a sensory object as a perceptual whole even if various aspects of the object are occluded, obscured by the background, or are not present at all. The visual system can easily detect coherencies in an object's local stimulus features, and is able to link, intensify and isolate them. These capabilities of grouping, mutual facilitation, and figure/ground separation require neural mechanisms for self-organization that are able to construct reliable and unique percepts out of ambiguous sensory signals. The mechanisms have to be highly flexible in order to cope with the immense variety of local feature combinations in natural scenes.

Perceptual 'association fields'

Our visual system can associate parts of a figure into a perceptual whole if a minimal feature contrast of the figure against the background is present, and if certain rules of gestalt properties are fulfilled. It is, for example, easy for us to recognize the triangles in Fig. 1 even though they are presented only as parts. Such type of feature association does not require previous knowledge of the figure. The association from its parts is a pre-attentively ('bottom-up') acting process. In contrast, the Dalmatian dog in Fig. 2 can be discovered fast and easily only if one knows this picture already. Here, the difficulty of preattentive feature association is due to the complexity of the figure, and its discovery is made more difficult still by the similarity of the features of dog and background. Feature association and recognition, in this case, requires visual memory or other 'top-down' support.

Psychophysicists have become increasingly interested in visual feature associations during the last years, paricularly in preattentive mechanisms. Their interest was stimulated by the discovery of oscillations in the visual cortex by neuroscientists and their hypothesis of 'feature association by synchronization' (Eckhorn et al., 1988; Gray and Singer, 1989). A good example of perceptual feature association that is related to neurophysiological findings is the work of Field and coworkers (Field et al., 1993) who found evidence for 'local association fields' for contour formation among strings of Gabor elements (fig. 3). Associations among local line (Gabor) elements into contours were induced within elongated 'association fields' around the Gabor elements only if neighboring elements had similar orientations and were

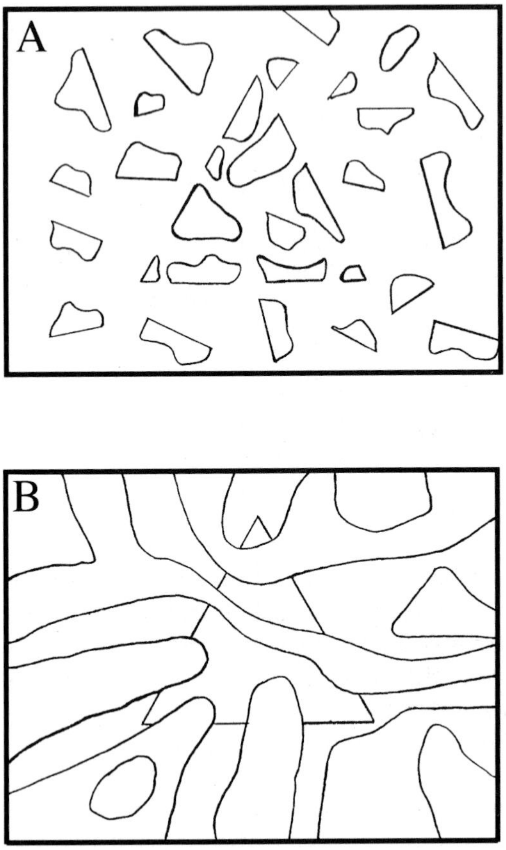

Fig. 1. Association of a figure of aligned elements by the human visual system. (A) The triangle is easily recognized even though the 'background' consists of similar elements. (B) The visible elements of the partly occluded triangle are easily associated into a figure.

aligned on first-order curvatures. These findings parallel observations in cat and monkey visual cortex that led to the definition of 'association fields of local populations of visual neurons' (Eckhorn et al., 1990b; discussed below).

Are feature associations supported by neural synchronization?

Neural mechanisms of associations have been discussed for a long time. For the visual system the main question is how the local cortical feature detectors, characterized by their receptive field (RF) properties, are combined such that their various properties are associated into coherent perceptual events. The 'synchronization hypothesis' seems a promising approach: it states that those neurons participate in the representation of a visual object whose activities engage in a common synchronized state in response to stimulation by that object (Reitboeck, 1983). The synchronization hypothesis attracted attention when stimulus-specific synchronized oscillations of 35–90 Hz were found in the visual cortex of anesthetized cats (Eckhorn et al., 1988; Gray et al., 1989, 1990) and awake monkeys (Freeman and van Dijk, 1987; Kreiter and Singer, 1992; Eckhorn et al., 1993b).

Recording synchronized activities simultaneously from several visual cortical areas

Locally, coding in the visual cortex can be characterized by receptive field (RF) properties of single neurons. Globally, features of a visual object are represented in a distributed fashion in several cortical areas. If we want to investigate the neural mechanisms that associate the distributed local feature representations into a global coherent percept we have to record neural activities in parallel from several cortical areas. This was done in our Marburg group in the typical experimental situation shown in Fig. 4. Fiber-microelectrodes were inserted in cat visual cortex areas 17, 18 and 19 or in monkey V1 and V2. The signal correlations between recording positions with overlapping and neighboring receptive fields in the same and in different cortical areas can, in this setup, systematically be varied and related to various receptive field properties.

Figure 5 shows the three different signal types that were generally extracted from the extracellular broad band recording (Eckhorn, 1992). The synchronized signal components of local cortical cell populations were of particular interest with respect to the 'feature-linking' hypothesis. Such a signal is the local slow wave field potential (LFP; 12–120 Hz band pass). LFP provide an estimate of the average activities of postsynaptic signals on dendrites and somata near the electrode tip (Mitzdorf, 1987). It is an averaged signal that

Fig. 2. Association and detection of a figure by the human visual system with help of 'top-down' mechanisms. Figure and ground are composed of similar elements, and the figure is not a simple visual object (dog). Its recognition requires help, e.g. from visual memory or by advice (Dalmatian dog with its head to the left and its tail to the right side) (modified from James, 1966).

mainly contains the synchronized components at the inputs of the local neural population near the electrode tip (range < 1 mm) because the synchronized components superimpose to relatively high amplitudes while statistically independent signals average out.

The synchronized components at the outputs of a local population were obtained by multiple unit recording activities (MUA) from the same microelectrodes that recorded LFP. MUA is spatially more confined (range < 150 μm), and gives an estimate of the average spike activity near the electrode tip. High MUA amplitudes occur preferentially if the outputs of the local population are synchronized. Single unit activity (SUA) was also recorded by the same microelectrodes by using conventional spike amplitude window discrimination. The capability of recording synchronized components of input (LFP) and output (MUA) signals of local populations enabled us to study their interactions at higher signal-to-noise ratios than is possible using single-unit spike trains alone.

Results of stimulus specific synchronizations in cat and monkey visual cortex

In the following paragraphs results are presented that were obtained from the visual cortex of anesthetized cat and an awake monkey. The data are

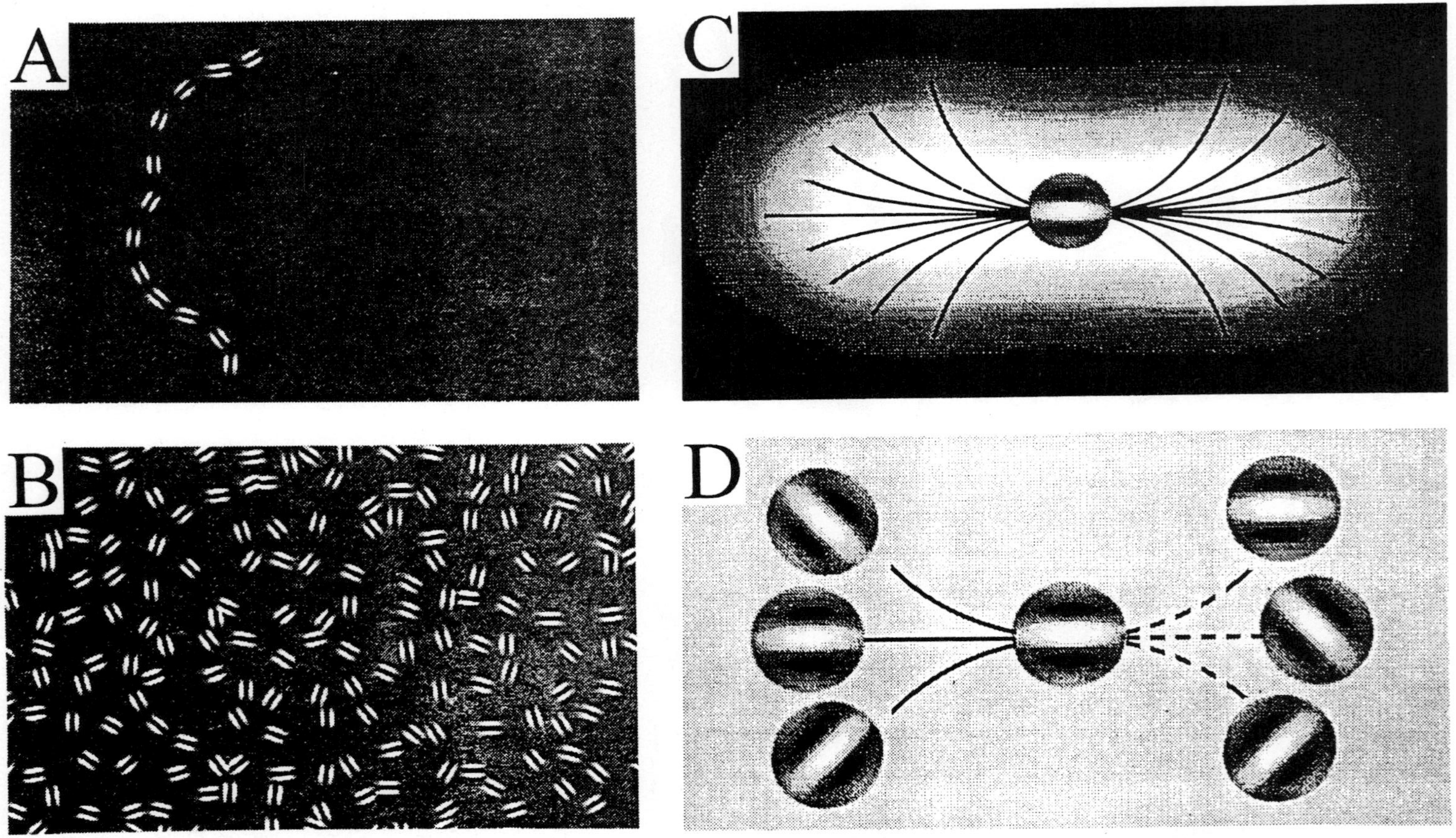

Fig. 3 Contour integration by the human visual system: evidence for a local 'association field'. (A) Path of elements the subject must detect when (B) embedded in an array of randomly oriented elements. (C) The 'association field', it represents the rules by which the elements in the path are associated and segregated from the background (the brighter the shading of the association field the stronger the associative forces). (D) The curves represent the specific rules of alignment. Grouping occurs only when the orientation of elements conforms to first-order curves (left side); elements like those at the right will not be associated (modified from Field et al., 1993).

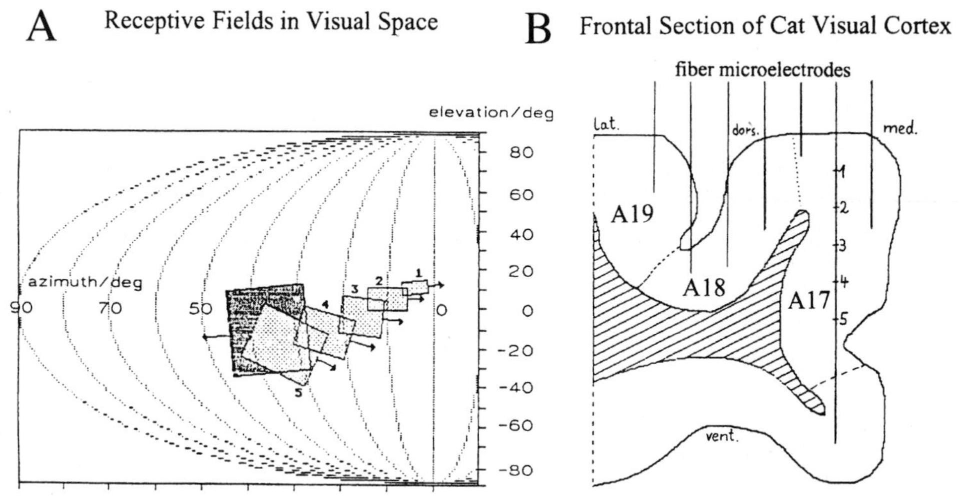

Fig. 4. Schema of experimental situation in multiple electrode recordings from several cortical areas. (A) Receptive fields in the left visual hemisphere. (B) Frontal section of cat visual cortex, areas 17, 18 and 19. The numbers at the tips of the schematically drawn fiber electrodes correspond to the respective receptive field numbers in (A).

not separately presented and discussed because the effects observed in both species so far are not essentially different.

High-frequency oscillations in the visual cortex of the awake monkey

An example of the simultaneous recording of LFP, MUA and single unit spikes with a single microelectrode from the primary visual cortex V1 of an awake monkey is shown in Fig. 6. LFP and MUA had mainly low frequency components at moderate amplitudes while the moving light stimulus was outside the receptive fields (left sides of Figs. 6A, 6B, 6C). When the stimulus entered the receptive fields, high frequency LFP and MUA oscillations appeared (right side panels), indicating that many neurons are active synchronously. The single unit spike train in this example was barely involved in the rhythmic activity of the neighboring neurons. This is evident from the lack of a pronounced peak in the spectrum at the oscillation frequency, as is present for LFP and MUA (details of single neuron participation in oscillatory states are presented below). This example is typical for the visual cortex of cat and monkey because it shows that oscillations occur nearly exclusively during direct activation of the neurons within their receptive fields (Eckhorn et al., 1988; Gray and Singer, 1989; Kreiter and Singer, 1992; Eckhorn et al., 1993b).

Orientation specific oscillations in monkey visual cortex

The validity of the 'linking-by-synchronization' hypothesis requires that oscillations occur when the neurons are activated by their preferred stimuli. Such stimulus specificity was extensively shown for orientation and movement direction of a contrast border in anesthetized cats (e.g. Eckhorn et al., 1988, 1990a; Gray and Singer 1989; Gray et al., 1989, 1990; Engel et al., 1990, 1991a,b). An example of a characteristic for the direction of stimulus movement that was measured for the oscillatory components in monkey primary visual cortex is Fig. 7. The oscillatory components of LFP and MUA were generally more sharply tuned (black characteristics) than the respective integrated responses (dotted characteristics; which show the conventional measure). This finding of narrower directional characteristics for oscillatory

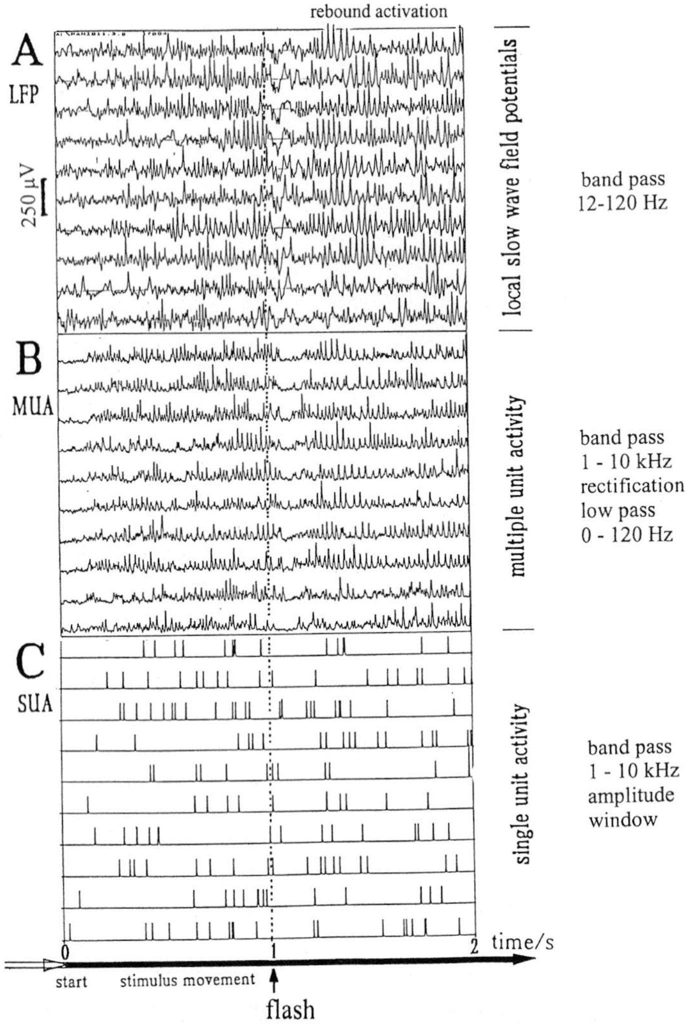

Fig. 5. Three types of neural signals from each micro-electrode. (A) Local slow wave (12–120 Hz) field potentials (LFP). (B) Multiple unit spike activities (MUA). (C) Single unit spike trains. All signals were recorded by a single electrode in response to identical stimulus repetitions. A whole-field grating stimulus (1 cycle/degree) started moving at $t = 0$ and continued at constant velocity (2°/sec) throughout the shown sweep to $t = 2$ sec. At about $t = 1$ sec a diffuse short photoflash was applied at the stimulation screen. It caused inhibition of oscillatory events (visible in LFP and MUA), and induced after a delay of 100–200 msec oscillations at higher amplitudes as before the flash. (Reproduced from Eckhorn et al., 1992.)

signals is supportive for the 'synchronization hypothesis' because it shows that the synchronized population signals code the directional properties of the stimulus better than the respective conventionally measured population responses.

Spatial profile of cortical coherence in cat visual cortex

The 'synchronization hypothesis' also requires that neighboring parts of the same visual object cause synchronized signals in the neurons representing these parts. This could be tested experi-

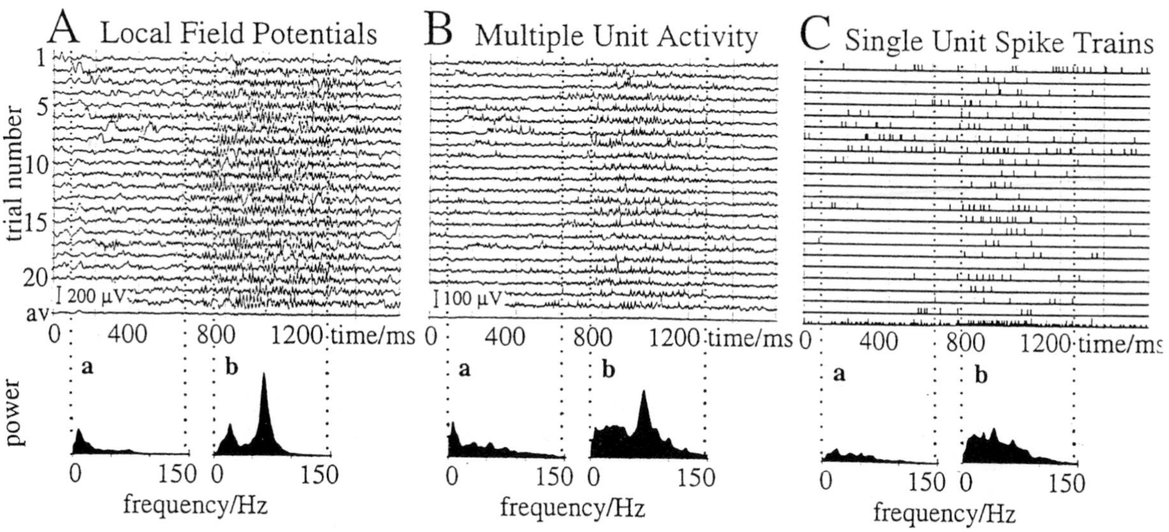

Fig. 6. High-frequency (60–90 Hz) oscillations in primary visual cortex of an awake monkey. Oscillatory activity during stimulation. Stimulus: light bar (2.47 × 0.12°) appearing at $t = 0$ and sweeping at 1°sec for 2 sec in the direction preferred by the neurones at the recording position. (A) Local field potentials (10–100 Hz). (B) Multiple unit activity (1–10 kHz, rectified, low-passed at 150 Hz). (C) Single cell spike trains (spectra were calculated after folding of the spikes by a Gaussian). Upper panels: 22 signal courses in response to identical stimulus repetitions, recorded by a single electrode. Lowermost traces (av) show the averages of the 22 responses (same amplitude scale). Lower panels: average power spectra calculated from the 22 single response epochs marked by dotted lines (arbitrary power scales). a: 512 msec data epochs before stimulation, b: 512 msec epochs during stimulation of the RFs. Note the strong high frequency LFP and MUA oscillations and its weak appearance in SUA. Note also, that in the average traces (av) the oscillations are not present because they are not phase-locked to the stimulus (modified from Eckhorn et al., 1993).

mentally with a spatially extensive stimulus and parallel recordings that were distributed across the 'visual map' of a single cortical area (Fig. 8). Brosch and coworkers found in cat area 17 and 18 that the average spatial coherence declined with cortical distance with a space constant of about 4 mm, corresponding to about 12° visual angle (Brosch et al., 1991). At a distance of 6 mm coherence was down to 20% of its initial value. The decline of coherence between oscillatory events was generally due to the increasing phase jitter with distance. It was not due to a spatial reduction in the oscillation amplitudes because these were, on average, similar in different cortical positions due to the large field stimulation. Closer inspection of the data in Fig. 8 revealed that the large variation of coherence at a single recording separation was mainly due to differences in receptive field properties. Paired recording positions with similar receptive field preferences had, on average, considerably higher values of coherence than recording pairs with differing properties.

The spatial decline of coherence indicates that oscillations do not occur in an ideally synchronized way within the range of representation of a larger visual object. According to the 'feature association hypothesis', this means that the coding of spatial continuity of a visual object by local groups of neurons in area 17 of the cat is restricted to a range of about 12–15° visual angle. However, such spatial confinement in the definition of continuity by a local group does not imply that it is impossible to define a larger figure on the basis of correlated activities of neurons within area 17. Coding of continuity would already be possible if the different neural subgroups that actually represent an object each have some of

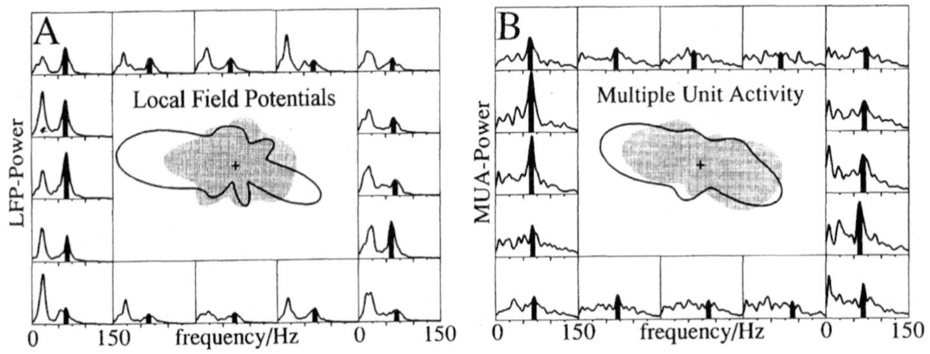

Fig. 7. High frequency oscillations in V1 of an awake monkey are orientation/direction specific. Directional tuning of oscillatory components in two different types of population activities. Simultaneous recordings of MUA and LFP from the same electrode. Stimulus: as for Fig. 6; movement in 16 different directions at 1°/sec, each direction ten times presented in random order. Outer panels: average power spectra of 512 msec response epochs during which the stimulus crossed the RFs (arbitrary power scales, identical for all panels). Black bars denote the range ± 4 Hz around the actual spectral maximum between 50 and 150 Hz for each stimulus direction (note that these dominant frequencies were very similar at different stimulus directions in cases where the oscillatory power was high). Central panel: direction tuning characteristic of oscillatory components in response to different stimulus movement directions. Amplitudes are proportional to the area of the black bars in the outer spectra. Center shading: direction tuning characteristic of the entire response power. The direction characteristics were smoothed by a sinc-interpolation. (A) Local field potentials. (B) Multiple unit activity. Note the similarity of the MUA and LFP tuning curves of the high frequency oscillatory components and MUA overall power and their similarity with the appearance of 20 Hz LFP components in contrast to the broad tuning of LFP overall power (modified from Eckhorn et al., 1993).

their signal components synchronized with other subgroups of the same assembly. This spatial confinement of synchronized oscillations parallels, for example, the restricted area of the association fields that were induced by Gabor line elements in the psychophysical experiments described above (Field et al., 1993). However, in those experiments the perception of contours of local elements is not confined to the 'local association field' of a single stimulus element.

Spatial relations of synchronized oscillations among different visual cortex areas

The presence of stimulus specific synchronization between different visual cortical areas is another prediction of the 'feature association hy-

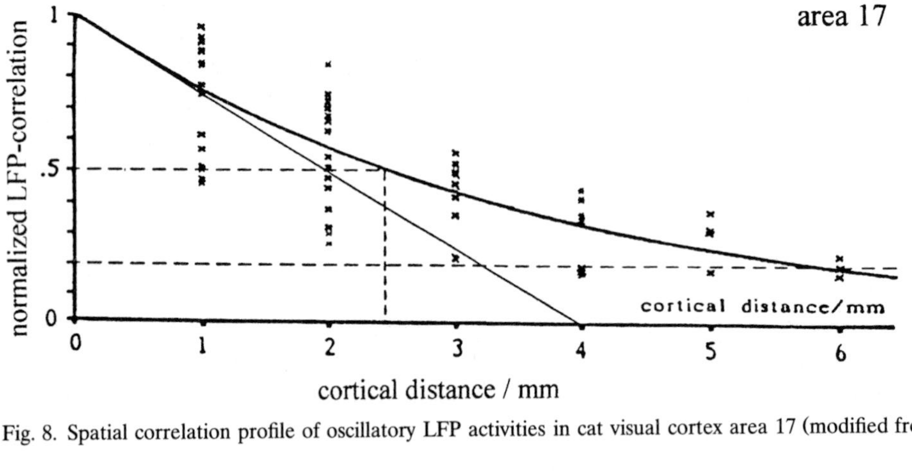

Fig. 8. Spatial correlation profile of oscillatory LFP activities in cat visual cortex area 17 (modified from Brosch et al., 1991).

pothesis'. Inter-areal synchronization is expected because different features from the same position of a visual object and spatially dispersed features of the same extensive object have to be associated by synchronization. Inter-areal synchronization of oscillations was indeed observed (including positions in which the receptive fields do not overlap) (Eckhorn et al., 1988, 1990a; Engel et al., 1991b). An example of synchronization between recording pairs from areas 17 and 18 of the cat at various distances is given in Fig. 9. Highest degrees of inter-areal synchronization were present at positions with overlap of the receptive fields, and up to separations of about 10° visual angle. Synchronization in these LFP recordings was present up to 15–20° visual angle (on average with a moderate level of confidence). This means in terms of the 'synchronization hypothesis', that feature associations are supported over a broad visual range by signals occurring synchronized in different cortical areas (here areas 17 and 18).

Synchronization of oscillations in different cortical areas depends on stimulus orintation and movement direction

The broad range of synchronization indices at a given distance of receptive fields in Fig. 9 can, again, be explained by the differences in preferred stimulus properties, because pair recordings were made between neurons of any type of receptive field properties. The more similarities in receptive field properties, the more probable were high degrees of synchronization among given neurons. An example of stimulus-specific synchronization among two cortical areas (A17 and 18 of the cat) is shown in Fig. 10: movement of a large

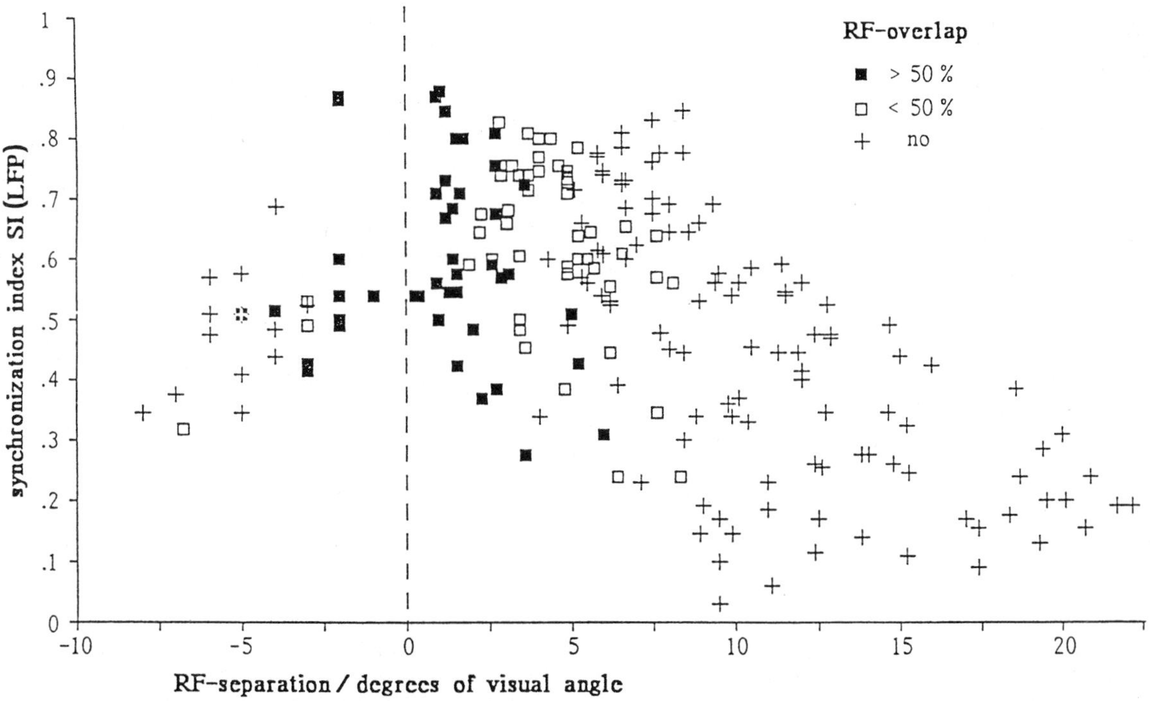

Fig. 9. Dependence of area 17–18 synchronization index of local field potentials on the separarion of the receptive field centers. Different symbols indicate the three classes of RF overlap and separation, respectively (see insert). RF separation was measured from the area 18 reference RF positions. Positive values indicate area 17 RF positions in peripheral retinal direction relative to the area 18 RF, while negative values indicate RF positions nearer to the central retinal position (modified from Eckhorn et al., 1990a).

grating in one direction synchronized neurons in the recording pair that preferred this direction, whereas the others were hardly activated or not at all (Fig. 10A). Change of the stimulus movement to the orthogonal direction synchronized the another A17–18 pair of neurons, while the previously synchronized pair became inactive.

Visual area 18 can dominate the synchronization with area 17

The stimulus-specific synchronizations that were observed between cat areas 17 and 18 (Eckhorn et al., 1990a) had another interesting specificity. They occurred with highest amplitudes and probability if the stimulus movement direction matched that of the area 18 neurons (e.g. Figs. 10, 11). This might be interpreted as a domination of area 18 neurons over those in area 17 (Fig. 11). As area 18 neurons in the cat have larger receptive fields, and prefer broader ranges of stimulus movement velocities in comparison with neurons in area 17, one might argue that area 18 of the cat strongly supports associations of the finer area 17 features by virtue of its coarser ones. Further speculations lead to parallels in psychophysical observations, in which dominance of coarse over fine spatial features is often found in feature association tasks.

Oscillatory events in different cortical areas occur at zero phase lag

In the cat, average phase differences between different positions of the same cortical area, and among different cortical areas (areas 17, 18, 19, LS) were approximately zero ($\leq$ 1 msec) (Eckhorn et al., 1988, 1990a, 1992; Gray et al., 1989; Engel et al., 1990, 1991a,b). Such in-phase oscillations were recently also demonstrated by us between visual areas V1 and V2 of an non-anesthetized monkey (Eckhorn et al., 1993a). Figure 12 shows

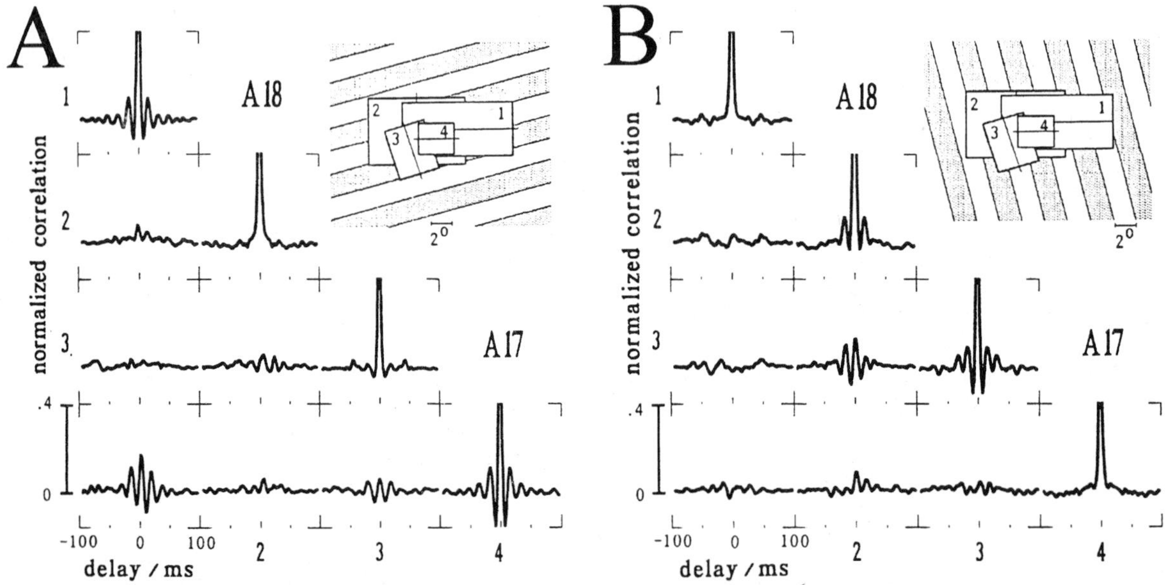

Fig. 10. Area 17–18 synchronization is direction specific. Example for the synchronization between area 17–18 spike activities SI (MUA) and its dependence on the orientation preferences of the neurons. Simultaneous recordings from two electrodes in area 17 (3,4) and two in area 18 (1,2). Autocorrelograms (clipped at 0.4) are plotted on the diagonals (upper left to lower right), cross-correlograms in the respective off-diagonal positions. The inserts show the RFs and (moving) grating stimuli drawn to scale. (A) Correlograms obtained with moving grating oriented near horizontally and (B) near vertically. Electrode 1: RF type not determined; el.2 and 3: simple-like RFs; el. 4: complex-like RF. Note the oscillatory modulations (autocor.) and synchronizations (crosscor.) of those recording pairs where the neurons are stimulated near their preferred orientations (A: 1 and 4; B: 2 and 3). Same stimulation as for Fig. 4 (modified from Eckhorn et al., 1990a).

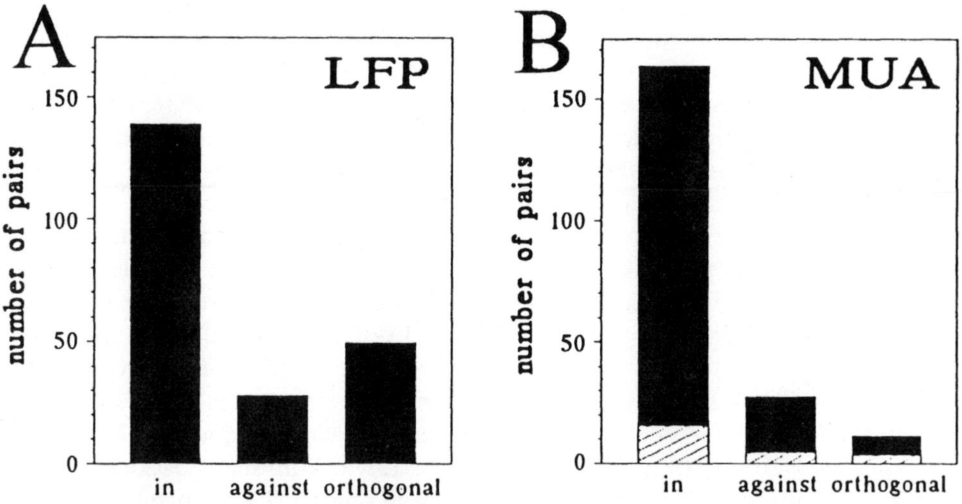

Fig. 11. Numbers of area 17–18 recording pairs in which synchronization was maximal in/against/orthogonal to/the preferred stimulus direction of the area 18 site. (A) Data for MUA. The hatched areas indicate the number of cases where only those recording positions from area 17 were selected in which the neurons had an orthogonal orientation preference (to the area 18 neurons) and were direction and orientation selective. (B) With LFP (modified from Eckhorn et al., 1990a).

an example and average values of V1–V2 phase differences. They were, in the data recorded so far, narrowly distributed around zero (Fig. 12A). Phase differences of individual recording pairs, however, had slightly broader distributions (SD = 3.2 msec).

The zero phase difference among V1 and V2 is of particular interest, because these visual areas are serially arranged with respect to the afferent visual stream, and because V1–V2 conduction delays are several milliseconds. A possible explanation of zero phase delay would be a common oscillatory input source to V1 and V2. It is more probable, however, that zero phase synchroniza-

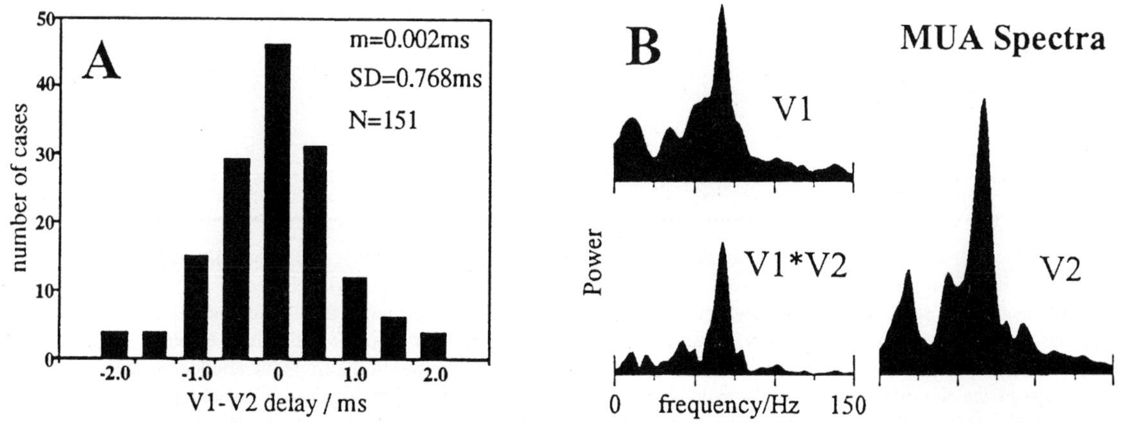

Fig. 12. Phase-locked high frequency oscillations between visual cortical areas V1 and V2 of an awake monkey have zero phase difference. (A) Distribution of average phase differences between oscillatory events in V1 and V2. (B) Example of power spectra and cross-power spectra of V1 and V2 multiple unit activities. Experimental parameters as for Fig. 6 (from Eckhorn et al., 1993).

tion is mainly due to temporally symmetric feedback interactions among the cortical areas (Eckhorn 1991, 1992; Engel et al., 1991a; Gerstner and Van Hemmen, 1993). In conclusion, associations of features the representations of which are dispersed among different cortical areas, could be efficiently supported by oscillations at zero phase difference because primarily the synchronized signals support one another at the spike encoder of a neuron while uncorrelated activities are transmitted with much less efficiency.

Single neurons are differently involved in stimulus-specific oscillations in cat visual cortex

The neural mechanisms and structures that generate and synchronize fast cortical oscillations have not yet been identified. We were therefore trying to clarify how and when single cortical neurons participate in oscillatory activities of remote as well as nearby neural populations (Eckhorn and Obermueller, 1993). We asked, in addition, if different types of participation can occur, and if they depend on visual stimulation and receptive field properties.

Three states of single cell participation in oscillations can be distinguished in spike-triggered averages of LFP or MUA from the same electrode (Fig. 13): (1) rhythmical states are characterized by the occurrence of relatively regular spike intervals at frequencies of 35–80 Hz, and these rhythms are correlated with LFP and MUA oscillations; (2) lock-in states lacking rhythmic components in single-cell spike patterns, although many spikes are phase-coupled with LFP or MUA

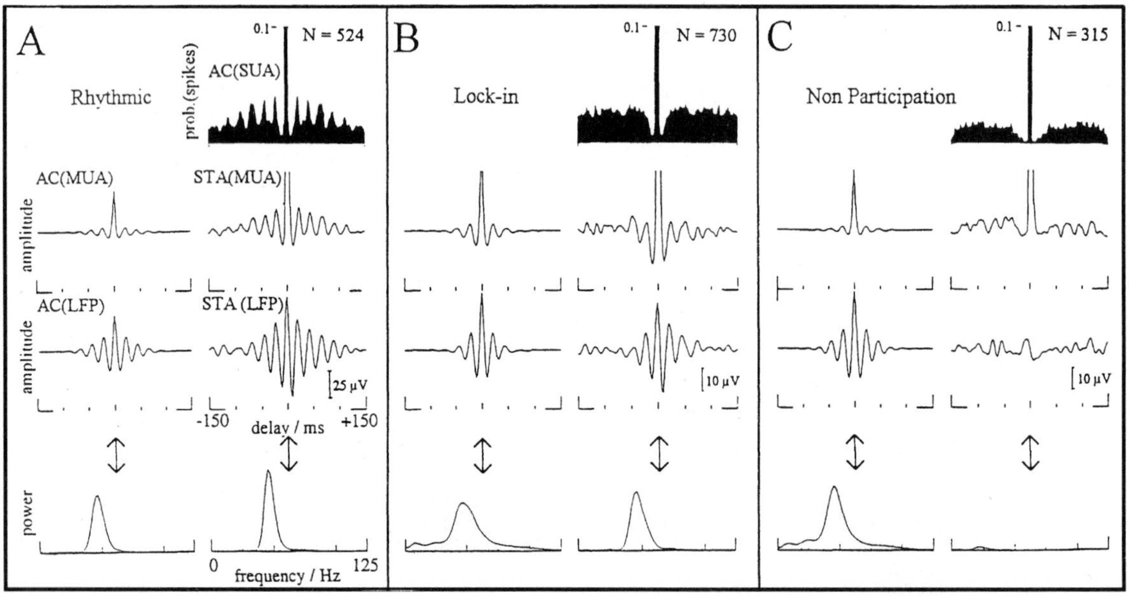

Fig. 13. Three different states of single cell couplings with oscillatory population activities. (A) Rhythmic, (B) lock-in, and (C) non-participation states of three different neurons. Each recording triple is from a single electrode. AC: auto- correlation histograms of single cell spikes (SUA), of multiple unit activity (MUA), and of local field potential (LFP). STA denotes spike-triggered averages of multiple unit activity or local field potentials. *N*: number of single cell spikes, equal to the number of averages in the STAs. According to the classification, STAs have oscillatory modulations in the rhythmic and lock-in states, and lack such modulation in the non-participation state. Note that in the rhythmic state (A) the single cell correlogram (top) is clearly modulated at 44 Hz, while in the lock-in (B) and the non-participation states (C) rhythmic modulations in the range 35–80 Hz are not visible (by definition). Lowest row of panels: power spectra for the above row of correlograms, calculated by Fast-Fourier transformation. The dotted lines in the power spectra of panel C indicate the respective confidence levels (four times background variance) (modified from Eckhorn and Obermueller, 1993).

oscillations; (3) during non-participation states, LFP or MUA oscillations are present but single-cell spike trains are neither 'rhythmic' nor phase coupled to these oscillations. Stimulus manipulations (ranging from 'optimal' to 'sub-optimal' for the generation of oscillations) often led to systematic transitions between these states (i.e. from rhythmic, to lock-in, to non-participation). Single-cell spike coupling was generally associated with negative peaks in LFP oscillations, irrespective of the cortical separation of single cell and population signals (0–6 mm). These results suggest that oscillatory cortical population activities are not only supported by local and distant neurons which display rhythmic spike patterns, but also by those showing irregular patterns in

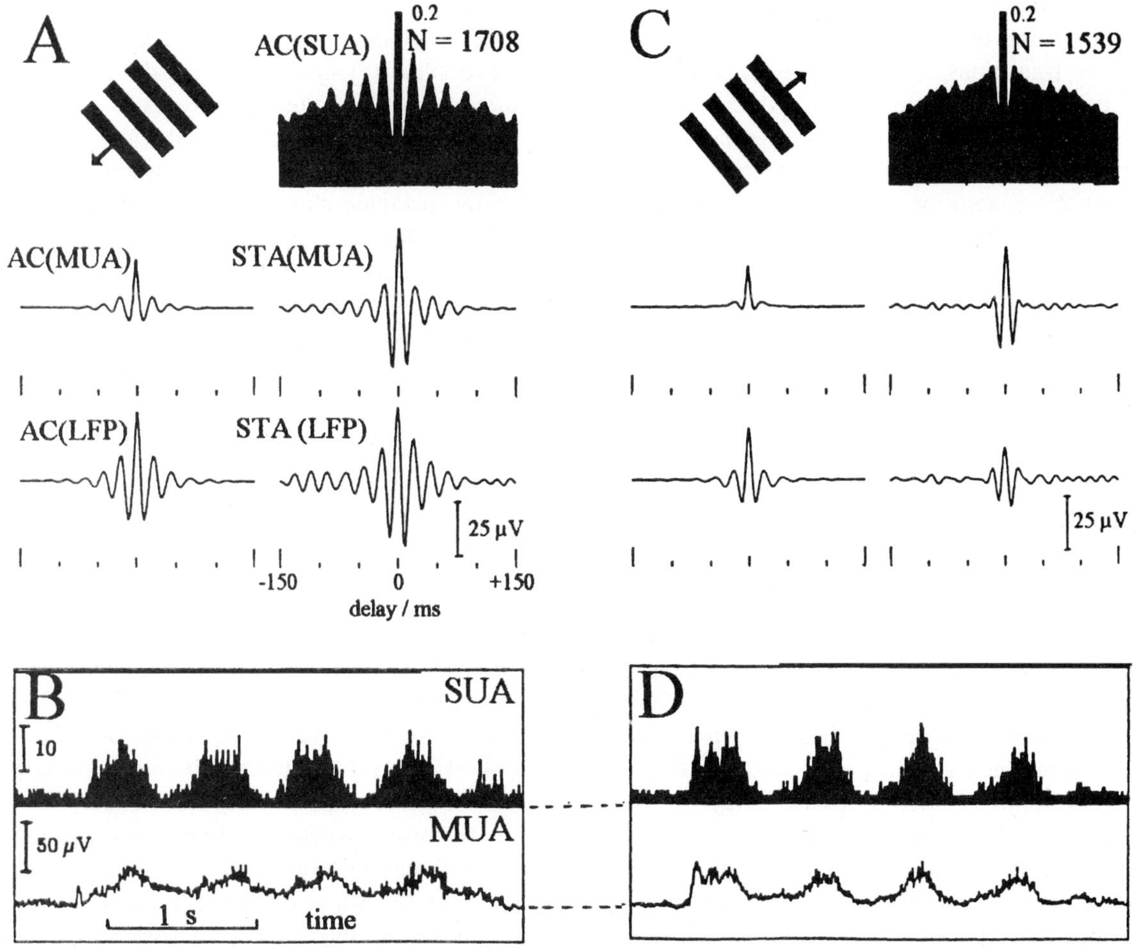

Fig. 14. Transition from rhythmic to lock-in state with reversing stimulus movement direction. Simple cell from area 17, sensitive for stimulus orientation but not for movement direction (defined by the number of spikes per stimulus). Stimulus: square wave grating of 0.7 cycles/°, moving at 2.4 °/sec for 3.5 sec across the receptive fields at preferred orientation. A and C are correlograms as in Fig. 13, both from the same recording position at two reverse stimulus movement directions (indicated by arrows at the grating schema). B and D show average single unit and multiple unit activities in response to 14 identical stimulus repetitions. Note the similar spike activation levels in A and B versus C and D while the simple cell changes its state from rhythmic (A) to lock-in (C). It is also noteworthy that the oscillatory modulation of the multiple unit activities and local field potentials are lower (hence significant) in the lock-in than in the rhythmic state in this example (modified from Eckhorn and Obermueller, 1993).

which some of the action potentials are phase-locked to oscillatory field events.

State transitions of single cell coupling with oscillatory population activities depend on stimulation and receptive field properties. Figure 14 is an example of changing a strong rhythmic state (A) into a lock-in state (B) by reversing the movement direction of a grating stimulus. Such reversals of stimulus movement direction cause often transitions from rhythmic to lock-in states and from lock-in to non-participation states.

Summarizing the influences of RF properties on state transitions, we have found that state transitions in the direction from rhythmic to lock-in to non-participation states often occurred by changing a stimulus from 'optimal' to 'sub-optimal' for the generation of oscillatory population activity. State changes in the reverse direction rarely occurred under these conditions. Here the term 'optimal' versus 'sub-optimal' is used with respect to the stimulus preferences of the neurons (tested were orientation, direction and velocity of a moving stimulus, binocular disparity, and temporal modulation frequency).

Two components of cortical synchronization: rhythmic and non-rhythmic

If internally generated rhythmic synchronization plays a role in perceptual feature-linking relations, one may assume that other types of stimulus related synchronizations probably also play a role. Most important in this respect are synchronizations that are due to sudden retinal image shifts. Such stimuli evoke stimulus-locked synchronizations at short response delays in the visual cortex (Eckhorn et al., 1990a). Rhythmic synchronizations, on the other hand, often occur at longer delays and last longer. It would be of advantage in temporally critical situations for such stimulus-locked responses to be able to 'overwrite' presently ongoing rhythmic synchronizations, in order to obtain an updated view of the visual scene. Such behavior has indeed been seen in the cat's visual cortex (Fig. 15): stimulus-locked cortical synchronizations could inhibit oscillatory activities immediately if the stimulus was strong and fast enough (Eckhorn and Schanze, 1991; Kruse et al., 1991). The occurrence of stimulus-locked response components in the cortex can be explained by the simultaneous firing, due to sudden image changes, of those neurons synchronously that have similar receptive fields and are stimulated simultaneously by similar features of the visual object (for simulations of stimulus-locked synchronizations in laterally coupled neural networks see Pabst et al., 1989).

In order to study interactions between externally driven (stimulus-locked) and internally induced (oscillatory) signals in some detail, we applied stimuli that elicited both types of activities (Kruse et al., 1991). Strong oscillations (40–90 Hz) were preferentially induced in our experiments by sustained slowly moving grating or bar stimuli that did not evoke brisk transient responses in cortical neurons. Externally driven (stimulus-locked) responses occurred in the range 1–40 Hz, and they were evoked in our experiments by (1) sudden stimulus movements (jerks), (2) random movements with variable amplitudes, and (3) light flashes (Fig. 15). Gradual increases in stimulus-locked response amplitudes were paralleled by gradual decreases in the amplitudes and occurrence probability of oscillatory events (Fig. 16): the stronger the stimulus-locked response components the stronger the inhibition of oscillatory events. After periods of inhibition, oscillatory events usually occurred with enhanced amplitudes and probabilities, compared with situations where no stimulus-locked responses interfered with their generation. It can therefore be concluded that strong stimulus-locked responses dominate cortical synchronization. After a certain 'relaxation' time, however, when the evoked stimulus response has partly worn off, a process of self-organization might lead to oscillatory synchronizations. Similar behavior was observed in neural network models (Eckhorn et al., 1990b; Arndt, 1993).

Periods of strong stimulus-locked and oscillatory stimulus-induced synchronizations might characterize extreme processing states of the visual cortex. In natural vision, short stimulus shifts are often followed by phases with more stationary retinal images, e.g. in saccade-fixation sequences or when a visual object suddenly moves and stops again. In both visual situations, the primary stimulus-locked responses that occur simultaneously in many cortical neurons might help to signal relatively crudely, but fast, the 'when?', 'where?' and 'what?' of an ongoing visual event. After sudden shifts of an object, or after saccades, post-inhibitory rebound activations might be useful for the sensitive generation of oscillatory synchronizations in just those parts of the cortical network in which the postsynaptic influence of the linking connections is still present (Eckhorn et al., 1990b).

We conclude that stimulus-locked and oscillatory synchronizations occur in alternation during natural vision if the stimulus-locked responses are strong. However, both components are simultaneously present whenever sustained and moderately transient stimulation drives the neurons. Both types of synchronized activities might, therefore, be capable of contributing to the perceptual integration in proportion to their actual amplitudes.

Feature associations might also be supported by coupled oscillations among different frequencies

High-frequency oscillations in cat and monkey visual cortex were found to occur often together with multiple spectral peaks at different frequen-

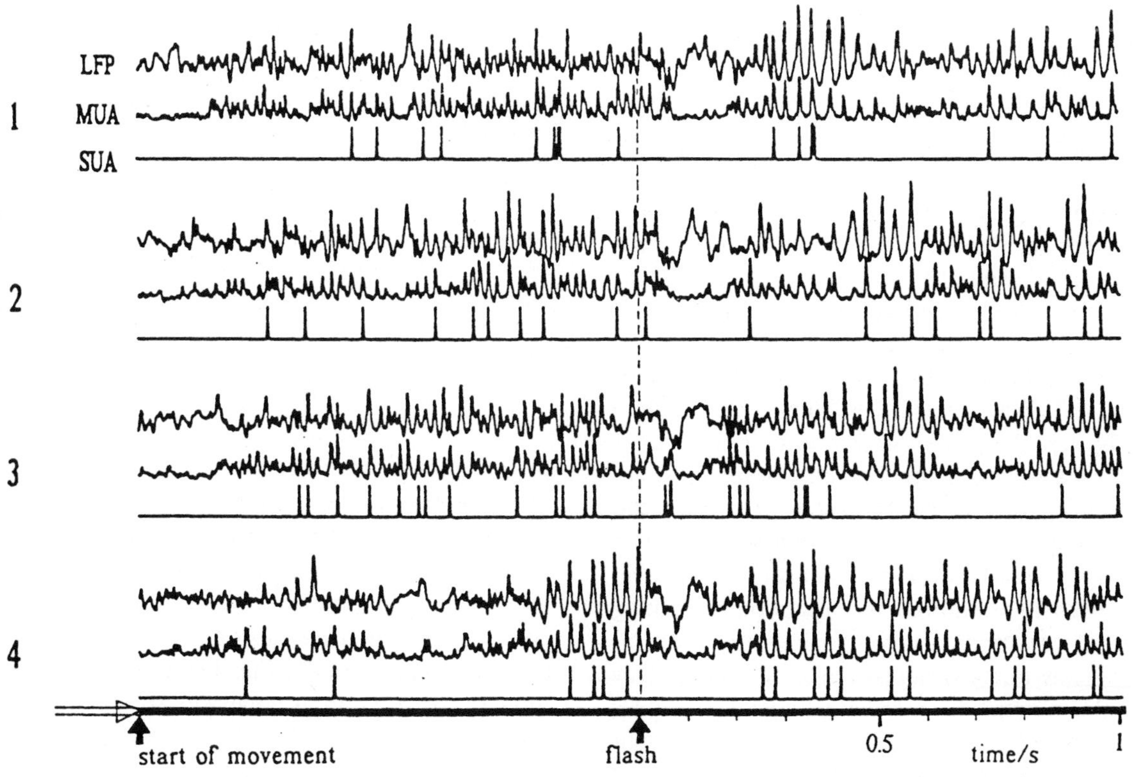

Fig. 15. Rhythmic signals in cat visual cortex induced by a moving stimulus (and inhibited by a flash). Responses to four identical stimulus repetitions of three types of neuronal signals recorded in parallel with the same microelectrode. Stimulus: grating of 0.25 cycles/degree moving at $v = 4.5$ deg/sec. Arrows below mark beginning of stimulus movement (left) and instance of a light flash (middle) that was given while the grating continued to move. Note the 200 msec post flash depressions of LFP high frequency components and of MUA and single unit spike discharges (from Eckhorn and Schanze, 1991).

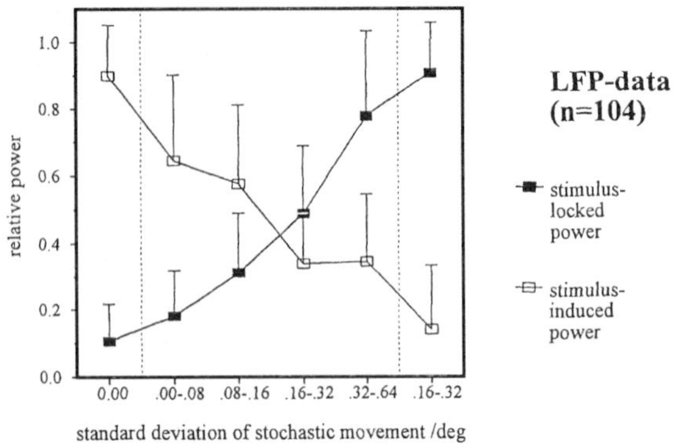

Fig. 16. Stimulus-locked response components gradually inhibit stimulus-induced oscillations. Local field potential recordings from cat visual cortex. Stimulus: grating, moving with a random component superimposed on a constant slow velocity. Open symbols: average normalized response power of the high-frequency oscillatory components; filled symbols: normalized power of the stimulus-locked components. Amplitudes (standard deviation) of the random stimulus movement are indicated at the abscissa. Each value is the average of $N = 104$ single curves (modified from Kruse et al., 1991).

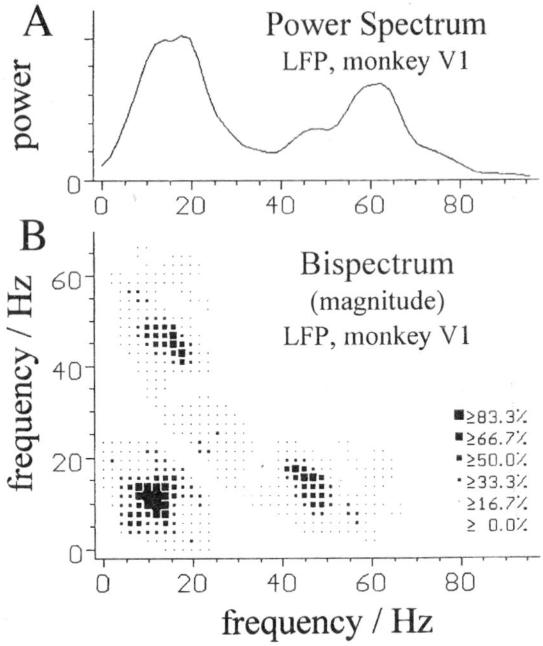

Fig. 17. Phase coupling between oscillations at different (incommensurate) frequencies. (A) Typical power spectrum of a local field potential recording from the visual cortex (V1) of an awake monkey during stimulation of the receptive fields at the recording position. Note the peaks at low and high frequencies. (B) Normalized amplitudes of the respective bispectrum. The black dots indicate frequency pairs that are phase coupled. In this example this is the case within the low frequency band (7–15 Hz) and between low and high frequency components (strong coupling, e.g. around 11 and 46 Hz (= 57 Hz) (modified from Schanze and Eckhorn, 1993).

cies. It is interesting for the hypothesis of 'feature association by synchronization' whether the features to be linked have to do so at the same frequency or whether phase coupling at different frequencies might also be able to support it. Coupling of relatively low-frequency oscillations with those in the high-frequency range are of particular interest because the former have been monitored for many decades in human and animal EEGs, and because low-frequency waves have been attributed to a broad variety of higher brain functions (for example the theta activity of the hippocampus for memory formation: for an overview, see Basar, 1980). Thomas Schanze in my group has followed this question by asking whether the frequency components recorded intracranially in the visual cortex of cat and monkey are phase-coupled, in order to reveal processes that previously were assumed to be independent (e.g. coupling between rhythms in the alpha and gamma frequency ranges of EEG) (Schanze and Eckhorn, 1993). Such phase-coupling among different frequencies cannot be detected in the conventional power spectrum, but the formalism of bi-spectrum is able to extract them if the correlations are due to quadratic phase couplings (Nikias and Raghuveer, 1987). Bi- and power spectra were calculated of multiple-unit activities (MUA) and local field potentials (LFP) for the visual cortex of anesthetized cats (A17, 18 and 19), and an awake monkey (V1; Fig. 17). For the cat and monkey cortical areas analyzed so far, the bispectra often were stimulus dependent and, in about 60% of the cases, revealed significant quadratic phase couplings of oscillations between two or several pairs of frequencies: (1) within the gamma-frequency band (30–90 Hz), (2) between gamma and low frequency (1–30 Hz) rhythms, and (3) among rhythms at lower frequencies (1–30 Hz).

We conclude that phase-coupling of oscillatory events may not be restricted to high frequency (35–90 Hz) events at a common frequency. It suggested that rhythmic processes at different time scales can be coupled in the visual system and, according to the 'association hypothesis', are thus candidate signals for the support of multimodal feature associations in sensory and association cortices.

Models and concepts of feature associations

Model networks with feeding and linking synapses

Based on the observed cortical synchronization effects the 'Marburg Biophysics Group' simulated such effects using neural network models (e.g. Eckhorn et al., 1990b). The model neuron consists of leaky integrators at its synapses, and a dynamic threshold mechanism with negative feedback (Fig. 18B). Despite these common features it has an important property for models of the visual system. Rapid synchronization of 'spike' activities is ensured in groups of model neurons that are coupled via 'linking' inputs, because linking $L(t)$ inputs act modulatory on the feeding $F(t)$ inputs: $F(t)L(t) + 1$. While the feeding input signals carry the local image coding properties of the model network (corresponding to receptive field properties of real visual neurons) the linking signals mediate the phases of the model neurons' activities. Independence in the local coding properties of coupled model neurons is guaranteed by the used amplitude modulation (additively acting coupling connections would, instead, lead to a superposition of the local coding properties, i.e. they would be 'smeared out') (Eckhorn et al., 1990b).

The neural networks used in the present simulations consist of either one or two layers of 'visual neurons' (Figs. 19–21). A typical dynamic response of a network is shown in the example of a single layer model (Fig. 19). A patch of increased intensity moves with constant velocity across the inputs of the single layer. It is obvious that, in such dynamic spatio-temporal stimulus situations, neurons at the region boundaries must either join or leave the synchronized assembly. The relatively constant extent of the synchronized assembly following the moving stimulus is clearly visible, although a high level of stochastic background activity was applied to the inputs.

In the simulation for Fig. 20, a second layer with feedback onto layer 1 is included. Each layer-2 neuron projects back to linking inputs of four layer-1 neurons. Thus, the two activated layer-1 regions are transiently linked by the 'higher visual map' according to their broader 'receptive fields'. Neurons in the two 'stimulated regions' in layer-1 (a and b; Fig. 20B), although separated by a gap of spontaneously active neurons, respond with phase-locked rhythmic burst discharges. The rhythmicities are quantitatively characterized by the autocorrelation functions of single neurons in the activated regions a and b in the network, while their synchronization is characterized by the a–b cross-correlation (Fig. 20C). Without layer 2-to-1 feedback the neurons in the two 'stimulated' patches a and b (Fig. 20B) would generate synchronized oscillatations within each patch that are independent of each other in their phases.

Synchronized activities like those in the model of Fig. 20 might be interpreted as serving the definition of spatial and temporal continuity in an object's 'neural represen- tation' throughout the synchronization period.

Spatial correlation contrast is improved by divergent negative feedback from higher to lower neural layer

The simulations with the above described neural network models were often dominated by two problems: (1) the difficulty of finding the correct values of the model parameters necessary for the expected effects because they had to be chosen within a small range, and (2) the low signal-to-noise ratio between the signals coding the 'figure' and those representing the 'ground' (measured as a spatial 'correlation contrast'). These unfavorable properties of the simulated neural networks are surely not present in real neural networks, so

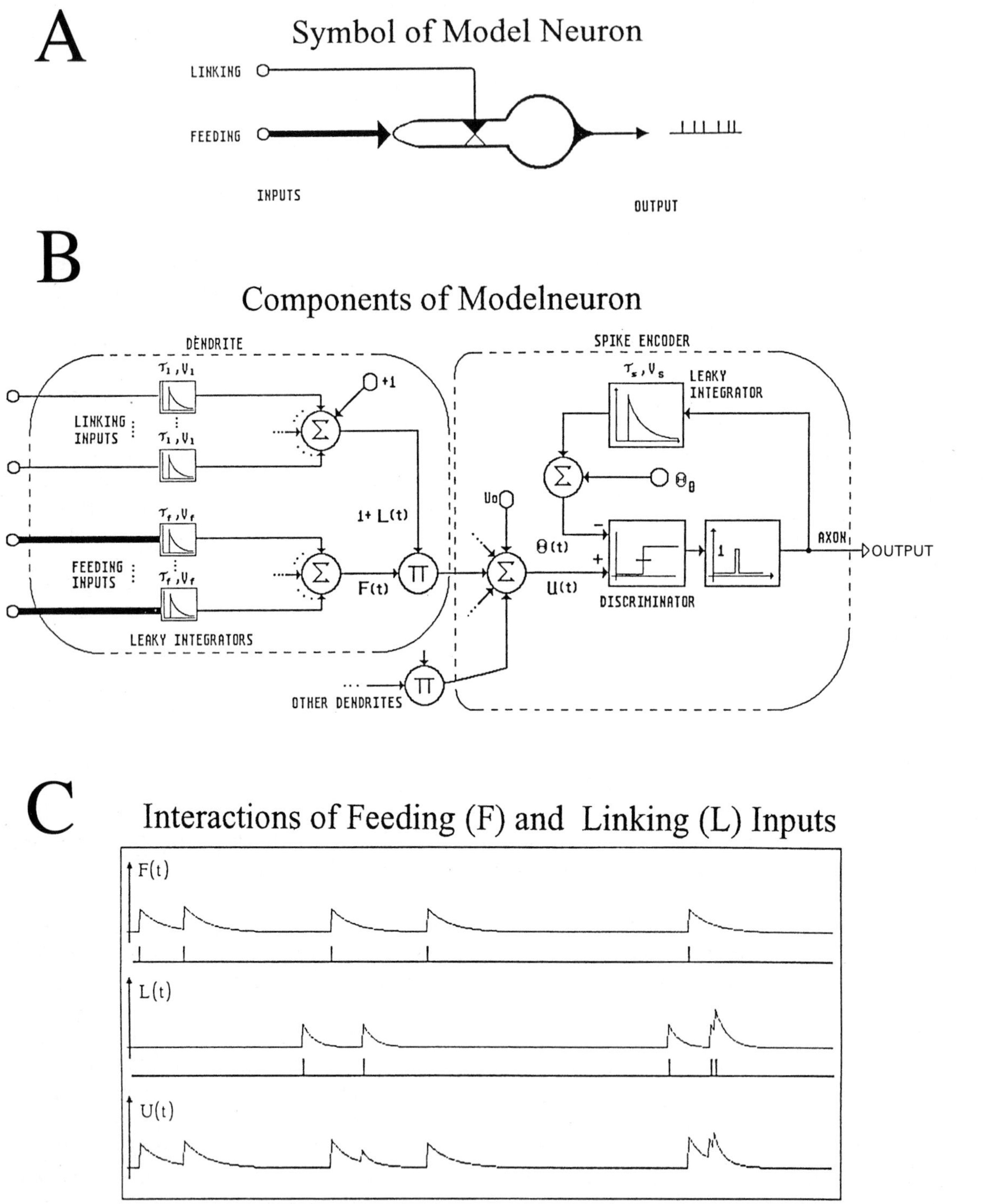

Fig. 18. Model neuron with dynamic synapses, two types of inputs (feeding and linking) and a dynamic spike encoder (modified from Eckhorn et al., 1990).

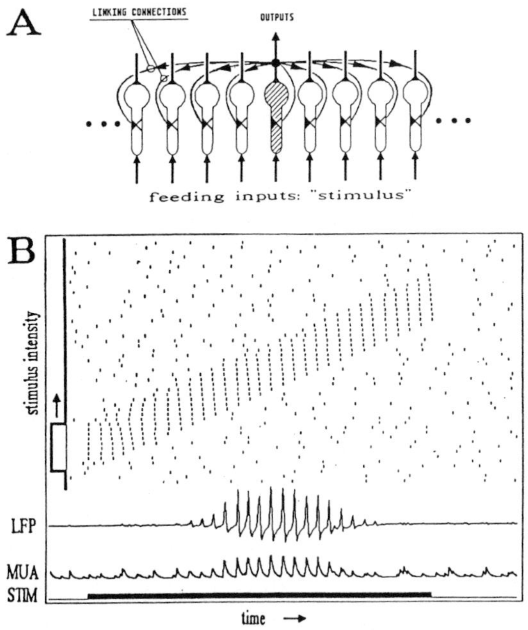

Fig. 19. Single layer linking network can code spatio-temporal continuity by synchronized oscillations. Stimulus: enhanced feeding input at 10 neurones moves at constant velocity. The black horizontal bar (B) indicates the on-duration of the stimulus. (A) One-dimensional layer of 50 model neurons, linking connections are drawn for a single neuron only. Thick lines: feeding connections; thin lines: linking connections. (B) Spikes at the outputs of the neurons are marked by a dash at the times of their occurrence. LFP and MUA are the respective spatially averaged dendritic and output signals from the model network (modified from Eckhorn et al., 1989).

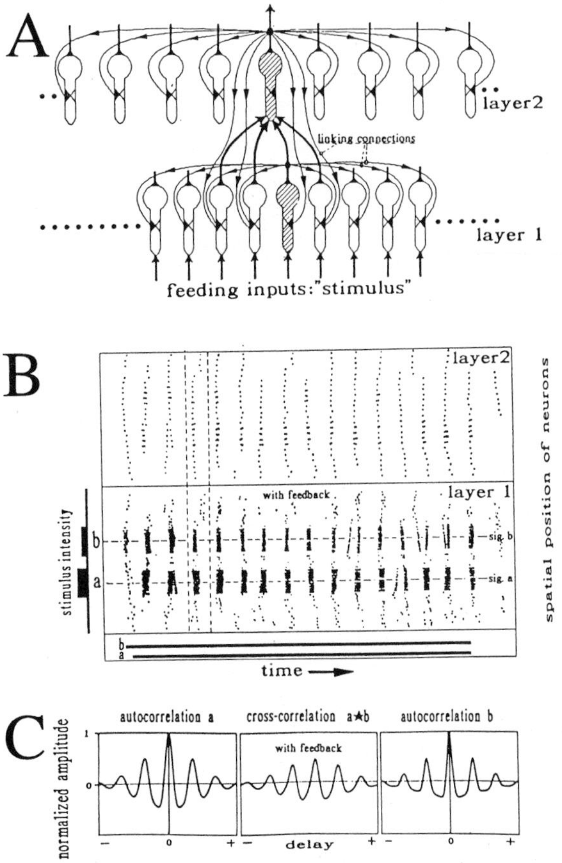

Fig. 20. Feedback associative (linking) connections from a higher (2) layer of model neurones synchronize spatially separated oscillatory groups in the lower (1) layer. Both layers are identically organized (as in Fig. 19), i.e. they have linking connections to the four nearest neighbors (modified from Eckhorn et al., 1990).

that an important and general property of real networks needed to be included: localized inhibitory feedback connections. These were implemented in the model by an activity control (inhibitory feeding connections) from the higher to the lower layer (shaded area in Fig. 21A) with a spatial divergence factor that was three times broader than the divergence of the feedback linking connections. Figure 21B shows the result of a simulation in which the strength of the feedback activity control was raised from zero to high values. The control function acted from every layer-2 neuron in a broad but topographically confined range on layer-1 neurons. This feedback is primarily active in layer-1 regions, where neurons are synchronized, because the coherence leads to high levels of activity in layer-2 neurons and, thus, reduces the activities in the range of the feedback inhibition. This inhibition, however, mainly affects the spontaneous activity of layer-1 neurons surrounding the synchronized population. The synchronized spikes in the region of stimulation are less inhibited because they facilitate their activities mutually, via the lateral linking connections.

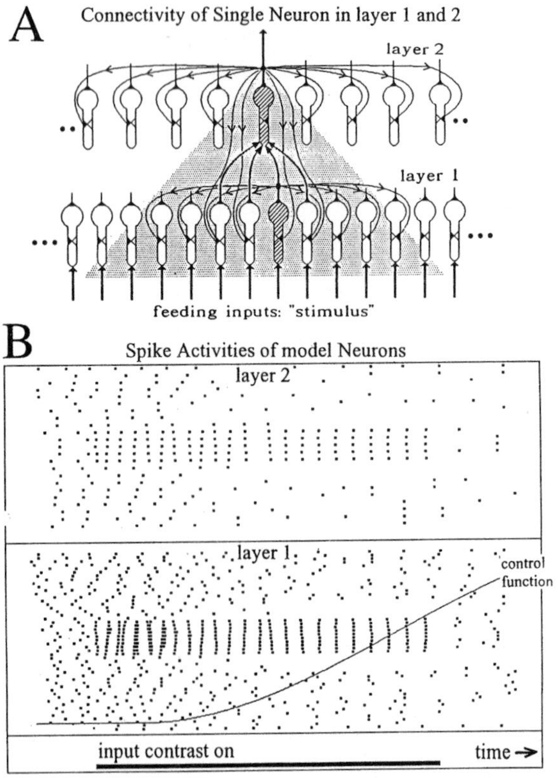

Fig. 21. Two-layer model with feedback and top-down activity control. An additional negative feeback control keeps the two layer network (same as in Fig. 20) in an appropriate working range for high contrast region definitions. (A) Network used for the simulation in B. The negative feedback is acting from each layer-2 neuron on feeding inputs of neurons in the shaded range of layer-1. These inhibitory inputs to layer-1 neurons have relatively long time constants (= 30 msec), and their coupling strengths are spatially distributed according to a Gaussian. (B) Simulation results. The line superimposed on the layer 1 spike activities (control function) indicates the strength of the negative layer 1-to-2 feedback, increasing from left to right (modified from Eckhorn et al., 1992).

It can therefore be concluded that top-down activity control via topographically arranged (diverging) projections increases the spatial 'correlation contrast', thereby improving the separability of a figure from the background. The same feedback inhibition also makes the synchronization effects more robust to changes in parameter values.

Visual pattern recognition by neural network models with linking connections

The Marburg group has used the model neuron, with its special linking connections, to model several interesting aspects of visual function. One example is feature association in a visuotopically arranged layer by a dynamic associative memory utilizing the same type of dynamic neurons (Arndt et al., 1991). The associative memory is already capable of supporting synchronized oscillations in the visuotopic layer if it is partly activated by a visual object which had been previously 'stored' in the associative memory. Both parts of the network that represent the actual 'visual object' then engage in a common synchronized oscillatory state.

Another example is a model that could 'recognize' illusory contours (Stoecker et al., 1991). For example , two layers of oriented line detectors and end-stop cells in a linking network topography suffice to represent, by synchronized oscillations in the respective line detectors, the illusorily perceived sides of a Caniza triangle, though physically not present.

Visual association fields of local cortical populations

In order to bring mechanisms of synchronization among neurons representing a visual object into correspondence with perceptual capabilities of feature associations, the concept of the 'linking or association field' of a local neural assembly was introduced (Eckhorn et al., 1990b). The linking field extended the concept of single-cell receptive fields (local coding) to neural ensemble coding. The association field of a local assembly of visual neurons is defined as that area in visual space where appropriate local stimulus features induce synchronization of activities within that assembly. Association fields are constituted both by the RF properties of the respective assemblies and by the properties of their linking interconnections (feed-forward, lateral, feedback). This implies that the association field of a given group of visual neurons is generally much broader than the super-imposed receptive fields of the neurons that constitute the 'association field assembly'.

Summary and conclusions

It was postulated that the perceived association of visual features is based on the synchronization of those neural signals that are activated by a coherent visual object. Two types of synchronized cortical signals were found by us in cat and monkey visual cortex, and were proposed as candidates for feature association: (1) stimulus-locked signals, evoked by transient retinal stimulation, and typically non-rhythmic; (2) oscillatory signals, induced by sustained stimuli, and typically not locked in their oscillation phases to stimulus events. Both types of signals can occur synchronously in those neurons that are activated by a common stimulus. Synchronized activities were found in paired recordings within vertical cortex columns, in separate columns of the same cortical area, and even between different cortical areas or hemispheres. The average phase difference between such common oscillatory events was typically close to zero ($<$ 1 msec mean $\pm$ 2 msec S.D.). For the dependence of synchronization from stimulus and receptive field properties, a preliminary 'rule' can be given: the coherence of fast oscillations in separate cortical assemblies depends inversely on the 'coding distance' between the assemblies' RF properties, but directly on the degree of overlap between the assemblies' respective coding properies and the features of a common stimulus. This means that oscillatory events in any two assemblies, in the same or in different cortical areas or hemispheres, are more closely correlated the more similar are their receptive field properties, and the better a common stimulus activates the assemblies simultaneously. Our results can explain some neural mechanisms of perceptual feature-linking, including mutual enhancement among similar, spatially and temporally dispersed features, definitions of spatial and temporal continuity, scene segmentation, and figure-ground discrimination. We further propose that mutual enhancement and synchronization of cell activities are general principles of temporal coding by assemblies, that are also used within and among other sensory modalities as well as between cortical sensory and motor systems.

Acknowledgements

I am thankful to Professor H.J. Reitboeck, the head of our Marburg Biophysics Department, for his support in developing both concepts and the multiple-electrode instrumentation. This article could not have been written without the numerous discussions with my colleagues and without their extensive help in experiments, data acquisition and processing, as credited in the text and figure captions. Expert help in care and preparation of our cats and monkeys and help in experimental techniques came from U. Thomas, J.H. Wagner and W. Lenz. The financial support by the Deutsche Forschungsgemeinschaft is also greatly acknowledged (Ec 53/4, Ec 53/6 and Ec 53/7).

References

Arndt, M. (1993) Representation of spatial and temporal continuity by synchronization of neural signals (in German). Doctoral Thesis, Philipps-University, Marburg, Germany.

Arndt, M., Dicke, P., Erb, M. and Reitboeck, H.J. (1991) A multilayered, physiology-oriented neural network model that combines feature-linking with associative memory. In: N. Elsner and H. Penzlin (Eds.), *Synapse, Transmission, Modulation,* Thieme Verlag, Stuttgart, New York, 587 pp.

Basar, E. (1980) *EEG-Brain Dynamics,* Elsevier North-Holland Biomedical Press, Amsterdam, New York, Oxford.

Brosch, M., Bauer, R., Schanze, T. and Eckhorn, R. (1991) Stimulus-induced oscillatory events and their spatial correlation profiles in cat visual cortex. *Eur. J. Neurosci.,* Suppl. 4: 54,1235.

Eckhorn, R. (1991) Stimulus-evoked synchronizations in the visual cortex: Linking of local features into global figures? In: J. Krüger (Ed.), *Neural Cooperativity,* Springer Series in Synergetics, Springer-Verlag, Berlin, Heidelberg, New York, pp. 184–224.

Eckhorn, R. (1992) Principles of global visual processing of local features can be investigated with parallel single-cell- and group-recordings from the visual cortex. In: A. Aertsen and V. Braitenberg (Eds.), *Information Processing in the*

Cortex, Springer-Verlag, Berlin, Heidelberg, New York, pp. 385–420.

Eckhorn, R. and Obermueller, A. (1993) Single neurons are differently involved in stimulus-specific oscillations in cat visual cortex. *Exp. Brain Res.,* in press.

Eckhorn, R. and Schanze, T. (1991) Possible neural mechanisms of feature linking in the visual system: stimulus-locked and stimulus-induced synchronizations. In: A. Babloyantz (Ed.), *Self-Organization, Emerging Properties and Learning,* Plenum Press, New York, pp. 63–80.

Eckhorn, R., Bauer, R., Jordan, W., Brosch, M., Kruse, W., Munk, M. and Reitboeck, H.J. (1988) Coherent oscillations: a mechanism of feature linking in the visual cortex? Multiple electrode and correlation analysis in the cat. *Biol. Cybern.,* 60: 121–130.

Eckhorn, R., Reitboeck, H.J., Arndt, M. and Dicke, P. (1989) A neural network for feature linking via synchronous activity: results from cat visual cortex and from simulations. In: R.M.J. Cotterill (Ed.), *Models of Brain Function,* Cambridge University Press, New York, pp. 255–272.

Eckhorn, R., Brosch, M., Salem, W. and Bauer, R. (1990a) Cooperativity between cat area 17 and 18 revealed with signal correlations and HRP. In: N. Elsner and G. Roth (Eds.), *Brain and Perception,* Thieme, Stuttgart, New York, p. 237.

Eckhorn, R., Reitboeck, H.J., Arndt, M. and Dicke, P. (1990b) Feature linking among distributed assemblies: Simulations and results from cat visual cortex. *Neural Computat.,* 2: 293–306.

Eckhorn, R., Schanze, T., Brosch, M., Salem, W. and Bauer, R. (1992a) Stimulus-specific synchronizations in cat visual cortex: multiple microelectrode and correlation studies from several cortical areas. In: E. Basar and T.H. Bullock (Eds.), *Induced Rhythms in the Brain,* Brain Dynamics Series, Birkhäuser, Boston, Basel, Berlin, pp. 47–82.

Eckhorn, R., Dicke, P., Arndt, M. and Reitboeck, H.J. (1992b) Feature linking of visual features by stimulus-related synchronizations of model neurons. In: E. Basar and T.H. Bullock (Eds.), *Induced Rhythms in the Brain,* Brain Dynamics Series, Birkhäuser, Boston, Basel, Berlin, pp. 397–416.

Eckhorn, R., Frien, A., Bauer, R., Kehr, H. and Woelbern, T. (1993a) Phase-locked high frequency oscillations between visual cortical areas V1 and V2 of an awake monkey. In: N. Elsner and M. Heisenberg (Eds.), *Gene, Brain, Behaviour,* Thieme Verlag, Stuttgart, New York, p. 444.

Eckhorn, R., Frien, A., Bauer, R., Woelbern, T. and Kehr, H. (1993b) High frequency (60–90 Hz) oscillations in primary visual cortex of an awake monkey. *NeuroReport,* 4: 243–246.

Engel, A.K., König, P., Gray, C.M. and Singer, W. (1990) Stimulus-dependent neuronal oscillations in cat visual cortex: inter-columnar interaction as determined by cross-correlation analysis., *Eur. J. Neurosci.,* 2: 588–606.

Engel, A.K., König, P., Kreiter, A.K. and Singer, W. (1991a) Interhemispheric synchronization of oscillatory neuronal responses in cat visual cortex. *Science,* 252: 1177–1179.

Engel, A.K., Kreiter, A.K., König, P. and Singer, W. (1991b) Synchronization of oscillatory neural responses between striate and extrastriate visual cortical areas of the cat. *Proc. Natl. Acad. Sci. USA,* 88: 6048–6052.

Field, D.J., Hayes, A. and Hess, R.F. (1993) Contour integration by the human visual system: evidence for a local 'association field'. *Vision Res.,* 33: 173–193.

Freeman, W. and van Dijk, B. W. (1987) Spatial patterns of visual cortical fast EEG during conditioned reflex in a rhesus monkey. *Brain Res.,* 422: 267–276.

Gray, C.M. and Singer, W. (1989) Stimulus-specific neuronal oscillations in orientation columns of cat visual cortex. *Proc. Natl. Acad. Sci. USA,* 86: 1698–1702.

Gray, C.M., König, P., Engel, A.K. and Singer, W. (1989). Oscillatory responses in cat visual cortex exhibit inter-columnar synchronization which reflects global stimulus properties. Nature, 338: 334–337.

Gray, C.M., Engel, A.K., König, P. and Singer, W. (1990) Stimulus-dependent neuronal oscillations in cat visual cortex: receptive field properties and feature dependence. *Eur. J. Neurosci.,* 2: 607–619.

Kreiter, A.K. and Singer, W. (1992) *Eur. J. Neurosci.,* 4: 369–375.

Kruse, W., Eckhorn, R. and Schanze, T. (1991) Two modes of processing and their interactions in cat visual cortex: stimulus-induced oscillatory and stimulus-locked synchronizations. In: N. Elsner and H. Penzlin (Eds.), *Synapse, Transmission, Modulation,* Thieme Verlag, Stuttgart, New York, p. 217.

Mitzdorf, U. (1987) Properties of the evoked potential generators: current source-density analysis of visually evoked potentials in the cat cortex. *Int. J. Neurosci.,* 33: 33–59.

Nikias, C.L. and Raghuveer, M.R. (1987) Bispectrum estimation: a digital signal processing framework. *Proc. IEEE,* 75: 869–891.

Pabst, M., Reitboeck, H.J. and Eckhorn, R. (1989) A model of pre-attentive region definition in visual patterns. In: R.M.J. Cotterill (Ed.), *Models of Brain Function,* Cambridge University Press, pp. 137–150.

Reitboeck, H.J. (1983) *IEE Syst. Man. Cybernet.,* 13: 676–682.

Schanze, T. and Eckhorn, R. (1993) Phase coupling between oscillations at different frequencies in the visual cortex of cat and monkey. In: N. Elsner and M.H. Heisenberg (Eds.), *Gene, Brain, Behaviour,* Thieme Verlag, Stuttgart, New York, p. 442.

Stoecker, M., Eckhorn, R. and Reitboeck, H.J. (1991) Oscillatory synchronizations in neural networks: responses to stimuli that induce the perception of subjective contours in humans. In: N. Elsner and H. Penzlin (Eds.) *Synapse, Transmission, Modulation,* Thieme Verlag, Stuttgart, New York, p. 216.

J. van Pelt, M.A. Corner H.B.M. Uylings and F.H. Lopes da Silva (Eds.)
Progress in Brain Research, Vol 102

CHAPTER 29

Modelling the cerebellar Purkinje cell: experiments in computo

Erik de Schutter*

Division of Biology, California Institute of Technology, Pasadena, CA 91125, U.S.A.

Introduction

The cerebellar Purkinje cell is among the largest and most complex neuron in the mammalian brain. The 150,000 to 175,000 granule cell inputs received by each Purkinje cell (Harvey and Napper, 1991) constitute the most massive synaptic convergence found on any neuron in the brain (Shepherd, 1990). Purkinje cells are also distinguished by high densities of Ca^{2+} channels on the dendrite and by a complex apparatus controlling cytoplasmic Ca^{2+} concentrations, e.g. with cytoplasmic Ca^{2+} stores possessing IP_3 and ryanodine receptors (Brorson et al., 1991), metabotropic receptors (Llano et al., 1991; Staub et al., 1992) and Ca^{2+} inflow through the Ca^{2+} channels (Hockberger et al., 1989; Lev-Ram et al., 1992). Numerous physiological (Llinás and Sugimori, 1992) and biochemical studies (Ito, 1984) have provided the neuroscience community with a wealth of details on the firing properties of Purkinje cells in vitro and on the identity and kinetics of synaptic and ionic channels, but an integrated view of how all these components interact and determine Purkinje cell responses in vivo is lacking. Such understanding is necessary because, as the only output neuron of the cerebellar cortex, it is essential to comprehend Purkinje cell function to understand the computations performed by the cerebellum.

Detailed computer models of neurons have become important tools for investigating how dendritic morphology and membrane biophysics interact in a complex neuron (Segev, 1992). These models allow one to fit all pieces of the puzzle together and see how they interact both locally and globally over time. The Purkinje cell has been the subject of several modelling efforts, but most of these models have explored only its passive electrical properties (Llinás and Nicholson, 1976; Shelton, 1985; Rapp et al., 1992). The few Purkinje cell models which have included active voltage dependent conductances have not studied the effects of these active properties on synaptic integration (Pellionisz and Llinás, 1977; Bush and Sejnowski, 1991).

In this short review I will summarize how a detailed Purkinje cell model was built, emphasizing why strict procedures can produce an accurate model, and describe some results based on in computo experiments with this model. For a more complete description of the model and other simulation results the reader should consult a number of recent papers (De Schutter and Bower, 1994a,b,c).

*Present address: Born Bunge Foundation, University of Antwerp, B2610 Antwerp, Belgium.

Methodology

How to build a realistic neuronal model

Compartmental modelling is a perfect tool for experimental neurobiologists as it involves little real mathematical work. Because all the equations describing the electrical (and sometimes also biochemical) events in the cell are discretized in space and time, the same set of equations can be used for any neuron (Rall, 1989). Several sophisticated software packages are available to simulate such compartmental models (De Schutter, 1992). The model presented in this paper was implemented with the GENESIS software (Wilson et al., 1989).

Because the equations used for compartmental modelling are fixed, building a model is reduced to the search for an appropriate set of parameter values. In practice it is impossible to measure all the parameters necessary for an active membrane model in a single neuron. Such a large range of ionic channels has been identified in neurons (Llinás, 1988) that even large experimental groups can no longer collect all the data by themselves (see for example McCormick and Huguenard, 1992). One usually needs to combine experimental data from a variety of sources, often from the same type of neuron in different experimental animals and from neurons in different brain regions in the same animal (De Schutter and Bower, 1994a). It is generally possible to obtain detailed morphological data although, apart from one notable exception (Weitzman et al., 1992), compartmental models were based on light microscopy reconstructions which did not include spines and other small structures. However, errors introduced by incomplete morphological reconstructions are a small problem in comparison with finding complete voltage clamp data on all the ionic channels present in a particular neuron. No model has yet been published where the data were complete enough to give an accurate representation of all the membrane channels. In all cases some of the equations describing channels were either highly simplified (usually for Ca^{2+}-activated channels, e.g. Yamada et al., 1989) or obtained from quite different preparations (Lytton and Sejnowski, 1991; Traub et al., 1991; McCormick and Huguenard, 1992). However, as both better experimental techniques (usually based on patch clamping) and the motivation to improve existing models continue to expand our knowledge of ionic and synaptic channels, this problem may soon be solved for several neuron types.

Finally, the density of the various ionic and synaptic channels is in almost all cases unknown. The measured maximal conductances are quite variable in experimental preparations (McCormick and Huguenard, 1992) and are influenced by a multitude of experimental manipulations (such as high ionic concentrations, isolation or culture of cells). In practice, voltage clamp experiments and single channel recordings mainly provide data about the presence and kinetics of particular channels, but the density of these channels remains a free parameter in the model. Calcium-imaging data have also been used to investigate the detailed distribution of channels, but, as many other factors can also influence these measurements, they are probably not conclusive (De Schutter and Bower, 1994a).

To summarize, in order to build a compartmental model one has to collect experimental data both on cellular morphology and on the distribution and kinetics of membrane channels. The channel density can then be determined by a fitting process (Bhalla and Bower, 1993), where the densities that cause the model to reproduce the normal firing properties of the neuron are assumed to be close to the real values. Most models also require computation of Ca^{2+} concentration, mainly to simulate inactivation of Ca^{2+} channels and activation of Ca^{2+}-activated K^+ channels. In a few cases, Ca^{2+} concentration has been modelled as accurately as possible (Yamada et al., 1989; Sala and Hernandez-Cruz, 1990), but usually highly simplified one-shell models are used (Traub, 1992).

The Purkinje cell model

Although the model has been described in great detail in the literature (De Schutter and Bower,

1994a,b), it seems useful in the context of this review to recapitulate its main features. Figure 1 gives an overview of all the components of the model; it also demonstrates how one can achieve a high degree of morphological and electrophysiological complexity by using compartmental models. The detailed dendritic geometry of the cell (based on morphological data provided by Rapp et al., (1992)) was replicated by 1600 electrically distinct compartments (Rall, 1962, 1989). Each compartment corresponds to one of the three electrical circuits shown in Fig. 1. These circuits differ with respect to the ionic channels present. As suggested by experimental data (Llinás and Sugimori, 1992), channel distributions in the model are not uniform but, rather, are distributed with the same density in each of three domains (the soma, the main dendrite and the rest of the dendrite, including spiny and smooth dendrites). Additionally, it was necessary to compute changes in Ca^{2+} concentration caused by inflow through Ca^{2+} channels. In the present model Ca^{2+} concentrations were computed only in a single submembrane shell with exponential decay (Traub, 1982).

Ten different types of voltage dependent channels previously shown to be present in Purkinje cells were modelled, 8021 channels in total. Channel kinetics were simulated using Hodgkin-Huxley-like (Hodgkin and Huxley, 1952) equations based on Purkinje cell specific voltage clamp data or, when necessary, on data from other vertebrate neurons. The soma possessed fast and persistent Na^+ channels (Gähwiler and Llano, 1989; French et al., 1990; Kay et al., 1990), low threshold (T-type) Ca^{2+} channels (Kaneda et al., 1990), a delayed rectifier, an A-current, non-inactivating K^+ channels (Hirano and Hagiwari, 1989; Gruol et al., 1991) and an anomalous rectifier (Crepel and Penit-Soria, 1986). The dendritic membrane included P-type and T-type Ca^{2+} channels (Regan, 1991), two different Ca^{2+}-activated K^+ channels (BK and K2) (Gruol et al., 1991) and a non-inactivating K^+ channel. The P-type Ca^{2+} channel is a high-threshold, very slowly inactivating channel, first described in the Purkinje cell (Llinás et al., 1989). In the model, the P channel constituted about 90% of the total Ca^{2+} conductance (Mintz et al., 1992). While there is some experimental evidence that Ca^{2+} release from internal Ca^{2+} stores plays a role in Purkinje cell responsiveness (Llano et al., 1991), such effects have not been incorporated into the model.

Synaptic inputs

To explore the effects of synaptic activation, synaptic channels were added to the Purkinje cell model without changing any of the other parameters. Granule cell excitatory synaptic inputs were modelled as a 0.7 nS AMPA type conductance (Garthwaite and Beaumont, 1989; Farrant and Cull-Candy, 1991), applied to one passive spine (Harris and Stevens, 1988), located on each spiny branch compartment. Spines were modelled only when parallel fiber inputs were supplied to the model; the membrane surface representing the other spines was collapsed into the membrane (Holmes and Woody, 1989; Rapp et al., 1992) of spiny compartments. Usually 1474 spines were modelled, representing approximately 1% of the number of spines found on real Purkinje cells (Harvey and Napper, 1991). For random asynchronous inputs the missing spines can be compensated for by increasing the firing rate of each synapse (Rapp et al., 1992; De Schutter and Bower, 1994a). An asynchronous firing rate of 10 Hz in the model would thus correspond to an average firing rate of about 0.1 Hz for real parallel fibers. All synaptic input firing rates mentioned are unscaled.

Climbing fiber inputs were also modelled as AMPA conductances (Knöpfel et al., 1990), distributed over the main and smooth dendrite (Palay and Chan-Palay, 1974). These synapses were fired in a volley, starting on the proximal dendrite and proceeding distally.

Both basket cell and stellate cell inhibition is mediated by $GABA_A$-receptors (Ito, 1984; Vincent et al., 1992). The kinetics for the synaptic conductance were based on recordings in pyramidal neurons of the hippocampus (Ropert et al., 1990). Basket cell synapses were placed on the

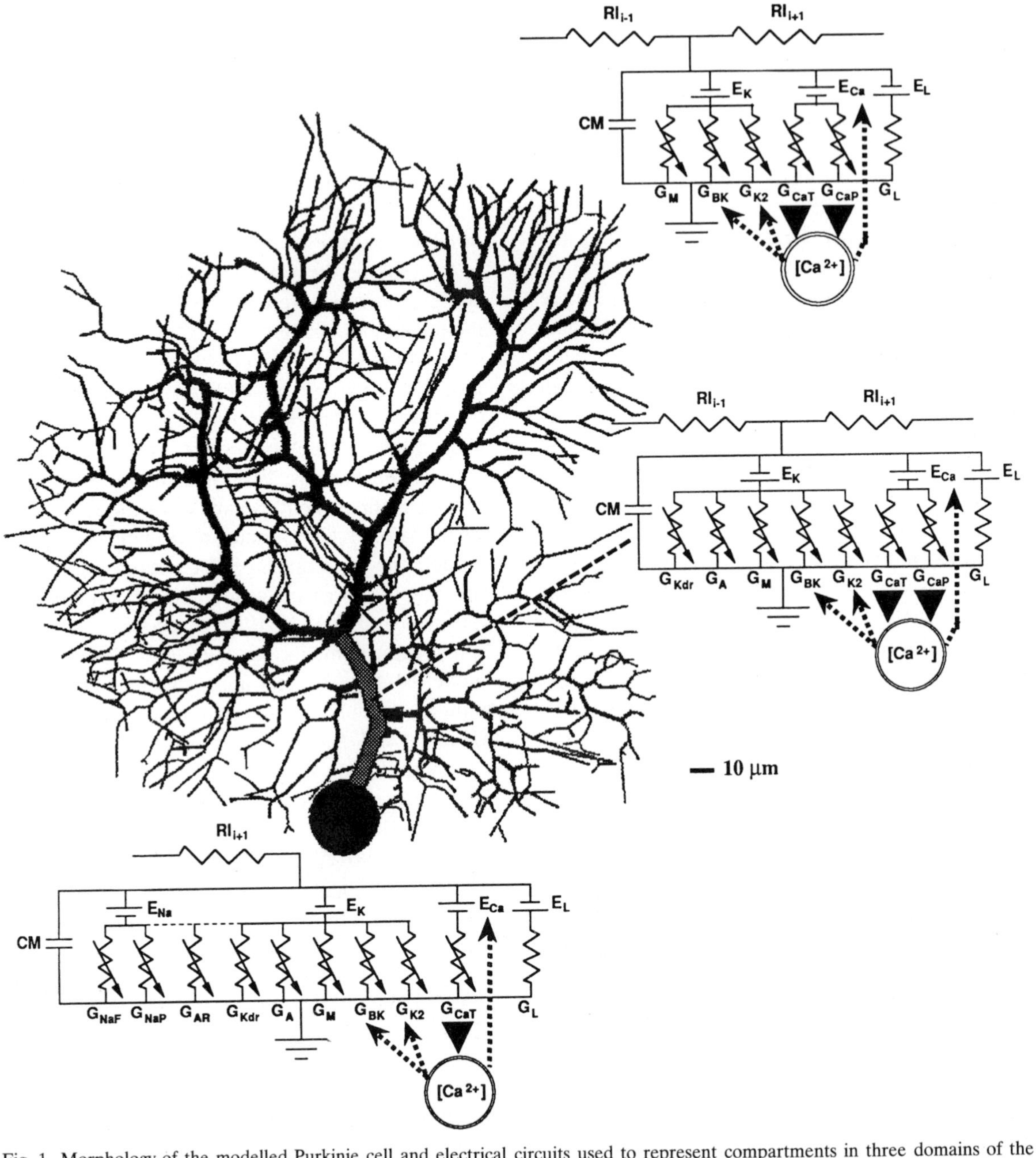

Fig. 1. Morphology of the modelled Purkinje cell and electrical circuits used to represent compartments in three domains of the model. The soma is modelled by the lower circuit, the main dendrite (colored gray) by the middle one and the rest of the dendrite by the upper one. Each circuit contains a membrane capacitance (CM), a leak conductance (GL), and a cytoplasmic resistance (RI) linking the compartment to other compartments. The number and types of ionic channels is different for each of the circuit types. Each channel is represented as a battery (the reversal potential) and a variable resistor (the voltage dependent conductance). The Ca^{2+} concentration, computed in a single submembrane shell, and its influence on the Ca^{2+} Nernst potential and on the Ca^{2+}-activation of K^+ channels is also represented.

soma and main dendrite and were fired synchronously. Stellate cell synapses were placed on all the smooth and spiny dendritic compartments, and were activated asynchronously following a Poisson distribution.

Simulation results

Simulation of current injections

Since the morphology of the cell and the kinetics of the ten different ionic channels were determined by experimental data, the only remaining free parameters in the model were the channel densities and Ca^{2+}-removal kinetics. These parameters were adjusted until the model replicated published in vitro physiological responses to current injections in the soma and dendrites (Llinás and Sugimori, 1980a,b; Hounsgaard and Midtgaard, 1988). The model shows the typical Purkinje cell firing of somatic sodium spikes during low amplitude current injection (Fig. 2) (0.5 nA), while dendritic calcium spikes appear with higher intensity currents (2.0 nA). Note also the delay in onset of firing during the 0.5 nA current injection (Fig. 2).

These simulations are described in greater detail in De Schutter and Bower (1994a), which also examines the contribution of different ionic channels to these firing patterns. Such an analysis is important, because the mere presence of channels does not establish that they participate in specific cell properties. For example, while it is generally accepted that the P channel is responsible for the fast dendritic Ca^{2+} spikes (Llinás et al., 1989), it has been suggested that the generation of Ca^{2+} plateau potentials often seen in the dendrites requires the slower kinetics of a T channel (Fortier et al., 1991). However, Llinás and Sugimori (1992) have argued that the P channel might also be capable of producing these prolonged potentials, providing it with a dual role. This was exactly what happened in the model, suggesting that the T channel is not involved in plateau generation.

Simulation of synaptic inputs

After constructing a complex neuronal model, it is important to see if this model can reproduce well-defined physiological responses, different from the properties used to initially tune the model. Purkinje cell responses to synaptic inputs (Ito, 1984) were used to test the basic response

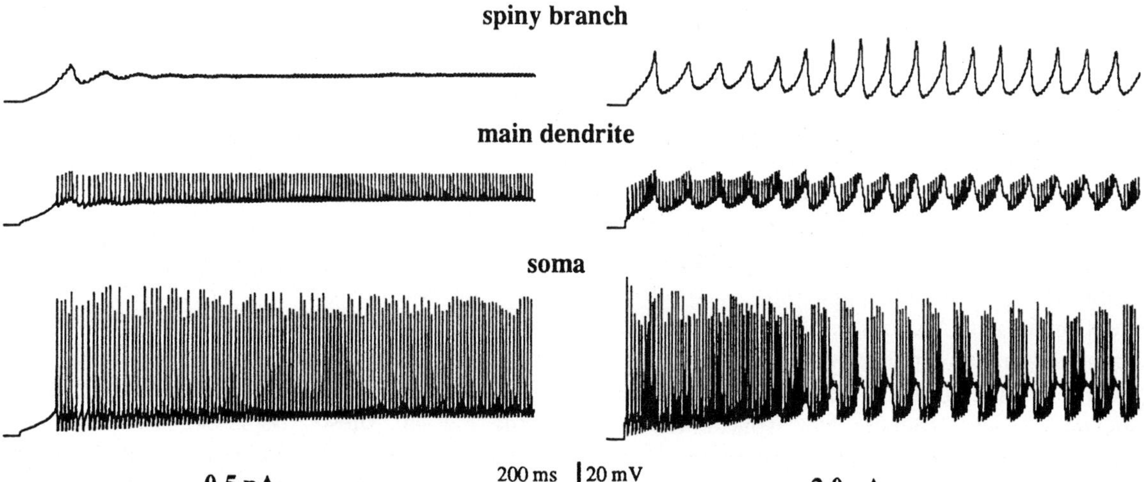

0.5 nA **2.0 nA**

Fig. 2. Simulation of steady current injections in the soma of the size indicated. Membrane potential as it was recorded in the soma, main dendrite and a distal spiny dendrite of the computer model.

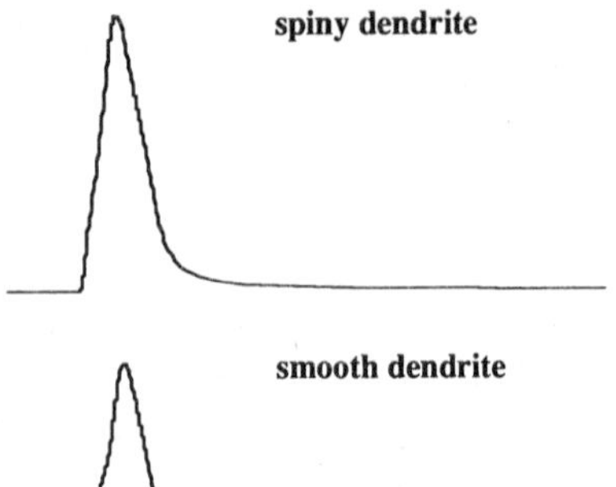

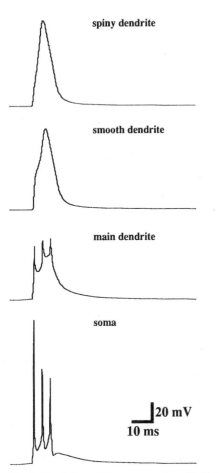

Fig. 3. Simulation of the response to a climbing fiber synaptic input. Complex spike as it was recorded in the indicated locations of the computer model.

properties of the model, two examples of which will be shown here; other simulations are described in De Schutter and Bower (1994b).

Figure 3 shows the simulated response to a climbing fiber input in different compartments of the model. This stimulus generates a complex spike (Llinás and Nicholson, 1976): the strong, distributed synaptic input activates dendritic Ca^{2+} channels, thus causing a large dendritic spike along with several somatic Na^{+} spikes. In the model, only P channels contributed to the dendritic depolarization, and, as reported in the literature (Knöpfel et al., 1991; Miyakawa et al., 1992), P channels were activated also in the spiny dendrites, outside of the region of synaptic contact. In Fig. 3 one can see clearly that the depolarization in the smooth dendrite precedes that in the spiny dendrite, and consists of two components, first a depolarization caused by the synaptic conductance, followed by the Ca^{2+} spike. These simulations also replicated the dual reversal potential of the complex spike, as described by Llinás and Nicholson (1976).

In the presence of asynchronous granule cell and stellate cell inputs, the modelled cell fired somatic action potentials at rates varying from 1 Hz to 200 Hz. The model was quite sensitive to the frequency of excitatory inputs in the normal range of spontaneous firing of Purkinje cells in vivo, i.e. 30–100 Hz (Murphy and Sabah, 1970). The effects of stellate cell inhibition on the simulated Purkinje cell response curve are analyzed in De Schutter and Bower (1994b). The remainder of this review will present unpublished results on the interaction in the Purkinje cell model of synchronous and asynchronous inputs from granule cells. The possible origin of such inputs in the cerebellar cortex will be described in the discussion. However, the input source does not affect single cell simulations, and this is a useful paradigm because it generates data comparable with most in vivo experimental recordings of Purkinje cells, as is illustrated in Fig. 4.

Figure 4A shows examples of the basic simulation protocol. The asynchronous drive made the model fire simple spikes at a rate of about 60 Hz. At a predetermined time, a small number of granule cell inputs fired synchronously, evoking an additional spike. A standard method for displaying in vivo recordings of the response of a particular neuron to a specific stimulus is the peri-stimulus histogram (PST). Figures 4B–4D compares simulated PSTs with an experimental one. Figure 4B shows the response to a synchronous activation of 100 granule cell inputs, while Fig. 4C shows the response to the same stimulus, followed 3 msec later by inhibitory inputs from basket cells. Figure 4D shows a PST generated from in vivo extracellular recordings of Purkinje cell responses in crus IIa of anesthetized

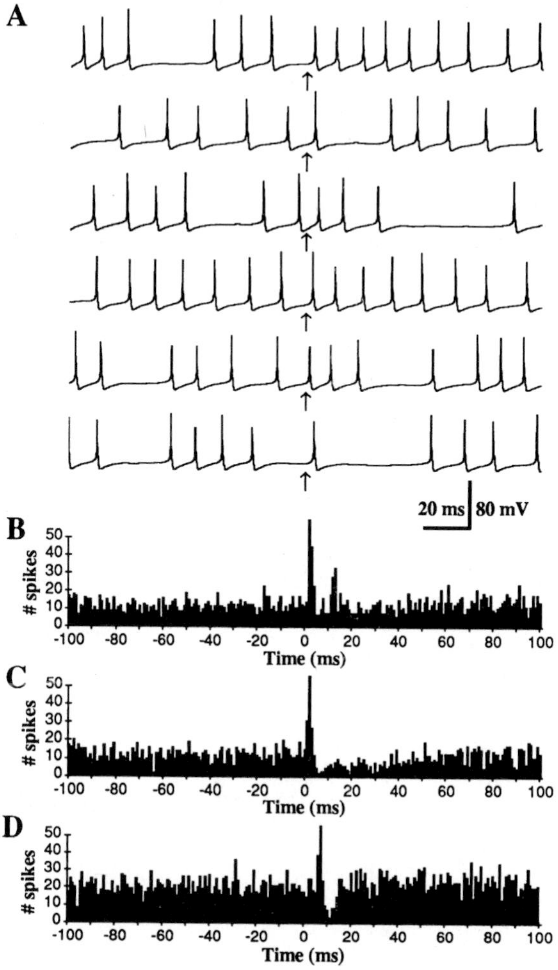

Fig. 4. Comparison of simulated and experimental response to synchronous excitation of a Purkinje cell. (A) examples of six simulation traces. Time of synchronous activation of 100 inputs is shown by arrow, asynchronous excitation at 28 Hz and inhibition at 1 Hz. (B) PST constructed from 200 simulations like the ones in A (C) PST of simulation of synchronous excitation, followed 3 msec later by synchronous inhibition from six basket cells. (D) PST from in vivo extracellular Purkinje cell recording (600 events) in crus IIa of the rat. Response to a facial tactile stimulus is shown. All PSTs have 1 msec bin widths.

rats to facial tactile stimuli, using standard methods (Bower and Kassel, 1990). The simulated data were not optimized so as to precisely simulate the experimental example of Fig. 4D. For example, there was a difference in the average firing frequency of the model in comparison with the experimental data.

More important is the similarity of the stimulus evoked response in both simulated and in vivo data. The input caused a sharp peak in the PSTs, which came sooner in the simulated data because direct synaptic activation was simulated. The in vivo data included delays caused by at least three additional synapses (at the sensory trigeminal ganglion neurons, at the trigeminal nucleus neurons, and at the granule cells). There was a small depression of spiking after the excitatory peak in the modelled PST without basket cell inhibition (Fig. 4B), which resulted from the activation of Ca^{2+}-activated K^{+} channels. A more pronounced depression is often seen in vivo (and also in the example in Fig. 4D), which is usually attributed to basket cell activity. When basket cell inhibition was added to the model (Fig. 4C), modelled and experimental PSTs looked much alike: the basket cell inhibition suppressed the small second peak in the PST of Fig. 4B. Interspike interval distributions (ISI) from the same set of experimental data were also very similar to those obtained from the simulations (De Schutter and Bower, 1994b). The fact that the model could faithfully reproduce experimental PSTs and ISIs inspires confidence in the realism of the simulated responses to granule cell synaptic inputs.

The response to synchronous granule cell inputs

As mentioned before, the Purkinje cell is distinguished by a huge number of converging granule cell inputs (Harvey and Napper, 1991). This leads to the interesting question of how sensitive this cell could be to individual inputs. Figure 4 already demonstrated that a relatively small number of synchronous inputs (100) was sufficient to cause a significant response in the PST. Figure 5 examines the relation between size of the simulated response and the number of inputs activated. Simulations were performed with a variable number of synchronously activated spines, evenly spread throughout the entire dendritic tree. The minimum number of synapses that needed to be activated in order to get a measurable response was 15–20. With such a small input a single spike was obtained in the PST (Fig. 5C) and when the number of inputs increased a second spike appeared. The second spike had a delay of more than 10 msec for 100–200 inputs but, when 400 or more inputs were active, the model fired doublets or even triplets. This corresponded to a small Ca^{2+} burst, which caused Ca^{2+}-activated K^+ currents to hyperpolarize the cell afterwards.

There are several ways in which one can measure the size of the response in a PST. Figure 5 compares the amplitude of the peak of the response (B) with a count of the average number of spikes in a 10 msec bin after the stimulus (A) (the 10 msec bin is stained black in the PSTs in Fig.

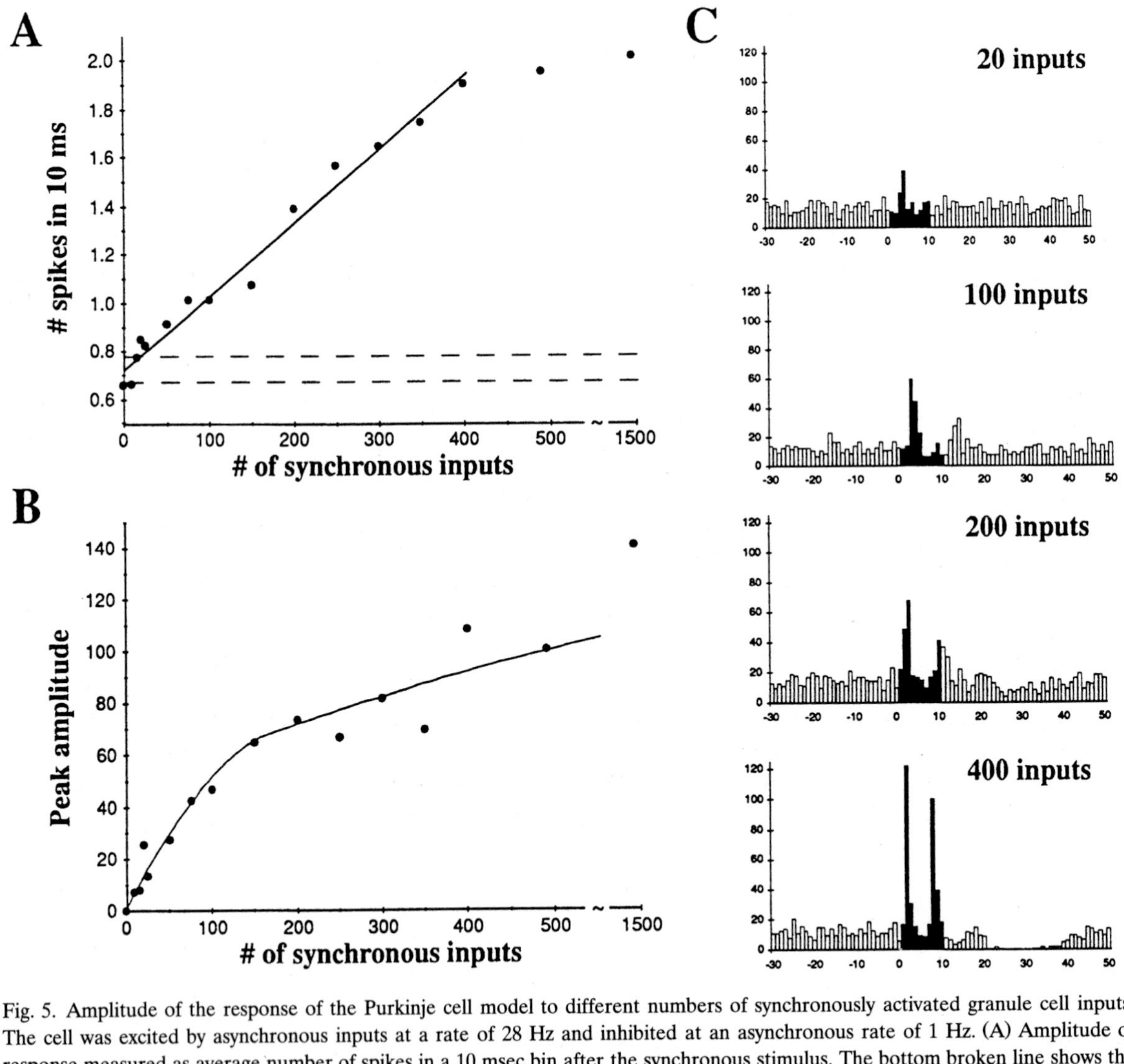

Fig. 5. Amplitude of the response of the Purkinje cell model to different numbers of synchronously activated granule cell inputs. The cell was excited by asynchronous inputs at a rate of 28 Hz and inhibited at an asynchronous rate of 1 Hz. (A) Amplitude of response measured as average number of spikes in a 10 msec bin after the synchronous stimulus. The bottom broken line shows the mean 10 msec bin counts before the synchronous stimulus, the top broken line is the mean +2 standard deviations. (B) Amplitude of the response measured as peak amplitude in the PST. (C) Example of PSTs obtained during synchronous activation of 20, 100, 200 or 400 granule cell inputs. The 10 msec bin used to compute the response shown in A is shaded black. PSTs are the average of 200 episodes, 1 msec bins.

5C). The average number of spikes within 10 msec after the stimulus was a better measure of the response: the relation was linear up to about 400 co-activated inputs. When more inputs were activated the response increased much more slowly. The peak amplitude did not show such a linear relation, except for 80 inputs or less, because it saturated (i.e. the model would always fire a spike 2 msec after the input in each trial) and was not sensitive to the presence of a second spike. Because 10-msec bins are often used to measure Purkinje cell activity in vivo (Sasaki et al., 1989; Marple-Horvat and Stein, 1990), the number of spikes in a 10 msec bin has been used as the standard measure of the responsiveness in the model (De Schutter and Bower, 1994c).

Attenuation of postsynaptic potentials by the spine neck

Granule cell axons make synapses on spines, which were modelled as passive membrane compartments. The Purkinje cell spine is relatively thick and stubby, so one can predict that a sizeable amount of the synaptic charge will flow into the dendritic shaft. An example is shown in Fig. 6, which shows the membrane potentials in a spine head and the underlying dendritic shaft during asynchronous excitatory inputs and a single synchronous input. This record was obtained in a distal dendrite, so that somatic action potentials were not noticeable.

When the synapse on the spine head was inactive, the membrane potential in the spine head was identical to that in the dendritic shaft. The short spine neck did not impede fast conduction of potential changes into the spine head. Note that because of asynchronous inputs on spines in other parts of the dendritic tree, the membrane potential varied continuously between about −58 and −48 mV.

A single input on the spine head caused a synaptic potential of 2.7–3.3 mV in the spine head, and one of 0.9–2.3 mV in the underlying dendritic shaft, an attenuation of 37–70% of the amplitude (average 54%). The synchronous activation of 100 spines caused a much larger synaptic potential (11.9 mV) in both the spine head and the dendritic shaft. There was no attenuation of the EPSP for synchronous inputs. This was caused by the active properties of the dendritic membrane. The synchronous input was large enough to activate the P channels, thus causing a large depolarization of the whole dendrite which conducted well into the spine head. The EPSP in the spine head after the synchronous input was a mixture of depolarization caused initially by synaptic current (the EPSP starts sooner in the spine head than in the dendrite) (Fig. 6A) and subsequently by current flowing in from the dendrite after activation of P channels. Note that the EPSP for the synchronous input was broader than for single asynchronous inputs, reflecting the activation of the P channels which also caused the increase in Ca^{2+} concentration in the dendritic shaft (Fig. 6C).

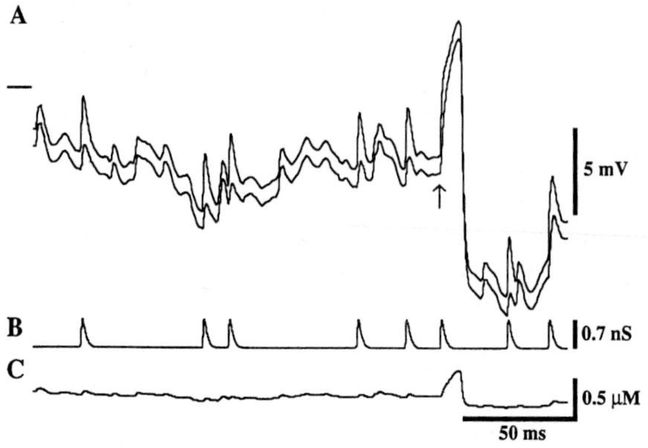

Fig. 6. Attenuation of postsynaptic potentials by the spine neck. (A) Membrane potential in the spine head (upper trace) and underlying dendritic shaft. The two traces have been offset by 1 mV from each other, the bar at the left indicates −50 mV for the spine head. The arrow indicates synchronous activation of 100 inputs, including one on the spine shown. (B) Excitatory synaptic conductance (asynchronous frequency 24 Hz) in the spine shown in A. (C) Ca^{2+} concentration in the dendritic shaft shown in A.

Discussion

Experiments in computo

This chapter reviews the simulation results obtained using a detailed model of a Purkinje cell. While a complete analysis of the biological realism of this model is beyond the scope of this chapter (see De Schutter and Bower 1994a,b), several examples were shown of simulated responses that are quite comparable with experimental data (Figs. 2–4). Of particular importance in this regard were the simulations of synaptic inputs, like the complex spike shown in Fig. 3, because the model had not been tuned specifically to generate these responses. These simulations were therefore true tests of the predictive power of the model.

Any model is a necessarily limited representation of reality (Bower and Koch, 1992), but compartmental models can be such accurate representations of neurons that one may compare them with an experimental preparation. In fact, though models are considered by many biologists to be inferior to experiments, laboratory preparations are very often limited representations of reality too. Because of constraints imposed by experimental procedures, investigations are sometimes performed on severely reduced systems, making the extrapolation back to the intact organism obscure. For example, there is a lot of evidence that the electrotonic properties of neurons in slice are quite different from those in vivo (Bernander et al., 1991; De Schutter and Bower, 1994a). Experiments are often performed under non-physiological conditions, such as pharmacological block of channels or high concentrations of ions in vitro, and the use of anesthetics in vivo, which can significantly affect the quantities being measured. An example is the use of Ca^{2+} ions to replace Ca^{2+} in voltage clamp investigations of Ca^{2+} currents, which results in significant shifts in the voltage dependency of the Ca^{2+} channels (Hille, 1991).

Of course, experiments are necessary to advance our scientific understanding of neuronal function, and are an essential ingredient in building realistic models. Intrinsically, however, there is no reason why the experimental method would be superior to simulation studies, as has been quite clear for elementary particle physics. An important reason why models have become essential to the progress of neuroscience is that they are no longer simply convenient descriptions of a complex reality, comparable with Newton's law of gravity. The modelling approach described here has actually an opposite goal because it tries to capture as much of the real complexity as possible. As such, the model provides a synthesis of a large body of experimental data. This integration of data in itself has become an important issue in neuroscience (Huerta et al., 1993).

But complex models also allow one to go a step further, as has been demonstrated in this and other papers (Lytton and Sejnowski, 1991; Traub et al., 1991): they can be used as the 'perfect' experimental preparation, because all simulated components of the cell are present in their 'native' form and can interact freely. Furthermore, one has continuous access to all parameters and complete control over the inputs. This allows investigation of neuronal properties and mechanisms which are presently not accessible to experimental procedures. An example is the response properties to synaptic inputs of Purkinje cells in vivo, which have been partially described in this chapter. While it is relatively straightforward to record such responses in vivo, there is no way even to roughly estimate the real excitatory and inhibitory synaptic drive onto the neuron in the living animal.

How much can one trust results obtained from computer modelling? This of course depends largely on the quality of the 'preparation', i.e. the model. Therefore, it is extremely important to use a careful and systematic approach in building a model along the lines described in this chapter. The data set and simulations used in building the model (current injections in slice) were clearly distinguished from those used subsequently to test the model and make predictions (synaptic inputs). This avoids one of the major pitfalls in simulation studies, that they cannot be tested because they were fitted to a complete experimental data set. At the same time it prevents accidental inclusion in the model of preconceived ideas on how the Purkinje cell integrates synaptic inputs. This is an important, often underestimated, issue (Bower, 1991). If hypothetical concepts of how a system operates intrude into decisions made during the model construction process, one may end up with a 'demonstration model' that establishes the feasibility of these theories, but has little predictive power.

To summarize, complex models of neurons and nervous systems can be used to investigate the properties of these systems in ways comparable with experiments in living systems. It has been suggested that this computational approach in neuroscience is complementary to theory and experiment (Koch and Segev, 1989). However, it might be more useful to consider such computations as a new experimental approach, instead of as a new branch of neuroscience. Hopefully, this will facilitate smooth interactions with experimentalists (Bower and Koch, 1992) and, in some cases, even the integration of simulation into other experimental methods (Sharp et al., 1993).

Response to small granule cell inputs

Under physiological conditions of stimulation, for example with tactile stimuli in vivo (Shambes et al., 1978; Bower and Kassel, 1990), Purkinje cells probably receive both asynchronous and synchronous excitatory stimuli. Asynchronous inputs arrive mainly over the parallel fibers and are due to spontaneous mossy fiber inputs (Ito, 1984) and to dispersion of initially synchronous signals caused by the variable propagation speeds along the parallel fibers (Bernard and Axelrad, 1991). The ascending branch of the granule cell axon also makes synaptic contacts on Purkinje cells (Llinás, 1982) and may be the main source of synchronous inputs. Because receptive fields in the granule cell layer are organized as small patches (Shambes et al., 1978), which are activated by peripheral stimuli in short bursts, granule cells underlying a particular Purkinje cell tend to fire together and can thus deliver synchronous inputs to that cell (Bower and Woolston, 1983).

The Purkinje cell model could respond to stimulation by a rather small number of synchronous granule cell inputs (about 20). This counters the common view that very many parallel fibers need to be activated to cause a Purkinje cell to fire a simple spike. Marr (1969) suggested on theoretical grounds that at least 500 inputs, and probably more, would be necessary to generate a Purkinje cell response. Sabah (1971) suggested that for maximizing reliability of computation in the cerebellum, the threshold should be set at 3000–6000 input lines. Similarly, Rapp et al. (1994) showed, in their analysis of a passive membrane model of the same Purkinje cell, that co-activation of 800 excitatory synapses resulted in a somatic EPSP with an amplitude of only 12.5 mV, barely enough to cross the threshold for action potential discharge.

These authors based their analysis on predicted responses to one large input in a silent Purkinje cell. In the approach presented here, asynchronous excitation and inhibition caused the model to fire continuously, similar to Purkinje cells in vivo (Fig. 4) (De Schutter and Bower 1994b), and responses to additional synchronous inputs were examined. The asynchronous excitatory drive increases the synchronous input by only a small amount. In Fig. 5 an excitation of 28 Hz was provided; assuming a scaling of 100 for asynchronous inputs (see Methodology) this corresponds to 0.28 Hz on 150,000 input lines. Thus, on average only 42 inputs would fire within the same millisecond as the synchronous input. While 42 is fairly large compared with the 20 input minimum for eliciting a response (Fig. 5), it is small compared with the 100–200 inputs which might represent the size of the signal coming over the ascending branches of a set of granule cell axons contacting the same Purkinje cell. A second spike was present in the simulated PSTs generated by such synchronous inputs (Figs. 4B, 5C). A second spike is unusual in in vivo recordings of Purkinje cells (Bower and Woolston, 1983); however, when basket cell inhibition was added to the model, the responses became very similar to experimental recordings (Figs. 4C, 4D). This suggests that in the living animal basket cell inhibition may truncate the response of Purkinje cells to stimuli.

The sensitivity of the Purkinje cell model to small synchronous inputs is caused by a sub-spiking threshold activation of P channels which amplifies small synchronous inputs (De Schutter and Bower, 1994c). The rise in Ca^{2+} concentration caused by this P channel activation, can be observed in Fig. 6C. The amplification has quite complex effects, as distal inputs are amplified more than proximal ones, effectively negating the importance of dendritic location of synaptic inputs (De Schutter and Bower, 1994c). The active properties of the dendrite are essential for obtaining this excitability, as a passive membrane model of the same neuron leads to entirely different conclusions (Rapp et al., 1994).

Recently, recordings of the Purkinje cell response to activation of single granule cells in cerebellar slice preparations suggested that 50 granule cell inputs would be sufficient to cause a single simple spike (Barbour, 1993). While this experimental evidence confirms one of the predictions made by the model (Fig. 5), this author did not address the question of the possible role of postsynaptic mechanisms.

The use of passive membrane spines

Finally, it is appropriate to discuss explicitly one of the simplifications used in the model, namely that the dendritic spines were modelled as passive compartments. Based on immunohistochemical studies, it has been claimed that P channels are present on Purkinje cell spine heads (Hillman et al., 1991), which would make these spines active. I had several good reasons for not including active membrane in the spine heads. First of all, adding a P channel and the associated Ca^{2+} concentration to every spine would have increased the size of the model considerably, thus reducing simulation speed. More important, however, is the fact that such addition would introduce new, badly constrained, parameters into the model. Nobody has any idea about the density of P channels on a spine. Even now that the size of the EPSP in response to a single parallel fiber input is known (Barbour, 1993), it is impossible to distinguish between activation of P channels in the spine head and in the dendrite. Finally, the intrasomatic current injections used to tune the model are not sensitive to differences caused by the presence of channels on the dendritic shaft as opposed to the spine heads. As a consequence, the conductance of any given channel in the spine heads has been collapsed into the corresponding conductance of the dendrite.

P channels in spine heads could have local effects, however, such as amplification of synaptic inputs (Miller et al., 1985; Segev and Rall, 1988); these would not be present in this model. The simulation data in Fig. 6 suggest that such local effects were negligible. Single synaptic inputs were too small to activate P channels in the spine head, as the resulting EPSP did not cross the -40 mV threshold of activation for P channels (Regan, 1991; Usowicz et al., 1992). Because the 0.7 nS maximum conductance used for a single granule cell input in the model might be rather large (Cull-Candy and Usowicz, 1989), it is probable that a single synaptic input could never activate P channels present in a spine head.

Synchronous synaptic inputs did activate P channels (Fig. 6), but this obviously involves summation of EPSPs in the dendrite itself (Shepherd and Brayton, 1987). Adding P channels to the spine heads would not have changed the response to synchronous inputs because the membrane potential was almost identical in spine head and dendrite during these inputs (Fig. 6) and, consequently, P channel activation would be identical in both locations.

Segev and Rall (1988) have suggested that excitable channels may be present on spine heads, because the same depolarization of the dendrite in response to a synaptic input can then be achieved with less channels than would be the case if they are on the dendritic shaft. Because less channels are needed, Segev and Rall (1988) called putting excitable channels on the spine heads more 'economical'. However, multiple spines need to be co-activated in their model also in order to obtain a sizeable depolarization of the dendrite. It is probable that P channels in Purkinje cell spine heads operate according to such a regime, i.e. the channels might potentiate synchronous, but not single inputs.

Summary

Detailed compartmental models of neurons are useful tools for investigating neuronal properties and mechanisms that are not accessible to experimental procedures. If a rigorous approach is used in building the model, simulation studies can be as valuable as laboratory experimentation. As such, modelling becomes an additional method for exploring the function of neurons and nervous systems.

As an example, a complex compartmental model with active dendritic membrane of a Purkinje cell is described. The response properties of the model to parallel fiber inputs were investigated. The model fired simple spikes in patterns comparable with those recorded from Purkinje cells in vivo. Synchronous activation of only 20 granule cell inputs was sufficient to generate a measurable response in simulated peri-stimulus histograms. This sensitivity to small excitatory inputs was caused by P-type Ca^{2+} channels in the dendritic membrane. Such P channels may also be present in the spine heads. Simulations suggest, however, that Ca^{2+} channels in spine heads cannot be activated by single parallel fiber inputs.

Acknowledgments

I thank M. Rapp, I. Segev and Y. Yarom for providing the morphological reconstruction of the Purkinje cell. This work was done in the laboratory of J.M. Bower at Caltech. J.H. Thompson provided the experimental data shown in Fig. 4. Results shown in Figs. 4 and 5 were obtained with the Intel Touchstone Delta System operated by Caltech on behalf of the Concurrent Supercomputing Consortium. Finally, I thank the reviewers for extensive stylistic improvements. This research was supported by Fogarty Fellowship F05 TW04368.

References

Barbour, B. (1993) Synaptic currents evoked in Purkinje cells by stimulating individual granule cells. *Neuron,* 11: 759–769.

Bhalla, U.S. and Bower, J.M. (1993) Exploring parameter space in detailed single neuron models: simulations of the mitral and granule cells of the olfactory bulb. *J. Neurophysiol.,* 6: 1948–1965.

Bernard, C. and Axelrad, H. (1991) Propagation of parallel fiber volleys in the cerebellar cortex: a computer simulation. *Brain Res.,* 565: 195–208.

Bernander, Ö, Douglas, R.J., Martin, K.A.C. and Koch, C. (1991) Synaptic background activity influences spatiotemporal integration in single pyramidal cells. *Proc. Natl. Acad. Sci. USA,* 88: 11569–11573.

Bower, J.M. (1991) Reverse engineering the nervous system: an anatomical, physiological, and computer based approach. In: S. Zornetzer, J. Davis and C. Lau, (Eds.), *An Introduction to Neural and Electronic Networks,* Academic Press, New York, pp. 3–24.

Bower, J.M. and Kassel, J. (1990) Variability in tactile projection patterns to cerebellar folia crus IIA of the Norway rat. *J. Comp. Neurol.,* 302: 768–778.

Bower, J.M. and Koch, C. (1992) Experimentalists and modellers: can we all just get along? *Trends Neurosci.,* 15: 458–461.

Bower, J.M. and Woolston, D.C. (1983) Congruence of spatial organization of tactile projections to granule cell and Purkinje cell layers of cerebellar hemispheres of the albino rat: vertical organization of cerebellar cortex. *J. Neurophysiol.,* 49: 745–766.

Brorson, J.R., Bleakman, D., Gibbons, S.J. and Miller, R.J. (1991) The properties of intracellular calcium stores in cultured rat cerebellar neurons. *J. Neurosci.,* 11: 4024–4043.

Bush, P.C. and Sejnowski, T.J. (1991) Simulations of a reconstructed cerebellar Purkinje cell based on simplified channel kinetics. *Neural Comput.,* 3: 321–332.

Crepel, F. and Penit-Soria, (1986) J. Inward rectification and low threshold calcium conductance in rat cerebellar Purkinje cells. An in vitro study. *J. Physiol. (London),* 372: 1–23.

Cull-Candy, S.G. and Usowicz, M.M. (1989) On the multiple-conductance single channels activated by excitatory amino acids in large cerebellar neurones of the rat. *J. Physiol. (London),* 415: 555–582.

De Schutter, E. (1992) A consumer guide to neuronal modelling software. *Trends Neurosci.,* 15: 462–464.

De Schutter, E. and Bower, J.M. (1994a) An active membrane model of the cerebellar Purkinje cell: I. Simulation of current clamps in slice. *J. Neurophysiol.,* 71: 375–400.

De Schutter, E. and Bower, J.M. (1994b) An active membrane model of the cerebellar Purkinje cell: II. Simulation of synaptic responses. *J. Neurophysiol.,* 71: 401–409.

De Schutter, E. and Bower, J.M. (1994c) Simulated responses of cerebellar Purkinje cell are independent of the dendritic location of granule cell synaptic inputs. *Proc. Natl. Acad. Sci. USA,* 91: 4736–4741.

Farrant, M. and Cull-Candy, S. G. (1991) Excitatory amino

acid receptor-channels in Purkinje cells in thin cerebllar slices. *Proc. R. Soc. London Ser. B,* 244: 179–184,.

Fortier, P.A., Tremblay, J.P., Rafrafi, J. and Hawkes, R. (1991) A monoclonal antibody to conotoxin reveals the distribution of a subset of calcium channels in the rat cerebellar cortex. *Mol. Brain Res.,* 9: 209–215.

French, C.R., Sah, P., Buckett, K.J. and Gage, P.W. (1990) A voltage-dependent persistent sodium current in mammalian hippocampal neurons. *J. Gen. Physiol.,* 95: 1139–1157.

Gähwiler, B.H. and Llano, I. (1989) Sodium and potassium conductances in somatic membranes of rat Purkinje cells from organotypic cerebellar cultures. *J. Physiol. (London),* 417: 105–122.

Garthwaite, J. and Beaumont, P.S. (1989) Excitatory amino acid receptors in the parallel fibre pathway in rat cerebellar slices. *Neurosci. Lett.,* 107: 151–156.

Gruol, D.L., Jacquin, T. and Yool, A.J. (1991) Single-channel K^+ currents recorded from the somatic and dendritic regions of cerebellar Purkinje neurons in culture. *J. Neurosci.,* 11: 1002–1015.

Harris, K.M. and Stevens, J.K. (1988) Dendritic spine of rat cerebellar Purkinje cells: serial electron microscopy with reference to their biophysical characteristics. *J. Neurosci.,* 8: 4455–4469.

Harvey, R.J. and Napper, R.M.A. (1991) Quantitative studies of the mammalian cerebellum. *Prog. Neurobiol.,* 36: 437–463.

Hille, B. (1991) *Ionic Channels of Excitable Membranes,* Sinauer, Sunderland, pp. 457–461.

Hillman, D., Chen, S., Aung, T.T., Cherksey, B., Sugimori, M. and Llinás, R.R. (1991) Localization of P-type calcium channels in the central nervous system. *Proc. Natl. Acad. Sci. USA,* 88: 7076–7080.

Hirano, T. and Hagiwara, S. (1989) Kinetics and distribution of voltage-gated Ca, Na and K channels on the somata of rat cerebellar Purkinje cells. *Pfluegers Arch.,* 413: 463–469.

Hockberger, P.E., Tseng, H.Y. and Connor, J.A. (1989) Fura-2 measurements of cultured rat Purkinje neurons show dendritic localization of Ca^{2+} influx. *J. Neurosci.,* 9: 2272–2284.

Hodgkin, A.L. and Huxley, A.F. (1952) A quantitative description of membrane current and its application to conduction and excitation in nerve. *J. Physiol. (London),* 117: 500–544.

Holmes, W.R. and Woody, C.D. (1989) Effects of uniform and non-uniform synaptic Tactivation-distributionsU on the cable properties of modelled cortical pyramidal neurons. *Brain Res.,* 505: 12–22.

Hounsgaard, J. and Midtgaard, J. (1988) Intrinsic determinants of firing patterns in Purkinje cells of the turtle cerebellum in vitro. *J. Physiol. (London),* 402: 731–749.

Huerta, M.F., Koslow, S.H. and Leshner, A.I. (1993) The Human Brain Project: an international resource. *Trends Neurosci.,* 16: 436–438.

Ito, M. (1984) *The Cerebellum and Neural Control,* Raven Press, New York, pp. 580.

Kaneda, M., Wakamori, M., Ito, M. and Akaike, N. (1990) Low-threshold calcium current in isolated Purkinje cell bodies of rat cerebellum. *J. Neurophysiol.,* 63: 1046–1051.

Kay, A.R., Sugimori, M. and Llinás, R.R. (1990) Voltage clamp analysis of a persistent TTX-sensitive Na currrent in cerebellar Purkinje cells. *Abstr. Soc. Neurosci.,* 16: 182.

Knöpfel, T., Audinat, E. and Gähwiler, B.H. (1990) Climbing fibre responses in olivo-cerebellar slice cultures. I. Microelectrode recordings from Purkinje cells. *Eur. J. Neurosci.,* 2: 726–732.

Knöpfel, T., Audinat, E. and Gähwiler, B.H. (1991) Climbing fibre responses in olivo-cerebellar slice cultures. II. Dynamics of cytosolic calcium in Purkinje cells. *Eur. J. Neurosci.,* 3: 343–348.

Koch, C. and Segev, I. (1989) Introduction. In: C. Koch and I. Segev (Eds.), *Methods in Neuronal Modelling: from Synapses to Networks,* MIT Press, Cambridge, MA, pp. 1–8.

Lev-Ram, V., Miyakawa, H., Lasser-Ross, N. and Ross, W.N. (1992) Calcium transients in cerebellar Purkinje neurons evoked by intracellular stimulation. *J. Neurophysiol.,* 68: 1167–1177.

Llano, I., Dreessen, J., Kano, M. and Konnerth, A. (1991) Intradendritic release of calcium induced by glutamate in cerebellar Purkinje cells. *Neuron,* 7: 577– 583.

Llinás, R.R. (1982) Radial connectivity in the cerebellar cortex: a novel view regarding the functional organization of the molecular layer. *Exp. Brain Res.,* Suppl. 6: 189– 194.

Llinás, R.R. (1988) The intrinsic electrophysiological properties of mammalian neurons: insights into central nervous-system function. *Science,* 242: 1654–1664.

Llinás, R.R. and Nicholson, C. (1976) Reversal properties of climbing fiber potential in cat Purkinje cells: an example of a distributed synapse. *J. Neurophysiol.,* 39: 311– 323.

Llinás, R.R. and Sugimori, M. (1980a) Electrophysiological properties of in vitro Purkinje cell somata in mammalian cerebellar slices. *J. Physiol. (London),* 305: 171–195.

Llinás, R.R. and Sugimori, M. (1980b) Electrophysiological properties of in vitro Purkinje cell dendrites in mammalian cerebellar slices. *J. Physiol. (London),* 305: 197–213.

Llinás, R.R. and Sugimori, M. (1992) The electrophysiology of the cerebellar Purkinje cell revisited. In: R.R. Llinás and C. Sotelo (Eds.), *The Cerebellum Revisited,* Springer-Verlag, Berlin, pp. 167–181.

Llinás, R.R., Sugimori, M. and Cherksey, B. (1989) Voltage-dependent calcium conductances in mammalian neurons: the P channel. *Ann. N.Y. Acad. Sci.,* 560: 103–111.

Lytton, W.W. and Sejnowski, T.J. (1991) Simulations of cortical pyramidal neurons synchronized by inhibitory interneurons. *J. Neurophysiol.,* 66: 1059–1079.

Marple-Horvat, D.E. and Stein, J.F. (1990) Neuronal activity in the lateral cerebellum of trained monkeys, related to visual stimuli or eye movements. *J. Physiol. (London),* 428: 595–614.

Marr, D.A. (1969) A theory of cerebellar cortex. *J. Physiol. (London),* 202: 437–470.

McCormick, D.A. and Huguenard, J.R. (1992) A model of the electrophysiological properties of thalamocortical relay neurons. *J. Neurophysiol.,* 68: 1384–1400.

Miller, J.P., Rall, W. and Rinzel, J. (1985) Synaptic amplification by active membrane in dendritic spines. *Brain Res.,* 325: 325–330.

Miyakawa, H., Lev-Ram, V., Lasser-Ross, N. and Ross, W.N. (1992) Calcium transients evoked by climbing fiber synaptic inputs in guinea pig cerebellar Purkinje neurons. *J. Neurophysiol.,* 68: 1178–1189.

Mintz, I.M., Venema, V.J., Swiderek, K.M., Lee, T.D., Bean, B.P. and Adams, M. E. (1992) P-type calcium channels blocked by the spider toxin omega-Aga-IVA. *Nature,* 355: 827–829.

Murphy, J.T. and Sabah, N.H. (1970) Spontaneous firing of cerebellar Purkinje cells in decerebrate and barbiturate anesthesized cats. *Brain Res.,* 17: 515– 519.

Palay, S.L. and Chan-Palay, V. (1974) *Cerebellar Cortex,* Springer-Verlag, New York, pp. 34–36.

Pellionisz, A. and Llinás, R.R. (1977) A computer model of cerebellar Purkinje cells. *Neuroscience,* 2: 37–48.

Rall, W. (1962) Theory of physiological properties of dendrites. *Ann. N.Y. Acad. Sci.,* 96: 1071–1092.

Rall, W. (1989) Cable theory for dendritic neurons. In C. Koch and I. Segev (Eds.), *Methods in Neuronal Modelling: from Synapses to Networks,* MIT Press, Cambridge, MA, pp. 9–62.

Rapp, M., Yarom, Y. and Segev, I. (1992) The impact of parallel fiber background activity on the cable properties of cerebellar Purkinje cells. *Neural Comput.,* 4: 518–533.

Rapp, M., Segev, I. and Yarom, Y. (1994) Physiology, morphology and detailed passive models of cerebellar Purkinje cells. *J. Physiol. (London),* 471: 87–99.

Regan, L.J. (1991) Voltage-dependent calcium currents in Purkinje cells from rat cerebellar vermis. *J. Neurosci.,* 11: 2259–2269.

Ropert, N., Miles, R. and Korn, H. (1990) Characteristics of miniature inhibitory postsynaptic currents in CA1 pyramidal neurones of rat hippocampus. *J. Physiol. (London),* 428: 707–722,.

Sabah, N.H. (1971) Reliability of computation in the cerebellum. *Biophys. J.,* 11: 429–445.

Sala, F. and Hernandez-Cruz, A. (1990) Calcium diffusion modelling in a spherical neuron: relevance of buffering properties. *Biophys. J.,* 57: 313–324.

Sasaki, K., Bower, J.M. and Llinás, R.R. (1989) Multiple Purkinje cell recording in rodent cerebellar cortex. *Eur. J. Neurosci.,* 1: 572–586.

Segev, I. (1992) Single neuron models: oversimple, complex and reduced. *Trends Neurosci.,* 15: 414–421.

Segev, I. and Rall, W. (1988) Computational study of an excitable dendritic spine. *J. Neurophysiol.,* 60: 499–523.

Shambes, G.M., Gibson, J.M. and Welker, W. (1978) Fractured somatotopy in granule cell tactile areas of rat cerebellar hemispheres revealed by micromapping. *Brain Behav. Evol.,* 15: 94–140.

Sharp, A.A., ONeil, M.B., Abbott, L.F. and Marder E. (1993) The dynamic clamp: artificial conductances in biological neurons. *Trends Neurosci.,* 16: 389–394.

Shelton, D.P. (1985) Membrane resistivity estimated for the Purkinje neuron by means of a passive computer model. *Neuroscience,* 14: 111–131.

Shepherd, G.M. (1990) *The Synaptic Organization of the Brain,* Oxford University Press, New York, pp. 3–31.

Shepherd, G.M. and Brayton, R.K. (1987) Logic operations are properties of computer-simulated interactions between excitable dendritic spines. *Neuroscience,* 21: 151–165.

Staub, C., Vranesic, I. and Knöpfel, T. (1992) Responses to metabotropic glutamate receptor activation of cerebellar Purkinje cells: induction of an inward current. *Eur. J. Neurosci.,* 4: 832–839.

Traub, R.D. (1982) Simulation of intrinsic bursting in CA3 hippocampal neurons. *Neuroscience,* 7: 1233–1242.

Traub, R.D., Wong, R.K.S., Miles, R. and Michelson, H. (1991) A model of a CA3 hippocampal pyramidal neuron incorporating voltage-clamp data on intrinsic conductances. *J. Neurophysiol.,* 66: 635–650.

Usowicz, M.M., Sugimori, M., Cherksey, B. and Llinás, R.R. (1992) Characterization of P-type calcium channels in cerebellar Purkinje cells. *Abstr. Soc. Neurosci.,* 18: 974.

Vincent, P., Armstrong, C.M. and Marty, A. (1992) Inhibitory synaptic currents in rat cerebellar Purkinje cells: modulation by postsynaptic depolarization. *J. Physiol. (London),* 456: 453–471.

Weitzman, D., Reuveni, I., White, E.L. and Gutnick, M.J. (1992) IPSPS control the time frame for neocortical conditioning: computer-simulation of a completely reconstructed spiny stellate neuron. *Eur. J. Neurosci.,* Suppl. 5: 281.

Wilson, M.A., Bhalla, U.S., Uhley, J.D. and Bower, J.M. (1989) GENESIS: a system for simulating neural networks. In: D. Touretzky (Ed.), *Advances in Neural Information Processing Systems,* Morgan Kaufmann, San Mateo, CA, pp. 485–492.

Yamada, W.M., Koch, C. and Adams, P.R. (1989) Multiple channels and calcium dynamics. In: C. Koch and I. Segev (Eds.), *Methods in Neuronal Modelling: from Synapses to Networks,* MIT Press, Cambridge, MA, pp. 97–133.

Subject Index